T0331487

MODERN TUNNELING SCIENCE AND TECHNOLOGY

PROCEEDINGS OF THE INTERNATIONAL SYMPOSIUM ON MODERN TUNNELING SCIENCE AND TECHNOLOGY (IS-KYOTO 2001) / KYOTO / JAPAN / 30 OCTOBER—1 NOVEMBER 2001

Modern Tunneling Science and Technology

Edited by

T.Adachi, K.Tateyama & M.Kimura
Department of Civil Engineering, Kyoto University, Japan

VOLUME 1

A.A.BALKEMA PUBLISHERS   LISSE / ABINGDON / EXTON (PA) / TOKYO

Published by: A.A.Balkema, a member of Swets & Zeitlinger Publishers
www.balkema.nl and www.szp.swets.nl

For the complete set of two volumes, ISBN 90 2651 860 9
For Volume 1, ISBN 90 2651 861 7
For Volume 2, ISBN 90 2651 862 5

Printed in the Netherlands

Modern Tunneling Science and Technology, Adachi et al (eds), © 2001 Swets & Zeitlinger, ISBN 90 2651 860 9

Table of contents

Preface   XI

Acknowledgements   XIII

Organizing committees   XV

Keynote Lectures

Safety aspects of the Lötschberg railway tunnel & Mont-Blanc road tunnel   3
F. Vuilleumier, A. Weatherill, Y. Trottet, B. Crausaz

On the fundamental mechanisms of tunneling   9
T. Tamura

Flexible response to difficult ground and controlled tunneling technology   21
H. Wagner

Design method of shield tunnel -present status and subjects-   29
Y. Koyama

The rehabilitation of tunnel structures   39
H. A. Russell

Underground space networks in the 21st Century –   55
new infrastructure with modal shift technology and geotechnology
T. Hanamura

Mechanism of tunneling

Optimal shape of underground structure   69
T. Tamura, J. Saito

An approximate solution for the stresses on a rigid tunnel   73
A. Verruijt

Surface displacements due to pressure modifications induced by tunnels   77
M. El Tani

Short and long-term load conditions for tunnels in low permeability ground   83
in the framework of the convergence-confinement method
A. Graziani, R. Ribacchi

3-D FEM analysis for layered rock considering anisotropy of shear strength   89
Y.J. Zhang, Y.P. Liu, C.Y. Xiong

A consideration on apparent cohesion of unsaturated sandy soil   95
R. Kitamura, K. Sako

Identification of the soil parameters in the analysis of a shallow tunnel   101
C.M.C. Moreira, J. Almeida e Sousa

A simple technique to improve the prediction of surface displacement profiles 105
due to shallow tunnel construction
A. Burghignoli, A. Magliocchetti, S. Miliziano, F.M. Soccodato

Numerical modeling of a nonlinear deformational behavior of a tunnel with shallow depth 111
S. Akutagawa, T. Kitani, K. Matsumoto, S. Mizoguchi

Prediction of ground settlements due to tunneling in clayey soils 115
using advanced constitutive soil models: a numerical study
A. Burghignoli, S. Miliziano, F.M. Soccodato

Analysis of three-dimensional effect of tunnel face stability on sandy ground 121
with a clay layer by the three-dimensional rigid plastic Finite Element Method
S. Konishi, T. Tamura

Hydraulic conductivity change around tunnel due to excavation 127
A. Kobayashi, M. Nishida, K. Hosono, T. Fujita

Displacement measurement in a two-dimensional tunneling experiment 133
N. Dolzhenko, Ph. Mathieu

Centrifugal model test on the clay arching 137
B. Wang, S. Zhou, Q. Gong, L. Yang

Arching effects on the deformation of sandy ground induced by tunneling 141
T. Honda, T. Hibino, A. Takahashi, J. Kuwano

Numerical simulation of centrifuge test of trapdoor by 3-D FEM 147
T. Adachi, F. Oka, M. Kimura, K. Kishida, M. Kikumoto, T. Takeda, F. Zhang

Earth pressure and ground movements due to tunneling-model tests and analyses 153
T. Nakai, D. Yamaguchi, H.M. Shahin, T. Kurimoto, M.M. Farias

Experimental study on the distribution of earth pressure through three-dimensional trapdoor tests 159
T. Adachi, M. Kimura, K. Kishida

Behaviour of a tunnel face reinforced by bolts: Influence of the soil/bolt interface 165
D. Dias, Y. Bourdeau, R. Kastner

3-D energy damage model for bolted rockmass and its application in a tunnel engineering 171
Q.Y. Zhang, P.Y. Lu, W.S. Zhu

Centrifuge model test of tunnel face reinforcement by bolting 177
H. Kamata, H. Mashimo

Development of MGF method based on the evaluation of forepiling supporting mechanism 183
Y. Kitamoto, K. Date, T. Yamamoto, K. Hibiya, H. Ohta

Experimental study on tunneling in the ground with inclined layers and its simulation 189
S.H. Park, T. Adachi, M. Kimura, K. Kishida, M. Kikumoto

Study on the strength enhancement theory of surrounding rock by bolting 195
C. Hou, N. Zhang, X. Li, P. Gou

An experimental study on the nonuniform tunnel lining 199
N. Yingyongrattanakul, T. Adachi, K. Tateyama, M. Kurahashi

Field measurements, monitoring, geophysics and site characterization

Seismic prospecting in underground excavation to investigate the behavior of surrounding bedrock 207
H. Sasao, K. Kaneko, A. Hirata, M. Yamazoe, Y. Obara, N. Nakamura

Evaluation of the geological condition ahead of the tunnel face 213
by geostatistical techniques using TBM driving data
T. Yamamoto, S. Shirasagi, S. Yamamoto, Y. Mito, K. Aoki

Application of CSAMT monitoring method and observational construction management to mountain tunnel 219
T. Fukui, T. Hirai, T. Morioka, A. Suga, T. Matsui

Nondestructive survey of shallow underground by electromagnetic method 225
T. Katayama, K. Ozaki, Y. Yoshizu, E. Kurata, T. Kozato, Y. Ashida

A study on the evaluation of in-situ rock conditions by the tomographic method 231
in tunnel geological investigations
S, Miki, H. Shiroma, K. Inoue, K. Nakagawa

Systems for forward prediction of geological condition ahead of the tunnel face 237
T. Yamamoto, S. Shirasagi, M. Inou, K. Aoki

Deep excavation in Bangkok -characterization, measurement and prediction 243
S. Shibuya, S.B. Tamrakar, T. Mitachi

Rock response during construction of cavern station at Bagatza (Metro Bilbao) 249
J. Madinaveitia

Excavation monitoring of a large cross section tunnel underpassing an existing railway 253
R. Sasaki, T. Takayama, M. Tsukada, M. Kimura, S. Torii, H. Nakagaki

Evaluation of load on a shield tunnel lining in gravel 259
H. Mashimo, T. Ishimura

Tunnel convergence measurement using vision metrology 265
S. Miura, S. Hattori, K. Akimoto, Y. Ohnishi

Sequential application of several survey systems in tunnelling for ground classifications 269
M. Okamura, T. Hara, T. Kimura, K. Ishiyama, T. Hirano

Application of several electric resistivities of rock masses for tunnel supporting design 273
M. Nakamura, H. Kusumi, E. Kondo

Development and application of seismic reflection survey looking ahead 277
of tunnel face using hydraulic impactor
T. Kato, H. Murayama, T. Yanai, M. Murayama, N. Shimizu

Application of CCD photogrammetry system to measurement of tunnel wall movement 281
due to parallel tunnel excavation
Y. Ohnishi, H. Ohtsu, S. Nishiyama, N. Okada, T. Seya, Y. Yoshida, T. Nakai, M. Ryu

Semi-automated tunnel measurement by vision metrology using coded-targets 285
S. Hattori, K. Akimoto, Y. Ohnishi, S. Miura

Development of vision metrology combined with auxiliary observations for tunnel profile measurement 289
K. Akimoto, S. Hattori, S. Miura, Y. Ohnishi

Assessment of grouting effect in rockmass using borehole jack tests 293
G. Zhang, J. Chen, S. Wang, Y. Li

Rock stress measurement using the compact conical-ended borehole overcoring (CCBO) technique 297
Y. Ishiguro, Y. Obara, K. Sugawara

Blast vibration monitoring and control of twin tunnels with small spacing 303
C. Wu, H. Liu

The monitoring result and consideration of vertical closed tunnels 307
K. Umeda, S. Koyama, T. Omichi, H. Sakamoto, K. Okaichi

Study on the tunnel load generation mechanism of twin tunnel 311
M. Yoshinaga, H. Kasamatsu, K. Ogawa, S. Ito, S. Azetaka, H. Tezuka

Determination of the horizontal subgrade reaction coefficient for the backside 317
of shoring systems in clayey soil
H. Nakamura, H. Suzuki

Maintenance of tunnel structures

Detection of cracks on tunnel concrete lining with electric conductible paint 325
T. Okada, S. Konishi, T. Mohri, K. Tateyama

Soundness investigation of tunnel concrete by core sampling 329
H. Saito, T. Uebayashi, N. Tasoko

Comprehensive safety inspections of Sanyo Shinkansen tunnels 333
T. Kondo, M. Ichida

A new non-destructive testing method for crack and its application to tunnel structure 337
T. Nakamura, N. Kawamura, Y. Hattori, K. Egawa, J. Wu

Development of new grouting material for tunnel rehabilitation 343
T. Asakura, S. Kohno, T. Kiuchi

A fundamental study on dynamic response of tunnel reinforcement structure for high speed railway 347
T. Kameda, T. Nishioka

Damage to mountain tunnels by earthquake and its mechanism 351
T. Asakura, S. Matsuoka, K. Yashiro, Y. Shiba, T. Ôya

The tunnelling at the crush zone in the Sanbagawa Metamorphic Belt 357
Y. Yoshida

Maintenance of the undersea section of the Seikan tunnel 363
M. Ikuma

Design and construction of mountain tunnels

A proposal of new rock mass classification for tunneling 371
W. Akagi, T. Ito, H. Shiroma, A. Sano, M. Shinji, T. Nishi, K. Nakagawa

Rock mass classifications study with the neural network theory 379
P.G.C. Lins, T.B. Celestino

Rock support design for tunnelling in Nepal Himalayan region 385
(with a case study of Khimti I Hydropower Project, 60MW)
G.L. Shrestha

Assessment of susceptibility of rock bursting in tunnelling in hard rocks 391
Ö. Aydan, M. Geni°, T. Akagi, T. Kawamoto

Design of large scale road tunnel based on the behavior of discontinuous rock mass -Ritto Tunnel- 397
Y. Sakayama, H. Niida, T. Ohnishi, Y. Tanaka, S. Inosaka

Behavior of mountain tunnel with large cross section in earth ground 403
N. Tomisawa, T. Matsui, M. Hino

Simplified behavior models of tunnel faces supported by shotcrete and bolts 407
L. Cosciotti, A. Lembo Fazio, D. Boldini, A. Graziani

Simulation analysis on deformation of soft rockmass due to excavation of tunnel 413
W. Zhu, S. Li, S.C. Li, Y. Zhang, S. Wang, W. Chen

Behavior of tunnels built in sedimentary rocks with joints dominant in one direction 419
T. Koyama, S. Nanbu, Y. Suzuki, Y. Tasaka, K. Kudo

Research of observational method on the groundwater in the tunnel approach crossing 425
Y. Ohnishi, H. Ohtsu, H. Ishihara, N. Okamoto, T. Yasuda, K. Takahashi

Main considerations on UDEC modeling of tunnel excavations and supports 433
S.G. Chen, H.L. Ong, K.H. Tan, C.E. Tan

Analysis on tunnel lining deformation and effect of countermeasures for earth pressure 439
T. Asakura, Y. Kojima, K. Yashiro, H. Shiroma, K. Wakana

Theory and practice of tunnel lining design 445
N.S. Bulychev, N.N. Fotieva

Experimental study on static behavior of road tunnel lining 451
H. Mashimo, N. Isago, H. Shiroma, K. Baba

Applicability of steel fiber reinforced high-strength shotcrete to a squeezing tunnel 457
M. Hisatake, T. Shibuya

Use of electric gradient to increse shotcrete early age strength 463
D.A. Ferreira, T.B. Celestino

Influence of invert construction procedure on the deformation and internal force of tunnel lining 469
H. Zhu, X. Liu, B. Ye, J. Liu

Efficient TBM driving by observational construction management 475
H. Niida, F. Kamada, T. Nishizono, H. Shigenaga, S. Mutaguti, Y. Miyajima, Y. Sawamura

Study on the application of a large TBM to Hida highway tunnel 481
K. Miura, M. Kawakita, T. Yamada, N. Sano, K. Ryoke

Constructing a tunnel through the embankment of national road No.4 using the AGF method   487
M. Miwa, M. Ogasawara

Ground reinforcement for a tunnel in weathered soil layer beneath Han riverbed in Korea   493
T. Yang, J. Woo, S. Lee

Stability and water leakage of hard rock subsea tunnels   497
B. Nilsen, A. Plastrøm

Development of long face reinforcement method with GFRP tubes   503
Y. Mitarashi, T. Matsuo, T. Okamoto, T. Tsuji, T. Haba, T. Okabe

Study on effect and evaluation of auxiliary methods for mountain tunneling in weak ground   509
T. Fukui, T. Hirai, Y. Kawamura, S. Nishimura, Y. Mitarashi, S. Azetaka, H. Tezuka, T. Matsui

Numerical evaluation of environmental impact on surrounding ground water by tunnel excavation   515
Y. Ohnishi, H. Ohtsu, H. Okai, M. Saga, T. Nakai, K. Takahashi

Back analysis for the bolted gate by coupling method of FEM and BEM   521
Y. Tan, X. Han, T. Wang, Z. Ma

Application of back analysis in assessing the stability of an Indian tunnel   525
A. Swarup, S. Akutagawa

Predicting surface settlement at shallow tunnels using Gompertz's curve as a stress release curve   529
T. Suzuki, T. Domon, K. Nishimura

Deformation and stability analysis of rectangular tunnel in soft rock ground
using a strain softening type elasto-plastic model   533
T. Adachi, F. Oka, T. Kodaka, J. Takato

Study on tunnel design and construction method for New Tomei-Meishin Expressway   539
I. Suzuki, H. Shiroma, T. Ito, S. Kaise

Construction work of Yamba tunnel TBM test section   545
K. Takagi, M. Tsukada, Y. Watabe, H. Iizuka

External water pressure on lining of tunnels in mountain area   551
Y.T. Zhang

Slope failure at tunnel entrance due to excavation and its countermeasure   557
A. Yashima, A. Matsumoto, K. Tanabe

Design and construction of tunnel through counterweight fill   563
H.G. Lee, T. Suzuki, K. Ookubo

Causes of Primary Crusher Conveyor Tunnel Failure in Sar Cheshmeh Copper Mine in Iran   569
M.M. Toufigh, M.E. Mirabedini

Study on the construction and design method of bolting support in coal roadway   573
N. Zhang, J. Bai, J. Zhou, C. Min, S. Hai

Rapid excavation for small section tunnel using TBM   577
H. Namura, H. Imaoka, T. Takamichi, Y. Kobayashi

The use of neurofuzzy modeling for performance prediction of tunnel boring machines   583
P.A. Bruines

Construction of Kamosaka tunnel by the NARAI excavation system   589
H. Haga, N. Takahashi, S. Morishima, H. Kamiyama

Development of low noise and vibration tunneling methods using slot by single hole continuous drilling   593
T. Noma, T. Tsuchiya

Experimental study on rock cutting by use of actual size disc cutter with round tip   599
H. Takahashi, T. Sato, H. Yamanaka, K. Kaneko, K. Sugawara

Establishment of ventilation design method for new Tomei-Meishin Expressway Tunnels   605
T. Iwasaki, K. Takekuni, T. Otsu, M. Yamada

Modern Tunneling Science and Technology, Adachi et al (eds), © 2001 Swets & Zeitlinger, ISBN 90 2651 860 9

Preface

During the 20th Century, great progress was made in the science and the technology. It goes without saying that remarkable advances were also made in the civil engineering technology. Civil engineering technology has been in existence, in some form or another, since the beginning of the human race. For example, the Akashi Channel Bridge and the Tokyo Bay Aqua-Line Tunnel were constructed making full use of the latest technology and are said to fundamentally support human life. These structures have left an indelible mark on the history of the civil engineering in Japan. Among the incredible landmark tunnels of the 20th Century are the Tanna Tunnel and the Seikan Tunnel, both in Japan, and the Euro Tunnel, in Europe; all three are representative tunnel structures of the 20th Century. The Tanna Tunnel, constructed at the beginning of the 20th Century, is a railway tunnel which was excavated by overcoming collapses and flooding due to the high pressure of groundwater. The Seikan Tunnel, an undersea tunnel and the longest tunnel in the world, connects the main island of Honshu to the northern island of Hokkaido. Construction of the Seikan Tunnel was a great challenge due to the terrible natural conditions which were encountered. It took twenty-four years to complete this tunnel. The Euro Tunnel, also an undersea tunnel, made full use of the technology by employing the tunnel boring machine. Construction of the Euro Tunnel brought about remarkable progress in the application of this tunnel boring machine. Moreover, as the Euro tunnel connects the United Kingdom with France, it is not only a remarkable landmark in the history of civil engineering, but its construction was also an epoch-making event in the history of Europe. The Euro tunnel is thought to be one of the most historical and monumental works in the 20th Century. As I have mentioned, remarkable advances have been made in tunneling technology thanks to the wisdom and the efforts of our predecessors. It is the wisdom and the efforts of our predecessors that have laid the foundation for the present-day civil engineering technology.

Now, in the 21st Century, the human race seems to have reached a turning point in regards to many issues. And, there also seems to be a natural turning point in the field of civil engineering. We civil engineers must consider the appointed tasks in the present-day civil engineering technology which involve not conquering nature, but creating a symbiotic relationship between the human race and our precious natural environment. In the mountainous country of Japan, where we must deal with various natural landform conditions and the responsibility of protecting the environment, concern over the use of underground space is increasing. Following this concern is the actual increase in underground construction. For the sake of urban renewal and the protection of the environmental in urban areas, laws that contribute to the creation and the effective use of underground space have been enacted and are being enforced. The effective uses of underground space presents a remarkable contribution to environmental protection as well as to the sustenance of human life itself. Recently, for the same reasons, European countries have also begun to carry out the effective use of underground space.

Due to the above-mentioned circumstances, there is an ever-increasing need for tunnel construction throughout the world. In particular, for the preparation of progressive traffic networks, the demand for the constructions of long-distance and larger-faced tunnels in on the rise. A growing tendency in the construction of tunnels and the creation of underground space now involves counteracting the problem of chronic traffic congestion and maintaining a comfortable ride for all. As part of a way to achieve these goals in Japan, the construction of another new highway between Tokyo and Osaka (the second Tomei and Meishin Highways) has begun. Due to natural landform restrictions and the fact that urban areas must be avoided, the route for this project will pass through mountainous areas, and therefore, the ratio of tunnel and bridge structures will be higher than for other existing highways. We are obliged to construct longer and larger-faced tunnels than have ever been previously constructed. In Europe, a new railway route, the Alp Transit, will be constructed. With this new route, railway service between Zurich and Milan will be one hour shorter than with the present service. The tunnel construction for the Alp Transit will have to overcome a higher confining pressure than has ever been tested. An effective geological prognosis is the key to successfully completing the tunnel construction safely and speedily. It is thought that the higher European tunneling technology, attained though the experience of constructing the Euro Tunnel, will help lead to the great success of the tunnel construction for the Alp Transit. In Southeast Asia, on the other hand, problems related to the defective infrastructure and chronic traffic congestion is ever-present worries in over-populated urban area. A rapid transit system for mass transportation and as an effective lifeline, is an urgent topic for that area of the world. Therefore, it is thought that positive discussions should be held and actions should be taken to achieve the effective use of underground space.

The construction of tunnels and the use of underground space not only help to create and preserve the environment, but they also can protect human beings from disasters. My present hope is that we engineers can contribute to creating a 21st Century society which attempts to maintain a good balance between human life and nature. Almost 120 years ago, here in Kyoto, a canal project, the Lake Biwa – Kyoto Canal, was performed which helped lead to the modernization of Japan. The first tunnel, whose length was 2,436 m, was excavated almost entirely by manpower, as part of this canal project. Since there was no such thing as modern technology in the field of civil engineering in Japan at that time, it is thought that the construction of the canals and the tunnels for this project was very difficult. In his graduation thesis, Professor Sakuro Tanabe proposed to lead the water from Lake Biwa to Kyoto City, his proposal was accepted and realized. The construction of canals and tunnels from Lake Biwa to Kyoto brought a stable supply of water to the area and enabled the generation of electric power. As a result, factories were opened for operation and Kyoto was the first place in Japan to have electric train. Therefore, we can say that a new environment for human life was attained because of that canal project. The canals and their tunnels still work well today and have benefited human life in Kyoto over the years. Tying this information together, we realize that there is quite a historical significance in having Kyoto be the site of this international conference where we will discuss the science and technology of tunneling for the 21st Century.

At IS-Kyoto 2001, we will not only present our ideas on design and construction technologies for tunnels, but our discussions will also include topics related to planning, geological and environmental investigations, as well as the maintenance and the longevity of tunnels. Addressing such issues as how to plan for the use of underground space, how to create effective designs for tunnels and underground structures, and how to sustain those tunnels and underground structures are important keys to successful discussions here at IS-Kyoto 2001. It is my sincere wish that remarkable results will be came out of IS-Kyoto 2001, and that results will contribute to the advancement of tunneling technology in the 21st Century and will help create a foundation for a sustainable society between the human race and nature.

Toshihisa Adachi
Chairman of the Organizing Committee, IS-Kyoto 2001
Professor, Department of Civil Engineering, Kyoto University, Japan

October, 2001

Modern Tunneling Science and Technology, Adachi et al (eds), © 2001 Swets & Zeitlinger, ISBN 90 2651 860 9

ACKNOWLEDGEMENTS

MANUSCRIPT REVIEWERS

The editors are grateful to the following persons who helped to review the manuscripts and hence greatly assisted in improving the overall technical standard and presentation of the papers in this proceeding:

H. Akagi	Y. Ishizuka	T. Kyoya	N. Shimizu
S. Akutagawa	Y. Jiang	H. Mashimo	M. Sugimoto
Y. Ashida	K. Kagawa	T. Matsuo	J. Takemura
T. Esaki	K. Kimura	Y. Mitarashi	T. Tamano
R. Fukagawa	M. Kimura	T. Miyake	K. Tateyama
T. Hagiwara	A. Kobayashi	K. Nishimura	J. Tohda
T. Hanamura	T. Kobayashi	S. Nishio	T. Yamabe
T. Hashimoto	T. Kodaka	J. Ohtani	A. Yashima
K. Hibiya	K. Kojima	H. Ohtsu	N. Yoshida
M. Hisatake	H. Komine	S. Ohtsuka	F. Zhang
Y. Ichikawa	S. Konishi	F. Oka	
A. Iizuka	Y. Koyama	S. Sakajo	

Modern Tunneling Science and Technology, Adachi et al (eds), © 2001 Swets & Zeitlinger, ISBN 90 2651 860 9

ORGANIZING COMMITTEES

Prof. T. Adachi
Chairman

Dr. M. Kimura
Secretary

MEMBERS

T. Chishaki	O. Kusakabe	Y. Ohnishi
T. Hanamura	S. Kuwahara	F. Oka
T. Hashimoto	T. Matsuda	K. Ono
K. Hibiya	T. Matsui	S. Sakurai
Y. Ikeda	M. Matsuo	Y. Shinomiya
K. Kamemura	K. Miura	T. Tamura
H. Kawata	K. Miyaguchi	M. Tezuka
A. Koizumi	T. Mizutani	H. Tsuji
T. Konda	T. Nishioka	
H. Kurihara	H. Ochiai	

INTERNATIONAL ADVISORY COMMITTEE

A. S. Balasubramaniam, *Thailand*	A. Haak, *Germany*	K. Ono, *Japan*
W. E. Bamford, *Australia*	K. Kovari, *Switzerland*	L. Ribeiro e Sousa, *Portugal*
N. Barton, *Norway*	S. L. Lee, *Singapore*	S. Sakurai, *Japan*
T. L. Brekke, *U.S.A*	A. Verruijt, *Netherland*	S. Valliappan, *Australia*
E. Broch, *Norway*	S. R. Lee, *Korea*	W. Zhu, *China*
T. B. Celestino, *Brazil*	R. J. Mair, *U.K.*	
H. H. Einstein, *U.S.A*	Y. Ohnishi, *Japan*	

SUB-ORGANIZING COMMITTEE

S. Akutagawa	K. Kishida	K. Nishimura
T. Ashida	T. Kitamura	H. Otsu
M. Imai	T. Kodaka	K. Tateyama
M. Kazama	S. Konishi	
M. Kimura	M. Mimura	

Keynote Lectures

Modern Tunneling Science and Technology, Adachi et al (eds), © 2001 Swets & Zeitlinger, ISBN 90 2651 860 9

Safety Aspects of the Lötschberg Railway Tunnel & Mont-Blanc Road Tunnel

F. Vuilleumier
Dr. of Eng., Director, IGWS, Brigue, Switzerland

A. Weatherill, Y. Trottet & B. Crausaz
Bonnard & Gardel Consulting Engineers Ltd, Lausanne, Switzerland

ABSTRACT: After the serious accidents, which happened recently in tunnels, most of the countries have established "Task Forces" in order to evaluate the safety of existing tunnels and to establish new safety measures. Based on two actual examples, the new safety measures are presented in this paper on a practical view for the first and on a theoretical view for the second.

1 INTRODUCTION

The serious accidents, which happened in the Mont-Blanc and Tauern tunnels in 1999, have highlighted the intrinsic difficulties of road traffic in tunnels. Most of the countries who have tunnels as part of their national infrastructures have established special "Task Forces". These have decreed directives and instructions concerning traffic in road and rail tunnels.

The road works undertaken in the framework of renovation and improvement of the Mont-Blanc Tunnel (11.6 km long) are intended to ensure its safety and are presented in the first part of this paper. In the second part the safety aspects of the Lötschberg rail tunnel project are analysed (35 km long). Construction was begun in 1998 and will be completed sometime in 2007.

2 THE MONT-BLANC ROAD TUNNEL

2.1 *Presentation and Characteristics of the Tunnel*

The Mont-Blanc tunnel, which represents a major road artery between France and Italy, is situated under the Mont-Blanc massif, the roof of Western Europe. At 11.6 km, the Mont-Blanc tunnel was the longest road tunnel in the world at the time of its completion in 1965.

The traffic in this single tube, two lanes and relatively small cross section tunnel can be very heavy. The traffic is characterised by a great deal of asymmetry (in both directions) and a high percentage of heavy goods vehicles. The amount of light vehicle traffic has increased by a factor of 2 while the amount of heavy goods vehicles has increased by a

factor 17. These figures show the importance of the tunnel for trade between France and Italy.

Vehicles loaded with dangerous goods are not allowed to use the tunnel. As shown by the disastrous 1999 fire, this ban has not been enough to prevent the occurrence of major accidents.

2.2 *Reinstatement of the Mont-Blanc Road Tunnel*

Following the March 1999 incident and subsequent to the court inquiry, the objectives of reinstatement have been defined as the repair of the damages caused by the accident and the installation of fittings and equipment but foremost to establish a global concept, which will ensure the tunnel's safety.

In the case of an accident or some other emergency, the safety measures aim at achieving the following objectives:
- Detecting abnormal situations and ensuring communication with tunnel users
- Providing protection and evacuation routes for tunnel users as well as access to rescuers
- Helping tunnel users to protect themselves and to fight against fire

The equipment acquired and the arrangements that have been made are described hereafter.

2.2.1 *Pits and turning spaces*
In both directions a garage is situated every 600 m allowing heavy goods vehicles to stop. Also every 600 m a turning space allows maintenance and rescue vehicles to operate in the tunnel.

2.2.2 *Refuges, fire-fighting facilities and escape routes*
In order to cater for the victims of an accident or emergency into a safe place before being able to

evacuate them, the refuges are situated on one side of the tunnel only and are spaced at intervals of 300 m. Their lay-out has been designed so as to protect occupants from the direct atmosphere of the tunnel by means of an airlock, situated between the tunnel and the refuge. These refuges are ventilated through fresh air ducts and put under light overpressure thereby imposing an air flux, which flows from the refuge into the tunnel. They are equipped with telephones, closed circuit TV cameras and public address systems. Water supplies, which are regularly replenished, have been placed in the refuges. In case of a fire in front of the refuge, the temperature inside it should not exceed 35 °C.

Located at the centre of the tunnel, fire-fighting facilities have been built at a place, which once harboured a turning space. Fire-fighters are always present, thus reducing the time they need to be on hand in the event of any emergency. The fire-fighting facilities are equipped with computer network terminals, telephones, closed circuit TV, radio as well as one heavyweight, and one lightweight fire truck. They are directly connected to the evacuation shaft.

From outside the tunnel, it is possible for rescuers to gain access to the refuges and evacuate the victims. The escape route is situated in an underground ventilation shaft. It is wide enough for one person to pass through easily. In the event of a fire in which this escape route were used, ventilation would be reduced to a minimum. At each end of the shaft airlocks have been installed to permit evacuation at station level. At the refuge points, stairs have been built which provide access to each individual refuge. The escape route has lighting along the entirety of its length and is equipped with emergency telephones. A motorised evacuation vehicle has been provided to facilitate injured or disabled victims of an accident or emergency.

2.2.3 *Safety enclaves, anti-fire enclaves and network*

Within the tunnel, safety enclaves have been placed alternately at intervals of approximately 100m. They are equipped with an emergency telephone, fire extinguishers (with sensors that detect when they have been removed) and electric sockets for rescue services, They are also equipped with glass doors (with sensors to detect when they have been opened) and are clearly indicated by means of specific signs. The anti-fire network is made up of anti-fire enclaves located every 150 m with a emergency fire pillar every 300 m on the north skewback (direction Italy-France) at the left of the refuges.

2.2.4 *Ventilation*

2.2.4.1 Fresh air ventilation
Fresh air is uniformly distributed through each of the eight ventilation sections. Each of these fans, connected to its specific fresh air gallery, has been built to provide 82.5 m³/s at maximum power.

Sanitary ventilation should permit the air quality to be maintained at a certain level within the tunnel and should prevent the air at any point in the tunnel from becoming polluted past a specific level.

2.2.4.2 Fire ventilation
The fire ventilation system is a key element in the new safety concept of the tunnel and has been redesigned in order to achieve the two following objectives: massive smoke extraction in a limited length domain surrounding the position of a fire and active control of the longitudinal velocity (smoke propagation management).

These objectives are intended to allow tunnel users to ensure their own safety in facilitating recognition and attainment of refuges and, upon reaching the scene, allowing rescuers to better combat the fire.

The concept chosen tends towards that of a semi-transversal ventilation system. The extraction capacity of 150 m³/s (using 7 smoke flues) is achieved by:
- 2 ventilators for vitiated air at each end and one for emergencies
- 4 relaying ventilators in the straight part of the extraction conduit
- Smoke flues every 100 m. with registers.

Control of the longitudinal velocity of air extraction within the tunnel is achieved thanks to 76 accelerators placed along the tunnel's roof. Their purpose is to achieve a longitudinal velocity of zero at the centre of the fire, helping to keep the smoke stratified.

2.2.5 *Closed circuit TV monitors, automatic incident sensors, radio-communication system and heat sensors*

Closed circuit TV monitors allow all refuges and pits in the tunnel to be monitored, and are supported by cameras placed half way up the walls allowing surveillance even in case of a fire. 150 cameras have been placed in the tunnel (one every 150 m on each skewback) to allow complete and unbroken surveillance of each section of tunnel. Automatic incident sensors permit the detection of any kind of incident (deceleration, stopping in a tunnel section, driving in the wrong lane, over-taking, speeding, driving too close to other vehicles, pedestrians, fires, losing of a load etc.) in the tunnel and uses the images provided by the cameras in any section of the tunnel.

The radio-communication system allows rescue workers from different services to communicate with each other as well as informing tunnel users of the situation and giving them instructions in several languages (12 different FM channels).

The heat sensors detect any rise in temperature within the tunnel, along the ceiling, within the refuges and the pits. An algorithm calculates the presumed extent of the smoke.

2.2.6 *Electricity, check points and network*

All the electronic equipment in the tunnel complex can be operated by either of the two tunnel ends, both under normal circumstances and in case of fire. In the event that both tunnel end supplies should fail simultaneously (French and Italian power supply) twin redundant inverters would come on line and provide emergency power for one hour for safety functions such as lighting within de current section, within the refuges, within the escape route, the GTC, telephone network, closed-circuit TV monitor system, the sound installation, signals and smoke extractors.

At each end of the tunnel and at the fire fighting facility a checkpoint has been set up. Two dedicated emergency teams (one at each end) are permanently on call. The principle responsibility to intervene resides with one or the other team, with the second team providing back-up if needed. All the IT systems and network platforms within the tunnel are redundant.

3 THE LÖTSCHBERG RAIL BASE TUNNEL

3.1 *Presentation and Characteristics of the Tunnel*

3.1.1 *Switzerland, a country of transit*

Switzerland has always been an important transportation junction and transit country in the middle of Europe despite the obstacle imposed on it by the Alps. In these last thirty years, the transportation of goods between the North and the South of Europe has increased by a factor of six.

3.1.2 *The consequences in Switzerland*

With the overloaded transportation network (rail and road), the environment is badly affected by noise and exhaust fumes. The quality of life and road safety are decreasing along the major highways. The saturation limits have long been exceeded.

3.1.3 *The AlpTransit*

Switzerland has been able to convince the European Community of the need to pursue a coherent transportation policy. With the Transit Agreements, the combined rail and road traffic also includes Europe. The role of the Alps is preserved for tourism and for the protection of the environment, because trains do not emit so many noxious substances. AlpTransit, the new railway link, will make the Alps easily transversable. This project is based on four constituent elements:
- The key element will be the new and additional railway route Arth-Goldau – Lugano with the Gotthard (57 km) and Monte-Ceneri (13 km) base tunnels
- The base railway route from Frutigen to the Rhône valley (Lötschberg Tunnel, 34.6 km) will

complete the picture in order to avoid a concentration of traffic on the Gotthard
- The Simplon route connected to the French TGV (Macon - Geneva) will connect the French part of Switzerland to the northern part of Italy and France
- The traffic routes east of Switzerland will be improved (Zurich – St-Gall)

3.1.4 *The Lötschberg Base Tunnel*

The Lötschberg base tunnel will be built to accept different classes of train, from goods transportation to passenger trains. The tunnel will also accommodate high speed trains and therefore will be designed for a maximal speed of 250 km/h.

Basically the tunnel is a two-tube tunnel comprising a rail tunnel East and West. But in the first phase the two-way tunnel will be open from Raron (South Portal) to the north of the emergency station in Ferden. The remaining part of the tunnel northwards will be a one-way tunnel as far as train operation is concerned (except at the North Portal in Frutigen).

3.2 *Safety Concept of the Tunnel*

Despite railways being a statistically safe form of transportation (due to guidance per rails, professional drivers, etc.) rail accidents may still happen. Even if accidents occur less often in tunnels (e.g. no level crossings, which cause most of the train-related accidents) the severity of the accidents increase seriously due to the tunnel configuration and the importance attributed to such accidents by the media. In Switzerland, the last accidents with fatalities occurred in 1971 and in 1932. The safety concept of the Lötschberg tunnel should bring a substantial improvement compared with the existing lines of the rail network. With a partial two-tube tunnel, access via two galleries, as well as a window gallery and reconnaissance gallery, an essential improvement in rail tunnel safety should be achieved.

3.2.1 *Safety Philosophy*

Safety is considered basically as the characteristic of the quality, which has to be guaranteed throughout its life period. It is based on the following four elements:
- Protection objectives
- Danger analysis and evaluation of risk
- Safety measures
- Transposition
  The first objective laid down is the "protection of the endangered person" i.e. passengers, personal, etc.

The second objective is the "protection of the natural environment, the constructions and technical installations and their use".

These objectives are considered to be obtained when:

- Principle 1: all necessary measures have been met in relation to the potential danger.
- Principle 2: all necessary measures have been met in relation to the technical and scientific standards which are applicable and the relevant circumstances.

The operator of the railway is responsible for its installations and the safety measures. In relation to the opening of the railway network for "free access", international jurisdiction takes on new importance. Protection objectives and safety measures play a very important role to the operator.

3.2.2 Methodology of the Risk Assessment

As part of the complex process of erecting a long railway tunnel, a safety analysis and concept serve the future operator to guaranty a high level of safety. The procedure consists of two steps:

1. A Qualitative Safety Analysis in order to limit the range of accident scenarios and related concepts, and to identify appropriate safety measures for the tunnel or its sections. It should also define the appropriate laws and applicable national and international norms and regulations.
2. A Quantitative Risk Analysis is used on the relevant hazard scenarios derived from step 1 only where this previous step was not able to define appropriate measures. Using international and domestic statistical data (available in Switzerland from the CFF and the BLS) and considering the effects of the Safety measures, each event is quantified with the frequency of its occurrence and the extent of its consequences.

In order to better define the remaining risks after having taken all essential preventive and curative measures, (so-called "residual risk"), the representation used is the frequency-consequences diagram, which allows for optimisation of the safety measures by means of applying specific assessment criteria and evaluating the acceptability of the risk in order to adjust said assessment criteria.

The diagram is subdivided into three ranges:
- Overall railway risk: unacceptable for new railway lines
- ALARP (As Low As Reasonably Practicable) Criteria: overall railway risk
  Irrelevant overall railway risk

In the area limited by the unacceptable risk and the irrelevant risk, the risk is to be reduced as far as is technically and operationally possible and should be contained within reasonable limits.

The risk assessment has been carried out taking the following critical events into consideration:
- Industrial accidents
- Accidents involving injury (passengers, others)
- Fire (all type of trains)
- Derailments
- Collisions

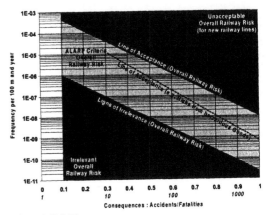

Figure 1. F-C Diagram.

- Losses of hazardous goods
- Operation failures
- Terrorism / violence

3.2.3 Objectives

From the results of the risk analysis, several objectives have been identified. A new tunnel cannot be designed without adopting the following measures:
- Measures to prevents incidents (appropriate conveys, railway in good order)
- Measures to reduce fatalities and damages
- Measures to improve self-rescue opportunities
- Measures to improve the possibilities of external rescue

The measures will also affect the rolling materials, but here, as the tunnel will be open to free access, requirements on materials aboard trains cannot be guaranteed. Due to the length of the tunnel and the difficulties involved for rescuers to get to the scene of the accident, self-rescue measures are essential during the initial phase. The safety measures should also ensure that rescuers are able to accomplish their duties.

3.2.4 Self-rescue

The event sequences, which have occurred in the past in tunnels, show that a great priority has to be attributed to self-rescue measures, particularly during fire. Given the great velocity at which a fire spreads and due to the high temperature, reduction of oxygen concentration, loss of visibility and propagation of toxic fumes, those persons who are able to escape rapidly have a real chance of getting themselves to safety on their own.

Therefore, the following aspects have to be considered:
- Length of escape paths and indicators thereof
- Equipment along the escape route (lighting, ventilation)
- Communication infrastructure (telephone, radio, etc.)

6

3.2.5 *Rescue by Third Party*

In order to facilitate the job of external rescuers, some measures have to be considered such as:

- Accessibility of tunnel portal
- Accessibility to the scene of the accident (extraction of and fresh air supply)
- Rapidity of intervention (training, planing, etc.)

3.2.6 *Service and Emergency Stations*

The safety measures will prevent a train already on fire from entering into the tunnel. Should this occur or in the event that a train should catch fire in tunnel, experience has revealed that the train will continue to roll a considerable distance (tens of kilometres). Therefore, it is necessary to reliably estimate the location in the tunnel where a train on fire will come to a standstill. Two stations, the service station Mitholz (which can later be transformed into to an emergency station) and the emergency station Ferden situated at about 20 km from the opposite portal, will greatly facilitate self-rescue, evacuation and intervention of the rescuers. Of course, it stands to reason that there can be no certainty that trains will be able to stop at these stations. It is therefore necessary that any safety concept include scenarios for a train stopping along the entire length of the tunnel.

The service station Mitholz will serve as an evacuation station in the eventuality that a train on fire stops in the station. A flight of steps the length of the station (about 440 m.) will cater for the victims of an emergency, who will then move to the reconnaissance gallery Kandertal. From there, they will pass through the locks and be evacuated by buses

Ventilation will be raised to its maximum capacity in order to ensure a safe environment for the self-rescue of the passengers from the service station to the reconnaissance gallery.

The emergency station Ferden is made up of two rail stations, one per tunnel (East and West) with a length of 450 m. A flight of steps will enable the victims of an accident to use one of the six escape routes spaced at intervals of 85 m and which lead to a protected and ventilated zone between the two tunnels. Any injured travellers can be picked up by ambulance, which can enter the tunnel from the Ferden dip gallery and reach this zone via the lock under the ventilation station of Ferden. The evacuation of any victims will follow by train via the safe tunnel.

The emergency phase ventilation will provide an air flow from the ventilation station Ferden in order to provide fresh air to the escape zone and the safe tunnel as well as to the tunnel where the train is on fire by passing through the escape routes. Any victims fleeing the scene will thus have a safe escape route free of fumes.

Fumes are extracted via seven dampers, one of which will be opened nearest to the position of the fire, so as to ensure optimal extraction of any fumes as well as maximum visibility in the tunnel. The fumes are extracted by the ventilation station Fystertellä through the ventilation shaft.

3.2.7 *Safety Concept in case of Fire Outside (South) of the Emergency Station*

In case of a burning train which has come to a stop south of the emergency station (i.e. in the twin tunnel), the evacuation of passengers shall occur through communication branches spaced every 333 m. nearest the location where the train has stopped. Another train will take the victims to the exterior of the tunnel. The ventilation system will go into emergency mode, the station Ferden will blow its maximum air capacity at the same time that the station Fystertellä will begin extracting at its maximum air capacity. This should create a pressure difference between the safety zone and the burning tunnel in order to prevent the fumes from hampering any escape efforts through the communication branches.

Coherent planing is needed in order to provide for the safe passage of trains, to allow the rescue train access to the tunnel and to prevent any other trains from entering the tunnel during an emergency. Planning is also required in order to organise rescue operations.

4 CONCLUSION

Two tunnels, both differing strongly from one another (one for road traffic and in the process of being renovated, the other for rail traffic and currently being constructed) have been the subject of study. Despite their differences, numerous similarities exist:

- The consequences of what might at first appear to be an incident of a relatively harmless nature can rapidly take on dramatic proportions
- In the case of fire, the most dreaded incident, the smoke and heat can rapidly cause the scene and the surrounding area to become a potentially fatal place
- The problems encountered when evacuating tunnel users and getting rescue workers to the scene of the accident are very similar

In this last case scenario, the time it takes rescue workers to get to the scene is often longer than life expectancy in an emergency situation. Victims of an accident or emergency must be able, in an initial phase, to take charge of their own rescue (self-rescue).

The route along which evacuation is to take place or along which rescue workers are to be dispatched must not be located in the affected tunnel. A shaft or other access route must provide access to the scene of the accident as well as allowing victims to be evacuated.

Tunnel workers and fire-fighters must be subject to regular drills, which must be as realistic as possible. Such exercises are designed to test the workings of the tunnel and their use by tunnel workers. They allow one to recognise shortcomings and undertake corrective action in order to solve them.

Past experience and experiences currently being acquired in the area of tunnel security aim at constructing or improving tunnel infrastructure in order to:

– Detect abnormal situations and rapidly inform tunnel workers and users of any danger
– Provide protection and facilitate evacuation of tunnel users and access of rescue workers
– Optimise preparations in view of the possibility of fire

The tunnels studied put into practice the maximum number of measures for obtaining the highest security objectives. Nevertheless the risk factor will never be reduced to zero even though we must do everything in our power to reduce the risks to the greatest extent possible.

Modern Tunneling Science and Technology, Adachi et al (eds), © 2001 Swets & Zeitlinger, ISBN 90 2651 860 9

On the fundamental mechanism of tunneling

Takeshi Tamura
Kyoto University, Kyoto, Japan

ABSTRACT: There have been developed a lot of methods which are quite useful for the excavation of tunnels in various types of ground conditions. A typical one is the so-called NATM for the mountainous tunnels where shotcreting and rockbolting are considered to be the main support against the earth pressure around tunnels. It is, however, noted that no understandable theories for it are found out to explain the performance of the interaction between the tunnel and the surrounding ground from the view point of applied mechanics. A trial will be made in this report to investigate the fundamental mechanism of the mountainous tunnel excavation by considering results of simple experiments and numerical calculation.

1 INTRODUCTION

The underground structures such as tunnels are kept in a stable state under the mechanical interaction with the surrounding ground. A relatively thin structure seems to support the heavy overburden in the deep mountainous tunnel as well as in the urban tunnel. This can be interpreted by the fact that a suitable distribution of earth pressure appears around tunnels as the result of the above interaction but the quantitative evaluation of such an earth pressure is quite difficult since it is very sensitive to the mutual displacement of the tunnel structure and the ground. A full understanding and an effective utilization of this interaction, however, lead to a rational design and a safe construction of tunnels. Therefore it is essential to clarify the mechanism of the earth pressure generation of the underground structure when their stability is discussed.

Kovari(1995) criticized the concept of the NATM by pointing out that the definition of the NATM and its terminology are quite ambiguous and that the fundamental ideas advocated by the founders of the NATM are nothing more than those in the old and traditional tunneling methods. His criticism is partly to the point since no model of the NATM was proposed by its founders to explain the mechanical performance of the NATM. Before the concept of the NATM was popular in the tunnel engineers, however, Peck(1963) explained clearly the effect of the interaction between the ground and the tunnel lining by showing a simple imaginary experiment. He stressed that a flexible lining plays an important role to make the stress distribution around the tunnel almost isotropic by being deformed according to the initial distribution in the ground. Peck's idea is fundamental and essential to the study of the modern tunneling. The whole content of this report follows Peck's interpretation and tries to discuss about it more concretely and quantitatively by mentioning the results of experiments and numerical analyses.

From this point of view, firstly in this report, two major issues of tunneling are introduced through a model experiment of tunnel excavation, secondly the trap door experiment and the loading experiment on a thin paper shell are illustrated to explain the above two issues and finally simple numerical analyses are done to confirm them.

2 NATM

The term of the NATM is popular in some countries such as Japan but it is neither authorized nor acknowledged worldwidely. This is, for instance, because the definition itself of the NATM is ambiguous as Kovari pointed out. Hereinafter in the present report, however, the NATM is defined tentatively to be a tunneling method where the shotcreting is used as the primary support followed by the concrete lining. The width of shotcrete lining, in general, is some 25cm without steel reinforcement which is relatively flexible compared with some 50cm thickness of concrete lining Therefore the tunnel support structure in the NATM is

Photo 1 Tunnel excavation experiment

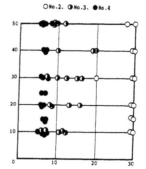

Figure 1 Overburden v.s. pull-out distance

	thickness (mm)
No.1	0.180
No.2	0.058
No.3	0.050
No.4	0.045

usually considered to be both of the shotcreting and the concrete lining as well as the surrounding ground itself. The shotcreting is done just after the tunnel excavation is completed and it is accompanied by rockbolting in most cases. In the current method of tunnel engineering, the shotcrete support and the concrete lining are designed independently. Namely, the shotcrete is designed to carry all of loads which are yielded by the tunnel excavation while the concrete lining is designed to carry the other loads which appear after the excavation. The loads to the concrete lining sometimes are assumed to be similar ones to those of the shotcreting with additional loads such as the underground water pressure. No interaction between the shotcrete support and the concrete lining is considered. In other words, the concrete lining seems designed to twofold guarantee the safety of the tunnel structure. Actually no forces are measured after the completion of the concrete lining in many cases.

It is obvious that the design of the shotcrete lining is considered to be the most important in the NATM when the tunnel stability at excavation is discussed.

3 MODEL EXPERIMENT OF TUNNEL EX-CAVATION

Adachi et al (1985) performed a model experiment of tunnel excavation in the sand ground as shown in Photo 1. A steel frame box (100cm × 50cm × 30cm) is used to contain the model ground. After depositing in it the lower sand layer up to around the half level of the box height as shown in Photo 1, a brass tube of 20cm in diameter wrapped with a sheet of thin (tracing) paper is set by fixing the both ends to the front (akril) and back(steel) walls. (In Photo 1, a number of paper strips are seen to be attached which model the rockbolting. But the cases without the rockbolting are investigated here.) The upper half of sand layer is further filled

up to the prescribed level corresponding with the overburden height. This is the initial state of the experiment of tunnel excavation where a thin paper shell prelining is supported by a brass tube to prevent it from collapsing.

The excavation process is simulated by pulling out the brass tube horizontally from the model ground by leaving a thin paper of shell structure in the sand ground. If the brass tube would come out completely of the 30cm thickness of sand ground without collapsing, a tunnel supported by a paper shell is said to be kept standing. It is evident that the sand ground can be stable with the aid of the paper shell since sand used in the experiment is totally in a dry state. A series of experiments were done with various types of tracing paper and different overburden heights.

The results of experiments are shown in Fig.1 where the lateral axis means the pull-out distance of the brass tube before collapsing and the vertical axis means the overburden height of covering sand, respectively. It is understood from Fig.1 that the sand ground collapses in the process of excavation if too thin tracing paper is adopted. On the other hand, it is remarkable that the brass tube can be pulled out completely through the 30cm thickness of the sand ground when a slightly thick tracing paper is used. It is still possible to completely pull out the brass tube even for the cases of the low overburden heights if a little thicker paper is used. The fact that the thin paper shell with almost no bending rigidity can bear the earth pressure of the sand ground gives us a hint to consider the stability of the mountainous tunnel supported by the shotcreting. In the following sections, adequate view points for the above experiment of tunnel excavation are introduced to discuss the mechanism of tunnel stability.

4   TWO MAJOR ISSUES OF TUNNELING

The tunnel excavation experiment introduced in the previous section shows us several important ideas to study the mechanical performance of the NATM. The following two major points will be discussed.

1)Loosened earth pressure

2)Characteristics of cylindrical flexible shell

The former is concerned with the magnitude of the vertical earth pressure on the tunnel. The vertical earth pressure in the initial state before the tunnel excavation is usually proportional to the depth of the tunnel. But its magnitude after tunnel excavation is much smaller than the initial value. This is the so-called loosened earth pressure. The quantitative evaluation of the loosened earth pressure was given by Terzaghi(1943).

The latter is concerned with the bearing capacity of the cylindrical thin shell which is subject to the normal load along its generator. The key with this is the interaction between the shell and the surrounding medium. The thick shell has a larger bearing capacity than the thin one when they are standing alone. But when they are surrounded by another medium, the thin shell may have a larger bearing capacity due to the mutual interaction.

The above two issues will be discussed in detail in 5 and 6, respectively, by showing the results of the experiment and the numerical simulation.

5   LOOSENED EARTH PRESSURE

The earth pressure on the tunnel structure is firstly considered. One of peculiar conditions of the underground structure is that the magnitude of external force is unknown in advance. This distinguishes it from the so-called upper structures such as bridges. In the case of bridges, both of static loading and live loading applied to them are definitely given quantitatively. But most of forces of the underground structure are difficult to be given. For instance, even the initial state of stress in the ground before construction is not clear. The vertical stress in the shallow ground can be well estimated by the unit weight of the ground and the depth. On the other hand, the lateral stress is not easy to be predicted than the vertical one. One of the practical ways for the evaluation of the lateral stress is to assume the coefficient of the lateral earth pressure which is usually set to be around 0.5. In the deep ground, the vertical stress may be smaller than the overburden and the lateral stress is sometimes larger than the vertical one.

Hence the first difficulty in the tunnel engineering is that almost no exact information on the initial stress is available. What is more complicated

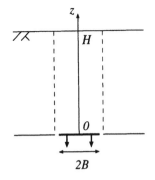

Figure 2 Terzaghi's theory

is that the initial stress is not directly related with the loading forces on the tunnel structure in general. If no displacement occurs in the ground by assuming the ideal tunnel excavation without disturbing the ground and the sufficiently large rigidity of the tunnel structure, then the initial stress is considered to be converted directly into the loading force on the tunnel structure. The actual ground as well as the tunnel structure, however, deforms mutually to change the interactive force between them. This seems to make complicated the evaluation of loading magnitude on the tunnel.

5.1   Terzaghi's theory

Fortunately there is a very convenient theory to predict the magnitude of the vertical load accompanied by tunnel excavation. This is called the theory of the loosened earth pressure on the trap door proposed by Terzaghi(1943) who assumes that the loosened ground region around the tunnel is formed to reduce the magnitude of the vertical stress to a remarkable extent. Referring to Fig.2, Terzaghi's formula is written as:

$$\overline{\sigma_z(z)} = \frac{\gamma B}{K_0 \tan \phi} \left[1 - \exp \left\{ \frac{K_0 \tan \phi}{B} (z - H) \right\} \right] \tag{1}$$

where $\overline{\sigma_z(z)}$ is the average value of loosened vertical stress at the level z, γ is the unit weight of ground material, B is the half width of the trap door, K_0 is the earth pressure coefficient, ϕ is the internal friction angle and H is the overburden height, respectively. It is noted that K_0 is not identical to the usual coefficient of lateral earth pressure which is used to evaluate the horizontal stress in the initial state. K_0 in Eq.(1) is the ratio of the magnitude of the horizontal stress to the average value of loosened vertical stress $\overline{\sigma_z(z)}$ in the loosened state after some amount of downward displacement of the trap door is given. In this loosened state, the column region of soil above the trap door goes downward and the soil of its both sides is going to topple laterally to the column region. This deformation mode forms a kind of flow caused by the downward displacement of the trap door. As a result, the lateral earth pressure on the trap door will increase to some extent. Therefore K_0 must be larger than that in the initial state. Actually Terzaghi recommends K_0 to be set equal to be around 1.

Eq.(1) can be obtained as the solution of a simple differential equation of equilibrium of the forces which is derived assuming the vertical slip lines passing through the both ends of the trap door as shown in Fig.2. The average magnitude of the vertical stress on the trap door is represented by

$$\overline{\sigma_z(0)} = \frac{\gamma B}{K_0 \tan \phi} \left[1 - \exp \left\{ \frac{-K_0 H \tan \phi}{B} \right\} \right] \quad (2)$$

since the trap door is located at the level $z = 0$. It monotonously increases with the overburden height H and approaches the following limit value σ_f when H becomes large enough.

$$\sigma_f = \frac{\gamma B}{K_0 \tan \phi} \quad (3)$$

It is interesting to note that the value of $B/(K_0 \tan \phi)$ is less than $2B$ when $K_0 = 1$ and $\phi = \pi/6$. This means that σ_f is smaller than the overburden stress corresponding to $2B$. More precisely, the total loosened load on the trap door is less than the self weight of ground material of the square region as shown in Fig.3.

Fig.4 shows the value of the normalized loosened stress $\overline{\sigma_z(0)}/\sigma_f$ on the trap door as the function of the normalized overburden height $H/(2B)$ when $\phi = \pi/6$ and $K_0 = 1$. The thin line in Fig.4 denotes the initial vertical stress which linearly increases with the overburden height H. When the normalized overburden height $H/(2B)$ is less than 0.2, the loosened vertical stress on the trap door is almost the same as the initial vertical stress. On the contrary, $\overline{\sigma_z(0)}$ is very close to its limit value σ_f when $H/(2B)$ is larger than 3.

The essential feature of the loosened earth pressure is as follows. The earth pressure on the trap door in the relatively deep ground is sharply decreased from the initial overburden stress down to the value represented by Eq.(2) when the trap door is lowered. Therefore even when a tunnel is excavated in the deep ground, the earth pressure on it is around σ_f if the tunnel excavation process is modeled similar to the trap door phenomenon.

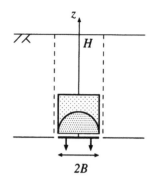

Figure 3 Load on the trap door

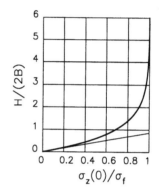

Figure 4 Loosened earth pressure on the trap door v.s. overburden

This reduction of the earth pressure is a consequence of the generation of the pressure relief zone around the tunnel which is easy to be formed when the tunnel is deeply located in the mountain. This corresponds to the experimental observation in 3 that the brass tube can be pulled out longer in the deeper deposit case. The interaction between the ground and the support plays an important role to produce the pressure relief zone.

It is usually said that Eq.(2) slightly overestimates the loosened earth pressure measured by the trap door experiment or calculated by the numerical analysis. The loosened earth pressure is easily reproduced in the laboratory by using the trap door experiment which will be explained below.

5.2 Trap door experiment

Adachi et al(1994) performed the trap door experiment by using the artificial ground of aluminum bars. Fig.5 shows the schematic view of experimental setting. The upper part of frame is filled with a number of aluminum bars of which radii

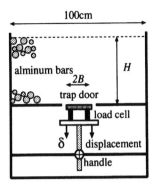

Figure 5 Trap door experiment

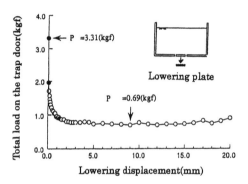

Figure 6 Displacement v.s. trap door pressure

The abscissa denotes the vertical displacement of the rigid plate at the bottom while the ordinate denotes the total earth pressure. The earth pressure on the trap door will suddenly change to the almost steady value. The critical earth pressure in the present paper is defined as the minimum value since such a definition is considered to be the simplest.

Fig.7 shows the total results of experiments and analyses. This figure includes the solutions by Terzaghi(1943) and by the rigid plastic finite element method(Tamura et al, 1984, 1987, 1990) with the associated and non-associated flows. It can be understood that the magnitude of the loosened earth pressure is almost proportional to the depth of the trap door location when it is located in the shallow ground. On the contrary, it converges to some value when it is located in the deep ground. It would appear that Terzaghi's solution Eq.(2) predicts a slightly larger values than those obtained in the experiment and the numerical analysis. Therefore the total load on the trap door is better estimated to be almost the same as the self-weight of

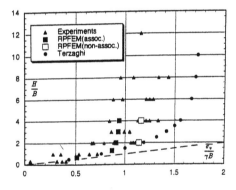

Figure 7 Results of trap door experiment

are 1.6mm or 3mm with 100mm of length. Their mixing ratio in weight is 3:2. Each bar is deposited horizontally in parallel to the viewing direction. The middle part of the bottom of the upper frame is made of a separated rigid plate of the length $2B(=10\text{cm})$ which is movable downward by controlling the handle through the screw gear. The initial values of the earth (i.e., aluminum) pressure just after deposited were almost the same as the overburden pressure which is calculated dividing the total weight of aluminum bars by the horizontal length of the frame (100cm). Lowering the rigid plate from the initial level makes the pressure on it decreased in the loosened state. The total earth pressure P on the rigid plate was measured at steps of the vertical displacement δ by a couple of load cells located below of the rigid plate. Experiments were performed for the cases of several values of the height H of the deposit.

Fig.6 illustrates typical data measured in the case of $H=30\text{cm}$ when the trap door is lowered.

the ground in the shape of a semi-circle of diameter 2B as shown in Fig.3.

6 CHARACTERISTICS OF CYLINDRICAL FLEXIBLE SHELL

The importance of the flexibility of the support or the shotcrete lining is discussed here. The thin paper shell in the tunnel excavation experiment mentioned in 3 can deform easily if some amount of load is applied to it. Such a flexible structure is classified to be unstable when it is subject to the non-uniform load. If the structural stability is required, a much more rigid material should be selected. As shown in the tunnel excavation experiment, however, the thin paper shell is proved to play a role of support. This seems to lead a contradiction concerning the flexibility of support. Therefore a fundamental question is arisen on whether the tunnel support should be rigid or flexible. The answer to this question is also found in consideration on the mechanical interaction between the support structure and the surrounding ground.

The ring model is often used to explain the role of shotcrete in the NATM. In most cases, it is discussed as the one-dimensional (axi-symmetric) model. But the two-dimensional consideration is truly required to understand the meaning of flexibility of thin paper shell since it contracts in the vertical direction and extends in the horizontal direction under the vertically dominant loading as shown in Fig.8. The ratio of contraction to extension of the flexible circular ring is almost 1. It is essential to note that the tunnel support stays in the ground and behaves interactively. This extension is the key to understand the stability of

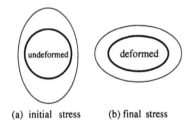

(a) initial stress   (b) final stress

Figure 8 Stress distributions around support

the tunnel support since it arises the passive earth pressure laterally to the tunnel structure. Particularly in the case of dense sand, a sufficient amount of passive earth pressure can be given to the tunnel structure. At the same time, the support deforms in the vertical direction in such a manner that the earth pressure is reduced down to the active earth pressure which is equivalent to the loosened earth pressure. As denoted just above, the flexible support in the ground will be subject to the active pressure in the vertical direction and to the passive earth pressure in the horizontal direction. This interaction between the support and the ground makes finally the stress state around the tunnel to be almost isotropic which causes a large amount of axial force with a small amount of bending moment. Hence the thin support structure can be in the stable state without collapsing.

What is the most important in this point is the flexibility of the tunnel support. If it is totally rigid, it bears the whole loading by itself without any deformation and the stress state in the ground is kept to be near the initial state. But an enough amount of thickness is necessary to make the magnitude of bending stress less than allowable value in this case. On the other hand if it is flexible enough, it deform itself to make the vertical stress loosened and the stress state around the support becomes almost isotropic which causes almost no bending moment. It results in the large bearing capacity of the thin flexible shell in the sand ground. The fundamental idea explained above was firstly proposed by Peck(1969) who considered an imaginary experiment similar to Fig.8.

6.1 Mechanical behavior of this paper shell

A very simple experiment is performed by Morinaga(1998) to see the mechanical behavior of the paper shell which is one of the models of the thin tunnel support. The paper used here is that for usual photocopy. Photos 2 and 3 show the results of a series of loading on the thin paper shell of 10cm in diameter.

The paper shell in Photo 2 has only two bottom-supporting points. On the other hand, the paper shell in Photo 3 has two side-supporting points in addition to the two bottom-supporting points. On the top on each circular shell, 5 of 1 yen Japanese coins(=1g each) are successively piled up. The deformations in Photo 2 are much larger than that in Photo 3. After 5 coins are put on, the paper shell without the side supports shows so large deformation that a negative curvature occurs at the top of the shell. But no large deformation can be observed in Photo 3 even if 5 coins are loading.

It is apparent that the difference between them does not come from the comparison of bending rigidity but comes from the effect of two side supports since they give the reaction forces when the paper shell is going to extend laterally. This reaction forces reduce the magnitude of deformation and the bending moment of the paper shell. Although the situations of the paper shells in this experiment and the tunnel excavation experiment mentioned in 2 are slightly different, essentially the same mechanism occurs in both cases since the vertical contraction yields the lateral reaction forces to reduce the bending moment in the paper shell. The difference between two sets of photos appears clearer if the bending rigidity is smaller. This means that the effectiveness of the lateral reaction forces is remarkable for thinner papers.

There are small pictures under the photos. They illustrate the results of a simple numerical simulation whose details will be explained in the next section. If each pair of photo and figure is examined, the accuracy of the simulation method in this study is understood.

6.2 Numerical simulation

Fig.9 shows the model of the numerical simulation where the line element is assumed to be rigid whose ends are connected through the bending spring (Tamura et al, 1998). In this figure, the number of elements $m(= 2m')$ is seen to be 4 but $40(m = 20, m' = 10)$ of equal-length elements arranged in a semi-circular form are used in the analysis. Since the line element are treated undeformable, no change in length of each bar occurs under the axial force likely to the beam model in structural mechanics. Therefore all of deformation in this model is attributed to the bending at the joints.

The bending spring constant K_θ is determined so that the deflection of the beam center in Fig.10 be the same as that in structural mechanics. At

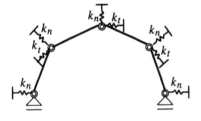

Figure 9 Simulation model

first, the deflection δ at the center of the straight beam of length $L(= ml)$ is obtained when a concentrated load P is applied. The bending moment M_j at the $j(\le m' + 1)$-th joint from the left end is

$$M_j = \frac{PL}{4} \times \frac{j-1}{m'}. \tag{4}$$

Hence the total complementary energy stored in this structure amounts to

$$U = \frac{1}{2K_\theta} \left\{ 2 \times \sum_{j=1}^{m'} M_j^2 + M_{m'}^2 \right\}$$

$$= \frac{1}{2K_\theta} \times \left(\frac{PL}{4}\right)^2 \times \frac{2m'^2 + 1}{3m'}. \tag{5}$$

The deflection δ at the center of the beam is calculated by using the Castigliano's theorem as follows.

$$\delta = \frac{dU}{dP} = \frac{m^2 + 2}{48m} \frac{PL^2}{K_\theta} \tag{6}$$

On the other hand, the solution by structural mechanics is $PL^3/48EI$ where E is the Young's modulus, I is the second moment of the section, respectively. Assuming the beam has the height t and a unit width in the direction perpendicular to the figure, $I = t^3/12$ is obtained. Comparing the both solutions, the following relationship between the two models is derived.

$$K_\theta = \frac{m^2 + 2}{12m} \frac{Et^3}{L} = \frac{1}{12} \left(1 + \frac{2}{m^2}\right) \frac{Et^3}{l} \tag{7}$$

When m is as large as 20 for instance, K_θ is approximated as

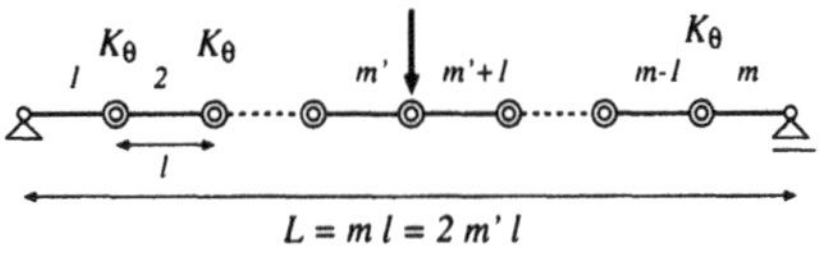

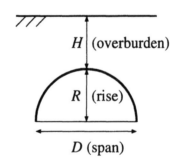

Figure 10 Calculation of K_θ

H (overburden)

R (rise)

D (span)

Figure 11 H, R ande D for analysis

$$K_\theta = \frac{Et^3}{12l} \tag{8}$$

The numerical analysis was done in the previous section by using this model. The tunnel excavation simulation is also numerically calculated by using this model attached by the ground springs to take into consideration the ground reaction. For simplicity, the same normal spring constant for the ground is assumed for both of contraction and ex-

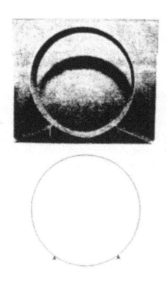

(a) initial state

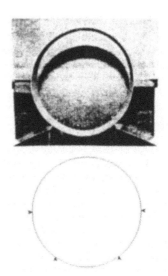

(a) initial state

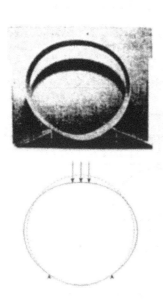

(b) 1g of loading

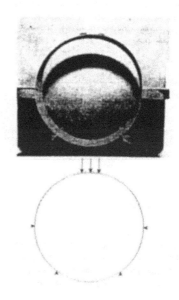

(b) 1g of loading

Photo 2 Paper shell with 2 supports

Photo 3 Paper shell with 4 supports

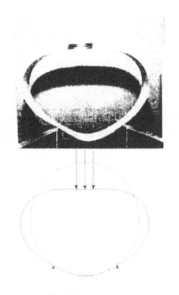

(c) 3g of loading

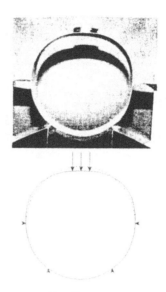

(c) 3g of loading

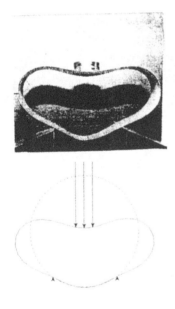

(d) 5g of loading

(d) 5g of loading

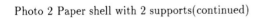

Photo 2 Paper shell with 2 supports(continued)

Photo 3 Paper shell with 4 supports(continued)

tension while the tangential spring has only one third of the normal spring constant. The vertical loading corresponding to the loosened earth pressure which is assumed here to be equivalent to H/D=1 and the lateral loading derived by the coefficient lateral earth pressure are applied to this model to simulate the tunnel excavation. The parameters in the calculation are as follows:

1)Tunnel diameter (span) : D=10m (fixed)
2)Loosened earth pressure: H
3)Coefficient of lateral earth pressure: K
4)Ground reaction : k
5)Width of shotcrete lining : t

The linear analysis is done only for the upper half of the tunnel.

Fig.12 illustrates the results in the case of $H = 10$m, $K = 0.5, k = 1000$tf/m^3 and $t = 25$cm which is taken as the standard version in this report. The figures show the distributions of axial force, bending moment, fringe stresses, loading vectors

Fig.13 illustrates the results when the thickness of the shotcrete lining t is set to be 50cm. Still the axial force is almost uniform but the bending moment became large to make the minimum stress tensile which should be avoided in the concrete structure.

Fig.14 illustrates the when the thickness of the shotcrete lining t is set to be 10cm. It is understood that the axial force is insensitive to the thickness of the shotcreting but that the bending moment is totally reduced. Since the cross section becomes small, the maximum (compressive) stress is increased. Therefore the shotcrete support has to be thin not to make the minimum stress tensile and to be thick to make the maximum stress small. If the thickness of the shotcreting be much thinner, the bending moment will almost disappear as is shown in Fig.15 where the thickness of the shotcrete lining t is set to be 1cm. An extreme case of the thin thickness of shotcreteing corresponds to the thin paper shell in the tunnel excavation experiment in

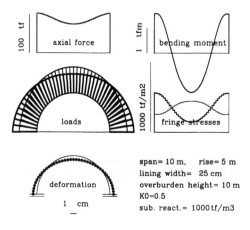

Figure 12 Standard version

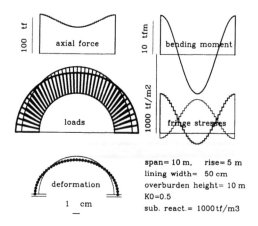

Figure 13 Case of thicker lining

on the shell and deformation mode, respectively. Each distribution reveals the value along the angular coordinate and the positive normal stress means compressive. It is observed that the loading pattern on the tunnel support has changed from the anisotropic initial state to the almost uniform one due to the effect of the ground reaction as Peck pointed out. As consequences, the axial force are almost uniform and a small value of bending moment occurs. (The magnitude of the bending moment becomes more than 100 times large if no ground spring is assumed.) In other words, the role of the reaction from the surrounding ground is remarkable to reduce the bending moment of the tunnel support.

3. When the thin support is considered, however, the buckling and the compressive failure may occur since the compressive stress become quite large.

Fig.16 illustrates the results when the ground reaction k is set to be 500tf/m^3 while Fig.17 shows the results when the ground reaction k is set to be 5000tf/m^3. It is understood that the large ground reaction makes the bending moment small and keeps the shotcrete in the stable state. The larger value of the ground reaction makes everything better.

Fig.18 illustrates the results when the ground reaction is effective only for the compressive forces. In other words, the springs subject to tensile forces are removed through the iterative calculation. This

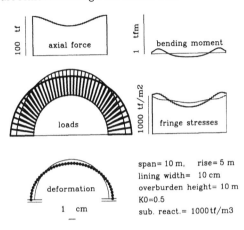

span= 10 m,   rise= 5 m
lining width=   10 cm
overburden height= 10 m
K0=0.5
sub. react.= 1000 tf/m3

Figure 14 Case of thin lining

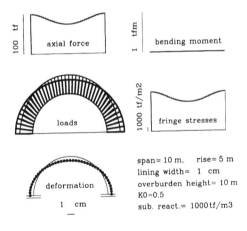

span= 10 m,   rise= 5 m
lining width=   1 cm
overburden height= 10 m
K0=0.5
sub. react.= 1000 tf/m3

Figure 15 Case of extremely thin lining

results in a slightly large amount of bending moment around the crown of the lining.

7 CONCLUSIONS

In this report, the following two major issues are introduced through the tunnel excavation experiment by Adachi et al(1985).

(1) The vertical load is reduced from the initial overburden stress into the loosened stress associ-

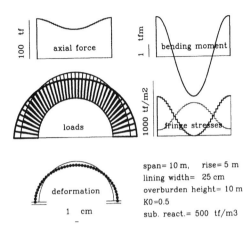

span= 10 m,   rise= 5 m
lining width= 25 cm
overburden height= 10 m
K0=0.5
sub. react.= 500 tf/m3

Figure 16 Case of weak ground reaction

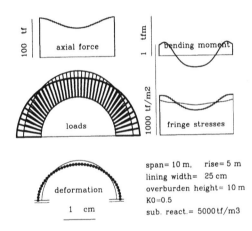

span= 10 m,   rise= 5 m
lining width=   25 cm
overburden height= 10 m
K0=0.5
sub. react.= 5000 tf/m3

Figure 17 Case of strong ground reaction

ated with the vertical displacement of the support.

(2) The lateral load is increased from the initial stress up to the passive earth pressure due to the lateral expansion of the flexible support.

The combination of the two issues explains that the stress state around the tunnel becomes almost isotropic and the thin tunnel support can be stable if the compressible failure is avoided. Both of them are related with the evaluation of the earth pressure under the interaction between the tunnel support structure and the ground, which is still difficult to be captured quantitatively. Furthermore the stability problem of the tunnel face is left to be more investigated since a kind of three-dimensional consideration is required. But it is true that the modern tunnel engineering cannot proceed any-

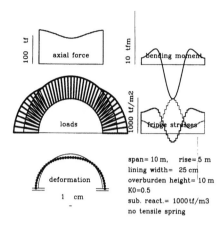

span= 10 m,   rise= 5 m
lining width= 25 cm
overburden height= 10 m
K0=0.5
sub. react.= 1000 tf/m3
no tensile spring

Figure 18 Case of no tensile spring

more without deep discussion on these issues. In this sense, more rational and economical design methods which include the theoretical background as well as the construction process are heavily expected.

REFERENCES

Adachi, T., Tamura, T. and Yashima, A. 1985. Experimental study on thin flexible tunnel support system, Proc. JSCE, No.358/III-3, pp.47-52.
Adachi, T., Kimura, M., Tamura, T. and Aramaki, S. 1994. Experimental and analytical studies of earth pressure, Computer Methods and Advances in Geomechanics, pp.2417-2422.
Kovari, K. 1994. Erroneous concepts behind NATM, Tunnel and Tunnelling, Nov.
Morinaga, H. 1998. A sensitivity analysis of flexible support by rigid-bar spring model, Graduation thesis of Department of Transportation Engineering, Kyoto University.
Peck, R.B. 1969. Deep Excavation and tunneling in soft ground, 7th Int. Conf. Soil Mech. And Found. Eng., State of the Arts Volume, pp.225-290.
Tamura, T., Kobayashi, S. and Sumi, T. 1984. Limit analysis of soil structure by rigid plastic finite element method, Soils and Foundations, Vol.24, No.1, pp.34-42.
Tamura, T., Kobayashi, S. and Sumi, T. 1987. Rigid-plastic finite element method for frictional materials, Soils and Foundations, Vol.27, No.3, pp.1-12.
Tamura, T. 1990. Rigid-plastic Finite Element Method in Geotechnical Engineering. Computational Plasticity, pp.135-164.
Tamura, T., Adachi, T., Umeda, M. and Okabe, T. 1998. On the tunnel support mechanism of shotcreting, Proc. JSCE, No.603/III-44, pp.11-20,(in Japanese).
Terzaghi, K. 1943. Theoretical Soil Mechanics, John Wiley and Sons.

Modern Tunneling Science and Technology, Adachi et al (eds), © 2001 Swets & Zeitlinger, ISBN 90 2651 860 9

Flexible Response to Difficult Ground and

Controlled Tunnelling Technology

H. Wagner, PhD, PE
D2 Consult Linz, AUSTRIA

ABSTRACT: Tunnel construction is a complex interaction between structural design concepts and flexible response to ground reactions in advanced tunnelling technology. The concept of structural design for tunnels is of increasing importance in advanced tunnel construction technology. There are different approaches in design modeling taking into account cross section geometry and overburden.

The size of tunnels represents one essential input among others. In order to view into reproducable conditions, this paper deals with two specific examples of transportation tunnels being located in Europe and South America. However, they could be located anywhere else. Such tunnels can be situated in shallow, shallow deep and deep conditions of overburden.

Shallow tunnels (Example 1) require the ability to take a greater number of different loading conditions, using stiff and rigid lining concepts. Deep tunnels (Example 2) do relate to loading conditions on the lining showing extended deformation behaviour. Deep tunnels have an essential need for implementation of flexible responses due to observations.

At both of the two examples there happened to be failures respectively collapses of the tunnel lining and the surrounding ground. This offered the chance to study the failure mechanisms and to learn for the future. There were no fatalities involved in both cases.

Shallow deep tunnels show benefits in the design both from the shallow tunneling and the deep tunneling philosophy. They are designed to use as a minimum requirement standard support measures. Depending on the ground they need to be supplemented by additional support measures as envisaged in the design under the requirements of standard safety needs during all constructions stages.

1 INTRODUCTION

In construction there is a need for handling the installation of standard and additional support by e.g. decision matrix. The decision factor must be defined, objective and measurable.

Through coordinated implementation of geomechanical measurements in the contract document, and measurement results in construction, the decision specifically for additional support measures such forming the flexible response.

In each individual contract, the designer has to decide on the criteria for the design of the decision matrix, controlling ground reactions by flexible response through standard and additional support measures for advanced tunnelling technology.

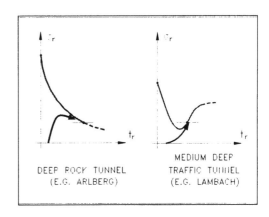

Figure 1. Simplified characteristic curve related to tunnel overburden.

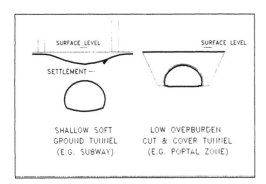

Figure 2. Construction method related to shallow tunneling.

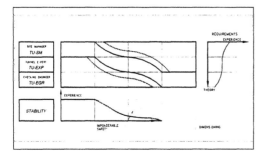

Figure 3. Tunnel safety and stability path

Tunnels could be located deep, · the overburden exceeds 200 meters and could have as much as up to 2.000 meters ·, or shallow (medium) deep with overburden exceeding 2 tunnel diameters up to less than 200 meters, or shallow with overburden starting at half of a tunnel diameter up to 2 tunnel diameters. Tunnel with less than half a diameter are usually built by using cut & cover methods.

There are different professional requirements related to the location of the tunnel. Construction practice is dominating deep rock tunnels, while more theoretical knowledge is needed for shallow tunnels. There is a complimentary relation especially when it comes to medium deep tunnels.

In regard to the stability of the tunnel structure there is a comparable relation between experience and calculation.

For deep tunnels experience is governing construction progress and success, while calculations sometimes at hydrostatic conditions are governing shallow tunnels.

Regarding shallow (medium) deep tunnels there is an area where calculations as well as experience have to be closer interrelated in the future.

2 SHALLOW TUNNELS (EXAMPLE 1)

The tunnel shown in Figure 4 is featured by big cross section (120 m²), and cohesionless and cohesive soft material with little overburden.

collapse during side bench niche excavation (1993).

Due to unfavourable conditions caused by geomechanic weaknesses superimposed by discontinuity of the lining due to niche excavation, a failure created the need for increasing safety by arranging of additional support.

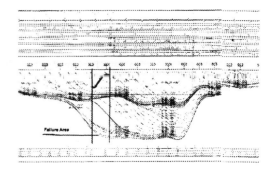

Figure 5. (Example 1) Measurement results, showing daily progress as reflected by individual curves. Failure area is indicated by unsteadiness.

In medium deep (shallow deep) tunnel locations deformations have been measured with less sensitivity when compared with shallow tunnelling.

In order to increase safety in construction numerous measures have been introduced related to design, construction and geotechnic. By introducing deformation criteria to determine additional support measures it was confirmed, that measurements are of continuous importance for future tunnel driving.

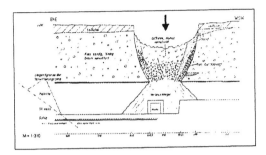

Figure 6. (Example 1) Cross section with geology in collapsed area (see also figure 4)

The need for increase of safety was targeted to

- Improved arching effects in both longitudinal and transversal direction,

- Prevention of lining discontinuities due to interrupted flow of forces,

- Improved composition effects between lining and underground leading to an active con-tribution for stability,

- Elongation of free stand-up time of bench by reduction of active bench height,

- Reduction of effective loads through reduc-tion of crown settlement,

- Reduction of risk for failure of side walls, and

- Improvement of mechanical behaviour of the underground for increasing of reserves of bearing capacity.

Following measures have been actively contributing to increase the level of safety.

- Reduction of ring closure distance (from 8 m to 5.5 m) resulting in reduced ring closure time,

- Excavation of niches with subsequent niche support installation decoupled from regular cross section excavation,

- Installation of systematic anchoring e.g. in 2 rows in the crown and 3 rows in the bench during each round,

- Increased thickness of shotcrete lining,

- Excavation in side wall and bench in parts as well as additional support measures at bench face,

- Deep foundation of crown footings to reduce crown settlements with injection piles fol-lowing deformation criteria,

- Soil nailing of deep sliding joint for im-provement of stability of side walls, and

- Improvement of mechanical behaviour of the ground by injection to increase bearing ca-pacity reserves.

Using all of these measures, the overall stability of the tunnel could be increased from 1.0 to 1.7.

3 DEEP TUNNELS (EXAMPLE 2)

3.1 Tunnelling Criteria

Since approximately July 21, 2000 tunnel deforma-tion measurement have indicated significant roof settlements in the area of an enlarged parking niche cross section.

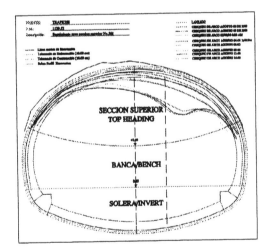

Figure 7. (Example 2) Measurement results, show-ing periodic increase of deformations of the lining geometry. Ground collapse has been expected while pressure relieved.

Overburden has exceeded 500 m. Nevertheless the contractor, being alarmed because of continuing roof settlements, did proceed with the works of excava-

23

tion and support installation, increasing the numbers of anchor bolts.

3.2 Sequence of Collapse

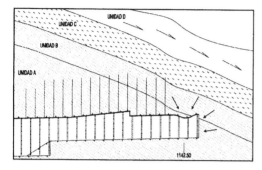

Figure 8. (Example 2) Initial crown drift deformation due to real rock pressure combined with water pressure, starting at entering at decomposed rock.

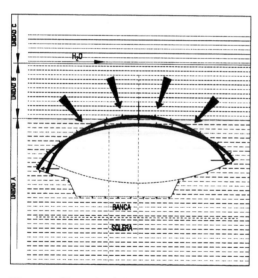

Figure 9. (Example 2) Initial failure mechanism. Pressure increase near tunnel face – roof settlement (27 cm) and crown footing divergency (90 cm).

On August 5, 2000 roof settlement has been measured in the range of maximum 1.2 meters whereas heavy infiltration of gravely material together with water amounted to approximately 500 m³ in 2 phases.

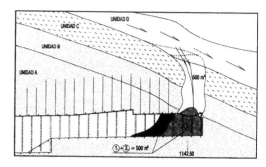

Figure 10. (Example 2) Subsequent loose material chimneying into crown drift in two separated collapse events.

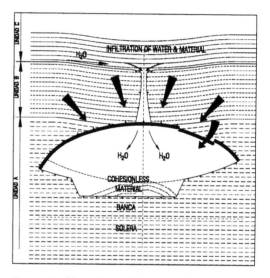

Figure 11. (Example 2) Subsequent failure mechanism. Bending, cracking and infiltration of water and loose material

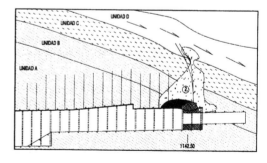

Figure 12. (Example 2) After reperfilation of deformed cross section driving of pilot tunnel through critical section.

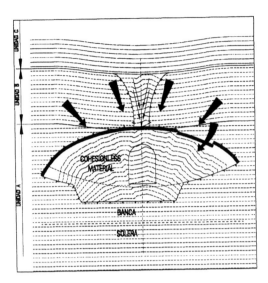

Figure 13. (Example 2) Crown drift, filled with material from ground failure and rehabilitation of deformed lining by reperfilation.

It became evident that on top of the roof an impermeable layer has broken, thus opening a vertical chimney like access to the cohesion less water saturated gravely material on top of the impermeable layer.

3.3 *Lessons Learned*

Figure 14. Tunnel face of crown drift with pilot drift and forepoling. Pilot drift at centre of crown drift served the purpose of investigation and stress relieve.

Following geotechnical and contractual lessons have been learned from driving NATM tunnels through difficult and faulty zones.

- Observation of measurement result did allow for preparation of failure.

- Horizontal measurement of enlarged niche cross section indicated divergencies.

- Bending of steel arches indicated deep cracking of impermeable layer above tunnel roof due to tension forces.

- Cracked rock allowed for increasing amount of water plus gravely material.

- Progressive ground failure happened between August 6 and August 18/19, 2000.

- Area of failure was in critical stability stage, whereas 51 m of unsupported tunnel invert was left open behind the face.

- Top heading stability should benefit from closed ring conditions with rock bolts.

- Excavation works at the tunnel face should be temporarily stopped until observation of measurement indicate stop of deformations.

- Final cross section excavation with structural invert should be installed as close as possible to the face.

- Excavation should continue with temporary top heading and drainage borings until full crown cross section is in sound ground.

4 SHALLOW DEEP TUNNELS

In order to proceed in design and construction of Tunnels it is necessary to define threshold values of deformations. Such deformations do have major importance in the decision making of support measures. Deformations could be related both to surface deformations and deformations within the tunnel.

The total measurable deformations contain individual deformations resulting from individual loading conditions.

It is obvious, that the more loading conditions are effecting a tunnel the more incremental deformations the tunnel will experience.

Besides the deformation parameter due to loading conditions there are several other parameters influencing deformations, which represent different influences in regard to their time dependent tendency respectively their absolute value.

Such parameters are the soil parameters, construction material parameters and most important the construction concept itself. From studying the influence of these parameters it may become necessary to modify the construction concept.

Beyond the fact that calculations are necessary, the experience of the designer is of utmost importance. Definition of threshold values, representing critical deformations is feasible. Deformations in the design stage are underlining the importance of geomechanical measurements during construction.

4.1 Deformation Threshold Value

In order to give both the resident engineer and the site managers an indication, from where on measured deformation are considered to require additional support, finite element calculations need to be performed during final design.

A rough estimate on an average value of overburden and differing between tunnel diameters for regular and for enlarged cross sections show different values, ranging from 5 to 10 cm.

In literature this threshold values are named critical deformations in the sense of the European understanding, whereas shear failure safety factor is on the very safe side, in most cases better than 2,0.

In the sense of American understanding, critical deformations are understood to have for instance shear failure safety factor equal to 1,0, which practically would mean, that the tunnel stability is close to collapse. However, as the term critical deformation has been introduced into the international literature for tunnelling it is recommended to stay with this term.

4.2 Surface Settlements

In shallow tunneling, surface settlements are forming a major concern. Contractors are taking this concern serious in addressing surface settlements specifically in the submittal of documents for the final proposal.

Surface settlements could become an issue due to deformations, measured within the tunnel. Such deformations could be caused among others by shifting of primary into secondary stress conditions.

There could be 2 and more such shiftings of primary to secondary to tertiary etc. stress transfers. It is obvious, that such stress transfers should be kept to a minimum in order to also minimize stress transfer related deformations.

Subsequently to deformations in the tunnel, excavation causes volume loss at the face of the shallow tunnel, usually in a range of 1,5 to 2,0 %. From measured results within projects of comparable size and geology, the development of deformations at different levels in between the tunnel roof and the surface should be studied. Experiences should be implemented for surface and ground settlement evaluation.

4.3 Installation of Support

Standard support measures are to be installed all along the length of the tunnel. Means and methods should be defined and documented. It should be demonstrated when and how additional support measures respectively contingency support measures will be installed.

It is necessary to base the decision for measures beyond standard support on objective geomechanic measurements.

It has been proposed, to cover regular expected ground conditions with standard support measures, not exceeding $1,0 \times d_{crit}$. d_{crit} represents a threshold value which is on the very safe side, for the purpose of defining the value requiring additional support measures.

If this value is understood to be a value, representing the condition close to the tunnel collapse, it would become necessary to define a different value which would be lower than the one defined in the tunnel literature.

4.4 Decision Algorithm Graph

In order to explain the practicality of when and where and how different categories of support measures in relation to deformations, measured in the tunnel, will be applied, the decision algorithm graph has been developed.

Actual deformation measurement starts at the tunnel face immediately after excavation. In real tunnel construction it will not be possible to measure deformations related from stress relieve ahead of the tunnel face.

For actual measurement, advanced measurement methods will be used based on laser beam reflection.

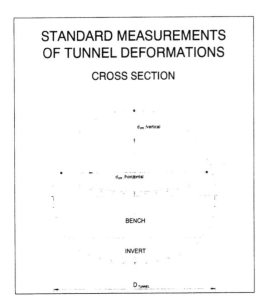

Figure 15. Example of time related tunnel deformation for evaluation of support.

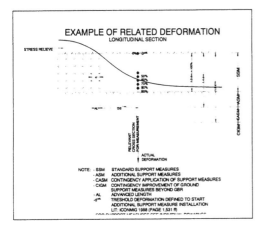

Figure 16. Longitudinal section-interaction of actual time and location related deformations to different tunnel support measure categories.

There will be series of curves, as this deformation is directly related to time. Actual amount of stress relieve deformation will be defined upon completion of back analysis and after completion of the test section.

Programmed days and weeks after excavation and installation of support, measurements will be taken frequently.

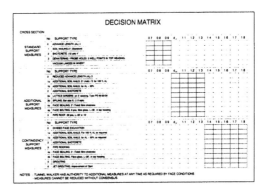

Figure 17. Three stage decision matrix for increasing application of different support measures related to increasing values of critical deformation factor.

These measurements will result in types of fibre curves, and should result finally in an approximation, where deformation increments will tend to become 0.

At this stage the fibre curves will become identical. Geomechanical measurements are therefore the only reliable instrument to judge the stability of the tunnel.

5 CONCLUSIONS

The defined threshold value respectively the critical deformation is controlling the freedom of decision making. It depends on the actual behaviour of the interaction between lining and underground, such satisfying needs in respect to safe tunnel lining design and construction. It is easy to use upon, defining the limits of stability.

Tunnel deformations developed within the proposed limits are considered to be adequate in regard to standard support measures.

Standard support measures should be sufficient to support the loading conditions in general. In addition to the standard support measures there is a need for flexible response with additional support measures.

Such measures must be estimated in advance of construction to the best of the combined knowledge of the engineering geologist and the structural consultant.

Flexible response measures must be identified prior to finalization of contract documents. With the flexible response measures, the tunnel will be under control.

The designer is asked to define the need for standard and additional lining support.

It should be specially indicated, why, when, where, and how, the additional support will be implemented in final construction.

6 LITERATURE

1. G. Vavrovsky / N. Ayaydin, "Bedeutung der vortriebsorientierten Auswertung geotechnischer Messungen im oberflächennahen Tunnelbau", Special Edition of Research + Praxis, 1988.

2. H. Wagner / A. Schulter, "Geonumerical computations for the determination of critical deformations in shallow tunnelling". International Conference on Numerical Methods in Geomechanics – ICONMIG, Austria 1988.

3. H. Wagner, Tunnel Expertise, Lambach Tunnel, Highspeed Railway Vienna-Salzburg, un-published report, 1992.

4. H. Wagner, "Civil Engineering for Urban Development and Renewal", International Sym-posium of Japan Society of Civil Engineers, Yokohama 1994.

5. H. Wagner / M. Srb, "Safety in Tunnels", Lambach Tunnel, Research Report SITU, 1995.

6. H. Wagner / D. Kolic, "Method of Statement", Seattle Metro Project, unpublished report, 2000.

7. G. Caro / H. Wagner, "Observational Methods for Buenavista Tunnel in Colombia – Review of Geotechnical and Contractual Practices", ITA Conference Milan 2001.

Modern Tunneling Science and Technology, Adachi et al (eds), © 2001 Swets & Zeitlinger, ISBN 90 2651 860 9

Design Method of Shield Tunnel   -Present Status and Subjects-

Y. Koyama
Railway Technical Research Institute

ABSTRACT: Recently, techniques of shield tunnel construction have been progressing remarkably and conquering many severe conditions. However, it is the present state that design of segment only seems to be on the starting point of progressing from the traditional method based on experiences, and it cannot follow in progress of constructional techniques. On account of the well-balanced progress between construction and design in shield tunneling method, an innovation of design technique is required. This paper shows an outline of the shield tunnel technique at present, behavior of shield tunnels in site and subjects for its design.

1 INTRODUCTION

The shield tunnel constructing method was invented in the 19th century and made rapid progress in the latter half of the 20th century. The open face and hand mining with air compression pressure were used for tunnelling at the initial stage. After that, a mud slurry method was developed in the former half of the 1970s. An earth pressure balanced method was also put into practical use in the latter half of the 1970s. Then, it became possible to keep the face stability in the super soft clay ground, which had not been possible even by using a supplementary method. Besides, in the same period, a new grouting material, which can harden immediately and keep plasticity, was developed for back-fill grouting. A new tail seal, which consists of a wire brush and special high viscose grease, was developed. Then, it became possible by the shield tunnelling method not only to excavate tunnels safely, but also to limit ground movements.

But, it is demanded recently to construct tunnels under severe conditions and develop further a new technique. As the design technique cannot appropriately reflect the construction technique, however the traditional method is still used. Then, an innovation of design technique is required. This paper shows an outline of the shield tunnel technique at present, movements of shield tunnels in site and subjects for its design.

2 PRESENT STATUS OF THE TECHNIQUE FOR SHIELD TUNNELLING

2.1 *Simple Circular Shield Tunnelling Method*

As construction circumstances are becoming severer in Japan, it is required to solve various problems. Under the circumstances, a number of technologies have been developing. To understand the development of the shield tunneling technology, it may be interesting to review its maximum diameter, water pressure, depth and minimum curve radius. Regarding the tunnel size, a shield tunnel with a diameter of about 15 meters has been constructed. As for the water pressure, a shield tunnel has been constructed for a water pressure of 0.9 MPa. In regard to the sharp curve, a tunnel with a curve radius of 50 meters has been constructed, even under high water pressure.

Moreover, as discussed below, extraordinary-shape shields have been developed. It is possible and performed increasingly to excavate a specific shape simultaneously for the entire cross section by means of these type shields.

2.2 *Multi-Circular Face Shield Method (MF SHIELD)*

The MF shield, utilizing the advantages of the slurry type shield, was developed to construct flat-shaped tunnels. To avoid the interference between cutter faces when they rotate, they are set at different positions in the longitudinal direction (Fig.2.1). A

Double-circular face type (Matsumoto,Y., 1994A).

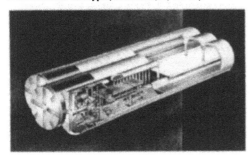

Triple-circular face type (Matsumoto,Y., 1994B)

Figure 2.1. Multi-circular Face Shield.

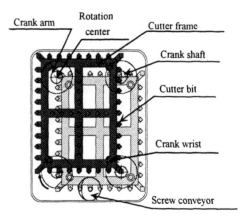

Figure 2.2. Principle of excavation by the DPLEX. (Kashima et al. 1997).

double-circular face type was developed first. Then, triple-circular face type shields have been developed to construct train stations in complicated underground spaces. In addition to the above, a four-circular-face type was developed to construct subway stations, whose shape is more complicated than that of triple-circular face type tunnels.

2.3 Double O-Tube Shield (DOT SHIELD)

The DOT shield is an earth pressure balanced type shield with the same purpose of MF shield. Two spoke cutter wheels of the shield rotate in an opposite direction on the same plane, and as not to come in contact, the rotation control is done.

2.4 Developing Parallel Link Excavation Shield (DPLEX Shield)

The DPLEX shield is quite a new method, which is modified to a large extent from the conventional earth pressure balanced type shield.

The shield employs multiple rotating shafts to which cranks are fit at the right angle (Fig.2.2). A cutter frame equipped with a number of cutter bits is universally coupled to the ends of rotating shafts. As the rotating shafts turn, the cutter frame starts a par-

allel link motion, which makes it possible to cut a tunnel with a cross section analogous to the shape formed by the cutter bits.

Advantages of the shield are not only a reduction of cutter torque due to short shafts, but also the applicability of unusual shapes.

3 PRESENT STATUS OF DESIGN METHOD FOR SHIELD TUNNELS

3.1 Present Status of Design Method for Shield Tunnels

In designing segments, the stress in each parts of segment is usually calculated with a design model, by taking account of the structural property of segmental lining and interaction between the ground and lining, so that the stress does not exceed the allowable limit. Hence, to design a segment, we need to reasonably model the segment structure and assume its interaction with the ground. There are a number of design models proposed so far. This paper presents model of interaction with the ground.

3.2 Interaction model for simple circular tunnels

Structural member dimensions and material strength have been determined by calculating the section forces by using a tunnel-ground interaction model. This procedure is adopted in various countries. We have a number of tangible interaction models, which can be categorized into two types. One is a method to use finite element analysis and the other is a method to use beam elements representing tunnel lining, which carries the earth pressure, water pressure and self-weight. While the former is not widely applied, different models proposed so far for the latter.

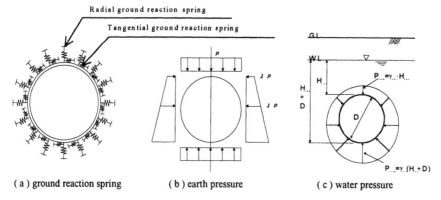

(a) ground reaction spring   (b) earth pressure   (c) water pressure

Figure 3.1 Full-circumferential ground spring model (RTRI, 1997).

These models are developed for the open type shield for manual excavation. The models can be sub-categorized into the following three types.

1) The tunnel is assumed to be made of a rigid material and the ground reaction force is determined independent of the tunnel deformation caused by active loads.
2) The reaction force is determined in consideration of the tunnel deformation caused by active loads.
3) In addition to the condition in 2), the shape is simplified.

Among Japanese tunnel engineers, for shield tunnel designing, the above 1) is called the Conventional Model, and 2) is called the Full-circumferential Spring Model (Fig.3.1). This section discusses the latter.

There are two modeling methods for the earth and pore water pressures. First, the modeling does not separate the two pressures, but second, the modeling is separated into the two parts. In general, the former is applied to the ground of low permeability such as clayey soil, and the other is applied to the ground of high permeability such as sandy soil. The earth pressure is divided into two parts, vertical and horizontal pressures.

1) Vertical earth pressure

There are two cases in assuming the vertical earth pressure acting on the upper part of segmental ring, one assumed to be a full overburden and the other a reduced one to take into account the soil shear strength. For both pressures, a uniform rectangle shape distribution is assumed. The latter is calculated by applying the formula for Terzaghi's loosening earth pressure. The upward earth pressure from the tunnel bottom is assumed to be the same as downward earth pressure in the upper part in magnitude and distribution.

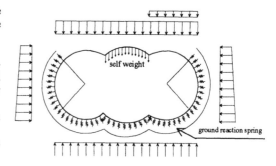

Figure 3.2 Design model for triple-circular tunnel (after Koyama et al. 1986).

2) Horizontal earth pressure

The horizontal earth pressure is assumed to have a trapezoidal distribution that increases into the direction of depth. Hence, the horizontal earth pressure at the top of the ring is derived from the vertical earth pressure multiplied by the coefficient of horizontal earth pressure. The vertical earth pressure increases into the direction of depth, but it should slightly be mitigated by the contribution of ground shear strength.

3) Ground reaction force

The ground reaction force is assumed to have a value that corresponds to tunnel deformation and displacement and modeled as ground springs located along the whole periphery of the tunnel.

4) Water pressure

The water pressure on the tunnel is assumed to act into the direction to the center of the ring, which increases in direction of depth from the ground water level.

31

3.3 Interaction model for designing shields of extraordinary shapes

The designing of the shield of extraordinary shapes is based on the above-mentioned methodology. For multi-circular face shields, the Full-circumferential Spring Model is generally used.

By taking account of their unusual shape, however, a bias load is considered although it is not considered for the design of simple circular shields (see Figure 3. 2). In applying the vertical downward earth pressure for example, a non-uniform load distribution is considered on either side, or at the center of shield. This strengthens the center column that carries severer conditions than those for circular shape segments in contact with the ground. Furthermore, the coefficient of ground reaction force is partially reduced to take account of ground irregularity. This is to increase the load-bearing capacity of the segment.

4 ACTUAL CONDITIONS OF EARTH PRESSURE AND WATER PRESSURE AFFECTING SHIELD TUNNELS

4.1 Present Status of Design Method for Shield Tunnels

Among a number of shield tunnels constructed so far, there are practically few cases where the loads affecting segments and occurring strain are measured at closed type shield tunnels. Generally speaking, the earth pressure is measured by means of earth pressure meters placed behind the segments. The pressure measured in this way includes both the effective earth pressure and water pressure. In following sections, the pressure measured by means of earth pressure meters is called the earth pressure or the total earth pressure. On the other hand, there are a number of cases where the water pressure is measured by means of the measurement of ground water flowing into bore holes. The bore holes are constructed from grouting holes in segments through the hardened grouting material after the segment lining goes out of the shield and stabilizes.

This chapter focuses on the examples of the measurement of earth pressure and water pressure, and describes the earth pressure and water pressure in connection with design methods, regarding them as what expresses some characteristics of the loads that are actually affecting the tunnels.

4.2 Long-term loads affecting simple circular shield tunnels

4.2.1 Measurement results at clayey layers

This paragraph shows the cases of earth pressure and water pressure measurement at tunnels in a clayey

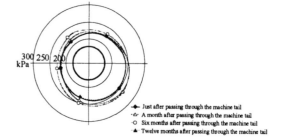

Figure 4.1. Distribution of earth pressure into the circumferential direction (in a alluvium) (Ariizumi et al. 1998).

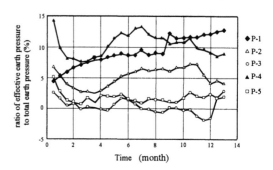

Figure 4.2. Changes in the ratio of effective earth pressure to total earth pressure over time (in a alluvium) (Ariizumi et al. 1998).

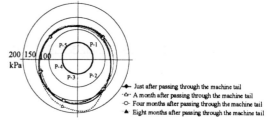

Figure 4.3. Distribution of earth pressure into the circumferential direction (in a diluvium) (Ariizumi et al. 1998).

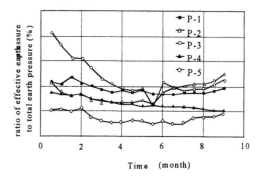

Figure 4.4. Changes in the ratio of effective earth pressure to total earth pressure over time (in a diluvium) (Ariizumi et al. 1998).

alluvium and a clayey diluvium. These two tunnels are constructed by using slurry shields. The external diameter and overburden of one tunnel are 3.95 m and 22 m, respectively, and those of the other tunnel are 4.95 m and 13 m.

Figure 4.1 and Figure 4.2 shows the distribution of earth pressure into the circumferential direction and changes in the ratio of effective earth pressure to total earth pressure over time in an alluvium, respectively. In the same way, those in a diluvium are shown in Figure 4.3 and Figure 4.4.

At both tunnels, the earth pressure gradually changes for a long period. This is thought to result from the changes in atmospheric temperature.

From these measurement results, the following hypotheses can be described on the earth pressure and water pressure affecting for a long period.
1) The effective earth pressure is never equal to zero at tunnels in clayey alluviums although it is small.
2) The effective earth pressure is almost equal to zero at tunnels in clayey diluviums.
3) The distribution of total earth pressure is extremely asymmetric at tunnels in alluviums where small effective earth pressure acts.
4) Relatively uniform earth pressure acts on tunnels in a diluvium where the effective earth pressure is small.

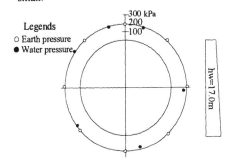

Figure 4.5. Distribution of earth pressure into the circumferential direction (in a alluvium) (Ito et al. 1991).

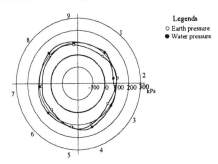

Figure 4.6. Distribution of earth pressure into the circumferential direction (in a diluvium) (Koyama et al. 1995).

4.2.2 Measurement results at sandy layers
This paragraph shows the cases of earth pressure and water pressure measurement at tunnels in a sandy alluvium and a sandy diluvium. These two tunnels are constructed by using slurry shields. The external diameter and overburden of one tunnel are 5.10 m and 17 m ,respectively, and those of the other tunnel are 9.80 m and 19 m.

Figure 4.5 shows the distribution of earth pressure and water pressure into the circumferential direction in an alluvium. In the same way, that in a diluvium is shown in Figure 4.6.

From these measurement results, the following can be described concerning the earth pressure and water pressure acting on tunnels for a long period.
1) The effective earth pressure is 10 to 20 percent of the total earth pressure and small at tunnels in sandy alluviums.
2) The water pressure is dominant in the earth pressure and the effective earth pressure is equal to zero in sandy diluviums.
3) The distribution of the total earth pressure is nearly symmetric.

In the case of tunnels in a diluvium, a model to assume that tunnels are subjected to buoyancy and pressed down by clayey layers can explain the shape of distribution of sectional force.

4.2.3 Measurement results at gravel bed
This paragraph discusses the cases of the measurement of one tunnel in a gravel alluvium and two tunnels in a gravel diluvium. These three tunnels are constructed by means of slurry shields. The external diameters of these tunnels are 3.35m, 6.2m and 4.75m, respectively, and overburdens are 25 m, 10m and 12 m, respectively.

The distribution of earth pressure and water pressure in the circumferential direction in an alluvium is shown in Figure 4.7, and that in a diluvium is shown in Figure 4.8 and Figure 4.9.

From these measurement results, the following can be described concerning the earth pressure and water pressure acting on tunnels for a long period.
1) The effective earth pressure is about 30 percent of the total earth pressure at tunnels in gravel alluviums though the water pressure is dominant.
2) If the ground water level is high at tunnels in gravel diluviums, only the water pressure acts on tunnels and the effective water pressure is equal to zero.
3) If the ground water level is low at tunnels in gravel diluviums, large effective erath pressure acts on the upper part of tunnels. On the other hand, the effective earth pressure acting on the lower part of tunnels is small.
4) In case the effective earth pressure is small, the distribution of the total earth pressure is uniform. On the other hand, if the effective earth pressure is

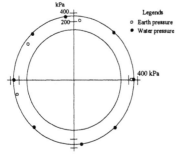

Figure 4.7 Distribution of earth pressure into the circumferential direction (in an alluvium) (Koyama,2000A).

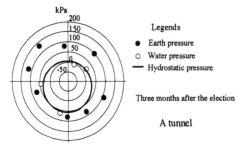

Figure 4.8. Distribution of earth pressure into the circumferential direction (in a diluvium) (Fujii et al. 2000).

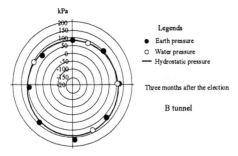

Figure 4.9. Distribution of earth pressure into the circumferential direction (in a diluvium) (Fujii et al. 2000).

dominant, the distribution of the total earth pressure is uneven to some extent.

4.3 Long-term loads acting on extraordinary shape shield tunnels

4.3.1 Multi-circular face shield

This section discusses the case of the measurement at a triple-circular tunnel constructed by using a slurry shield shown Figure 2.1. The tunnel is located in the strata of clay, sand and gravel. Its size is 15.6 m in width and the overburden is 21 m. Figure 4.10 compares the measured earth pressure and design load.

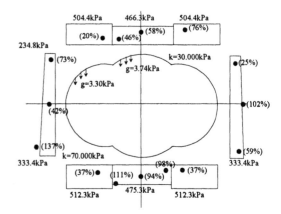

• The ratio of measured earth pressure to design load

Figure 4.10. Comparison between the measured earth pressure and design load (Okado et al. 2000).

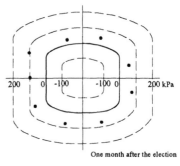

Figure 4.11. Distribution of earth pressure (Kashima et al. 1997).

From this measurement result, the following can be described concerning the earth pressure and water pressure acting on tunnels for a long period.
1) The earth pressure acting on multi-circular face shield tunnels is less uniform than that acting on simple circular shield tunnels.
2) The effective earth pressure is larger than the water pressure.
3) The distribution of the earth pressure looks as if tunnels have wholly turned around.

4.3.2 Rectangular shield

This section discusses the case of the measurement of a rectangular shield tunnel constructed by means of DPLEX shield. The tunnel is located in a diluvium and its size is 4.2 m in width and the overburden is 3 m. The measured earth pressure is shown in Figure 4.11.

From this measurement results, the distribution of the earth pressure acting on tunnels for a long period is as uniform as that acting on multi-circular face shield tunnels.

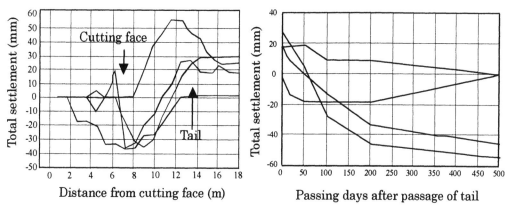

Figure 4.12. Distribution of earth pressure into the circumferential direction (in a diluvium) (Hirata, 1989).

4.4 *Factors influencing the loads acting on tunnels*

4.4.1 *Relationship between ground displacement and earth pressure*

The earth pressure acting on a tunnel for a long period is considered to depend on the strain of soil near the surface of tunnel. Actual strain conditions of soil near the tunnels are estimated from ground displacement near the tunnel in this section. Figure 4.12 shows the measurement results during the construction of subway shield tunnels in Osaka. In all cases, the values of measurement vary widely although the ground conditions are similar (in clayey alluviums). The results of measurement generally show a tendency of settlement, but there are some cases where they shows a tendency of rise. This suggests that the strain of soil just above the tunnel can either extend or shrink and that the earth pressure acting on the tunnel crown can also either increase or decrease when compared with the situation before the excavation of the tunnel.

There are no examples of detailed measurement around tunnels. The strain of ground near the tunnel can be considered to vary widely, depending on the operational control of shield, the way of backfill grouting and other conditions. The scattering of strain can be considered to cause the heterogeneity of distribution of earth pressure affecting tunnels.

4.4.2 *Influence of backfill grouting*

An influence of backfill grouting on shield tunnels has been confirmed by indoor and other tests, but an interesting test has been carried out to confirm it by using an actual shield tunnel. The test was carried out at an earth pressure balanced shield tunnel whose external diameter is 5.30 m. The test was carried out in two cases (Table 4.1).

Table 4.1. Design of back-fill grouting (Nishizawa et al. 1996).

	Case1	Case2
Design grouting pressure	150 kPa	50 kPa
Design grout ratio	139 %	100 %
Backfilling material	Standard mixture	Lean mixture
Grouting method	Usual method	Stop grouting when grouting pressure becomes bigger than design pressure

The backfill grout was injected according to the ordinary maintenance method in one case and the grouting pressure was lower and the volume was smaller than those by the ordinary maintenance method in the other case. The earth pressure and water pressure affecting the tunnel is measured in the test. The test is carried out at two sections of the tunnel in the clayey diluvium and sandy diluvium to grasp the influence of soil condition. These results are shown in Figure 4.13 and Figure 4.14.

The followings were obtained from these results.
1) If the backfill grouting pressure is low, the earth pressure is very small and its distribution is uniform in all ground conditions.
2) If the backfill grouting pressure is high, the earth pressure is high and its distribution is uneven in all ground conditions.
3) The earth pressure caused by the backfill grouting pressure just before the execution tends to remain for a long period.

In the actual execution of shield tunnel, the backfill grouting is expected to fill up the gaps between the segments and ground and to stabilize the tunnel.

35

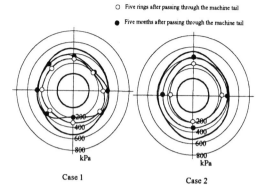

Case 1   Case 2

Figure 4.13. Distribution of earth pressure into the circumferential direction in the clayey diluvium (Nishizawa et al. 1995).

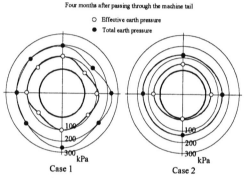

Case 1   Case 2

Figure 4.14. Distribution of earth pressure into the circumferential direction in the sandy diluvium (Nishizawa et al. 1996)

At the same time, the backfill grouting gives a measure to add high pressure to the cutting face in order to reduce the settlement of ground. As the segments receive the reaction force caused by the backfill grouting, the segments are considered to receive larger loads than those considered at design. Consequently, in case the backfill grouting is carried out at high pressure and causes local destruction of ground around the tunnel, the final distribution of earth pressure is assumed to be uneven. This is indicated from the indoor tests of backfill grouting (Koyama, 2000B).

4.4.3 *Influence of operational control*

Figure 4.15 shows the results of measurement of changes in the horizontal and vertical diameters of the same tunnel. Although the ground condition is almost uniform, there are some sections where deformation modes vary from the long sideways mode to the mode with the shorter side at the top. It is impossible to explain the differences of deformation modes by using a model where the earth pressure and water pressure act on the tunnel. It is understood that the modes with the shorter side at the top occur

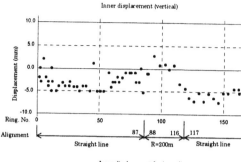

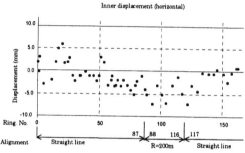

Figure 4.15. Inner displacement of segment (Example of measurement) (TEPCO, 1998).

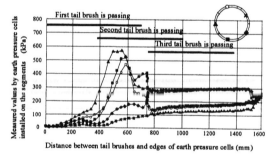

Figure 4.16. Influence of tail brushes on the segments (Ariizumi et al. 1999).

at curved line sections by comparing this deformation data with execution data. It is reasonable to think that this modes result from the conflict between the segmental rings and shield tails. That is to say, the operational control remarkably influences the loads acting on tunnels and subsequent deformation.

4.4.4 *Influence of impermanent loads in execution*

The earth pressure and water pressure is mainly discussed above. The loads acting on the segments become maximum when the shield machine is driven. Figure 4.16 shows the examples of the measurement by using the earth pressure cells installed on the segments to grasp the influence of tail brushes on the segments. This result shows that the pressure becomes maximum in a passage of shield brush and two

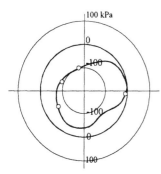

100 kPa

Figure 4.17. Result of measurement of the pressure acting on the segment (in a sharp curve) (Okada et al. 1996).

times as large as that into the ground. However, this pressure doesn't remain and only impertinently affects the segments

Figure 4.17 shows the result of measurement of the pressure acting on the segment when the shield is driven along a sharp curve. From the result of measurement, high partial pressure is known to act on the segments as the shield is driven. However, the pressure is known to be impermanent and disappear as the shield stops. A part of the partial pressure remains in the tunnel and controls the condition of tunnel deformation as mentioned above.

The impermanent pressure acting on the segments as mentioned above doesn't affect stability of the tunnels for a long period directly. However, the impermanent pressure causes cracks and chips of segments and a leakage of water, and then a quality of segments ,consequently, become low.

5 ISSUES FOR SEGMENT DESIGN AND R&D TREND

For the designing of segments, the following two hypotheses should hold when we adopt a design methodology based on the above-mentioned long-term earth pressure and water pressure.

1) Tunnel deformation calculated in designing a tunnel is subject to the earth pressure (including the ground reaction force) and water pressure acting on the tunnel.
2) Tunnel deformation before the tunnel reaches the long-term stable condition is smaller than that under the stable condition.

As shown in the former section, however, the above two points do not necessarily hold in actuality. Hence, the followings are conceivable.

1) Tunnels deform during construction due to various feature other than the earth pressure and water pressure, and the temporary loads remain for long period. It is thought that the reason of this phenomenon is the existence of back-fill grouting between the segment and ground.

2) If the effect of operational loads of shield machine does not exist under normal conditions of construction, the tunnel would be under an ultimately stable condition while subjected only to the earth and water pressure, but the earth pressure always disperses to a considerable extent. This fact seems to show that the ground deformation during construction depends not only on the ground condition but also on the construction condition. It can also be assumed that intensive ground failures are due to the back-fill grouting.

From the above facts, we can conclude that the segment designing requires detailed construction conditions. Since the information for this purpose is limited when we start designing, however it is impossible to clarify the construction conditions. Hence, during shield tunneling, it is not permitted to exceed the deformation calculated based on the assumed earth and water pressure.

At present, the excessive deformation is not observed during construction in the above-mentioned practice. But there will be uncertainties in keeping the quality of construction in the future, as in the past, under severe tunneling conditions in deeper or greater cross-section shields.

As the discussions in this paper focused on the design of cross-section, the effect of operational jack force was also refereed to only in relation to cross-section. Because the jack force directly acts into the tunnel axis direction, it is important to evaluate its effect in the longitudinal direction.

To maintain the same quality of shield tunnel as that in the past, under severer conditions, it is important that the constructional loads properly be included in design. To realize this idea, it is necessary to develop a simple measuring system for the behavior of tunnel and ground during construction, collect the measured data and accumulate experience. It is also important to develop a methodology which reflects actual practices.

6 CONCLUDIONG REMARKS

Shield tunnel construction technique is remarkably progressing and solving a number of difficult problems in Japan. While novel technologies are recently adopted in the segment design, we are just at the starting point to depart from the empirical method without sufficient following the progress of construction technologies.

It is obvious that the above mentioned situation is a barrier in adopting new shield tunneling methods. To ensure well-balanced progress both in construction and designing shield tunneling, we are required to solve these problems.

REFERENCES

Ariizumi,T. , Okadome,K. & Nagaya,J. 1998. Result and Analysis on Site Measurement of Load Acting on the Shield Tunnel. Proceedings of Tunnel Engineering: Vol.8, pp.367-372: JSCE. (in Japanese)

Ariizumi,T.,Okadome,K., Igarashi,H. & Nagaya,J. 1999. Investigation of the Load Acting on Shield Segments during Tunneling. Proceedings of Tunnel Engineering: Vol.9, pp.271-276: JSCE. (in Japanese)

Fujii,K., Mashimo,H. & Ishimura,T. 2000. Load Acting on Shield Tunnel in Gravel Ground. Proceedings of Tunnel Engineering, JSCE: Vol.10, pp.257-262. (in Japanese)

Hirata,T. 1989. Study on Ground Movement Caused by Shield Driving and Improvement of Shield Tunneling Method. Doctoral Thesis for Kyoto University: pp.100-110. (in Japanese)

Itoh,H. & Saito,M. 1991. Result and Consideration of Field Measurement Acting on the Shield Tunnel. Proceedings of the 46st annual conference of the JSCE: 3, pp.160-161: JSCE. (in Japanese)

Kashima,Y. & Kondo,N. 1997. Construction of Extremely Close-set Tunnels Using the Muddy Soil Pressure Balanced Rectangular Shield Method. Proceedings of World Tunnel Congress '97: p.255-260: Balkema.

Koyama,Y., Kato,M. & Shimizu,M. 1986. Design of a Multi-circular Shield Tunnel. Proceedings of the 41st annual conference of the JSCE: 3, pp.801-802: JSCE. (in Japanese)

Koyama,Y., Okano,N., Shimizu,M., Fujiki,I. & Yoneshima,K. 1995. In-situ Measurement and Consideration on Shield Tunnel in Diluvium Deposit. Proceedings of Tunnel Engineering: Vol.5, pp.385-390: JSCE. (in Japanese)

Koyama,Y. 2000A. Study on the Improvement of Design Method of Segments for Shield-driven Tunnels. RTRI REPORT: Special No.33, pp.80-83: RTRI. (in Japanese)

Koyama,Y. 2000B. Study on the Improvement of Design Method of Segments for Shield-driven Tunnels. RTRI REPORT: Special No.33, pp.47-55: RTRI. (in Japanese)

Matsumoto,Y. 1994A. The Most Novel Shield Tunnel Technologies: p.119: Nikkei BP. (in Japanese)

Matsumoto,Y. 1994B. The Most Novel Shield Tunnel Technologies: p.133: Nikkei BP. (in Japanese)

Nishizawa,K., Shiotani,T., Tsutiya,K., Mima,K., Hashimoto,T. & Nagaya,J. 1995. The Field Measurement of the Earth Pressure on the Segments in the Diluvium Sand. Proceedings of the 50st annual conference of the JSCE: 3, pp.1318-1319: JSCE. (in Japanese)

Nishizawa,K., Shiotani,T., Tsutiya,K., Mima,K., Hashimoto,T. & Nagaya,J. 1996. The Field Measurement of the Earth Pressure on the Segments in the Stiff Clay. Proceedings of the 31st Japan national conference on geotechnical engineering: pp.2289-2290: The Japan Geotechnical Society. (in Japanese)

Okada,H., Kishi,T., Yoshida,R., Sugishima,T. & Onawa,H. 1996. A Study on Behavior of Shield Tunnel Lining with Large Section and Large Depth at Sharply Curved Portion. Proceedings of Tunnel Engineering: Vol.6, pp.399-404 : JSCE. (in Japanese)

Okado,N., Yahagi,S., Oishi,K. & Komatsu,Y. 2000. Measurement Results and Discussions of a Three-centered Shield Tunnel with Different Diameters. Proceedings of Tunnel Engineering: Vol.10, pp.281-286: JSCE. (in Japanese)

Railway Technical Research Institute (RTRI). 1997. Design Standard for Railway Structures (Shield-driven Tunnel): pp.157-158: Maruzen. (in Japanese)

Tokyo Electric Power Co., Ltd. (TEPCO). 1998. Report on Improvement of Shield Tunneling. (in Japanese)

Modern Tunneling Science and Technology, Adachi et al (eds), © 2001 Swets & Zeitlinger, ISBN 90 2651 860 9

The Rehabilitation of Tunnel Structures

Henry A. Russell, P.E.
Parsons Brinckerhoff Quade & Douglas, Inc., Boston, Massachusetts, USA

ABSTRACT: Parsons Brinckerhoff has been a leader in the rehabilitation of underground structures and in particular highway and rail tunnels for over 20 years. Over that period of time Parsons Brinckerhoff has developed detailed procedures for the inspection and rehabilitation of tunnel structures and their associated systems. Most underground facilities that require rehabilitation are located in urban areas and are a major element of the infrastructure. The rehabilitation of these facilities requires that the facility be maintained in service during the repair/restoration work. As a result of these requirements, special procedures have been developed to facilitate the inspection, design and implementation of the repairs. These procedures require the use of specialty products, construction methods and in most cases prequalification of the contractor. This paper will discuss the procedures used for the inspection, design requirements, and in particular the use of specialty materials, construction practice, and current methodology using case histories to illustrate successful underground rehabilitation projects.

1 INTRODUCTION

Highway, rail and fluid conveyance tunnels are vital components of the metropolitan areas that they serve. These systems represent enormous long-term public investment by the community's municipalities and regions that they serve. Without the use of tunnels and their subsystems vehicular commerce, public transportation and the transfer of water and wastewater in a region would be paralyzed. Many of these systems are aging and are nearing the end of their original design life. Numerous operators of these systems have embarked on a systematic program of inspection, prioritization and the rehabilitation of these tunnels to extend their service life well into the 21-century. The rehabilitation of these underground structures is usually triggered by an event that has caused an interruption in service, or a failure of components in the underground system that has provided inefficient operation of the underground facility. The regions daily dependence on these underground facilities requires that the inspection and rehabilitation of the facilities be performed during periods of limited shutdowns. This inability to have possession of the site for long periods of time necessitates a different design approach, the repairs must be able to be performed in a limited time, be suitable for service within a very limed time and provide for long-term

economic service. For example in the case of highway and transit tunnels the site is generally available during off peak hours usually at night, and requires the facility to be placed back in service within four to six hours. Fluid conveyance (water and wastewater) tunnels generally may be placed out of service for longer periods of time, since they usually have redundant alternative conveyance systems. I will discuss the parameters for the inspection and rehabilitation of the more difficult vehicular and rail tunnels, with an emphasis on the structural inspection and rehabilitation of the structures.

2 TUNNEL INSPECTION

The purpose of a tunnel inspection is to observe and document current physical conditions. A tunnel condition survey is most useful if clearly conducted using established standardized criteria for determining the quality of the tunnel elements inspected. This information is to be documented for use in establishing the prioritization for the repairs and to establish a long-term data base for the operator of the tunnel to observe the long-term performance of the tunnel, and the prevention of future catastrophic structural or system failures that would cause the tunnel to be closed for long periods of time.

These tunnel surveys are usually conducted during periods of limited use during non-revenue hours; those hours when transit and highway tunnels may be shut down without creating unnecessary burdens on the system users and its operators.

2.1 Inspection Personnel Staffing

The effective management of a tunnel inspection and rehabilitation project requires the selection of appropriate staff for many varied functions, each requiring certain minimum standards of experience. The required staff varies with the magnitude of the program and the time constraints imposed by the construction contract or the operation of the facility. In general, a tunnel inspection program usually consists of specialists in structural, electrical, and mechanical, ventilation tunnel design and operation. Each of these disciplines require that the personnel be led by a Senior Technical Engineer/Specialist experienced in the design and operation of the tunnel and its systems. The function of the senior personnel is to develop the criteria and supervise the day-to-day inspection and provide quality control and quality assurance for the project.

2.2 Detailed Inspection of the tunnel

The detailed inspection of a tunnel is performed as follows:
- Collection of existing data
- Development of a Project Safety Plan
- Project Schedule
- Coordination of Field Operations
- Preliminary Inspection
- Database Management Procedures
- Inspection Parameter Selection
- Survey Control

2.21 Collect Existing Data

The collection of existing data is key in providing a working knowledge of the original design and construction of the tunnel system. This working knowledge is only obtained through exhaustive review of all existing plans and specifications, including original plans and all subsequent modifications to the tunnel. This information provides the basis for interpreting the inspection data as it pertains to observed tunnel defects.

The plans and specifications are usually obtained from the owner of the tunnel and are often supplemented with construction photographs. It is also useful to contact historical societies and in the case of transit facilities transit enthusiasts clubs for information. All of this material must be systematically catalogued and filed for rapid retrieval.

2.22 Project Safety Plan

The inspection of tunnels is often performed in areas where a potential for hazardous gasses, concrete/rock falls and other operational conditions may place inspection personnel at risk. A Project Safety Plan is required to protect inspection personnel and must be developed with the cooperation of the tunnel operator and identify the hazards of the workplace. This safety training is essential for the safe efficient and rapid inspection of the tunnels and its associated structures.

2.23 Project Schedule

The project schedule is developed as a function of the scope of work within the contract. The schedule must be developed by personnel experienced in tunnel inspection and rehabilitation with consideration made for available staff, hours of access and contract milestones. The schedule should be sufficiently flexible to allow for modifications based on the progress of the inspection, and special unforeseen operational requirements. Typical inspection progress, shown in Table 1 is a typical guideline for the structural inspection, using a three-person team.

2.24 Coordination of Field Operations

The coordination of the field operations requires close day-to-day contact with the operations and maintenance personnel of the tunnel to be inspected. Rail systems in particular require specific train orders for personnel to be present on the right-of-way. Transit systems are very safety conscious and require strict adherence to specific safety procedures by individuals on the systems right-of-way Usually these requirements consist of notification to rail traffic of the presence of personnel and the use of trained flagmen for protection of the inspection team.

2.25 Preliminary Tunnel Inspection

The preliminary inspection of the tunnel by senior personnel is a key element in effective inspection of the tunnel structure. This inspection is performed by selected members of the of the project team who have extensive experience in the inspection or design

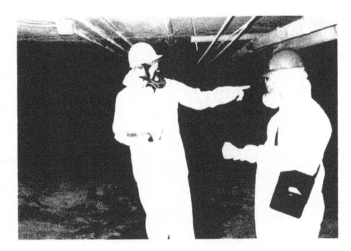

Figure 1 Inspection Team in hazardous location wearing protective equipment

Table 1. Tunnel Inspection Rates

Tunnel Type	Inspection Rate (M/hr)			
Arch				
Single Cell	75- 120	53	38	22
Double Cell	60 – 90	45	30	15
Circular				
Single Cell	83 - 136	61	46	22
Double Cell	75 - 106	54	46	15
Arch				
Single Cell	53 - 121	46	38	15
Double Cell	45 – 60	38	75	10

Notes:
1. Rates based on 3- person teams
2. Tunnel Condition (percentage of surface wall area)
- Good = 10 % defects
- Average = 20=30 % defects
- Fair = 30-60% defects
- Poor = More than 60% defects

Figure 2 Hand-held data recorder

Table 2. Standard Structural Inspection Codes

Symbol	Identification
Concrete Spalls	
S-1	Concrete spall less than less than 5 cm
S-2	Concrete spall to reinforcing steel
S-3	Concrete spall behind Reinforcing steel
S-4	Special Concrete spall (requires sketch)
Concrete Delaminations	
D	Delamination
Concrete Cracks	
C-1	Less than 3 mm
C-2	3 mm – 8 mm
C-3	8 mm – 12 mm
C-4	greater than 12 mm
Concrete Joints	
J-1	Joint less than 3 mm
J-2	3 mm – 8 mm
J-3	8 mm – 12 mm
J-4	Greater than 12 mm
Reinforcing Steel	
R-1	Surface rust
R-2	Loss of section (%)
R-3	Out of Plane H= Horiz., V= Vert.
R-4	Broken
R-5	Buckled
R-6	Other (requires sketch)
Framing Steel	
F-1	Surface rust
F-2	Loss of section
F-3	Out of Plane H=Horiz. V= Vert.
F-4	Broken
F-5	Buckled
F-6	Other (requires sketch)
Steel Liner Plate	
SP-1	Surface rust
SP-2	Loss of section (%)
SP-3	Out of plane H= Horiz. V=Vert.

SP-4	Broken
SP-5	Bucked
SP-6	Other (requires Sketch)
Steel Liner Plate Flanges	
FL-1	Surface Rust
FL-2	Loss of Section (%)
FL-3	Out of plane H= Horiz., V=Vert.
FL-4	Other (requires sketch)
Precast Concrete Liner Segments	
SC – 1	Concrete spalls less than less than 54mm
SC-2	Concrete spall to reinforcing steel
SC-3	Concrete spall behind reinforcing steel
SC-4	Other (requires sketch)
SCD	Delamination
Bolt Connections	
B-1	Surface rust
B-2	Loss of section (%)
B-3	Out of plane H= Horiz., V= Vert.
B-4	Broken
B-5	Buckled
B-6	Other (requires sketch)
B-7	Bolt missing
Tunnel Brick	
BR-1	Loss of section, depth= mm, area = M
BR-2	Loss of mortar SM
BR-3	Other (requires sketch)
Dimension Stone	
DR-1	Loose stone, area =SM
DR-2	Open joints, area = SM
DR-3	Failed, area = SM
DR-3	Other (requires notes)
Tunnel Moisture	
D	Dry
PM	Past Moisture
GS	Glistening Surface
F	Flowing, (cc/day)
M	Moist

of tunnel structures. Typically this inspection team should consist of a structural engineer, geotechnical engineer, materials specialist, and a specialist in any other special discipline (i.e. electrical, mechanical engineers) required by the scope of work. This inspection is performed at an average rate of 2 km per hour and it's purpose is to:

- Observe the condition of the tunnel and it's subsystems.
- Make recommendations for prioritization of elements to be inspected.
- Provide guidance for the development of the project schedule.
- Assist in development of repairs.

2.26 Data Management Procedures

Management of the inspection data requires the development of a system for the flow of the information form the field to the office and back to the field for a quality assurance check. This orderly transfer of the information is essential for the efficient documentation of the data and the required quality assurance necessary for a successful project. Typically the data is collected with the use of hand held data recorders of the non-progressive type and is downloaded on a daily basis to a central computer located in the project office. The data is recorded in a database utilizing Microsoft EXCEL as the data base program. Excel was selected due to its ease in operation and the widespread accessibility to owners, and its ability to be incorporated into CADD drawings.

In order to use a database program for the management of the data standardized inspection parameters have been developed. The coding of the information, necessary for the consistency of reporting also aids to assure quality control by providing guidelines for the inspection personnel and standardizing visual observations. Table 2 illustrates some typical standard coding for cataloguing tunnel defects.

2.26 Survey Control

All condition (inspection) surveys require a definitive baseline for location (survey) purposes. This requirement is the same for the inspection of tunnels. Generally most tunnel systems have an established survey baseline. These survey baselines were established for the original construction or in the case of rail tunnels for the maintenance of the track. The base line is usually identified on the walls of the tunnels or at the portals. In rail tunnels one must be cautious since the baseline is laid out at the centerline of the track and will not always match a base line established for the structure due to changes in track alignment.

The tunnel inspection must be tied to the existing baseline stationing system used by the owner/ operator of the tunnel for the following reasons:

- Allows inspection data to be used for long-term monitoring of the tunnel structure by the owners/ engineering/ maintenance staff.
- Rapid start-up of inspection teams.
- Reduction for project costs and confusion.

In addition to locating the identified defect along the alignment it is necessary to identify it in relation to its position in the structure. To accomplish this location in the tunnel, the limits of the walls roof, and invert must be delineated for conformity as shown in Figure 1. Circular tunnels are divided into 30-degree sections clockwise from the tunnel crown. This delineation is always performed looking up station on the established survey baseline.

3 TUNNEL CONDITION SURVEY

The inspection of a tunnel is best performed by a three-person technical team. The inspection always proceeds up station along the established survey baseline. The team's personnel are assigned sectors, limits of the tunnel cross-section for inspection as shown in the previous section. The roof and crown of the tunnel are generally inspected visually unless serious defects are noted and the area is inspected by sounding with hammers or by the use of non-destructive testing, using portable truck mounted scaffolding.

3.1 Simultaneous Inspection

As the inspection progresses, two members are each responsible for the inspection of assigned areas of half of the tunnel, with the third member used as a recorder. It is not always necessary to assign a third member of the team to act as a recorder, but is more efficient way to inspect more tunnel per actual hours of inspection. Unless otherwise required all of the elements of the tunnel should be inspected simultaneously and coded as shown in the previous section. This simultaneous inspection allows for rapid inspection and efficient use of the limited hours for the tunnel inspection. It is also necessary because these elements function in conjunction with each other; they must be analyzed and evaluated as an all-inclusive unit rather than a series of small independent elements.

The inspection is performed in 50-meter increments for each cell of the tunnel. A 50-meter fiberglass tape or roller wheel is used to station the 50-meter section, from 0 to 50 meters. If a roller wheel is used the third member of the tram uses the wheel to locate the team, otherwise the location is read from the tape. The location within the tunnel is identified as shown in Figure 2. At this time the defect is recorded, using the codes for standard inspection parameters. The recording of the data may be performed using the traditional pen and paper system or most commonly on large projects the use of a computer based data recorder. All coding is as provided in Table 2. For example, a concrete delamination with sever cracking, on the right wall (moving up station) at station 121+56 is recorded as follows:

Station	Location	Identification	Remarks
121+56	R	D	severely cracked

3.2 *Calibration of Inspection Crews*

It is necessary to standardize observations of the inspection teams. The standardization/calibration of the crews is to be performed at the outset of the field inspection. Having a "dry run" of a 50-meter segment of the tunnel performs the calibration. This "dry run" is to be an independent inspection of the tunnel section by a senior member of the staff and by the inspection team/s who will perform the actual collection of the data. The two inspections will then be compared for accuracy, and additional training

will be performed at this time, if required, and an additional check section for inspection calibration may be required. This method of calibrating the inspection team/s has proven to be very successful in assuring similar observations among all personnel involved in the collection of field data.

3.3 *Special Testing*

In locations where structural analysis is required, detailed information as to the makeup of the materials or the presence of certain elements in the

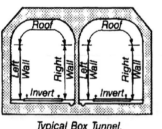

Typical Box Tunnel,
Double Cell

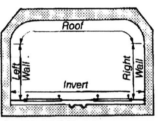

Typical Box Tunnel,
Single Cell

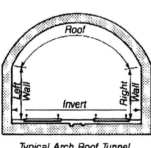

Typical Arch Roof Tunnel,
Single Cell

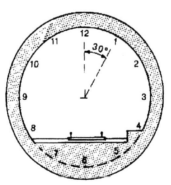

Typical Circular Cell Tunnel

Figure 3 Tunnel cross sections for inspection

TABLE 3. Special Tests

In-Situ Concrete Strength Special Tests

Type	Test Method	Property Measured	Region Tested
Semi-Destructive			
	Pull Out Test	Indirect Shear/Tensile Strength	Surface Zone
	Break-Off tests	Flexural Strength	Surface Zone
	Winsor Probe	Penetration Resistance	Surface Zone
	Tescon Probe	Stress-Strain Relationships	Internal Zone
	Test Cores	Strength	Internal Zone
	Cartbonization	Depth of CO_2	Surface Zone
	Petrographic Analysis	Composition	Internal Zone
Non-Destructive			
	Rebound Hammer	Resilience	Internal Zone
	Ground Penetrating Radar	General Condition	Internal Zone
	Refraction Survey	Micro Cracking	Internal Zone
	Resistivity	Steel Corrosion	Internal Zone
	Ultrasonic	Elastic Modulus	Surface and
	Pulse Velocity	Density/using Poisson's Ratio	Internal Zone
	Thermography	General Condition	Surface Zone

Steel Tunnel Elements

	Test Method (US)	Property Measured
Destructive		
	ASTM E8	Strength
	ASTM E10	Hardness
	ASTM E18	Hardness
	ASTM A370	Strength of Steel Element
	ASTM E390	Quality of Welds
Non- Destructive		
	ASTM A325	Torque of Bolted Connections
	ASTM 490	Bolt Connections
	Radiography	Quality of welds

structural unit that will require special testing. Special tests are those tests, which are utilized to determine certain characteristics of the materials, which comprise the tunnel system. Table 4 illustrates typical special testing commonly used to determine the condition of the structure.

These testing techniques are based on the measurement of different material properties such as thermal conductivity, (thermography), electrical conductivity, and direct current (electromagnetic, galvanic electrical and ground penetrating radar), and the velocity of a stress wave within the material, (sonic/ultrasonic). Thermal and electrical properties are dependent on the moisture content of the concrete substrate. All of these methods require some physical testing such as concrete cores to calibrate the measurements and rely heavily on the experience of the analyst. However, nondestructive sonic/ultrasonic wave measurements have been successful in determining the properties of the tunnel liner by measuring the deformation moduli value using Young's bulk shear and unconfined compressive strength. This system is particularly effective to determine the presence of micro-cracking and large areas of honeycombing of the concrete liner. Ground penetrating radar, (GPR) identifies the presence of metal reinforcing, voids, and water within the liner or other concrete components of the tunnel.

45

TABLE 4. Geophysical Test Method Comparison

Test Method	Cracks Macro/ Micro		Chemical Alteration	Delamination	Reinforcing Steel	Moduli Values	Strength
Sonic/ Ultrasonic, Compressional Sear, Velocity & Attenuation	X	X	X	X		X	X
Sonic/Ultrasonic Reflection/Resonance				X		O	X
Radar (GPR)	O		O	O	X		
Thermography	O		O	O			

Notes:
"X" indicated best-suited method of for identifying item; "O" indicates questionable results dependent on conditions; No comment: means not generally used for identifying item

Thermography (infra-red) and Resistivity are not particularly successful for use in determining the condition of tunnel structural components. Thermography generally will not penetrate deep with the concrete element and the accuracy is questionable if the substrate is at the same ambient temperature as the air within the tunnel. Reisitivity has been used for many years with limited success by measuring the potential between two fixed points on the liner. This system has difficult in providing accurate measurements in electrical transit tunnels and in tunnels with active cathodic protection systems. Table 4 illustrates a comparison between various non-destructive test methods.

3.4 Evaluation of the Data

The culmination of a survey is to determine the status of the tunnel elements inspected. This determination is based on a review of the original design and the performance of that design. The condition survey will also provide valuable information for future use by the owner/ operator of the tunnel system to manage the capital and maintenance budgets.

The inspection data must be preserved in an easy to understand format for future use and comparisons. This is best performed in a simple database format with spreadsheets used to document the factual information collected. In addition to presenting the data in a logical format a portion of the final report must contain an analysis of the observed conditions, and recommendations for future inspections frequency and the repair and maintenance of the tunnel system.

In addition this data maybe sorted for use in future rehabilitation contracts. The data is commonly sorted by station and summarized by defect on the contract documents. In addition to illustrating the quantities of the tunnel defects to be repaired, cross sections of the tunnel are provided along with a location plan. This easy to read format provides important information to the contractor, including horizontal and vertical clearances within the tunnel and the actual quantities of the work to be performed.

4 Case Study

The Sumner/ Callahan Tunnel Project Boston, Massachusetts USA

4.1 Introduction

The Massachusetts Turnpike Authority, (MTA), operates two cross-harbor tunnels in the city of Boston Massachusetts, USA. These tunnels connect downtown Boston with East Boston and Logan International Airport. The tunnel system consists of twin two lane circular tunnels approximately 1,661.5 meters in length. Each tunnel has approximately 50,000 vehicle trips per day, with the majority of the trips occurring between the hours of 0500 hrs and 2300 hrs. These tunnels are an integral link in the infrastructure of Boston and as such have not been shutdown or had extensive maintenance since the 1960's. In 1989 the MTA initiated a comprehensive program to rehabilitate these tunnels, due to minor structural failures of the tunnel ceiling and degradation for the roadway slabs due to chloride contamination as a result of road salting outside of the tunnels. Parsons Brinckerhoff in association with Sverdrup Corporation was retained as joint venture to catalogue the tunnel defects and to perform as a Construction Manager for the rehabilitation of the tunnels. Parsons Brinckerhoff was in the lead in this joint venture in the performance of this work. The MTA required that all of the work be performed during off-peak hours between the hours of 2300hrs and 0500 hours on a daily basis with no work performed during special holiday periods where holiday traffic at the airport was intense and other periods of high traffic volume due to special events within the city.

4.2 Tunnel Description

4.21 Sumner Tunnel

The Sumner Tunnel was opened to traffic in 1934 and is a two lane semi-transverse circular tunnel with bi-directional traffic. The tunnel roadway slab was constructed over the fresh air duct and a ceiling was installed over the roadway for the exhaust duct. The interior finish of the tunnel was a ceramic tile with the walls and tunnel ceiling being coated with the tile. The exhaust air was carried out by the means of an exhaust duct constructed over the roadway slab. The vertical clearance within the tunnel was 4.15 meters and each lane is 3.38 meters in width. A one-meter walkway was constructed on the west side to allow for operational personnel to enter the tunnel. The tunnel was constructed under compressed air and with the temporary lining being segmental steel liner plate. The liner was nine meters in diameter. With the final liner being cast-in-place concrete, 50 cm in thickness.

In 1960 the ceiling in the tunnel was removed and replaced with a cast-in-place concrete ceiling. The ceiling was removed because of extensive cracking due to improper placement. At the same time as the ceiling was replaced the granite cobblestone roadway wearing course was removed and replaced with bituminous concrete, and the signalization was modified to change the traffic flow to unidirectional.

4.22 Callahan Tunnel

The Callahan Tunnel was constructed in 1960 due to the increased need for an additional tunnel to serve the growing Logan International Airport. This tunnel was constructed to be unidirectional with the Sumner Tunnel to be converted to unidirectional traffic after the opening of the Callahan Tunnel. The tunnel is semi-transverse with a vertical clearance of 4.15 meters and two lanes of 3.68 meters in width. A one-meter walkway was constructed on the east side of the tunnel for the use of operational personnel. The tunnel was constructed by the use of compressed air as in the Sumner Tunnel. The tunnels primary liner is high strength steel with the final liner being concrete infill of the liner of 25 cm in thickness. The tunnel finish is ceramic tile on the walls and ceiling. The concrete roadway slab is supported by steel floor beams approximately 75 cm on centers.

4.23 Inspection Program

In 1989 a comprehensive inspection program was instituted to inspect all structural mechanical and electrical elements of the tunnel system. The inspection was carried out in two phases: The preliminary phase occurred during the first week of the program. In this phase, senior tunnel, electrical, geotechnical, and mechanical engineers familiar with the design of tunnel systems viewed in depth the tunnels system and identified items to be inspected in depth and other ancillary items to be catalogued. The Senior Inspection team after conferencing developed a series of standard codes for our database to properly categorize the elements identified and their condition. Once the parameters were developed, a meeting was held with the owner to identify any special elements they wanted to be inspected and these elements were incorporated into the database for inspection. The detailed inspection was carried out using two-three person teams of each specialty, (i.e. structural, mechanical, electrical etc.). The data was collected during the nighttime hours and entered into a database on a daily basis.

Once the data was collated the information was prioritized to develop a five year program to systematically correct the tunnel system defects. The prioritization of the work was to develop construction contracts and to assist the owner (MTA) in the development of their capital and maintenance programs. The following criteria were used to evaluate the system components:

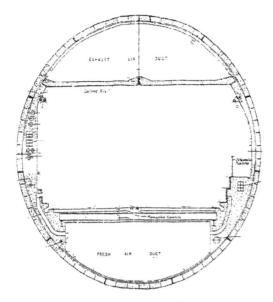

Figure 4. Typical tunnel cross section

- Immediate (repairs to be performed within 90 days)

- Priority (repairs to be conducted within one to three years)
  P-1 Repairs within one year
  P-2 Repairs within two years
  P-3 Repairs within three years

- Routine (repairs to be conducted within three to five years or earlier if funded and convenient)

  P-4 Routine in nature not necessary for system function

P-5   Routine   tasks,   normal maintenance
N   No work required

Based on this evaluation, it was determined that the most critical repairs were as follows:

- The replacement of the tunnel ceiling in the Callahan Tunnel.
- The restoration of deteriorated structural concrete in the tunnel liner of both tunnels.
- The replacement of the tunnel ceiling in the Sumner Tunnel.

4.25 Callahan Tunnel Ceiling Replacement

The Callahan tunnel ceiling was a cast-in-place concrete ceiling separating the roadway from the exhaust duct. The ceiling was finished on the under side with ceramic tile set in a mortar bed. Over the years the washing of the tunnel and the accumulation of diesel exhaust had created a mild sulfuric and carbonic acid, which attacked the mortar creating a debonding of the ceiling tile. In 1989 nine vehicles were damaged from falling tile, so it was determined that this rehabilitation work was to take priority.

The MTA and Parsons Brinckerhoff reviewed the types of tunnel ceiling used in the United States and in Europe and evaluated each ceiling type for its long term performance. The types reviewed were:
- Painted/unpainted cast-in-place concrete
- Tiled finished cast-in-place concrete
- Painted deep dish steel pans with concrete infill
- Porcelain coated deep dish pans with concrete infill

Figure 5. Rotary head milling machine for grinding existing tunnel ceiling

48

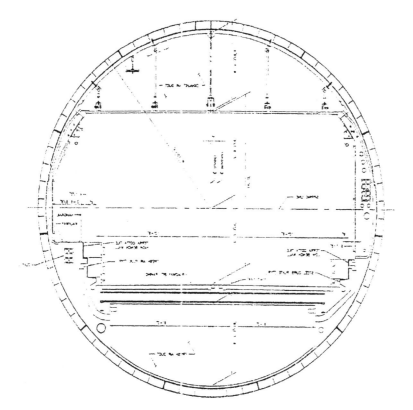

Figure 6. Typical cross section of new tunnel ceiling

- Aluminum deep dish with concrete infill
- Porcelain coated totally encapsulated steel panels with fire retardant infill

After careful evaluation it was decided that the most durable of the ceiling types in use was the porcelain coated totally encapsulated steel panels with a fire retardant infill. This type of panel was chosen due to its resistance to water tightness, resistance to rusting, high gloss and ease of installation.

A design was instituted with the ceiling being supported by five hangers of galvanized steel with all of the hardware to be stainless steel. The panels were to be installed in modules and could be installed within the four hour working window as directed by the owner.

4.26 Ceiling Installation

The existing 15cm cast-in-place ceiling was removed by the use of a concrete shear that would nibble the concrete ceiling and shear the reinforcing

at a rate of 33 meters a work shift. In the portal sections under the ventilation buildings the existing ceiling was ground using a special rotary head grinder to remove sufficient concrete for the installation of a porcelain panel. The debris was dropped into two large dump trucks and hauled to an offsite storage area for disposal during the day. During the preliminary construction it was noted that the soot on the top of the ceiling was contaminated with lead from the combustion of leaded gasoline (petrol) and, as such would make all of the debris for the ceiling demolition become a hazardous material and therefore increase the disposal cost by over 300%. In order to prevent this additional cost the ceiling work was stopped and a power wash and vacuum-cleaning program was instituted and the work was delayed one month for the cleanup. Once the cleanup was completed, the concrete from the ceiling was placed in an ordinary landfill.

The new ceiling was placed at a rate of 30-40 meters a shift and was installed approximately 50 meters behind the demolition work. The reason for this open space was to ensure that the two activities did not interfere with each other.

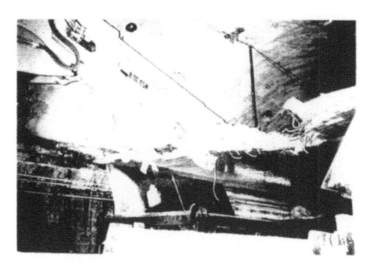

Figure 7. Concrete demolition of existing ceiling

Figure 8. New porcelain ceiling Callahan tunnel

Figure 9. Deteriorated tunnel liner concrete

The ceiling installation procedure was to set the hangers for approximately 30 meters at a time using polyester resin anchors for the anchorage in the lining of the tunnel. The hangers were set to grade using an electronic leveling device. Once set to grade a specially fabricated erection machine would lift panels that were the full width of the tunnel 6.76 meters wide by 4 meters long would be erected and

Figure 10. Concrete reinforcing; coated with zinc-rich coating

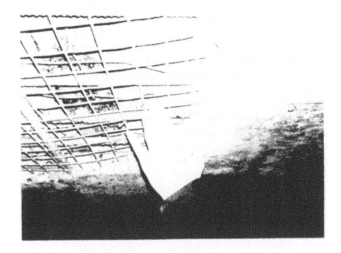

Figure 11. Welded wire mesh and coated reinforcing steel with guide wire for shotcreting

Figure 12. Application of prepackaged polymer-modified shotcrete

attached to the hangers by the use of pin connections. All connection bolts were double nutted and peened to secure them from vibration.

4.27 *Structural Concrete Repairs*

The roadway slab and the tunnel liner in the area of the fresh air duct below the roadway was found to be severely deteriorated as a result of road salts being carried into the tunnel from the vehicles and once, mixed with the drainage water in the tunnel caused severe chloride attack on the concrete and in some instances reduced the concrete to a gravel. Due to the limited time allotted for the repair work and the necessity to return the roadway and tunnel to traffic within 5 hours of the start of work, PB worked with Fosroc (now THORO in the U.S.) to develop a rapid setting high strength polymer modified shotcrete that would allow the work to progress without interfering with the revenue stream of the tunnel system. Based on field tests an existing product called Rederoc SP was modified to meet the needs of the project. The material was a dry spray polymer modified shotcrete that was shrinkage compensated and chloride resistant. The material would have an initial set within 20 minutes and had a compressive strength of 5000 psi (34.5MPa) @ 24 hours with an ultimate strength of 8000psi (55.2 MPa) at 28 days. The use of a polymer modified shotcrete also provided for minimum rebound and dust to be created, which would have caused severe problems for the operating ventilation system.

The concrete surfaces were prepared for the shotcrete by removing all of the deteriorated concrete, coating the reinforcing with a zinc rich coating and the addition of 10X10 cm welded wire mesh and placing the shotcrete. The spalls varied in depth from 4 cm to over 15 cm. The entire invert of the fresh air duct was covered with a flash coat (3 cm thick) for protection of the existing concrete lining from chloride attack. The flash coat was approximately 40,000 m^2. No shrinkage cracking was noted.

5 Conclusions and Lessons Learned

The inspection and rehabilitation of an operating tunnel system requires a different inspection and design approach. The inspection program must be developed to accommodate the owner's requirements and his operational schedule. Standardized codes and procedures for the identification of tunnel defects must be applied consistently and in a calibrated system. The data must be sorted and reviewed constantly during the inspection to ensure good quality assurance and accuracy. The design of the rehabilitation must be cognizant of the fact that the site is only available for a limited period of time, with little or no storage space. All materials must be transported into the tunnel on a daily basis and removed at the end of each shift. The materials used must be capable of being installed and in service in a short period of time and therefore specialized materials must be utilized. This use of specialized materials and a short work window causes the rehabilitation of an operating tunnel to be more costly than one that is closed for a long period of time.

In addition the work requires contractors who are knowledgeable in this type of work and all contracts should have clauses for pre-qualification to ensure project success. The owner, designer, and contractor must be flexible in regard to changes in site conditions and must work as a team to allow for modifications to the contract in a timely manner.

References:

ACI Committee 318, *Building Code Requirements for Reinforced Concrete*, American Concrete Institute 1983.

American Society for Testing and Materials, *Annual Book of Standards, Part 2,4,10,19, ASTM 1999.*
American Institute of Steel Construction, *Manual for Steel Construction,*, AISC, 1992.

Bikel, J.O., *The Tunnel Engineering Handbook*, Chapman &Hall New York, NY, 1993.

Lincoln Arc Welding Foundation, *Design of Welded Structures*, Lincoln Arc Welding Foundation 1999.

Massachusetts Turnpike Authority, *Callahan Tunnel Ceiling Contract No.0072*, Boston Massachusetts, 1991.

Massachusetts Turnpike Authority, *Sumner Tunnel Ceiling Contract No. 0073*, Boston, Massachusetts, 1994.

Matthews, R. *Tunnel Manual*, Report DOT-TST-76-40 Monitor 18, 1975.

Romanoff, *M Underground Corrosion*, U.S. Department of the Interior, National Bureau of Standards, 1957.

Russell, H.A. *The Inspection and Rehabilitation of Transit Tunnels,* William Barclay Parsons Fellowship Monograph 3 , 1987, New York, NY.

Voort, H.B. *Tunnel Instrumentation, Shotcrete for Ground Support*, Proceedings Engineering Foundation Conference, 1976.

Whitehurst , E.A., *Evaluation of Concrete Properties from Sonic Tests*, ACI Monograph No. 2 American Concrete Institute, 1996.

Modern Tunneling Science and Technology, Adachi et al (eds), © 2001 Swets & Zeitlinger, ISBN 90 2651 860 9

Underground Space Networks in the 21st Century:
New Infrastructure with Modal Shift Technology and Geotechnology

T. Hanamura
Okayama University, Okayama, Japan

ABSTRACT: The focal point of underground space use is the regeneration of city by relocating numerous city functions underground in order to reinforce, develop and improve urban infrastructure for the future enrichment of society. Underground space is no more optional space but necessary space for the city. In the 21st century, the underground space will burst into use in many parts of the world. Subways will become common transportation means in most of large cities even in the developing countries, and public utilities and lifelines do the same. The major feature that will emerge in the 21st century is the network use for infrastructure system. Networks will be strengthened in the transportation systems as well as telecommunication and electric power networks. Sewerage and drinking water systems will be networked among treatment plants and purification plants. Storm water reservoir tunnels will also be networked. In order to respond to this need, new technologies have to be developed. First is the network development of infrastructure with the new modal shift concepts and technologies. Modal shift concepts and technologies that have already been adopted in the freight transportation systems will be applied to the network of underground infrastructures. Second is the methodological use of deep underground with geo-technology. Third is the geo-technical development to use deep and large underground space. Geo-technological issues are the design and construction technologies to use deep underground and to create the large caverns and tunnels.

1 HISTORICAL OVERVIEW OF UNDERGROUND SPACE USE

1.1 *The Start of Modern Use of Underground Space in 19th Century*

The modern use of underground space started practically in Europe in the mid-19th century. They were a sewerage system in Paris and an underground railway system for a subway line in London. Paris sewerage system was developed by Baron Haussmann, Prefect of the Seine, and the engineer Eugene Belgrand in 1850's. Paris sewerage lines had been constructed under almost all streets. This system has become a big asset for Parisian, for it accommodated later on drinking water lines, electric power lines, telecommunication lines, etc. in the sewer tunnels and functioned as multi-purpose tunnels.

First underground railway system was opened in London in 1863. First subway system used steam locomotives at the beginning of service. It caused serious air pollution problems together with other steam engines. However soon after, the electric engine had been introduced and applied to the subway system.

Both sewerage and subway systems influenced lots on European countries and USA in the early 20th century and have developed throughout the world in the 20th century.

1.2 *Underground Space Use in 20th century*

Underground space has been extensively developed and used in the 20th century in the urban areas of the cities. They are underground traffic systems of railway and road, public utilities and lifelines of drinking and sewage water, electricity, gas and telecommunication, and human activity space as underground shopping mall, car park and pedestrian passage.

1.2.1 *Underground Space Development in Advanced Countries in the First Half of 20th Century.*

Underground space use was limited mostly in advanced countries of Europe and US in this period.

Sewage systems had applied and developed to most of large cities in Europe and US. Public utilities and lifelines had also been developed in many cities using underground. However subways had been limited only in the big cities. Big cities needed mass transport-means in the city. Outside the advanced countries described above, only Tokyo, Osaka in Japan and Buenos Aires in Argentine had subway systems in this period. Together with subway development, underground shopping malls and passages had been constructed and developed.

1.2.2 *Urban Congestion Problems and Underground Space Developments*

The 20th century can be called the century of industrial society with mass production and high efficiency as its goals. Developed countries succeeded in achieving the mass-production capacity of an industrial society. Advanced countries have developed underground facilities extensively such as subways, roads and highways, sewerage, drinking water, electric power and telecommunication lines, shopping malls, and other facilities. Industrialization with mass production system has spread to the developing countries. Because of the development of worldwide economy and efficient transportation system for the passenger and freight, people and goods have been migrating very quickly throughout the world. Economic liberalization pushes industries to find less expensive labors and materials in worldwide market. This economic movement pushes cities of each nation to be industrialized and urbanized. Hence, the most of big cities in the world have faced urban congestion problems. Take the traffic congestion for instance; almost every city experiences this problem. This is particularly serious in developing countries, which experience a heavy migration of their population towards the larger cities as a result of the push-out effect of local poverty.

1.2.3 *Underground Space Use from Advanced Countries to Developing Countries*

Though the underground space use had started at first to find more comfortable solutions in large cities in developed countries, it has expanded throughout the world including developing countries to confront urban congestion problems. A new subway system was opened in Calcutta, India, in 1995 and a water supply tunnel for people and agriculture was opened in Manabi, Ecuador in 2000. In order to alleviate congestion problems, it is a must to use urban space more efficiently in three dimensions. Underground public utilities and lifelines have been widely used in many cities in the world. Underground space is no more optional space in developed countries only, but necessary space for the city even in developing countries.

2 UNDERGROUND SPACE USE FOR INFRASTRUCTURES

2.1 *Subway as a Key Transportation System in the Large City*

The New York subway system has almost one hundred year history. NY Subways function as both suburban railways and the city MRT (Mass Rapid Transit). Each weekday 4.3 million people use NY subways and about 1.3 billion passengers a year. Largest annual rider-ship is Moscow subway of 3.2 billion, and follows Tokyo of 2.7, Seoul of 1.6, Mexico of 1.3 and NY of 1.3 billions (New York City Transit 2001). And then Paris, Osaka, London, Hong Kong and St. Petersburg follow. Recently city of Los Angels has operated subways and is still constructing new lines.

Some US and European cities that use subways are less than one million in population. However in Asia, former USSR and South America, subways run only in big cities with a population of more than one million. At this moment subways in Mainland China are in operation in 5 cities and under construction in 3 cities. In Brazil, 7 cities are in operation and one is under construction. Thus subway has become key transportation system in large cities in the world. It has spread in big cities even in developing countries.

Therefore subways can be concluded to be inevitable in big cities for the large capacity of transportation that cannot cover by automobiles only.

2.2 *Underground Space for Freight Transport*

There are a few freight transport systems by using underground space. One is the postal train system in London. This has been used for more than 70 years. Train carries postal cargo among postal stations with two cars a train and four containers a car. Though the handling system is not modern enough from the present standard of technology, the underground train system works for postal cargo transportation.

The road passes underneath the Forum Des Halles in Paris and there is an unloading platform under the building. Schiphol airport in the Netherlands use underground space for baggage handling and transporting. Under the city center of Dallas there are three underground truck terminals, that have numbers of platforms of loading and unloading goods from trucks to some buildings around the place, and vice versa. Master plan for the city center of Dallas defined three levels as aboveground level for cars, 1st level of underground for pedestrian and 2nd level for trucks and goods handling. Disney World in Orland, USA has a complete underground delivery system, which transport merchandise and daily necessaries through the tunnel among a distribution center, 22 shops and other facilities such as rest rooms.

Channel tunnel between France and England has a function of tunnel for car transport by shuttle train. Cars coming to Folkestone station from British highways go into the double decked car train and when train arrives at the Calais station cars go out of car train and go into French Highways. In Europe some of long tunnels have such functions as shortcut tunnels for car transport by shuttle trains. The Simplon tunnel between Switzerland and Italy is one of them. The shuttle train plays a role of car transport.

2.3 Underground Space Use for Road and Highways

Roads and highways normally use aboveground space due to the need for exhaust fumes dissipation into the air. In special cases, however, roads and highways must utilize underground space in and around cities. One common use of underground space is for the underpasses of street and railway crossings. Other uses include utilizing underground space for environmental considerations.

The Central Artery of Highways in Boston is a typical example of using underground space for this purpose. An elevated highway goes down into a tunnel through downtown Boston and creates surface parcels. The surface is reserved for parks and open space.

Highways in Tokyo also run underground. Underground highways help to alleviate traffic congestion on the surface and some of the noise from the vehicles. In Metropolitan Tokyo, the Shinjuku Ring Road Line of the Metropolitan Expressway and the Ken-O-Dou, the outer ring road connecting inter-city highways outside of Tokyo, use underground space extensively.

2.4 Underground Space Use for Sewerage and Storm Water Reservoir & Tunnels

The sewerage system is a basic use of underground as described in chapter one. Sewerage systems are becoming complex and innovated. Singapore has just started to construct a mega infrastructure project called "Deep Tunnel Sewerage System" (DTSS). By replacing the existing system of sewage treatment works, the new DTSS links deep tunnel sewers to two new wastewater treatment plants located in southern coastal areas. Sewer water will be conveyed by gravity via deep tunnels to the treatment plants and the treated effluent will then be discharged through deep-sea outfalls (DTSS 2001). Special feature of this project is that Singapore government requisitioned 100-year durability. The DTSS has three faces of technology innovations that are the use of deep underground, the network

use of sewerage systems, and the long life durability of 100 years.

The storm-water reservoir for sewage or river is another kind of underground infrastructure, which has purpose of disaster prevention and environmental protection from storm-water disaster in the urban area. The Tunnel and Reservoir Plan (TARP), being put to use in Chicago, is really a unique concept for preventing disasters and for protecting the environment by storing water in tunnels. Before the TARP comes out, the combined sewage from homes and industries, along with storm runoff drainage, had to spill directly into rivers and canals as combined sewage overflow, when the combined sewage exceeded the capacity of the treatment plants. Untreated sewage, diluted with storm runoff, bypassed treatment plants and began to pollute lakes, rivers, and streams and caused flooding. Under the TARP plan, huge underground tunnels have been excavated underneath the city of Chicago, starting in the 1970s, to intercept combined sewage overflow and convey it to large storage reservoirs in the tunnels. And, for instance, after a storm has subsided, the overflowed water can be conveyed to treatment plants for cleaning before it is sent to a waterway. The U.S. Environmental Protection Agency provided nearly 75% of the funding for this project. The reservoir tunnels are about 10 m in diameter and have been bored in limestone rock from 60 to 100 m below the ground by a Tunneling Boring Machine (TBM). TARP projects continue to be conducted up to the present time, and approximately 85% of the tunnels have been completed. As each section of the tunnel lines is finished, it is opened for service. The entire TARP project is scheduled to be completed in 2004. The concept has been adopted in other U.S. cities, such as Milwaukee.

The TARP concept has influenced the handling of storm water in Japan. In order to avoid the inundation from flooding due to a heavy rainfall, the storm water can overflow and be stored in tunnels. After the storm has subsided, the overflow can then be conveyed to waterways. They are referred to as underground rivers in Japan, and many such facilities have been constructed in Osaka and Tokyo.

2.5 Other Infrastructure such as Multi-Purpose Conduits and Tunnels

Underground installation of electric power lines to buildings and houses are common in the city in Europe and in North America. Underground installation of electric cables is good for landscape. This installation system has been adopted in the central districts of big cities in Japan. High voltage power lines such as 1 million volts are inevitable to run underground from the safety, heat and magnetic insulations. As IT technologies develop, telecommunication cables, especially optical fiber cables are installed underground. Though lots of private or leased lines use underground for one purpose, cables of different kinds, drinking water lines and sewer lines use together multi-purpose conduits or tunnels. In Japan multi-purpose conduits or tunnels are commonly used under the street. Optical fibers are now installed in sewer lines as Paris Sewer tunnels housed other lines in the past.

3 UNDERGROUND SPACE USE FOR HUMAN ACTIVITIES

3.1 Enrichment of City Environment with Underground Space Use

Underground space will offer a comfortable environment in the city, under severe and hazardous climatic conditions like cold, heat, wind and rain. Underground space will offer convenient connections between blocks of cities through underground facilities. They can be connected at any depth and thus create three-dimensional use. This is quite different from high-rise buildings, which cannot connect easily in any floors of buildings.

Many people associate underground space with darkness, humidity and confinement and claustrophobia. However, those fears will disappear very easily, once people create and experience a nice underground environment. Elements for comfortable environment in underground space are lightings, ventilation & air conditioning, space size and interior designs, safety measures against fire protection including fire extinguishing and evacuation, and safety walks during night, acquisition of aboveground climatic information etc. Large underground atrium spaces with high ceilings will provide comfortable environment for those people using these underground areas. This large-scale underground space is impressive to the beholder and provides people with feelings of security and peace. It embraces people with a feeling of maternal warmth, which is expressed as the great mother feeling. Underground Olympic Ice Hockey Arena in Gjøvik, Norway has given strong impacts of underground warmth on many people worldwide.

Underground space is not only for convenient connections in the city, but also for the improvement in the aboveground environment for human life. The relocation of numerous city functions underground will reinforce, develop and improve urban infrastructure and activate aboveground by creating more open and human space. If the railway, highway, electric power or telecommunication line goes underground, landscape improves and greenery and open-space increases aboveground. The underground space use will enrich the city environment.

3.2 Typical Use of Underground for Human Activities

3.2.1 Underground Shopping Malls

Underground Shopping Malls have been constructed in many parts of the world in the 20th century. Forum Des Halles in Paris, Eton Center in Toronto, Place Ville-Marie in Montreal, the Gallery in Philadelphia and Queen's Square in Yokohama are some typical examples among them. These shopping malls have good connections to subways or intercity railways, car parks in underground, nearby other shopping malls with convenient walkways, convenient passages to nearby buildings for office workers. Take Montreal underground network for instance, it is called as "The underground city of Montreal". The main part of this underground city is the pedestrian network, which is accessible from subway stations, streets, via lobbies of the nearby buildings and their basements. This creates the large underground environment of indoor city that spread 30 kilometers of corridors, squares and commercial malls. Queen's Square Yokohama is a newly built terminal, shopping and office complex which uses underground and aboveground space at the same time and create a three dimensional urban human space. The atrium space will connect to the underground railway system directly in a few years. When the subway line is completed, it will house the huge atrium of 45m in height, which penetrate through 5th floor aboveground level to the 5th floor underground subway line level. Subway trains and station can be seen from the 5th floor aboveground level, and oppositely subway passengers can look up at the large atrium from the train or the platform. Subway level is approximately 30m below surface. The atrium is designed to be one body by connecting subway, underground shopping area and aboveground shopping mall. Forum Des Halles is another fine example, which creates huge underground building-complex of sunken garden type and conveniently connects to the subways and commuter railways under the buildings. The underground complex houses, not only shopping malls, but also museum, theater and swimming pool and others. It is a gigantic underground commercial and cultural complex.

All these underground shopping complexes conveniently connect to other facilities and make underground networks.

3.2.2 Underground Car- and Bicycle-Parks

3.2.2.1 Underground Car-Parks and the Network

There are lots of underground car parks. Some of them are inside the buildings. There are some of unique independent underground car parks. Large-scale underground car parks have been constructed in tunnels and in vertical shafts. One of them is a tunnel type of triple decked car park in Landsberg, Germany, which was constructed in large NATM tunnel of 18.9m wide and 16.4m high. Another is a unique helical car park under the Opera House in Sydney, in which helical parking lots are situated in the annular type of cylindrical and vertical shaft of

75 m in outer diameter and 60m deep below the sea level. Idea was from helical parking buildings in Chicago. Similar type of underground car parks in vertical shaft is also in Cesena, Italy. In Tokyo a unique car park was constructed and opened, which is situated in the canal under the water.

Most of car parks are independently located. However there are some underground car-park network systems in Japan. Here the private and public car parks are networked by the underground roads. One is in Hiroshima. Another is a comparatively large-scale network system in Tokyo, which has two main roads having around 400 m in one way and has two connection loops underground. Road system has completed and the car park system will open in around 2003 after the completion of nearby building complexes construction. Cars can choose either of four big car parks through driving the road in the area without disturbing aboveground traffic.

3.2.2.2 Underground Bicycle-Parks and the Bicycle-Park & Ride System

The underground bicycle parking area has been gaining attention in Japan. Commuters use bicycles from home to the nearest station. The number of possessor of bicycles amounted to 73 millions approximately which is roughly 0.6 bicycles per person in Japan. The small size motorcycles of engine capacity less than 50 cubic centimeters reached around 10 millions. In most cases suburban stations are inundated with bicycles and small motorcycles. In order to store these bicycles in order, bicycle-parking areas have been constructed in Japan. The numbers of bicycle parks reached around 9,800 and storing capacity reached 3,6 millions in Japan. If there is enough area in front of the station, bicycle-parking places are arranged nearby aboveground. However the area in front of the station has land values as the terminal places for cars, buses and buildings for stores in the neighborhood. Therefore municipalities and private companies are constructing bicycle parks above- and underground nearby the station, or on or under the railways. Most of bicycle parks are constructed aboveground. However the underground bicycle parks are optionally constructed in the busier stations. Underground bicycle parks are rapidly increasing. The Kita Asaka Bicycle Park in Tokyo is one of examples constructed and operated by the city.

The park and ride system is popular and functions well in the US. However densely populated countries like Japan, the car park needs large area and cannot arrange large place in front of the station. In stead bicycle parks can be easily arranged at the station. Commuters normally go to the nearest suburban station either by bus, by bicycle, on foot, or by car that is called "kiss and ride" and car goes back home after kissing the station by the partner's driving. Bicycles play important roles for commuting

from home to the station in Japan. Since bicycles are environmentally friendly, the bicycle-park & ride system is getting popular in Japan. This bicycle-park & ride system also functions an efficient modal shift in the suburban stations.

3.2.3 *Underground Passages*

Unique underground passages are existed. O'Hare Airport has unique underground passage between the terminal building and the island of gates to the airplanes. It has rainbow colored electronic illuminations on the side walls and neon tubes on ceiling that synchronize with music and moving belt lines on the floor. It is really beautiful and passengers enjoy moving into the islands. There are lots of nice pedestrian crossing under the intersection or crossing of the roads and highways. Some of examples are in London, Vienna, Montreal, Hiroshima, Seoul, etc. These underground passages connect to the underground shopping malls and make the network of underground passages or pedestrian walkways and shopping malls.

4 NETWORKING AND LINKAGE

4.1 *Importance of Networking and Linkage*

4.1.1 *Increase in Convenient Connections for Peoples' Movement*

Underground Space will offer convenient connections among underground and aboveground facilities. Underground passages connect each block of the city and people. Take the pedestrian walk networks for instance, it gives free and safe space for those people to move, and connects to the communities once separated by the roads in the car society. Networking among subway stations, shopping malls, parking areas of both cars and bicycles and pedestrian passages will activate people's movement. People feel reluctant to climb up and down by stairs against gravitational force. Therefore the networking on the same level is preferable. If there is a difference of levels among spaces, like the levels of subway platform, concourse, shopping mall, underground car or bicycle parks, some installations are necessary. Escalators and elevators are essential installations for easy way to reach different levels. Some ideas are necessary for the space to provide the feeling with comfort, liveliness or safety.

4.1.2 *Improvement on Traffic Congestion Aboveground by Transportation Linkage*

The linkage for transportation primarily aims to alleviate aboveground traffic congestion. Underground road and parking network systems will decrease the aboveground traffic jams and solve parking problems as well. Underground bus road in Seattle verifies this functions clearly. Aboveground roads are mostly used for cars and jams are reduced. Underground space provide convenient connecting points between the railway systems of subway and the road systems of bus and taxi. Smooth flow of people can be induced through the use of underground space. Underground road networks or car parks networks also provide efficient loading and unloading places which are directly connected with buildings. Japan has initiated this underground linkage of car-parks.

4.1.3 *Improvement in Efficiency by Decentralization and Distribution of Infrastructure with Network*

In order to maintain and develop urban functions more effectively in a large city, infrastructure should be distributed and networked among the densely populated areas, This concept is the decentralization and distribution of infrastructure with networks, Take energy for instance, electric power lines are connected by high voltage cables between substations. Power lines are connected among power comstations and independent power producers' (IPP) plants in order to compensate each other and to make most use of effective powers from different origins. The district heating and cooling system also can be connected between the main plant and cogenerating plants located in each building. Sewage systems connect the final treatment plant and intermediate plant located in the local area. The large scale treatment, disposal and energy plants play major roles as the main functions in the city which can be connected and networked to the sub-systems in each local area to compensate and to attain high efficiency in total in the city. These sub-systems and networking are suitably placed in underground. Underground space linkage for infrastructure like energy and lifelines activate entire city to be efficient in total.

4.2 *Improvement in Overall Efficiency of the Entire City*

In order to increase the overall efficiency in the entire city, it is necessary to find optimum solution to use numbers of facilities in the network as efficient as possible in the entire city. Electricity for instance use numbers of power plants to use as efficiently as possible by networking power plants through power lines to compensate each other and make the maximum power at the time of peak use. Therefore the aim of networking is to search for and then perform the most efficient methods in the entire city for physical transportation, sewage, garbage and refuse collection and disposal, and treatment as a city management. Underground space for railways,

motorways or waters is another option for networking in the city management.

5 NETWORKING WITH MODAL SHIFT CONCEPTS AND TECHNOLOGIES

Modal shift concepts and technologies have been applied to the freight transportation systems between trunk lines for inter-city transport and peripheral lines for intra-city transport, such that ships, railways or large trucks & lorries have been used for trunk lines and the small size of city trucks have been used for door-to-door; end users lines. In the future even the sewerage lines will adopt network systems with modal shift concepts to transport sewage from one treatment plant to another. They need new technologies for conveying and handling transporting materials effectively from one mode to another.

5.1 *Subway as One Mode for Transportation in the City*

The underground metro, the subway is one of options in the railway systems. The railway systems can be categorized as rails in inter-city, intra-region and intra-city.

High-speed railway is a typical artery line for inter-city or even inter-continent rail, such as Shinkansen line in Japan, TGV in France, etc. The high-speed railway has verified its competitiveness of trains with the motorcars or aircrafts in some long distance say 600 km between Tokyo and Osaka in Japan. For this high-speed rail, underground space is only space in tunnels. Euro channel tunnel is one of them. In future underground space will be used even for high-speed rails in large cities if there is no surface area available.

In the regional area of mega-city, or metropolitan area, the commuter train connects between the suburban area and the city terminal or center. Take US for instance, the commuter train is looked over again such as the Altamont Commuter Express in California in order to reduce and smooth out the highway traffic jam. This commuter line offers special transportation services for elderly and disabled persons, creates safer routes for bicycling and walking, and provides cities with money for pothole repairs for highways (Altamont Commuter Express 2001). The commuter train normally uses surface rails, but sometimes needs to go underground near the city centers. As railways come to the center of cities, they gradually tend to use underground from the difficulties of surface land availability and the landscape concerns.

In the city if it is necessary to transport large capacity of commuters, train systems are very effective. Most common railway systems in the city region are Mass Rapid Transit (MRT) and Light Rail Transit (LRT). MRT and LRT use normally railways as both elevated and underground systems. The subway is normally included in MRT, but sometimes included in LRT too. The short distance transportation is normally done by surface transportation systems such as tramcars, buses and other means including bicycles. In the developing countries, jeepneys for instance, small motor carriers, or rickshaws are used for transportation means in the metropolitan region and city centers.

The subway system of underground railway is normally selected as MRT in the city area where it is difficult to find surface or elevated space for railway systems. In the suburban hometown, commuters or passengers at first use cars, buses or bicycles besides walking from home to railway station, and then use the suburban railway as commuter trains. Next they transfer from suburban railway to MRT for subway mostly or LRT, in the city terminals. When they arrived by MRT or LRT at the nearest station to office or shopping place, they take trams, bus, taxi, bicycles or walks to their destination. This means commuters and passengers select different modes of transport-means to find most efficient and comfortable ways to their destinations. People have to shift modes of transportation without consciousness. In transportation modes, the subway plays a vital role as MRT and sometimes LRT in the city.

Railway is a line facility, but station is a nodal point facility. A modal shift point between railway systems is a connecting station as a nodal point. The modal shift for passengers should be supported by mechanical systems. Escalators and elevators are inevitable transport means at the connecting station. Sometimes moving walks are necessary where the connecting length to walk is long. Design for layout of connecting passages is important to have smooth flow of passengers at the connecting station. Park and Ride systems have also been developed for use with cars and bicycles. Underground is also used for car or bicycle parks at the railway station. These car or bicycle parks also play important roles for modal shift application between rail system and other vehicle systems. When you change modes at the nodal point between railway and other transport system, some special facilities are necessary. They are railway stations, connecting passages, bus terminals, taxi stands, car parks, bicycle parks, etc.

5.2 Underground Physical Transportation System with Modal Shift Technology

The modal shift concept comes from traffic engineering, such as the modal shift of transporting goods by railway and maritime transport systems to city trucks that carry the goods within a city. The modal shift idea has developed with the mechanical systems of storing and handling goods, such as con-

tainer systems. In the near future, a new modal shift for transporting goods will be introduced in large cities with the technology for mechanical handling systems. Goods, transported inter-continent or inter-city by means of maritime, railway, and/or truck or lorry transportation systems, will go into cities with the different mode of transportation. The modal shift should be supported by mechanical handling system. The cargo container systems have to change between transporting modes. Large cargo changes into small cargo containers, which have smaller-scale capacities.

The underground physical transportation system will be employed in the near future because of traffic jam and air pollution by heavy lorries and trucks aboveground. Inter-city transports are to be supported by maritime, railway or large trucks or lorries. However in the city aboveground space are sometimes fully used and no space available for large capacity transport-means to move. Underground delivery systems become feasible. They are good for environment from noise, air pollution and traffic jams. Underground delivery systems as physical transportation system connect maritime ports, large railway stations, and suburban depots of trailer or truck terminals with wholesale markets or distribution centers in the cities through tunnels. New physical transportation systems, that is, more mechanically or electronically controlled than the train system of postal parcels in London, will be introduced in some large cities in the near future. Electric railway system or wheel carrier system will be developed with mechanical handling systems. Civil engineers have to take care the tunnel and terminal cavern construction based on the geological evaluation and to collaborate interdisciplinary with mechanical handling and other field of engineers.

Garbage or refuse will be transported by underground tunnels. Incineration plants and final garbage disposal places are networked by garbage transport tunnels. This is efficient in city management and environmentally excellent from traffic jam, noise, smells and landscape.

5.3 Underground Sewerage, Drinking Water and Storm-Water Reservoir Tunnel with Modal Shift Technology

Sewerage water lines will be networked to a number of treatment plants in order to provide the best service to customers and the most efficient use of treatment plants throughout the entire city. Among networking ideas, Singapore has just introduced networking in DTSS. Modal shift and networking of sewerage tunnels need new technologies for transporting waters from one place to any other place. Drinking water lines are networked with several purification plants to find a uniform supply in the city zone.

The TARP for constructing storage tunnels for combined sewage and storm water is another application of a modal shift in storm water management. River water or combined sewage water, accumulating during a heavy rainstorm, is stored in the reservoir tunnels. After the storm has subsided, the stored sewage water is conveyed to treatment plants, in order to avoid water pollution and to prevent flooding from the area; it is then conveyed to rivers or the sea. The mode of the water shifts from streams to overflow and then to stored water. Preserving water from pollution and preventing it from flooding have triggered technological developments by applying the modal management of storm water.

Networking of reservoir tunnels is another challenge to find optimum solutions in the storm water management. Water should be transported from one tunnel to another. They need new technologies for pressurized tunnel and conduit systems with pump stations for storing and distributing the sewage waters and check-valves for the prevention of back flows.

5.4 Networking For an Optimal Efficiency in the Entire City

If you want to increase the total efficiency, you have to increase overall efficiency. Though it is important to increase efficiency of each system, effort in each system has limitation. In order to increase the total efficiency of an entire system, combination of concentration and distribution of capacity has to be considered in between or among systems, that are designated as modes. Especially it is important for capacity to transfer smoothly among systems. Modal shift ideas become important to distribute say passengers' capacity from one system to another at the connection point in the region. The modes of the inter-city, intra-region and intra-city transportation have to be selected to have most efficient ways to handle the capacity transfer in entire modes. Subway system is one of the optional modes of the transportation management as a mass rapid transit. Underground car or bicycle park system also functions as the optional mode of the transportation. Storm-water reservoir tunnel is another mode to store the storm-water overflowed from the river and to flow water to the sea or river after a storm has subsided. Therefore systematical improvement of an entire efficiency by using modal shift technology is the matter of concerns in the transportation and storm-water management. The use of underground space network should be managed normally by the city to get a high-efficiency in transporting capacity of passengers, goods, or waters.

In order to increase the overall efficiency in the entire city, it is necessary to find optimum solution to use numbers of facilities in the network as efficient as possible in the entire city. Electricity for instance use numbers of power plants to use as efficiently as possible by networking power plants by power lines to compensate each other and make the maximum power at the time of peak use. Networking is another option for the city management. Underground space for railways, motorways or waters function as networking in the city management.

Since the modal shift and networking is closely related, it is important to find the optimum solution in the network by applying the modal technology.

6 GEO-TECHNICAL DEVELOPMENT

6.1 Deep Underground Use

6.1.1 Need for Using Deep Underground

Some typical ways in Japan, in which deep underground space is being used, include a tunnel for high-voltage electric power lines at a depth of 66 m in Osaka, a conduit for a gas pipeline at a depth of 60 m in Yokohama, and a sewer line tunnel at a depth of 57 m in Tokyo. When you look at data from around the world regarding the use of deep underground space, a subway station for Moscow Metro at a depth of 100 m, a subway station for Washington, D.C. Metro at a depth of 60 m, a New York water supply tunnel at a depth of 150 m and its vertical shaft at a depth of 200 m, and the Chicago TARP tunnel at a depth of between 61 and 107 m are noteworthy.

Focusing on three major cities in Japan, namely, Tokyo, Osaka, and Nagoya, shallower underground space has already become congested with various underground structures like subways, electric power lines, gas pipes, telecommunication lines, water supply and sewer lines, and many other tunnels and pipelines.

A new law called "the special measures act for public use of deep underground space" has been enacted in Japan. This new law came into force on April 1, 2001. The deep underground is specified as the depth that is not normally used for construction basements, normally 40 m below the surface, or 10 m below the surface of the bearing layers of foundation piles. The main purpose of this law is to regulate the use of deep underground space in appropriate and rational manners for public purposes. Benefits are expected to include the smooth implementation of public projects, the selection of desirable routes which would shorten construction periods and reduce costs, the systematic use of deep underground space by cities, the creation of more and safer structures from earthquakes, the reduction of noise and vibrations on the ground surface, and the preservation of the landscape above the ground. The above-mentioned act in Japan enables government to permit the use of deep underground space without prior compensation to landowners unless they experience damage or loss. Since deep underground space is normally not used for private purposes, permission to use this space for public projects normally does not result in losses requiring compensation.

Since the construction cost is high in deep underground, only important facility can be assigned in deep underground. The deeper the line goes underground, the larger the transporting capacity of the line should increase. This is mainly come from economic evaluation. Therefore the deep line should be the trunk line.

This is similar to the blood vessel systems in the body. The main artery and vein of the blood vessel are deep in the body, and the capillary tubes of blood vessel are near the skin. This is mainly come from the protection from injury.

Facilities in deep underground and the main artery and vein of the blood vessel have similarity that important organ or facility should be kept deep inside, the capacity of the flow should be larger, and that the safety should be ensured in deeper place.

6.1.2 *Technology for Using Deep Underground Space*

The difference between using deep underground space in Japan and using it in Europe or the U.S.A. is the geological conditions. The geology in Japan consists of soil, which is soft mostly from the plains of estuaries or the part of reclaimed land. In Europe and the U.S.A., on the other hand, the geology is mostly rocks.

In order to design underground structures, earth pressure levels have to be obtained as the external force to the structures. In most cases, Rankine earth pressures are adopted in which the earth pressure increases linearly with depth. Is this true in deep underground? We need more data that effect on the design for deep underground. At shallower depths, the earth pressure does not strongly effect on the design because of the ambiguity of the safety factors in civil engineering structures. At the same time, water pressure levels are not necessarily hydrostatic any more in deep underground because of the water flow through the soil layer. In order to make the economical design, it is necessary to evaluate the earth and the water pressures more precisely. This means that pressure evaluations become the important issues for deep underground technology.

6.2 *Technical Development of Tunnel and Cavern for Underground Space Use*

6.2.1 *Need for Capacity Increase and Large Tunnel or Cavern*

Population growth and consequent urban congestion need the capacity increase of transportation and infrastructure. Therefore capacity rates have to be increased in transportation and infrastructure system. Since the discharge rate can be expressed as multiplication of the velocity by the sectional area. The velocity can be increased by the speed-up of flow such as the speed-up of railway or road. High-speed railway or highway systems have to be sought. It is also important for highway or road system to avoid traffic jams; this means keeping the velocity high enough is to keep the flow capacity. Another factor to increase the capacity is to have larger sectional area. If you have large numbers of truck lines or lanes in railway or highway, you can get large capacity to pass those vehicles through the way. Advanced highways have large numbers of lanes and hence wide, so that they can have large capacity of vehicles flow.

In underground, the tunnel should be smooth enough in alignment to have smooth flow to avoid the speed reduction, and should be section-wise large enough to have large numbers of truck lanes in one tunnel. In railway system, high-speed rail is the crucial issue in inter-city or inter continental railways. However in highways, the crucial issues are to increase numbers of lanes and hence to widen the tunnel width. Therefore the tunnel should be wide and large enough. One of the big targets for the tunnel engineering is the increase of the diameter of the tunnel. One single shield tunnel has attained 14 m in the diameter and mountain type of tunnel is getting bigger and bigger with the special support like umbrella method. These technological efforts to increase the diameter of tunnel will continue in the future.

6.2.2 *Technological Assignments for Future Tunnel*

6.2.2.1 Shield Tunnel

Shield tunnels are being constructed with larger diameters because they must accommodate a large capacity of road traffic. Undersea and under-river tunnels have adopted large shield tunnels with diameters around 14 m. Some of them are the undersea road tunnel of the Trans Tokyo Bay (TTB) Highway in Japan (JTA 1992) whose outer diameter is 14.14 m, and the under-river tunnel of the Elbe Road Tube in Hamburg, Germany whose outer diameter is 13.75 m (Zell et al. 2000).

The TTB Highway was opened in 1997, and the 4th Elbe Tunnel Tube will complete in 2003. The Metropolitan Expressway in Tokyo also goes underground to alleviate surface traffic congestion and to reduce noise to nearby residents. This expressway employs both an open-cut tunnel and a shield tunnel. The shield tunnel has an outer diameter of 13 m.

There are lots of storm water tunnels in Tokyo and Osaka that have diameters in the range of 10 to 12 m. An outer discharge channel called the underground river in the Tokyo Metropolitan Area will be completed in 2001. It will connect and network some rivers to prevent them from flooding (JTA 2000). Shield tunnels are used with an inside diameter of 10.6 m and a depth of around 60 m below the surface in Tokyo. In Otsu, Japan, the same kind of discharge channel tunnel is under construction and some of parts will open in 2004, which has dimensions with an inside diameter of 10.8 m, outer of 12.64 m and a depth of around 40 m below the surface. These underground rivers will function as discharge channels as well as storm water reservoirs for storing floodwaters and conveying them to non-flooding rivers.

Future target of diameter is 20 m in shield type of tunnel. This is because in urban area mountain tunnel of NATM has a diameter around 20 m such as Mt. Baker Ridge Tunnel in Seattle and Ome Tunnel in Tokyo.

6.2.2.2 Mountain Type Tunnel of NATM

Besides the shield tunnel, Mountain type tunnel of NATM is applied under soil or soft rock conditions in urban area. This method is referred to as a "City NATM" or an "Urban NATM". The urban NATM is common in Japan as well as in Europe. The Ome highway tunnel in Tokyo has a section of 18 to 19 m in height and 14 to 15 m in width. It is a double deck tunnel for vehicle lanes and was completed in August 2001. This NATM tunnel adopted the umbrella method for reinforcement with roof pipe supports.

One of other examples is the Underground Car Park in Landsberg, Germany. The tunnel was excavated to be 16.4 m in height and 18.9 m in width. Another is a storm water reservoir tunnel in Kawasaki, Japan, which was also constructed using urban NATM technology in a soft rock environment. The bored tunnel section has an oval shape with a height of 18.2 m and a width of 17.0 m.

The urban NATM uses technologies such as the CD (Central Diaphragm) method for temporary support by steel rib walls coated with shotcrete, the umbrella method with roof pipe supports, etc.

Some of tunnels have special reinforcements besides NATM technologies. A large tunnel constructed near the ground surface in Milan is an example of what is called the cellular arch method. A cellular arch was formed by roof arched concrete pipes, 2 m in diameter each, and concrete arch ribs to support the pipes; this created the large bored tunnel section with a width of 28.8 m and a height of 20 m. The Mt. Baker Ridge Tunnel in Seattle is a double-deck road tunnel with an inner diameter of around 20 m; the outer shell structure is a concrete filled multi-drift tunnels (Parker 1990).

The larger tunnels described above have cross sections of around 220 to 260 m2. This means that the tunnel is equivalent to a circular tunnel of 17 to 18 m in diameter.

6.2.2.3 Large Scale Outer Shell Reinforcements

The revolutionary idea of reinforcing the outer shell was initiated by the Mt. Baker Ridge Tunnel in Seattle in which small tunnel drifts were used as elements of the continuous outer shell. The cellular arch method, employed in Milan, was another kind of outer shell reinforcement that used large pipes around the roof (Lunardi 1990).

Modified idea for a large tunnel or a cavern is the Geo-Dome method. The outer shell was reinforced by a continuous spiral tunnel (Hanamura and Yamaguchi 1992). Another example of a large-scale outer shell structure is the MMST method reinforced by numbers of outer-shell rectangular shield tunnels and forms large rectangular-shaped sections of around 700-800 m2. The conceptual new ideas to reinforce the outer shell include the Whalebone and the Sardine-Bone Methods (JTA 1998). These efforts to produce large cavern will continue.

7 CLOSING

In 20th century underground space was used mostly by advanced countries. However in the 21st century, the underground space will burst into use in many parts of the world.

Key factors for underground space development are networking and modal shift concept and technologies among underground facilities. Interdisciplinary effort is necessary among city dwellers, civil engineers and mechanical engineers.

Subways will become common transport-means in most of large cities even in the developing countries, and public utilities and lifelines do the same. Bicycle–Park and Ride system will develop among densely populated countries.

Physical transportation of goods and garbage will be transported underground and networked to find the optimum solution in the city management. Sewerage and storm-water reservoir systems will spread into developing countries and will be networked to find optimum solution in advanced countries as well.

Deep underground use will develop in big cities in advanced countries. In order to have large capacity of traffic flow and to raise environmental concerns, underground road system will expand in congested area of large cities. Tunnels and caverns will become large in size to accommodate the large traffic capacity especially in road and highways. Shield tunnel will target to have 20 m in diameters from the social need. Mountain type of urban tunnel will become gradually large by outer-shell reinforcement.

Underground space use can be liberated to all parts of world in 21st century.

8 REFERENCE:

1) Altamont Commuter Express (2001), http://www.acerail.com/
2) AUA (American Underground Association) 2001, Project of the Month; Tren Urbano, web site
3) Desai, D.B., K. Rossler and H. Wagner 2000, Large station design for Tren Urbano's Minillas Extension, North American Tunneling ' 00, Balkema, Rotterdam : 239-248
4) DTSS(2001),http://www.pub.gov.sg/dtss.html
5) Hanamura, T. and K. Yamaguchi 1992, Joint R&D Program For Underground Space Use Between Private Industries and Government in Japan, Proc. ICUSESS 92 at 5th Int'nl Conf. on Underground Space and Earth Sheltered Structures, Delft pp.170-179
6) JTA (Japan Tunnelling Association) 1992, Tunnelling Activities in Japan 1992, JTA
7) JTA (Japan Tunnelling Association) 1998, Tunnelling Activities in Japan 1998, JTA
8) JTA (Japan Tunnelling Association) 2000, Tunnelling Activities in Japan 2000, JTA
9) Lunardi P. 1990. The Cellular Arch Method: Technical Solution for the Construction of the Milan Railway's Venezia Station. Tunnelling and Underground Space Technology Volume 5, Number 4: 351-356.
10) New York City Transit (2001), http://www.mta.nyc.ny.us./nyct/facts/ffintro.htm
11) Parker, H.W. 1990. Concept and construction of the Mount Baker Ridge Tunnel. Tunnels and Tunnelling, March: 57-58.
12) Zell, Dyckerhoff and Widmann 2000, The elbe Road Tunnels in Hamburg-Three generations of tunnelling, http://www.ita-aites.org/tribune9/fog2.html

1. Mechanism of tunneling

Modern Tunneling Science and Technology, Adachi et al (eds), © 2001 Swets & Zeitlinger, ISBN 90 2651 860 9

Optimal shape of underground structure

Takeshi Tamura & Jun Saito
Kyoto University, Kyoto, Japan

ABSTRACT: The most typical and popular shape of tunnels is considered to be circular. The circular shape of cavity can perform in a stabilized manner under a wide range of initial stresses since the vertical contraction of the circular tunnel yields a similar amount of horizontal expansion to make the surrounding stress state around the tunnel almost isotropic. This mechanism is the fundamental principle of the NATM (New Austrian Tunneling Method) and based upon the tacit assumption that the tunnel is located with a enough overburden larger than its diameter at least. It is, however, a still controversial issue to determine the optimal shape of the tunnel or the cavity under a given initial stress field from the mechanical point of a view. In this paper, the optimal shape of the underground structure is studied by using a simple mathematical formulation.

1 INTRODUCTION

One of the most interesting problems in the application of optimization technique to tunnel engineering is the shape determination of underground structures. Thus far various theories for the shape optimization problem of elastic structure have been published, which are mostly based upon a variational principle by considering the variation of the domain (coordinates) as well as the main variables (displacements). However few studies can be found in the literature which cover the whole formulation process from the optimality criteria to the numerical procedure. In the present paper, an optimization problem will be considered where the shape of elastic underground structures with the minimum compliance is sought for. The theoretical idea is almost the same as that of Banichuk(1983) but a more precise formulation, following Tamura(1992), to get optimality criteria and their weak forms for the finite element method will be shown. Finally the results of a simple example and its application to the tunnel shape problem will be explained.

2 VARIATIONAL PRINCIPLE

Let $\sigma_{ij}, \varepsilon_{ij}$ and f_i be the stress, the strain and the body force, respectively. As shown in Fig.1, a distribution of traction T_i is applied to a part of boundary S_σ (stress boundary) of the whole do-

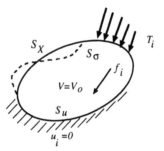

Figure.1 Boundary conditions and moving boundary.

main V. Then the total potential energy I of an elastic structure is defined as

$$I = \int_V \left(\frac{1}{2}\sigma_{ij}\varepsilon_{ij} - f_i u_i \right) dV - \int_{S_\sigma} T_i u_i dS. \quad (1)$$

Through the first variation of I which considers the change of domain configuration on S_X, the following conditions for the optimal shape of the domain can be obtained.

$$\frac{1}{2}\sigma_{ij}\varepsilon_{ij} - f_i u_i = \lambda \quad \text{(constant)} \quad S_X \quad (2)$$

with the usual equations of equilibrium and boundary conditions:

$$\sigma_{ij,j} + f_i = 0 \quad \text{in} \quad V \quad (3)$$

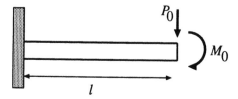

Figure.2 Cantilever subject to loads.

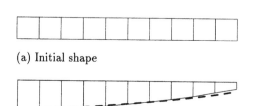

(a) Initial shape

(solid:present method, dotted:classical method)

(b) Optimal shape

Figure.3 Initial and optimal shapes.

$$T_i = \sigma_{ij} n_j \qquad \text{on } S_\sigma \qquad (4)$$

It is noted that the total potential energy I in Eq.(1) under Eqs.(3) and (4) can be shown to be equal to the negative value of the stored elastic energy:

$$I = -\frac{1}{2} \int_V \sigma_{ij} \varepsilon_{ij} dV = -(\text{Stored Energy}). \quad (5)$$

Therefore Eqs.(2), (3) and (4) with the constraint condition:

$$V = V_0 \qquad (6)$$

lead to the optimal shape of the elastic structure which stores the minimum elastic energy among all structure with the same volume V_0 under the given external forces.

3 A SIMPLE EXAMPLE

As a simple example of numerical analysis of the shape optimization, the design of bottom surface of a cantilever without body force will be considered since its close form solution is easily obtained within the frame of the classical beam theory. In the present study, a unit magnitude of vertical force is applied at the top-right node and also a unit magnitude of bending moment is applied at the right end as shown in Fig.2. This applied bending moment is to make the bending moment along the beam non-zero. The displacement is fixed at the top-left node and the lateral displacement is prevented at another node on the left end. Poisson's ratio is assumed to be 0.333. The initial shape of beam is rectangular which is divided into a number of the 4-node isoparametric elements. Fig.3(b)shows the results of numerical analysis as well as the solution of the beam theory. As is seen from this figure, the validity of the present formulation is confirmed.

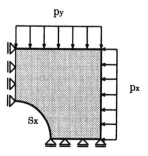

Figure.4 Cavity under external forces.

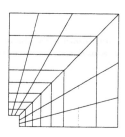

(a) Initial shape

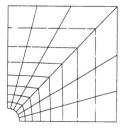

(b) Optimal shape

Figure.5 Initial and optimal shapes of cavity.

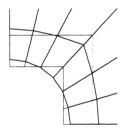

Figure.6 Shape around cavity.

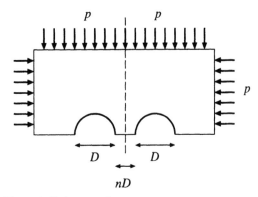

Figure.7 Twin tunnels.

4 OPTOMAL SHAPE OF UNDERGROUND

4.1 *Single cavity*

Fig.4 shows the domain with boundary conditions for the shape optimization of a single cavity. A quarter sector is analyzed due to the horizontal and vertical symmetries. A pair of distributed uniform load p_x and p_y are applied to the right and top boundaries, respectively. A traction-free curved boundary S_X denotes the cavity of which shape is optimized under the sense of the minimum compliance stated before. The area of cavity is unchanged in the course of optimization as the constrained condition. It is expected that the circular shape of cavity is obtained if the isotropic stress condition $p_x = p_y$. Fig.5(a) and (b) shows the initial and final shapes of the cavity. Fig.6 illustrates the close-up view of cavity from which almost complete circular shape is confirmed to be the optimal solution to this fundamental problem.

A number of examples are tried under the anisotropic stress condition $p_x \neq p_y$. According to the stress ratio p_y/p_x, several elliptic shape of cavities are resulted although no figures are present here.

4.2 *Twin tunnels*

The twin tunneling is quite often used and a lot of its design methods are proposed. A typical argument around the twin tunnels is on the evaluation of the earth pressure applied to the center pillar between the two cavities since the stability of the twin tunnels are strongly affected by the mechanical performance of the center pillar. There has not been found the researches on the shape of twin tunnels since almost all of twin tunnels are designed to be circular. Of course, non-circular tunnels are not preferable from a view point of construction process. But it is interesting to understand what kind of shape is recommended through consideration of minimum storage of elastic energy (or compliance). Furthermore the mechanical influence of the width

of the center pillar on the optimal shape is another issue to be considered.

Fig.7 shows the schematic view of region with boundary conditions for the shape optimization of twin tunnels. The right half is analyzed due to the horizontal symmetry. An isotropically distributed uniform load p is applied to both of the horizontal and vertical boundaries. A traction-free curved boundaries denote the cavities of the twin tunnels of which areas are unchanged in the optimization. The width of the center pillar nD is a variable in the following results where D denotes the diameter in the initial circular shape of each tunnel as shown in Fig.7. Fig.8(a) shows the initial finite element mesh in the case of $n = 1$ and Fig.8(b) illustrates the close-up view near the tunnel where the thin and thick lines correspond to the initial and final shape of the cavity, respectively. It is understood that the optimal shape seems to be almost circular in this case and that the center pillar of width D is not considered to be so mechanically influential to each cavity. Figs.9 and 10 correspond to the cases of $n = 0.5$ and $n = 0.25$, respectively. As shown in Fig.10(b), the shape of cavity deforms top-leftward from the initial circular shape as if the single tunnel is subject to the vertically dominant external stresses.

5 CONCLUSIONS

A numerical study is presented to seek for the optimal shape of the underground structures by using the variational principle of elasticity. Its validity is confirmed through a simple example of structural mechanics. As a practical problem, the optimal shape of the twin tunnels is considered and it is found that the optimal shape of such tunnels are rationally deformed from the circular one when the distance of the two tunnels become smaller than the tunnel diameter. In other words, the twin tun-

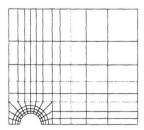

(a) Initial shape

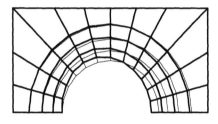

(b) Optimal shape

Figure.8 Twin tunnels with $n = 1$.

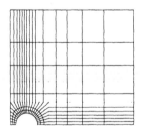

(a) Initial shape

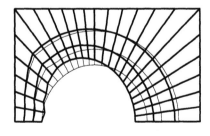

(b) Optimal shape

Figure.9 Twin tunnels with $n = 0.5$.

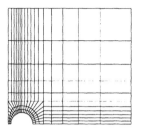

(a) Initial shape

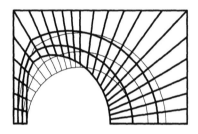

(b) Optimal shape

Figure.10 Twin tunnels with $n = 0.25$.

nels closer than their diameter are mechanically influenced each other.

REFERENCES

Banichuk, N.V. 1983. Problems and methods of optimal structural design, Plenum Press, 1983.

Tamura, T. 1992. A fundamental study on shape optimization of continuum, Proc. Korea-Japan Joint Seminar on Structural Optimization, pp.119-127, 1992.

Modern Tunneling Science and Technology, Adachi et al (eds), © 2001 Swets & Zeitlinger, ISBN 90 2651 860 9

An approximate solution for the stresses on a rigid tunnel

A. Verruijt
Delft University of Technology, Delft, Netherlands

ABSTRACT: A simple approximate expression for the distribution of normal stresses and shear stresses on a rigid circular tunnel is presented. The tunnel is supposed to be constructed in a homogeneous soil, below the groundwater table. Results are compared with an exact analytic solution for a rigid circular tunnel in an elastic half plane. The approximate solution appears to be accurate for a small or deep tunnel, up to values of the cover depth equal to the radius of the tunnel.

1 THE PROBLEM

1.1 Introduction

The design of a tunnel lining requires the analysis of the stresses acting upon it from the surrounding soil. These may be dependent upon the deformation of the lining, and upon the construction method. In this analysis the initial stresses in the soil play an essential role.

A simple method of construction is that a circular hole is bored in the ground, by a tunneling machine, in which a circular tunnel lining is constructed, usually from concrete elements. This results in a tunnel that is slightly smaller than the bored hole. The gap between the concrete lining and the soil is often filled with injected grout. In this paper all the deformations involved in such a procedure are neglected: it is assumed that the shape of the tunnel is exactly equal to the shape of the soil body that has been removed, and the tunnel itself is assumed to be completely rigid.

For a coupled analysis of the soil-structure interaction problem for a deformable tunnel lining, the finite element method (Lee & Rowe, 1991; Vermeer & Bonnier, 1991) seems to be the most promising method, although an exact analytic solution may be possible for the problem of a circular tunnel embedded in an elastic half plane, using Muskhelishvili's complex variable method (Muskhelishvili, 1953; Verruijt, 1998). Such analytic solutions may be useful for the validation of numerical solutions.

The very first step in an analytic or numerical method of analysis of tunnel behaviour is the determination of the initial stresses, before the deformation of the tunnel. Although this may seem to be an almost trivial problem, it is not, because the weight of the tunnel in general is unequal to the weight of the removed soil mass. This means that the state of stress should include a contribution due to the application of a force equal to the difference of the weight of the tunnel and the weight of the original soil, acting upon a rigid circular inclusion in the soil. For an elastic material this is a problem similar to the problem of a circular cavity in a half plane (Mindlin, 1939; Verruijt & Booker, 2000). An approximation to that solution is presented here.

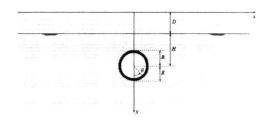

Figure 1. Tunnel in half plane, below groundwater.

1.2 Basic assumptions

It is assumed that the tunnel is constructed in a homogeneous soil, with a horizontal top surface, and a horizontal groundwater table at a depth D. The tunnel is of circular shape, with radius R, located at a depth H below the groundwater table (Fig. 1). The unit weight of the dry soil is γ_d, the unit weight of the saturated soil is γ_s, and the unit weight of the pore water is γ_w. The weight of the tunnel will be expressed by its average unit weight γ_t (the total weight of the tunnel divided by its total area πR^2). It is postulated that the stress distribution at the soil-tunnel interface should at least satisfy the conditions that it ensures equilibrium of the tunnel, and that it is in agreement with classical soil mechanics results for the limiting cases of a tunnel of very small radius, and a tunnel weight equal to the weight of the excavated soil. The validity of Terzaghi's effective stress principle is also assumed, with the pore water pressures being determined by the location of the groundwater table. The ratio of the horizontal to the vertical stresses in the initial state, prior to the construction of the tunnel, is denoted by K_0. This is assumed to be a given constant.

2 STRESS DISTRIBUTION

2.1 Initial stresses

Before the construction of the tunnel, the vertical total stresses in points on the future tunnel circumference are

$$\sigma_{yy} = \gamma_d D + \gamma_s (H + R \cos \theta),$$

or

$$\sigma_{yy} = \gamma_d D + \gamma_s (H - R) + \gamma_s R(1 + \cos \theta). \quad (1)$$

Written in this form the first term describes the effect of the weight of the dry soil, the second term represents the weight of the saturated soil above the crest of the tunnel, and the third term represents the variable stress level in the points on the boundary of the future tunnel, as a function of the angle θ.

The pore pressures are

$$p = \gamma_w (H + R \cos \theta). \quad (2)$$

Thus the vertical effective stresses are

$$\sigma'_{yy} = \gamma_d D + (\gamma_s - \gamma_w)(H + R \cos \theta), \quad (3)$$

and the horizontal effective stresses are

$$\sigma'_{xx} = K_0 \gamma_d D + K_0 (\gamma_s - \gamma_w)(H + R \cos \theta). \quad (4)$$

Adding the pore pressure gives the horizontal total stresses,

$$\sigma_{xx} = K_0 \gamma_d D + [\gamma_w + K_0 (\gamma_s - \gamma_w)](H + R \cos \theta). \quad (5)$$

These stresses simply represent a stress distribution due to the weight of the overlying material, with a uniform lateral stress coefficient K_0.

2.2 Stresses after constructing the tunnel

It is now assumed that after construction of the tunnel the horizontal stresses σ_{xx} do not change, and that the vertical stresses above the tunnel remain

equal to the weight of the overlying material. In order to satisfy the condition of equilibrium of the tunnel it is now assumed that in the expression (1) for the vertical total stresses only the parameter γ_s in the third term is modified to γ_t, the average unit weight of the tunnel,

$$\sigma_{yy} = \gamma_d D + \gamma_s (H - R) + \gamma_t R(1 + \cos \theta). \quad (6)$$

The horizontal total stresses remain the same,

$$\sigma_{xx} = K_0 \gamma_d D + [\gamma_w + K_0 (\gamma_s - \gamma_w)](H + R \cos \theta). \quad (7)$$

Together with the assumption that before and after the construction of the tunnel the shear stress

$$\sigma_{xy} = 0, \quad (8)$$

this is the stress distribution that is postulated as an approximation to the true stress distribution.

The resulting forces of the stress distribution are

$$F_x = \int_0^{2\pi} \sigma_{xx} \sin \theta R d\theta, \quad (9)$$

$$F_y = \int_0^{2\pi} \sigma_{yy} \cos \theta R d\theta. \quad (10)$$

Substitution of equations (6) – (8) into (9) and (10) shows that $F_x = 0$ and $F_y = -\gamma_t \pi R^2$, which is the weight of the tunnel. It appears that the given stress distribution indeed satisfies the condition that its resulting force corresponds to the total weight of tunnel.

It may be assumed that the stress distribution is a proper first approximation if the shape of the final tunnel is equal to the shape of the soil body that has been removed, because the only difference is the weight of the material inside the circle. For the limiting case that the tunnel is just as heavy as the excavated material the stress distribution reduces to the original one, and if the radius of the tunnel $R = 0$ it also reduces to the undisturbed stress distribution. It should be recognized, however, that effects such as ground loss due to a difference between the tunnel diameter and the diameter of the cavity, and deformation of the tunnel lining, are excluded.

2.3 Stresses in polar coordinates

For the design of the tunnel lining it is most convenient to express the stresses acting on the tunnel lining in the form of a radial normal stress and a tangential shear stress. In this case, with $\sigma_{xy} = 0$, the radial normal stress and the tangential shear stress are related to the stresses in Cartesian coordinates by the equations

$$\sigma_{rr} = \sigma_{xx} \sin^2 \theta + \sigma_{yy} \cos^2 \theta, \quad (11)$$

$$\sigma_{rt} = (\sigma_{xx} - \sigma_{yy}) \sin \theta \cos \theta. \quad (12)$$

74

With (6) – (8) this gives

$$\sigma_{rr} = A_0 + A_1 \cos\theta + A_2 \cos 2\theta + A_3 \cos 3\theta, \quad (13)$$

$$\sigma_{rt} = B_1 \sin\theta + B_2 \sin 2\theta + B_3 \sin 3\theta, \quad (14)$$

where

$$A_0 = \tfrac{1}{2}(1+K_0)\gamma_d D + \tfrac{1}{2}[\gamma_s + \gamma_w + K_0(\gamma_s - \gamma_w)]H$$
$$+ \tfrac{1}{2}(\gamma_t - \gamma_s)R, \quad (15)$$

$$A_1 = \tfrac{1}{4}[3\gamma_t + \gamma_w + K_0(\gamma_s - \gamma_w)]R, \quad (16)$$

$$A_2 = \tfrac{1}{2}(1-K_0)\gamma_d D + \tfrac{1}{2}(1-K_0)(\gamma_s - \gamma_w)H$$
$$+ \tfrac{1}{2}(\gamma_t - \gamma_s)R, \quad (17)$$

$$A_3 = \tfrac{1}{4}[\gamma_t - \gamma_w - K_0(\gamma_s - \gamma_w)], \quad (18)$$

$$B_1 = -\tfrac{1}{4}[\gamma_t - \gamma_w - K_0(\gamma_s - \gamma_w)]R, \quad (19)$$

$$B_2 = -\tfrac{1}{2}(1-K_0)\gamma_d D - \tfrac{1}{2}(1-K_0)(\gamma_s - \gamma_w)H$$
$$- \tfrac{1}{2}(\gamma_t - \gamma_s)R, \quad (20)$$

$$B_3 = -\tfrac{1}{4}[\gamma_t - \gamma_w - K_0(\gamma_s - \gamma_w)]R. \quad (21)$$

These stresses have been expressed in the form of a Fourier series, because this is probably the appropriate form for the design of the tunnel lining.

For certain applications the Cartesian components of the surface traction on the tunnel lining may be of interest. These can be obtained using the transformation formulas $t_x = -\sigma_{xx} \sin\theta$ and $t_y = -\sigma_{yy} \cos\theta$ (where it has been noted that $\sigma_{xy} = 0$). In the form of a Fourier series expansion the resulting expressions are

$$t_x = C_1 \sin\theta + C_2 \sin 2\theta, \quad (22)$$

$$t_y = D_0 + D_1 \cos\theta + D_2 \cos 2\theta, \quad (23)$$

where

$$C_1 = -K_0\gamma_d D + [\gamma_w + K_0(\gamma_s - \gamma_w)]H, \quad (24)$$

$$C_2 = \tfrac{1}{2}[\gamma_w + K_0(\gamma_s - \gamma_w)]R, \quad (25)$$

$$D_0 = -\tfrac{1}{2}\gamma_t R, \quad (26)$$

$$D_1 = -\gamma_d D - \gamma_s H - (\gamma_t - \gamma_s)R, \quad (27)$$

$$D_2 = -\tfrac{1}{2}\gamma_t R. \quad (28)$$

It is the first term in equation (23) that leads to the resultant force $F_y = -\gamma_t \pi R^2$ when integrating the surface tractions along the tunnel surface. All other terms are periodic, and therefore do not contribute to a resultant force.

3 COMPARISON WITH ANALYTIC SOLUTION

3.1 Analytic solution

For a homogeneous linear elastic material the problem considered in this paper admits an exact analytic solution, using complex potentials and conformal mapping (Muskhelishvili, 1953). For similar problems, referring to an elastic half plane with a circular stress free cavity, the solutions have been given by Verruijt (1998) and Verruijt & Booker (2000). The solution for the case of a rigid inclusion loaded by a given force may be obtained in a similar fashion, see also Verruijt (1997). Details are omitted here.

3.2 Examples

In this section some examples will be given. In all examples the distinction between dry and saturated soil is eliminated, by assuming that $D = 0$ and $\gamma_w = 0$. This also implies that there is no difference between effective and total stresses. The relevant soil parameters are K_0 and Poisson's ratio v. A first example is shown in Figure 2, for the case that $v = 0$, $K_0 = 0.5$, $R/H = 1/3$. The figure shows the analytical solution for the vertical traction, and its approximation by the formula (23).

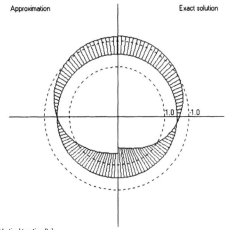

Approximation   Exact solution

Vertical traction (ty)

Figure 2. Vertical traction on tunnel lining, $v = 0$, $K_0 = 0.5$, $R/H = 0.3333$.

It appears from Figure 1 that the approximation in this case, in which the cover of soil $H - R$ above the tunnel is equal to the diameter $2R$, is rather good. For smaller or deeper tunnels the approximation appears to be even better, both for the vertical and the horizontal tractions.

As a further demonstration of the accuracy of the approximation the horizontal traction is shown in Figure 3, for the same case.

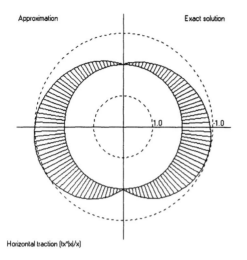

Figure 3. Horizontal traction on tunnel lining, $\nu = 0.0$, $K_0 = 0.5$, $R/H = 0.3333$.

In this case the approximation is very good, indicating that the installation of the tunnel has practically no influence on the horizontal tractions.

As further illustrations of the applicability of the approximation the horizontal and vertical tractions for the case of a large tunnel ($R/H = 0.8$) are shown in Figures 4 and 5.

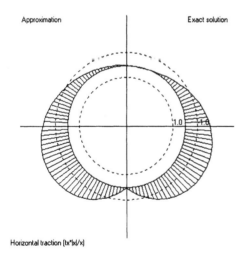

Figure 4. Horizontal traction on tunnel lining, $\nu = 0.5$, $K_0 = 1.0$, $R/H = 0.8$.

It appears that even for such a large tunnel the approximation is not very inaccurate. Further computations (not shown here) indicate that the influence of the soil parameters ν and K_0 on the results is rather

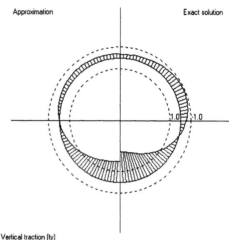

Figure 5. Verical traction on tunnel lining, $\nu = 0.5$, $K_0 = 1.0$, $R/H = 0.8$.

small, except for the influence of K_0 on the absolute magnitude of the horizontal tractions, of course.

4 CONCLUSION

It has been shown that a reasonably accurate approximation for the stresses on a rigid circular tunnel can be obtained by a simple generalization of the undisturbed stress distribution. The accuracy of the approximation is rather good if the diameter of the tunnel is smaller than the cover depth.

5 REFERENCES

Lee, K.M. & Rowe, R.K. 1991. An analysis of three-dimensional ground movements: the Thunder Bay tunnel. *Canadian Geot. J. 28*: 25-41.

Mindlin, R.D. 1939. Stress distribution around a tunnel. *Proc. ASCE 21*: 619-642.

Muskhelishvili, N.I. 1953. *Some Basic Problems of the Mathematical Theory of Elasticity.* Translated from the Russian by J.R.M. Radok. Groningen: Noordhoff.

Vermeer P.A. & Bonnier, P.G. 1991. Pile settlements due to tunnelling. *Proc. 10th Eur. Conf. Soil Mech. and Found. Eng., Florence.* Rotterdam: Balkema, 869-872.

Verruijt, A. 1997. A complex variable solution for a deforming circular tunnel in an elastic half-plane. *J. Numer. Methods Geomecanics 21*: 77-89.

Verruijt, A. 1998. Deformations of an elastic half plane with a circular cavity. *Int. J. Solids Structures 21*: 2795-2804.

Verruijt, A. & Booker, J.R. 2000. Complex variable analysis of Mindlin's tunnel problem. D.W. Smith & J.P. Carter (eds.), *Developments in Theoretical Geomecanics; Proc. Booker memorial symp., Sydney, 16-17 November 2000.* Rotterdam: Balkema, 3-22.

Modern Tunneling Science and Technology, Adachi et al (eds), © 2001 Swets & Zeitlinger, ISBN 90 2651 860 9

Surface displacements due to pressure modifications induced by tunnels

Mohamed El Tani
Lombardi Eng. Ltd., Minusio-Locarno, Switzerland

ABSTRACT: Tunnel excavation in an aquifer rock mass modifies water pressure which generates deformations and surface displacements. The amplitude and shape of surface displacements are related to a large number of parameters which characterise the rock mass, the aquifer, the tunnel and their interactions. A weak procedure is used in this paper to simplify the basic equations. The simplified equations serve to predict surface displacement behaviour for varying anisotropy ratios, rock mass dimensions and tunnel properties.

1 INTRODUCTION

Tunnel excavation produces different kinds of ground movement whose amplitude depends on the generating effect and its location. The amplitude of surface displacements which are produced by volume loss, such as tunnel convergence or opening collapse, decrease with their depth location. Surface displacements which are produced by pressure modifications increase with tunnel depth. From the different generating effects of surface displacements, those which are due to water may become predominant for tunnels excavated at great depth. Pressure modifications bring about subsidence and horizontal displacements which are of primary importance for the monitoring instances of surface and underground structures. Certain structures, such as arch dams, are sensitive to horizontal displacements which may widen or narrow the valleys they are closing. In urban areas, both vertical and horizontal deformations are needed to assess the potential exposure risk to buildings from damage charts.

Many concepts are involved when simulating ground displacements induced by pressure modifications. Some of these are the space delimitation of the phenomena, the boundary behaviour and the rock mass properties. This paper partially tackles these concepts. It will attempt to clarify the parameters role needed for these concepts and find their relation to surface displacements based on a simplified set of equations.

The basic equations can be simplified using numerical techniques or physical observations. In this paper a numerical technique is used. A partial application of the finite element method is used to eliminate one space dimension from the basic equations. This is called a weak approximation and the resulting equations are called semi discrete. This technique is applied considering differing behaviour of the rock mass base, which may slide or be rigid. For each behaviour, a set of semi discrete equations is obtained and may serve in evaluating different effects on the amplitude and shape of surface displacements, such as the rock mass dimensions, anisotropy ratios and the tunnel depth and geometry.

2 BASIC EQUATIONS

Induced pressures and displacements which are brought about by an excavated tunnel may be classed, in most cases, as 2D phenomena. This is a highly limiting hypothesis to which the following is added; the pressures brought about by the tunnel excavation are known. They are, in the case of linear behaviour, independent of displacements and may be calculated separately. Displacements follow pressure modifications instantaneously when inertia and viscosity are ignored. This will simplify the basic equations and enables the accent to be put on the 2D induced displacements.

The basic equations are the equilibrium equations

$$\frac{\partial \sigma_y}{\partial y} + \frac{\partial \tau}{\partial z} = 0 \qquad [2.1]$$

$$\frac{\partial \tau}{\partial y} + \frac{\partial \sigma_z}{\partial z} = 0 \qquad [2.2]$$

and the elastic relationships coupled to the Fillunger-Terzaghi porous material effective stress or the Dalton-Truesdell mixture partial stress (zero order flux stress)

$$\sigma_y = E_y(\varepsilon_y + \lambda_{yz}\varepsilon_z) - p \qquad [2.3]$$

$$\sigma_z = E_z(\varepsilon_z + \lambda_{zy}\varepsilon_y) - p \qquad [2.4]$$

$$\tau = 2G_{yz}\varepsilon_{yz} \qquad [2.5]$$

where σ_y, σ_z and τ are the horizontal, vertical and shear components of the total stress tensor, ε_y, ε_z and ε_{yz} are the horizontal, vertical and shear components of the strain tensor, p is water pressure, E_y, E_z and G_{yz} are the elastic and shear modulus which are positive and λ_{yz} and λ_{zy} are crossed effect coefficients which verify $E_y\lambda_{yz}=E_z\lambda_{zy}$ and $\lambda_{yz}\lambda_{zy} \leq 1$.

The elastic coefficients E_y, E_z, λ_{yz}, λ_{zy} and G_{yz} are those of a 2D orthotropic material which fit major elastic continuous models for rocks and soils. Many models are proposed for a direct evaluation of the elastic coefficients for jointed rocks or porous soils. These models describe a gross response behaviour of a representative element containing a large number of joints for a rock or a large number of void inclusions for a soil. Here it is supposed that in the case of a rock mass there are at least two intersecting sets of fissures.

Generally, measured or computed elastic coefficients are those of 3D elasticity and a conversion is necessary for 2D elasticity. The last five columns of Table 1 give the 2D elastic coefficients in terms of 3D cubic and tetragonal coefficients for both deformation modes, which are called plane deformation and plane stress.

Strain definitions complete the set of equations and relations [2.1] to [2.5] and are

$$\varepsilon_y = \frac{\partial v}{\partial y} \qquad \varepsilon_z = \frac{\partial w}{\partial z} \qquad 2\varepsilon_{yz} = \frac{\partial v}{\partial z} + \frac{\partial w}{\partial y} \qquad [2.6]$$

where v and w are the horizontal and vertical displacements. oy and oz are the horizontal and vertical axes.

3 THE TUNNEL VERTICAL AXIS

The tunnel vertical axis has special properties. In the low shear model, it is the location of the inflexion point for the horizontal displacement, the location of the pressure modification maximum value and the location of the vertical displacements maximum value. It will become a sliding vertical axis if it is an axis of symmetry. The tunnel vertical axis behaves as a boundary even if it is not a material one. It is natural to adopt it as such since its position is indisputable. The choice of the position of the other vertical and horizontal boundaries of the rock mass remains. These should result from geological surveys. However, if their positions are unknown, they must figure as parameters during the entire analyses of the strain effects on surface structures.

Table 1. Relations between 3D elastic coefficients and 2D elastic coefficients.

Material	Coefficients*	Mode	E_y	E_z	λ_{yz}	λ_{zy}	G_{yz}
Cubic**	E v G	Plane deformation	$\dfrac{(1-v)E}{(1+v)(1-2v)}$	$\dfrac{(1-v)E}{(1+v)(1-2v)}$	$\dfrac{v}{1-v}$	$\dfrac{v}{1-v}$	G
		Plane stress	$\dfrac{E}{(1+v)(1-v)}$	$\dfrac{E}{(1+v)(1-v)}$	v	v	G
Tetragonal***	E v G E' v' G'	Plane deformation	$\dfrac{(E'-v'^2 E)E}{(1+v)[(1-v)E'-2v'^2 E]}$	$\dfrac{(1-v)E'^2}{(1-v)E'-2v'^2 E}$	$\dfrac{v'(1+v)E'}{E'-v'^2 E}$	$\dfrac{v'E}{(1-v)E'}$	G'
		Plane stress	$\dfrac{EE'}{E'-v'^2 E}$	$\dfrac{E'^2}{E'-v'^2 E}$	v'	$\dfrac{v'E}{E'}$	G'

* 3D elastic coefficients of an orthotropic material are the elastic modulus: E_1,E_2,E_3, Poisson coefficients: v_{12},v_{13},v_{23} and shear modulus: G_{12},G_{13},G_{23}.

** Cubic material is a special orthotropic material which has three independant coefficients defined by: $E_1=E_2=E_3\equiv E$, $v_{12}=v_{13}=v_{23}\equiv v$, $G_{12}=G_{13}=G_{23}\equiv G$. Isotopic elasticity is a special cubic material for wich $G=E/2(1+v)$.

*** Tetragonal material: $E_1=E_2\equiv E$, $E_3\equiv E'$, $v_{12}\equiv v$, $v_{13}\equiv v_{23}\equiv v'$, $G_{12}\equiv G$, $G_{13}\equiv G_{23}\equiv G'$. Transverse isotropy is a special case for which $G=E/2(1+v)$. Saint Venant material is a special case for which $1/G'=(1+v)/E+(1+v')/E'$.

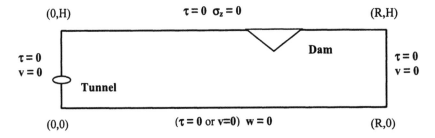

Figure 1. Schematic representation of the rock mass with a tunnel and a dam.

All of the preceding comments suggest the consideration of a rock mass bordered by the vertical tunnel axis from the beginning. The tunnel vertical axis will also be considered as an axis of symmetry to allow the horizontal displacement to be zero there. The rock mass and boundary conditions are shown in Figure 1. The upper surface is unloaded. Two conditions are possible at the base:
- The base may slide on its support and in this case the shear stress is zero.
- The base cannot move horizontally and the horizontal displacement is zero there.
These antagonist conditions represent the limiting cases of possible behaviour of the base and they lead to two different evolutions of the mass. They will be differentiated and called the "sliding base" and the "rigid base". Concerning the final and right boundary, sliding conditions are considered.

4 RIGID BASE MASS

The rigid base mass does not lend itself to a global analysis of an analytical character. A detailed analysis would be possible with the aid of a bi-harmonic analysis or using the integral method with Green functions. These are two possibilities that would be interesting to explore but for the moment it is through a semi-weak approach that the rigid base model yields some useful information. The semi-weak method is a partial application of the finite element method which reduces the equations to new, simplified and approximate equations. This method is described in the Appendix and is applied to both mass models: the rigid and sliding base.

4.1 The rigid base mass in the weak formulation

The simplified equations that are obtained by applying the weak formulation to the rigid base

model are Equations [A9] and [A10] in the Appendix. They only take into account surface displacements. It is, therefore, possible to use the following notations without risk of confusion:
- $v(y)$ for $v(y,H)$ which is the horizontal surface displacement value.
- $w(y)$ for $w(y,H)$ which is the vertical surface displacement value.

The simplified equations of the rigid base mass are coupled but, luckily, they separate when the crossed effect coefficient is such that $E_y\lambda_{yz}$ is equal to G_{yz} and, in this case, they assume the following forms:

$$\frac{d^2v}{dy^2} - \frac{3G_{yz}}{E_y}\frac{v}{H^2} = \frac{3}{HE_y}\int_0^H \frac{z}{H}\frac{\partial p}{\partial y}dz \qquad [4.1]$$

$$\frac{d^2w}{dy^2} - \frac{3E_z}{G_{yz}}\frac{w}{H^2} = \frac{3}{HG_{yz}}\int_0^H \frac{z}{H}\frac{\partial p}{\partial z}dz \qquad [4.2]$$

4.2 Uncoupled steady state displacements

Final pressure can be expressed in an integral form, which does not really help to extract surface displacements from [4.1] and [4.2] in a simple parametric form. Remembering that pressure is a harmonic function, the task becomes easier by decomposing pressure into a sum of harmonic functions of the form $e^{-\beta y}\cos(\alpha z)$; $\alpha/\beta = (k_y/k_z)^{1/2}$ and k_y and k_z are the horizontal and vertical permeabilities. There is still a simpler form, which is the weak form; a semi weak pressure may be obtained by applying the same formalism described in the Appendix to the pressure governing equation. The semi discrete triangular shaped pressure in an infinite strip (R is infinite in Figure 1) of height H is

$$p(y,z) = \Delta p(1 - \frac{z}{H})e^{-\sqrt{\frac{3k_z}{k_y}}\frac{y}{H}} \qquad [4.3]$$

79

in which Δp is determined from pressure values. It will be chosen such that [4.3] is closest to pressure modifications at the tunnel vertical axis. For a rapid and practical estimation of Δp, both sides of [4.3] are integrated considering H as twice the tunnel depth and p(y,z) is replaced by its integral approximation. This leads to

$$\Delta p = \frac{3\gamma \ln 3}{2\pi} \frac{Q}{\sqrt{k_y k_z}} \qquad [4.4]$$

Q is the volume of water which flows into the tunnel per unit time and unit tunnel length and is related to k_y and k_z and to h_r, h and r being the resultant head, tunnel depth and radius. γ is the specific weight.

Substituting [4.3] into [4.1] and [4.2], uncoupled surface displacements induced by pressure modifications in an infinite strip are obtained as

$$v = \frac{H\Delta p}{2\sqrt{3}E_y} \sqrt{\frac{k_z}{k_y}} \frac{e^{-\sqrt{\frac{3k_z}{k_y}}\frac{y}{H}} - e^{-\sqrt{\frac{3G_{yz}}{E_y}}\frac{y}{H}}}{G_{yz}/E_y - k_z/k_y} \qquad [4.5]$$

$$w = \frac{H\Delta p}{2G_{yz}} \frac{e^{-\sqrt{\frac{3k_z}{k_y}}\frac{y}{H}} - \sqrt{k_z G_{yz}/k_y E_z}\, e^{-\sqrt{\frac{3E_z}{G_{yz}}}\frac{y}{H}}}{E_z/G_{yz} - k_z/k_y} \qquad [4.6]$$

Boundary conditions are v(0)= 0 and dw(0)/dy =0. The latter is the remainder of the condition of zero shear on the tunnel vertical axis when the horizontal displacement is constant or zero.

4.3 Displacement properties

Displacement signs are identical to the pressure modification. Δp is negative for an excavated tunnel. In this case surface points will settle and move horizontally towards the tunnel axis. If Δp is positive, surface points will heave and move outwards. The amplitude of the surface horizontal displacement increases from zero at the tunnel vertical axis to attain a maximum and decrease back to zero. The maximum horizontal displacement amplitude is obtained at

$$\frac{y_{v\_max}}{H} = \frac{1}{2\sqrt{3}} \frac{\ln \dfrac{k_z}{k_y}\dfrac{E_y}{G_{yz}}}{\sqrt{\dfrac{k_z}{k_y}} - \sqrt{\dfrac{G_{yz}}{E_y}}} \qquad [4.7]$$

Vertical surface displacement amplitude is maximum at the origin and decreases to zero moving outwards. Its inflexion point is attained at

$$\frac{y_{w\_inflexion}}{H} = \frac{1}{2\sqrt{3}} \frac{\ln \dfrac{k_z}{k_y}\dfrac{G_{yz}}{E_z}}{\sqrt{\dfrac{k_z}{k_y}} - \sqrt{\dfrac{E_z}{G_{yz}}}} \qquad [4.8]$$

y_{v_max} and $y_{w\_inflexion}$ locations depend on the outer boundary which has been, for practical purposes, moved to infinity. They might be re-written for a bounded rock mass, possibly using a harmonic pressure decomposition. This is needed to refine the prediction concerning surface displacements and the position of their particular points.

As an example, if the surface structure is an arch dam and is located between 0 and y_{v_max}, its abutments will move toward each other. If the arch dam is located beyond y_{v_max}, the abutments will move in opposite directions. If y_{v_max} is in the middle of the dam, the latter will be shifted toward the tunnel axis but not deformed horizontally.

4.4 Isotropic displacements

The de-coupling relation $E_y\lambda_{yz} = G_{yz}$ in the case of isotropic behaviour is satisfied with a unique value of Poisson's coefficient in each of the deformation modes. ν is 1/4 for plane deformation and 1/3 for plane stress (Table 1). G_{yz} is noted G and E/G is $2(1+\nu)$; E is the 3D isotropic elastic modulus. Cross coefficients λ_{yz} and λ_{zy} are both equal to 1/3 and are independent of the deformation mode. Ratios G_{yz}/E_y, G_{yz}/E_z and k_z/k_y are equal to 1/3, 1/3 and 1, respectively. Inserting these values in [4.5] and [4.6] leads to surface displacements of an isotropic infinite rigid base strip. Their amplitudes are shown in Figure 2.

5 CONCLUSION

The semi discrete weak model has demonstrated its usefulness in isolating parameters of the water-ground-tunnel interaction and looking at their individual effects. An important point has not been treated; it is the precision and application range of the weak equations. Tunnel excavation consolidates the rock mass and then drains the aquifer, lowering the water table. Displacements in this paper reflect the consolidating part, since the water table has been implicitly assumed to hold jointly to the surface.

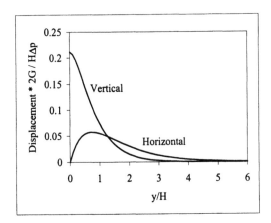

Figure 2. Amplitudes of uncoupled steady state surface displacements of an isotropic infinite rigid base strip.

REFERENCES

Attewell, P.B. & Woodman, J.P. 1982. Predicting the dynamics of ground settlement and its derivitives caused by tunnelling in soil. *Ground Engineering*, 15, 13-16, 18-20, 22 and 36.

Berry, D.S. 1960. An elastic treatment of ground movement due to mining-I. Isotropic ground. *J. Mech. Phys. Solids*, 8, 280-292

Berry, D.S. & Sales, T.W. 1961. An elastic treatment of ground movement due to mining-II. Transversely isotropic ground. *J. Mech. Phys. Solids*, 9, 52-62.

Boscardin, M.D. & Cording E.J. 1989. Building response to excavation-induced settlement. *ASCE, Journal of Geotechnical Eng.*, 115, 1-21.

Burland, J.B. 1997. Assessment of risk damage to buildings due to tunnelling and excavation. *1st Int. Conf. Earthtquake Geot. Eng. ,IS-Tokyo 95*, 1189-1201. Balkema.

Celestino, T.B., Ferreira, A.A. & Re, G. 2000. Shallow tunnel excavation safety evaluation based on ground distorsion measurements, *AITS-ITA 2000, Durban*, 79-83, SAIMM.

Dhatt, G. & Touzot, G. 1984. *Une présentation de la méthode des éléments finis*. Maloine, Paris.

Dormieux, L., De Buhan, P. & Leca, E. 1992. Estimation par une méthode variationnelle en élasticité des déformation lors du creusement d'un tunnel. *Revue Française de Géotechnique*, 59, 15-32.

El Tani, M. 1999. Water inflow into tunnels. *Proc. World Tunnel Cong. ITA 1999 , Oslo*, 61-70. Balkema.

El Tani, M. 2001. Tunnel induced displacements in a low shear rock aquifer. *Proc. WTC ITA 2001, Milan*, 4 p, Pàtron.

Fillunger, P. 1913. The uplift in dams (in German). *Österreich. Wochenschrift öffentlichen Baudienst 19*, 532-535, 552-555, 556-570, 586-593.

Hugh, T. 1987. *The finite element method*. Pretince Hall, New Jersey.

Ingham, B.D. & Kelmanson, M.A. 1984. Boundary integral equations. Analyses of singular, potential and biharmonic problems. Lecture notes in Engineering. Springer-Verlag.

Kachanov, I., Tsikrov, I. & Shafiro, B. 1994. Effective moduli of solids with cavities of various shapes. *Appl. Mech. Rev.* 47, 151-174.

Loganathan, N. & Poulos, H.G. 1998. Analytic prediction for tunneling-induced ground movements in clays. *J. Geo. & Geoenv. Eng.*, 124, 846-856.

Lombardi, G. 1992. The F.E.S. rock mass model, Part I, *Dam Engineering*, III, 46-76.

Lombardi, G. 1999. Tassements de massifs rocheux au-dessus de tunnels. *Vorerkundung und Prognose der Basis-tunnels am Gotthard und am Lötschberg, Zürich*, 269-278, Balkema.

Lurie, S. A. & Vasiliev V.V. 1995. *The biharmonic problem of the theory of elasticity*. Gordon and Breach.

Morland, L. W. 1974. Elastic response of regularly jointed media. *Geophys. J. R. Astr. Soc.*, 8, 435-446.

O'Reilly, M.P. & New, B.M. 1982. Settlements above tunnels in the United Kingdom - their magnitude and prediction. *3rd Int. Sym. , Tunnelling '82, IMM London*, 173-181.

Peck, R.B. 1969. Deep excavations and tunneling in soft ground. *State of the art volume. 7th Int. Conf. Soil Mech. Foundation Eng, Mexico*, 225-290.

Pougatsch, H. 1990. Le barrage de Zeuzier. Rétrospective d'un événement particulier. *Eau, Energie, Air, 82, 9*, 195-208.

Rat, M. 1973. Ecoulement et répartition des pressions interstitielles autour des tunnels. *Bull. Liaison du Laboratoire des Ponts et Chaussées*, 68, 109-124.

Sagaseta, C. 1987. Analysis of undrained soil deformation due to ground loss. *Géotechnique*, 37, 301-320.

Singh, G. 1973. Continuum characterization of jointed rock masses, Part I, *Int. J. Rock Mech. Min. Sci.*, 10, 311-335.

Tal A. & Dagan J. 1983. Flow toward storage tunnels beneath water table 1. Two dimensional flow. *Water Resource Researc*, 19, 241-249.

Ting, T. 1995. *Anisotropic elasticity*. Oxford University Press.

Terzaghi, K. 1924. The theory of hydrodynamic stresses and its geotechnical applications (in German). *Proc. Int. Cong. Applied Mechanics, Delft*, 288-294.

Truesdell, C. 1957. Sulle basi della termomeccanica. *Atti della Accademia Nazionale dei Lincei. Serie Ottava. Rendiconti*, XXII, 33-38, 158-166.

Verruijt, A. & Booker, J.R. 1996. Surface settlements due to deformation of a tunnel in an elastic half plane. *Géotechnique*, 46, 753-756.

Verruijt, A. 1998. Deformations of a half plane with a circular cavity. *Int. J. Solids Structures*, 35, 2795-2804.

REFERENCES ATTRIBUTION

Section 1. Volume loss; elastic subsidence in an analytical form: Berry (1960), Berry et al. (1961), Sagaseta (1981), Verruijt et al. (1996) and Verruijt (1998); subsidence based on the Gauss function: Peck (1969), Attewell et al. (1982) and O'Reilly et al. (1982); subsidence based on the Yield function: Celestino et al. (2000); semi empirical subsidence using the Gauss function: Dormieux et al. (1992) and Loganathan et al. (1998). Elastic subsidence and tunnel's induced pressure: El Tani (2001). Tunnels and dams: Pougatsch (1990) and Lombardi (1999). Tunnels and buildings: Boscardin et al. (1989) and Burland (1997).

Section 2. Pressure induced by tunnels: Rat (1972), Tal et al. (1972) and El Tani (1999). Effective stress: Fillunger (1913) and Terzaghi (1924). Partial stress in mixtures: Truesdell (1957). Jointed rock modulus: Singh (1973), Morland (1974) and Lombardi (1992). Porous material modulus: Kachanov et al. (1994). Monograph on anisotropic elasticity: Ting (1991).

Section 3. Low shear model: El Tani (2001).

Section 4. Monograph on biharmonic elasticity: Lurie et al. (1995). Monograph on biharmonic integral using Green functions: Ingham et al. (1984). Pressure integral form and water inflow in tunnels: El Tani (1999).

Appendix. Monographs on finite elements: Dhatt et al. (1984) and Hugh (1987).

APPENDIX: THE SEMI-DISCRETE WEAK FORMULATION

The semi-weak formulation consists of a partial weighted integration of the basic equations. They are made discrete in one space direction and analytical in the other.

A1 *The semi weak procedure*

A set of nodal lines z_i and functions $\phi_i(z)$ are chosen with the following properties:

$$\phi_i(z_j) = \delta_{ij} ; \quad i,j \geq 1 \qquad [A1]$$

where δ_{ij} is Kronecker's delta. The functions ϕ_i are used as interpolation functions for the displacements:

$$v(y,z) = \sum_i \phi_i(z) v(y,z_i) \qquad [A2]$$

$$w(y,z) = \sum_i \phi_i(z) w(y,z_i) \qquad [A3]$$

The first component of the equilibrium equation [2.1] is multiplied by the functions $\phi_i(z)$ whenever $v(y,z_i)$ is non zero. The second component of the equilibrium equation [2.2] is multiplied by the functions $\phi_i(z)$ whenever $w(y,z_i)$ is non zero. The resulting products are integrated over a vertical line. Taking into account the conditions at the base and surface shown in Figure 1, the integration leads to

$$\int_0^H [\phi_i \frac{\partial \sigma_y}{\partial y} - \tau \frac{d\phi_i}{dz}]dz = 0 ; \quad i \geq 1 \text{ and } v(y,z_i) \neq 0 \qquad [A4]$$

$$\int_0^H [\phi_i \frac{\partial \tau}{\partial y} - \sigma_z \frac{d\phi_i}{dz}]dz = 0 ; \quad i \geq 1 \text{ and } w(y,z_i) \neq 0 \qquad [A5]$$

The final weak equations are obtained by inserting the interpolations [A2] and [A3] into strains [2.6] which are taken into the material relations [2.3] to [2.5] which are then inserted in [A4] and [A5].

A2 *Application : Sliding base mass*

The linear interpolation functions over the segment [0,H] are chosen. There are two functions which are $\phi_1(z) = 1-z/H$ and $\phi_2(z) = z/H$. The nodal lines z_1 and z_2 are 0 and H. Only the horizontal and vertical displacements at the base and at the surface are involved. Thus, it is possible to use the following notations without risk of confusion:

- w(y) for w(y,H) which is the value of the vertical displacement at the surface.
- v(y) for (v(y,0)+v(y,H))/2 which is the mean value of the horizontal displacement.
- Δ(y) for v(y,H)-v(y,0) which is the difference between the surface displacements and those at the base.

Applying the procedure given in section A1, the following equations may be obtained after some manipulations:

$$\frac{d^2 v}{dy^2} + \frac{\Lambda}{HE_y} \frac{dw}{dy} = \frac{1}{HE_y} \int_0^H \frac{\partial p}{\partial y} dz \qquad [A6]$$

$$\frac{d^2 \Delta}{dy^2} - \frac{12G_{yz}}{E_y} \frac{\Delta}{H^2} - \frac{6G_{yz}}{E_y H} \frac{dw}{dy} = \frac{6}{HE_y} \int_0^H (\frac{2z}{H}-1) \frac{\partial p}{\partial y} dz \qquad [A7]$$

$$\frac{d^2 w}{dy^2} - \frac{3E_z}{G_{yz}} \frac{w}{H^2} + \frac{3}{2H} \frac{d\Delta}{dy} - \frac{3\Lambda}{G_{yz}H} \frac{dv}{dy} = \frac{3}{G_{yz}H} \int_0^H \frac{z}{H} \frac{\partial p}{\partial z} dz \qquad [A8]$$

where $\Lambda = E_y \lambda_{yz} = E_z \lambda_{zy}$. The boundary conditions are $v(0)=v(R)=\Delta(0)=\Delta(R)=dw/dy(0)=dw/dy(R)=0$.

A3 *Application : Rigid base mass*

The same nodal lines z_1 and z_2 and the same interpolation functions ϕ_1 and ϕ_2 from section A2 are used. Here, only v(y,H) and w(y,H) are involved. The following notation is used:
- v(y) for v(y,H) which is the surface horizontal displacement.
- w(y) for w(y,H) which is the surface vertical displacement.

Applying the procedure in A1 leads to the following equations:

$$\frac{d^2 v}{dy^2} - \frac{3G_{yz}}{E_y} \frac{v}{H^2} + \frac{3(\Lambda - G_{yz})}{2E_y H} \frac{dw}{dy} = \frac{3}{HE_y} \int_0^H \frac{z}{H} \frac{\partial p}{\partial y} dz \qquad [A9]$$

$$\frac{d^2 w}{dy^2} - \frac{3E_z}{G_{yz}} \frac{w}{H^2} + \frac{3(G_{yz} - \Lambda)}{2G_{yz}H} \frac{dv}{dy} = \frac{3}{HG_{yz}} \int_0^H \frac{z}{H} \frac{\partial p}{\partial z} dz \qquad [A10]$$

where $\Lambda = E_y \lambda_{yz} = E_z \lambda_{zy}$. The boundary conditions are $v(0)=v(R)=dw/dy(0)= dw/dy(R)=0$.

Modern Tunneling Science and Technology, Adachi et al (eds), © 2001 Swets & Zeitlinger, ISBN 90 2651 860 9

Short and long-term load conditions for tunnels in low permeability ground in the framework of the convergence-confinement method

A.Graziani & R.Ribacchi
University of Rome "La Sapienza", Italy

ABSTRACT: The "convergence-confinement" method, based on the simplified scheme of a circular tunnel under isotropic in situ stress, represents a useful tool for analyzing ground-lining interaction in preliminary design stages. In the paper, the stress-flow conditions around a tunnel driven through a saturated elastic or elasto-plastic medium are investigated. As common problems in soil mechanics, the stability and the deformation of a structure can be assessed under short- and long-term conditions. The same approach has sometimes been adopted in tunnel design by utilizing the characteristic curve of the tunnel evaluated separately for undrained and steady-state drained conditions. The applicability of such a simplified method is discussed by comparing it to the results of the fully-coupled model.

1 INTRODUCTION

Driving a tunnel in a water-bearing ground modifies the total stress regime as well as the water flow and pore pressure distribution around the excavation.

If the permeability of the ground is low, the excavation of the tunnel and the installation of the lining can be analyzed in practice as a time-independent process, which takes place in undrained conditions, before any consolidation phenomenon is initiated.

Pore pressure changes within the ground, generated in the short-term situation (t=0), and the modified boundary conditions for flow after lining installation, determine the development of a flow-deformation time-dependent coupled process. The load on the lining may markedly change during the consolidation process until the long-term (t=∞) steady-state flow conditions are attained.

Ground-lining interaction is usually evaluated in the preliminary design stage by the "convergence-confinement" method, based on the simplified scheme of a circular tunnel under isotropic in situ stress.

The problem is solved by assuming plane strain conditions and by simulating the effect of excavation advance by progressively relieving the initial load S applied on the tunnel wall, which in most deep tunnel cases can be assumed to be isotropic and equal to the mean overburden stress at the depth of the tunnel axis.

Some basic stress-diffusion solutions for fluid-saturated elastic porous media have been worked out by various authors in the last decades.

Rice & Cleary (1976) gave some explicit solutions for the special cases of short-term and steady-state flow conditions around a hole in an infinite medium.

Carter & Booker (1982) and Detournay & Cheng (1988) examined the coupled stress-flow transient when a lining is installed before any deformation of the ground occurs. This assumption may be reasonable only for the special case in practice where the lining is jacked into the clay ahead of the excavation face.

To analyze practical situations characterized by more general boundary conditions and material behavior, resort can be made to numerical models. As an example the work by Ohtsu et al. (1999) may be quoted.

In this paper some simplified uncoupled solutions for elastic and elasto-plastic ground (Lembo-Fazio & Ribacchi 1984) are assessed by comparing them with coupled numerical analyses performed by the finite differences FLAC code (Itasca 1998).

2 ELASTIC POROUS MEDIUM

The case of a circular tunnel (radius a) driven in a porous medium with uniform initial total stress S and pore pressure u_o is considered (Figure 1).

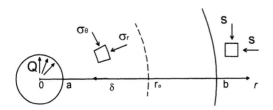

Figure 1. Problem definition.

The axisymmetric problem at hand exhibits some special features: relieving the internal pressure S (which simulates the excavation in undrained conditions) does not change the mean stress $p = (\sigma_r + \sigma_\theta)/2$ in any point around the tunnel; therefore no excess pore pressure Δu is generated at $t = 0$. Moreover, Rice & Cleary (1976) have shown that for an infinite domain $(b/a \rightarrow \infty)$ the pore pressure distribution during the transient flow phase $(t > 0)$ is simply governed by a homogeneous diffusion equation:

$$\frac{\partial^2 u}{\partial r^2} + \frac{1}{r}\frac{\partial u}{\partial r} = c_V \frac{\partial u}{\partial t} \qquad (1)$$

where c_V is the consolidation coefficient given by

$$c_V = \frac{2kGB^2(1-\nu)(1+\nu_u)^2}{9(\nu_u - \nu)(1-\nu_u)\gamma_w} \qquad (2)$$

where k is the permeability of the medium, G the shear modulus, B the Skempton parameter, ν and ν_u

the Poisson coefficient respectively for drained and undrained conditions.

The solution $u(r,t)$ of the governing equation (1) is a function only of two normalized variables: the scaled radius $\rho = r/a$ and the time factor $T = c_V t/a^2$.

For the typical case of soft rock tunneling, considered here, solid grain compressibility can be neglected, and B and ν_u are always very close to the limit value 1 and 0.5 which hold for incompressible fluids (soil mechanics case).

Rehbinder (1996) has clearly pointed out the marked influence of fictitious outer boundaries used in numerical models (Fig. 1) to approximate a real infinite domain stress-flow problem. A number of numerical experiments have demonstrated that a good agreement with exact solutions can be achieved by the following assumptions:

(i)   the outer boundary of the model is located at a distance r = b, with $b/a \geq 50$;

(ii)   at r = b, the radial stress σ_r is assumed to be equal to the stress S for $r \rightarrow \infty$, while displacement is free;

(iii)   the domain (a,b) is divided into two sub-zones by a limit radius r_0: an internal zone (a, r_0) where the pore pressure can vary following the perturbation imposed at the tunnel wall; an external zone (r_0, b) where the pore pressure remains undisturbed, equal to the initial value u_0.

Unlike the Detournay & Cheng (1988) solution, which can only represent the case of an indefinitely time-increasing drainage effect due to tunnel excavation, assumption (iii) allows an asymptotic steady state for pore pressure distribution to be

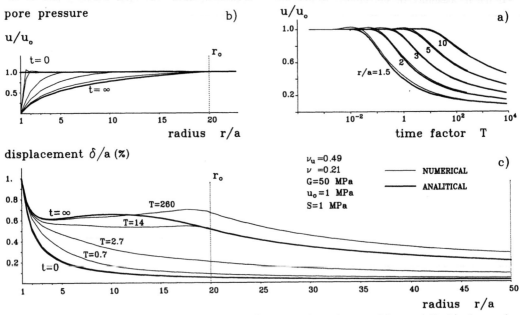

Figure 2. Tunnel in an elastic porous medium: a) pore pressure history at various points around the tunnel; b), c) isochrones of pore pressure and displacement variation with scaled radius.

reached in a finite time. The latter scheme can better represent many real situations where long-term equilibrium conditions are actually reached thanks to a natural recharge from the water-bearing stratum, reservoir or from rainfall, if any.

In this case the long-term drained conditions ($t = \infty$) are characterized by a pore pressure profile (Figure 2b) which is a logarithmic function of the radial distance

$$u(r, \infty) = u_a + (u_o - u_a)\frac{\ln(r/a)}{\ln(r_o/a)} \quad , \quad a \le r \le r_o \quad (3)$$

For the analyses presented here, the boundary of the flow zone has always been assumed to be $r_o/a = 20$ and the pore pressure at the tunnel wall $u_a = 0$.

The pore pressure perturbation propagates from the tunnel wall inside the rock mass: Figure 2a compares the pore pressure history at various points around the tunnel afforded by numerical and analytical coupled models. The grid used for the numerical model is composed of 100 axisymmetric elements, with an outer boundary of $r = b = 50a$.

Figure 2c shows the isochrones of displacement around the tunnel: the numerical results perfectly fit the analytical solutions for the limit case of undrained ($t = 0$) and drained ($t = \infty$) conditions, at least in the vicinity of the tunnel wall ($r/a < 10$).

It is worth noting that consolidation phenomena lead to an increase in displacements only at points inside the rock mass, while the convergence δ_a of the tunnel wall remains exactly the same as for the short-term undrained conditions:

$$\delta_a = (S - Q)\, a\, /2G \quad (4)$$

which in turn is the same as for a tunnel in a completely dry porous medium. Q stands for the total pressure applied to the wall, if a support system is present.

The axi-symmetric problem of a circular tunnel in a linear-elastic porous rock mass, notwithstanding its great interest as a basic conceptual model and a benchmark for numerical models, proves to be trivial as far as the ground-lining interaction is concerned.

The load Q acting upon the lining is not affected by consolidation phenomena, but depends only on the installation method and distance from the face; it can therefore be determined on the basis of the characteristic curve (4).

3 ELASTO-PLASTIC POROUS MEDIUM

The development of plastic deformations around the tunnel strongly affects the solid matrix-fluid interaction and, ultimately, also the load acting on the lining during the transient following a rapid undrained excavation.

Before analyzing the fully coupled hydraulic-mechanical process triggered by the relief of internal total pressure Q, it is worth considering the characteristic curves $\delta_a(Q)$ of the tunnel evaluated separately for the two limit cases of undrained ($t = 0$) and drained ($t = \infty$) conditions.

The yielding behavior of the solid matrix is represented by a simple ideal-plastic model, with Mohr-Coulomb strength parameters c', φ' and non-associated flow rule. The specific material parameters assumed in the analyses (see Table 1) are representative of a medium-hard clay.

Table 1. Material and geometric properties of tunnel.

G	33. MPa	ν	0.25
S	1 MPa	ν_u	0.49
u_o	0.4 MPa	c'	0
k	10^{-9} m/s	φ'	30°

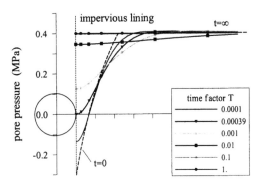

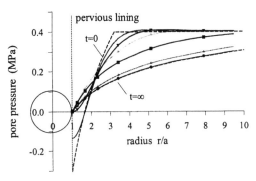

Figure 3. Isochrones of pore pressure profiles around a pervious or impervious tunnel.

3.1 *Short-term conditions*

For a saturated Mohr-Coulomb material with no dilation ($\psi = 0$), the undrained yielding limit can be expressed in the total stress space by the undrained

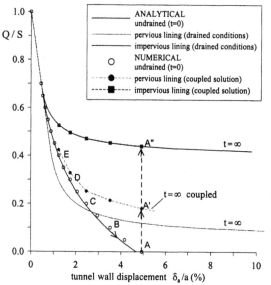

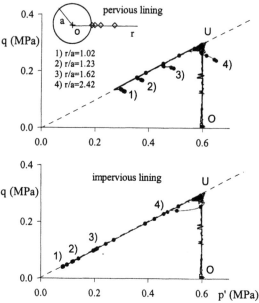

Figure 4. Characteristic curves of the tunnel for the short and long-term conditions obtained by different methods.

Figure 5. Effective stress paths around a tunnel in elasto-plastic ground.

cohesion c_u. For the problem at hand, c_u is independent of r/a and equal to $(S-u_o) \cdot \sin\varphi$.

On the basis of the aforementioned assumptions, a simple closed form solution can be obtained (Salençon 1969) for a tunnel excavated in undrained conditions (soil mechanics case):

$$\frac{\delta(r)}{a} = \frac{c_u}{2G} \exp\left(\frac{S-Q}{c_u} - 1\right) \cdot \frac{a}{r} \qquad (5)$$

Excess pore pressure is predicted only within the plastic zone, and has the following expression:

$$\Delta u(r) = -S + Q + c_u[1 - 2 \cdot \ln(r/a)] \qquad (6)$$

Relationship (6) indicates a general decrease in pore pressure, which near the tunnel wall can also attain negative values, as in the case of Figure 3 (curve for t = 0).

The analytical solution is perfectly matched by the numerical model, which takes the pore-fluid explicitly into account and operates with a yielding condition in terms of effective stress.

The characteristic curve of the undrained tunnel is plotted in Figure 4, together with numerical results for a set of decreasing values of stabilizing pressure Q.

3.2 Long-term conditions

Following the conventional soil mechanics approach, the long-term situation of the tunnel, when the excess pore pressure has dissipated and steady-flow conditions have been established, could be assessed on the basis of an uncoupled hydro-mechanical analysis.

The pore pressure distribution for t = ∞ depends only on boundary conditions: the set of hydraulic conditions corresponding to the pore pressure profile (3) is deemed sufficiently general to represent most practical cases.

Lembo-Fazio & Ribacchi (1984) derived a closed form solution for drained conditions for t = ∞, by assuming the pore pressure distribution of relationship (3). The extension of the plastic zone (plastic radius R) obtained by this approach is:

$$\frac{R}{a} = \left\{ \frac{\left[S - u_R - \frac{(1-2v)}{2(1-v)}(u_o - u_R) + c' \cdot \cotg\varphi' \right](1 - \sin\varphi') - h}{Q - u_a + c' \cdot \cotg\varphi' - h} \right\}^{\frac{1}{N-1}}$$

$$h = \frac{u_o - u_a}{(N-1) \cdot \ln(r_o/a)} \quad , \quad N = \frac{1 + \sin\varphi'}{1 - \sin\varphi'} \qquad (7)$$

where the parameter h represents the effect of the flow gradient on the plastic zone; u_R is the pore pressure at the plastic radius R which must also satisfy equation (3).

An examination of relationship (7) shows that pore pressure influence is equivalent to a reduction in both initial stress (which is favorable to stability) and rock mass cohesion (which is obviously unfavorable). Therefore, the extension of the plastic zone can either be lower or greater with respect to the condition where groundwater is absent. Relationship (7) also shows that for the stability of a tunnel located below the water table a minimum support pressure is necessary, and is given by

$$Q_{cr} = u_a + h - c \cot g\varphi \qquad (8)$$

The expression of the displacement function $\delta(r)$, which is rather cumbersome, is reported in APPENDIX.

The characteristic curves $\delta_a(Q)$ of the tunnel, for the two limit cases of a perfectly tight ($u_a = u_o$) and completely pervious ($u_a = 0$) tunnel wall, are plotted in Figure 4. The two curves are asymptotic to the critical value Q_{cr}, given by (8), which, when scaled with respect to S, is equal to 0.4 and 0.07, respectively.

3.3 *Fully-coupled stress-flow model*

The issue still to be discussed is whether or not the long-term situation predicted by the uncoupled analysis matches the results of a fully-coupled model.

The loading path assumed for the coupled numerical analyses starts from the undrained curve (points A, B, C, D, E in Figure 4) at different initial support pressure, it then follows the support characteristic line up to the point corresponding to the long term load $Q(t = \infty)$. For the typical relative stiffness of a closed ring concrete lining, the support line is almost vertical.

The two limit hydraulic conditions of a perfectly impervious and pervious lining have been considered.

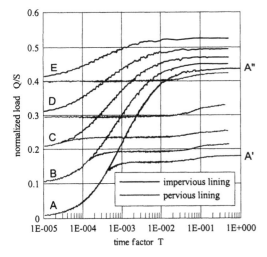

Figure 6. History of lining load during consolidation for various initial loads at the time of installation (see also Figure 3).

While in the impervious case the final load is exactly the same as that predicted by the simplified uncoupled solution, in the pervious case the coupled model gives a significantly higher load than the uncoupled solution (Figure 4). The reason for such different behavior can be better understood if the stress-paths near the tunnel wall during the consolidation phase are monitored (Figure 5).

While in the impervious case the plastic zone formed during undrained excavation expands, and the stress-path always lies on the plastic limit, in the pervious case all the points around the tunnel undergo elastic unloading during the last phase of the consolidation process. This kind of behavior makes the exact solution stress-path dependent and therefore different from the uncoupled solution.

Moreover, only the coupled model can give an estimate of the time required to reach the final steady state conditions. Figure 3 shows the isochrones of the pore pressure at selected times: the steady state conditions are practically attained for a time factor T close to 1.

The histories of lining load Q are plotted in Figure 6: the curves marked as A, B, ... correspond to different initial loads in undrained conditions, as indicated in Figure 4.

4 APPLICATIONS

The closed form solution (5) has been successfully applied by Mair & Taylor (1992) to the prediction of undrained deformation around tunnels driven in London clay and Boom hard clay.

The mean values of short–term loads for tunnels in London clay is about 20-30% of the overburden load (that is Q/S = 0.2-0.3).

The purpose of this paper was to propose a more comprehensive model which could encompass both the short and long-term conditions for a tunnel excavated under the water table. It is the aim of the authors to review the observed behavior of deep tunnels in different clayey grounds in the framework of the proposed simplified models.

One of the open questions which deserves more careful examination of real cases to be answered is what are the more realistic hydraulic boundary conditions at the extrados of the lining. The characteristic curves of Figure 4 highlight that the permeability of the lining is the key parameter to asses the long-term load.

Notice that material parameters assumed in the example of Figure 4 could also be suited to the case of deep tunnels in London clay.

Data gathered by Barrat et al. (1994) suggest that, for tunnels in London clay at depth of 25-30m, the long-term load is about 60% of the overburden. A typical load history measured in London Underground at Regents Park is shown in Figure 7.

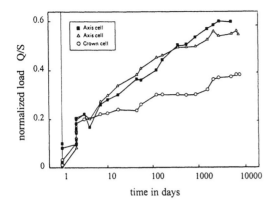

Figure 7. Long-term measurements of lining load in London clay (Regents Park tunnel, adapted from Barrat et al. 1994).

This empirical finding can only be explained if the lining effectively acts as an impervious barrier, in contrast to the common opinion according to which the tunnel acts as a drain (Ward & Pender 1981).

Some recent measurements reported by Gourvenec et al. (1999) confirm that only local seepage occurs towards the tunnel, with pore pressures corresponding to the far field being reached within some 1.5m from the tunnel wall (radius a = 2m).

5 CONCLUSIONS

Coupled analyses of the consolidation process around a circular tunnel have shown that:
- for an elastic medium, the lining-ground interaction is restricted to the undrained deformation phase because the consolidation phenomena do not modify the convergence of the tunnel walls;
- for an elasto-plastic medium the conventional approach provides correct predictions of the long-term load only for the special case of a perfectly tight lining.

In many real cases, when the tunnel acts as a drain inside the ground, the long-term load on the lining predicted by the conventional approach can be remarkably underestimated.

These findings may be a reasonable explanation for the gradual increase of load and some lining failures observed after years of the operation of tunnels.

REFERENCES

Ward, W.H. & M.J. Pender 1981. Tunnelling in soft ground – General report. *10th ICSMFE, Stockholm*, 4: 261-275. Rotterdam: Balkema.

Rice, J.R. & M.P. Cleary 1976. Some basic stress diffusion solutions for fluid-saturated elastic porous media with compressible constituents. *Reviews of Geophysics and space physics*, 14, 2: 227-241.

Carter, J.P. & J.R. Booker 1982. Elastic consolidation around a deep circular tunnel. *Int. J. Solids Structures*, 18, 12: 1059-1074.

Detournay, E. & A.H. Cheng 1988. Poroelastic response of a borehole in a non-hydrostatic stress field. *Int. J. Rock Mech. Min. Sci.*, 25, 3: 171-182.

Rehbinder, G. 1996. Influence of fictitious outer boundaries on the solution of external field problems. *In Coupled thermo-hydro-mechanical processes of fractured media, Developments in Geotechnical Engineering*, 79, 231-243.

Lembo-Fazio, A. & R. Ribacchi 1984. Influence of seepage on tunnel stability. *Int. Conf. Design and performance of underground excavations, Cambridge*, ISRM/BGS, 21, 173-181.

Ohtsu, H., Y. Onhishi, H. Taki & K. Kamemura 1999. A study on problems associated with finite element excavation analysis by the stress-flow coupled method. *Int. J. Num. Anal. Meth. Geomech.*, 23: 1473-1492.

Itasca 1998. FLAC, User's Manual, Version 3.40. *Itasca Consulting Group Inc., Minneapolis, Minnesota.*

Salençon, J. 1969. Contraction quasi-statique d'une cavité à symmétrie sphérique ou cylindrique dans un milieu élastoplastique. *Ann. Ponts et Chaussées*, 4: 231-236.

Mair, R.J. & R.N. Taylor 1992. Prediction of clay behaviour around tunnels using plasticity solutions. *Wroth Mem. Symp., Predictive Soil Mech., Oxford*, 449-463.

Barrat, D.A., M.P. O'Reilly & J. Temporal 1994. Long-term measurements of loads on tunnel linings in overconsolidated clay. *Tunnelling '94, IMM*, 469-481. Chapman & Hall.

Gourvenec, S.M., M.D. Bolton, K. Soga & M.W. Gui 1999. Field investigation of long term loading on an old tunnel in London Clay. *Int. Symp. Geotech. Aspects of Underground Construction in Soft Ground, Tokyo*, 219-224.

APPENDIX

Analytical expression of displacement function for long-term drained conditions (uncoupled model):

$$\frac{\delta(r)}{a} \cdot 2G = \left[(S - u_o)\sin\varphi + (u_o - u_R)\frac{1-2v+\sin\varphi}{2(1-v)} \right] \cdot \frac{(R/a)^{K+1}}{(r/a)^K} +$$

$$+ \left[(S - u_o + c \cdot \cotg\varphi)(1-2v) - h\frac{1+NK - v(K+1)(N+1)}{K+1} \right] \cdot \left[\frac{(R/a)^{K+1}}{(r/a)^K} - (r/a) \right] +$$

$$- \left[(Q - u_a + c \cdot \cotg\varphi - h)\frac{1+NK - v(K+1)(N+1)}{N+K} \right] \cdot \left[\frac{(R/a)^{N+K}}{(r/a)^K} - (r/a)^N \right]$$

where:

$$h = \frac{u_a - u_g}{(N-1) \cdot \ln(r_o/a)} \quad , \quad N = \frac{1+\sin\varphi}{1-\sin\varphi} \quad , \quad K = \frac{1+\sin\psi}{1-\sin\psi}$$

Modern Tunneling Science and Technology, Adachi et al (eds), © 2001 Swets & Zeitlinger, ISBN 90 2651 860 9

3-D FEM Analysis for Layered Rock Considering Anisotropy of Shear Strength

Zhang Yujun
Institute of Rock and Soil Mechanics, The Chinese Academy of Sciences, Wuhan 430071, China

Liu Yiping
Institute of Physics and Mathematics, The Chinese Academy of Sciences, Wuhan 430071, China

Xiong chuanyi
Institute of Rock and Soil Mechanics, The Chinese Academy of Sciences, Wuhan 430071, China

ABSTRACT: A layered rock possesses obvious anisotropic mechanical behavior, and its shear strength should be a function of inclination with respect to the foliated planes. For this the first author suggested an empirical expression of C, ϕ for layered rock. Now this expression is compared with a test result made by the former researchers and a good agreement between both can be seen. On the basis of this the constitutive relationship of a transversely isotropic medium and Mohr-Coulomb's criterion in which C, ϕ vary with directions are employed, and a relative 3-D elasto-plastic FEM code is developed. Taking an underground opening as the calculation object, the numerical analyses are carried out by using the FEM code for two cases of transversely isotropic rock and isotropic rock respectively. Finally, some conclusions are obtained according to this research.

1 INTRODUCTION

It is well-known that a layered rock possesses obvious anisotropic behavior of deformation and strength, so compared with a homogeneous rock, its conditions of stability and failure are more complicated(Serrano A & Olalla C 1998). At present a layered rock can be treated theoretically as a transversely isotropic medium, its stress-strain relationship includes five constants. If Mohr-Coulomb's criterion is employed to describe the failure characteristics of a layered rock, there should be $\tau = C(\theta) + \sigma \, tg \, \phi(\theta)$, that is, here both cohesion and internal friction angle aren't constants, but are functions of the angle between the failure surface and the foliated plane (or the direction of maximum principal stress σ_1)(Smith B & Cheatham JR B 1980, Nova R.1980). For this reason, the first author suggested an empirical expression of C, ϕ for a layered rock. Now this expression is compared with the test results made by the earlier researchers and a good agreement between both can be seen. On the basis of this the constitutive relationship of a transversely isotropic medium and Mohr-Coulomb's criterion in which C, ϕ vary with directions are introduced, therefore a relative 3-D elasto-plastic FEM code is developed. Taking an underground opening as the calculation object, the numerical analyses are carried out for both cases of a transversely isotropic rock and an isotropic rock, respectively, and some conclusions are obtained according to this research.

2 ANISOTROPIC SHEAR STRENGTH OF A LAYERED ROCK

2.1 *Expression of C, ϕ for a layered rock*

Generally speaking, the shear strength of a layered rock S has a minimum value S_{min} and a maximum value S_{max} in directions parallel to and perpendicular to the foliated plane, respectively, and among them S varies with the θ angle with respect to the foliated plane according to some rules (Figure 1) as:

$$S = f(S_{min}, S_{max}, \theta) \qquad (1)$$

Jaeger also gave an expression of S(Jaeger J. C. 1960):

$$S = S_1 + S_2 \cos 2(\alpha - \beta) \qquad (2)$$

where: S_1, S_2 are constants; α, β are angles between the failure surface as well as the foliated plane and the direction of maximum principal stress σ_1, respectively.

As a kind of approximation, the first author suggested an empirical expression for C, ϕ as (Y. Zhang & Y. Tang 1999):

$$C = \frac{2\theta}{\pi}(C_{max} - C_{min}) + C_{min}$$
$$\phi = \frac{2\theta}{\pi}(\phi_{max} - \phi_{min}) + \phi_{min} \qquad (0 \le \theta \le \frac{\pi}{2}) \qquad (3)$$

where: C_{min}, ϕ_{min} and C_{max}, ϕ_{max} are the cohesion and the internal friction angles parallel to and perpendicular to the foliated plane separately.

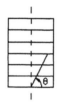

Figure 1. Inclination of plane with C, Φ concerned θ with respect to foliated plane.

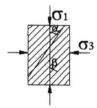

Figure 2. Inclination α , β of the failure plane and foliate plane with respect to σ_1.

2.2 Comparison with earlier test result

Here the test results made by Attewell et al.(Attewell B, Sandford R.1974)is used. For Penrhyn slate through changing the inclination β of the foliated plane with respect to the axis of maximum principal stress, Attewell et al. obtained the C, Φ values for different β magnitudes under the condition of triaxial compression . A curve-fit formula of C, Φ acquired by Attewell et al. is stated as:

$$C = C_1 - C_2 \cos 2(\gamma - \beta)$$
$$\phi = \phi_1 - \phi_2 \cos 2(\delta - \beta)$$
(4)

where: C_1, C_2 and ϕ_1, ϕ_2 are constants; γ, δ and β are inclinations of the failure surface and the foliated plane with respect to σ_1.

Because Attewell et al. didn't write the concrete values of C_1, C_2, γ and ϕ_1, ϕ_2, δ in(Attewell B & Sandford R. 1974), again the authors make a curve-fitting on the basis of Atwell's results and produce:

$$C = 4.30 - 2.28 \cos 2(40° - \beta)$$
$$\phi = 39.5° - 11.3° \cos 2(44° - \beta)$$
(5)

From equation (5) we have that:

$$C_{min} = C_1 - C_2 = 2.20, \quad C_{max} = C_1 + C_2 = 6.58$$

$$\phi_{min} = \phi_1 - \phi_2 = 28.2°, \quad \phi_{max} = \phi_1 + \phi_2 = 50.8°$$

Thus:

$$C = \frac{2\theta}{\pi}(6.58 - 2.02) + 2.02$$
$$\phi = \frac{2\theta}{\pi}(50.8° - 28.2°) + 28.2°$$
(6)

Now let β =90 ° ,that is, the foliated plane is perpendicular to the axis of σ_1 and change γ, δ and θ in expressions (4), (6) from 0° to 90° , the relative curves for C, Φ are demonstrated in Figure 3. From this it can been seen that both expressions (4), (6) have a good agreement.

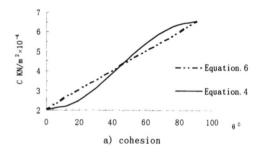

a) cohesion

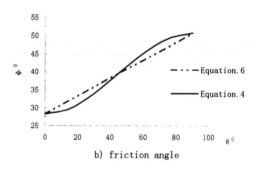

b) friction angle

Figure 3. Relationship beteen shear strength and θ .

3 PRINCIPLES OF THE 3-D FEM

3.1 Constitutive relationship

For the transversely isotropic rock shown in Figure 4, the foliated plane is taken as x´ y´ plane, and the normal direction of the foliated plane as z´ axis, then the elastic matrix in the local coordinate system x´y´z´ is written as(Zhou Weiyuan.1990).Where: M= $E_1/ (1+\mu_1) (1-\mu_1-2n\mu_2^2)$; n=$E_2/E_1$, $G_1=E_1/2 (1+\mu_1)$; E_1, μ_1 are the elastic modulus and Poisson's ration within the foliated plane, and E_2, G_2, μ_2 are the elastic modulus, the shear modulus and

$$[D'] = M \begin{bmatrix} 1-n\mu_2^2 & & & & & \\ \mu_1+n\mu_2^2 & 1-n\mu_2^2 & & & \text{Symmetry} & \\ \mu_2(1+\mu_1) & \mu_2(1+\mu_1) & \dfrac{1-\mu_1^2}{n} & & & \\ 0 & 0 & 0 & \dfrac{G_1}{M} & & \\ 0 & 0 & 0 & 0 & \dfrac{G_2}{M} & \\ 0 & 0 & 0 & 0 & 0 & \dfrac{G_2}{M} \end{bmatrix} \quad (7)$$

Poisson's ratio in the direction perpendicular to the foliated plane.

Stresses, strains are transformed from the local coordinate system $x'y'z'$ into the global coordinate system xyz, we get:

$$\{\sigma\} = [D] \cdot \{\varepsilon\} \qquad (8)$$

where:

$$[D] = [L] \cdot [D'] \cdot [L]^T$$

$$[L] = \begin{bmatrix} l_1^2 & l_2^2 & l_3^2 & 2l_1l_2 & 2l_2l_3 & 2l_3l_1 \\ m_1^2 & m_2^2 & m_3^2 & 2m_1m_2 & 2m_2m_3 & 2m_3m_1 \\ n_1^2 & n_2^2 & n_3^2 & 2n_1n_2 & 2n_2n_3 & 2n_3n_1 \\ l_1m_1 & l_2m_2 & l_3m_3 & l_1m_2+l_2m_1 & l_2m_3+l_3m_2 & l_3m_1+l_1m_3 \\ m_1n_1 & m_2n_2 & m_3n_3 & m_1n_2+m_2n_1 & m_2n_3+m_3n_2 & m_3n_1+m_1n_3 \\ n_1l_1 & n_2l_2 & n_3l_3 & n_1l_2+n_2l_1 & n_2l_3+n_3l_2 & n_3l_1+n_1l_3 \end{bmatrix}$$

$l_i, m_i, n_i (i=1,2,3)$ are cosins of the angles between the axes of the local and global coordinate systems.

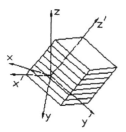

Figure 4. Local and global coordinate systems.

3.2 Failure criterion

For the layered rock and stress state presented in Figure 5, if Mohr-Coulomb's criterion is applied, a shear surface should occur in $\sigma_1 - \sigma_3$ plane, we have:

$$\tau = C(\theta) + \sigma\, tg\,\phi(\theta) \qquad (9)$$

where:

$$\tau = \frac{1}{2}(\sigma_1 - \sigma_3)\sin 2\rho \qquad (10)$$

$$\sigma = \frac{1}{2}(\sigma_1 + \sigma_3) + \frac{1}{2}(\sigma_1 - \sigma_3)\cos 2\rho \qquad (11)$$

Substituting equation (10) and equation (11) into equation (9) leads a yield function as:

$$F = \frac{1}{2}(\sigma_1 - \sigma_3)\sin 2\rho - C(\theta)$$
$$- \frac{1}{2}[(\sigma_1 + \sigma_3) + (\sigma_1 - \sigma_3)\cos 2\rho]tg\phi(\theta) \qquad (12)$$

After a transformation, equation (12) becomes(Owen D.R.J, Hinton E. 1980):

$$F = \sqrt{J_2'}[(\sin 2\rho - \cos 2\rho\, tg\phi(\theta))\cos\eta$$
$$+ \sqrt{3}tg\phi(\theta)\sin\eta] - \frac{1}{3}tg\phi(\theta)J_1 - C(\theta) \qquad (13)$$

where:

$$J_1 = \sigma_x + \sigma_y + \sigma_z$$

$$J_2' = \frac{1}{2}(\sigma_x'^2 + \sigma_y'^2 + \sigma_z'^2) + \tau_{xy}^2 + \tau_{yz}^2 + \tau_{xz}^2$$

$$\sin 3\eta = -\frac{3\sqrt{3}}{2} \cdot \frac{J_3}{(J_2')^{3/2}}$$

$$J_3 = \sigma_x'\sigma_y'\sigma_z' + 2\tau_{xy}\tau_{yz}\tau_{xz} - \sigma_x'\tau_{yz}^2 - \sigma_y'\tau_{xz}^2 - \sigma_z'\tau_{xy}^2$$

When the angle ρ between the failure surface and σ_3 axis and the relative $C(\theta)$, $\phi(\theta)$ are known, the plastic matrix can be computed as:

$$[D_p] = \frac{[D]\left\{\dfrac{\partial F}{\partial\sigma}\right\}\left\{\dfrac{\partial F}{\partial\sigma}\right\}^T[D]}{\left\{\dfrac{\partial F}{\partial\sigma}\right\}^T[D]\left\{\dfrac{\partial F}{\partial\sigma}\right\} + A} \qquad (14)$$

where: A is a hardening function.

It is necessary to point out: if the rock is an isotropic one, its C, ϕ are constants, so there is

$$\rho = \frac{\pi}{4} + \frac{\phi}{2} \text{(Brady B.H.G \& Brown E.T. 1990)when}$$

the failure happens; but here the rock is a transversely isotropic one, C, ϕ are functions of the orientations with respect to the foliated plane, so the angle ρ and the relative C, ϕ can't be determined explicitly when the rock breaks. For this reason the authors put forward a search-trial method to find the right as follows.

In the local coordinate system shown in Figure 6, it is assumed that a shear failure surface has been produced, and its normal cosins are l_x, m_x, n_x. As the normal is within $\sigma_1 - \sigma_3$ plane, the angles between it and axes are T_1, T_2 and 0, respectively, so this results in:

$$A = l_1 l_x + m_1 m_x + n_1 n_x$$

$$B = l_3 l_x + m_3 m_x + n_3 n_x$$

$$0 = l_2 l_x + m_2 m_x + n_2 n_x \qquad (15)$$

where: $A = \cos T_1$, $B = \cos T_2$.

Solving for the normal cosins from equation (15), we have:

$$m_x = \frac{Aa_2 - a_1 B}{b_1 a_2 - a_1 b_2}$$

$$l_x = \frac{A - b_1 m_x}{a_1} \qquad (16)$$

$$n_x = -\left(\frac{l_2}{n_2} l_x - \frac{m_2}{n_2} m_x \right)$$

where

$$a_1 = l_1 - \frac{n_1 l_2}{n_2}, \quad b_1 = m_1 - \frac{m_2 n_1}{n_2}$$

$$a_2 = l_3 - \frac{n_3 l_2}{n_2}, \quad b_2 = m_3 - \frac{m_2 n_3}{n_2}$$

n_x is the cosin of angle between the normal of the shear failure surface and z' axis (rotationally elastic symmetry axis), so it leads

$$\theta = \cos^{-1}(n_x) \qquad (17)$$

Then according to the relationship between C, ϕ and θ in equation (3), the relative C, ϕ may be calculated. But usually the $\{l_x, m_x, n_x\}$ are unknowns beforehand, for this purpose a search-trial method is employed, that is, let angle ρ change from $0°$ to $90°$, and for every ρ_f the $A_i = \cos T_{1i}$, $B_i = \cos T_{2i}$ are known, so the n_{xi} and θ_i can be determined, therefore the relative C_i and ϕ_i are obtained.

Using $f_i = \tau - C_i - \sigma tg\phi_i$ to find out the angle ρ_f corresponding to the maximum value of f_i, the surface with ρ_f is the real or potential shear failure surface to be evaluated. This work can be done by the computer easily.

4 CALCULATON EXAMPLES

It is assumed that the longitudinal axis of a tunnel, whose orientation is $\beta_0 = 0°$, is the z axis of a global coordinate system, and the strike and dipping angle of the foliated plane of a layered rock around the tunnel are $\beta = 45°$, $\alpha = 45°$, respectively, so we have (Yu Xuefu et al. 1983):

$$l_1 = \sin(\beta - \beta_0), \quad m_1 = 0, \quad n_1 = -\cos(\beta - \beta_0)$$

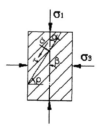

Figure 5. Stress state and foliated planes.

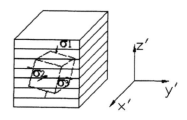

Figure 6. Layered rock and x' y' z' coordinate system.

$$l_2 = \cos(\beta - \beta_0)\sqrt{1 - \sin^2 \alpha}, \quad m_2 = \sin \alpha,$$

$$n_2 = \sin(\beta - \beta_0)\sqrt{1 - \sin^2 \alpha}$$

$$l_3 = \cos(\beta - \beta_0)\sqrt{1 - \cos^2 \alpha}, \quad m_3 = \cos \alpha,$$

$$n_3 = \sin(\beta - \beta_0)\sqrt{1 - \cos^2 \alpha}$$

The cross-section of the tunnel is a rectangular one with a size of 10m × 10m. The thickness of the overburden is 20m. For the cases of taking the rock as a transversely isotropic one and an isotropic one respectively, the numerical analyses by using the 3-D elasto-plastic FEM code developed by the authors are carried out. The mechanical parameters of rock and the mesh for FEM are shown in Table 1 and Figure 7.

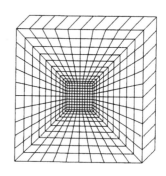

Figure 7. FEM mesh.

Table 1. Mechanical parameters of rock.

Rock type	Elastic modulus		Poisson's ratio		Shear modulus G(MPa)	Density γ (KN/m³)	Cohesion		Friction angle	
	E_1 (MPa)	E_2 (MPa)	μ_1	μ_2			C_1(MPa)	C_2(MPa)	$\Phi_1(°)$	$\Phi_2(°)$
Isotropy*	150.0	150.0	0.3	0.3	57.7	25.0	0.22	0.22	19.8	19.8
Transverse isotropy	200.0	100.0	0.2	0.4	40.0	25.0	0.11	0.33	14.2	25.4

* The mechanical parameters are obtained by averaging those of the transversely isotropic one.

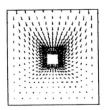

a) displacement vectors

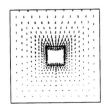

a) displacement vectors

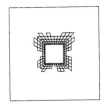

b) plastic zones

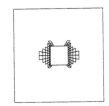

b) plastic zones

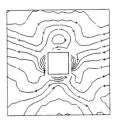

c) σ_x contours (10^{-1}Mpa)

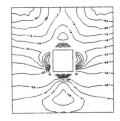

c) σ_x contours (10^{-1}Mpa)

d) σ_y contours (10^{-1}Mpa)

Figure 8. Calclation case of transversely isotropic rock.

d) σ_y contours (10^{-1}Mpa)

Figure 9. Calclation case of isotropic rock.

Table 2. Displacements at some points on opening boundary. cm

Rock type		Middle point of roof		Middle point of bottom		Middle point of left wall		Middle point of right wall	
		ux	uy	ux	uy	ux	uy	ux	uy
Isotropy		0.0	-6.7	0.0	3.7	1.2	-2.0	-1.2	-2.0
Transverse isotropy	Front face	-0.4	-10.9	0.3	4.2	6.2	-2.1	-7.4	-2.7
	Back face	-0.4	-9.8	0.2	4.7	6.7	-3.0	-6.9	-3.6

* ux is the horizontal displacement and it points towards the left for a positive sign; uy is the vertical displacement and it points upwards for a positive sign.

The displacement vectors, plastic zones and contours of stresses in the plane perpendicular to the z axis σ_x, σ_y obtained from the calculations are presented in Table 2 , Figure 8-9. From these data it can be seen that when the rock is a transversely isotropic one, all the distributions of the displacements, plastic zones and σ_x, σ_y contours in the surrounding rock are non-axisymmetric about the tunnel's vertical axis, and this is very different from that case of isotropic rock. So we know: when a layered rock has very different mechanical properties in the directions parallel to and perpendicular to the foliated plane separately, and its occurrence shows certain geometrical relationship with the excavation factors (for example, with the tunnel's axial line), the dynamic state of the surrounding rock is more complicated compared with the case of isotropic rock. So if a layered rock is met in the practices of rock mechanics and engineering, it is necessary to treat seriously the anisotropy of mechanical behavior of the rock.

5 CONCLUSIONS

1 The authors compare the empirical expression of C, Φ for layered rock suggested by themselves with a test result made by former researchers, and see that both have a good agreement. This shows the empirical expression is reliable.

2 Describe the principles of 3-D elasto-plastic FEM analysis in which a layered rock is taken as a transversely isotropic one and the suggested empirical expression of C, Φ is used. Among them the important thing is to adopt a search-trial method to find the shear failure plane (or weakest shear plane) as well as the relative angle ρ, C and Φ.

3 Through the calculation examples it can be seen: when a layered rock possesses obvious anisotropic mechanical behavior and its occurrence presents certain geometrical relationship with the excavation factors (for example, a tunnel's axial line), the dynamic state of surrounding rock is more completed compared with the case of isotropic rock. So this feature should be considered in numerical analysis and construction of a project.

** Acknowledgement- The financial support is provided by the Chinese National Science Foundation under Grant No.59879027.

REFERENCES

Attewell B. and Sandford R. Intrinsic shear strength of a brittle, anisotropic rock-I: Experimental and mechanical interpretation. Int. J. Rock Mech. Min. Sci., 1974, 11(11): 423-430.

Brady B.H.G. and Brown E.T. Rock mechanics for underground mining Coal Industry Press. Beijing, 1990,83-84 (in Chinese, translated by Feng Shuren et al.).

Jaeger J. C. Shear failure of anisotropic rock.Geol. Mag. 1960, 97: 65-72.

Nova R. The failure of transversely isotropic rocks in triaxial compression.Int. J. Rock Mech. Min. Sci.,1980, 17(6): 325-332.

Owen D.R.J. and Hinton E. Finite elements in plasticity: Theory and practice.Pineridge Press Limited, West Cross, Swansea U.K. 1980, 215-231.

Serrano A. and Olalla C. Ultimate bearing capacity of an anisotropic discontinuous rock mass, Part I: Basic modes of failure.Int. J. Rock Mech. Min. Sci., 1998, 35(3): 301-324.

Smith B. and Cheatham JR B. Anisotropic compacting yield condition applied to porous limestone.Int. J. Rock Mech. Min. Sci., 1980, 17(3): 159-165.

Y. Zhang. and Y. Tang. 2-D FEM analysis for an underground opening considering the strength-anisotropy of the layered rock mass.Chinese Journal of Geotechnical Engineering, 1999, 21(3):307~310(in Chinese).

Yu Xuefu et al. Stability Analysis for Surrounding Rock Masses of Underground Projects Coal Industry Press. Beijing, 1983,505-508 (in Chinese).

Zhou Weiyuan. Higher Rock Mechanics.Water Conservancy and Power Press. Beijing, 1990,116-119 (in Chinese).

Modern Tunneling Science and Technology, Adachi et al (eds), © 2001 Swets & Zeitlinger, ISBN 90 2651 860 9

A consideration on apparent cohesion of unsaturated sandy soil

R. Kitamura
Prof. of Kagoshima Univ., Japan

K. Sako
Graduate Student of Kagoshima Univ., Japan

ABSTRACT: The inter-particle force due to meniscus at a contact point of particles is investigated and the relationship between inter-particle force, surface tension and suction is used to derive the relation between the inter-particle force and apparent cohesion for unsaturated sandy soil. The soil-water characteristic curve calculated by Kitamura's model for unsaturated sandy soil is also used to relate the apparent cohesion to suction. Furthermore the inter-particle force is related to the apparent cohesion based on the mechanical and probabilistic consideration on the scale of soil particle size. Finally the relationship between the suction and apparent cohesion is calculated and applied to the stability analysis of tunnel heading in the sandy ground.

1 INTRODUCTION

In the performance of the tunneling the horizontal earth pressure at the tunnel heading is one of the important factors to be taken into account. In the sandy ground some calculation methods of horizontal earth pressure at the tunnel heading were proposed by Murayama et al. (1966) and Tamura et al. (1999). The horizontal earth pressure at the tunnel heading deeply related to the apparent cohesion of unsaturated soil. Therefore the estimation of apparent cohesion of unsaturated soil is important to perform the tunneling efficiently. On the other hand Kitamura et al. (1998) proposed the numerical model called Kitamura's model to calculate the relation between suction, unsaturated-saturated permeability coefficient and water content for unsaturated sandy soils.

In this paper the inter-particle force due to meniscus at a contact point of particles is investigated and the relationship between inter-particle force, surface tension and suction is used to derive the relation between the inter-particle force and apparent cohesion for unsaturated sandy soil. The soil-water characteristic curve calculated by Kitamura's model for unsaturated sandy soil is also used to relate the apparent cohesion to suction. Furthermore the inter-particle force is related to the apparent cohesion based on the mechanical and probabilistic consideration on the scale of soil particle size. Finally the relationship between the suction and apparent cohesion is calculated and the results are applied to estimate the horizontal earth pressure at the tunnel heading in the performance of tunneling.

2 DERIVATION OF COHESIONAL COMPONENT FROM INTER-PARTICLE FORCE

The inter-particle force between two adjacent particles is generated by the surface tension of pore water in soil mass as shown in Fig.1. The inter-particle force F_i can be expressed by the following equation (Karube et al. (1978)).

$$F_i = 2 \cdot \pi \cdot r' \cdot T_s + \pi \cdot r'^2 \cdot s_u \qquad (1)$$

where F_i: inter-particle force;
T_s: surface tension; s_u: suction ($= u_a - u_w$);
a and r': radius of curvature of meniscus in Fig.1.

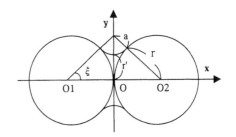

Figure 1. Inter-particle force between two particles due to surface tension.

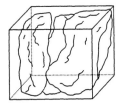

Fig.2(a) Soil particles in element

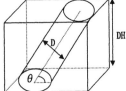

Fig.2(b) Pipe and other impermeable parts

Figure 2. Modeling of soil particles and voids.

Karube et al. (1978) also related the water content to r' in Eq.(1). Eq. (1) includes the surface tension and suction in the right side. The surface tension is one of the common physical quantities and its value can be easily obtained from the table. Therefore the suction must be related to the water content to estimate the inter-particle force at a contact point of soil particles.

Kitamura et al.(1998) proposed a numerical model for seepage behavior of pore water in unsaturated soil. Figure 2(a) shows an element with a few soil particles. This situation is modeled by Fig.2(b), in which the voids and soil particles are modeled by the pipe with a diameter D and an inclination angle θ, and the other impermeable parts respectively. The void ratio e, suction s_u, volumetric water content W_v and unsaturated-saturated permeability coefficient k can be derived by using some mechanical and probabilistic considerations in soil particle size as follows.

$$e = \int_0^x \int_{-\frac{\pi}{2}}^{\frac{\pi}{2}} \frac{V_p}{V_e - V_p} \cdot P_d(D) \cdot P_c(\theta) d\theta dD \qquad (2)$$

$$s_u = \gamma_w \cdot h_c = \frac{4 \cdot T_s \cdot \cos\alpha}{d} \qquad (3)$$

$$W_v = \frac{e(d)}{1+e} = \frac{1}{1+e} \int_0^d \int_{-\frac{\pi}{2}}^{\frac{\pi}{2}} \frac{V_p}{V_e - V_p} \cdot P_d(D) \cdot P_c(\theta) d\theta dD \qquad (4)$$

$$k = \int_0^d \int_{-\frac{\pi}{2}}^{\frac{\pi}{2}} \frac{\gamma_w \cdot D^3 \cdot \pi \cdot \sin\theta}{128 \cdot \mu \cdot \left[\frac{D}{\sin\theta} + \frac{DH}{\tan\theta} \right]} \cdot P_d(D) \cdot P_c(\theta) d\theta dD \qquad (5)$$

where V_p: volume of pipe in Fig.2(b);
V_e: volume of element in Fig.2(b);
$P_d(D)$: probability density function of D;
$P_c(\theta)$: probability density function of θ;

W_v: volumetric water content;
α: contact angle between the pipe and water;
γ_w: unit weight of water;
d: maximum diameter of pipe filled with water.

Substituting Eq.(3) into Eq.(1), the inter-particle force can be obtained.

The apparent cohesion c can be assumed to be the summation of two components, i.e., cohesion due to suction and others such as interlocking and physical-chemical actions, as follows.

$$c = c_0 + c_1 \qquad (6)$$

where c: apparent cohesion;
c_0: component of cohesion due to interlocking, physical-chemical actions and so on;
c_1: component of cohesion due to suction.

Figure 3 shows the cross section of soil mass. In the situation as shown in Fig.3 the following equation can be derived for the number of particles per unit volume and the void ratio.

$$(1+e) \cdot Nv \cdot \left(\frac{4 \cdot \pi \cdot r^3}{3} \right) = 1 \qquad (7)$$

where N_v: number of particles per unit volume;
e: void ratio;
r: radius of sphere.

And the number of soil particles per unit area is

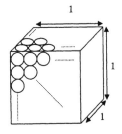

Figure 3. Cross section of soil mass.

Shearing force

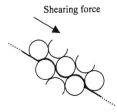

Figure 4. Potential slip plane in soil mass.

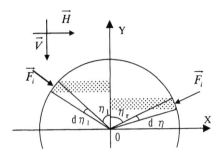

Figure 5. Equilibrium of force on the surface of a particle.

obtained by the following equation.

$$Np = Nv \cdot 2 \cdot r = \frac{1}{1+e} \cdot \frac{3}{2 \cdot \pi \cdot r^2} \qquad (8)$$

where Np: number of particle per unit area.

In the slope the shearing force is generated and consequently the potential slip plane is supposed as shown in Fig.4. Field (1963) proposed an empirical equation to relate the number of contact points per particle as follows.

$$Ca = \frac{12}{1+e} \qquad (9)$$

where Ca: coordinate number (number of contact points per particle).

Using Eqs.(8) and (9) the number of contact point (coordination number) per unit area of potential slip plane can be expressed as follows.

$$Nc = \frac{Ca}{2} \cdot Np \qquad (10)$$

where   Nc: number of contact points per unit area.

Figure 5 shows a semi-sphere particle taken out of the potential slip plane in Fig.4. The probability density function of contact points on the narrow belt designated by the dark zone is introduced to obtain the number of contact points at the narrow belt. Then the following equation is derived.

$$dN = N_c \cdot D(\eta_l) \cdot ds_l + N_c \cdot D(\eta_r) \cdot ds_r \qquad (11)$$

where D(η_l), D(η_r): weight function of number of contact points on the narrow belt with angle η_l and η_r; η_l, η_r: contact angle; ds: area of the narrow belt in Fig.5.

Supposing that the number of contact points is maximum at a contact angle $\pi/2$ and minimum at a contact angle zero, and that it changes linearly, the following equation can be derived as the weight function of contact points.

$$
\begin{cases}
D(\eta_l) = -\dfrac{2}{\pi-2} \cdot \eta_l + \dfrac{\pi}{\pi-2} & \left(0 \le \eta_l \le \dfrac{\pi}{2}\right) \\[4mm]
 & \qquad\qquad\qquad (12) \\[2mm]
D(\eta_r) = -\dfrac{2}{\pi-2} \cdot \eta_r + \dfrac{\pi}{\pi-2} & \left(0 \le \eta_r \le \dfrac{\pi}{2}\right)
\end{cases}
$$

Referring to Fig.5, the following equation can be derived from the equilibrium of forces in the vertical and horizontal direction.

$$\overline{H}' = \overline{H} + \int (\overline{F_i} \cdot \sin\eta_l) dN_l - \int (\overline{F_i} \cdot \sin\eta_r) dN_r + c_h \qquad (13)$$

$$\overline{V}' = \overline{V} + \int (\overline{F_i} \cdot \cos\eta_l) dN_l + \int (\overline{F_i} \cdot \cos\eta_r) dN_r + c_v \qquad (14)$$

where $\overline{H}'$: horizontal resultant force in the narrow belt with angle η in Fig.5;

$\overline{V}'$: vertical resultant force in the narrow belt with angle η in Fig.5;

$\overline{H}$: average resultant force of horizontal direction acting on a particle;

$\overline{V}$: average resultant force of vertical direction acting on a particle;

$\overline{F_i}$: inter-particle force at a contact point;

c_h: horizontal component of c_0;

c_v: vertical component of c_0.

Applying the law of friction at a contact point, the following equation can be derived by using Eqs.(13) and (14).

$$\frac{\overline{H}'}{\overline{V}'} = \frac{\overline{H} + \int (\overline{F_i} \cdot \sin\eta_l) dN_l - \int (\overline{F_i} \cdot \sin\eta_r) dN_r + c_h}{\overline{V} + \int (\overline{F_i} \cdot \cos\eta_l) dN_l + \int (\overline{F_i} \cdot \cos\eta_r) dN_r + c_v} = \tan\delta \qquad (15)$$

where δ : average frictional angle between particles.

Changing force to stress (force per unit area), the following equation is obtained.

$$\tau = \overrightarrow{F_i} \cdot \tan\delta \cdot \left\{ \int \cos\eta_l \, dN_l \big/ S + \int \cos\eta_r \, dN_r \big/ S \right\}$$

$$- \overrightarrow{F_i} \cdot \left\{ \int \sin\eta_l \, dN_l \big/ S - \int \sin\eta_r \, dN_r \big/ S \right\}$$

$$+ \left(c_v / S \cdot \tan\delta - c_h / S \right) + \sigma \cdot \tan\delta \qquad (16)$$

where τ : shear stress;
      σ : normal stress;
      S: area of shear slip plane in Fig.4.

Supposing that Eq.(14) corresponds to the equation of Mohr-Coulomb failure criteria($\tau = c + \sigma \tan\phi$), the following equations are obtained.

$$\phi = \delta \qquad (17)$$

$$c = c_0 + c_1 = \left(c_v / S \cdot \tan\delta - c_h / S \right)$$

$$+ \overrightarrow{F_i} \cdot \tan\delta \cdot \left\{ \int \cos\eta_l \, dN_l \big/ S + \int \cos\eta_r \, dN_r \big/ S \right\}$$

$$- \overrightarrow{F_i} \cdot \left\{ \int \sin\eta_l \, dN_l \big/ S - \int \sin\eta_r \, dN_r \big/ S \right\} \qquad (18)$$

where ϕ : internal friction angle;
      c: apparent cohesion;
      c_0: component of cohesion due to interlocking, physical-chemical actions and so on.

$$c_0 = c_v / S \cdot \tan\delta - c_h / S \qquad (19)$$

     c_1: component of cohesion due to suction.

$$c_1 = \overrightarrow{F_i} \cdot \tan\delta \cdot \left\{ \int \cos\eta_l \, dN_l \big/ S + \int \cos\eta_r \, dN_r \big/ S \right\}$$

$$- \overrightarrow{F_i} \cdot \left\{ \int \sin\eta_l \, dN_l \big/ S - \int \sin\eta_r \, dN_r \big/ S \right\} \qquad (20)$$

Using Eqs.(11) and (12), Eq.(20) can be

Table1.   Values used for calculation.

Sample	Shirasu
Density of soil particle (g/cm³)	2.37
Surface tension (N/m)	73.48*10⁻³
Coefficient of viscosity (Pa · s)	1.138*10⁻³
Number of division	180
Lowest height of probability density function of θ : ζ_c	0.159
Initial value of void ratio	1.30
Initial value of apparent cohesion (kPa)	10.0
Internal friction angle (°)	38.0

transformed as follows.

$$c_1 = \frac{\pi}{4 \cdot (\pi - 2)} \cdot \overrightarrow{F_i} \cdot N_c \cdot \tan\phi \qquad (21)$$

3 NUMERICAL EXPERIMENT

Table 1 shows the values used in the calculation. These values are supposed to be Shirasu that is local sandy soil in Kagoshima Prefecture. The numerical experiments are shown in Figs. 6, 7 and 8.

4 APPLICATION TO ESTIMATION OF HORIZONTAL EARTH PRESSURE AT TUNNEL HEADING

Tamura et al. (1999) proposed the simple estimation method of horizontal earth pressure at tunnel heading based on Fig.9 and derived the following equations.

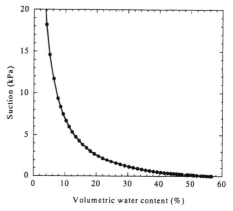

Figure 6. relation between volumetric water content and suction.

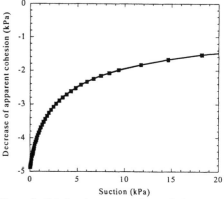

Figure 7. Relation between suction and decrease of apparent cohesion due to suction Δc_1.

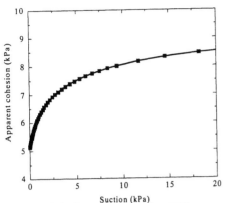

Figure 8. Relation between suction and apparent cohesion.

$$P = \left\{ \frac{1}{2} \gamma D \tan \alpha + \sigma_v (\tan \alpha - \tan \phi) - 3c \right\} D \tan \alpha \quad (22)$$

$$\sigma_v = \frac{\gamma B - 2c}{2K_0 \cdot \tan \phi} (1 - e^{-\frac{2K_0 H}{B} \tan \phi}) \quad (23)$$

where P: resultant force of horizontal earth pressure;
 σ_v: vertical overburden stress;
 c: apparent cohesion;
 ϕ: internal friction angle;
 B: width of lowering floor;
 γ: unit weight of soil;
 K_0: coefficient of earth pressure at rest(=1);
 α: inclination angle of slip plane;
 H: height of overburden;
 W: weight of slip soil mass.

The concrete values of calculation are listed in Table 2. The numerical result is shown in Fig.10.

5 CONCLUSIONS

The apparent cohesion is derived based on some consideration on the scale of soil particle size, in which Kitamura's model and two particle model are applied to estimate the inter-particle force. The calculated apparent cohesion was applied to Tamura's equation to estimate the horizontal earth pressure at heading. It is found that the horizontal earth pressure at heading, which is a function of the water content under unsaturated condition, can quantitatively be estimated. The proposed method may be promising to estimate the horizontal earth pressure at heading under unsaturated condition for sandy tunnel.

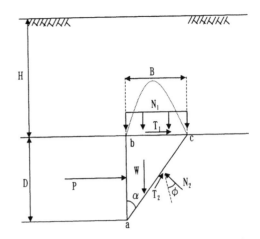

Figure 9. simple estimation method of horizontal earth pressure at tunnel heading proposed by Tamura et al. .

Table 2. The concrete values of calculation.

Height of heading D(m)	2.0
Internal friction angle ($^\circ$)	38.0
Coefficient of earth pressure at rest K_0	1.0
Height of overburden H(m)	10.0
Density of soil particle ρ_s (g/cm^3)	2.37
Void ratio e	1.3

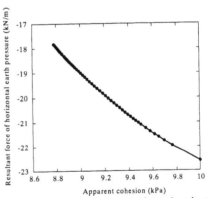

Figure 10. Relations apparent cohesion and resultant force of horizontal earth pressure.

ACKNOWLEDGEMENT

This research was supported by the Grant-in-Aid of Scientific Researches (Project No.09555153, No.10650490 and No.12792009) of the Ministry of Education.

REFERENCES

Field, W.G.: Towards the statistical deformation of a granular mass, Proc. 4th A. and N. Conf. of

SMFE, pp.143-148, 1963.

Karube, D., Namura, K., Morita, N. and Iwasai, T.:
Fundamental study of the stress-strain behavior of
an unsaturated soil, J. of Geotechnical
Engineering, JSCE, No.269, pp.105-119, 1978 (in
Japanese).

Kitamura, R., Fukuhara, S., Uemura, K., Kisanuki, J.
and Seyama, M.: A numerical model for seepage
through unsaturated soil, Soils and Foundations,
Vol.38, No.4, pp.261-265, 1998.

Murayama, S., Endo, M. and Hashiba, T.:
A geotechnical consideration on the tunnel
performance of shield machine, Proc. 1st Japan
National Conf. On Geotechnical Engineering,
pp.75-79, 1966 (in Japanese).

Tamura, T., Adachi, T., Konishi, S. and Tsuji, T.:
Stability analysis of tunnel heading by rigid-
plastic finite element method, J of Geotechnical
Engineering, JSCE, No.638/III-49, pp.301-310,
1999 (in Japanese).

Modern Tunneling Science and Technology, Adachi et al (eds), © 2001 Swets & Zeitlinger, ISBN 90 2651 860 9

Identification of the soil parameters in the analysis of a shallow tunnel

C. M. C. Moreira
Instituto Superior de Engenharia de Coimbra, Portugal

J. Almeida e Sousa
Faculdade de Ciências e Tecnologia, University of Coimbra, Portugal

ABSTRACT: The identification of the deformability modulus, the shear modulus and the coefficient of earth pressure at rest of the ground involved by the opening of a shallow tunnel is presented. The identification of these parameters is obtained by solution of the inverse problem, using field observations. The algorithm employed uses the finite element method. The analysis is 2D and assumes a plain strain state. The 3D effect of the tunnel face advance is simulated using the convergence-confinement method. A linear elastic behaviour is assumed for all the materials. The support is considered isotropic, while the ground is also considered with transverse-isotropic behaviour, attempting to simulate the inherent anisotropy that most of the natural soil deposits exhibit. Some relationships between parameters are derived in order to achieve the numerical solutions that better fit the observations carried out. Finally, the identified parameters are compared with those derived from the characterisation tests.

1 INTRODUCTION

The solution of a large number of the problems that presently concern geotechnical practice comprehends numerical modelling. To simulate the mechanisms involved by the geotechnical structures in an increasingly realistic way, a wide range of constitutive laws have been formulated to describe the behaviour of the various materials. However, the reliability of the results depends not only on the sensible choice of an appropriate model, based on the available geological information, but also on the selection of the most suitable values for the parameters of the adopted constitutive law. As the estimation of these parameters is covered by an inaccuracy degree that can be very high in many cases, the attempts to evaluate or correct them based on field measurements of some variables, particularly displacements, becomes more and more precious for geotechnical engineering.

The identification of some of the geotechnical parameters of the ground involved by the opening of a shallow tunnel is performed in this paper, by resolution of the so called inverse problem. It is the tunnel of the temporary terminus of the Alameda II underground station of Lisbon Metro.

The algorithm for resolution of the inverse problem was implemented by using the finite element method according to the traditional methodology (Villalba 1987). In simple terms, its structure can be described by the following sequence:

1. Formulation of a numerical model of linear elastic behaviour, in some cases isotropic and in others with anisotropy;

2. Definition of an objective function based on the maximum likelihood criterion;

3. Minimisation of the objective function using the Gauss-Newton algorithm.

2 DESCRIPTION AND OBSERVATION OF THE TUNNEL

The tunnel in reference is 165 m long and involved an excavation volume of about 70 m^3 per meter length. Site investigation revealed over the entire length of the tunnel, the existence of a landfill topsoil laying over two strata characteristic of the Lisbon marine Miocene period, composed by overconsolidated materials (Marques et al. 1997).

The landfill, 2.5 to 3.0 m in thickness, is very heterogeneous and comprises essentially sandy clays. Under that landfill appears a layer containing sands basically fine, normally micaceous, silty clayey and presenting gravel intercalations. It is the 'Estefânia' sands stratum with thickness between 11 and 12 m. Finally there is the 'Prazeres' clays layer where the tunnel is mostly excavated, and which is mainly composed by very stiff sedimentary fine silty clays.

The geotechnical characterisation of these materials was performed by in situ tests – SPT, permeability and pressiometric, both Ménard and self-boring –

and laboratory tests, specially of identification, classification and determination of the principal physical properties, oedometric and undrained consolidated triaxial (Marques 1998, Sousa Coutinho et al. 1996).

The tunnel was constructed in agreement with the New Austrian Tunnelling Method. For each advance cycle, the vault and sides excavation was processed in three stages of 1m, leaving a central core that worked as support for the front. The invert was excavated in 2.4 m stages.

In the primary lining were used steel ribs and sprayed concrete in two layers until a 0.2 m thickness was reached. The final lining, 0.4 m thick, was executed in concrete after the complete excavation of the tunnel.

The observation plan to analyse the structure behaviour allowed the measurement of surface settlements in several sections. Their distribution in the cross section named S69, aim of this work, was approached by a Gauss curve where the settlement u_v in any point at a distance of x meters of the symmetry axis can be evaluated by:

$$u_v = 8.9 e^{-x^2/414.7} \text{ (mm)} \tag{1}$$

3 METHODOLOGY OF THE ANALYSIS

This study is an attempt to identify the geotechnical parameters of the 'Prazeres' clays, which best fits the Gauss curve deduced from the surface settlements observed. The analyses were performed in two dimensions with plain strain conditions and assuming undrained behaviour for the clays. Figure 1 shows the discretization and profile of the ground in the section S69 under consideration.

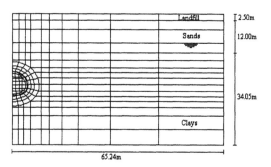

Figure 1. Finite element mesh for the cross section S69.

The three-dimensional stress and strain states that the opening of a tunnel always produces in the vicinity of the excavation face were considered by simulating the construction in two phases. The first one tried to represent the deformation occurring before the support installation by relieving the nodal forces

equivalent to the initial stress state in the tunnel contour applying a relieve factor α. The second simulates the deformation after the support installation, by activating the elements that represent the support and relieving the remaining percentage of the initial forces in the tunnel contour. The relieve factor α adopted was 0.5, which was obtained by confronting the results of a three dimension calculation, trying to reproduce the constructive process effectively followed in the vicinity of the section S69, with those of the plain strain analysis.

The initial stress state was admitted to be geostatic. Initially, all the ground materials were assumed to have isotropic linear elastic behaviour. This is justified because if the soil strength parameters derived from the tests were considered, particularly those about the undrained strength of the clays, there are almost no ground zones plastically yielding (Almeida e Sousa 1998). Table 1 shows the elastic parameters, the unit weights and the coefficients of earth pressure at rest used in soil characterisation.

Table 1. Mechanical properties adopted in the analysis.

Soil	γ kN/m^3	K_0	E', E_u MPa		v, v_u
Landfill	18.0	0.5	20		0.25
Sands	20.5	0.8	80	for z<6 m	0.25
			80+8.6(z-6)	for z≥6 m	
Clays	21.0	*	*		0.49

*To identify

With the purpose of simulating the inherent anisotropy that most natural soils exhibit, some calculations were also performed assuming a transverse-isotropic behaviour for the 'Prazeres' clays. As the influence of the anisotropy related to the deformability modulus (E_h/E_v) in the behaviour of the ground adjacent to the opening is very small (Almeida e Sousa 1998, Lee et al. 1989), only the anisotropy concerning the shear modulus in any vertical plan (G_{vh}/E_v) was considered.

All the calculations were done assuming an isotropic linear elastic behaviour for the support with E=10 GPa and v=0.2.

4 IDENTIFICATION OF PARAMETERS

In the analysis where it was assumed an isotropic behaviour for the strata an attempt was made to identify K_0 and E of the clays. These parameters were chosen because they are those that most significantly condition the behaviour of the ground mass and also because the existence of a large number of tests, in situ and in laboratory, from which they were estimated, allows a conclusion about the suitability of the model used.

The deformability modulus, as the test results showed its growth in depth, was defined assuming an increasing rate of m with depth and a value E_0 for the top of the clay layer, such as:

$$E = E_0 + m(z - 14.5) \quad (MPa) \qquad (2)$$

The most efficient strategy in the identification process was the application of the Gauss-Newton minimisation algorithm simultaneously with a sensibility analysis. This allowed in certain phases of the process the readjustment of the parameters, based on the search path already observed, in order to reach the goal more rapidly.

The greatest difficulty during the search was the dependency between E_0 and m, because it produces a collection of solutions physically acceptable instead of a unique one. Thus, it was chosen to set m and identify only E_0 and K_0. Then, pairs (E_0, K_0) were identified for several values of m, all of them producing solutions similar to that represented in Figure 2, correspondent to $m=4$. Table 2 synthesises the results of the situations analysed.

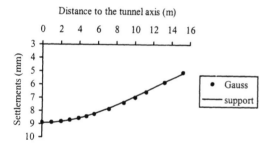

Distance to the tunnel axis (m)

Figure 2. Settlements observed and evaluated with $m=4$.

Table 2. Identified geotechnical parameters of the clays.

Case	m	E_0	K_0
	Mpa/m	MPa	
1	0	108.6	0.432
2	2	90.2	0.508
3	4	72.3	0.579
4	6	55.5	0.647
5	8	39.8	0.716
6	9	32.4	0.751
7	10	25.3	0.789
8	12	12.4	0.861

The previous table shows the existence of a high degree of interdependence between the three parameters analysed. The variations of E_0 and K_0 with increasing m are shown in Figure 3. The variation of K_0 with m can be represented univocally by:

$$K_0 = 0.432 + 0.036m \qquad (3)$$

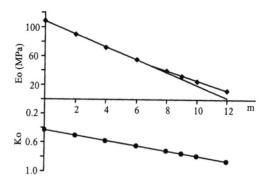

Figure 3. Relationship between the parameters m, E_0 and K_0 numerically evaluated.

On what concerns E_0, the corresponding variation with m cannot be obtained in a singular manner. Instead, there is a 'two-linear' dependence between those parameters represented by:

$$E_0 = 108.6 - 8.833m \quad (MPa) \quad \text{for } m \leq 7 \qquad (4)$$

and

$$E_0 = 94.6 - 6.850m \quad (MPa) \quad \text{for } m > 7 \qquad (5)$$

Each of these relations reveals a collection of pairs (E_0, m) that simultaneously with the corresponding values of K_0 lead to the solution to the problem under analysis.

The deformability moduli determined by some of the geotechnical characterisation tests performed are confronted in Figure 4 with those obtained in the analysis. Besides the deformability modulus evaluated by the Ménard pressiometer tests, the secant modulus corresponding to axial strains of 0,1, 0,3 and 0,5%, derived from extension and compression triaxial tests, are also indicated. Note that these last moduli were corrected in order to eliminate the effects of the disturbance caused by sampling (Almeida e Sousa 1998).

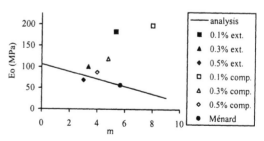

Figure 4. Comparison between E_0 and m estimated by the tests and numerically evaluated.

103

As the figure emphasises, it is possible to conclude that the solution to the problem can be well adjusted using the parameters E_0 and m evaluated by the Ménard pressiometer tests. Also the secant deformability modulus corresponding to strains of 0.5% (compression) and 0.3 and 0.5% (extension) seem to allow the solution to be fairly adjusted.

However, the success of this adjustment depends unconditionally on the sensible choice of a convenient value for K_0. According to (3) the value of this parameter would have to be about 0.55 for the secant modulus and 0.64 for the pressiometric tests modulus, values which are near the inferior limit of the range obtained by the tests with the self-boring pressiometer ($0.57 < K_0 < 1.33$).

From the above it can be inferred that the isotropic linear elastic model is quite adequate to the problem analysed, as long as a value close to the minimum of the tests is adopted for K_0.

The analysis assuming a transverse-isotropic behaviour for the clays began by setting E equal to the value obtained by the Ménard pressiometer tests and then tried to identify both K_0 and the ratio between the shear and deformability modulus ($n = G_{vh}/E_v$) that better fit the solution. Once again, that was not possible because of the existence of a large collection of solutions instead a unique one. So, several values of K_0 were set and the corresponding value of n was identified to each of them. The results are shown in Figure 5. It can be seen that the variation of n with K_0 is represented by:

$$n = 0.480 - 0.217K_0 \qquad (6)$$

If K_0 is assigned the value 0.87, which corresponds to the average value evaluated by the self-boring pressiometer tests and which is close to the values empirically estimated from the overconsolidation ratio determined by the tests performed, the above expression allows the conclusion that n equal to 0.29 enables a good agreement between the curves of the surface settlements evaluated and observed.

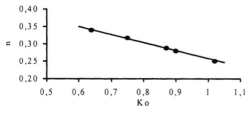

Figure 5. Relationship between n and K_0 that fits the solution, with E derived from Ménard tests.

5 CONCLUSIONS

Besides the good functioning of the algorithm implemented, it is important to emphasise on what concerns the geotechnical parameters of the ground involving the Alameda II terminus tunnel of Lisbon Metro, the following:

1. There is a high degree of interdependency between the deformability modulus, the coefficient of earth pressure at rest and the shear modulus, which leads to a collection of physically acceptable solutions, instead of a unique one;

2. The deformability modulus obtained from tests with the Ménard pressiometer and some of the secant modulus derived from the triaxial tests are in harmony with values that can allow an adequate solution;

3. The isotropic linear elastic model proves to be adequate to the problem if the value adopted for K_0 approaches the minimum obtained from tests;

4. Assigning K_0 values close to the medium values obtained in the tests imposes the adoption for the shear modulus of a value inferior to the corresponding isotropic one.

REFERENCES

Almeida e Sousa, J. 1998. Tunnels in earth massives – Behaviour and numerical modelling. (In Portuguese). *Ph. D. Thesis*, FCT University of Coimbra, Portugal.

Lee, K. M. & R K.Rowe 1989. Deformations caused by surface loading and tunnelling: the role of elastic anisotropy. *Géotechnique*, 39:125-149.

Marques, F. 1998. Analysis of the behaviour of a tunnel opened in the Miocene formations of Lisbon. (In Portuguese). *Master Thesis*, FCT University of Coimbra, Portugal.

Marques, F., D. D. Langton, R. J. Furtado & J. Almeida e Sousa 1997. Evaluation of the deformability and consolidation features of the Prazeres clays. (In Portuguese). *6° Congresso Nacional de Geotecnia*, Lisbon, 1:179-188.

Sousa Coutinho, A.G.F. & Marques, M. & Costa, A. & Veiga, J. 1996. Pressiometric tests in the zone of Alameda station. Works, interpretation and results of the tests. (In Portuguese). *Relatórios 221/96 a 225/96* – NF, LNEC, Lisbon.

Villalba, A. 1987. Identification of geotechnical parameters. Application to the excavation of tunnels. (In Spanish). *Ph. D. Thesis*, E. T. S. D'Enginyers de Camins, Canals i Ports de Barcelona.

Modern Tunneling Science and Technology, Adachi et al (eds), © 2001 Swets & Zeitlinger, ISBN 90 2651 860 9

A simple technique to improve the prediction of surface displacement profiles due to shallow tunnel construction

A. Burghignoli, A. Magliocchetti, S. Miliziano & F.M. Soccodato
Dept. of Structural and Geotechnical Engineering, University of Rome 'La Sapienza', Italy

ABSTRACT: The usual techniques for the simulation of a shallow tunnel excavation in two-dimensional numerical analyses often produce relatively wide and flat surface settlement profiles, even when highly non-linear and sophisticated elasto-plastic soil models are used. In this paper, the first results of a new approximate procedure for the simulation of the effects of the excavation is proposed. The technique is based on a differential reduction of the vertical and horizontal components of the geostatic stresses initially acting on the tunnel boundary. A conventional and simple elasto-plastic *Mohr-Coulomb* soil model was adopted because of its wide application in engineering practice. The results obtained from the numerical study showed that the proposed procedure is able to furnish surface settlement profiles in agreement with the experimental-based inverted normal (Gaussian) distribution form, right from the very small strain levels. The differential stress reduction relationships were found to be dependent from the geometry of the problem and from the soil parameters: thus, a trial and error procedure must be followed in order to obtain the appropriate 'differential reduction rules'.

1 INTRODUCTION

When dealing with shallow tunneling in an urban environment, the prediction of the effects related to the tunnel construction on existing buildings resting on ground surface plays a major role. A first approach to the problem consists in neglecting the influence of the presence of the building on the shape of the settlement profile: in this case, by assuming a value for the ground loss and a typical shape for the settlement profile in a free field conditions (e.g. the normal distribution form), the charts proposed by Burland (1995) may be used in order to evaluate the expected damage of the building. In this way, an overestimation of damage is often obtained, because the stiffness of the building strongly modifies the shape of the free field settlement profile, reducing both angular distortions and horizontal strains. Thus, the numerical analysis of the tunnel-soil-building interaction represents a necessary step in order to obtain a realistic evaluation of damage (Addenbrooke & Potts 1996).

Being the computational costs of a fully three-dimensional (3_D) analysis still very high, conventional two-dimensional (2_D) analyses under plane strain conditions are usually carried out. When a 2_D approach is followed, a proper procedure for the simulation of the effects of the excavation must be firstly defined. The procedure should be able to give rise to a displacement field in a quantitative agreement with the in-situ measurements: it's well known that, in a free field condition, an inverted normal (Gaussian) distribution form has been proved to successfully match most of the experimental data.

The techniques for a 2_D simulation of the tunnel excavation usually adopted (e.g. the proportional reduction of the geostatic stresses initially acting on the tunnel boundary or the imposition of a displacement field to the tunnel boundary) often produce relatively wide and flat surface settlement profiles, even when highly non linear and sophisticated soil models are used (e.g. Gunn 1993; Addenbrooke et al. 1997; Burghignoli et al. 2001). Even more importantly, the horizontal strain profile, which greatly affects the damage of buildings, is significantly underestimated by the above quoted techniques.

In this paper, a new approximate procedure for the simulation of the effects of tunnel excavation, in a 2_D approach of the problem, is proposed in order to obtain a reasonable numerical prediction of the surface displacement (both vertical and horizontal) profiles. Attention is focused on short term movements (tunnels in clayey soils, undrained conditions). The results of a number of numerical analyses carried out following the new procedure will be presented and compared with those obtained following some of the available techniques for the simulation of the effects of tunnel construction.

2 CONVENTIONAL TECHNIQUES

When dealing with shield tunneling, 2_D analyses are usually carried out by following two numerical techniques in order to simulate both the tail void (gap) closure and the 3_D extrusion effects at the tunnel face: *i*) initial stresses reduction or *ii*) displacements imposition to the tunnel boundary.

The method of the *proportional stress reduction* is the most used (Negro & de Queiroz, 2000): the effects of the excavation are simulated by a progressive reduction of the stresses initially acting on the tunnel boundary. The vertical and the horizontal components of the geostatic forces are decreased by the same factor and the amount of unloading is usually defined by comparing the obtained ground loss (i.e. the volume in excess of the notional excavated volume) with the designed (or the observed) one. This technique may be considered as an application of the convergence-confinement methods derived for the rock-lining interaction analysis of deep tunnels (Panet 1995).

The imposition of a displacement field to the tunnel boundary (*imposed convergence* technique) is the alternative method of tunnel analysis. The imposed tunnel convergence may be uniform and (in a closed-form solution in an elastic half plane and in undrained conditions has been proposed by Verrjut & Booker 1996) or may be oval-shaped. Figure 1b shows the shape of tunnel convergence proposed by Loganathan & Poulos (1998) in generalizing the closed-form solution derived by Verrjut & Booker (1996); a slightly different tunnel convergence, derived from the studies carried out by Rowe & Kack (1983), is shown in Figure 1c.

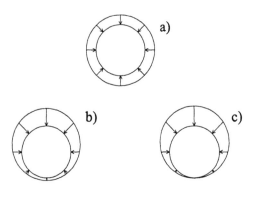

Figure 1. Different imposed tunnel convergence: a) uniform; b) oval-shaped (Loganathan & Poulos 1998); c) oval-shaped (after Rowe & Kack 1983).

As in the stress reduction method, the ground loss is once again the parameter controlling the magnitude of the tunnel convergence. Following Loganathan & Poulos (1998), tail void closure (physical gap), 3_D effect at the tunnel face (extrusion) and quality of workmanship may be regarded as the main factors controlling the amount of the ground loss.

3 THE PROPOSED TECHNIQUE

The approximate procedure proposed in this paper belongs to the stress reduction methods. It consists in a *differential* reduction of the vertical and horizontal forces initially acting on the tunnel boundary.

By denoting ps_{vb} (ps_{vu}) and ps_{hb} (ps_{hl}) the percentages of unloading of the vertical (v) and horizontal (h) components of the geostatic forces initially acting on the upper (u) (i.e. from the spring to the crown) and on the lower (l) (i.e. from the spring to the invert) half-circumference of the tunnel, the following relationship defines, for a given unloading step, the current value of the (reduced) force component acting on the tunnel boundary:

$$F_{ij} = (1 - ps_{ij}) \cdot F_{i0} \qquad (1)$$

where the subscripts i can be v or h, j can be u or l and 0 refers to the initial geostatic value.

In this way, the reduction of each nodal force is also associated with a rotation of the force itself. Furthermore, the method of proportional stress reduction is fully recovered by assuming a unique and constant value for ps_{ij}.

It is worth noting that the proposed technique, as well as any other conventional technique, should be regarded as an expedient in order to obtain reasonable results from a 2_D analysis of a 3_D real problem rather than a close simulation of what happens at the tunnel boundaries.

4 NUMERICAL ANALYSES

The numerical analyses were carried out by using the finite difference code FLAC (Itasca, 1996) under plane strain and undrained conditions. The geometry of the problem, the finite difference grid and the assumed boundary conditions are shown in Figure 2. The grid is conveniently mapped in order to allow for the introduction (via interfaces and elements) of a building resting on ground surface.

Two sets of numerical analyses were carried out by considering two different covers (H=7 and 12 m) of a 6 m diameter tunnel.

A simple elasto-perfectly plastic *Mohr-Coulomb* soil model was used because of its wide use in engineering practice. Two types of soils, with different stiffness parameters, were considered. The values of the parameters, typical of a medium to stiff clayey soil, are reported in Table 1; the Young' modulus E' has been assumed as linearly increasing with depth.

Figure 3 shows the vertical and horizontal surface displacements obtained following some of the conventional methods for the simulation of the effects of the excavation. The results are relative to a ground loss V_L equal to 1.5% of the notional excavated volume per unit length of the tunnel.

Vertical settlements are normalized by the maximum settlements S_{ymax} obtained from an inverted normal (Gaussian) distribution form also reported in the Figure 3a. A typical value of the settlement trough parameter $i=0.5 \cdot Z_0$ (O'Reilly & New 1982), Z_0 being the depth of the tunnel axis, was used in order to evaluate S_{ymax} according to the following relationship:

$$V_S = \sqrt{2\pi} \cdot i \cdot S_{y\,max} \qquad (2)$$

where the ground loss into the tunnel V_S equals, in undrained conditions, the volume of the settlement trough per unit length. Furthermore, the usual hypothesis of ground surface movements directed towards the tunnel axis was assumed in order to derive a reference surface horizontal displacement profile.

The vertical and horizontal surface displacements predicted by the Gaussian distribution are thus described by the following relationship:

$$S_y = S_{y\,max} \cdot e^{\left(-\frac{x^2}{2 \cdot i^2}\right)} \qquad (3)$$

and

$$S_x = \frac{x}{Z_0} \cdot S_y \qquad (4)$$

where x is the distance from the tunnel center-line, i the horizontal offset from the tunnel center-line to the inflection point of the settlement trough and S_{ymax} the maximum surface settlement for x=0. The maximum horizontal displacement S_{xmax} obtained from eqns. (3) and (4) was used in order to normalize the horizontal surface displacement profile.

Figures 3 show that the method of uniform stress reduction furnishes surface displacement profiles significantly different from those predicted by the Gaussian distribution. A similar result was also obtained by imposing to the tunnel boundary the oval-shaped convergence proposed by Loganathan & Poulos (1998). In this case, an elastic constitutive model, with the same parameters shown in Table 1, was adopted according to the hypothesis of the Authors. The poor predictions obtained in this numerical study as compared to the closed-form solutions proposed by the Authors should be attributed to the effects the rigid bottom boundary and of the variable stiffness profile assumed.

A better prediction was obtained with the method suggested by Rowe & Kack (1983). However, differential settlements, angular distortions and horizontal strains are clearly underestimated when compared with the Gaussian distributions.

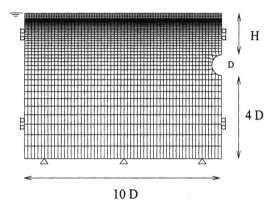

Figure 2. The problem analysed.

Table 1. Soil parameters adopted in the analyses.

Soil	γ (kN/m³)	c' (kPa)	φ' (°)	E' (MPa)	ν'
A	20	15	28	$6.7 \Rightarrow 20.0$	0.25
B	20	15	28	$33.3 \Rightarrow 46.7$	0.25

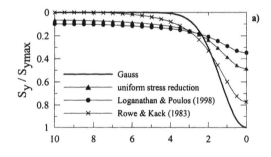

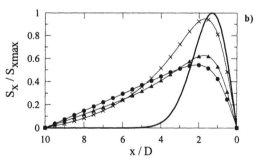

Figure 3. Normalized surface displacements obtained using conventional techniques (soil type A, tunnel with H=12 m, V_L=1.4%): a) vertical settlements; b) horizontal displacements.

107

Figures 4 report the results obtained following the differential stress reduction technique proposed in this paper. In this analyses, the same law relating vertical and horizontal forces unloading was used for the upper (i.e. from the spring to the crown) and the lower (i.e. from the spring to the invert) part of the tunnel half-circumference (Fig 4a).

Figure 4b, for the two soil types studied, shows the variation of the ground loss as a function of the unloading percentage of the geostatic vertical forces which may be considered as the driving variable of the analysis. The normalized vertical and horizontal surface displacements are reported in Figures 4c and 4d, respectively. Two percentages of ground loss are shown in order to study the influence of the progressive developments of plastic strain: an almost elastic behavior of soil is observed for a ground loss equals to 0.4%, while significant zones of plastic straining appear for V_L=2.5%.

A remarkable agreement between numerical results and Gaussian distributions, independently of the amount of ground loss was obtained. In this case, the law of differential stress reduction is characterized by a zero unloading of the horizontal component of the forces when an almost elastic soil response is observed (ps_v<30%); then, a progressively increasing percentage of reduction of the horizontal component of the geostatic forces is applied (Fig. 4a).

Satisfactory results were also obtained for the other geometry of the problem studied (tunnel with H=7 m). In this analysis it was necessary to introduce a different rule of stress reduction for the forces initially acting on the upper and on the lower part of the tunnel circumference (Fig. 5a) in order to match, for each value of the ground loss, the Gaussian distribution. Furthermore, it may be noted that, as compared with the previous analysis, a reduction of the horizontal component of the geostatic forces is necessary right from a relatively small amount of unloading (i.e. in the elastic range).

The set of numerical analyses just above reported indicates that the suitable unloading relationship between vertical and horizontal components of the geostatic forces are dependent on the mechanical properties of the soil as well as on the problem geometry (tunnel diameter and cover). A trial and error procedure must be thus followed. However, a preliminary assessment of the stresses-reduction laws is briefly outlined in the following. Very small percentages of horizontal unloading should be adopted in order to match the Gaussian distribution forms when small amount of ground losses are considered (i.e. when the soil response is almost elastic). With increasing plastic strain developments, the predicted displacement profiles rapidly depart from the expected distribution: the percentages of horizontal unloading must be increased.

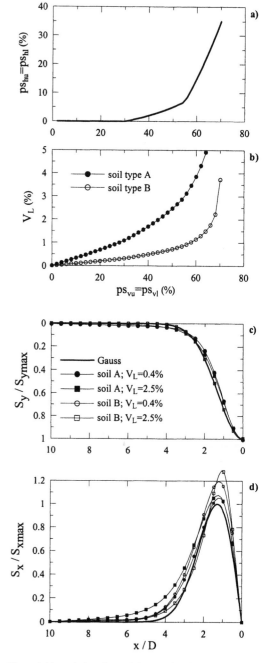

Figure 4. Numerical analyses of the tunnel with H=12 m: a) differential reduction relationship; b) ground loss as a function of unloading percentage of vertical components of the geostatic forces; c) normalized surface settlements; d) normalized surface horizontal displacements.

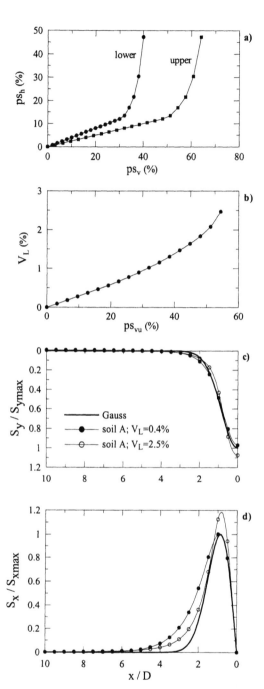

Figure 5. Numerical analyses of the tunnel with H=7 m, soil type A: a) differential reduction relationships; b) ground loss as a function of unloading percentage of the vertical components of the geostatic forces acting on the upper (from the spring to the crown) circumference of the tunnel; c) normalized surface settlements; d) normalized surface horizontal displacements.

5 CONCLUSIONS

In this paper, a new approximate technique is proposed to simulate the excavation of a shallow tunnel by a 2_D numerical analysis. Attention was paid to the prediction of undrained surface displacement profiles.

The proposed technique consists in a differential reduction of the vertical and horizontal components of the geostatic stresses initially acting on the tunnel boundary.

A number of numerical analyses were carried out by adopting a simple elasto-perfectly plastic soil model. Two different problem geometry with two sets of elastic soil parameters were considered.

A remarkable agreement between numerical predictions and the experimental-based inverted normal Gaussian distribution form was obtained.

The suitable relationships governing the unloading of the vertical and horizontal components of the geostatic forces were found dependent on the mechanical properties of the soil as well as on the problem geometry (tunnel diameter and cover). A trial and error procedure must be thus followed in order to define the appropriate 'stress reduction rules'.

ACKNOWNLODGMENT

This work has been developed under the financial support of MURST, National Research Project: *Gallerie in Condizioni Difficili*.

REFERENCES

Addenbrooke, T.I. & Potts, D.M. 1996. The influence of existing surface structure on the ground movements due to tunnelling. In Mair & Taylor (eds). *Geotechnical Aspects of Underground Construction in Soft Ground; Proc. intern. symp., London, 15-17 April 1996.* Rotterdam: Balkema.

Addenbrooke, T.I., Potts, D.M. & Puzrin, A.M. 1997. The influence of pre-failure soil stiffness on the numerical analysis of tunnel construction. *Geotechnique* 47(2): 693-712.

Burghignoli, A., Miliziano, S. & Soccodato, F.M. 2001. Prediction of ground settlements due to tunneling in clayey soils using advanced constitutive soil model: a numerical study. This Symposium.

Burland, J.B. 1995. Assessment of risk of damage to buildings due to tunnelling and excavation. Proc Int. Conf. Earthquake Geotechnic. IS-Tokio 95.

Gunn, M.J. 1993. The prediction of surface settlement profiles due to tunneling. In Houlsby & Schofield (eds), *Predictive Soil Mechanics. Proc. Wroth memorial symp., Oxford, 27-29 July 1992.* London: Thomas Telford.

Itasca Consulting Group 1996. FLAC (Fast Lagrangian Analysis of Continua) version 3.4. User' manuals. Minneapolis.

Loganathan, N. & Poulos, H.G. 1998. Analytical prediction for tunnelling-induced ground movements in clays. *Journal of Geotechnical and Geoenviromental Engineering ASCE* 124(9): 846-856.

Negro, A. & de Queiroz, P.I.B. 2000. Prediction and performance: a review of numerical analyses for tunnels. In Kusakabe, Fujita & Miyakazi (eds). *Geotechnical Aspects of Underground Construction in Soft Ground; Proc. intern. symp., Tokyo, 19-21 July 1999*. Rotterdam: Balkema.

O'Reilly, M.P. & New, B.M. 1982. Settlements above tunnels in the United Kingdom – their magnitude and effects. *Tunneling 82*, IMM, London, 173-181.

Panet, M. 1995. *Le calcul des tunnels par la méthode convergence-confinement*. Paris: Presses Ponts et Chausées.

Rowe, R.K. & Kack, G.J. 1983. A theoretical examination of the settlements induced by tunnelling: four case histories. *Canadian Geotechnical Journal* 20: 299-314.

Verruijt, A. & Booker, J.R. 1996. Surface settlements due to deformation of a tunnel in an elastic half plane. *Geotechnique* 46(4): 753-756.

Modern Tunneling Science and Technology, Adachi et al (eds), © 2001 Swets & Zeitlinger, ISBN 90 2651 860 9

Numerical modeling of a nonlinear deformational behavior of a tunnel with shallow depth

S. Akutagawa
Department of Architecture and Civil Engineering, Kobe University, JAPAN

T. Kitani
Pacific Consultants, Tokyo, JAPAN

K. Matsumoto
Omoto Gumi, Okayama, JAPAN

S. Mizoguchi
Graduate School of Science and Technology, Kobe University, JAPAN

ABSTRACT: A new finite element analysis procedure is proposed for simulation of deformational behavior around a shallow tunnel. The method incorporates reduction of shear stiffness, as well as strain softening effects of a given material. An illustrative example is shown in which the proposed method is applied to simulate an excavation process of a tunnel made of aluminum bars. Results obtained indicated that surface settlement, ground reaction curve, minimum supporting pressure, and formation of kinematic collapse are in good agreement with experimental results. The typical deformational mechanism observed in the experiment is also seen in the field, which is hopefully modeled with accuracy by the proposed numerical procedure.

1 INTRODUCTION

Deformational mechanism of an urban NATM tunnel at shallow depth is often characterized by a set of unique kinematic movements in subsidence profile and crown settlement, etc. Computational tools available, so far, for design purposes are used to predict deformational behavior around tunnel, interaction between support structures and ground. However, results from most attempts might not be satisfactory in that those unique deformational behaviors of nonlinear nature are not properly represented.

This paper proposes an improved computational scheme by which characteristic deformational behavior of a shallow tunnel is properly modeled. The new computational procedure, incorporating strain-induced anisotropy and strain softening, was applied to simulate a tunnel collapse experiment performed in laboratory using aluminum bars. The results obtained showed that the new computational scheme could represent 1) surface subsidence and crown settlement, 2) shear bands developing from tunnel shoulder, 3) formation of an unstable zone (the primary zone) above crown and an induced sliding zone (the secondary zone) beyond, which were all observed in experiments.

2 DEFORMATION IN SHALLOW TUNNELS

Deformational behavior around a shallow tunnel is often characterized by formation of shear bands developing from tunnel shoulder reaching, sometimes, to the ground surface. Figure 1 shows a strain distribution derived from the results of displacement measurements taken from a subway tunnel in Washington D.C. (Hansmire and Cording, 1985).

One possible explanation of this deformational behavior may be best stated with a help of an illustration given in Figure 2. Region-A, surrounded by slip plane *k-k*, is regarded as a potentially unstable zone which may displace downward at the lack of

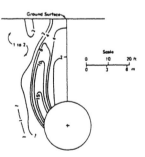

Figure 1. Strain distribution around a subway tunnel (after Hansmire and Cording, 1985).

frictional support along *k-k* planes. What is separating region-A from the surrounding is shear band *a* formed along *k-k* line with some thickness, as region A slides downward. The adjacent region B follows the movement of region A, leading to the formation of another shear band *b*. The direction of shear band *b* is related to $45° + \phi/2$ (ϕ: friction angle) and often coincides with what is called a boundary line of zone influenced by excavation.

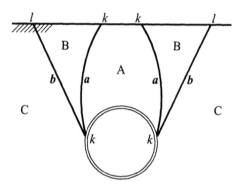

Figure 2. Typical deformational mechanism around a shallow tunnel.

Regions A and B correspond to the primary and secondary zones of deformational behavior pointed out earlier by Murayama et al (1969, 1971) in the series of trap door experiments. Confirming the presence of these zones is equivalent to acknowledging formation of shear bands *a* and *b*, which may not be a desirable practice in view of minimizing deformation during construction of shallow tunnels. However, it is regarded very important that a reliable method be established in order to reveal nonlinear deformational mechanism and identify the state of deformation with reference to an ultimate state, which is of current interest in the new design practice.

3 NUMERICAL PROCEDURE

In the framework of applying general numerical analysis tools, such as finite element methods, there have been series of approaches taken for simulation of tunnel excavation. Adachi et al (1985) made use of classical slip line theory to define geometrical distribution of joint elements for modeling shallow tunnel excavation. Okuda et al (1999) applied a back analysis procedure to identify deformational mechanism, in which anisotropic damage parameter *m* was employed. A strain softening analysis was

conducted by Sterpi (1999) in which strength parameters (cohesion and friction angle) were lowered immediately after the initiation of plastic yielding. This approach was applied for the interpretation of field measurements by Gioda and Locatelli (1999) who succeeded to simulate the actual excavation procedure with accuracy. These attempts incorporate some of the key factors that must be taken into consideration for modeling shallow tunnel excavation. However, there still is shortage in modeling capability which is expected to cope with development of shear bands, formation of primary and secondary zones and etc.

By reviewing the pervious works, the authors concluded that the essential features to be taken into the numerical procedure would be reduction of shear stiffness and strength parameters after yielding (namely, strain softening). Following is a brief summary of the procedure employed in this work (Matsumoto, 2000). A fundamental constitutive relation between stress σ and strain ε is defined by an elasticity matrix D

$$D = \frac{E}{1-v-2v^2} \begin{bmatrix} 1-v & v & 0 \\ v & 1-v & 0 \\ 0 & 0 & m(1-v-2v^2) \end{bmatrix} \quad (1)$$

where $\sigma = D\varepsilon$ holds. E and v stands for Young's modulus and Poisson's ratio, respectively. The anisotropy parameter m is defined as

$$m = m_e - (m_e - m_r)[1 - Exp\{-100\alpha(\gamma - \gamma_c)\}] \quad (2)$$

where m_e is the initial value of m, m_r is the residual value, α is a constant, γ is shear strain, γ_c is the shear strain at the onset of yielding.

The constitutive relationship is defined for conjugate slip plane direction ($45° \pm \phi/2$) and transformed back to the global coordinate system. Strength parameters, namely cohesion *c* and friction angle ϕ are reduced from the moment of initiation of yielding to residual values, as indicated in Figure 3. This implies that the admissible space for stress is gradually shrunk as strain-softening process takes place. Any excess stress, which is computed on the transformed coordinate system based on slip plane direction, outside an updated failure envelop is converted into unbalanced forces that are compensated for in an iterative algorithm.

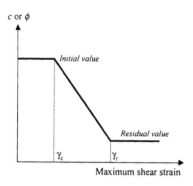

Figure 3. Reduction of strength parameters.

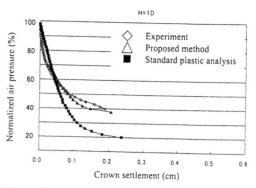

Figure 4. Ground reaction curves from computation and experiment.

4 RESULTS

The proposed analysis method was applied for simulation of a laboratory tunnel excavation experiment (Akutagawa et al, 1998). The experiment uses a model tunnel with a diameter of 15cm. The tunnel is made of air bags surrounded by ground material represented by aluminum bars of different diameters (1.6 and 3mm) having length of 5cm. Excavation is modeled by reducing air pressure from an initial value until a tunnel collapses, while deforming aluminum bars are monitored to record displacements.

For example, some results are shown here for the case in which an overburden height H was taken equal to the tunnel diameter D. Figure 4 shows crown settlements plotted against normalized air pressure, when the input parameters were chosen as seen in Table 1. It is seen that the settlement-air pressure curve (ground reaction curve) computed by the proposed method agrees well with the experimental results; whereas the results from a standard plastic analysis, without considering strain softening, show unsatisfactory match.

Table 1. Input parameters.

Young's modulus		530 KPa
Cohesion	Initial	0.2548 KPa
	Residual	0.0 KPa
Friction angle	Initial	30.7°
	Residual	23.2°
Increment of maximum shear strain during which strength parameters drop		0.0025
Residual value of m		0.0001
Constant α		1.0
Poisson's ratio		0.333
Unit weight		21.36 KN/m³

Normalized ground settlement curve is shown in Figure 5 again in comparison to the results from a standard plastic analysis and the experiment. It is clearly seen that the transition from a concave to convex shape of settlement profile, which is relevant to the formation of shear bands, is properly modeled by the proposed method.

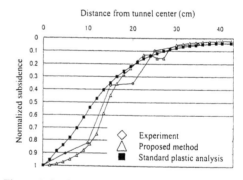

Figure 5. Surface subsidence profiles.

Lastly, distribution of maximum shear strain is shown for several stages of excavation process in Figure 6. The strain distribution at collapse is also shown in Figure 7 for experimental results. According to Figure 6, a shear band started to develop when the stress release ratio was approximately 46% and then branched into two bands later. The vertical one on the left would be regarded as shear band *a* in Figure 2, dividing region A from B. The inclined one represents shear band *b* that identifies the boundary between regions B and C. Both shear bands then reached to the surface at last, leading to the ultimate collapse of the tunnel. The air pressure at which the final collapse was defined also agreed well with the experimental results.

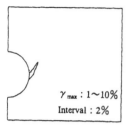

(a) Stress release ratio 46%

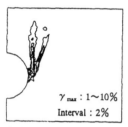

(b) Stress release ratio 62%

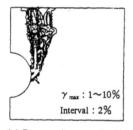

(c) Stress release ratio 64%

Figure 6. Transition of maximum shear strain distribution.

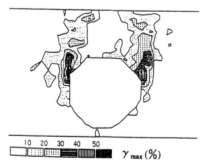

Figure 7. Strain distribution from experiment at collapse

5 CONCLUSION

A nonlinear finite element analysis procedure was proposed for modeling a deformational behavior, which is unique to tunnels with shallow depth. An objective was to point out the importance of modeling a nonlinear nature of the deformational mechanism for obtaining a better understanding of design load on tunnel linings and its relation to kinematics of the surrounding ground. The results obtained showed that modeling of a ground behavior, which is essentially of nonlinear nature, by an elastic or elastic-perfectly plastic approach, leads to incorrect understanding of the deformational mechanism. The proposed approach produced strain distribution, deformational mechanism, surface settlement profile, ground reaction curve, which were in good agreement with the results of the model test. Use of the proposed approach would enable a better understanding of deformational characteristics of the ground medium not only in identifying local plastic zones, but also in revealing kinematics of movement of blocks formed between slip planes, or shear bands. This makes a good starting point for optimizing ground support for reducing surface settlement, considering a particular nature of the deformational mechanism of shallow tunnels.

REFERENCES

Hansmire, W. H. & E. J. Cording 1985. Soil tunnel test section : Case history summary, *Journal of Geotechnical Engineering*, ASCE, 111(11):1301-1320.

Murayama, S. & H. Matsuoka 1969. On the settlement of granular media caused by the local yielding in the media, *Proceedings of JSCE*, 172: 31-41. (in Japanese)

Murayama, S. & H. Matsuoka 1971. Earth pressure on tunnels in sandy ground, *Proceedings of the JSCE*, 187: 95-108. (in Japanese)

Adachi, T., T. Tamura & A. Yashima 1985. Behavior and simulation of sandy ground tunnel, *Proceedings of the JSCE*, 358(III-3): 129-136. (in Japanese)

Okuda, M., T. Abe & S. Sakurai 1999. Nonlinear analysis of a shallow tunnel, *Journal of Geotechnical Engineering*, JSCE, 638(III-49): 383-388. (in Japanese)

Sterpi, D. 1999. An analysis of geotechnical problems involving strain softening effects, *International Journal for Numerical and Analytical Methods in Geomechanics*, Vol.23, 1427-1454.

Gioda, G. & L. Locatelli 1999. Back analysis of the measurements performed during the excavation of a shallow tunnel in sand, *International Journal for Numerical and Analytical Methods in Geomechanics*, Vol. 23, 1407-1425.

Matsumoto, K. 2000. Fundamental investigation on design pressure of tunnels, *Masters thesis*, Graduate School of Science and Technology, Kobe University, Japan. (in Japanese)

Akutagawa, S., T. Kitani, Y. Abe & S. Sakurai 1998. A consideration on tunnel pressure derived from the Terzaghi's formula based on an equilibrium assumption in a limit state, *Proceedings of Tunnel Engineering*, JSCE, 8:95-100. (in Japanese)

Modern Tunneling Science and Technology, Adachi et al (eds), © 2001 Swets & Zeitlinger, ISBN 90 2651 860 9

Prediction of ground settlements due to tunneling in clayey soils using advanced constitutive soil models: a numerical study

A. Burghignoli, S. Miliziano & F.M. Soccodato
Dept. of Structural and Geotechnical Engineering, University of Rome 'La Sapienza', Italy

ABSTRACT: In this paper the main results of a numerical study on the shape of the surface settlement profiles related to shallow tunnel construction in cemented clayey soils are presented. The analyses were carried out under plane strain and undrained conditions, based on a constitutive soil model proposed recently in the literature. Despite the high non linear description of the mechanical behavior of the soil embedded in the model, only a slight improvement in the prediction of the settlement trough was obtained, especially when comparing the settlement profiles obtained from the numerical analyses with those predicted by the inverted normal (Gaussian) distribution form which reasonably matches most of the experimental data. The analysis of the results seems to indicate that an important role in producing differences between numerical predictions and observed performances is played by the three-dimensional nature of the problem.

1 INTRODUCTION

A satisfactory prediction of the ground movements related to the construction of geotechnical engineering structures is often dependent on, among other factors, the constitutive model adopted to simulate the mechanical behavior of the soil. In particular, it is well known the important role played by the non-linearity of the soil behavior, right from the very small strain levels which are typical of the working loading conditions of the structures.

When dealing with cemented soils, the observed reduction of the stiffness and strength of the soil due to the progressive bonds degradation plays an additional role that often cannot be neglected (Burghignoli et al., in press).

Though the construction of a shallow tunnel (and, hence, the numerical simulation) is a complex and three-dimensional problem, the use of non-linear soil models in conventional two-dimensional numerical analyses allowed to improve the prediction of the surface settlement profiles (e.g. Gunn 1993, Addenbrooke et al. 1997).

The aim of this study is to explore the potential of a highly non-linear constitutive model for natural clayey soils recently published in the literature (Rouainia & Muir Wood 2000). Attention is focused on the numerical prediction of the surface settlement profiles in a free-field condition.

In the following, after a brief description of the new model, the results of a number of numerical analyses will be presented and discussed. The analy-ses were carried out in the transverse plane of the tunnel (plane strain) and under undrained conditions. The effects of the excavation process were simulated following the usual technique of a proportional reduction of the geostatic stresses initially acting on the tunnel boundary.

The results obtained with the new model are compared with those obtained by using a simple and conventional elastic-perfectly plastic soil model and with the experimental-based inverted normal (Gaussian) distribution form.

2 THE CONSTITUTIVE MODEL ADOPTED

The kinematic hardening model for natural clays with loss of structure recently proposed by Rouainia & Muir-Wood (2000) has been used in this study. In the following, a brief description of the model is drawn; further and more detailed information can be found in the referenced paper.

The model, which will be referred as the RMW model, is based on an elasto-plastic formulation through a Bounding Surface plasticity approach (Dafalias 1986); a mixed isotropic and kinematic hardening rule is used. The model is characterized by three elliptical surfaces in the stress space (Fig. 1). The inner surface (the bubble) defines the borders of the elastic region. Isotropic hypoelasticity, with a bulk modulus linearly dependent on mean effective stress, is assumed. The kinematic hardening of the bubble (the bubble is dragged by the stress path in a plastic loading process) allows to maintain a certain memory of the recent stress history experienced by the material. The outer surface is the *structure* surface and it bounds the admissible stress states; the size and the position of this surface reflects the effects of the soil structure (e.g. degree of cementation bonding, anisotropy). The Bounding Surface approach, in which the plastic modulus is a function of a normalized distance between current stress and conjugate stress on the structure surface, allows to model the observed continuos decrease of soil stiffness with strain. With increasing development of plastic strain the structure surface shrinks, simulating the effects of progressive damage of soil structure, until it eventually reaches the third surface defined in the model: the *reference* surface is representative of a purely isotropic and volumetric hardening soil behavior, like in the *Modified Cam clay* model (Roscoe & Burland 1968).

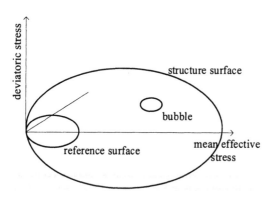

Figure 1. The surfaces of the model proposed by Rouainia & Muir Wood (2000).

The isotropic version of the model (i.e. with the structure surface centered along the isotropic stress axis) is characterized by 9 parameters and 3 hardening variables. Four parameters (M, λ, κ and ν) are identical to the well-known parameters of the *Cam clay* family models, even though, in this case, κ is conceptually related to the *true* (small strain) elastic behavior of the soil; B and ψ control the amount of plastic strain (Bounding Surface Plasticity), while A and k the additional developments of plastic strain related to the progressive bond degradation. The last parameter R defines the constant ratio between the sizes of the elastic nucleus and of the reference surface. The three hardening variables p'_c, r and α define the dimension of the reference surface, the dimension of the structure surface and the position of the center of the elastic nucleus, respectively.

The calibration of the model requires a significant effort: trial and error or optimization procedures are often necessary when using the results of laboratory tests in order to obtain the numerical values of the

parameters and the initial values of the hardening variables. Examples of the model calibration are already reported in Rouainia & Muir Wood (2000) and in Burghignoli et al. (in press) with reference to the experimental results obtained on a cemented clayey soil from Avezzano (Italy).

3 NUMERICAL ANALYSES

The geometry of the problem is shown in Figure 2: two diameters (D=5 and D=10 m) of the tunnel have been considered in the numerical study presented in this paper. The analyses were carried out by using the finite difference code FLAC (Itasca 1996) under plane strain and undrained conditions; the excavation process was simulated by adopting the usual technique of a proportional reduction of the forces initially acting on the tunnel boundary.

The relevant mechanical properties of the soil shown in Figure 2 are representative of Avezzano cemented clay. The variations with depth of the small strain shear modulus G_0, derived from in situ dynamic measurements, and of the undrained shear strength c_u are also reported in the figure.

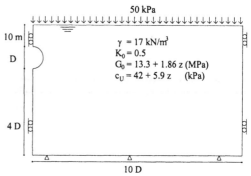

Figure 2. The problem analysed.

Table 1 lists the values of the RMW model parameters used in the analyses.

Table 1. Rouainia & Muir Wood (RMW) model parameters used in the analyses.

λ	0.09	B	10	p'_c (kPa)	42.9 + 6z
κ	0.0025	ψ	0.6	r	3.0
M	1.40	A	1.3	p'_α (kPa)*	32.1 + 4.5z
ν	0.125	k	0.5	q_α (kPa)*	22.5 + 3.2z
R	0.3				

* p'_α and q_α are the isotropic and deviatoric components of the hardening tensor α, respectively.

The results of a numerical simulation of an undrained triaxial compression test carried out on a soil element located at 15 m of depth (i.e. at the level of the tunnel axis) are shown in Figure 3. The main features of the mechanical behavior of cemented clayey soils (elastic threshold, continuos decay of the stiffness, progressive reduction of the strength and positive excess pore pressure development after peak) appear to be well captured by the model.

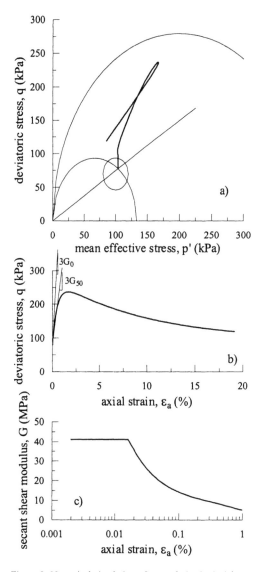

Figure 3. Numerical simulation of an undrained triaxial compression test: a) effective stress path and initial position of the three surfaces characterizing the RMW model; b) deviator stress versus axial strain curve; c) secant shear modulus versus axial strain.

Turning back to the numerical analyses of the tunnel, it was decided to compare the results obtained with the RMW model with those obtained adopting a conventional elasto-perfectly plastic model with the

Tresca yielding criterium. Even though the latter model required a total stress approach, it was possible to match the peak undrained strength predicted by the two models (both in compression and in extension stress paths). Two sets of numerical analyses were carried out with the simple model: the first one (EPH analyses) adopting the same small strain elastic stiffness (G_0) introduced in the RMW model; the second one (EPM analyses) adopting an *operational* elastic stiffness (G_{50}) conventionally defined at a level of strain corresponding to the 50% of soil strength (Fig. 3b).

The trend of the ground loss V_L (i.e. the volume in excess of the notional excavated volume per unit length expressed as a percentage of the notional excavated volume) with increasing reduction of the forces initially acting on the tunnel boundary are shown in Figure 4. It is possible to note that, in all the three analyses carried out, the tunnel with D=10 m collapses when about 90% of unloading is reached (Fig. 4a). This result confirms the overall equivalence of the soil strength predicted by the simple elasto-plastic and by the *advanced* model obtained with the appropriate selection of the relevant parameters. The tunnel with D=5 m withstands a complete reduction of the geostatic forces (Fig. 4b).

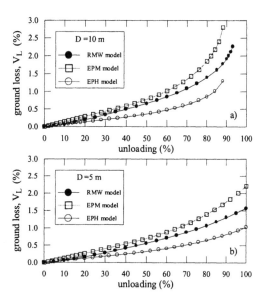

Figure 4. Ground loss as a function of the percentage of the reduction of the geostatic forces: a) 10 m diameter tunnel; b) 5 m diameter tunnel.

For a given value of the percentage of unloading, the ground loss and, hence, the settlements predicted by the three models are generally different. Only when a very small percentage of unloading is considered and the soil response is almost elastic, the EPH and the

117

RMW models are clearly equivalent, the elastic parameters adopted in the two model being the same.

Figure 5 reports the surface settlements associated with a 40% of unloading. It appears that, in this case, the maximum settlements obtained with the RMW and with the EPM model are comparable; however, at great distance from the tunnel center-line the settlements predicted by the RMW model are smaller and similar, as expected, with those predicted by the EPH model (small strain stiffness).

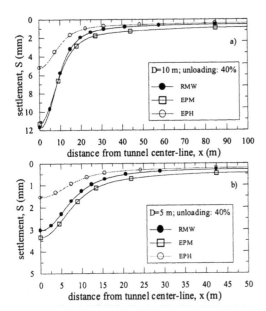

Figure 5. Surface settlement profiles obtained from the numerical analyses for a 40% of unloading: a) 10 m diameter tunnel; b) 5 m diameter tunnel.

4 ANALYSIS OF RESULTS

A convenient normalization of the results must be introduced in order to properly compare the shape of the surface settlement profiles. It is well known that the observed surface settlement profiles can be successfully described by an inverted normal (Gaussian) distribution form:

$$S = S_{max} \cdot e^{(-\frac{x^2}{2 \cdot i^2})} \qquad (1)$$

where x is the distance from the tunnel center-line, i the horizontal offset from the tunnel center-line to the inflection point of the settlement trough and S_{max} the (maximum) surface settlement for x=0.

In undrained conditions the volume of the ground loss into the tunnel V_S (i.e. in excess of the notional excavated volume) equals the volume of the settlement trough per unit length. The following equation holds true:

$$V_S = \sqrt{2\pi} \cdot i \cdot S_{max} \qquad (2)$$

With reference to tunnels constructed in clayey soils, the value of the parameter i may be estimated from the following empirical relationships (Peck 1969; O'Reilly & New 1982):

$$i = \frac{D}{2} \cdot \left(\frac{Z}{D}\right)^{0.8} \qquad (3)$$

and

$$i = 0.5 \cdot Z \qquad (4)$$

where D is the tunnel diameter and Z the depth of the tunnel axis.

Table 2 reports the values of the settlement trough parameter i directly derived from the results of numerical analyses, together with the i values predicted by eqns (3) and (4).

Table 2. Values of the settlement trough parameter i.

	D=10 m	D=5 m
Numerical analyses	6.9-7.1	5.5-5.7
Equation (3)	6.9	5.2
Equation (4)	7.5	6.3

The values of i derived from the numerical analyses were found only slightly dependent on the constitutive soil model adopted and on the amount of unloading; also, the values of i fall in the range predicted by the empirical relationships (3) and (4).

Introducing in the eqn (2) the values of i and V_S obtained from each numerical analysis it is possible to evaluate the maximum settlement S_{max} predicted by the normal distribution. The values of i and S_{max} have been used in order to normalize the distance from the tunnel center-line and the settlement troughs obtained from the numerical analyses, respectively. In this way, the settlement profiles have the same area and it is possible to compare their shapes. Attention is focused on three percentage of unloading (10, 40 and 70 %) in order to consider the effect of progressive development of plastic strain.

Figures 6a-6f report the normalized settlement troughs together with the normal distribution. In all cases, the settlement profiles are more 'flat' than the normal distribution and clearly underestimate the maximum settlement predicted by eqn (2). Figures 6a and 6d are relative to the 10 % of unloading. The curves are substantially coincident each other, despite of the different soil models and tunnel diameters: in fact, for this percentage of unloading, a predominant elastic behavior of the soil is expected (see Fig. 4). With increasing percentages of unloading, the normalized curves obtained with the RMW model define the higher maximum settlement and, consequently, the lower settlements at large distance from the tunnel (Figs. 6b and 6e).

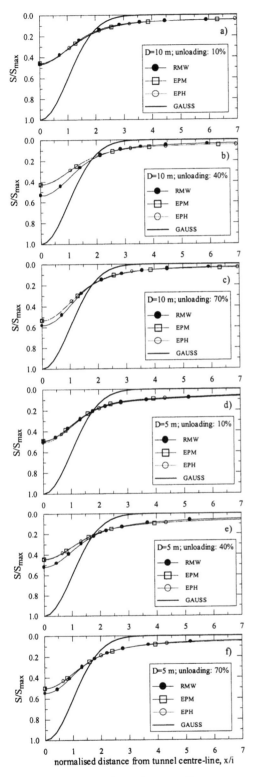

Figure 6. Normalized surface settlement profiles.

However, when looking at the normal distribution, it may be noted the little improvement of the prediction obtained with the RMW model as compared the results obtained with the more simple and conventional elasto-perfectly plastic models. Furthermore, when high values of the percentages of unloading are considered, the predictions obtained with the different models appear to be comparable (Figs. 6c and 6f): this fact may be attributed to the significant development and localization of plastic strain near the tunnel.

Results in agreement with those just above described have been obtained by other researchers following the same approach used in this study (two-dimensional undrained analysis and proportional reduction of the forces initially acting at the tunnel boundary) and with different non-linear soil models. For example, in Figure 7 the results obtained by Gunn (1993) using a non linear elastic model based on the experimental data at small strain level obtained in a triaxial apparatus by Jardine et al. (1985), and the results obtained by Addenbrooke et al. (1997) using the non linear elasto-plastic model proposed by Puzrin & Burland (1998) are reported in the normalized form. In the same figure is also reported an observed settlement trough (Standing et al., 1996); in particular, the numerical analysis carried out by Addenbrooke et al. (1997) refers to the tunnel monitored by Standings et al. (1996).

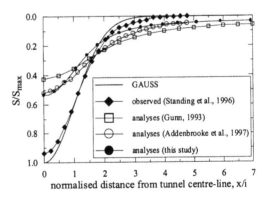

Figure 7. Comparison between numerical predictions and observed performances.

It is clearly apparent that all the soil models, even if highly non-linear, are not able to match the shape of the settlement trough associated with the normal distribution form to which the measured settlement profiles are generally very close. In detail, an underestimation of the maximum settlement of about 50% and a flat profile are obtained from the numerical analyses.

The differences between numerical predictions and observed performances may be attributed to two major facts: inadequate description of the mechanical behavior of the soil and inadequate modeling of the problem (geometry and simulation of the excavation process).

As for as the first aspect is concerned, many *advanced* constitutive soil models published in the literature in the last decades – even if their use is still confined to research purposes - appear to correctly reproduce the mechanical behavior of soils right from the small strain levels. In fact, it is worth noting the improvement of the numerical predictions obtained by adopting non-linear soil models not only when dealing with the simulation of laboratory tests but also when two-dimensional boundary values problems are considered. For example, the use of highly non linear soil models has greatly improved the prediction of the settlement field related to excavations (Simpson et al. 1979) or the prediction of the shape of the load-settlement curve of a shallow foundation in cemented clayey soils (Burghignoli et al., in press).

A similar improvement of the predictions has not been observed in the numerical analysis of tunnels: the results of this study, as well as the analysis of the results of other studies, seems to indicate that the differences between numerical prediction and observed performance should be mainly attributed to the use of a two-dimensional approach for a really three-dimensional problem.

5 CONCLUSIONS

In this paper the results of a numerical study aimed to the analysis of surface settlements profiles related to shallow tunnels construction in cemented clayey soils are reported.

The numerical analyses were carried out using an *advanced* soil model recently proposed in the literature which appears to correctly reproduce the main features of the mechanical behavior of natural cemented clayey soils, as the reduction of stiffness and strength associated to the progressive damage of cementation bonds.

The results of the numerical analyses are not satisfactory: the maximum settlement appears to be underestimated and the width of the settlement trough too large when compared to the shape of the settlement profiles generally observed and reasonably matched by an inverted normal distribution form.

Considering the high non-linearity characterizing the model and the satisfactory predictions of the model at the single element level, even at very small strain amplitude, the differences between numerical results and typical settlement profiles observed may be attributed to the effects of a two-dimensional modeling of the real three-dimensional problem.

However, because the computational costs of a complete three-dimensional analysis are still too

large, a reasonable prediction of the settlement profiles from a simplified two-dimensional analysis is of great importance, especially when dealing with the interaction between tunnel construction and existing buildings. To this aim, even if an improvement of the model (e.g. introducing the elastic anisotropy or changing the law of evolution of the plastic modulus) is always possible, the results of this work seem to indicate that a major effort should be addressed towards the study of techniques of numerical simulation of the excavation effects different from those generally used.

ACKNOWNLODGMENT

The Authors wish to thank Mario Cocciuti for his help in implementing the RMW model in the FLAC code. This work has been developed under the financial support of MURST, National Research Project: *Gallerie in Condizioni Difficili*.

REFERENCES

Addenbrooke, T.I., Potts, D.M. & Puzrin, A.M. 1997. The influence of pre-failure soil stiffness on the numerical analysis of tunnel construction. *Geotechnique* 47(2): 693-712.

Burghignoli, A., Cocciuti, M., Miliziano, S. & Soccodato, F.M. (in press). Evaluation of advanced constitutive modeling for cemented clayey soils: a case history. *Mathematical and computer modelling*.

Dafalias, Y.F. 1986. Bounding surface plasticity I: mathematical formulation and hypoplasticity. *J. Engng. Mech. ASCE* 112(9): 966-987.

Gunn, M.J. 1993. The prediction of surface settlement profiles due to tunneling. In Houlsby & Schofield (eds), *Predictive Soil Mechanics. Proc. Wroth memorial symp., Oxford, 27-29 July 1992.* London: Thomas Telford.

Jardine, R.J., Symes, M.J. & burland, J.B. 1985. Tee measurement of soil stiffness the triaxial apparatus. *Geotechnique* 34(3): 323-340.

Itasca Consulting Group 1996. FLAC (Fast Lagrangian Analysis of Continua) version 3.4. User' manuals. Minneapolis.

O'Reilly, M.P. & New, B.M. 1982. Settlements above tunnels in the United Kingdom – their magnitude and effects. *Tunneling 82*: 173-181.

Peck, R.B. 1969. Deep excavations and tunnelling in soft ground. *Proc. 7th Int. Conf. Soil Mech. and Found. Eng.,* State of the Art Volume: 226-290.

Puzrin, A.M. & Burland, J.B. 1998. Non-linear model of small strain behaviour of soil. *Geotechnique* 48(2): 217-233.

Roscoe, K.H. & Burland, J.B. 1968. On the generalised stress-strain behaviour of 'wet' clays. In *Engineering plasticity.* Cambridge University Press.

Rouainia, M. & Muir Wood, D. 2000. A kinematic hardening constitutive model for natural clays with loss of structure. *Geotechnique* 50(2): 153-164.

Simpson, B., O'Riordan, N.J., Longworth, T.I. & Burland, J.B. 1979. A computer model for the analysis of ground movement in London clay. *Geotechnique* 29(2): 149-175.

Standings, J.R., Nyren, R.J., Longworth, T.I. & Burland, J.B. 1996. The measurement of ground movements due to tunnelling at control sites along the Jubilee Line Extension. In Mair & Taylor (eds). *Geotechnical Aspects of Underground Construction in Soft Ground; Proc. intern. symp., London, 15-17 April 1996.* Rotterdam: Balkema.

Modern Tunneling Science and Technology, Adachi et al (eds), © 2001 Swets & Zeitlinger, ISBN 90 2651 860 9

Analysis of Three-dimensional Effect of Tunnel Face Stability on Sandy Ground with a Clay Layer by the Three-dimensional Rigid Plastic Finite Element Method

S. Konishi
Railway Technical Research Institute, Tokyo, Japan

T. Tamura
Kyoto University, Kyoto, Japan

ABSTRACT: Sandy ground with a clay layer is frequently encountered in tunnel construction work. However, there are few studies of tunnel face stability for alternate ground. The tunnel face stability is affected strongly by the three-dimensional effect. We evaluated the three-dimensional effect of the tunnel face stability for sandy layers with a clay layer, by applying three-dimensional Rigid Plastic Finite Element Method (3D-RPFEM) and two-dimensional Rigid Plastic Finite Element Method (2D-RPFEM). Special features of the method, which were cleared by results, are shown below. 1) 3D-RPFEM is more sensitive to a clay layer than 2D-RPFEM. 2) 3D-RPFEM expresses remarkably influence in the dilatancy at the upper part of the tunnel.

1 INTRODUCTION

1.1 *Introduction*

Ground with sandy and clay alternative layers is frequently encountered in tunnel construction work. However, there are few studies of tunnel face stability for alternate ground. On the other hand, tunnel face stability is affected by not only property of the ground and geometric shape in the cross section but also those and lining in the longitudinal section. To study the three-dimensional effect, we simulated model tests in the uniform sandy ground and sandy ground with a clay layer by the 3D-RPFEM and 2D-RPFEM respectively. And, we compared the results and evaluated three-dimensional effect of tunnel face stability in the sandy ground with a clay layer.

1.2 *Rigid plastic finite Element method*

The analysis method concerned with plastic is effective in estimation of the critical state, such as collapse of the tunnel face. Rigid Plastic Finite Element Method (RPFEM) is the limit analysis based on the upper bound theorem. Merits of RPFEM are shown below (Tamura 1984, 1987, 1999).

1) A elastic coefficient, which is useless under the critical state, is unnecessary.
2) The method can express well elastic flow near critical state, of which error becomes big with elasto-plastic model.
3) Initial stress is not required.

We applied the RPFEM for materials of the Drucker-Prager (or the Mohr-Coulomb) type under an assumption of the associated flow rule in order to consider effects of internal friction angle and dilatancy on the critical state of soil structure.

2 STUDY METHOD

We analyzed into tunnel face stability in the uniform sandy ground and sandy ground with a clay layer by 3D-RPFEM and 2D-RPFEM respectively and compared results. Conditions of analyses are same as those of experiment model tests (Konishi 2000). Main material properties and dimensions are shown below.

1) An Internal friction angle in a sandy layer is 41 degree.
2) Cohesion of a clay layer is 0.4kPa.
3) Unite weight of sandy is 14.91 kN/m^3. Unite weight of clay is 13.53 kN/m^3.
4) The tunnel face is 200mm in height and 100mm in width.
5) Overburden is 200mm in thickness.
6) The model ground is 800mm in length, 700mm in width and 600mm in height.
7) A clay layer is placed on the middle level of tunnel face. It is 25mm in thicknss.

Fig.1 (a) shows a ground model for 3D-analysis. A hollow place expresses a tunnel. A section of tunnel is semicircular in an experiment model. But, it is rectangular in the model for 3D-analyses due to simplification. The model has 1897 nodes and 1420 elements. Fig.1 (b) shows a ground model for 2D-analysis. It has the same shape as the lateral face of the 3D model.

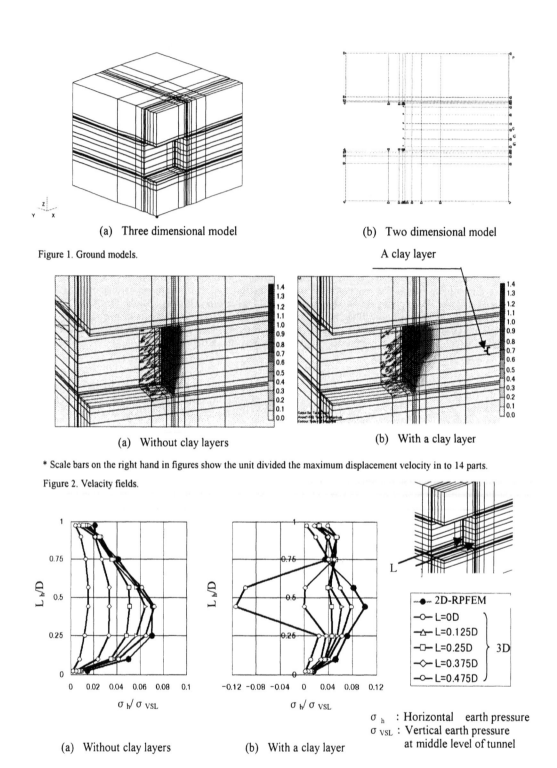

(a) Three dimensional model   (b) Two dimensional model

Figure 1. Ground models.

A clay layer

(a) Without clay layers   (b) With a clay layer

* Scale bars on the right hand in figures show the unit divided the maximum displacement velocity in to 14 parts.

Figure 2. Velacity fields.

2D-RPFEM
L=0D
L=0.125D
L=0.25D   3D
L=0.375D
L=0.475D

σ_h : Horizontal earth pressure
σ_{VSL} : Vertical earth pressure at middle level of tunnel

(a) Without clay layers   (b) With a clay layer

Figure 3. Horizontal earth pressures on faces.

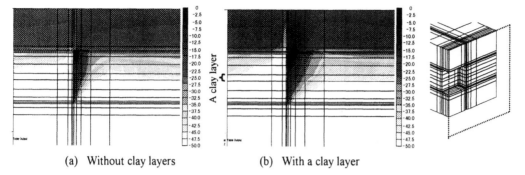

(a) Without clay layers   (b) With a clay layer

* Scale bars on the right hand in figures show the unit divided the maximum vertical earth pressure (=8.94kPa) into 20 parts.

Figure 4. Vertical earth pressure distribution (A longitudinal section cut at the tunnel center line).

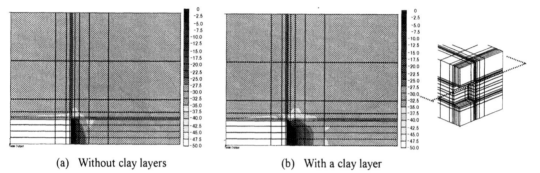

(a) Without clay layers   (b) With a clay layer

* Scale bars on the right hand in figures show the unit divided the maximum vertical earth pressure (=8.94kPa) into 20 parts.

Figure 5. Vertical earth pressure distribution (A horizontal section cut at top of the tunnel).

3 RESULTS

3.1 Velocity of displacement

Fig.2 shows velocity fields at face collapse obtained by analyses with 3D-RPFEM. Fig.2 (a) shows a result in sandy ground and Fig.2 (b) shows it in alternate ground. Those are expressed with the unit divided the maximum velocity into 14 parts. Velocity at upper middle part on the face is biggest in both cases. Velocity becomes smaller as near to circumference of a face on cross section and sliding line on longitudinal section. Fig.2 (b) shows zone of the fast velocity to widen on crossing and longitudinal directions due to a clay layer. By the result, it is known that loosening area extend three-dimensionally due to a clay layer.

3.2 Horizontal earth pressure

Fig.3 shows horizontal earth pressures on face. In the uniform sandy ground (Fig.3(a)), distributions obtained by analyses with 2D-RPFEM are the same shape. Distributions are convexity. Maximum value of earth pressure appears at the little low point from a middle level of the face. It is known that the position needs the largest power to hold collapsing face.

The shape become flat, as section approach to the side wall. In alternate ground, horizontal earth pressure in a clay layer projects into plus side on a center cross section. But, it projects into minus side, as section approach to the side wall.

3.3 Vertical earth pressure

Fig.4 (a) and (b) show vertical earth pressure on longitudinal sections. Those are expressed with the unit divided the maximum earth pressure (8.94kPa) into 20 parts. A zone in which the vertical earth pressure decreases widens clearly due to a clay layer. By the result, it is known that a damaged area on surface extent in alternate ground.

Fig.5 (a) and (b) show vertical earth pressures on horizontal sections. On the crown level, the vertical earth pressure decreases in a zone of collapse and increases in a range surrounding collapsed zone. The phenomenon is similar to results of drop door tests in two-dimension and due to re-distribution of stress (Murayama 1971). We suggested the three-dimensional distribution in figures for the first time.

A decreasing zone reduces suddenly near a side wall. The phenomenon in alternate ground is clearer than that in the uniform sandy ground. An increas-

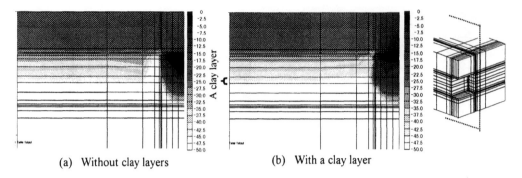

(a) Without clay layers   (b) With a clay layer

* Scale bars on the right hand in figures show the unit divided the maximum vertical earth pressure (=8.94kPa) into 20 parts.

Figure 6. Vertical earth pressure distribution (A cross section cut at 0.075D before face).

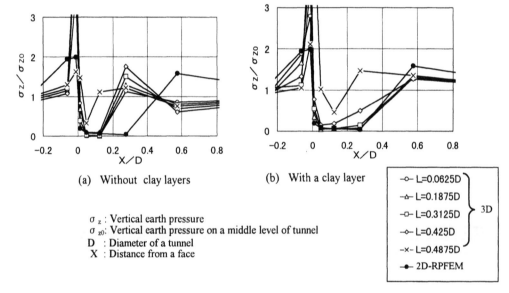

(a) Without clay layers   (b) With a clay layer

–○– L=0.0625D	⎫
–△– L=0.1875D	⎪
–□– L=0.3125D	⎬ 3D
–◇– L=0.425D	⎪
–×– L=0.4875D	⎭
–●– 2D-RPFEM	

σ_z : Vertical earth pressure
σ_{z0}: Vertical earth pressure on a middle level of tunnel
D   : Diameter of a tunnel
X   : Distance from a face

Figure 7. Vertical earth pressures at a top level of the tunnel on longitudinal sections.

ing range in alternate ground is wider than that in the uniform sandy ground.

Fig.6 (a) and (b) show vertical earth pressures on cross-sections. A position of the cross section is 0.075D (A mark D is diameter of tunnel: 200mm) before the face. In the uniform sandy ground, earth pressure at outside of side wall increases remarkably. But, in the sandy ground with a clay layer, increase of it is unremarkable. On the other hand, a decreasing zone widens to side wall due to a clay layer.

Those results show that the clay layer transmits loosening in a lower sandy layer to an upper sandy layer widely and weakly. The uniform sandy ground is sensitive to re-distribution of stress due to excavation. But, the sandy ground with a clay layer is insensitive.

Fig.7 (a) and (b) show vertical earth pressures at

top level of the tunnel on longitudinal sections. In a legend, a mark L suggests a distance from center of the tunnel. A zone due to 3D-RPFEM in which vertical earth pressures decrease is smaller than that due to 2D-RPFEM in the uniform sandy ground. Zones are similar, even if a section is away from the tunnel center line. But, it reduces rapidly near side wall in results of 3D-RPFEM. In alternate ground, decreasing zones (L=0.0625D, 0.175D, 0.2625D) due to 3D-RPFEM are similar to that due to 2D-RPFEM. But decreasing ranges on sections near side wall (L=0.425D, 0.4875D) due to 3D-RPFEM are smaller than that due to 2D-RPFEM.

Fig.8 shows vertical earth pressures at a top level of tunnel on cross-sections. X suggests a distance from tunnel face. Results in the uniform sandy ground are shown in Fig.8 (a). On a cross section near a lining edge (x=-0.025D), a vertical earth

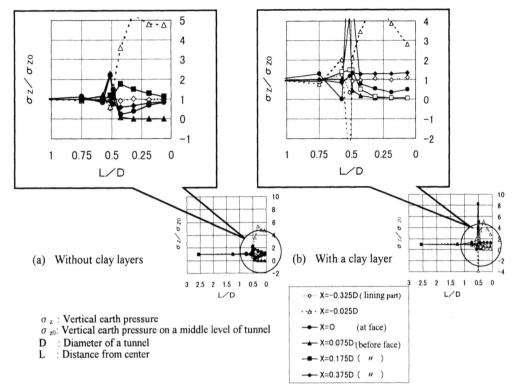

σ_z / σ_{z0}

L／D

1   0.75   0.5   0.25   0

(a) Without clay layers

L／D

3 2.5 2 1.5 1 0.5 0

(b) With a clay layer

L／D

1   0.75   0.5   0.25   0

L／D

3 2.5 2 1.5 1 0.5 0

σ_z : Vertical earth pressure
σ_{z0}: Vertical earth pressure on a middle level of tunnel
D   : Diameter of a tunnel
L   : Distance from center

····◇··· X=-0.325D (lining part)
·-△· X=-0.025D
──●── X=0   (at face)
──▲── X=0.075D (before face)
──■── X=0.175D (〃)
──◆── X=0.375D (〃)

Figure 8. Vertical earth pressures at the top level of tunnel on cross sections.

pressure increases largely on tunnel lining and does not increase so much on outside ground of side wall. On cross-sections at face and before face (x=0 and x=0.075D), a vertical earth pressure decreases on position in front of tunnel lining and increases on near position in front of a side wall. The special feature grows weaker, as a section goes away from face. By results, it is considered that next phenomenon's occur due to stress re-distribution.

1) A large vertical earth pressure (6 times as large as usual state) acts on a front edge of tunnel lining.
2) Value of a increasing earth pressure on ground before face is about 0.4 times as large as it on tunnel lining.

Results in the uniform sandy ground are shown in Fig.8 (b). Comparing to results in sandy ground, special features of results in alternate ground are below:

1) Values of the earth pressure in increasing zone (x=0.375D) are small.
2) Values of the earth pressure in increasing zone (x=0.375D) are equal.

3.4 Earth pressure on the tunnel face

Table 1 shows total earth pressures acting on faces due to 3D-RPFEM and 2D-RPFEM. Total earth pressure acting on a face suggests force, which is necessary to support a face. In the uniform sandy ground, a total earth pressure due to 3D-RPFEM is 0.69 times as large as that due to 2D-RPFEM. In sandy ground with a clay layer, it becomes 0.53 times. A result due to 3D-RPFEM show that total earth pressure in sandy ground with a clay layer is smaller than that in the uniform sandy ground. But, a result due to 2D-RPFEM show that total earth pressure in sandy ground with a clay layer is larger than that in the uniform sandy ground. By comparing results, next features are considered:

1) 3D-RPFEM is more sensitive to cohesion of a clay layer than 2D-RPFEM.
2) 3D-RPFEM is more sensitive to changes in the dilatancy at the upper part of a tunnel than 2D-RPFEM.

4 CONCLUSIONS

We have the following remarks as some conclusions of the present paper.

(1)   Velocity of displacement at upper middle part on a face is biggest in results of 3D-analysis. Zone of the fast velocity widens on crossing and longitudinal directions due to a clay layer.

(2)   In the uniform sandy ground, distribution of

Table 1. Total earth pressure acting on a tunnel face due to 3D-RPFEM and 2D-RPFEM.

Item	(a) Total earth pressure acting on face due to 2D-RPFEM (kPa)	(b) Total earth pressure acting on face due to 3D-RPFEM (kPa)	(b)/(a)
Uniform sandy ground	0.218	0.150	0.686
Sandy ground with a clay layer	0.261	0.138	0.529

horizontal earth pressures on face obtained by 2D-RPFEM and it obtained by 3D-RPFEM on a tunnel center section are the same shape. Distributions are convexity. The shape becomes flat, as the section approaches to the side wall. In sandy ground with a clay layer, horizontal earth pressure in a clay layer projects into plus side on a tunnel center cross section. But, it projects into minus side, as a section approaches to the side wall.

(3)   A zone, in which the vertical earth pressure decreases, widens on crossing and longitudinal directions due to a clay layer. A clay layer transmits loosening in a lower sandy layer to an upper sandy layer widely and weakly.

(4)   On a top level of tunnel, the vertical earth pressure decreases on a zone in front of face and increases on a part surrounding it due to re-distribution of stress.

(5)   A uniform sandy ground is sensitive to re-distribution of stress due to excavation. But, sandy ground with a clay layer is insensitive.

(6)   3D-analysis is more sensitive to the effects of clay layer than 2D-analysis. Results by 3D-analysis are more remarkable than that by 2D-analysis, concerning phenomenon's shown below.

1)   The zone, in which the vertical earth pressure decreases, extends larger on a clay layer.

2)   Increasing values of the vertical earth pressure in sandy ground with a clay layer are smaller than those in the uniform sandy ground due to the effect of the clay layer.

(7)   In the uniform sandy ground, a total horizontal earth pressure acting on faces due to 3D-analysis is 0.69 times as large as it due to 2D-analysis. In sandy ground with a clay layer, it becomes 0.53 times. Reason for it is considered that 3D-RPFEM is more sensitive to co-

hesion of a clay layer and changes in the dilatancy at the upper part of the tunnel than 2D-RPFEM.

REFERENCE

Tamura,T., Kobayashi,S. & Sumi,T. 1984 Limit analysis of soil structure by rigid plastic finite element method, Soils and Foundations, JSSMFE, Vol.24, No.1, pp34-42.

Tamura,T., Kobayashi,S. & Sumi,T. 1987 Rigid-plastic finite element method for frictional materials, Soils and Foundations, JSSMFE, Vol.27, No.3, pp1-12.

Tamura,T., Adachi,N., Konishi,S., & Tsuji,T. 1999 Rigid-plastic finite element method for frictional materials, Journal of geotechnical engineering, JSCE, No.638/III-49, pp301-310.

Konishi,S., Asakura,T., Tamura,T. & Tsuji,T. 2000 Evaluation of tunnel face stability for sand strata with clay layers, Journal of geotechnical engineering, JSCE, No.659/III-52, pp51-62.

Murayama,S., & Mtauoka,H. 1971 Evaluation of tunnel face stability for sand strata with clay layers, Journal of geotechnical engineering, JSCE, No.187, pp95-108.

Modern Tunneling Science and Technology, Adachi et al (eds), © 2001 Swets & Zeitlinger, ISBN 90 2651 860 9

Hydraulic conductivity change around tunnel due to excavation

A. Kobayashi, M. Nishida
Graduate School of Agricultural Science, Kyoto University, JAPAN

K. Hosono
Geoscience Research Laboratory, JAPAN

T. Fujita
Japan Nuclear Cycle Development Institute, JAPAN

ABSTRACT: To examine the mechanism of the change in permeability around a tunnel due to excavation, two in-situ tests are simulated by the coupled mechanical and hydraulic model with a non-linear elastic constitutive law for single fracture and anisotropic properties equivalent to the fractured rock mass. Through the comparison between measured and calculated results, the mechanism is estimated. It is concluded from the examination that the behavior of the horizontal fractures has a large effect on the reduction of the hydraulic conductivity, and the deformation in the drift direction also has an effect on the change in the hydraulic conductivity. Moreover, it is found that the fractures contributing to the mechanical behavior may be limited.

1 INTRODUCTION

Pore water pressure in the rock mass has an effect on the drainage-way and the thickness of lining of tunneling design. The disturbed zone around the tunnel due to excavation has a large influence on the pore pressure distribution in the rock mass. If the permeability becomes large at the zone, the inflow rate comes to be large and significant drainage-way system may be necessary. If the drainage system cannot have enough drainage capacity, the degradation of lining may be caused by high water pressure. Furthermore, this zone has an important role for the waste disposal tunnel. The change in permeability around the tunnel has a large influence on the performance assessment of the repository. Thus, the change in permeability around tunnel due to excavation should be examined correctly before construction.

To do this, the mechanism of coupled mechanical and hydraulic behavior should be understood well and its effect has to be considered into the numerical model. In this paper, the continuum model considering the non-linear mechanical characteristics of fractures and anisotropic properties of fracture system is introduced and the application to two in-situ tests is presented.

The equivalent continuum model to the fractured medium is introduced by combining the crack tensor theory proposed by Oda (1986) and the constitutive model by Barton and Bandis (1985; BB model). This model can realize the anisotropy of the media by crack tensor theory and the non-linear mechanical and hydraulic behavior by BB model. The model is incorporated into the coupled FE approach. Excavation process can be simulated.

To validate the model and examine the behavior around the tunnel due to excavation, two in-situ excavation tests are simulated by the model. Both test tunnels are adjoining and excavated in different directions. Change in permeability and deformation are measured at each test and the comparison with numerical results is carried out. Through the comparison, the behavior in the rock mass is examined.

2 THEORY

The Barton-Bandis (BB) model represents the change in the normal and shear stiffness due to the stress change by using *JRC* (Joint Roughness Coefficient) and *JCS* (Joint Compression Strength). By using the JRC_n^I and JCS_n^I (MPa) which is the mean values for each set I, the initial normal stiffness of joint, K_{nI} (MPa/mm), is calculated for each set I,

$$K_{ni}^I = -7.15 + 1.75 JRC_n^I + 0.02 \left(JCS_n^I / a_j^I \right) \quad (1)$$

where a_j(mm) is the initial joint aperture given by

$$a_j^I = \frac{JRC_n^I}{5} \left(0.2 \frac{UCS}{JCS_n^I} - 0.1 \right) \quad (2)$$

where *UCS* is the uniaxial compression strength of rock.

The normal stress, σ_n^I, to each set is calculated by $\sigma_{ij} n_i^I n_j^I$ in which σ_{ij} is the calculated stress of each element, and n_i^I is the mean unit normal vector of each set. The normal stiffness of the joint is revised by using the normal stress at each time step.

$$K_n^I = K_{ni}^I \left[1 - \frac{\sigma'_n}{V_m K_{ni}^I + \sigma'_n} \right]^{-1} \tag{3}$$

and the shear stiffness is obtained from

$$K_s^I = \frac{100}{L^I} \sigma_n^I \tan \left[JRC_n^I \log_{10} \left(\frac{JCS_n^I}{\sigma'_n} \right) + \phi'_r \right] \tag{4}$$

V_m in the equation (3) is assumed to be the same as a_j^I of the equation (2). L^I is the mean fracture length of the set (mm) and ϕ_r is the residual friction angle.

By using $h^I = K_n^I L^I$, $g^I = K_s^I L^I$, the elastic compliance C_{ijkl} and hydraulic conductivity tensor k_{ij} are calculated by

$$C_{ijkl} = \sum_I \left(\frac{1}{h^I} - \frac{1}{g^I} \right) F_{ijkl}^I + \frac{1}{4g^I} \left(\delta_{ik} F_{jl}^I + \delta_{jk} F_{il}^I + \delta_{il} F_{jk}^I + \delta_{jl} F_{ik}^I \right) \tag{5}$$

$$k_{ij} = \sum_I \frac{1}{12} e^{I3} \left(P_{kk}^I \delta_{ij} - P_{ij}^I \right) \tag{6}$$

where the summation is carried out for sets. δ_{ij} is the Kronecker delta. F_{ijkl}, F_{ij} and P_{ij} are obtained from

$$F_{ijkl}^I = \rho^I \frac{\pi}{4} L^{I3} n_i^I n_j^I n_k^I n_l^I \tag{7}$$

$$F_{ij}^I = \rho^I \frac{\pi}{4} L^{I3} n_i^I n_j^I \tag{8}$$

$$P_{ij}^I = \rho^I \frac{\pi}{4} L^{I2} n_i^I n_j^I \tag{9}$$

where ρ^I is the fracture density of set I.

The values of F_{ijkl}, F_{ij} and P_{ij} are not changed through the analysis process. The increment of the aperture of the set is given by

$$\Delta V^I = \frac{\Delta \sigma_n^I}{K_n^I} \tag{10}$$

The new aperture of the set is revised by

$$e^I = e_{int}^I - \Delta V^I \tag{11}$$

in which e_{int}^I is estimated from

$$e_{int}^I = \sqrt[3]{\frac{18 \times \mu_l \times K_{mean}}{9.8 \times P_{KK}^I}} \tag{12}$$

where μ_l is the viscosity of water. K_{mean} is the mean hydraulic conductivity.

By using the above nonlinear equivalent model, the following coupled equations are solved by the three-dimensional finite element method. The continuity equation is given by

$$\left\{ \frac{\rho_l g k_{ij}}{\mu_l} h_{,j} \right\}_{,i} - \rho_{l0} n S r \rho_l g \beta_P \frac{\partial h}{\partial t} - \rho_l \frac{\partial \theta}{\partial \psi} \frac{\partial h}{\partial t} - \rho_l Sr \frac{\partial u_{i,i}}{\partial t} = 0 \tag{13}$$

where ρ_l is the density of water, g is the gravitational acceleration. n is the porosity, S_r is the degree of saturation, β_P is the compressibility of water, u_l is the displacement vector, h is the total head and t is the time.

The equilibrium equations are written as

$$\left[\frac{1}{2} T_{ijkl}^{-1} (u_{k,l} + u_{l,k}) + \chi T_{ijkl}^{-1} C_{kl} \rho_l h \right] + \rho b_i = 0 \tag{14}$$

where $T_{ijkl} = (M_{ijkl} + C_{ijkl})$, in which M_{ijkl} is the elastic compliance of the rock matrix. $C_{ij} = C_{ijkl} \delta_{kl}$.

3 IN-SITU TEST AND CONDITIONS

To verify the applicability of the model and examine the real behavior, the in-situ excavation tests are analyzed with the above model. The in-situ test was carried out at Kamaishi mine located in north-east of Japan. The test is carried out within the Kurihashi granodiorite that is one of the granitic bodies of early Cretaceous age. Figure 1 shows the schematic view of the test site. The overburden thickness at the site is about 260 m. When KD-88 is excavated by the normal blasting method, the hydraulic conductivity test is carried out with KE-3, the rock stress is measured with KE-2, and the strain is observed with KE-1. At KE-3, the borehole jack test is also carried out to measure the change in the deformability. KD-89 is excavated by the smooth blasting method. Before and after excavation, the hydraulic conductivity test is carried out at KE-6, 7, which exists at 1 and 3 m from the drift wall of KD-89, the rock stress is also measured at KE-4, and the strain is observed at KE-5. All boreholes exist on the lateral side of the excavation drift. As shown in Table 1, the strike of the main fractures is east and dip of the most fractures is over 70°. The direction of the KD-88 drift is perpendicular to the strike of dominant fractures, while that of the KD-89 drift is parallel to the strike. Table 2 shows the initial earth stress and mechanical properties. The data of Tables 1 and 2 are measured at the location A shown in Figure1. As shown in

128

both tables, the mechanical properties are isotropic, while the stiffness matrix becomes anisotropic by the crack tensor. The hydraulic conductivity tensor is also anisotropic, while the initial aperture is the same for all sets.

Figure 2 shows the FE mesh used in the calculation. The shape of the drift of KD-88 is U-shaped, while that of KD-89 is rectangular. The excavation is carried out at 5m/day by removing the elements of the drift. The excavated drift wall has zero water pressure. The outer boundary has a zero deformation condition in normal direction to the planes and fixed total head condition. The prescribed total head at boundary is 20m, which is a measured value at the test site. In the case of the excavation of KD-88, x-axis coincides with north and y-axis is west. For KD-89, x-axis is east and y-axis is north. Thus, the components of the crack tensor are different between both cases. The direction of the initial earth pressure is also different for each case. The difference between excavation methods is, however, not considered in the simulation. The equivalent nodal stress to the initial earth pressure is unloaded at the excavated drift wall.

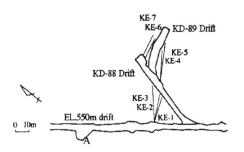

Figure 1. Schematic view of test site.

Table 1. Fracture geometry information.

No.set I	1	2	3	4	5
Dip direction(°)	316.8	248.9	178.5	179.7	353.1
Dip(°)	70.1	88.5	15.8	76.2	76.2
Density(Num./m³)	0.846	0.743	0.404	0.625	0.515
Length(m)	1.5	1.65	1.86	2.18	1.81
JRC	8.83				
JCS(MPa)	104.52				

Table 2. Mechanical properties.

Int. stress (MPa)	29.3	7.6	2.9
Dip direction of int. stress	342°	248°	107°
Dip angle of int. stress	13°	17°	68°
E (GPa)	29.0		
Poisson's ratio	0.3		
UCS (MPa)	124		

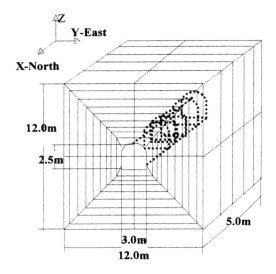

(a) KD-88

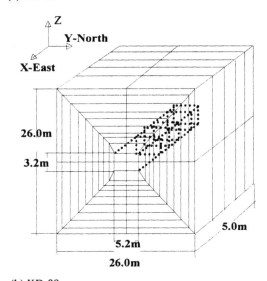

(b) KD-89

Figure 2. Finite element mesh.

4 PREVIOUS STUDY

The values of JRC and JCS are examined with the core samples obtained at the location near the excavation drift, which is indicated by point A in Figure 1. JCS is inferred from the Schmidt hammer test and JRC is examined by the shear test. The number of the samples is 30. Thus, it is difficult to identify those values for each fracture set. The values shown in Table 1 are the average ones of all samples. As the first examination, the average values are used with the same finite element mesh having a circle drift section for both drifts (Kobayashi, et al. 2000). The results showed that the mechanical behavior was realized well by the above model, while the hydraulic conductivity change due to excavation was difficult to represent. The measured change in hydraulic conductivity is indicated in Table 3. The rate of change is also shown in the table. The positive rate means the decrease of hydraulic conductivity. After excavation, the hydraulic conductivity is reduced so much at KD-88, while the change in hydraulic conductivity is not so clear at KD-89. One location has an increased hydraulic conductivity and the others have a decreased one. At KD-89, the hydraulic tests were carried out at the different excavation steps with the borehole parallel to the drift. Thus, multiple results are obtained at the same distance from the drift wall of KD-89. It is also found the change in hydraulic conductivity cannot be recognized at the distance of 2 m from the drift wall of KD-88, while a clear change is seen at the distance of 3m at KD-89.

Table 3. Measure change in hydraulic conductivity.

(x 10^{-9} m/s)

KD-88			
Distance	Before	After	Rate*
0.85m	0.241	0.0497	0.79
1m	1.63	0.441	0.73
2m	2.21	2.28	-0.03
KD-89			
Distance	Before	After	Rate*
1m	0.0108	0.715	-65.20
1m	54.9	15.6	0.72
1m	0.115	0.105	0.09
3m	0.0285	0.0215	0.25
3m	2.57	2.17	0.16

* (Before-After)/After

5 CALIBRATION RESULTS

As mentioned above, the values of JRC and JCS as shown in Table 1 are used for all sets at the previous calculation. In this paper, JRC and JCS are calibrated to realize the change at KD-88, and then. the analysis for KD-89 is carried out by using the same parameter values. After trial and error, by using the values shown in Table 4, the results shown in Table 5 is obtained. The previous results, in which the same JRC and JCS are used for all sets, are also given in the table. In the model, the initial hydraulic conductivity is anisotropic but homogeneous.

As shown in Table 5, while the hydraulic conduc-

Table 4. JRC and JCS values.

Set	1	2	3	4	5
JRC	20	20	20	3	20
JCS(MPa)	124	124	124	89	124

Table 5. Calculated hydraulic conductivity change.

(x 10^{-9} m/s)

		KD-88		
Distance		K_{xx}	K_{yy}	K_{zz}
	Initial	1.22	1.85	1.93
0.85m	Prev.*	3.11	3.4	4.38
	Rate	-1.55	-0.84	-1.27
	Pres.**	1.75	1.54	2.89
	Rate	-0.43	0.17	-0.50
1m	Prev.	3.27	3.61	4.58
	Rate	-1.68	-0.95	-1.37
	Pres.	1.62	1.54	2.71
	Rate	-0.33	0.17	-0.40
2m	Prev.	3.2	3.91	4.37
	Rate	-1.62	-1.11	-1.26
	Pres.	1.29	1.65	2.16
	Rate	-0.06	0.11	-0.12
		KD-89		
	Initial	1.85	1.22	1.93
1m	Prev.	7.81	2.86	7.14
	Rate	-5.40	-0.55	-2.70
	Pres.	1.32	0.71	1.71
	Rate	-0.08	0.62	0.11
3m	Prev.	3.48	1.91	3.43
	Rate	-1.85	-0.03	-0.78
	Pres.	1.26	0.749	1.78
	Rate	-0.03	0.60	0.08

* Previous model with the same JRC and JCS for all sets
** Present model with JRC and JCS shown in Table 4

tivities in all directions come to be large due to excavation in the previous results, the hydraulic conductivity in y direction, which is horizontal direction, becomes small for both drifts at the present model. This is because the horizontal fractures are closed due to increase of normal stress and has a decreased hydraulic conductivity. The dominant fracture sets at the test site are steep as shown in Table 1, and the number of the horizontal fracture set is one. Thus, if the homogeneous distribution of JRC and JCS for all sets is assumed as the previous calculation, the hydraulic conductivity becomes large by the effect of the aperture opening of the steep fracture sets. Thus, the present model has more deformable fracture in the horizontal direction than the steep directions. The behavior of the steep fractures is influenced by that of the horizontal fractures.

The hydraulic conductivity in the other directions at KD-88 becomes large, while that in z direction at KD-89 is reduced. Moreover, the rate of change at KD-89 larger than the one at KD-88 can be seen in the calculated results as well as the measured results. In particular, the change in x direction, i.e., the drift direction, is large. On the other hand, the change in hydraulic conductivity after excavation becomes small drastically at the distance larger than 2 m at KD-88 similarly to the measured results. In particular, the change in x direction becomes small. This large change at the far distance from the drift wall of KD-89 is caused from the behavior of the steep fractures parallel to the drift. In the analyses, the steep fractures have a high stiffness and the horizontal one has a low stiffness calculated by JRC and JCS

shown in Table 4. Thus, the horizontal fracture is deformable than the steep ones. The effect of the deformation of the horizontal fracture on the steep fractures is very important for the entire behavior at this test site. This also means the importance of 3-D analysis.

Moreover, the effect of the direction of earth pressure has an important role on the results. At the test site, the maximum pressure works in the north direction. Since the drift of KD-89 is perpendicular to the north direction, the unloading due to excavation is larger than that of KD-88. This introduces the reduction of normal stress on the steep fractures parallel to the drift of KD-89. On the other hand, the drift direction of KD-88 coincides with the maximum pressure direction. This is one of the reasons of

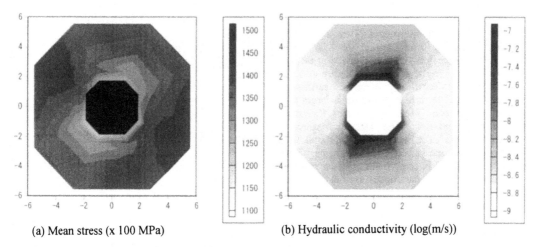

(a) Mean stress (x 100 MPa)   (b) Hydraulic conductivity (log(m/s))

Figure 3. Profile of mean stress and hydraulic conductivity at KD-88.

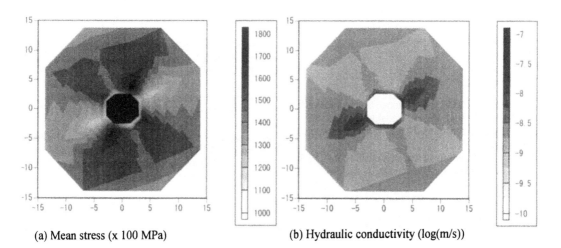

(a) Mean stress (x 100 MPa)   (b) Hydraulic conductivity (log(m/s))

Figure 4. Profile of mean stress and hydraulic conductivity of KD-89.

the small deformation at the place far from the drift of KD-88 and the large hydraulic change at the place far from the drift of KD-89.

Figures 3 and 4 show the profiles of the mean stress and mean hydraulic conductivity of KD-88 and KD-89. As the above discussion, the stress field and deformation of fractures induced by that give the change of hydraulic conductivity distribution. Thus, it is difficult to identify the clear relation between stress components and hydraulic conductivity distribution as shown in the previous our study (Kobayashi, et. al., 2000). It is, however, seen as shown in Figures 3 and 4 that the mean stress distribution has a good correlation with the hydraulic conductivity. The mean stress is related to the volumetric change. In the figures, the positive stress is the compressive one. Thus, it is found that the low mean stress coincides with the high hydraulic conductivity. The mean hydraulic conductivity at the initial state is 1.67×10^{-9} m/s (= -8.8 in log(m/s)). The initial distribution is homogeneous. Thus, it is found that the upper and lower parts of the KD-88 drift have a high permeability and the horizontal sides of the KD-89 drift become a high permeable part.

Table 6 shows the change in vertical earth pressure at KD-88. The measured change is larger than the calculated one. Table 7 indicates the strain caused by excavation. The measured strain is smaller than the calculated one. It is found from those results that the stiffness of the rock mass is estimated to be small in comparison with the real one. However, the fracture parameters used in the model, i.e., JRC, JCS, are set at the maximum values for the steep fractures. Thus, if the stiffness is considered to be larger than the present model, the information on fracture density and fracture length should be modified to be small values. The data shown in Table 1 is, however, obtained from the in-situ geological investigation. Thus, it is inferred that the fractures contributing to the mechanical behavior is limited and the number of fractures important from the mechanical aspect is small in fact.

6 CONCLUSIONS

The hydraulic conductivity change due to excavation is examined through the measured and calculated results. It can be concluded from the examination as followings;

1) The reduction of the hydraulic conductivity is occurred at the lateral side of the drift in the case where the dominant fractures are perpendicular to the drift direction. On the other hand, in the case that the dominant fractures are parallel to the drift, the increase of hydraulic conductivity may be occasionally occurred.

2) The above change in hydraulic conductivity is caused from the behavior of fracture system.

The steep fractures work to increase the hydraulic conductivity of the rock mass, while the horizontal fractures contribute to the reduction of it.

3) The deformation in the drift direction has a large effect on the change in hydraulic conductivity. Thus, the three-dimensional analysis should be necessary to examine the mechanical and hydraulic coupling problem.

4) The earth pressure also has a large effect on the results. Thus, the three-dimensional consideration of the earth pressure should be incorporated into the examination.

5) The stiffness of the rock mass is effected by the mechanical and geometric characteristics of the fractures. The fractures contributing to the mechanical behavior may be limited.

Table 6. Earth pressure change at KD-88 (MPa).

KD-88		
Distance	Measured	Calculated
1 m	4.7	0.97
2.5 m	5	0.13

Table 7. Strain due to excavation (μ).

KD-88			
Direction	Drift	Vertical	Horizontal
Measured	-75	220	-190
Calculated	-310	608	-287
KD-89			
Measured	-30	-120	-350
Calculated	-54	-909	-897

REFERENCES

Barton, N.R., Bandis, S.C. and Bakhter, K. 1985. Strength, Deformation and Conductivity Coupling of Rock Joints, *Int. J.Rock M.M.S.G.A*, 22,3, 121-140.

Oda, M. 1986. An Equivalent Continuum Model for Coupled Stresses and Fluid Flow Analysis in Jointed Rock Masses, *Water Resour. Res.*, 22,13, 1845-1856,

Kobayashi, A., Hosono, K., Fujita, T. and Chijimatsu, M. 2000. Coupled hydraulic and mechanical model with nonlinear elasticity of fractured rock mass, *GeoEng2000*

Modern Tunneling Science and Technology, Adachi et al (eds), © 2001 Swets & Zeitlinger, ISBN 90 2651 860 9

Displacement measurement in a two-dimensional tunneling experiment.

N. Dolzhenko
URGC, Géotechnique, INSA de Lyon, France

Ph Mathieu
URGC, Géotechnique, INSA de Lyon, France

ABSTRACT: Ground movements are inevitably caused by bored tunnel construction in soft ground. In urban areas their potential effects on buildings are an important in the construction of tunnels.
This paper investigates experimental study of a circular tunnel with the analogue Taylor-Schneebeli material consisting of metallic rods of 3 to 5 mm diameter. This work is to obtain a better simulation of soil movements due to construction of shallow tunnels, dug with pressurized shield tunneling boring machine, inducing in the soil mass, successive deconfining, reconfining cycles. These different cycles are due to overcutting, to the conicity of the boring machine, to grouting and to grout consolidation. For these reasons, we decided to reproduce these different phases with a small-scale two-dimensional model.

1 INTRODUCTION

Ground movements are caused by bored tunnel construction in soft ground. In urban areas their potential effects on buildings are an important in the construction of tunnels.

In order to better include the behavior of the soil around and on the ground tunnel surface, the several authors carry out experiments on the small-scale model. Many experimental research works analyzed the deformation in transverse sections of tunnels (Marshall [96], Mair [93]). It is now possible to obtain accurate measurements of deformations with digital camera. In our work we are studying the small-scale model, for this reason we have used digital image techniques, complete displacement and strain fields were obtained within the entire soil mass.

2 EXPERIMENTAL TEST

2.1 Small scale experimental model

The experimental model of tunneling has a cylindrical form, which is 200 mm in diameter. It consists of several metal parts allowing a variation of diameter of approximately 12 mm while preserving its cylindrical form. The weight of the tunnel is 14kg, which corresponds to the weight of the analogue soil.

The material consists a mixture of metallic rods of 3, 4 and 5 mm diameter, 60mm long and represents a two-dimensional analogue soil.

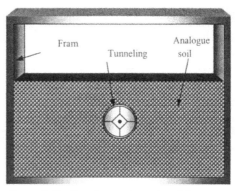

Figure 1. Small scale experimental model.

The specific weight of the mixture is: $\gamma=65$ kN/m^3. With an angle of friction $\varphi=23°$, cohesion $c = 0$, void ratio $e = 0.206$. Our material mechanical properties are close to dense sand.

The frame is 2 m of width and 1.5 m of height (Fig. 1) are made of structural steel sections.

The displacement field in the soil mass is obtained from digital pictures using correlation techniques (software Sifasoft, Mguil_Touchal & al. [2]). It is based on the correlation of the distribution of the level of gray (which exists naturally in the small rods of Schneebeli) between two digital images. The accuracy of measurements is about 0.1mm.

The image processing consist of the following phases:

♦ Imagery using the digital camera.

- ◆ Calculation of the displacements field with software Sifasoft.
- ◆ Calculation of the strain field using software Deplac (Al Abram and Mathieu[98]).

2.2 *Deconfining and reconfining phases.*

The digging of tunnel with pressurized shield tunneling boring machine, induce in the soil mass, successive deconfining, reconfining cycles of the ground.

According to experimental research done by Ollier [97], the three significant phases during the advance of the tunneling are as follows :
1) Digging with overcut, to conicity of the boring machine (reduction in the diameter of the tunneling).
2) Injection of the grout in the annular space behind the segments (increase in diameter).
3) Consolidation of the grout (second reduction in diameter).

In order to realize our search works, we decided to reproduce, with a small-scale model, these different phases in order to measure the induced soil movements. By using the digital camera we carried out a succession of digital images corresponding to the phases of digging.

We defined the loading plot. A schematical diagram of the test is shown on the figure 2:

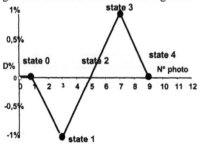

Figure 2. Loading plot.

The deconfining phase (State0-1) corresponds to reduction of the initial diameter of tunneling ($\Delta D=-\Delta D_0$), The reconfining phase (State1-3) corresponds to a growth of diameter ($\Delta D=+2\Delta D_0$) of tunnel and the final phase (State 3-4) is the second decreasing cycle ($\Delta D=-\Delta D_0$), where $\Delta D_0 = 1\%D$ (D=200mm).

3 GROUND MOVEMENTS

3.1 *Displacement field*

We present the results of displacement field of the state 0-1 and of state 2-3, corresponding to the same absolute variation of diameter. The figures 3 and 4 show the iso-displacement field. We note that the displacement fields are not quite symmetrical for

both cases compared to the vertical axis of tunnel, due to slight horizontal movements of the center of tunnel. The width of iso-displacements field in the surface of state2-3 (Fig. 4) almost is 1,5 times broader than state 0-1 (Fig. 3). The iso-displacement curves are push closer together at the tunnel level.

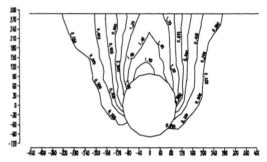

Figure 3. Displacement fields (State 0-1).

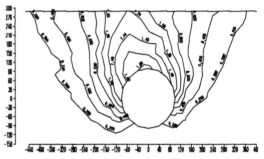

Figure 4. Displacement fields (State 2-3).

3.2 *Strain fields*

The maximum shear strains $(\varepsilon_1-\varepsilon_2)/2$ (where ε_1 is major principal strain, ε_2 is minor principal strain in absolute value) for the state 0-1 and state 2-3 are very similar and shown in figure 5. The main shear strains are concentrated in two strips localized on the two sides of tunnel. The maximal strain value is around the tunnel and strains reduce towards the surface.

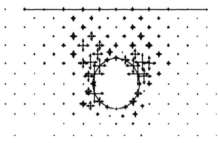

Figure 5. Maximum shear strain for the state 0-1 and state 2-3.

134

3.3 Volumetric strain

The volumetric strain ($\varepsilon_v=\varepsilon_1-\varepsilon_2$) is presented in figures 6 and 7. If $\varepsilon_v>0$ the mass of soil is contraction, if $\varepsilon_v<0$ the mass of soil is dilatance. We note that the analogue soil for the state 0-1 is contraction on the sides of tunnel and in the surface. The rest of the ground shows a dilatance behavior above the tunnel.

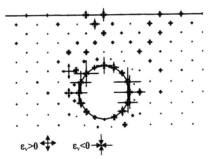

$\varepsilon_v>0$ ✛ $\varepsilon_v<0$ ✖

Figure 6. Volumetric strain. State 0-1.

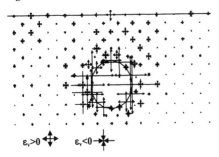

$\varepsilon_v>0$ ✛ $\varepsilon_v<0$ ✖

Figure 7. Volumetric strain. State 2-3.

On the other hand, the volume change for the increasing phase state 2-3 (Fig.7) is opposite to state 0-1. The zones which were in contraction for the $\Delta D = -\Delta D_0$, in case of $\Delta D = +\Delta D_0$ these zones are becoming dilated, and the rest of soil is in contraction.

The presented results agreed with many experimental researches on the small scale model. (Rogers [92]).

4 GROUND SURFACE MOVEMENTS

4.1 Ground surface parameters.

Following the analysis of many case histories, Peck [69] concluded that the surface settlement profile was well described with the Gauss formula:

$$S_x = S_{max} \exp\left[\frac{-x^2}{2*i^2}\right]$$

Where :
- S_x is the settlement at distance x
- x is the horizontal distance from the center line
- S_{max} is the maximum settlement over the tunnel centerline

- i is the horizontal distance from the tunnel centerline to the point of inflexion (figure 8):

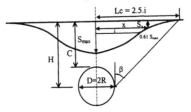

Figure 8. Parameters of settlement.

4.2 Experiment.

In order to analyze the surface settlement, we decided to compare different states with the same variation of diameter.

We investigated the effect of the different phases on typical value (S_{max}, S_{crown}, i, V_s/V_t) examine their influence on the ground movements produced. We can observe that curves are similar in form to Peck formula and almost symmetrical compared to the tunnel centerline.

4.3 Ground surface movement curves

➢ State 0-1 and state 3-4

Figure 9 illustrates the ground surface settlements for two states, caused by the deconfining operations. Plotting the vertical displacement showed us that the slight differences between the two curves.

The curve for the state 3-4 is larger and deeper to the state 0-1. The ground loss of state 3-4 is 1.1 times larger to state 0-1. We note that the same variation of the diameter of tunnel in successive phases does not provoke the same ground surface movements, due to down movement a base of tunnel.

The diagram shown in Figure 10 illustrates the settlement changes for the two confining states.

➢ State 1-2, state 2-3 and state 0-3

State 1-2, 2-3 and 0-3 have the same value of confinement, but this results from the experiment curves

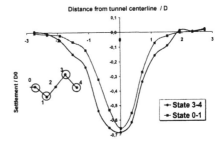

Figure 9. Surface settlements. State 0-1 and state 3-4.

show the differences between the curves is probably due to settlement irreversible of analogue soil mass. The maximum displacement value in the surface for the state 1-2 and state 0-3 is almost similar, but the ground loss value is larger for the state 0-3.

The ground surface value for the State 2-3 is $1.33D_0$ larger to state 1-2. Consequently, the volume loss for the State 1-2 is about twice smaller compared to the volume loss for second State 2-3.

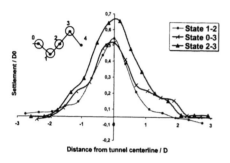

Figure 10. Surface mouvements.

➢ State 0-2, State 2-4 and State 0-4.

Equally, we investigated the surface movement when the diameter of tunnel finds again his initial value. Three phases are presenting in the figure 11.

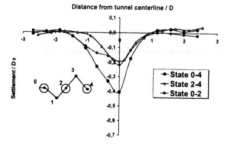

Figure 11. Surface settlements.

Residual settlement (State 0-2 and State 2-4) is in the order of 1/3 of settlement of previous phases. We note that the distribution of settlement ground movement for the state 0-4 almost twice deep to state 0-2 or 2-4, therefore the soil masse pack down. The ground settlement of state 0-4 is sum of value of state 0-2 and state 2-4.

5 CONCLUSION

The field measurements obtained by small-scale model presented in this paper can be used to predict the behaviour of future similar tunnel constructions. The experiment work shows the results for change from 1% of diameter of tunnel. The previous results drive us to the following conclusions:

♦ It is possible to determine the displacement field in the whole soil mass in such an experiment. The results can be used to validate a computation technique. These two approaches will be compared in another paper to appear soon.

♦ In case of increasing diameter phase, the surface displacement profile is twice wider than in case of decreasing phase.

♦ The same variation of the diameter of tunnel does not provoke the same ground surface movements due the tunnel movements.

REFERENCE

Al Abram, I. (1998). Etude sur modèle réduit bidimensionnel du champ de déplacement induit par le creusement d'un tunnel à faible profondeur. Thesis, INSA de Lyon.
Al Abram, I. Mathieu, Ph. & Kaster, R. (1998). Mesure de déplacements par Imagerie Numérique dans un modèle réduit simulant le creusement du tunnel. *Collique photomécanique* 98, Marne la Vallée, 56-72.
Mguil-Touchal, S. et al (1996). Mesure de champs de déplacements et de déformations par corrélation d'images numériques. *Proceeding of the Mécamat'96 - Mécanisme and Mechanics of the Large stains*, Aussoie-France, 179-182.
Marshall, M.A. et al. (1996). Movements and stress changes in London Clay due to the construction of pipe jack. *"Geotechnical Aspects of Underground Construction in Soft Ground."* Mair R.J & Taylor R.N. (end), Rotterdam: A.A. Balkema, 719-724. ISBN 0727719165
Mair, R.J. & Taylor, R.N. Prediction og clay behaviour around tunnels using plasticity solutions. *"Predictive soil mechanics"* Wroth P., Houslby G.T., … (eds), London: Thomas Telford, 449-463. ISBN 0727719165
Peck, R. B. (1969). Deep excavations and tunneling in soft ground. *Proceeding of the 7th International Conference on Soil Mechanism References* Foundation Engineering. Mexico, vol.3, 225-290.
O'Reilly, M.P. & New B.M. (1982). Settlements above tunnels in the United Kingdom – their magnitudes and prediction. *Tunneling' 82,* London,173-181.
Ollier, C. (1997). Etude expérimental de l'interaction sol machine lors du creusement d'un tunnel peu profond par tunnelier a pression de boue. Thesis, INSA de Lyon.
Rogers, C.D.F. & Yonan, S.J.S. (1992). Experimental study of a jacked pipeline in sand. *Tunnels & Tunneling*, June, 35-38.

Modern Tunneling Science and Technology, Adachi et al (eds), © 2001 Swets & Zeitlinger, ISBN 90 2651 860 9

Centrifugal model test on the clay arching

W. Binglong, Z. Shunhua, G. Quanmei& Y. Longcai
Dept. of Railway and Architecture, TongJi University, China

ABSTRACT: This paper takes a sewage tunnel as an example, which crosses the Huangpu River in Shanghai. The pipe of the tunnel is made of prefabricated reinforced concrete and the outside diameter is $\Phi 2500mm$. The centrifugal model tests simulate the slump caused by the unbraced excavation of single clay hole and double holes. The effect of arching is analyzed through displacement field. The deformation during the braced excavation of double holes is analyzed and the superimposed soil pressure over the braced structure is measured.

The experiment results show: during the excavation of single unbraced hole, the earth arch will emerge automatically. The arch is 5.1m high. While the excavation of the double unbraced holes, the thickness of superimposed soil is at least 7.0m. When the tunnel is braced, the arch will not emerge.

1 INTRODUCTION

The effect of soil arching exists universally in the nature as a kind of mechanic phenomenon. More and more attention has been put to the effect of earth arching in some large projects that involve the excavation. Project costs can be saved by making use of the effect of earth arching (Handy L, 1985; Wu Zhishu, 1995).

In recent years, lots of underground engineering projects have been built in Shanghai. In this sewerage project, one or two parallel sewage pipes were used in many segments. The pipe is made of prefabricated reinforced concrete. Its outside diameter is $\Phi 2500mm$. The net space between two pipes is 1.28m. They cross the second soil layer on which is the first layer (mucky clay). The major physical and mechanical properties are given in Table 1. Whether the layer II can be arched and the effect of arching can be made use of so as to decrease the cost of construction is very important.

2 THE DESIGN OF THE CENTRIFUGAL MODEL

The curve of stress-strain relation for all points in the centrifugal model is the same as the one of stress--strain relation for all points in the prototype. When making the model, if the geometrical size of the prototype is diminished N times while the acceleration of gravity amplified N times, the stress of the soil in the model is the same as that in the prototype. The soil in the model is undistorbed soil sample from the site. It is mixed, remolded in the agitator, and consolidated by layers in the centrifuge. The simulation of soil is to control the water content and strength index of soil.

The deformation of pipe system is affected by the rigidity of the pipes. The deformation of pipes affects the pressure caused by super-imposed soil. The simulation of pipes is based on its bending deformation capacity under soil pressure. The models are designed according to the equivalent rigidity standard (Zhou Shunhua, 1992). The

Table 1. Major physical and mechanical properties of the stratum.

Soil strata	Soil name	Water content ω (%)	Void ratio e	Unit weight γ (KN/m³)	Cohesion C (KPa)*	Angel of internal friction $\psi(^{0})$*	Plasticity index I_p
I	Mucky clay	45.3	1.226	17.3	10.5	9.0	19.0
II	clay	33.4	0.940	18.8	24.0	12.0	20.1

* is the result of direct shear test

relation between the prototype and the model must satisfy the following equations:

$$\frac{E_p I_p}{E_m I_m} = C_E C_L^3 \qquad (1)$$

$$C_L = \frac{L_p}{L_m} = N \qquad (2)$$

$$I = t^3 / 12 \qquad (3)$$

Where:
E = elastic modulus of pipe material
I = the second moment of inertia for pipe material of unit length
C_E = similitude coefficient of elastic modulus
C_L = geometrical similitude coefficient
N = the model ratio
t = thickness of pipe wall.
Subscripts "p" and "m" represent prototype and model respectively.

Through computation, the model ratio is determined to be 65. Steel pipes are used to simulate the bracing structure. The steel pipes are 2mm thick. The outer diameter is 38mm.

3 THE SIMULATION AND THE ANALYSIS OF RESULTS

The enough thickness of superimposed layer and certain deformation of the arch are the important conditions for arching. The test can determine the conditions causing the arching, the shape and range of loosening area. The excavation of every hole is finished at the same time.

3.1 The test of single-hole

Before the experiment, the soil is consolidated by layers to simulate the layers I and II until physics and mechanical properties of simulated soil approximates those of the field soil. The excavation of single hole is finished at one time. During the tests, a photo is taken every time, the acceleration of gravity increases 5g and closed circuit TV is used to monitor the test.

3.1.1 Tests of the first group of single holes

Fig.1 shows the simulation. The layer II is 9.1m thick. The thickness of superimposed soil of layer II on the top of the single hole is 5.2m , and layer I on II is 9.2m. The total thickness of the simulated soil is 18.3m.

Fig.2 shows the relation between acceleration and vertical displacement of soil on the single hole. During the experiment, when the N ranges from 5g to 25g, the displacement of top of the hole is slight.

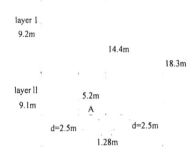

Figure 1. The simulations of first and second group of tests (single and double holes).

When N=30g, the displacement begins to be obvious, and the radial displacement emerges from the top of hole to earth's face.layer I II

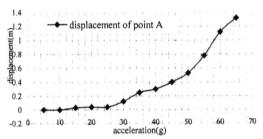

Figure 2. The relation between vertical displacement on the top of the hole and acceleration .

The displacement of all points inclines mostly toward the hole. With the increment of the acceleration and displacement, sliding plane emerges in the soil and develops gradually. At last, a steady earth arch forms. The arch is 5.1m high, which is shown in the Fig.3.

Figure 3. Tests of first group of single hole (65g).

When the acceleration ranges from zero to 65g, the deformation of soil under the bottom of arch is obvious. There is only comparatively obvious deformation near the hole in the soil between the top and the bottom of the hole.

3.2 *Tests of double holes*

The second group of tests simulates the case in which the double holes cross the soil of layer II, Fig.1 shows that another hole is excavated symmetrically on the basis of first group of single hole. When N=30g, the bigger deformation appears on the right side of the hole. The vertical displacement is 0.325m. When N=65g, the diameter of the hole reduces to half. The vertical displacement on the top of the hole is 1.3m. There is no obvious displacement at the bottom of hole. The displacement of points above the bottom of the hole basically develops toward the hole. The vertical central axis between two holes is the symmetric axis of displacement. The groove emerges on the border between two layers. The bottom of groove is placed in the symmetric axis of two holes. Beyond the distance of 5.2m from the symmetric axis, the vertical displacement of border between layers I and II is approximately equal to zero. Crack in the soil can be found. The combination of left and right crack is just like a symbol "()". The bottom of the "()" is just on the bottom of hole. The top of the "()" is just on the border between layers II and I. Because layer I is very soft, cracks can not develop upward any more. The excavation of the hole affects the superimposed layer, stable arch will not emerge in the layer II which is 5.2 m thick upon the top of hole. It is shown in Fig.5. The third group of tests is done to determine the necessary thickness of superimposed soil in layer II, which is important to the arching.

Enough thickness is necessary conditions to form stable natural arch of double holes in the soil of layer II. The two holes are excavated together and finished at the same time. The thickness of layer II is 13.6m. The thickness from the top of holes to border of two layers is 8.6 m. The thickness of layer I is 7.9 m. The simulation is shown in Fig.4. The results of test show that natural arch of double holes emerges in layer II. The natural arch of double holes is 7.0 m high.

4 TESTS OF THE HOLES BRACED BY PIPE SLICES

The fourth groups of tests simulate the case in which the hole is braced by pipe slices. Steel pipes are used with the outside diameter of 38mm to simulate prefabricated reinforced concrete pipe with the diameter of 2500mm. Two earth pressure cells are

Figure 4. The simulation of the third group of tests.

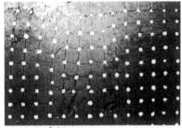

Figure 5. Test of the second group of double-hole (65g).

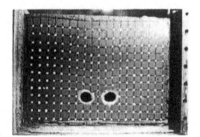

Figure 6. Test of the third group of double holes (65g).

placed just on the top of pipes to measure the variation of soil pressure. The place and embedment of two parallel sewage pipes is the same as those in the second group. The thickness from the top of sewage pipes to the border of two soil layers is 5.2 m and the thickness of layer I is 9.2 m, which is shown in Fig. 1. During the tests, the displacement is so slight that it can be neglected. The measured pressure at the top of pipes is approximately equal to the computed stress caused by the own weight of the soil. It shows that arching can't form in clay with pipe slices braced.

4 CONCLUSIONS

(1) Enough thickness of superimposed soil and certain deformation are important conditions for arching.

(2) The stable natural arch emerges in the layer II. During the excavation of single hole, the height of arch is 5.1m.

(3) In the layer II, the excavation of double holes needs the thickness of superimposed layer II to be at least 7.0 m .

(4) When there is bracing structure, the earth pressure at the top of the hole is basically equal to the own weight of the superimposed soil. In this case, arching can not emerge.

REFERENCES

Handy L. The arch in soil arching ASCE. 1985. 111(3). 302~318.

Wu, Zhishu. The study of arching shape in soil, Geotechnical Mechanics and Engineering. *Dalian University of Natural Sciences and Engineering:* 1995. 404~409.

Zhou Shunhua. The study on principle of similitude on the centrifugal model tests, *Shanghai Tiedao University*, 1992(4), 20~22.

Modern Tunneling Science and Technology, Adachi et al (eds), © 2001 Swets & Zeitlinger, ISBN 90 2651 860 9

Arching effects on the deformation of sandy ground induced by tunneling

T. Honda
Dept. of Civil Engineering, University of Tokyo, Tokyo, Japan

T. Hibino
Dept. of Civil Engineering, Tokyo Institute of Technology, Tokyo, Japan

A. Takahashi
Dept. of International Development Engineering, Tokyo Institute of Technology, Tokyo, Japan

J. Kuwano
Dept. of Civil Engineering, Tokyo Institute of Technology, Tokyo, Japan

ABSTRACT: Centrifuge model tests on shallow tunnels in sand were carried out to investigate ground deformation caused by tunneling. An image processing system was introduced to measure displacements of the targets accurately which were attached in the model to observe ground deformation. This system can measure displacements of the targets with the accuracy of 0.1 mm or better in these tests. Strains in the model are calculated from them. Through the observation of ground deformation, it was found that arching in the ground was formed by dilation with shear deformation and was not formed until some deformation developed, that the deformation when arching was formed depended on the density of sand around a tunnel, and that an arch was made even in a shallow tunnel.

1 INTRODUCTION

Recent excessive land use within urban areas has lead to an increase in the number of tunneling projects for service and transportation purposes. This trend will be accelerated in the future. Underground construction in soft ground and in a land adjacent to other structures will be necessary. However tunnel excavation inevitably cause ground deformation and may affect existing structures near the tunnel. It is important to predict ground deformation and take measures to it. Many researchers have studied predicting ground deformation, for instance settlement troughs near the ground surface have empirically proposed by Peck (1969), O'Reilly & New (1982), Mair et al. (1993) and Kuwano et al. (1999). Settlement troughs proposed from measurement data in situ and model tests have been classified by soil material and a ratio of tunnel cover to tunnel diameter. If the tunneling is thought to be made under undrained condition such as in clay, the ground deformation can be empirically predicted by estimating a volume loss at the excavation. In granular materials, it is known that the volume loss at the surface is smaller with the deeper tunnel cover. This phenomenon would result from arching in ground. In general, arching is known to have effects to reduce earth pressure on a tunnel wall. Many researchers have been extensively studied the effects of arching on earth pressure by measuring earth pressure and bending moment of tunnel walls in situ and model tests. However the arching effects on ground deformation have not been investigated enough because it is dif-

ficult to measure deformation in the ground. In the model tests, photographs and X-ray have been used to measure co-ordinates of targets in model ground. However the accuracy of these methods is not high enough to observe the strain accurately, for the co-ordinates of targets are determined from a manual process. As a new technique, image processing systems with CCDTV cameras have been recently used to measure displacements precisely in model grounds (Taylor et al. 1998). This system can capture images continuously and measure the co-ordinate of targets automatically.

In this study, an image processing system with a CCDTV camera was introduced to measure the ground deformations accurately in centrifuge model tests which were carried out to investigate ground deformation caused by tunneling. This image processing system made it possible to measure not only displacements in the model but also ground deformation such as shear strains, principal strains and volumetric strain by using the displacements of the targets arranged as a grid. From the observation of ground deformation, deformation mechanism of sandy ground around tunnel was investigated.

2 TEST PROCEDURE

2.1 Centrifuge model test

A series of plane strain centrifuge model tests was carried out. Figure 1 illustrates a schematic diagram of centrifuge model. Dry Toyoura sand with relative density of about 80 % was used as the model ground

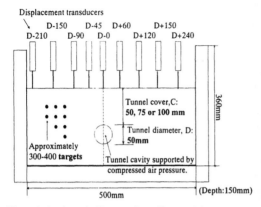

Figure 1. A schematic diagram of centrifuge model.

Figure 2. A image captured from CCDTV camera (TS-2: at failure).

Table 1. Property of Toyoura sand.

Gs	2.645		
D_{60} (mm)	0.21	D_{50} (mm)	0.20
D_{30} (mm)	0.19	D_{10} (mm)	0.18
e_{max}	0.972	e_{min}	0.605

Table 2. Test program and summary of results.

No.	C/D	Pluviation	Dr %	σ_{TF} kPa	$S_{max,F}$ mm
TS-1	2.0	Vertical	77.9	---*	0.45
TS-2	1.5	Vertical	82.8	7.4	1.23
TS-3	1.5	Horizontal	78.5	2.8	0.50
TS-4	1.0	Vertical	78.4	6.8	0.79
TS-5	1.0	Horizontal	82.9	5.2	0.46

* TS-1 did not collapse since tunnel support pressure did not decrease below 10.1 kPa for the trouble of test apparatus.

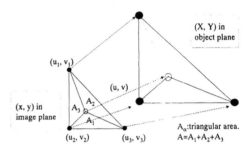

Figure3. Shift of co-ordinates from image plane to object plane.

(Table 1). Model tunnel had a diameter of 50 mm. It was lined with a rubber bag and a thin plastic sheet with thickness of 0.06 mm which is used for OHP transparency. Air pressure was supplied in the rubber bag to keep the tunnel against the weight of sand around and the sheet was rolled around the rubber bag to prevent the tunnel from inflating bigger than 50 mm. Model grounds were prepared by air pluviation method from above a container ("Vertical" in Table 2) or from the front of it with removing the front window ("Horizontal"). Model tunnels were set with tunnel cover of 100, 75 or 50 mm. Influence of the methods of model ground preparation and the thickness of tunnel cover on ground deformation were investigated. Test program is shown in Table 2.

The model was subjected to a centrifuge acceleration of 100 G. Air pressure was supplied to the tunnel cavity to balance the weight of overburden with the increase of centrifuge acceleration. Upon reaching equilibrium, air pressure was reduced to simulate the excavation process at a rate of 20 kPa/min. Settlements at the ground surface were measured by displacement transducers, and tunnel support pressure was measured by a pore pressure transducer. Pattern of displacements and strains in the ground

were determined from images of targets obtained by CCDTV camera which was set at the front of the container.

2.2 Image processing system

Figure 2 shows an image captured by CCDTV camera. Two types of targets were prepared, i.e. those for reference and for displacement measurement. The reference targets were buried in the Perspex window. The measurement targets were painted on the thin rubber membrane to be a grid of 10mm. This membrane was placed between the front vertical face of sand and Perspex window with silicone oil for lubrication. Pins were glued on the membrane behind the targets to ensure the targets moving together with the model ground.

Image Processing System which was introduced in these tests consists of two processes. The first is to measure the co-ordinates of targets on an image plane. The second is to shift the co-ordinates on the image plan to those on the object plane of the model ground. The former was developed in Geotechnical Engineering Research Centre, City University in UK. The captured images consist of 768×576 pixels. The center co-ordinates of targets on the image plane can be calculated with a resolution of 1/10 pixel by fitting a certain curve to brightness of pixels in a target. The observation area of the model ground captured by CCDTV camera was about 320 × 240 mm, the size of a pixel was about 0.4-0.5 mm. Accordingly, the co-ordinates of targets in the model were calculated with the resolution of 0.04-0.05 mm. The co-ordinates in the image plan are shifted to the co-ordinates in the object plane as equation (1).

$$X = x + u, \quad Y = y + v \tag{1}$$

where x and y are co-ordinates on the image plane, X and Y are ones in objective plane, u and v are for displacement adjustments. Adjusting displacements are calculated with 3 reference targets surrounding the measurement targets by equation (2).

$$u = \sum_{\alpha=1}^{3} \frac{A_\alpha}{A} \times u_\alpha, \quad v = \sum_{\alpha=1}^{3} \frac{A_\alpha}{A} \times v_\alpha \tag{2}$$

where u_α and v_α are adjusting displacements of the reference target, A_α / A are area co-ordinates (Figure 3). From the co-ordinates of the targets, displacements were obtained and strains were calculated from the displacements of the targets by using isoparametric element with 4 nodes.

The accuracy of the co-ordinates obtained from the above processes was lower than the resolution of 0.04-0.05 mm, because image data were slightly disturbed by a noise when the data was sent to outside of centrifuge model test apparatus. However, the accuracy was 0.1mm or better in displacement and less than 1% in strains of the grid of 10 mm. It was possible to catch ground deformation before a failure of the tunnel.

3 RESULTS AND DISSCUSSION

3.1 Settlement curves

Summary of the results in centrifuge model tests is shown in Table 2, where σ_{TF} is a tunnel support pressure and $S_{max,F}$ is a ground surface settlement above a tunnel axis at failure. The failure was determined by the temporary increase in the tunnel support pressure, because the collapse of the tunnel wall caused momentary increase of pressure which was beyond discharge ability of an air regulator. TS-1 did not collapse since tunnel support pressure did not decrease below 10.1 kPa for a trouble of test apparatus and was stopped on the way. $S_{max,F}$ in TS-1 is the final settlement at the σ_T of 10.1kPa.

Figures 4 and 5 show settlements at the surface above a tunnel axis, S_{max}, and at the tunnel crown, S_c. Settlement curves at the surface were almost same until σ_T of about 10 kPa except for TS-2, and then TS-2 and 4 whose sand was deposited vertically from above the container induced sudden large settlement until a failure. On the other hand, TS-3 and 5 whose sand was deposited horizontally from the front of the container induced less settlement. Similarly, settlement curves at the crown in Figure 5

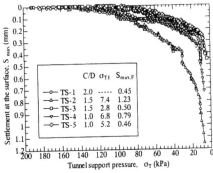

Figure 4. The surface settlements with the reduction of tunnel support pressure.

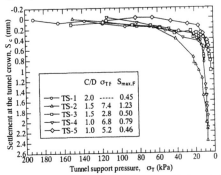

Figure 5. Settlements at the tunnel crown with the reduction of tunnel support pressure.

showed almost same trends. In TS-2, settlements at the surface and tunnel crown gradually developed from early stage of the reduction in tunnel support pressure and large settlements were observed at the failure. It was found from the volumetric strain in the model with the increase of centrifuge acceleration, that TS-2 had loose sand area around the tunnel, and that the loose sands caused large settlements. This is discussed more in 3.2. It was also found that the tunnel support pressure at failure was much lower than overburden pressure and were almost same in all the tests, though tunnel covers were different. It was presumed that the difference in the initial density of the ground due to the direction of sedimentation influenced the tunnel support pressure at failure rather than the difference in the tunnel cover, since σ_{TF} in TS-3 and 5 were lower than in TS-2 and TS-4. Moreover, judging from settlement curves to tunnel support pressure which was not normalized by overburden pressure, it is assumed that tunnel cover hardly influence settlement curves before σ_T of about 10 kPa.

3.2 Strain distribution

Distributions of the maximum shear strain at failure of TS-2 - 5 are shown in Figures 6-9. Some scatter

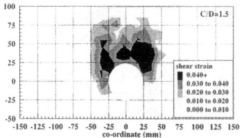

Figure 6. Distribution of the maximum shear strain at failure; $S_{max,F}$ = 1.24 mm and σ_{TF} = 7.4 kPa in TS-2.

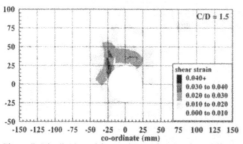

Figure 7. Distribution of the maximum shear strain at failure; $S_{max,F}$ = 0.50 mm and σ_{TF} = 2.8 kPa in TS-3.

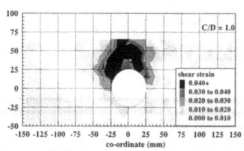

Figure 8. Distribution of the maximum shear strain at failure; $S_{max,F}$ = 0.79 mm and σ_{TF} = 6.8 kPa in TS-4.

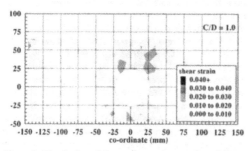

Figure 9. Distribution of the maximum shear strain at failure; $S_{max,F}$ = 0.46 mm and σ_{TF} = 5.2 kPa in TS-5.

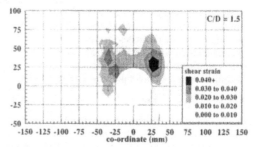

Figure 10(a). Distribution of the maximum shear strain; S_{max} = 0.91 mm and σ_T = 11.6 kPa in TS-2.

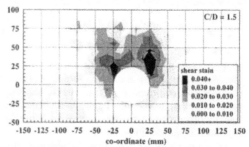

Figure 10(b). Distribution of the maximum shear strain; S_{max} = 1.04 mm and σ_T = 8.7 kPa in TS-2.

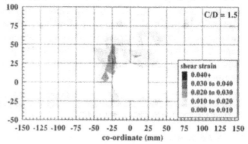

Figure 11(a). Distribution of the maximum shear strain; S_{max} = 0.40 mm and σ_T = 4.7 kPa in TS-3.

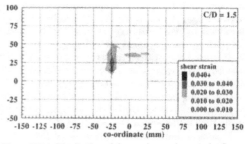

Figure 11(b). Distribution of the maximum shear strain; S_{max} = 0.45 mm and σ_T = 3.5 kPa in TS-3.

was seen in TS-5. This is probably because the quality of images in TS-5 was not sufficient. The shear strain concentrated on above a tunnel and was hardly observed below the tunnel axis in all the tests. TS-2 and 4 whose settlements were large induced the ap-

parent concentration of shear strain. Distributions of shear strain at failure were obviously different among the tests.

Variations of shear strain with the reduction of the tunnel support pressure in TS-2, 3 and 4 are shown

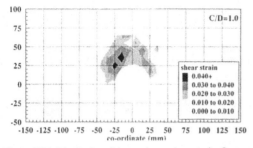

Figure 12(a). Distribution of the maximum shear strain; S_{max} = 0.58 mm and σ_T = 8.9 kPa in TS-4.

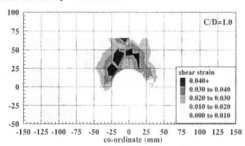

Figure 12(b). Distribution of the maximum shear strain; S_{max} = 0.65 mm and σ_T = 7.8 kPa in TS-4.

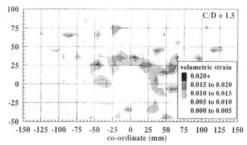

Figure 13. Volumetric strain with the increase of centrifuge acceleration in TS-2.

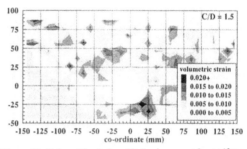

Figure 14. Volumetric strain with the increase of centrifuge acceleration in TS-3.

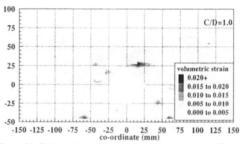

Figure 15. Volumetric strain with the increase of centrifuge acceleration in TS-4.

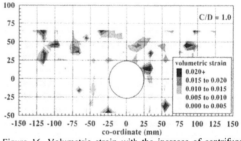

Figure 16. Volumetric strain with the increase of centrifuge acceleration in TS-5.

in Figures 10-12. Shear deformation in TS-3 first developed in the region of around 45° from the horizontal and then was transmitted along a tunnel wall to above a tunnel. The region on which shear strain concentrated formed a shape of arch along a tunnel wall. In TS-4, shear deformation similarly appeared in the region of around 45° from the horizontal first and then transmitted to above a tunnel. The arch was formed in the distant region from a tunnel wall. In TS-2, shear deformation also developed in the region of around 45° from the horizontal and then concentrated on the same region. This distribution of shear strain was similar to it in clay (Kuwano et al. 1999).

The reason why the different distributions of shear strain were observed was thought to be due to the initial conditions of the model grounds. To compare the initial conditions of the model grounds, volumetric strain with the increase of centrifuge acceleration from 1G to 100G were evaluated as shown in Figures 13–16. With the increase of centrifuge acceleration, a soil element in the model ground is compressed by the weight above it. If the model ground has homogeneous density, the volumetric strain should be larger with the depth. In other words, if the model ground is not homogeneous, volumetric strain should be larger locally in the region of loose sand. It is seen in Figure 13–16 that there are the regions of large volumetric strain around a tunnel wall. If sand was poured from above the container as in TS-2 and 4, the sand seemed to collide with the tun-

nel wall and reflect. Then, the region above the tunnel became looser and positive volumetric strain concentrated on above the tunnel. In TS-3,4 and 5, local volumetric strains around the tunnel were slightly observed but the sand above the tunnel was

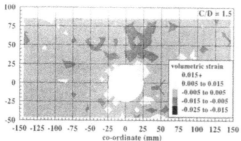

Figure 17. Distribution of volumetric strain at failure; $S_{max,F}$ = 1.24 mm and σ_{TF} = 7.4 kPa in TS-2.

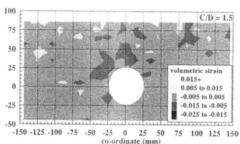

Figure 18. Distribution of volumetric strain at failure; $S_{max,F}$ = 0.5 mm and σ_{TF} = 2.8 kPa in TS-3.

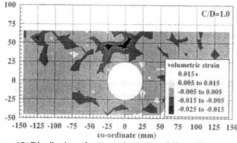

Figure 19. Distribution of volumetric strain at failure; $S_{max,F}$ = 0.79 mm and σ_{TF} = 6.8 kPa in TS-4.

not as loose as in TS-2. The reason why large settlements developed from the early stage in TS-2 will be due to loose sands above the tunnel. In addition, comparing shear strain at the failure and volumetric strain with the increase of centrifuge acceleration, shear strain concentrated on the region where the larger volumetric strain was observed. The following mechanism of ground deformation was presumed from the shear strain and volumetric strain distribution above. With the reduction in tunnel support pressure, shear deformation initiated at 45° from the horizontal. The dense sand in this region dilated by shear deformation. It resulted in the increase in the confinement of soil elements. Therefore, the progress of deformation in this region was prevented and the surface settlements were restricted. Then, the area of shear deformation shifted to above the tunnel. Repeating this process, the region of large shear

deformation became an arch as seen in Figures 7 and 8. Moreover, confining stress by the dilation was transmitted along the arch from above the tunnel wall to both sides of the tunnel. The weight above the tunnel was, therefore, thought to be supported by the arch. As the result, deformation appeared only inside of this arch and the settlements at the surface were restricted until this arch collapse at the failure. If the sand in the ground is loose, the above arch mechanism based on the soil dilation may not be seen as in TS-2.

Finally, volumetric strain distributions at failure were shown in Figures 17–19. Some dilatant volumetric strain was observed around a tunnel wall in TS-3 and around distant region from a tunnel wall in TS-4. These areas coincided with the region where the larger shear deformation was observed. In TS-2, contracted volumetric strain was observed around a tunnel wall where the loose sand existed.

4 CONCLUSION

From centrifuge model tests on a shallow tunnel in sand, the followings were found.

1. If the density of sand is homogeneous, the settlements at early stage of the reduction in tunnel support pressure hardly depend on the thickness of tunnel cover. They are rather influenced by the density of sand around the tunnel.

2. Shear deformation initiated in the region of 45° from the horizontal. Close to a failure, shear deformation concentrate on the same region in loose sand and large settlements are observed. In dense sand, shear deformation shifts to above the tunnel and forms a shape of arch and the settlements are not large.

5 REFERENCES

Kuwano, J., A. Takahashi, T. Honda and K. Miki 1999. Centrifuge investigation on deformations around tunnels in nailed clay. *Proc. Geotechnical Aspect of Underground Construction in soft Ground*, 205-210.

Mair, R.J., R.N. Taylor & A. Bracegirdle 1993. Subsurface settlement profiles above tunnels in clay. *Geotechnique 43(2)*, 315-320.

O'Reilly, M.P. and B.M. New 1982. Settlements above tunnels in United Kingdom – their magnitude of prediction. *Proc. Tunneling '82 Symposium. London, Institute of Mining and Metallurgy*, 173-181.

Peck, R.B. 1969. Deep excavations and tunneling in soft ground. *Proc. 7th International Conference Soil Mechanics and Foundation Engineering. State of the art volume*, 266-290.

Taylor, R.N., S. Robson, R.J. Grant, and J. Kuwano 1998. An image analysis system for determining plane and 3-D displacements in centrifuge models. *Proc. Int. conf. Centrifuge 98, Tokyo, Balkema, Rotterdam*, 73-78.

Modern Tunneling Science and Technology, Adachi et al (eds), © 2001 Swets & Zeitlinger, ISBN 90 2651 860 9

Numerical simulation of centrifuge test of trapdoor by 3-D FEM

T. Adachi, F. Oka, M. Kimura, K. Kishida, M. Kikumoto & T. Takeda
Department of Civil Engineering, Kyoto University, Kyoto, Japan

F. Zhang
Department of Civil Engineering, Gifu University, Gifu, Japan

ABSTRACT: In this paper, centrifuge tests that are conducted to investigate the mechanism of trapdoor in a sand ground are simulated with 3-D finite element analyses. In the centrifuge tests, a circle shaped trapdoor is adopted and different ratios of the overburden to the width of the trapdoor are considered. The mechanical behaviors of the trapdoor, such as the load-displacement relationship, the distribution of the stresses on the trapdoor and the deformation pattern of ground are carefully investigated. The 3-D finite element analyses are based on a strain softening constitutive model, which satisfies the uniqueness of the solution in an initial value problem on Valanis's sense (Valanis, 1985). It is found that the numerical analyses conducted in this paper can simulate the mechanism of trapdoor to a reasonable accuracy.

1 INTRODUCTION

In designing a tunnel, the most important facts that should be considered are how to evaluate the earth pressure acting on the lining of the tunnel and the deformation of the surrounding ground. Experiments with trapdoor are often performed in order to clarify the mechanical behavior of the ground and lining in a tunnel excavation, because of the simplicity of conducting a trapdoor test. Research on trapdoor tests in granular materials (Szechy, 1966) was reported in which the pressure upon the trapdoor drops rapidly to its minimum value but then turns to increasing again when the door is further lowered. Ladanyi and Hoyaux (1969) also carried out a series of model trapdoor tests in which the pressure on the trapdoor was measured and the displacement trajectories of the groundmass, which consisted of aluminum rods with a frictional angle of 25°~26°, were photographically recorded.

Adachi et al. (1997) conducted a trapdoor test with centrifuge machine, using a cylindrical model ground filled with dried sand. In the test, particular attention had been paid to the earth pressure acting on the trapdoor at different ratio of the overburden to the diameter of the trapdoor.

In this paper, a constitutive model proposed by Adachi and Oka (1995), which can describe both the strain hardening and the strain softening behavior, is adopted in the finite element analysis in which the

centrifuge tests conducted by Adachi et al. (1997) are simulated.

The purpose of the paper is to propose a suitable estimating method for the trapdoor behavior using finite element analyses based on a suitable elastoplastic constitutive model with strain softening. Based on the results obtained from the analysis conducted in this paper, it is possible to understand the mechanism of tunnel excavation.

2 CONSTITUTIVE MODEL WITH STRAIN SOFTENING

Adachi and Oka (1995) proposed an elastoplastic model with strain softening, using a strain measure expressed as

$$dz = (de_{ij} \, de_{ij})^{1/2} \qquad (1)$$

where dz is an incremental strain measure and de_{ij} is an incremental deviatoric strain tensor. The stress history tensor $d\sigma_{ij}^{*}$ is expressed by introducing a single exponential type of kernel function, namely,

$$\sigma_{ij}^{*} = \frac{1}{\tau} \int_{0}^{z} \exp(-(z-z')/\tau)\sigma_{ij}(z') \, dz \qquad (2)$$

where τ is a material parameter which expresses the retardation of stress with respect to the strain measure and σ_{ij} is the effective stress tensor. It is clear from the equation that the stress history tensor is not only related to the current stress but also the stress history condition.

The total strain increment tensor is composed of the elastic and plastic components:

$$d\varepsilon_{ij} = d\varepsilon_{ij}^e + d\varepsilon_{ij}^p \qquad (3)$$

The plastic strain increment is given by the non-associated flow rule as,

$$d\varepsilon_{ij}^p = H\left(\partial f_p / \partial \sigma_{ij}\right) df_y \qquad (4)$$

where f_p is the plastic potential function, f_y is the yield function and H is a positive function describing the strain hardening and strain softening characteristics. The subsequent yield function is defined by

$$f_y = \eta^* - \kappa = 0, \quad \eta^* = \sqrt{S_{ij}^* S_{ij}^*} / \sigma_m^* \qquad (5)$$

where S_{ij}^* is the deviatoric stress history tensor, σ_m^* is the mean stress history, κ is the strain hardening and softening parameter and is given by the following evolution equation:

$$d\kappa = d\gamma^p G'\left(M_f^* - \kappa\right)^2 / M_f^{*2}$$

$$d\gamma^p = \left(de_{ij}^p de_{ij}^p\right)^{1/2} \qquad (6)$$

In the case of proportional loading, it can be integrated as

$$\kappa = M_f^* G' \gamma^p / \left(M_f^* + G'\gamma^p\right)$$

$$\gamma^p = \int d\gamma^p \qquad (7)$$

where e_{ij}^p is a deviatoric plastic strain tensor. G' and M_f^* are the strain hardening-softening parameters.

For the yielding function defined in Equation 5, the following Prager condition should be satisfied,

$$df_y = d\eta^* - d\kappa = 0 \qquad (8)$$

The loading condition is given by the following relations:

$$d\varepsilon_{ij}^p \begin{cases} \neq 0 & \text{if } f_y = 0, \ df_y > 0 \ \text{loading} \\ = 0 & \text{if } f_y = 0, \ df_y = 0 \ \text{neutral} \\ = 0 & \text{if } f_y = 0, \ df_y < 0 \ \text{unloading} \end{cases} \qquad (9)$$

It is assumed that plastic potential function is expressed by the relation as

$$f_p = \bar{\eta} + \overline{M} \ln\left[(\sigma_m + b)/(\sigma_{mb} + b)\right] = 0$$

$$\bar{\eta} = \left(S_{ij} S_{ij} / (\sigma_m + b)^2\right)^{1/2} \qquad (10)$$

where S_{ij} is the deviatory stress tensor, σ_m is the mean stress and $\overline{M}$ is the parameter that controls the development of the volumetric strain. σ_{mb}, the plastic potential parameter, is determined by isotropic consolidation tests and takes the value of the pre-consolidated stress. b is the plastic potential parameter that represents the tensile strength of the material.

The following relation expresses a boundary surface, which defines the normally consolidated and over-consolidated region:

$$f_b = \bar{\eta} + \overline{M}_m \ln\left[(\sigma_m + b)/(\sigma_{mb} + b)\right] = 0 \qquad (11)$$

Based on this relation, the value of $\overline{M}$ in Equation 10 can be determined based on the boundary surface:

$$\begin{cases} \overline{M} = -\bar{\eta}/\ln\left[(\sigma_m + b)/(\sigma_{mb} + b)\right] & \text{if } f_b \leq 0 \\ \overline{M} = \overline{M}_m & \text{if } f_b > 0 \end{cases} \qquad (12)$$

where $\overline{M}_m$ is a material parameter. Combining Equations 4, 5, 6, 8, 10 and 12, the following equation for the plastic strain increment tensor can be derived:

$$d\varepsilon_{ij}^p = \Lambda \left[\frac{\bar{\eta}_{ij}}{\bar{\eta}} + \left(\overline{M} - \bar{\eta}\right)\frac{\delta_{ij}}{3}\right]\left[\frac{\eta_{kl}^*}{\eta^*} - \eta^* \frac{\delta_{kl}}{3}\right]\frac{d\sigma_{kl}^*}{\sigma_m^*} (13)$$

$$\Lambda = M_f^{*2} / G' / \left(M_f^* - \eta^*\right)^2$$

$$\eta_{ij} = S_{ij} / (\sigma_m + b) \qquad (14)$$

8 parameters are involved in the model and they can be determined with the conventional triaxial compression tests. Detail description of the determination of these parameters can be referred to the reference (Adachi and Oka, 1995)

The model includes a feature by which the uniqueness of the solution for Valanis's (1985) sense is satisfied in the initial value problem (Adachi and Oka, 1995). It also has the advantage that in finite element analyses based on the model, numerical results have very small mesh dependency.

3 CENTRIFUGE TEST OF TRAPDOOR

The centrifuge tests considered here were conducted under axial symmetrical condition. Figure 1 shows the experimental apparatus of the trapdoor tests. The apparatus has a cylindrical container, with a size of 38 cm in diameter and 30 cm in depth. A circle shaped trapdoor, with a diameter of 5 cm, is located at the center of the cylindrical container. Various kinds of ground, such as dried sand, saturated sand, saturated clay or the mixture of these materials, can be prepared within the container. In present case, dried Toyoura Standard Sand was used. After the sand is filled up in the container, the apparatus is then set up on a centrifuge machine.

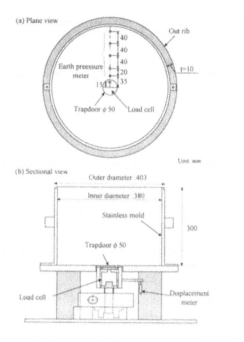

Figure 1. Outlet of trapdoor in Centrifuge tests (Adachi et al., 1997).

In the tests, in order to understand the deformation pattern of trapdoor in different overburden ratios, four kinds of grounds with different overburden ratios, that is H/D=0.5, 1.0, 1.5, 2.0, are tested. Here, H stands for the depth of the ground and D stands for the diameter of the trapdoor. The tests were conducted under 40g gravitational field. The material parameters of the sand are listed in Table 1.

Table 1. Material parameters of sand.

Young' modulus E (MPa)	15.68	Residual stress ratio M^*_f	0.90
Poisson Ratio v	0.333	b (kPa)	0.98
Density γ' (g/cm^3)	1.6	σ_{mb} (MPa)	1.96
Stress history parameter τ	0.002	M_m	0.80
Strain-softening parameter G'	300.0		

Apart from Young's modulus and Poisson's ratio, 6 extra parameters are necessary for the Adachi-Oka model. They are needed to be determined with conventional triaxial compression tests.

Figure 2 shows the comparison of the stress-strain-dilatancy relations between the theoretical and the test results of Toyoura Standard Sand whose material parameters are listed in Table 1, in triaxial extension test. It is known from the figure that the theoretical simulation agrees well the tested results.

4 FINITE ELEMENT ANALYSIS OF TRAPDOOR

The finite element analyses are conducted based on

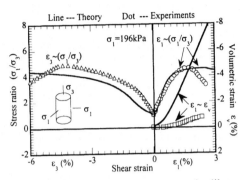

Figure 2. Simulation of the stress-strain-dilatancy relationship in triaxial extension test.

the Adachi-Oka model and the material parameters of the ground are listed in Table 1. According to the symmetric condition of the tests, it is possible to simulate the tests with 2-D axial symmetrical finite element analysis. Because the final aim of the research is to evaluate the mechanical behavior of the ground near a cutting face of a tunnel which is a typical 3-D problem, as a first step, the present 2-D problem is also considered in 3-D FEM.

Figure 3 shows the model used in the 3-D finite element analyses. According to its axial symmetrical condition, only a small part of the cylinder which occupies 5° of central angle is considered.

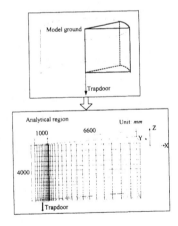

Figure 3. Calculation model.

Figure 4 shows the finite element mesh adopted in the analysis of the trapdoor (H/D=1.0). The numbers of node and 4-node isoparametric element are 2860 and 1344, respectively. For different overburden ratios, finite element meshes will be different. The high the ratio is, the larger the numbers of the node and the element of a mesh will be. The initial stress field of the ground is a gravitational field with a value of $K_0 = 0.5$.

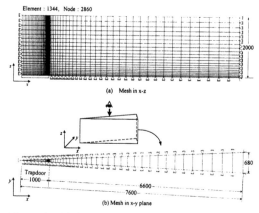

(a) Mesh in x-z

(b) Mesh in x-y plane

Figure 4.   Finite element Mesh (H/D=1.0).

In simulating the decent of trapdoor, a prescribed given displacements are applied along the trapdoor. In order to avoid a discontinuity of the descent at the corner of trapdoor, a transient area with a width of 40 mm is arranged in such a way the displacement along the transient area distributes linearly, as shown in Figure 5.

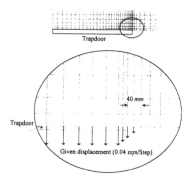

Figure 5. Given displacement at the corner of trapdoor.

Figure 6 shows the load-displacement relations of the trapdoor in different overburden ratios. It is found from the figure that in every case, the load drops rapidly to its minimum value at the beginning, and then keeps in a constant value even when the trapdoor is further lowered. This phenomenon can be linked to the stress-strain relation of the ground as shown in Figure 2. Before the shear stress reaches its peak value, the load acting upon the trapdoor will monotonically decrease. When the stress reaches the peak value, shear stress will no longer increase anymore, resulting in constant shear strength. The resistance of the ground then is remained constant, which on the consequently, results in an unchanged load no matter how the decent increase. There is a

tendency in both experiment and numerical analysis that the larger the overburden ratio is, the smaller the load acting on trapdoor will be. The numerical analysis can simulate the experimental results to some extent, especially the overall tendency of the load-displacement relationship.

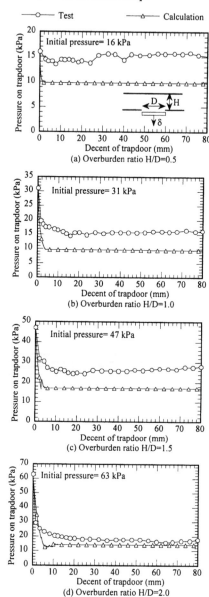

Figure 6.   Simulation of load-displacement relations of trapdoor in different overburden ratios.

Figure 7 shows the distribution of the earth pressure acting on the trapdoor and the bottom of ground. In either the cases of H/D=0.5 and H/D=2.0, the numerical analyses can simulate the experimental result to a reasonable accuracy. In both cases, there is a tensional region at the area right above the trapdoor. In reality, however, tensional region will never occur in a non-cohesive material when it is kept in dried condition. The reason why the discrepancy occurs is that a small value of tension (same as the value of b in Table 1) is allowed in the numerical analysis.

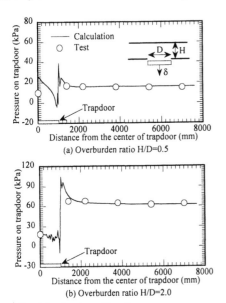

(a) Overburden ratio H/D=0.5

(b) Overburden ratio H/D=2.0

Figure 7. Distribution of earth pressures.

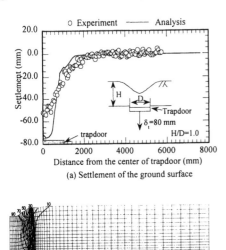

(a) Settlement of the ground surface

(b) Deformation of ground by numerical analysis

Figure 8. Deformation of ground (H/D=1.0).

Figure 8 shows the distribution of the displacements of the ground at the overburden ratio H/D=1.0. From Figure 8(a), it is known that the deformation pattern obtained from the numerical simulation agrees well with the result from the experiment. In Figure 8(b), the numbers represent the percentages of the settlements within the ground with respect to the decent of the trapdoor. Due to the formation a vertical shear band as shown in Figure 9, the ground above the trapdoor sinks on whole. While in the cases of H/D=1.5 and H/D=2.0, deformed area of the ground is only restricted to the area right above the trapdoor, with a height of 0.5B (not shown in the figure). It is known from the experiment and numerical analyses that on the condition that the deformed zone does not reach the surface of the ground, the arch effect provided by the ground can prevent the other area from further deformation.

Figure 9 shows changes in the distribution of shear strain ($\varepsilon=(2I_2)^{1/2}$, $2I_2=e_{ij}e_{ij}$) with the displacement of the trapdoor when the overburden ratio is H/D=0.5. It is found from the figure that at the beginning, a large shear strain zone occurs only in the vicinity of the corner. It develops upward along with the increase of the decent of the trapdoor. Finally a vertical shear band that reaches the surface of the trapdoor forms in the ground, which means that in shallow tunnel excavation, deformed zone may reach the surface of ground.

In the cases of overburden ratios H/D=1.5 and 2.0, the above-mentioned phenomenon does not occur. It is found from the Figure 10 (H/D= 2.0) that the area in which a large shear strain develops is only restricted to the vicinity of the trapdoor. Due to the formation of an arch within the ground above the trapdoor, a further decent of the trapdoor does not affect too much the ground above the arch. Because of the existence of the arch, the pressure acting on the trapdoor becomes small, as can be seen in the load-displacement relation shown in Figure 6.

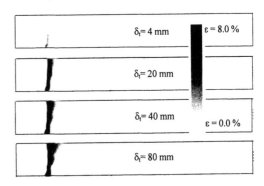

Figure 9   Distribution of shear strain ($\varepsilon = \sqrt{2I_2}$) at different decent of trapdoor (H/D=0.5)

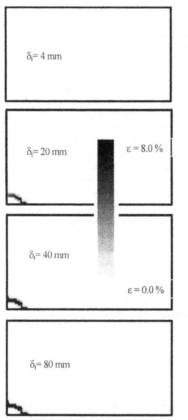

Figure 10. Distribution of shear strain ($\varepsilon = \sqrt{2I_2}$) at different decent of trapdoor (H/D=2.0).

5 CONCLUSION

The following conclusions can be given based on the present study:

① Compared to the load-displacement relationship of trapdoor by centrifuge test, numerical simulation always gives a larger reduction of the pressure acting on the trapdoor when the trapdoor is lowered. The difference between the test and analytical results, however, decreases when the overburden ratio increases. On the whole, the numerical analyses conducted in this paper can predict the test results to a reasonable accuracy.

② The deformation pattern of the ground may be totally different for different overburden ratio. In shallow ground, larger displacement may reach the surface of the ground due to the formation of a vertical shear band. In deep ground, however, displacement is only restricted to the area right above the decent trapdoor.

③ The earth pressure acting on the trapdoor and the bottom of the ground can be will simulated by the numerical analyses.

REFERENCES

Adachi T., Kimura M., Nishimura T., Koya N. & Kosaka K., 1997. Trapdoor experiment under Centrifugal conditions, Proc. of Int. Conf. on Deformation and Progressive Failure in Geomechanics (IS-Nagoya'97), pp.725-730.

Adachi T. and Oka F., 1995, An elasto-plastic constitutive model for soft rock with strain softening, Int. Jour. for Numerical and Analytical Methods in Geomechanics, Vol. 19, pp. 233-247.

Ladanyi B. and Hoyaux B., 1969, A study of trapdoor problem in a granular mass, Canadian Geotech. J., Vol.6, No.1, pp. 1-15.

Szechy K., 1966, The Art of Tunnel, Akademiai, Kiado Budapest, Printed in Hungary.

Valanis K. C., 1985, On the uniqueness of solution of the initial value problem in softening materials," J. Appl. Mech. 52, pp. 649-653.

Modern Tunneling Science and Technology, Adachi et al (eds), © 2001 Swets & Zeitlinger, ISBN 90 2651 860 9

Earth pressure and ground movements due to tunneling – model tests and analyses

T. Nakai, D. Yamaguchi, H. M. Shahin & T. Kurimoto
Nagoya Institute of Technology, Nagoya, Japan

M. M. Farias
University of Brasilia, Brasilia, Brazil

ABSTRACT: In order to investigate the fundamental mechanism of the generation of earth pressure and the ground movement due to tunneling, 2D and 3D trap door tests and the corresponding elastoplastic finite element analyses are carried out. The analytical results agree well with the observed earth pressure and ground movements. These results depend not only on the geometrical condition but also on the construction sequence. In this sense, it is shown that the results in 3D model tests and analysis are different from those in 2D model tests and analyses, and that 3D analysis is necessary for good predictions.

1 INTRODUCTION

2D and 3D model tests for trap door problems and the corresponding finite element analysis were carried out previously (Nakai, *et al.*, 1997). In model tests, stepwise downward displacements were imposed and only surface settlements were measured. However, the practical process of tunneling is more continuous and the earth pressure distributions in tunneling are as important as ground movements. New model tests and finite analyses in which the excavation sequence were considered were carried out later (Nakai *et al.*, 2000). In the present study, the device for trap door tests is improved to measure the earth pressure, and a more continuous tunnel excavation process is simulated in finite element analyses using finer meshes and employing a more sophisticated elastoplastic constitutive model for geomaterials. Carrying out these model tests and analyses, we will investigate the fundamental mechanism which generates the earth pressure and the ground movements in tunneling.

2 METHODS OF MODEL TESTS AND ANALYSES

Figure 1 shows a schematic diagram of the apparatus used in 2D model tests. The lowering part consists of 10 brass blocks (width of each block B=80mm), which are movable in upward and downward directions independently. A mass of aluminum rods is used to model the soil in the 2D tests. The mass is composed of two kinds of rods having diameters of 1.6 and 3.0mm and a length of 50mm are mixed

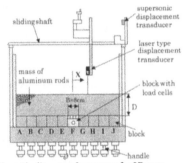

Figure 1.Schematic diagram of apparatus for 2D tests.

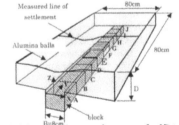

Figure 2. Schematic diagram of apparatus for 3D tests.

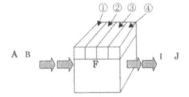

Figure 3. Outline of block with load cells.

with a ratio of 3:2 in weight, and has unit weight of $\gamma=21.4$kN/m^3 at the stress level of the model tests. Figure 2 shows a schematic diagram of the apparatus for the 3D model tests. We used a mass of alumina balls (unit weight at stress level of the model tests $\gamma=22.3$kN/m^3), which was obtained by mixing two kinds of balls (2mm and 3mm) of equal weight ratio. As shown in Figure 2, ten square brass blocks (80mm×80mm), which are the same as those used in 2D model tests in Figure 1, are buried along the centerline of the table. One of these ten brass blocks can be replaced by a cube having load-cells to measure the earth pressure. Three ratios of depth D of model ground to the width B of the movable block are employed in the model tests (D/B=0.5, 1.0 and 2.0). To measure the earth pressure distribution on the block, its upper part is divided into 4 small rectangular blocks, and the load cell is set up at the bottom of each divided block as shown in Figure 3. A laser type displacement transducer moves along a slide shaft over the model ground. The position of the laser type transducer is measured by a supersonic wave transducer. All the data of earth pressures and surface settlements are automatically collected by a data logger connected to a personal computer.

Figure 4 shows the finite element meshes used in the 2D and 3D analyses in the case of D/B=1.0. Analyses were carried out for the cases of D/B=0.5 and 2.0 as well as D/B=1.0, similar to the model tests. In the 3D analyses, half of the ground is analyzed, since the lowering blocks are symmetric with respect to the direction of excavation. Smooth boundary conditions are assumed in the lateral faces and bottom of the mesh. To simulate the lowering of the blocks in the numerical analyses, we impose the displacements at the nodal points, which correspond to the top of the lowering blocks in the model tests.

Table 1 shows the patterns of the experiments and the analyses. The 2D tests in which a downward displacement of d=4mm is imposed to the block F alone in Figure 1 are called Series I (2D sectional or single block excavation) and are intended to simulate the excavation of a cross section of the tunnel. In the experiment of this series, setting up the block with load cells at other positions as well and carrying out the same tests to measure the earth pressures not only on the lowering block but also around it, we can obtain the earth pressure distribution along cross section. Series II (2D block by block excavation) refers to the tests in which the downward displacement of d=4 mm is imposed block by block from A to J and measurements are performed in block F. The idea is to simulate the advance process along the longitudinal excavation direction. Series III (3D block by block excavation) represents the three-dimensional condition of Series II. However, practical tunneling process is more sequential than Series III in which excavation is done in every 80 mm along the excavation direction. The numerical analy-

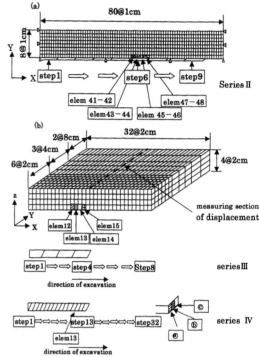

Figure 4. FEM meshes for 2D and 3D FEM and processes of lowering basements.

Table 1. Patterns of excavation

series	sutudy	type of excavation	D/B		
I	Experimental & numerical	2D single block	0.5	1.0	2.0
II		2D block by block	0.5	1.0	2.0
III		3D block by block	0.5	1.0	2.0
IV	Numerical	3D sequential	0.5	1.0	2.0

Table 2. Parameters of excavation analyzed

λ	0.008
κ	0.004
R_{cs}	1.8
β	1.2
e_0(P=1.0 × 98kPa)	0.3
a	1300

$e_{ini.}$(P=1.0 × 10^{-5} × 98kPa)	0.328

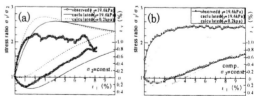

Figure 5. Observed and calculated stress-strain curves.
(a) biaxial test on aluminum rods mass
(b) triaxial test on alumina balls mass

ses of Series IV (3D sequential excavation) are then carried out to simulate the downward displacement of d=4 mm to the corresponding nodal points in Figure 4(b) from the left to the right sequentially. The process to impose the displacements to the nodal points in the analyses is illustrated at the bottom of Figure 4.

An elastoplastic constitutive model for sand, named subloading t_{ij} model (Nakai et al., 2001), is used in the finite element numerical analyses. This model can describe properly the following typical characteristics of sand, despite its small numbers of parameters:
(i) Influence of intermediate principal stress on the deformation and strength of sand.
(ii) Influence of stress path on the direction of plastic flow.
(iii) Negative and positive dilatancy.
(iv) Influence of density and/or confining pressure.

The dots in Figure 5(a) and (b) show the observed principal stress ratio: major principal strain: volumetric strain relations in biaxial test on aluminum rods mass and in triaxial test on alumina balls mass under constant minor principal stress (σ_x=19.6Kpa) respectively. The solid curves in Figure 5 are the calculated results corresponding to the observed ones for the masses of aluminum rods mass and alumina balls. Here, the same values of material parameters in Table 2 are used for both materials, since the strength and the deformation characteristics of these materials are not much different. The dotted curves in these figures are the calculated results in which the initial confining pressure is assumed two orders smaller in magnitude. This is because the initial confining pressure in model tests is much smaller than that in the biaxial and triaxial tests. We can see that the constitutive model describes strain softening behavior as well as the influence of the confining pressure. The initial stresses of the ground are calculated by simulating the self-weight consolidation in these analyses.

3 RESULTS AND DISCUSSIONS

3.1 Series I (2D single block excavation)

Figure 6 shows the observed profiles of surface settlements for the cases of D/B=0.5, 1.0 and 2.0. Figure 7 illustrates the computed profiles of surface settlements corresponding to the observed ones in Figure 6. In these figures, the vertical dotted straight lines indicate the centerline, and the horizontal axes represent the distance from the centerline. The prescribed displacement pattern of the lowering block is indicated at the bottom of each figure. It can be seen from Figure 6 that although the surface settlements at the center of the model is almost the same as the imposed displacement of the lowering block in the

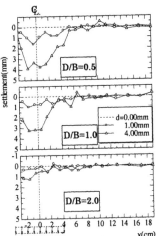

Figure 6. Observed profile of surface settlement: 2D single block.

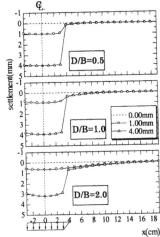

Figure 7. Computed profiles of surface settlement: 2D single block

shallow ground (D/B=0.5), it becomes smaller with the increase of the tunnel depth. Such tendency of the observed surface settlements is simulated by the numerical analyses, although the reduction of the computed settlement with the increase of tunnel depth is less than the observed.

Figures 8 and 9 show the observed and computed distributions of the earth pressures normalized by the initial vertical stress $\sigma_{z0}=\gamma D$ on the base level. Since the initial vertical stress is proportional to the depth D, the scale of vertical axis in each figure is proportional to D/B. We can see from these figures that the earth pressure on the lowering block decreases, while increasing in the adjacent one due to arching effect in the cases of D/B=1.0 and 2.0, as reported by the previous experimental studies (Murayama and Matsuoka, 1971; Adachi et al., 1994). For the cases of D/B=0.5 this change was not very significant, but this tendency becomes more remarkable for greater

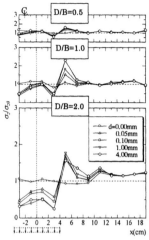

Figure 8. Observed distributions of earth pressure: 2D single block

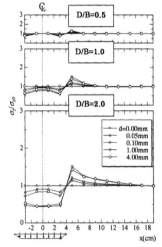

Figure 9. Computed distributions of earth pressure: 2D single block

depths. We can also notice that such change of the earth pressures occurs for very little displacement of the block.

3.2 Series II (2D block by block excavation)

Figure 10 shows the history of the observed earth pressures normalized by the initial earth pressures on block F in Figure 1. The letter in each column denotes the block to which the vertical displacement was imposed, and the horizontal axis in the column indicates the displacement of the corresponding lowering block. The symbols ① to ④ represent the earth pressures that are measured by the four load cells mounted in block F, as shown in Figure 3 - ① is the most backward, and ④ is the most forward. It can be seen that the earth pressure in block F decreases extremely during the lowering of the block

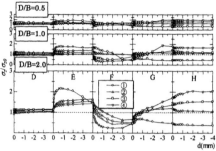

Figure 10. History of observed earth pressure: 2D block by block

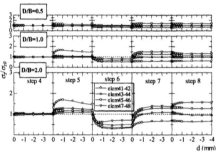

Figure 11. History of computed earth pressure: 2D block by block

itself in case of D/B=1.0 and 2.0, however it increases during the lowering of the previous and next blocks. This is due to the arching effect formed in the direction of excavation. Such arching effect is simulated by the computed results in Figure 11. Here, elements 41-42 to 47-48 in the finite element mesh correspond to the divided sections ① to ④ of block F in the model tests.

3.3 Series III (3D block by block excavation)

Figure 12 shows the observed profiles of the surface settlements of the model tests of Series III, at the end of the downward displacement of d=4 mm at the blocks A, E, F and J in Figure 2. Here, the measured section of surface settlement is along the line between the blocks E and F. Figure 13 is the computed surface settlements in Series III, where the steps 1, 3, 4, 5 and 8 in the computations correspond to the imposition of a downward displacement of 4 mm to the blocks B, D, E, F and I in the experiment, respectively. Step 4 is the condition where the excavation face reaches the measured line of settlements. There is a good agreement between the observed and computed profiles of settlements not only qualitatively but also quantitatively. We can also see from these figures that the surface settlement at the centerline for D/B=2.0 is less than half of the imposed displacement, while that in the case of D/B=0.5 it is almost the same as the imposed displacement.

156

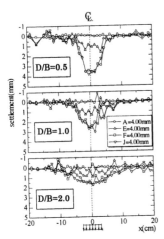

Figure 12 Observed profiles of surface settlements: 3D block by block

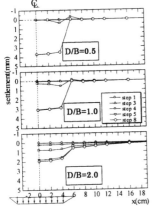

Figure 13 Computed profiles of surface settlements: 3D block by block

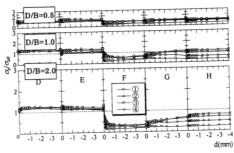

Figure 14. History of observed earth pressure: 3D block by block

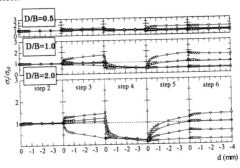

Figure 15. History of computed earth pressure: 3D block by block

Figure 14 shows the history of the observed earth pressure, arranged with respect to the same relations as in Figure 10 of Series II. The change of earth pressure with excavation advance is qualitatively similar to the results in Figure 10. However, the degree of its change is not so remarkable, because of the three-dimensional effect. Figure 15, which presents the computed history of earth pressure, shows the same trend as the observed one.

3.4 Series IV (3D sequential excavation)

To be precise, the excavation process in Series III is not sequential, because the downward displacement is imposed at intervals of 8cm to the lowering blocks. To simulate the construction process of tunneling more accurately, numerical analyses in which the downward displacement is sequentially imposed to the corresponding mesh nodes are carried out.

Figure 16 shows the computed surface settlements in this series. Here, the results at step 13 rep-

resent the surface settlements when the excavation face reaches the measured line, and the results at step 31 are the final surface settlements. Comparing Figure 16 with Figure 13, we can see that although the shapes of settlement profiles in Series IV are similar to those in Series III, the magnitudes of settlements in Series IV are smaller than those in Series III in every case.

Figure 17 shows the history of the computed earth pressure on the bottom part of the model. Here, ⓐ represents the earth pressure near the center of the lowering block, and ⓒ represents the earth pressure near the edge. We can see that the vertical earth pressures on the lowering blocks become almost zero at and near the face regardless of the depth of the ground and increase gradually after the face passes the measured line except for the very shallow ground ($D/B=0.5$). Also no remarkable increase of the earth pressures on the lowering blocks was observed before the face passage like Figure 15. Figure 18 shows the computed distributions of earth pressures in the direction of cross section at every step. We can see that there are arching effects not only in the direction of the cross section but also in the excavation direction. Thus, the earth pressures depend on the distance from the face, and their final distributions are also different from the two-dimensional ones in Figure 9.

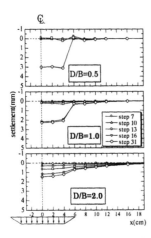

Figure 16 Computed profiles of surface settlements: 3D sequential

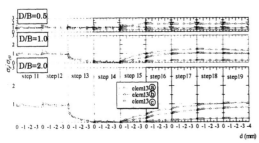

Figure 17. History of computed earth pressure: 3D sequential

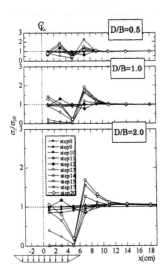

Figure 18. Computed distributions of earth pressure: 3D sequential

4 CONCLUSIONS

Experimental and numerical studied on 2D and 3D trap door problems have been done to investigate the surface settlements and the earth pressure due to tunneling. Throughout the present studies, the following conclusions are obtained:

(1) The surface settlements in sandy ground under 3D sequential excavation are smaller than those under 2D excavation and under block by block excavation, though there is not much difference in their shapes. It is then necessary to take into consideration the proper three-dimensional construction process for a precise prediction.

(2) Not only the arching effects in the direction of cross section but also the arching effects in the excavation direction influence on the earth pressure in tunneling. The earth pressure in tunneling should also be predicted as three-dimensional problems.

(3) Though the arching was not very significant for the case of D/B=0.5 in three-dimensional tunneling problems, this effect is remarkable for the cases of D/B=1.0 and 2.0. As a result, the vertical earth pressure at the excavation face is expected to be almost zero except for very shallow tunnel.

(4) 3D finite element analysis in which the elastoplastic stress-strain behavior of soil and the construction process are properly taken into consideration is a powerful tool for the prediction of ground movements and earth pressure in tunneling.

REFERENCES

Adachi, T., Tamura, T., Kimura, M. & Aramaki, S. 1994. Earth pressure distribution in trap door tests: *Proc. of 29th Japan National Conference of SMFE*, 3, 1989-1992 (in Japanese).

Murayama, S. & Matsuoka, H. 1971. Earth pressure on tunnels in sandy ground: *Proc. of JSCE*, 187: 95-108 (in Japanese).

Nakai, T., Hinokio, M., Hoshikawa, T., Yoshida, H. & Chowdhury, E.Q. 2001. Shear behavior of sand under monotonic and cyclic loadings and its elastoplastic modeling: *Proc. of 10th Int. Conf. on Computer Methods and Advances in Geomechnics*, Tucson, 2: 367-372.

Nakai, T., Matsubara, H., Kusunoki, S. & Farias, M.M. 2000. Effects of excavation sequence on the 3D settlement of shallow tunnels: *Proc. of Int. Sym. on Geotechnical Aspects of Underground Construction in Soft Ground*, Tokyo, 1: 403-408.

Nakai, T., Xu, L. & Yamazaki, H. 1997. 3D and 2D model tests and numerical analyses of settlements and earth pressure due to tunnel excavation: *Soils and Foundations*, 37(3): 31-42

Modern Tunneling Science and Technology, Adachi et al (eds), © 2001 Swets & Zeitlinger, ISBN 90 2651 860 9

Experimental study on the distribution of earth pressure through three-dimensional trapdoor tests

T. Adachi, M. Kimura & K. Kishida
Department of Civil Engineering, Kyoto University, Kyoto, Japan

ABSTRACT: In order to clarify the mechanical behavior in tunnel excavations, the authors develop and perform three-dimensional trapdoor experiments. The influence of excavations is considered in the experiments. Based on the results, the distribution of earth pressure is found to greatly affect the process of the excavations. It can be confirmed, therefore, that a higher level of earth pressure is working on a ground where loosened conditions occur due to a previously lowered trapdoor.

1 GENERAL INSTRUCTIONS

In mountain areas of Japan, the need to effectively use underground space has increased. In urban areas, traffic infrastructures as well as lifeline infrastructures have been utilized to develop underground space in relation to the available land, the landscape, and environmental preservations. When constructing underground caverns and the tunnels, therefore, more safety and assurance than presently exist should be required. This is because as underground structures become larger, the influence of pre-existing structures on and in the ground must be considered. The purpose of this research work is to grasp the mechanical behavior of excavated grounds and understand the importance of it.

In order to grasp the mechanical behavior of excavated grounds, the authors previously developed a two-dimensional trapdoor tester and carried out experiments with it (Adachi et al. 1994, 1995a, b, and 1997). In the present paper, the authors develop a three-dimensional trapdoor tester in order to discuss in detail the mechanical behavior of tunnel excavations and perform tests with it. Concerning the tunnel excavating process, earth pressure and ground surface settlement levels are measured. The mechanical behavior of tunnel excavations is discussed using the obtained data. In this paper, only the results of the earth pressure measurements will be described and discussed.

2 THREE-DIMENSIONAL TRAPDOOR EXPERIMENTS

2.1 Apparatus

Figure 1 shows the three dimensional trapdoor apparatus developed by the authors. Figure 2 presents the arrangement of six trapdoors and earth pressure gauges on the bottom plate. The chamber is $1090 \times 1090 \times 600$ *mm*. The model shown in Figure 2 is for the single tunnel excavations. Since the bottom of the plate is replaceable, tests can be performed for twin tunnel and crossroad tunnel models by employing other types of bottom plates. For all trapdoors, the width of the cross section is 150 *mm*. For trapdoor numbers 1 and 6, the length of the trapdoors is 200 *mm*, while it is 150 *mm* for trapdoor numbers 2, 3, 4, and 5. The shape of trapdoor numbers 2, 3, 4, and 5 is square. The lowering of the trapdoors is performed by control jacks which are introduced under the trapdoors. The loads applied to the trapdoors are measured by load cells (TCLP-50KA) which are also installed under the trapdoors. Earth pressure gauges, 25 *mm* in diameter (P325SV-02), are arranged so as to measure the detailed distributions of earth pressure on the trapdoors. In order to grasp the distributions of earth pressure around the trapdoors on the chamber, earth pressure gauges, 50 *mm* in diameter (KD-2F), are installed. At the top of the chamber, a surface settlement profiling system is set up, as shown in Figure 3. A non-contacted laser scanning micro sensor is adopted to profile the model ground surface, and the profiling data on the model ground surface can easily be obtained.

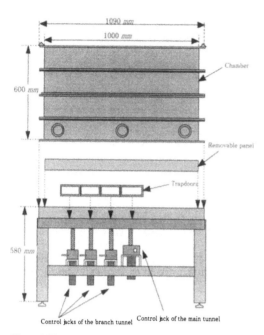

Figure 1. Three-dimensional trapdoor tester (side view).

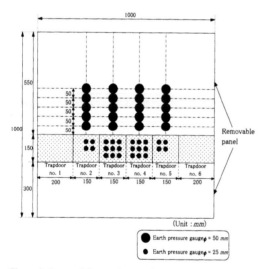

(Unit : *mm*)

● Earth pressure gauge ϕ = 50 *mm*
• Earth pressure gauge ϕ = 25 *mm*

Figure 2. Set up of the trapdoors and the earth pressure gauges on the removable panel.

2.2 Experimental conditions and patterns

Silica sand number 6 is used for the model ground. The physical properties of Silica sand number 6 and the properties of the ground are presented in Table 1.

In this research work, trapdoor experiments are carried out under three overburden heights, namely, $H = 150$ *mm* (1D), 300 *mm* (2D) and 600 *mm* (4D), respectively, where D is the width of the trapdoor, namely, 150 *mm*.

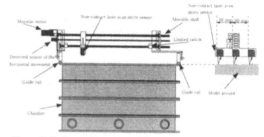

Figure 3. Surface settlement profiling system.

Table 1. The physical properties of Silica sand number 6 and the properties of the ground.

Physical properties		Properties of making ground	
Specific gravity	2.63	Relative density [%]	70.94
Maximum void ratio	1.03	Dry density [kN/m³]	14.72
Minimum void ratio	0.64	Internal friction angle	36°
		Void ratio	0.753

In order to suppose the tunnel excavation in this paper, the six trapdoors shown in Figure 2 are lowered one after another. The descending order of the trapdoors is trapdoor numbers 1, 2, 3, 4, 5, 6, and each trapdoor is lowered until it reaches a descending movement (δ_t) of 5.0 *mm*.

The earth pressure is measured at intervals of descending movement 0.05 *mm* in the range of descending movement (δ_t) from 0.0 to 0.2 *mm*, 0.1 *mm* in the range of δ_t from 0.2 to 1.0 *mm*, 0.5 *mm* in the range of δ_t from 1.0 to 3.0 *mm*, and then 1.0 *mm* in the range of δ_t from 3.0 to 5.0 *mm*, respectively. After lowering each trapdoor at a descending movement of 5.0 *mm*, the ground surface settlements are profiled.

3 EXPERIMENTAL RESULTS AND DISCUSSIONS

3.1 Earth pressure on the trapdoors

Figure 4 shows the relationship between the standardized earth pressure of each trapdoor and the descending movement, in which the standardized earth pressure is defined by dividing the measurement of earth pressure by the initial earth pressure. Regardless of the overburden height, the earth pressure applied to the trapdoors decreases with the lowering of the trapdoors and the decrease in earth pressure converges to a descending movement of 5.0 *mm* for all trapdoors. The convergent value of the standardized earth pressure becomes the same of all trapdoors at each overburden height. It is 0.2 at $H = 1D$, 0.1 at $H = 2D$, and 0.05 at $H = 4D$. With larger overburdens,

160

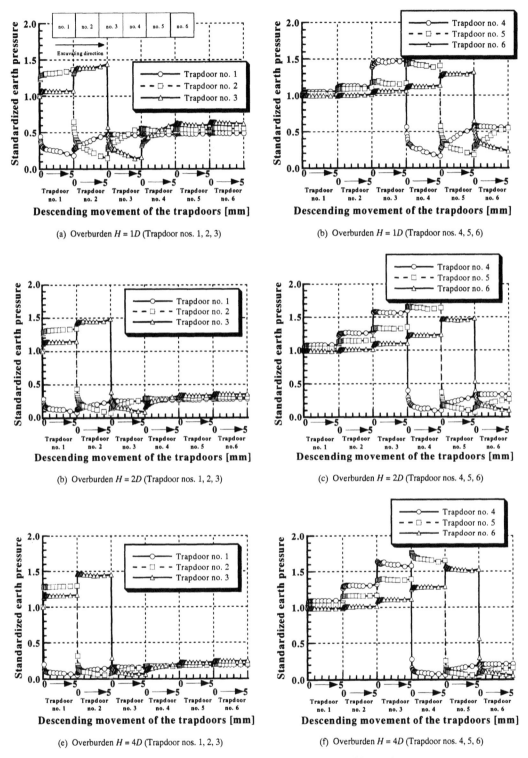

(a) Overburden $H = 1D$ (Trapdoor nos. 1, 2, 3)

(b) Overburden $H = 1D$ (Trapdoor nos. 4, 5, 6)

(b) Overburden $H = 2D$ (Trapdoor nos. 1, 2, 3)

(c) Overburden $H = 2D$ (Trapdoor nos. 4, 5, 6)

(e) Overburden $H = 4D$ (Trapdoor nos. 1, 2, 3)

(f) Overburden $H = 4D$ (Trapdoor nos. 4, 5, 6)

Figure 4. Relationship between standardized earth pressure and descending movement of the trapdoors.

161

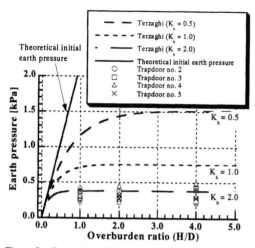

Figure 5. Comparison of theoretical and experimental earth pressure levels.

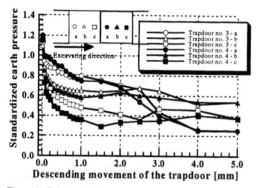

Figure 6. Relationship between standardized earth pressure and descending movement of the trapdoors in consideration of the location s of the pressure gauges on the trapdoors.

therefore, the reduction from the initial earth pressure appears remarkable, and it is confirmed that an arching action is in full play.

Before trapdoor numbers 2 ~ 6 themselves are lowered, their earth pressure levels gradually increase because they absorb the earth pressure which is occurring due to the lowering of the other trapdoors. In particular, earth pressure levels increase remarkably when an adjacent trapdoor is lowered. And, increments in earth pressure also appear with the descending trapdoors when they are located at more distant locations than an adjacent trapdoor. This tendency is clearly present with larger overburdens. On the other hand, after trapdoors numbers 1 ~ 5 themselves have been lowered, their earth pressure levels increase, absorbing the influence of the next trapdoor being lowered for each overburden. However, earth pressure levels hardly increase when trapdoors located far away are lowered.

Figure 5 shows the relationship between the minimum earth pressure of the trapdoors and the overburden ratio. It is also confirmed that a reduction in earth pressure is remarkably present with larger overburdens, and an arching action is in full play. Regardless of the overburden and the location of the trapdoors, minimum earth pressure levels present almost equivalent values and reach less than 0.5 kPa in each case. Based on Terzaghi's earth pressure theory (Terzaghi, 1943), the authors have described an earth pressure equation for three-dimensional trapdoor experiments as follows (Adachi et al., 1999):

$$\overline{\sigma_v} = \frac{\gamma x}{4K_h \cdot \tan\phi}\left[1 - \exp\left(-\frac{4}{x}K_h \cdot H \cdot \tan\phi\right)\right] (\phi \neq 0) \qquad (1)$$

where x, H, K_h, and ϕ are a side length of the trapdoors, the overburden, the coefficient of earth pressure, and the internal friction angle, respectively. Applying three values for K_h, namely, 0.5, 1.0, and 2.0, Figure 5 also shows the estimated earth pressure levels. Then, experimental results are plotted around an estimated curve at $K_h = 2.0$. In a sandy ground, however, the coefficient of earth pressure generally supports almost 0.5. In the present tests, the fact that the coefficient of earth pressure is 2.0 is barely considered. Therefore, Terzaghi's earth pressure theory hardly applies to the earth pressure obtained in these experiments.

Changes in standardized earth pressure levels along the center lines of trapdoors numbers 3 and 4 are shown in Figure 6, while the trapdoors themselves are being lowered. Figure 6 shows the results at 1D overburden height. In Figure 6, it is confirmed that the changes in standardized earth pressure differ depending on the distance of the pressure gauges from the centers of the trapdoors. When the earth pressure is located at a distance of 1/4D from the centers of the trapdoors in the direction of the excavation, it decreases until it reaches the descending movement of 1.0 mm, and then reaches the convergent state. The changes in earth pressure, when it is located at the centers of the trapdoors, also represent the same behavior as the pressure gauge at a location of 1/4D. However, the convergent value at the centers of the trapdoors is larger than that located at a distance of 1/4D from the centers of the trapdoors in the direction of the excavation. On the other hand, when the earth pressure is located at a distance of -1/4D from the centers of the trapdoors in the direction of the excavation, still decreases while the trapdoors are being lowered. On the front sides of the trapdoors, is the direction of the excavation, the arching action works and the ground becomes stable in an earlier stage of the descending process on the front of the trapdoors. The arching action does not appear on the back sides of the trapdoors, and it is found that the influence of the excavation expands. This is the reason why the ground has already achieved a loosened state when the trapdoors are lowered.

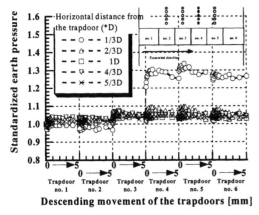

(a) Overburden H = 1D

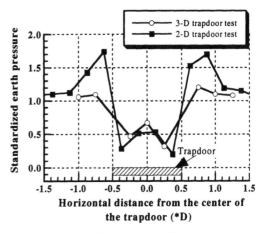

(a) Overburden H = 1D

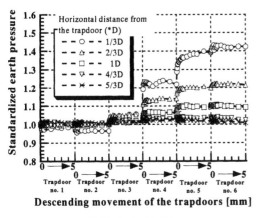

(b) Overburden H = 2D

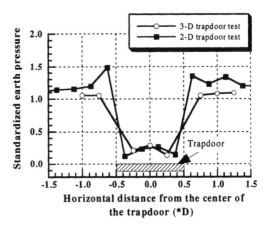

(b) Overburden H = 4D

Figure 8. Comparison of two-dimensional and three-dimensional trapdoor tests.

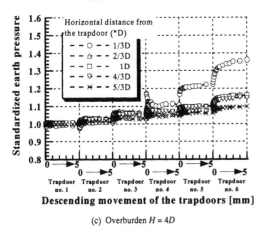

(c) Overburden H = 4D

Figure 7. Changes in earth pressure around the trapdoors.

3.2 Earth pressure acting around the trapdoors

The overburden heights of 1D, 2D, and 4D and the standardized earth pressure levels around the trapdoors are shown in Figure 7. The results have been measured by the pressure gauges which are located on an extended line from the center of trapdoor number 4 along the cross section. For each overburden, the earth pressure increases as the trapdoors are lowered. The increments in earth pressures have a close correlation to the distance from the pressure gauges to the descending trapdoors. In case of a short distance between the pressure gauges and the descending trapdoors, the increments in earth pressure increase. The largest change in earth pressure can easily be found when trapdoor number 4 is lowered.

In the case of the 1D overburden (Figure 7(a)), the largest change in earth pressure can be found at the location which is at a distance of 1/3D from the side of the trapdoor. Concerning the other pressure gauges located far away, no remarkable changes in earth pressure can be found. In the case of the 2D overburden (Figure 7(b)), the earth pressure located

163

within a distance of $1D$ from the trapdoor is found to increase. In the case of a $4D$ overburden (Figure 7(c)), all earth pressure gauges show increments. The earth pressure located at a distance of $1/3D$ from the trapdoor, in particular, increases rapidly and reaches the peak value with the lowering of trapdoor number 4. After reaching the peak, the earth pressure decreases. Then, it increases again with the lowering of trapdoor number 5. Therefore, it is thought that the region where trapdoor number 4 was lowered and the arching action occurred is spreading due to the lowering of trapdoor number 5."

3.3 Comparison of two-dimensional and three dimensional trapdoor experiments

In this section, the results of two-dimensional and three-dimensional trapdoor experiments are compared. The two-dimensional trapdoor experiments are performed using aluminum rod layers. The width of the trapdoors in the two-dimensional experiments is 100 *mm*, while it is 150 *mm* in the three-dimensional experiments. The results are shown after being standardized by the width of the trapdoors. Samples of the test results are shown in Figure 8. The results after the lowering of the trapdoors to $0.02D$ are presented in Figure 8. Under various overburdens, the standardized earth pressure applied on the trapdoors decreases remarkably in both the two- and the three-dimensional experiments, and the reduced standardized earth pressure levels are equivalent for the two- and the three-dimensional experiments. The standardized earth pressure levels appearing around the trapdoors increase. The standardized earth pressure levels in the two-dimensional experiments increase more than those in the three-dimensional experiments. This tendency does not depend on the overburden. In the case of the three-dimensional experiments, it is thought that all four sides can accept the earth pressure which occurs because of the descending trapdoor. On the other hand, only two sides (left and right) of the descending trapdoor can accept the earth pressure in the two-dimensional experiments. Therefore, the earth pressure working around the trapdoors would be over-estimated in the two-dimensional trapdoor experiments.

4 CONCLUSION

In considering the excavating process for tunnels, three-dimensional trapdoor experiments have been carried out. Based on the obtained data for the distribution of earth pressure, the authors have discussed the mechanical behavior of tunnel excavations. Since the ground already has a loosened condition due to the lowering of one trapdoor, earth pressure levels with have an influence on the next descending

trapdoor and will not be distributed equally around the trapdoor. It is confirmed that extra earth pressure acts in the direction of the loosened area because the loosened area can easily be deformed.

REFERENCES

Adachi, T., Tamura, T., Kimura, M., Aramaki, S.. 1994. Experimental and analytical studies of earth pressure. *Computer Methods and Advances in Geomechanics* : 70 – 75. Rotterdam: Balkema.

Adachi, T., Tamura, T., Kimura, M., Nishimura, T.. 1995a. Axisymmetric trap door tests on sand and cohesion soil. *Proceedings of the 30th Japan National Conference of Soil Mechanics and Foundation Engineering*: 1973 – 1976. (in Japanese)

Adachi, T., Tamura, T., Kimura, M., Koya, N.. 1995b. Mechanical behavior of twin trap door. *Proceedings of the 30th Japan National Conference of Soil Mechanics and Foundation Engineering*: 1977 – 1980. (in Japanese)

Adachi, T., Kimura, M., Kosaka, K., Koya, N.. 1997. Trap door tests on sand and cohesion soil in centrifuge. *Proceedings of the 32nd Japan National Conference on Geotechnical Engineering*: 2149 - 2150. (in Japanese)

Adachi, T., Kimura, M., Kishida, K., Kosaka, K., Sakayama, S.. 1999. The mechanical behavior of tunnel intersection through three dimensional trap door tests. *Journal of Geotechnical Engineering*. JSCE, No. 638/Ⅲ-49 : 285 – 299. (in Japanese)

Terzaghi, K.. 1943. *Theoretical Soil Mechanics*, John Wiley & Sons. New York. pp. 66 – 75.

Modern Tunneling Science and Technology, Adachi et al (eds), © 2001 Swets & Zeitlinger, ISBN 90 2651 860 9

Behaviour of a tunnel face reinforced by bolts : Influence of the soil/bolt interface

D. Dias, Y. Bourdeau & R. Kastner
INSA Lyon, Villeurbanne, France

ABSTRACT: The design of reinforced structures requires a perfect knowledge of the behaviour of the soil, the bolts and of the soil/bolt interface . To determine the frictional law between the soil and the bolt, full scale pull-out tests were performed on radial anchor bolts, equipped with strain gauges. The analytical method developed by [Bourdeau, 1994] allows to derive the local frictional law parameters (assumed constant along the rod) from the global relationship between the total load and the head displacement. This frictional law is introduced in a 3D tunnel model with reinforced face. The 3D numerical model provides an accurate geometrical description of the tunnel structure, such as the lining behind the face, as well as the interface law between bolts and ground. To observe the influence of the soil/bolt interface, we study the influence of several frictional laws in terms of face displacements and axial forces in the bolts.

1 INTRODUCTION

During tunnel construction, it is necessary to control the face stability and to restrain the settlements due to the excavation. Two techniques can be coupled : mechanical pre-cutting and face reinforcement by nailing. Following problems which occurred during the tunnel construction under Toulon City (France) when boring a red Permian geological layer, full scale pull-out tests were performed on radial anchor bolts, equipped with strain gauges (Figure 1&2).

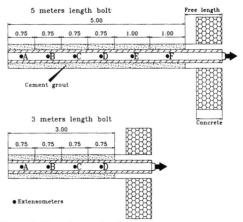

Figure 2. Experiment details.

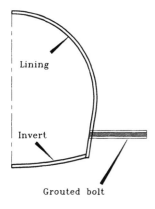

Figure 1. Experimentation

2 EXPERIMENTAL CONDITIONS

2.1 *Fiberglass rods characteristics*

The hollow fiberglass rod has a tensile strength between 500 and 600 MPa and a failure strain between 1 and 2 %.

This rod is placed in a boring of $\varnothing$ 100 mm, then sealed with CPA grout with a cement water ratio of 2. The sealing is realised in one step under low pressure.

The sealed part is isolated from the free part attached to the anchor by a separator of 0.8 m long. The ground properties are done by Dias et al. (1998a).

Table 1 gives elastic modulus value E, sections S and stiffness E.S for the fiberglass rods, the grout and the composite. These values result from the analysis done by Dias et al. (1998a).

Table 1. Mechanical characteristics.

	E	S	E . S
	MPa	m²	kN/m^2
Fiberglass rod	19000	$15.7.10^{-4}$	29830
Grout	2000	5.10^{-3}	10000
Composite			39830

2.2 Pull-out test

After a first tension value of 63 kN, the load was increased by steps of 26 kN (maintained constant for 15 minutes). The pull-out tests all ended with the failure of the rod. The failure load T_{TR} and the corresponding head displacement U_{TR} are given in Table 2 and Figure 3.

Table 2. Tests results.

	T_{TR}	U_{TR}	T_{TC}
	kN	mm	kN
3 m long rod	260	13.7	195 to 221
5 m long rod	370	22.7	300 to 326

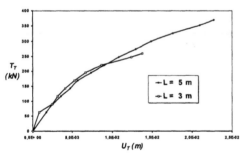

Figure 3. T_T - U_T curves.

The analysis of the creep displacements (Figure 4) obtained in 15 minutes shows that the rod failure occurs beyond creep load (an estimation of T_{TC} is given in Table 2).

The good agreement between U_T-T_T curves and creep curves up to a 221 kN load proves that the interface behavior is independent of the bolt length. Beyond this load, differences which occur required a more detailed analysis of the progressive mechanism of soil/composite interface friction (failure of the rod or of the grout).

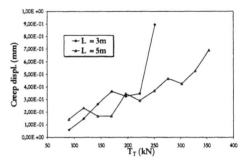

Figure 4. Creep curves.

3 INTERPRETATION

3.1 Local friction models

The global behavior of the bolt during a pull-out test results from its deformability on the one hand and the local relation between the friction load and the relative movement at the soil-bolt interface (Figure 5 with T tension, ε strain).

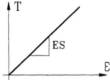

Figure 5. Tensile law.

In this study, three friction models were considered (figure 10). The first corresponds to the law suggested by Frank-Zhao for the piles, with a final stage preceded by two linear parts whose slopes are in a ratio equal to 5. In the second trilinear model, we does not fix a priori this ratio. Finally a bilinear simplified model was tested (Figure 6 with τ friction stress, qs constant level of friction stress, k, first slope and U local displacement), concretised by a fictitious threshold in which there is no soil-bolt displacement. The inclusion is characterised by a linear elastic behavior of stiffness ES.

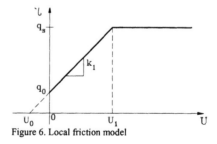

Figure 6. Local friction model

These various laws lead to an analytical description of global behaviour during a pull-out test of the bolt whose complex form is given by [Bourdeau, 1994 and Dias, 1998a]. These global models lead Bourdeau to distinguish two types of inclusions:

• The first known as "extensible" are characterised by the fact that when the q_s value (Figure 6) is reached at the rod head before any displacement of the rod end occurs.
• The second known as " Stiff " when rod end displacement occurs before the q_s value is reached at the head.

From the global analytical model and specific measurements, Bourdeau proposes methods of identification of the parameters. As an example, we give hereafter the step for the simplified model.

3.2 Evaluation of the friction model parameters

A graphical method to obtain the parameters of the local friction model has been proposed by Bourdeau et al. (1994) based on equation (1) and on an iterative process in which for the first step the second term of the right end side of (1) is neglected.

Equation (1) is used to find the local law assumed to be the same along the rod, without introducing any additional assumptions on the law shape.

$$\tau_T(U_T) = \frac{1}{p} \frac{T_T}{ES} \frac{dT_T}{dU_T} + \tau_Q \frac{dU_Q}{dU_T} \qquad (1)$$

Where T_T and U_T head force and displacement, p inclusion perimeter, τ_Q and U_Q tail end friction and displacement.

The second term of (1) vanishes when no displacement occurs at the tail end : that allows to build the τ - U curve till the constant level in case of "extensible" inclusions (the first step of the iteration process is sufficient). Iterative corrections are necessary in the other cases ([Dias, 1999b]).

An application of this methodology is presented in Figure 7.

For the 5 m long bolt, the τ - U (MNOP) relation calculated with only the first term of the formula (1) permits to reach the level value q_s inducing that the rod is "extensible" as foresaid. The decrease of the friction stress τ observed beyond the point O is due to the end bolt displacement that has been neglected in the first iteration. Nevertheless a q_s value can be inferred from this first step. An approximation of τ_Q can thus be introduced in (1) for the second iteration. This results in the MNOQ path. This process allows to reach back the initial maximum value, proving that there is no softening behavior and justifying the choice we made to interpret the results (Figure 6).

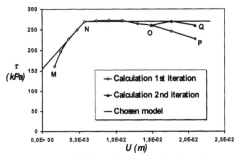

Figure 7. τ - U model for the 5 m long bolt (solution 1).

Table 3. τ - U model parameters.

	q_0 kPa	q_s kPa	k_1 kPa/m
3 m long rod	150	290	22500
5 m long rod	125	270	26000

Table 3 summarizes τ - U parameters used for the experimental curve simulations.

4 FRICTIONAL LOCAL LAW

Figure 10 presents the three local frictional laws selected, which simulate the T_T-U_T curve correctly. The values of the various parameters are presented in Table 5.

Table 5 . Frictional local law parameters.

	Properties	CASE 1	CASE 2	CASE 3
Bolt & Grout	E (MPa)	6.05E+09		
	S (m^2)	6.60E-03		
	E.S (kN)	3.99E+07		
		Frank-Zhao (n = 5)	Trilinear (Frank-Zhao with n = 40)	Bilinear (Toulon's work)
Interface	k_1 (kPa/m)	3.49E+07	3.18E+08	1.80E+09
	k_2 (kPa/m)	6.98E+06	7.95E+06	
	Bond strength (kN/ml)	87.9	84.8	86.4

It should be noted that these 3 laws are able to precisely simulate the relation between head load and displacement but that the Frank-Zhao model is less accurate for the simulation of the local loads in the bolt [Dias, 1999a].

5 APPLICATION TO A TUNNEL FACE

5.1 Data relating to calculations

In order to investigate the influence of the friction

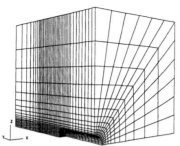

Figure 9. 3D mesh.

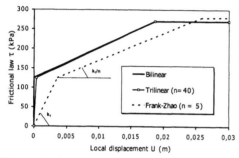

Figure 10. Frictional local laws retained.

Table 4. Soil mass, shotcrete lining and bolts properties [Dias, 1999b].

	Soil Mass : Red Permian	Support	Bolts
Behaviour	Mohr-Coulomb	Elastic	Elastic Plas-tic
Young's modulus (MPa)	300	10 000	20
Poisson's ratio	0.3	0.2	
Cohesion (kPa)	50		
Friction angle (de-grees)	20		
Dilatancy angle ψ (degrees)	0		
Initial state of stress (MPa)	0,8		
Section (m^2)	0		$1.4.10^{-3}$
Tension strength (t)	0.8		70

law, the analytical results are used and introduced into a numerical approach. The analysis has been carried out using a three-dimensional finite difference code (FLAC 3D), which allows to take into account the geometry of the tunnel, among others the lining behind the face and the non-linearities of the behavior. The tunnel is considered to be at great depth, and the initial stress field is supposed to be homogeneous and isotropic. As a consequence of the symmetry, only a quarter of the geometry is considered and each bolt, parallel to the tunnel axis, is modeled individually (Figure 9).

In the comparative study presented hereafter, the geometry as well as the geotechnical parameters considered, based on in situ measurements, are those adopted for the design of the cross-town tunnel in Toulon (France). The model is that of a circular tunnel of radius R= D/2 = 5.80m, in a deep and isotropic ground, modeled as an elastic-perfectly plastic material obeying Mohr Coulomb's yield criterion and its non associated flow rule. In the numerical model, the lining support behind the face (shotcrete) is linear elastic and is installed up to the tunnel face simultaneously with excavation step. This sequence of work is equivalent to pre-vaults executed ahead of the face. Geomechanical parameters for soil and shotcrete lining are as follows :

The bolts are modeled as linear elements of elastic-perfectly-plastic behavior with a fixed yield strength (Table 4).

Excavation has been simulated in 12 steps of 3 meters : at the beginning, each bolt has a length of ten times the tunnel diameter ($L_o= 10$ D). For each excavation step and for each bolt, a segment of three meters is removed from the model, therefore after n excavations we have bolts of length $L_n= (10D - 3n)$ in meters. The "degree" of reinforcement can however be considered constant despite this length reduction, since it has been shown that no significant increase of performance can be observed by increasing bolt length beyond 9 meters.

5.2 Results

The influence of various parameters is considered through the distribution of the bolt loads, the interface soil/bolt and the distribution of bolt displacements for the three laws selected.
- Frank-Zhao law corresponding to CASE 1.
- Trilinear law corresponding to CASE 2.
- Bilinear law without threshold corresponding to CASE 3.

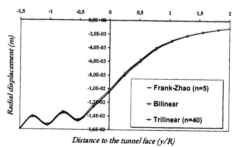

Figure 11. Radial displacement.

Figure 11 shows that the influence of the frictional law is not very significant on radial displacements (about 4 %). In order not to multiply the figures, we summarized in the table 6 the various quantities which have been compared. Calculation with the trilinear law is taken as reference in this comparison, because it corresponds to the solution nearest to the experimental results during pull-out tests.

Table 6. Difference between the several cases.

Frictional local law	Max. axial displacement	Max. bolt load	Axial volume
Trilinear (n=40)	Reference		
Frank-Zhao (n=5)	-2.9 %	6.4 %	-3.8 %
Bilinear	-6.2 %	11.2 %	-2.0 %

Only the maximum bolt loads seems to present notable variations (about 12 %) what nevertheless have a little influence on axial volumes and maximum axial displacements (variations about 6 %).

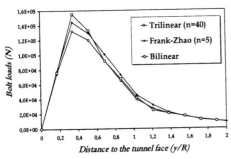

Figure 12. Medium load in the 52 bolts.

It is interesting to note that the shape and the values of the mobilised loads are overall very close (Figure 12), except differences located in the vicinity of the maximum. This can explain why the global behavior of the ground is very little affected.
Finally, it appears that the very simplifying assumption which we choose in the parametric study has a a minor influence on the total behavior of the ground, but can lead to differences of about 10 % on the maximum mobilized loads in the bolting.

5.3 Limits of the models in case of face bolting

The first strong assumption relates to the reversible behavior of the interface. This assumption is acceptable as long as the threshold q_s is not reached. On the other hand, when this threshold is reached it is certainly appropriate to take into account a nonreversible behavior. This phenomenon can be met in tunnels where the direction of mobilisation of a bolt evolves with the progress of the work, passing from a role of anchoring to that of a reinforcement applying of confinement loads at the tunnel face.

6 CONCLUSIONS

To determine the incidence of the frictional law between the soil and the bolts during our three-dimensional simulations, three laws have been considered :
• Two trilinear models
• A simplified bilinear model with threshold.

The global behavior observed during the extraction tests of bolts can be described in a correct way with all these models. On the other hand, the trilinear model appears more accurate to simulate the local behavior. The initial slope obtained by identification on this model is sufficiently high so that we can substitute to it, without notable loss of precision, the bilinear model with threshold.

If we consider the global behaviour during the digging of a tunnel, as well for the axial displacement of the face as for convergence, it appears that the different models leads at nearly identical results. The only notable differences relate to the mobilized loads in inclusions, where the variation on the maximum load did not exceeds, however, 18 %.

REFERENCES

Bourdeau, Y., Ogunro, V., Lareal, P. & Riondy, G. 1994. « Use of strain gages to predict soil-geotextile interaction in pull-out tests ». Fifth International Conference on Geotextiles, Geomembranes and Related Products, Singapore, 451-455.

Dias, D. & Bourdeau, Y. 1998a. « Comportement en ancrage de boulons en fibre de verre scellés au coulis de ciment ». Congrès Universitaire de Génie Civil, Reims-France, 8 p.

Dias, D. & Bourdeau, Y. 1998b. « Etude phénoménologique du comportement en ancrage de boulons scellés au coulis. ». Proceedings of A.I.G.I.E., Vancouver, 8 p.

Dias, D. & Bourdeau, Y. 1999a. « Influence of anchor length on grouted bolt behaviour ». Proceedings of the 5th International Symposium on Field Measurements in Geomechanics, Singapore, 4 p.

Dias, D. 1999b. « Renforcement du front de taille des tunnels par boulonnage – Etude numérique et application à un cas réel en site urbain ». Doctorate thesis, Insa Lyon, 320 p.

Modern Tunneling Science and Technology, Adachi et al (eds), © 2001 Swets & Zeitlinger, ISBN 90 2651 860 9

3-D energy damage model for bolted rockmass and its application in a tunnel engineering

Q.Y.Zhang   P.Y.Lu
Shenzhen Geological Bureau, People's Republic of China

W.S.Zhu
Institute of Rock and Soil Mechanics , The Chinese Academy of Sciences ,Wuhan, People's Republic of China

ABSTRACT: According to energy damage deformation mechanism of jointed rockmass, a 3-D energy damage model for intermittently jointed rockmass under initial damage, damage evolution and plastic damage deformation state is established in this paper. On the basis of this model, a cylindrical energy damage rock-bolt element model (the CEDRB model) is proposed to simulate effects of bolts reinforcement on the jointed rockmass. The proposed models have been applied to the stability analysis of Shenzhen Xiawan supply water tunnel under construction . The computed results of 3-D non-linear energy damage FEM effectively guide the engineering construction.

1 INTRODUCTION

A jointed rockmass contains a considerable number of intermittent joints and cracks which result in remarkable weakening and intensive anisotropy of the rockmass in its mechanical properties. In the case of large-sized discontinuities penetrating through the rockmass, the DEM technique and DDA method can be employed to simulate their mechanical properties (Cundall 1977,Shi 1988). The more difficult problem is how to simulate the intermittently distributed joints which are generally found in rockmass . For such joints, literature (Yang 1990) gives constitutive relation of elastoplastic damage based upon self-consistent method and literature (Xu 1992) gives constitutive relation of damage fracture using the principle of geometrical damage, however, these models can not simulate the damage evolution of jointed rockmass effectively .

This paper, according to the damage propagation process of 3-D crack and by applying the theories of energy damage mechanics and FEM computation method , studies the 3-D energy damage properties coupled with plastic damage deformation of intermittently jointed rockmass as well as spacial reinforcing effect of bolts in jointed rockmass , and a 3-D nonlinear energy damage FEM program has been developed based upon the models proposed in the present paper and successfully applied to the Shenzhen Xiawan water-supplying tunnel engineering.

2 ENERGY DAMAGE MODEL FOR JOINTED ROCKMASS

2.1 Energy damage evolution equation for jointed rockmass

Assuming that the joints and cracks in a rockmass are penny-shaped, the initial damage strain energy density ϕ_e can be given as following (Zhang 1998):

$$\phi_e = \frac{1}{2}\sigma_{ij}c^e_{ijkl}\sigma_{kl} + \frac{1}{E_0}\sum_{k=1}^{M}\{a^{(k)^3}\rho_v^{(k)}[G_1(1-C_v^{(k)})^2\sigma^{(k)^2}$$

$$+ G_2(1-C_s^{(k)})^2\tau^{(k)^2}]\} \tag{1}$$

Where σ is the applied stress tensor in a far-field ; C^e_{ijkl} is the elastic flexibility tensor of non-damaged rockmass or intact rock , as expressed in eq . (2)

$$C^e_{ijkl} = \frac{1+v_0}{2E_0}\left(\delta_{ik}\delta_{jl} + \delta_{jk}\delta_{il}\right) - \frac{v_0}{E_0}\delta_{ij}\delta_{kl} \tag{2}$$

where E_0, v_0 are the elastic modulus and Poison's ratio of non-damaged rockmass respectively;M is the number of prevailing joint sets in a unit volume of jointed rakmass ; $a^{(k)}$ is the average radius of the k th joint set in statistics ; $\rho_v^{(k)}$ is the average volume density of k th joint set;$C_v^{(k)}$ and $C_s^{(k)}$ are the pressure-transmitting and shearing-transmitting coefficients of the k th joint set respectively; σ , τ are the projections of the applied stress tensor along the normal and tangential directions of a joint plane respectively; and G_1, G_2 are the coefficients (Zhang 1998).From the theory of elasticity , we have initial damage flexibility tensor of the jointed rockmass by derivation of ϕ_e with respect to σ_{op} :

$$
\begin{aligned}
C_{opkl}^{od} = \frac{1}{E_0} \sum_{k=1}^{M} \{ & \rho_v^{(k)} a^{(k)^3} [2G_1(1-C_v^{(k)})^2 n_o^{(k)} n_p^{(k)} n_k^{(k)} n_l^{(k)} \\
& + \frac{1}{2} G_2(1-C_s^{(k)})^2 (\delta_{kp} n_o^{(k)} n_l^{(k)} + \delta_{ko} n_p^{(k)} n_l^{(k)} + \delta_{lo} n_p^{(k)} n_k^{(k)} \\
& + \delta_{lp} n_o^{(k)} n_k^{(k)} - 4 n_o^{(k)} n_p^{(k)} n_k^{(k)} n_l^{(k)})]\}
\end{aligned}
\tag{3}
$$

in which n_k, n_l, n_k', n_l' $(k,l,k',l' = 1,2,3)$ are the directional cosine of the unit vector of the joint plane .

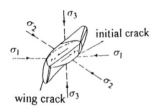

Figure 1. Propagation model of 3-D crack.

According to propagation mechanism of 3-D crack (Figure 1),we can get the flexibility tensor of damage evolution of jointed rockmass C_{opkl}^{ad} (Zhang et al.1999)

$$
\begin{aligned}
C_{opkl}^{ad} = \frac{1}{E_0} \sum_{K=1}^{M} \{ & \rho_v^{(k)} a^{(k)^2} [B_1^{(k)} n_o^{(k)} n_p^{(k)} n_k^{(k)} n_l^{(k)} \\
& + B_2^{(k)} (n_o^{(k)} n_l^{(k)} \delta_{kp} + n_p^{(k)} n_l^{(k)} \delta_{ko} \\
& + n_o^{(k)} n_k^{(k)} \delta_{lp} + n_p^{(k)} n_k^{(k)} \delta_{lo})]\}
\end{aligned}
\tag{4}
$$

Coefficients in eq.(4) are referred to Zhang et al.(1999).From the above, we have the equivalent elastic damage flexibility tensor of a jointed rockmass which takes the initial damage and damage evolution into account:

$$
C_{opkl}^{e-d} = C_{opkl}^e + C_{opkl}^{od} + C_{opkl}^{ad}
\tag{5}
$$

where C_{opkl}^e , C_{opkl}^{od} , C_{opkl}^{ad} are expressed as in eqs.(2), (3)and (4) respectively and C_{opkl}^{e-d} obviously meets symmetrical condition.

According to the theory of energy damage, we have the damage tensor representation under 3-D stress state , which reflects the anisotropic damage degree of a jointed rockmass:

$$
\omega_{ijkl} = I_{ijkl} - \left(C_{ijop}^{e-d} \right)^{-1} C_{opkl}^e = I_{ijkl} - E_{ijop}^{e-d} C_{opkl}^e
\tag{6}
$$

where I_{ijkl} is the four-dimentional unit tensor , E_{ijop}^{e-d} is the equivalent elastic damage stiffness tensor . Then the energy damage evolution equation of the jointed rockmass can be derived from eq.(6)

$$
\dot{\omega}_{ijkl} = - \frac{\partial E_{ijop}^{e-d}}{\partial \sigma_{mn}} C_{opkl}^e \dot{\sigma}_{mn} = F_{ijklmn} \dot{\sigma}_{mn}
\tag{7}
$$

2.2 Constitutive model of energy damage for jointed rockmass

Referring to literatures (Zhang et al.1999), We introduce the effective stress $\bar{\sigma}$ to reflect the coupled efficiency of damage and plastic deformation and establish the plastic damage surface equation of the rockmass in the effective stress space as following:

$$
f = f(\bar{\sigma}, q) = 0
\tag{8}
$$

According to the generalized orthogonal flow rule and the condition of plastic damage consistency $\dot{f} = 0$,We have 3-D constitutive equation of energy damage for jointed rockmass (Zhang et al.1999).

$$
\dot{\sigma}_{kl} = K_{klrs} \dot{\varepsilon}_{rs}
\tag{9}
$$

where K_{klrs} is the modulus tensor of energy damage for jointed rockmass, and whose numerical evaluation can be found in Zhang et al.(1999).

3 CYLINDRICAL ENERGY DAMAGE ROCK-BOLT ELEMENT MODEL

The bolts are represented by a cylindrical energy damage rock-bolt element (CEDRB element) to simulate reinforcement effect of bolting. Suppose that the bolt and jointed rockmass within a certain range comprise a CEDRB element (see Figure 2).A CEDRB element is buried in a rockmass to form a composite element of jointed rockmass with bolts, referred to as the BDR element in Figure 3.

The CEDRB element is assumed as orthogonal anisotropic damage medium. The contribution of CEDRB element to the overall stiffness matrix of the system, i.e., the additional stiffness matrix $[K_e]_{add}$, can be derived according to the interpolation theory of FEM and the theorem of virtual work:

$$[K_e]_{add} = [N]^T [K_e][N] \qquad (10)$$

where $[N]$ is the matrix of shape function , $[K_e]$ is the stiffness matrix of CEDRB element in the global coordinate system of xyz, having value of

$$[K_e] = [T]^T \{\int_v [B]^T [D_e][B]dv\}[T] \qquad (11)$$

in which $[T]$ is the stiffness conversion matrix , $[D_e]$ is the elastic energy damage matrix of the CEDRB element .

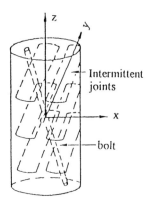

Figure 2. Cylindrical energy damage rock-bolt element model (The CEDRB model).

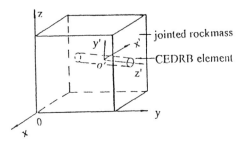

Figure 3. Bolted damaged rockmass element (The BDR element).

$$[D_e] = \begin{pmatrix} S_{11} & S_{12} & S_{13} & 0 & 0 & 0 \\ S_{21} & S_{22} & S_{23} & 0 & 0 & 0 \\ S_{31} & S_{32} & S_{33} & 0 & 0 & 0 \\ 0 & 0 & 0 & 2G_{yz} & 0 & 0 \\ 0 & 0 & 0 & 0 & 2G_{zx} & 0 \\ 0 & 0 & 0 & 0 & 0 & 2G_{xy} \end{pmatrix} \qquad (12)$$

Where S_{ij} ($i, j=1,2,3$), G_{yz}, G_{zx}, G_{xy} are elastic damage coefficients of the CEDRB element, they are evaluated numerically, see Zhang et al.(2000) for detail.

4 APPLICATION IN A TUNNEL ENGINEERING

4.1 General situation of engineering

Shenzhen Xiawan water-supplying tunnel is located at Buji town of ShenZhen. Its northern latitude is $20° 35'$, east longitude is $114° 11'$, Axis strike of the tunnel is $N 15° 30' W$. Southern cavity door of the tunnel is away from the Shenzhen reservoir about $700m$, northern cavity door of the tunnel is away from the Buji road about $300m$.

The whole length of the tunnel is $400m$. The tunnel section is $5m \times 5m$. The tunnel shape is archway of a city gate (see Figure 4). There to be two water-supplying pipes and a roadway in it .

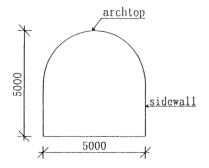

Figure 4. The cross section of the tunnel.

4.2 Mechanical parameters of surrounding rockmass ,geometric characteristics of joints and FEM computation scope

Rockmass surrounding the tunnel mainly consists of mixed granite.Mechanical parameters of surrounding rockmass is in table 1.There are three joint sets distributing in opening zone. Geometric

Table 1. Mechanical parameters of surrounding rockmass.

Rockmass	Density (KN/m^3)	Saturated compressive Strength (Mpa)	Saturated tensile strength (Mpa)	Poison's ratio	Young's modulus (Mpa)	shearing strength C (Ma)	tg ϕ
Completely Weathered Mixed granite	17.5	0.75	0.0	0.35	4.36	0.025	0.46
Strongly weathered Mixed granite	18.4	1.5	0.0	0.30	4.72	0.028	0.50
Slightly weathered Mixed granite	26.58	62.92	7.15	0.25	15465	2.0	0.85

Table 2. Geometric characteristics of joint sets in opening zone.

Joints	Strike $(°)$	Dip $(°)$	Dip angle $(°)$	Interval (m)	length (m)
Set 1	214	304	80	1.0	1.0
Set 2	65	155	41	0.4	2.0
Set 3	166	256	59	0.8	2.5

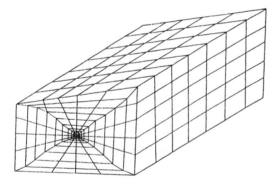

Figure 5. Schematic diagram of finite element mesh.

characteristics of joint sets is in table 2. Initial ground stress field of opening zone is obtained by back analysis .The installed bolts are made of indented steel bars of high-strength and their installing length is 2.5~3.5m with spacing of 1m*1m.

Finite element computation scope is three hundred metres long , one hundred and fifty metres wide and two hundred metres high. Schematic diagram of finite element mesh is Figure 5 .

4.3 Analysis for computed results

Radial stress of the tunnel is released and tangent stress of the tunnel is increased after excavation. Stress concentration appears in arch foot and sidewall foot . Deformation of the tunnel faces to inside of opening(see Figure 6). The maximum

computed displacement for arch top is 10.5mm, which is 8.7% larger than the measured displacement. The maximum computed displacement for sidewall is 19.7mm,which is 3.6% larger than the measured displacement. The computed displacements of the tunnel are basically coincident with the measured displacements of the tunnel and the computation error between them is less than 10% (see Figure 7), which indicates that the mechanical models proposed in this paper are rational and reliable .

Figure 8 shows that the distribution of damage evolution zone of opening after excavation. After the reinforcement by bolts, the distribution zone of damage evolution is much smaller than before the reinforcement by bolts (see Figure 9), showing significant improvement of the integrity, fracture toughness and bearing capacity of rockmass due to bolting.

According to above computed results, We have carried on systematically combined bolting and shotcreting at arch top, arch foot and sidewall of the tunnel and achieved satisfactory effect .

5 CONCLUSION

According to the bolted energy damage models proposed in this paper, We carry on stability analysis for Shenzhen Xiawan water-supplying tunnel during construction. The computed displacements are basically coincident with the measured displacements and the computed error between them is less them 10% . Xiawan tunnel is basically stable under excavation , but arch top , arch foot as well as sidewall of the tunnel must be systematically

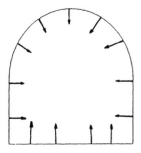

Figure 6. Deformation of opening after excavation

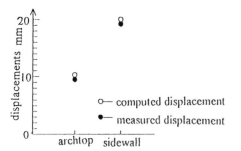

Figure 7. Comparison between the computed displacements and the measured displacements.

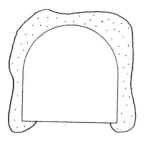

Figure 8. Distribution of damage evolution zone (dot area) before the bolting.

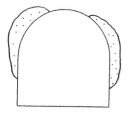

Figure 9. Distribution of damage evolution zone (dot area) after the bolting.

supported. Computed results show that the bolted energy damage models proposed in this paper well simulate mechanical deformation properties of jointed rockmass and reinforcement effect of bolts. The non-linear energy damage FEM results also show that the rock-bolt supporting can increase the integrity, fracture toughness and bearing capacity of the surrounding rockmass and decrease the damage evolution zone significantly.

Construction for Xiawan water-supplying tunnel was managed by first author. By adopting New Austrian Tunnelling Method, the tunnel construction was completed with high quality and high speed and the engineering quality has achieved excellent grade .

REFERENCES

Cundall,P.A.1977.Measurement and analysis of accelerations in rock slopes.*Ph.D dissertation*, Imperial College ,London.
Shi,G. H . 1988. Discontinuous deformation analysis ---a new numerical model for the statics and dynamics of block systems . *Ph.D dissertation* , Department of Civil Engineering , University of California ,Berkeley ,USA.
XU,J.N.& Zhu, W. S. 1992. The damage fracture analysis of jointed rockmass and its application in engineering. *Engineering Fracture Mechanics* . *43(2) :165-170.*
Yang,Y.Y.1990. A fracture damage model for Jointed rockmass and its application to the rock engineering. *Ph.D dissertation*, Qinghua university ,China.
Zhang, Q.Y.1998 . Research on 3-D damage fracture model for bolted intermittent multi-cracks rockmass and its numerical simulation with application in engineering . *Ph .D dissertation* , Institute of Rock & soil Mechanics , The Chinese Academy of sciences , Wuhan ,China.
Zhang, Q.Y. et al.1998. Analysis of elastoplastic damage for high jointed slope of the Three Gorges project shiplock during unloading due to excavation. *Chinese Journal of Hydraulic Engineering.(8):19-22.*
Zhang, Q .Y. & Zhu,W.S.1998.Elastoplastic damage constitutive model for jointed rockmass and its rock-bolt computation. *Chinese Journal of Geotechnical Engineering.20(6):90-95.*
Zhang,Q.Y. et al. 1999.Application of elastoplastic damage model in a large sized underground power house.*Chinese Journal of Rock Mechanics and Engineering.18(6):654-657.*
Zhang,Q.Y. & Xiang,W.1999.Constitutive model for energy damage of jointed rockmass and its engineering application.*Chinese Journal of Engineering Geology.7 (4):310-314.*

Zhang,Q.Y. & Xiang,W.2000.Application of 3-D elastoplastic damage model with bolts to a large-sized rockmass slope engineering.*Chinese J.Wuhan Univ.of Hydr. & Elec. Eng. 33(2):6-10*

Zhang,Q.Y. & Xiang,W. 2000.Application of 3-D elastoplastic damage model with bolts in XILUODU underground power house.

Chinese Journal of Computational Mechanics.17(4):475-482.

Modern Tunneling Science and Technology, Adachi et al (eds), ©2001 Swets & Zeitlinger, ISBN 90 2651 860 9

Centrifuge model test of tunnel face reinforcement by bolting

H.Kamata & H.Mashimo
Public Works Research Institute, Independent Administrative Institution, Tsukuba, Ibaraki, Japan

ABSTRACT: In order to clarify the effect of the typical auxiliary methods using bolting (face bolting, vertical pre-reinforcement bolting and forepoling) on tunnel face stability centrifugal model test was carried out with various arrangements of the bolts and lengths in sandy ground. Also, the experimental results were compared with the analytical results by DEM (Distinct Element Method). The experimental results show that each bolting has an optimum length and arrangement to act effectively and it is shown that the effect of the bolting obtained from the experimental results can be simulated by DEM.

1 INTRODUCTION

To excavate a tunnel safely by mountain tunneling method, keeping tunnel face stable is necessary. When the face stability cannot be assured at the bad ground condition auxiliary methods using bolting such as face bolting, forepoling, etc. are often used for tunnel face stability. However the design of these auxiliary methods, such as the optimum length, arrangement, stiffness of the bolts, etc. have not been established yet. Mashimo et al. (1998) have carried out the same centrifugal model tests and have proposed the way of evaluating the face stability in sandy ground. Concerning face bolting, R.AlHallak et al. (2000) have found that a face bolting is effective and it depends on the bolt density by centrifugal model tests using dry sand.

In this study, in order to clarify the effect of the typical auxiliary methods (face bolting, vertical pre-reinforcement bolting and forepoling) on the face stability, centrifugal model tests were carried out with various arrangements of bolts and lengths, and the results were compared with the analytical results by Distinct Element Method (DEM).

2 CENTRIFUGE MODEL TEST

2.1 *Test equipment and test procedure*

2.1.1 *Test equipment*

Figure 1 shows the cross section of the centrifugal model. A container with a transparent acrylic panel in front surface and dimensions of 140×500×400 mm was used. A model tunnel with a diameter D of 80mm made of half a cylindrical acrylic shell was used. 5mm thick and a semicircular aluminum plate was also installed to hold back the tunnel face in the container and Teflon sheet was adhered to the inner walls of the container in order to reduce friction.

2.1.2 *Test procedure*

The model ground with the model tunnel embedded was taken to the centrifuge equipment. When the centrifugal acceleration achieved to the fixed value, the aluminum plate of the tunnel was pulled out to release the stress at the face, and then the stability of the face was observed. In order to consider no span near the face, the model tunnel with the aluminum plate protruding of 0.1D to the ground from the shell was set.

Unsaturated Toyoura standard sand was used.

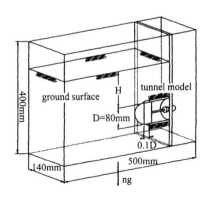

Figure 1. View of container with model tunnel.

Table 1 shows the material properties. The model ground was made by pouring and compacting every 2cm layer.

Bolts made of phosphor bronze and coated with sand for friction were used. Bolts were installed when the ground model was made. The overburden ratio H/D (the ratio of the overburden H to the tunnel diameter D) was fixed at 1.0 in all tests.

2.2 Results of model test and consideration

2.2.1 No reinforcement

Figure 2 shows the states of face failure at centrifugal acceleration of 25g and of 30g, respectively. In either case a slip surface originated near the bottom of the face and spread toward the top ahead of the face. The failure area reached approximately 0.2D to 0.3D toward the face. However, at 25g the failure area is closed approximately 0.4D above the crown of the tunnel (dome-shaped called in this paper), while at 30g the failure reached the ground surface.

2.2.2 Face bolting

Face bolts of 1.2mm in diameter were used. The tests were performed with various lengths of bolts (0.25D,0.5D,1.0D and 1.5D) and arrangements (in the full section , in the upper half section and in the lower half section) for each length. The tests were performed at 25g. Figure 3 shows the arrangement of the bolts.

Figure 4(a) shows the states of face failure where bolts were installed in the full section. It can be seen that the extent of failure area was minimized with the bolt length longer than 0.5D. In other words, 0.5D bolt length provided the same reinforcing ef-

fect as longer bolts. On the other hand, in the case of 0.25D, although the failure area above the crown was a little smaller than without reinforcement, the effect on face stability cannot be expected. The necessary fixing length for the bolt to act effectively may be insufficient in the case of 0.25D because face failure occurs to the extent approximately 0.2D to 0.3D from the face without reinforcement.

Figure 4(b) and 4(c) show the states of face failure where the bolts were installed in the lower half section and in the upper half section, respectively. In figure 4(b), some amount of effect on face stability was recognized at the lower part of face. However, the dome-shaped failure occurred above the crown for every length as well as without reinforcement. On the other hand, in figure 4(c), although small failure occurred at the lower part of face, the dome-shaped failure did not occur. This is due to the following. Failure occurrs from the lower part of face

Table 1. Properties of Material.

Specific gravity Gs	2.64
Unit weight $\gamma t(kN/m^3)$	15.1
Water content (%)	6.5
Cohesion Cd(Kpa)	4.6
Friction angle $\phi d(deg)$	34.5

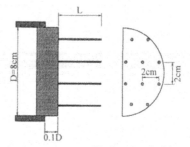

Figure 3. Arrangement of bolts (face bolting).

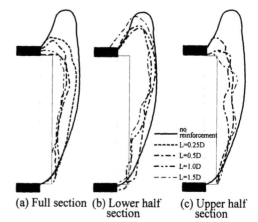

(a) Full section (b) Lower half (c) Upper half
section   section

Figure 4. Failure patters (face bolting).

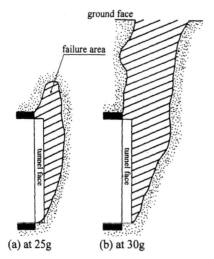

(a) at 25g   (b) at 30g

Figure 2. Failure pattern (no reinforcement) .

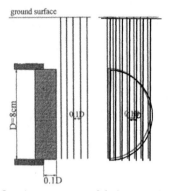

Figure 5. Arrangement of bolts (Vertical pre-reinforcement bolting).

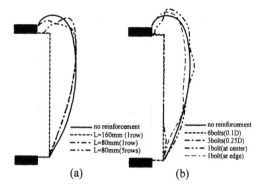

Figure 6. Failure patters (vertical pre-reinforcement bolting).

at the beginning, however the face bolts in the upper half section prevent the failure from extending further to the upper area. These results suggest that, in case the face bolting is adopted, installing them in the upper section is more effective and safety than in the lower section

The axial force and the bending moment of the bolts were measured in the case of the full section. The tensile force was generated along the overall length and the maximum value was approximately 1.0N at the center of the bolts. On the contrary, the bending moment hardly was generated except in the area near the face. The maximum value was little more than 0.03 N·mm. This indicates that the effect of face bolting depend mainly on an axial force.

2.2.3 *Vertical pre-reinforcement bolting*

Vertical pre-reinforcement bolts of 1.0mm in diameter were used. In order to investigate the influence of the length on face stability, two kinds of length were adopted; 160mm (from the ground surface to the tunnel invert) and 80mm (from the ground surface to the tunnel crown). As shown in Figure 5, bolts were installed at the 0.1D spacing along the cross section and along the longitudinal section.

Figure 6(a) shows the states of face failure. It can be seen that the face stability was assured where the bolts (L=160mm) was installed only in one row along the longitudinal section at a distance of 0.1D from the face. In the case of the bolts (L=80mm), failure occurred although the dome-shaped area above the crown was a little smaller than in the case of no reinforcement. In addition, the case of the bolts (L=80mm) in five rows along the longitudinal section was performed. The state of failure did not changed in comparison with in one row. As is evident from these results, bolt length is the important factor for face stability.

Further tests were carried out to obtain the information on the influence of the arrangement of bolts along the cross section. The number of row and the length are fixed at 1 and at 160mm, respectively, and

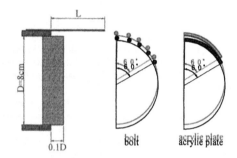

Figure 7. Arrangement of bolts (forepoling).

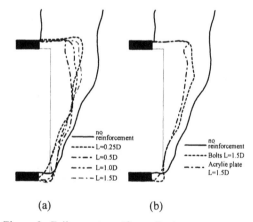

Figure 8. Failure patters (forepoling).

the spacing along the cross section was varied. Figure 6(b) shows the results. Although failure occurred with one bolt, face stability can be assured with more than three bolts.

From the above experimental results, it is found that vertical pre-reinforcement bolts reaching the tunnel invert is more effective on face stability than many short bolts ahead of the face, and it is also made clear that the optimum spacing of the bolt along the cross section exists.

2.2.4 *Forepoling*

Forepoling of 1.0mm in diameter were used. The length ahead of the face was varied 0.25D, 0.5D, 1.0D, and 1.5D. As shown in figure 7, bolts were installed at the range of 60 ° from the crown and at 1cm spacing. In addition, to investigate the influence of the forepoling stiffness, a test was performed using the forepoling made of acrylic plate. The tests were performed at 30g because of examining the effect not only on face stability but also on the restraint of settlement at the ground surface, where the failure reached ground surface without reinforcement (see figure 2 (a)).

Figure 8(a) shows the state of failure in the case of the forepoling. No evident differences were found among the tests in the failure shape and area. Concerning the effect on the restraint of settlement at the ground surface, it can be seen that the forepoling prevents the failure reaching the ground surface for every length. However, the effect on face stability cannot be seen. Figure 8(b) shows the results using the forepoling of acrylic plate. The results are almost the same as in figure 8(a).

Although forepoling can be expected to support the soil pressure induced by the loosen area ahead of face, it may have a limit to the effect on face stability in the bad ground condition. If face stability needed, it is necessary to adopt the other method such as enhancing the strength of face, dividing the face to be excavated at a time.

3 NUMERICAL ANALYSIS

3.1 *Analysis method*

Numerical simulation was carried out for comparison with the above experimental results. The behavior as particles may account for a significant portion of face stability analysis as well as that as a continuum. Therefore, Distinct Element Method (DEM) was used. The program code used is UDEC (Itasca 1996). The discontinuous medium (such as a jointed rock mass) is represented as a assembly of discrete blocks and large displacements along discontinues and rotations of blocks are allowed. Individual blocks can be treated as either rigid or deformable material. In addition, this program can simulate the reinforcement such as fully-grouted bolts. Therefore, it is possible to evaluate the effect of bolting on face stability.

The computed results depend greatly on block size and joint patterns. Therefore, the smaller the block size is made, the closer the model is to a real soil. However, a practical limit exists because the computational time and capacity increase with making block size small. Besides, the way of determining the parameters such as the stiffness, strength between the blocks has not been established yet.

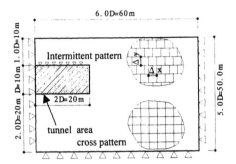

Figure 9. Numerical model.

Table 2. Parameters .

Sand (Block)	E(Mpa)	20
	ν	0.35
	γ t (KN/m³)	16
Discontinuities (Between blocks)	Kn (KN/m)	2.2 ×10⁵
	Ks (KN/m/m)	7.0× 10⁴
	C(Kpa)	1.0
	φ	0.0
Bolts (Cable Element)	E(Mpa)	2.06×10⁵
	Area (m²)	5.1×10⁻⁴
	Tensile and Compressive Strength(kN)	123
	Kbond(MN/m/m)	12
	Sbond(kN/m)	96

In spite of the uncertainties of determining the parameters, DEM may be more suitable than FEM (Finite Element Method) to evaluate the effect of bolting on face stability because DEM is thought to be able to judge the effect from whether the failure occurs or not as well as the centrifugal model test. First of all, the block size, joint patterns and parameters were investigated to simulate the failure pattern without reinforcement. Thereafter, under the same conditions, the effects of bolting on face stability were simulated with installing bolts.

Input parameters consist of a block elasticity (E), normal and shear stiffness (kh, kv) between blocks ,and joint strength (C,φ). For the simplification, kh kv and E were constant and C,φ were varied. The simulations were performed using the field scaled model in figure 9 and at 1g. Two types of joint patterns were adopted: intermittent and cross-continuous. Blocks of 0.5 × 0.5 m and of 0.25 × 0.25m also were adopted .The calculation steps are as follows. At the first step the stresses before tunnel construction are calculated. At the second step

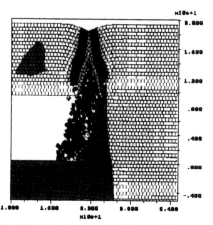

Figure 10. Block displacement without reinforcement
(A block size of 0.5 ×0.5m).

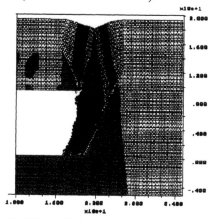

Figure 11. Block displacement without reinforcement (A block size of 0.25 ×0.25m) .

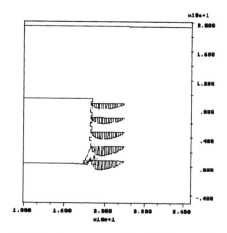

Figure 12. Velocity vector and axial forces with
face bolts of 0.5D.

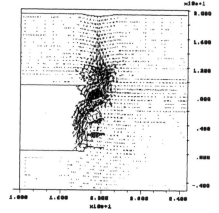

Figure 13. Velocity vector and axial forces with
face bolts of 0.25D.

stresses and deformations are calculated after excavation of the tunnel part at a time.

3.2 Analysis results and consideration

3.2.1 No reinforcement

Figure 10 shows displacements of blocks in the case
of intermittent joint pattern and blocks of 0.5 ×
0.5m. It can be seen that failure occurred. A slip surface was originated from the bottom of the face and
spread toward the top ahead of the face (at first
dome-shaped was formed and finally failure reached
the ground surface). The failure area reached approximately 0.3D from the face. These results are in
good agreement with the experimental (see figure 2).
However, it was not able to simulate such a equilibrium state as dome-shaped at 25g. In the case of
cross-continuous joint pattern, the face was extruded
from the center of the face and a slip surface reached
the ground surface at a stretch, which are different

from the experimental results. Concerning the influence of block size, figure 11 shows the results with
the intermittent joint pattern and blocks of 0.25 ×
0.25m .It can be seen that the failure pattern and area
are almost the same as that in figure10. This means
that if the block size is made small to some extent, it
is not necessary to reduce block size further to simulate the failure. Parametric analyses were performed
to investigate the relationship between parameters of
strength (c,φ) and face stability. In the case of c below 4kpa failure occurs where φ=0. In addition, in
the case that the diameter of the tunnel is 5.0m under
the same parameters, face stability is assured.

3.2.2 Face bolting

Intermittent joint pattern and blocks of 0.5 ×0.5m
were adopted hereafter to simulate the effect of bolting. Table 2 shows the parameters for the calculation. Cable element of UDEC generating only axial
force was used for the model of bolting. Figure 12

shows the result in the case of face bolts of 0.5D in length in five rows on the face. The calculation achieved equilibrium state finally and face stability were assured although a small failure occurred at the part of lower face. In the equilibrium state, velocity of a block turns out to be approximately zero. Tensile forces were generated along the overall length and had a maximum approximately halfway of the bolt. In the case of 1.0D, face stability was also assured. On the other hand, in the case of 0.25D the calculation was not able to achieve equilibrium state and finally failure occurred as shown in figure 13. These results are consistent with the experimental results. In order to make face bolt act effectively, it is important to install bolts of the necessary and sufficient length.

Furthermore calculations were made to examine the influence of position of bolts. Face stability was able to be assured with face bolts neither only in the upper half section nor only in the lower half section. Any evident differences were not observed between cases unlike the experimental results.

3.2.3 Vertical pre-reinforcement bolting

Vertical pre-reinforcement bolts reaching to the invert of the tunnel was able to prevent the failure as shown in figure 14. On the contrary, in the case of vertical pre-reinforcement bolts to the crown, failure occurred finally as shown figure 15. These results are also in good agreement with the experimental results. If examined in detail in Figure 14, the axial forces of the bolt were almost compressive, which is different from the result of face bolting. From this result it is found that the mechanism of reinforcement of vertical pre-reinforcement bolting might be different from that of face bolting.

4 CONCLUSION

This paper involved the performance of centrifugal model test of face stability, and analyzed the test results with DEM. The principal conclusions were obtained as follows.

1 Face bolting has a great effect on the face stability when the bolt length is longer than the 0.5D. Face bolting installed in the upper half section of tunnel face has more effect than that in the lower half section.

2 Vertical pre-reinforcement bolting reaching tunnel invert ahead of the tunnel face has a great effect on face stability

3 Forepoling has an effect to prevent the failure extending to the ground surface. However, it may be less effective on the face stability than other bolting in the bad ground condition.

4 The failure without reinforcement and the effect of bolting shown in the experiment can be simu-

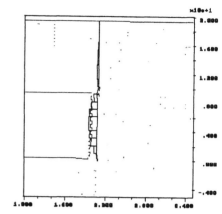

Figure 14. Velocity vector and axial forces with vertical pre-reinforcement bolts reaching to the invert.

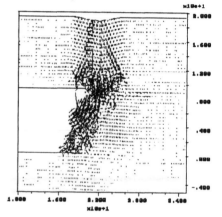

Figure 15. Velocity vector and axial forces with vertical pre-reinforcement bolts reaching to the crown.

lated using the model of the intermittent joint patterns and blocks of 0.5 × 0.5 m by DEM.

REFERENCES

H.Mashimo & M.Suzuki 1998. Stability conditions of tunnel face in sandy ground. *Centrifuge 98*: 721-725.Rotterdam: Balkema.

R.AlHallk et al. 2000. Experimental study of the stability of a tunnel face reinforced by bolts, *Geotechnical Aspect of Underground Construction in Soft Ground*: 65-68 Rotterdam: Balkema.

Modern Tunneling Science and Technology, Adachi et al (eds), © 2001 Swets & Zeitlinger, ISBN 90 2651 860 9

Development of MGF Method Based on the Evaluation of Forepiling Supporting Mechanism

Y. Kitamoto, K. Date, T. Yamamoto, & K. Hibiya
Civil Engineering Department, Kajima Technical Research Institute, Tokyo, Japan

H. Ohta
Dept. of Civil Eng., Tokyo Institute of Technology, Tokyo, Japan

ABSTRACT: The authors, investigating the rational and economical tunnel presupport and its design procedure, developed a new method to decrease surface settlement better than conventional one. In this paper, the advantage of the new method is verified and its supporting mechanism is evaluated by centrifugal model tests and the cylindrical shell theory.

1 INTRODUCTION

In recent years, in place of shield tunneling method, NATM has been frequently adopted in soils with little cohesion in urban areas mainly for the economic purposes. Also, as the cross section of NATM tunnel is getting larger, in even ground of worse geological conditions, the development of the technique to control the deformation is especially important.

While construction techniques are advancing, no rational design methods have yet been established. The development of a simple design method based on the understanding of supporting mechanism is heavily demanded.

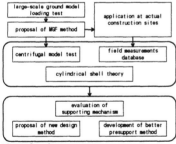

Figure1. Flow of study.

The presupport method using long steel pipes (referred to as the conventional method below) is generally selected from among existing presupports because of its great supporting effect in spite of high cost. As shown in Figure 1, the authors continued studies to develop a better presupport method providing greater supporting effect at lower cost

than the conventional one, and to propose a new design method based on the understanding of supporting mechanism.

This paper presents an outline and the results of field application of the presupport method using medium-length steel pipes (referred to as the MGF method below, MGF: Multi-Ground-Forepiling) (Yamamoto et al. 1999). Furthermore, the results of a centrifugal model test carried out to verify the supporting effect and to evaluate presupport mechanism and the evaluation of the results of a centrifugal test and field measurement based on the cylindrical shell theory are discussed.

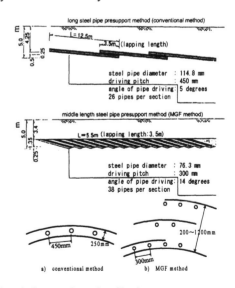

Figure2. Cross section and profile of presupports.

2 MGF METHOD AND THE RESULTS OF ITS APPLICATION

The MGF method is classified between conventional forepoling and forepiling methods. It is characterized by the driving of medium-length steel pipes of a diameter smaller than in the conventional method at greater angles (Fig. 2). The MGF method enables cost reduction and faster construction by using jumbos and integrated special pilot-and-reaming bits (referred to as special bits below). At the same time, it controls settlement more effectively than the conventional method. Its effectiveness has been verified by loading tests using large-scale models (Date et al. 2000). Table 1 lists the features of the MGF method.

At a railway tunnel construction site, the conventional and MGF methods were applied under almost similar geological and earth cover conditions for comparison of the measurements obtained (Fig. 2). Table 2 shows actual surface settlement data (average measurements). As shown in the table, surface settlement at the end of the excavation of the top heading and the bench was 48.3 mm and 15.4 mm in the sections where the conventional and MGF methods were adopted, respectively. That is, settlement in the section with the MGF method was held to one-third of that in the section with the conventional method to show that the effectiveness of the MGF method in controlling surface settlement was verified.

Table1. Features of MGF method.

Benefits	Description
Lower cost	-Construction by jumbos -Adoption of inexpensive medium length steel pipes, e.g. gas pipes -Guarantee of rod bit recovery owing to the use of medium-length steel pipes
Better quality	-Greater face stability and better control of ground deformation owing to the thickening of improved zone and the effect of steel pipes as diagonal bolts -Prevention of the falling of ground between pipes by close driving of steel pipes, and control of ground deformation by creating a shell -Greater tunnel stability owing to decrease of tunnel cross section
Shorter construction period	-Elimination of rod connection work as a result of use of medium length steel pipes -Smooth construction, and guarantee of rod bit recovery through the adoption of medium length steel pipes and special bits -Easier work in boulder stones or fracture zone as a result of use of small-diameter pipes
Greater safety	-Less demanding work at high places owing to the elimination of rod connection work

Table2. Surface settlement.

	Section with the conventional method	Section with MGF method
	mm	mm
At the end of top heading excavation	43.1	12.4
At the end of bench excavation	48.3	15.4

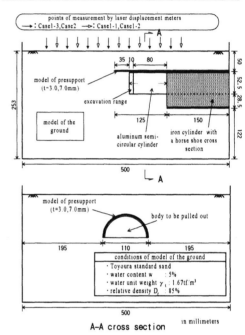

Figure3. Outline of tunnel excavation test model.

3 CENTRIFUGAL MODEL TEST

The surface settlement controlling effect of the MGF method in actual construction was attributed mostly to close driving of steel pipes creating a shell, and to the driving at greater angles producing a thicker presupport. In order to verify the superiority of the MGF method and evaluate the supporting mechanism of the presupport, therefore, centrifugal model tests were carried out in four cases.

3.1 Outline of model test

Figure 3 gives an outline of an apparatus for the tunnel excavation tests.

First, for the presupport, steel pipes and improved soil mass were assumed to be integrated into a shell and thus modeled by a shell made of resin (a modulus of elasticity: 100MPa). The shell was installed on a hard iron cylinder in view of the site condition (with relatively hard gravel and sandstone layers at the foot of the top heading).

Then, for tunneling, excavation of top heading from the installation of one support to that of the next support was adopted as the basic excavation step in the test and simulated by pulling out an aluminum semi-circular cylinder shown in Figure 3 in the field of a centrifugal force of 50 G at a rate of 20 mm/min for a length of 10 mm (about 1/10 the tunnel diameter).

Finally, as the model of the ground, sandy soil with little cohesion was used. That is, unsaturated Toyoura sand with a water content w of 5% was compacted to achieve a relative density D_r of 85% (unit weight γ_t of 16.7 kN/m³). As an earth cover, 0.5 diameter was adopted considering the field condition where the MGF method was applied. As a result, the earth cover was determined as 50 mm above the top of the hard iron cylinder.

In addition, the distribution of surface settlement due to the excavation of top heading was measured by laser displacement gauges installed along the tunnel axis as shown in Figure 3.

3.2 Cases of model test

Table 3 shows test cases. The conventional and MGF methods were classified as Cases 1 (Cases 1-1, 1-2 and 1-3) and 2, respectively.

First, the thickness of the presupport or shell was set at 3 and 7 mm for Cases 1 and 2, respectively according to that in actual construction.

Then, in relation to the length of presupport remaining in the ground, a condition considered most unstable (right before the next round of steel pipe driving) either in the conventional or MGF method was modeled. A length of 35 mm was determined as a standard value based on the general remaining length under the above condition. In the conventional method, however, the length longer than the determined value remains for a long time. A model was, therefore, also made for the condition with the longest length. Thus, the length was set at 95 mm in Case 1-2.

Finally, the shell was basically installed 180 degrees in the cross section of the semicircular heading. In Case 1-3, the shell was installed 120 degrees based on actual application to examine the effect of the range of the shell in the cross section of the heading on surface settlement, and to collect basic data for the case where the cylindrical shell theory was applied.

Moreover, in Case 2, strain gauges were installed as shown in Figure 4 to grasp changes in presupport strain due to the excavation of top heading. Excavation was carried out in three steps in a 30-mm length in Case2.

Table3. Cases of centrifugal model test.

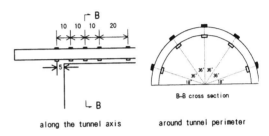

Case	Thickness*	Remaining length**	Range of the shell	Schematic cross section of presupport
Case1-1	t=3.0mm	35mm	180°	
Case1-2	t=3.0mm	95mm	180°	
Case1-3	t=3.0mm	35mm	120°	
Case2	t=7.0mm	35mm	180°	

* thickness of presupport
** length of presupport installed ahead of the face

Figure4. Points where strain gauges were installed (Case2).

along the tunnel axis   around tunnel perimeter

3.3 Results of test

(i) Distribution of moment

Figure 5 shows the distribution of moment along the tunnel axis. The figure gives the test results in Case2, and shows the moment increments calculated based on the strain measurements obtained in three steps of top heading excavation: (i) 0 to 10 mm, (ii) 10 to 20 mm, and (iii) 20 to 30 mm. The moment values were converted to those in the actual ground.

It is evident that moment distribution shifted with excavation, and that the maximum moment was achieved in the middle of the excavation range in each step.

(ii) Distribution of surface settlement

Figure 6 shows the distribution of surface settlement along the tunnel axis that occurred when a semi-circular cylinder was pulled out in centrifugal field of 50G for a length of 10 mm to simulate tunnel excavation.

In Cases 1-1 and 1-2, the maximum surface settlement and the trend of settlement are almost similar. This means that under this test condition increasing the length of a presupport remaining in the ground over 35mm does not increase surface settlement controlling effect. Furthermore, a comparison between Case 1-1 and 1-3 shows that a shell installed 180 degrees has greater settlement controlling effect than a shell installed 120 degrees.

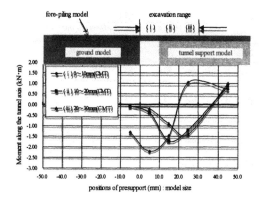

Figure5. Moment distribution along the tunnel axis (Case2).

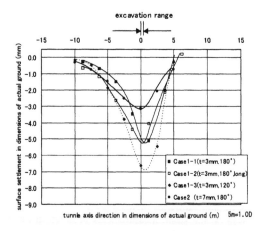

Figure6. Surface settlement distribution along the tunnel axis.

A comparison between Cases 1 and 2 (shell thickness: 3.0 mm, 7.0mm, respectively) shows that settlement is controlled much better in the latter case especially near the excavation range. Consequently, the MGF method is shown to have a greater settlement controlling effect than the conventional method.

4 EVALUATION BASED ON CYLINDRICAL SHELL THEORY

4.1 Outline of cylindrical shell theory

To establish the simple and rational design method, the authors have evaluated the supporting mechanism by the cylindrical shell theory that enables the evaluation of three-dimensional deformation behaviors (Kitamoto et al. 2000). According to the cylindrical shell theory, where load p is uniformly applied on a cylindrical shell in the direction of diameter and for a certain length l (defined as cylindrical loading for further discussion), displacement in the direction of diameter δ_x is expressed by equation (3.1) in the range $0<x<l$.

$$\delta_x = p\, r^2\, \{2 - e^{-\beta x}\cos\beta x - e^{-\beta(l-x)}\cos\beta(l-x)\}/(2\,E\,t) \quad (3.1)$$

where, r: radius of cylindrical shell, t: shell thickness, E: modulus of elasticity, $\beta = \{3(1-\nu^2)/(r^2 t^2)\}^{0.25}$, ν: Poisson's Ratio.

Then, the maximum displacement in the direction of diameter in the loading range δ_{max} can be expressed as shown below using equation (3.1).

$$\delta_{max} = p\, r^2\, \{1 - e^{-\beta(l/2)}\cos\beta(l/2)\}/(E\,t) \quad (3.2)$$

Moment M_x along the tunnel axis can be represented in the ranges $0< x <l$ and $x >l$ as shown below.

$$M_x = p\, \{e^{-\beta x}\sin\beta x - e^{-\beta(l-x)}\sin\beta(l-x)\}/(4\,\beta^2) \quad (3.3)$$
$$M_x = p\, \{e^{-\beta x}\sin\beta x - e^{-\beta(x-l)}\sin\beta(x-l)\}/(4\,\beta^2) \quad (3.4)$$

4.2 Evaluation of the results of centrifugal model test

(i) Distribution of moment

Figure 7 shows the moment distribution along the tunnel axis obtained by test and computation. The test results are moment increments during excavation for a length of 0 to 10 mm shown in Figure 5, and the computation results were obtained by equations (3.3) and (3.4). Both results are converted to those in actual ground.

For conditions for computation, total overburden load on the crown was adopted as uniformly distributed load in the direction of diameter p, and a length of 0.5 m equivalent to tunnel advance (equivalent to 10 mm in the model) was used as loading range l. The Poisson's Ratio is 0.49 based on the test conducted for the specimen. Figure 7 also shows the results of beam structure analysis (one-dimensional), which is generally used for design computation for presupports. The length of a beam in Figure 7 is defined as shown in Figure 8. Beam structure analysis was also made for a case where the beam length was set the tunnel advance (0.5m).

Figure 7 shows that only the results based on the cylindrical shell theory agreed well with the test results, on the contrary, the results of one-dimensional beam structure analysis agreed poorly with the test results. As a result, it is verified that the characteristics of a presupport can be modeled as a cylindrical shell structure with cylindrical loading.

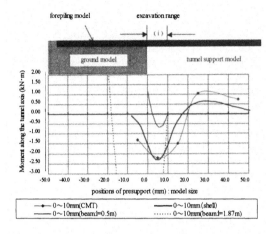

Figure7. Comparison in moment along the tunnel axis.

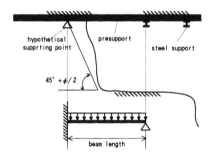

Figure8. Beam structure analysis.

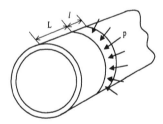

Figure9. Definition of L

(ii) *Distribution of surface settlement*

First, analysis of the agreement of the results of Cases 1-1 and 1-2 is made. As a prerequisite for equations (3.1) through (3.4) to be valid, the shell should have a sufficient length. That is, $L \geqq \pi / \beta$ (where L: length between the edge of loading range and the end of the cylinder, shown in Figure9.) needs to be satisfied. Computation in Case 1-1 results in L=35 mm, larger than π / β=32 mm. Thus, the prerequisite is satisfied. Where the length of presupport remaining in the ground has such a value as to satisfy the above prerequisite, settlement controlling effect is expected to remain unchanged.

Second, the maximum value of surface settlement w_{max} in the test was compared with the maximum value δ_{max} in the direction of diameter computed by equation (3.2). It is made on the basis of the fact that crown settlement and surface settlement right after the face is passed have a similar tendency and are of a similar amount in tunneling under shallow and little cohesive ground. The results of the above comparison are shown in Figure10. The thickness of the presupport is plotted along the horizontal axis (in the dimensions of the model), and the ratio between analysis and test values δ_{max} / w_{max} (referred to as peak value ratio below) along the vertical axis. The figure also shows the results of one-dimensional beam structure analysis adopted in Figure 7 for comparison.

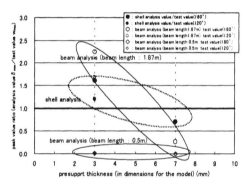

Figure10. Comparison between analysis and test values.

Figure10 shows that the peak value ratio based on the cylindrical shell theory is independent of presupport thickness and concentrates around 1.0. In the beam structure analysis, on the other hand, hardly any deformation occurs where the beam length is set at 0.5m. Even if the beam length is set at 1.87m, the peak value ratio fluctuates with the presupport thickness. This means that the former is the better method than the latter to evaluate the characteristics of presupport deformation

The results of evaluation based on the cylindrical shell theory, however, do not completely agree with test results. Even for cases where the shell is installed 180 degrees in the face, the peak value ratio fluctuates from lower to higher than 1.0 according to presupport thickness. This is because the evaluation of presupport deformation was basically aimed at identifying crown settlement as described earlier, because the improved zone did not form a complete cylinder, and because the shell made for presupport was not completely hollow. In the future, a study should be made about methods for evaluating parameters in the cylindrical shell theory according to various presupport conditions.

187

4.3 *Evaluation of the results of field construction*

Section 4.2 above showed the validity of the cylindrical shell theory under the condition where little settlement occurs at the foot of the top heading. Then, the field measurements obtained during construction as described in Section2 are evaluated based on the cylindrical theory. For field measurements, relationship between the distance from the face and surface settlement is available and used for fitting. It is known from the cylindrical shell theory that the relationship between the distance from the face and settlement can be obtained relatively easily by accumulating settlement per tunnel face advance (Kitamoto et al. 2000). The conditions set for calculation are shown in Table 4. The results of comparison with field measurements are shown in Figure11.

Table4. Calculation conditions.

	Conventional method	MGF method	Remarks
$p(kN/m^2)$	75		$p=\gamma_t \times total$ overburden (γ_t : 15kN/ m^3,overburden : 5m)
ν	0.3		
r (m)	5.7	5.4	MGF requires smaller increase of tunnel cross section
t (m)	0.25	0.7	Refer to Figure2.

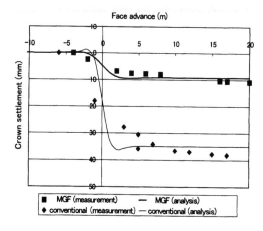

Figure11. Fitting of field measurements.

For the conventional method, the analytical value represents the affected range slightly smaller than the test value. For the MGF method, measurements are positioned along with the analytical curve, and thus, settlement characteristics with face advance are

well evaluated. For crown settlement, the ratio between MGF and conventional methods is well evaluated by the cylindrical shell theory. Modulus of deformation E obtained as an unknown by back calculation is about 300 N/mm^2, smaller than a general stiffness of the presupport consisting of steel pipes and improved mass. This is ascribable to the installation of the shell 120 to 180 degrees, and the failure to form an ideal shell shape of the presupport owing to chemical grouting condition. In the future, studies should be made to examine combined stiffness of an improved zone integrating steel pipes and improved mass.

5 CONCLUSIONS

-The effectiveness of thicker presupport in controlling settlement was verified by a centrifugal model test. The length of presupport remaining in the ground in the conventional method is found to have little influence on the settlement controlling effect. Thus, the new presupport method with medium-length steel pipes is found to be superior to the conventional method in controlling surface settlement.

-The results of the centrifugal model test were evaluated based on the cylindrical theory. As a result, the cylindrical theory is found to be more appropriate than one-dimensional beam structure analysis in explaining the characteristics of surface settlement where a shell-type presupport was installed.

-In addition, the actual construction data were evaluated based on the cylindrical theory. As a result, it was verified again that the cylindrical theory could represent the effect of presupport thickness on surface settlement.

REFERENCES

Yamamoto, Kitamoto, Date & Okamoto. 1999. Effect of the new forepiling method with middle length pipes.. Proceedings of Tunnel Engineering, JSCE (Japan Society of Civil Engineers). Vol. 9. 167-172. (in Japanese)

Date, Kitamoto, Yamamoto, Goto & Ohta. 2000. Behavior and deformation controlling effect of tunnel presupports. Proceedings of the 55th annual conference of JSCE. (in Japanese)

Kitamoto, Date, Yamamoto, Hibiya & Ohta. 2000. Simple methods for evaluating tunnel presupports. Proceedings of the 35th conference for Geotechnical Engineering. (in Japanese)

Modern Tunneling Science and Technology, Adachi et al (eds), © 2001 Swets & Zeitlinger, ISBN 90 2651 860 9

Experimental study on tunneling in the ground with inclined layers and its simulation

S. H. Park
Geostructure Group, Civil Eng. Div., KICT, Korea

T. Adachi, M. Kimura, K. Kishida, & M. Kikumoto
Dept. of Civil Eng., Kyoto University, Kyoto, Japan

ABSTRACT: A tunneling model tests, simulated by a trapdoor apparatus, are performed in the ground with inclined layers using aluminum blocks. The 60-degree formation with highly inclined layers shows the most significant feature on the non-symmetrical distribution of the earth pressure on the upper part of the trapdoor, while the 45-degree formation with moderately inclined layers shows that feature on the outer parts of the trapdoor. In order to verify the experimental results, a numerical simulation by the FE analysis is conducted using joint elements to explain the discontinuous behaviors of the model ground. By calculating the distributions of earth pressure and surface profiles with the trapdoor displacement, it is confirmed that the calculated results can reasonably explain the experimental results within lower displacements of about 1.00 mm.

1 INTRODUCTION

Nowadays, tunnels are constructed not only in sound and stable grounds, but also sometimes in unstable ground conditions composed of faults, fractured zones, bedding (stratification), and joints, etc. If a ground is homogeneous and isotropic, tunnel construction can be performed easily, rapidly and safely. However, it is supposed to frequently be accompanied by overbreaks and squeezing due to tunneling in a discontinuous ground conditions. It may cause the appearance of cavities in the tunnel surroundings and sinkholes in the ground surface. Moreover, large excavations in localized areas have been performed, it is known that loosening, load concentrations, erratic loads, and discontinuous behaviors of the ground may occur. From this point of view, tunneling model tests and its simulation using the FE method are performed in the discontinuous ground, in particular, with inclined layers.

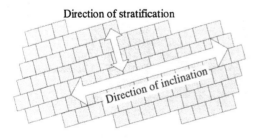

Figure 1. Modeling on the ground with inclined layers.

2 TUNNELING MODEL TESTS

Discontinuous conditions of the ground are modeled by making the inclined layers that have the same inclination and spacing between them. The model ground shows staggered geometry with two kinds of predominant discontinuous planes. As shown in Figure 1, one of the discontinuities is in the direction of the inclination of the layers of the continuous type, while the other is in the direction of the stratification of the discontinuous type that is perpendicular to the previous one.

2.1 *Testing apparatus*

In the present study, trapdoor apparatus (Adachi et al., 1994) is used to measure the distribution of earth pressure at the bottom of the model ground with the trapdoor displacement and the variations in earth pressure. Figure 2 shows the testing apparatus with 1075 mm in height, 1000 mm in horizontal length, and 150 mm in width. It is composed of forty supporting blocks, 2.45 cm in width, and each block has a load cell glued on the bottom of it. The tunneling process is simulated by lowering the supporting plate, namely, the trapdoor, to reduce the confining stress in the localized area. Trapdoor displacement can be applied by a control jack. Surface settlements are measured by a laser displacement system, which is shown in Figure 2. The horizontal location of the laser displacement sensor and the vertical displacements of the surface settlements are measured at the same time.

2.2 Ground materials

In order to simulate the model ground with inclined layers, aluminum rods and aluminum blocks, in which no cohesion between particles exists, are used. The aluminum rods, shown in Figure 3(a), are used as an aggregate in which two kinds of rods, with diameters of 1.6 mm and 3.0 mm and a length of 50 mm, are combined in a weight ratio of 3:2. The inclined-layer formation in the present study mainly consists of aluminum blocks, as shown in Figure 3(b). The aluminum blocks are divided according to shape into two kinds of rectangular bricks and three kinds of triangular prisms. Two kinds of bricks can make the ground a staggered geometry with two directions of discontinuities, as shown in Figure 1. The triangular prisms are used to form the angles of the inclined layers.

2.3 Testing parameters

Figure 4 exhibits the typical state of the ground conditions for a tunneling model test. The trapdoor has three widths, namely, D = 10 cm, 15 cm, and 20 cm. The overburden represents the height of the deposit of aluminum blocks H, namely, 0.5D, 1.0D, and 2.0D, with respect to the trapdoor width D. At the lower and the upper parts of the deposit of aluminum blocks, there are aggregates of aluminum rods with the height of 2.5 cm and 1.5 cm, respectively. In this paper, the experimental results are presented only for the inclination of the layers of θ = 30, 45, and 60 degrees formations.

3 EXPERIMENTAL RESULTS

3.1 Distribution profiles of earth pressure

Figures 5 exhibits examples of the distribution patterns of earth pressure in the ground with the trapdoor width of D = 20 cm. The overburden height of the ground is H = 20 cm. Each curve exhibits the distribution of earth pressure at the trapdoor displacements of δ_t = 1.00 mm according to the angle of the inclined layers of θ = 30, 45, and 60 degrees. The Y-axis represents the normalized earth pressure measured by forty load cells, while the X-axis is the horizontal distance from the center line of the trapdoor. The normalized earth pressure is the obtained value when the observed earth pressure is divided by the initial earth pressure. The shaded area indicates the location of the trapdoor.

From this figure, it is confirmed that the ground with inclined layers exhibits various patterns of earth pressure for the inclination of the layers. For the 30-degree formation, the distribution profiles of the earth pressure show symmetry about the center line of the trapdoor. On the other hand, non-symmetrical patterns of earth pressure about the

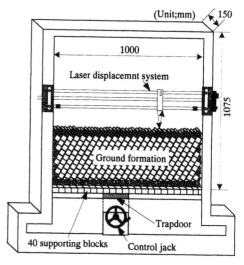

Figure 2. Trapdoor testing apparatus.

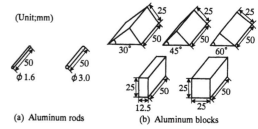

(a) Aluminum rods   (b) Aluminum blocks

Figure 3. Ground materials.

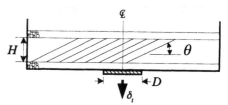

Figure 4. Typical state of the model ground.

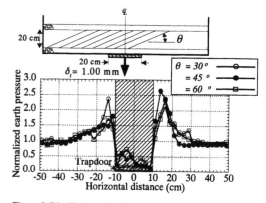

Figure 5. Distribution of earth pressure at δ_t = 1.00 mm with H = 20 cm.

190

center line of the trapdoor are confirmed in the 45- and the 60-degree formation. Non-symmetrical patterns of earth pressure are observed not only on the upper part of the trapdoor, but also on the outer parts of the trapdoor.

3.2 Non-symmetry on the distribution of earth pressure

To investigate the non-symmetry on the distribution of earth pressure, variations in earth pressure on the right and left sides of the trapdoor are examined centering on the center line of the trapdoor. The sections to be evaluated are divided into four parts, 10 cm each, which are shown in Figure 6. Namely, as for the upper part of the trapdoor, there are the upper left-hand part of the trapdoor and the upper right-hand part. As for the outer parts of the trapdoor, there are the outer left part of the trapdoor and the outer right part.

Figure 7 exhibits the variations in the normalized earth pressure on the upper parts of the trapdoor with overburden heights of $H = 20$ cm. Asymmetry of the earth pressure on the upper parts of the trapdoor is not so notable for the 30-degree formation. As for

the 45- and the 60-degree formations, however, there are obvious asymmetries according to the trapdoor displacements such that the normalized earth pressure of the upper left-hand part is higher than that of the upper right-hand part. The most significant asymmetry develops in the 60-degree formation for the two upper parts of the trapdoor.

Variations in the normalized earth pressure on the outer parts of the trapdoor are presented in Figure 8. From this figure, the 45-degree formation exhibits the most notable asymmetry on the outer parts of the trapdoor, in which the earth pressure acting on the outer right part is greater than that on outer left part.

3.3 Surface settlement profiles

Figures 9 exhibits an example of the surface settlement profiles for the trapdoor displacement of $\delta_t = 1.00$ mm with the overburden height of $H = 20$ cm according to the angles of the inclined layers of $\theta = 30$, 45, and 60 degrees. The X-axis represents the distance from the center line of the trapdoor, while the Y-axis represents the surface settlement values. From this figure, it is known that the settlement position and the amount of surface settlement differ according to the angles of the inclined layers. Surface settlement is predominantly observed in the upper left direction of the trapdoor in the 30-degree formation with lowly inclined layers. For the 45-degree formation with moderately inclined layers, surface settlement is predominantly observed in the upper right direction of the trapdoor. For the 60-degree formation with highly inclined layers, surface settlement is observed only in the upper right direction of the trapdoor. The surface settlement induces over 90% of the trapdoor displacement of $\delta_t = 1.00$ mm for the 60-degree formation. The surface settlement

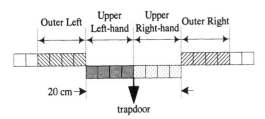

Figure 6. Asymmetrical distribution of earth pressure.

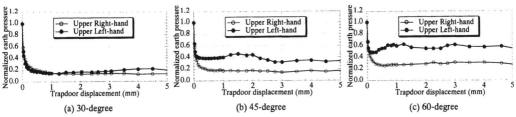

(a) 30-degree   (b) 45-degree   (c) 60-degree

Figure 7. Normalized earth pressure on the upper parts of the trapdoor with $H = 20$ cm.

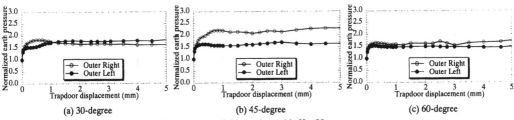

(a) 30-degree   (b) 45-degree   (c) 60-degree

Figure 8. Normalized earth pressure on the outer parts of the trapdoor with $H = 20$ cm.

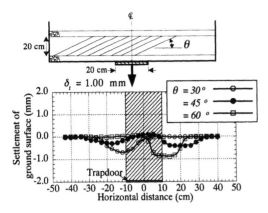

Figure 9. Surface profiles at $\delta_t = 1.00$ mm with $H = 20$ cm.

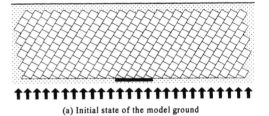

(a) Initial state of the model ground

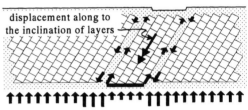

(b) In the early stage of the trapdoor displacement

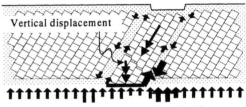

(c) As the trapdoor is lowered considerably

Figure 10. Asymmetrical distribution of earth pressure.

is very severe when ground deformation occurs along the direction of the inclination of the layers.

3.4 Investigation on the non-symmetrical distribution of earth pressure

When the tunneling model tests are carried out in the ground with inclined layers in this study, the most significant characteristic is the non-symmetrical distribution of earth pressure. In other words, the normalized earth pressure acting on the upper left-hand part of the trapdoor is higher than that of the upper right-hand part for the 45- and the 60-degree formations. This result is investigated by assuming the mechanical process of the model ground below.

Figure 10(a), (b), and (c) are the initial state of the model ground before the trapdoor is lowered, the early stage of the trapdoor displacement, and after the trapdoor is considerably lowered, respectively. It is reasonable to assume that the initial earth pressure acting along the bottom of the model ground has a uniform distribution since the horizontal surface is considered. As the trapdoor starts to be lowered, a mass deposit located between the sliding surfaces slides in the direction of the inclination of the layers. At the same time, frictional resistance is exerts on the sliding surfaces as shown in Figure 10(b). If the trapdoor is lowered slightly during the early stage of the trapdoor displacement, it can be thought that the frictional resistances on both sliding planes are almost identical. Induced resistances are to be shared on the both outer part of the trapdoor so that the increment in vertical earth pressure on both sides of the outer trapdoor might be identical with those. Subsequently, as shown in Figure 10(c), the trapdoor is lowered further and the mass deposit located between the sliding planes deforms remarkably. At this time, the lower part of the mass deposit causes it to fall down vertically because of its weight, while the upper part of the mass deposit slides in the direction of the inclination of layers, as seen during the early

stage of the trapdoor displacement. Since the lower part of the mass deposit falls down vertically, a separation or an exfoliation at a contact section between the inclined layers on the left-hand side of the mass deposit may occur. This separation causes the confining lateral pressure to decrease in that area. In contrast, there is a great increase in frictional resistance on the right-hand side of the mass deposit. In that respect, the normalized earth pressure acting on the upper left-hand part of the trapdoor, for which lateral earth pressure is reduced, shows greater loads than that of the upper right-hand part.

4 NUMERICAL SIMULATION

4.1 Outline of the analysis

Numerical analysis is conducted using inter-element slip model (IESM, Adachi et al., 1985) to explain the discontinuous behavior of the model ground. Figure 11 shows an example of FE mesh for the 60-degree formation used in this study. The aggregate of aluminum blocks, which is the main part of the model ground, is composed of quadrangular solid elements and joint elements with no thickness, whereupon the features of FE mesh are almost identical to those of the model ground. The deformation

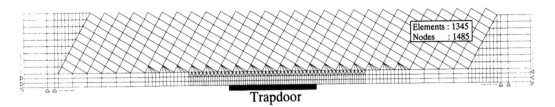

Trapdoor

Figure 11. An example of FE mesh for 60-degree formation.

of each aluminum block is disregarded, incorporating considerable high elastic Young's modulus and elasto-perfectly plastic behavior into the deformation of the joint elements. Aluminum rods deposits are installed on the right-hand and the left-hand sides, as well as beneath of the aggregate of aluminum blocks. The elasto-plastic behavior is incorporated into the aluminum rods deposits. The upper aggregate of aluminum rods used in the model tests is conveniently excluded from the numerical analyses. Table 1 summarized the material properties used in the analysis.

4.2 Comparisons with experimental results

Comparisons of analytical and the experimental results are conducted for the distribution patterns of earth pressure with trapdoor displacements. Figure 12 shows a comparison of the distribution patterns of earth pressure that were obtained at the lowest elements of the FE mesh with the trapdoor displacements of $\delta_t = 1.00$ mm and the overburden height of $H = 20$ cm. Non-symmetrical distributions of earth pressure on the upper part of the trapdoor are relatively confirmed for the 60-degree formation. For the outer parts of the trapdoor, on the other hand, the non-symmetrical distributions of earth pressure are relatively confirmed for the 45-degree formation although they are not so notable.

Figure 13 indicates the surface profiles at a trapdoor displacement of $\delta_t = 1.00$ mm for overburden height of $H = 20$ cm. Non-symmetrical profiles of the surface settlements around the center line of the trapdoor are confirmed. The 30-degree formation greatly shows its surface settlement on the upper left side of the trapdoor. This implies that the surface settlement is predominant in the direction of the stratification that is modeled in the experiments. As for the 45-degree formation, surface settlements are observed both on the upper right and the upper left sides of the trapdoor for which the extent of surface settlements is widest. As for the 60-degree formation, surface settlement is observed only on the upper right side of the trapdoor.

Table 1. Material properties.

Unit weight	aluminum rods	21.1 kN/m³
	aluminum blocks	26.4 kN/m³
Young's modulus	aluminum rods	49.1+9.8 σ_m kPa
	aluminum blocks	6.2×10^4 kPa
Poisson's ratio (rods/blocks)		0.33/0.20
Cohesive strength (rods/blocks)		0.0/0.0 kN/m²
Internal friction angle of the aggregate of aluminum rods		30°
Sliding friction angle of joint elements		20°
Coeff. of lateral earth pressure		1.0
Joint stiffness (Normal/Shear)		$150\sigma_{n0}/500\sigma_{n0}$

* σ_m ; Mean stress, σ_{n0} ; Initial normal stress

5 CONCLUSION

Mechanical behavior of the ground with inclined layers is examined by carrying out the tunneling model tests. Based on the experimental results, numerical analysis is conducted to explain the discontinuous behaviors of the model ground with incorporating the joint elements. The results are described as below.

1) The experimental results featured non-symmetrical distributions of earth pressure obtained by lowering the trapdoor. Non-symmetry was remarkably confirmed with the 60-degree formation for the upper part of the trapdoor and with the 45-degree formation for the outer parts.

2) Examinations were conducted with a comparison of the distribution of earth pressure on the bottom line of the ground and the surface settlements obtained through the experiments. It is confirmed that the calculated results can reasonably explain the experimental results within lower displacements of about 1.00 mm. For the surface settlements, in particular, the numerical results explain a good correspondence with the experimental results.

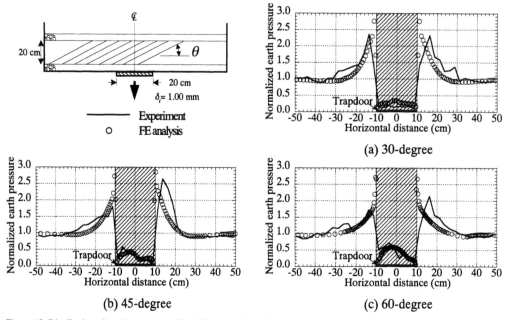

Figure 12. Distribution of earth pressure at $\delta_t = 1.00$ mm and $H = 20$ cm.

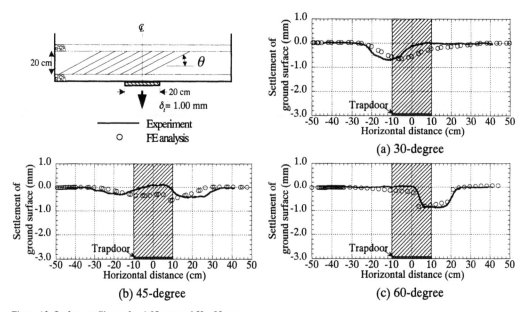

Figure 13. Surface profiles at $\delta_t = 1.00$ mm and $H = 20$ cm.

REFERENCES

Adachi T., Tamura T., Kimura M. & Aramaki S.(1994) : Experimental and Analytical Studies of Earth Pressure, *Computer Methods and Advances in Geomechanics, A.A.Balkema*, pp.70~75.

Adachi T., Tamura T., & Yashima A. (1985) : Behavior and simulation of sandy ground tunnel, *Proc. 11th Int. Conf. on Soil Mechanics and Foundation Engineering, San Francisco*, pp. 709~712.

Modern Tunneling Science and Technology, Adachi et al (eds), © 2001 Swets & Zeitlinger, ISBN 90 2651 860 9

Study on the strength enhancement theory of surrounding rock by bolting

C. Hou, N. Zhang & X. Li
China University of Mining and Technology, Xuzhou City, China
P. Gou
Jiaozuo Institute of Technology, Jiaozuo city, Henan province, China

Abstract: based on laboratory strength tests of mudstone and laboratory simulation tests of bolted rockmass, this paper puts forward the strength enhancement theory of surrounding rock, which aims at explaining the bolting mechanism of big-deformation roadways. The enhancement effect on surrounding rock is closely interrelated to the mean load density of bolting, so a combination of coupling numerical analysis and field measurement is applied to determine rational mean load density in practice.

1 PREFACE

Correct designing and applying of bolting support depends, to an extraordinary extent, on the guidance of a satisfactory theory. Traditional bolting theories, such as suspension theory, compound-beam theory and compound-arch theory, each can only explain the bolting mechanism from a certain angle. Thus their practical application is greatly limited. It is extremely so when great damage occurs in the surrounding rock of big-deformation roadways.

The big-deformation roadways fall into three basic types. The first is softrock roadways cut by joints and fractures, whose surrounding rock is loose, broken, nonisotropic, and stratified. The second is the roadways of high initial stress of rock, including deep shaft mine and great structural stress. The third is the roadways induced by mining activities, whose coefficient of stress concentration can be 3~5 times of the initial stress of rock. Fracture zone and broken zone usually occur in the surrounding rock of big-deformation roadways. A typical plastic zone (including broken zone) of roadway in coal seam is shown as figure 1. And as for: whether bolting can work well in the plastic zone? how can it work? and what is the mechanism of bolting? The strength enhancement theory of surrounding rock can give a good answer to them.

2 THE TEST OF MECHANICAL PROPERTIES OF ROCK AND MECHANISM OF BOLTING

What are employed in the experiment are the rigid servo facilities MTS81502S and samples gained from floor mudstone in Maomin Mining District.

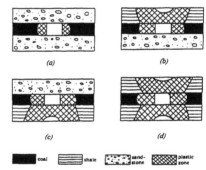

(a)   *(b)*

(c)   *(d)*

coal   shale   sand-stone   plastic zone

Figure1. the typical distributed sketches of plastic zone (including broken zone) of surrounding rock in gateway
(a) rocks around two walls with low strength
(b) rocks around two walls and roof with low strength
(c) rocks around two walls and floor with low strength
(d) rocks around gateway with low strength.

The overall stress-strain curves of the mudstone under various lateral pressures are shown in Figure 2. When the lateral pressure increases, the peak strength and residual strength increase. Obviously, the corresponding strength under tri-axial loading conditions is larger than that under uni-axial loading conditions.

The same goes for the surrounding rock of roadway. The strength of surrounding rock is not always the same when loading conditions are different. Its strength is bound to enhance with the increasing lateral pressure of rockmass. When bolts

Table1. The C and φ of bolted rockmass before being broken under various bolts density distribution.

The number of bolts /400cm²	0	2	3	4	5	6	8
C/MPa	0.3466	0.3568	0.3626	0.3677	0.3828	0.3773	0.3869
φ /MPa	31.51	31.53	33.51	35.57	37.14	38.8	40.4

Table2. The C* and φ* of bolted rockmass after being broken under various bolts density distribution.

The number of bolts /400cm²	0	2	3	4	5	6	8
C*/MPa	0.0168	0.0182	0.0183	0.184	0.0186	0.0194	0.021
φ* /MPa	31.51	31.53	33.51	35.57	37.24	38.8	40.4

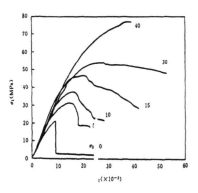

Figure2. The overall stress-strain curve of mudstone under various lateral pressures

are installed in the roof, the strength of bolted rockmass will be increased. If plastic zone and broken zone occur in the roof, the rock stress of these two zones will be in the post-peak curve section, and their strengths are post-peak strength and residual strength respectively. The lateral pressure induced by bolting similarly enhances the post-peak strength and residual strength of bolted rockmass. Meanwhile, the strength of bolted rockmass will increase with the increasing of mean load density.

3 THE STRENGTH ENHANCEMENT THEORY OF SURROUNDING ROCK

Now take a coal roadway of medium hard rock(uni-axial compressive strength is 15 ~20MPa) as an example, laboratory simulation tests are performed to explain strength enhancement of surrounding rock of roadway. The original bolted rock is $2.0 \times 2.0 \times 2.0 m^3$, and the model is $20 \times 20 \times 20 cm^3$. Bolts are laid out only on one of the surfaces. Loading tests are performed on unbolted rockmass and then on bolted ones with various numbers. The mean load density is adjusted with the number of bolts. Suppose the cohesive force of

bolted rockmass before and after being broken is C and C* respectively, the internal friction angle before and after being broken is φ and φ* respectively, the results can be shown in table1. and table 2.

It can be seen that the value of C, C*, φ, and φ* all become larger with the increase of bolting density. The enhancement of φ and φ* are even much bigger, while the changing law of which is the same. The change of C is very small, then the enhancement of C* is much more obvious and far beyond that of C.

The ratio between the strength of bolted rockmass and the strength of original rockmass is determined as enhancement coefficient.

The enhancement coefficient of peak strength of bolted rockmass is as follows

$$K_j = \sigma_1 / \sigma_c \qquad (1)$$

The enhancement coefficient of residual strength of bolted rockmass is as follows

$$K_c = \sigma^*_1 / \sigma^*_c \qquad (2)$$

Where: σ_1 = the peak strength of original rockmass; σ_c = the peak strength of bolted rockmass; σ^*_1 = the residual strength of original rockmass; σ^*_c = the residual strength of bolted rockmass.

The peak strength and residual strength of bolted rockmass are shown in Table 3. It can be seen that the strength of bolted rockmass have been enhanced when the strength of surrounding rock is in various curve sections. The enhancement coefficient of bolted rockmass will increase with the increase of the mean load density of bolting. The enhancement coefficient of residual strength of bolted rockmass is larger than that of the peak strength.

On the basis of regression analysis of the test results, the formulae of the peak strength of bolted rockmass(σ_b) and the residual strength of bolted rockmass(σ^*_b) are as follows:

$$\sigma_b = 0.4 + 15.89 \, \sigma'' + 2 \, C \, tan(45^0 + \phi /2)$$
$$(r = 0.968) \qquad (3)$$

Table3. The enhancement coefficient of bolted rockmass..

The number of bolts /400cm^2	Peak strength /MPa	Residual strength /MPa	K$_j$	K$_c$
0	1.65	0.525	1	1
2	1.725	0.5875	1.04	1.12
3	1.8315	0.625	1.11	1.19
4	1.9275	0.6675	1.168	1.271
5	2.075	0.7	1.258	1.333
6	2.17	0.75	1.315	1.429
8	2.275	0.82	1.379	1.562

$$\sigma\,^*_b =0.4+26.4\ \sigma\,'' +2\ \ C^* \ \tan(45^0+\ \phi\,^*\ /2)$$
$$(r=0.\,967) \qquad\qquad (4)$$

where: $\sigma\,''$ = the axial power per square given by bolt (mean load density)

According to formulae (3) and (4), the increase of σ_b and $\sigma\,^*_b$ are related to the increase of $\sigma\,''$, meanwhile the increase of $\sigma\,''$ leads to the increase of C, C*, ϕ, and ϕ^*, which further enhances the value of σ_b and $\sigma\,^*_b$. Here the increase of $\sigma\,^*_b$ is much more rapid than that of σ_b. In other words, the effect of bolting on the broken zone is more notable. So any increasing of mean load density($\sigma\,''$) may enhance the strength of surrounding rock, thus keep the roadway stable.

The bolting support of gob edge entry driving in 5318 fully mechanized top coal caving face can be used as a typical engineering example.

Based on field measurement, the method of coupling numerical analysis is applied to gain the relative curve of the roof subsidence and mean load density under the action of bolting (shown as fig.3). Many useful conclusions can be drawn from this figure. When the mean load density of bolting is less than 0.2 MPa, the roof subsidence will sharply increase, and the roof stability is rapidly worsened. When it is more than 0.3 MPa, the roof subsidence will decrease, but the effect of bolting is not distinct. So the rational mean load density should be 0.2~0.3 MPa. When it is the case that the mean load density of bolting is less than 0.2 MPa, the strengthened supporting methods should be applied, such as cable anchor and prop-support ahead of mining face.

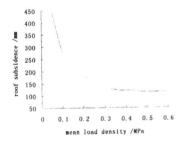

Figure3. the relative curve of the roof subsidence and the mean load density
(The initial anchorage is 12 kN).

4 CONCLUSIONS

1.The surrounding rock of most gateways is loose, and subject to strong mining activity, thus the plastic zone and broken zone will occur in the surrounding rock, whose strength is in post-peak strength state.

2. The increase of mean load density of bolting and any change arising from the increase of C, C*, ϕ, and ϕ^* of surrounding rock will enhance the peak strength and residual strength of bolted rockmass.

3. The enhancement theory of strength of surrounding rock can be widely applied to the bolting of gateways and soft rock roadways.

REFERENCES

Nianjie Ma & Chaojiong Hou. A research into plastic- zone of surrounding strata of gateway affected by mining abutment stress, Proceedings of the 31st U.S. symposium: 211~217. A.A. balkema /rotterdam/ brookfield/ 1990.

Chaojiong Hou, Yanan He & Xiao Li, A new method to control floor heave in hetrogeneous strata, Proceedings of the ninth international conference on computer methods and advances in geomechanics: 1519-1522, A.A. balkema/rotterdam/brookfield/1997.

Gale W.J. et al. Design Approach to Assess Coal Mine Roadway Stability and Support Requirements, VIII Australian Tunelling conference, 24.8.1993.

Modern Tunneling Science and Technology, Adachi et al (eds), © 2001 Swets & Zeitlinger, ISBN 90 2651 860 9

An experimental study on the nonuniform tunnel lining

N. Yingyongrattanakul, T. Adachi, K. Tateyama & M. Kurahashi
Department of Civil Engineering, Kyoto University, Kyoto, Japan

ABSTRACT: The elliptic and the horseshoe-shaped tunnels are considered to be more economical than the conventional circular tunnel because an amount of the excavated soil and the unnecessary space can be reduced. In this research, the model tests are carried out for investigating the effects of the tunnel shape, and the variation of the lining thickness on the tunnel stability. The tests are carried out on three types of the tunnel shape: circle, ellipse and horseshoe. For the elliptic and the horseshoe-shaped tunnel, the variations of the lining thickness are taken into consideration. The Styrofoam and silica sand are used to model tunnel lining and underground material respectively. The circumferential strains, generated at the surface of the lining, are measured after loading the tunnel by earth pressure. The axial force and the bending moment are calculated and demonstrated. The stability of each lining-thickness pattern is compared and the efficiency of increasing the tunnel stability by varying the lining thickness is discussed.

1 INTRODUCTION

Shield tunneling methods have become more and more popular in urban areas, because of the resulting small ground surface settlement, which minimizes potential damage to the nearby buildings and other civil infrastructure. Recently in Japan, the construction of the large cross-section tunnel by the shield tunneling methods is increasingly required. The large cross-section tunnel means the tunnel for the three-lane road that a width and an area are more than 16 m and 100 m^2 respectively. For this kind of tunnel, the conventional circular tunnel is considered to be unsuitable because a large shield machine is required; a large amount of the excavated soil is generated and lots of space is unutilized. In order to overcome these problems, the noncircular tunnels: elliptic or horseshoe-shaped tunnel, are taken into consideration instead.

There are many researchers have studied these problems, Hashimoto et al. (1996) conducted the two-way loading experiments on the circular, elliptic and rectangular tunnel and concluded that the bending moment becomes large when the tunnel becomes flat. The results also concluded that the bending moment is extremely concentrated at the corners of the rectangular tunnel so it is not appropriate for using this shape to construct a large tunnel. Sou et al. (1993) carried out the failure tests on the elliptic tunnels and concluded that the stress is concentrated at the spring line and can be reduced

by increasing the thickness at the spring line. However the mechanical behaviors of the elliptic and horseshoe-shaped tunnel under the gravity load have not been clarified. Thus the studies to investigate the mechanical behaviors and the design method of these kinds of tunnel shape are necessary.

In the case of same tunnel width and lining thickness, the noncircular tunnel has less stability than the circular one. In the conventional design method, the stability of the tunnel is improved by increasing the overall lining thickness. Increasing the lining thickness means rising up the construction cost in the same way. In this study, the concept of improving the tunnel stability by varying the lining thickness, whereas unchanging the overall lining area, is proposed.

The laboratory tests are conducted to compare both effects of the tunnel shape and the lining-thickness pattern on the stability of lining. Three types of tunnel shape, circular, elliptic and horseshoe-shaped tunnel are tested. The effects of the lining-thickness pattern are investigated by testing on the nonuniform-thickness lining for each type of the elliptic and horseshoe-shape tunnel. The strains, the axial forces and the bending moments generated on the tunnel lining are demonstrated. Finally, the efficiency of improving the stability of the noncircular tunnels is discussed.

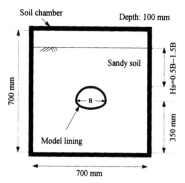

Figure 1. Laboratory test apparatus.

Table 1. Properties of the material used in the tests.

Material	Ground	Tunnel lining
	Silica sand	Styrofoam
Unit weight (kN/m³)	13.19	0.30
Relative density	60%	--
Young's Modulus (MPa)	50.00	15.24

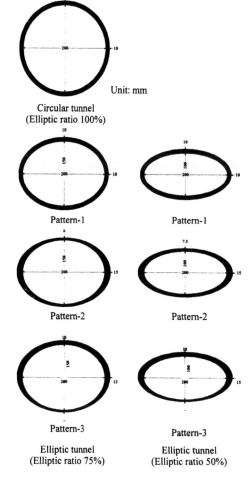

Figure 2. Model lining of the circular and the elliptic tunnel.

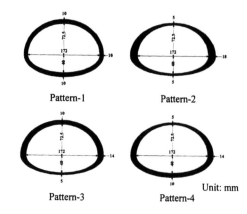

Figure 3. Model lining of the horseshoe-shaped tunnel.

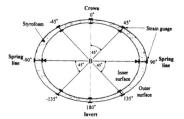

Figure 4. Positions of the strain gauges.

2 LABORATORY TEST DEVICE

A laboratory test device used in this study is shown in Figure 1. First the model lining is set at the position as shown in the figure, after that the steel chamber is filled up with the dry silica sand. In this test the silica sand No. 6 is used to produce the experiment ground and the model lining is made from the Styrofoam. The properties of the sand and the Styrofoam are shown in Table 1. The tests are carried out on three types of tunnel shape, circle, ellipse and horseshoe. The size and the shape of the model linings are shown in Figure 2 and 3. The strain gauges are attached symmetrically at the positions: 0° (crown), 45°, 90° (spring line), 135°, and 180° (invert), on both outer and inner sides of the lining surface, totally 16 gauges, as shown in Figure 4.

After filling the sand up to the designate height and leveling the soil surface, the circumferential strains generated at the lining surface are measured. The axial force and the bending moment of each lining section are calculated from the following equations:

$$N = EA\frac{(\varepsilon_i + \varepsilon_o)}{2} \tag{1}$$

$$M = 2\frac{EI}{d}\frac{(\varepsilon_i - \varepsilon_o)}{2} \tag{2}$$

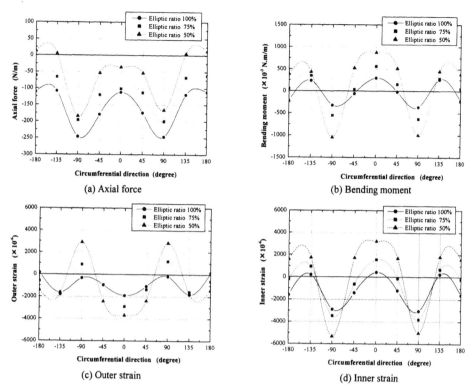

(a) Axial force

(b) Bending moment

(c) Outer strain

(d) Inner strain

Figure 5. Test results of the uniform thickness patterns of the circular and the elliptic tunnel at $H_s/B = 1.0$.

where N = axial force; M = bending moment; E = Young's Modulus; A = cross-sectional area, I = the moment of inertia; ε_i = the strain generated at the inner surface (inner strain); ε_o = the strain generated at the outer surface (outer strain), note that the minus sign of the strain means compressive strain.

3 RESULTS AND DISCUSSIONS OF THE ELLIPTIC TUNNEL

In this section, the test results of the circular and the elliptic model linings are presented and discussed. As shown in Figure 2, three tunnel shapes: elliptic ratio 50%, 75% and 100% (circle) are used, where the elliptic ratio is the ratio of width and height of the tunnel. The width of all tunnel shapes is same, 200 mm, to express the same utilizable capacity of the tunnel. Moreover in order to investigate the effects of the variation of the lining thickness on the lining stability, for the elliptic ratio 50% and 75% tunnels, three patterns of the thickness variation are used. For each tunnel shape, the area of tunnel lining for all patterns is equal. The tests are carried out on 7 patterns of the model lining at the cover to width ratio (H_s/B) equals 0.5, 0.75 and 1.0.

3.1 The effects of the tunnel shape

The results of the pattern-1 (uniform thickness) of the elliptic ratio 50%, 75% and 100% tunnels at H_s/B equals 1.0 are shown in Figure 5(a)~(d). The results show, when the tunnel shape becomes flat, the axial force becomes small while the bending moment becomes large. Thus it can be concluded that, for the circular tunnel, the axial force dominantly supports the earth pressure acting on the tunnel. In the case of the elliptic tunnel, the bending moment becomes dominant part to support the tunnel instead. From the results of the circumferential strain generated at the lining surface, when the tunnel becomes flat, the strain becomes large. For the circular tunnel, there is no remarkable tensile strain generated, while in the case of the elliptic tunnel, the tensile strain remarkably generated specially at the inner side of the crown, the invert and the outside of the spring line.

From the above results, it is no doubt to conclude that, in the case of uniform-thickness lining, the circular tunnel is the most stable shape. For the elliptic tunnel, the flatter tunnel is, the less stability tunnel becomes.

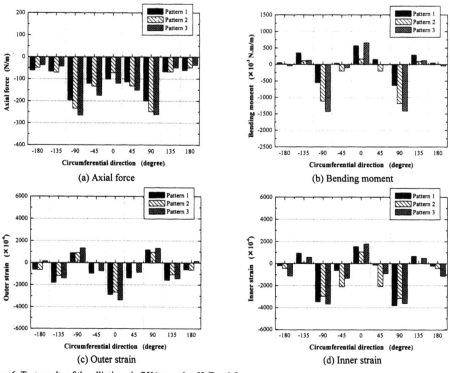

(a) Axial force

(b) Bending moment

(c) Outer strain

(d) Inner strain

Figure 6. Test results of the elliptic ratio 75% tunnel at $H_s/B = 1.0$.

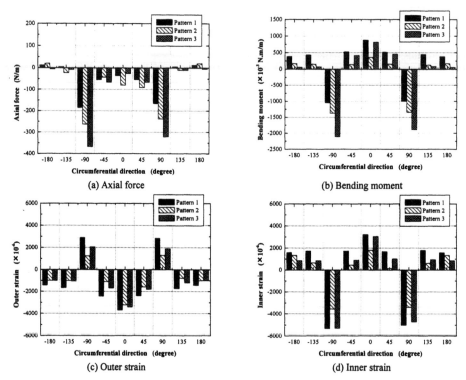

(a) Axial force

(b) Bending moment

(c) Outer strain

(d) Inner strain

Figure 7. Test results of the elliptic ratio 50% tunnel at $H_s/B = 1.0$.

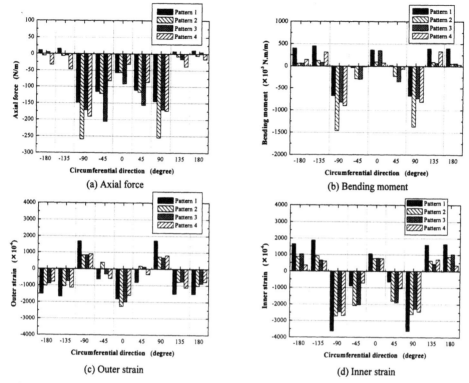

(a) Axial force

(b) Bending moment

(c) Outer strain

(d) Inner strain

Figure 8. Test results of the horseshoe-shaped tunnel at $H_s/B = 1.0$.

3.2 *The effects of the thickness pattern*

Figure 6(a)~(d) and Figure 7(a)~(d) show the comparisons results of three thickness patterns of the elliptic ratio 75% and 50% respectively. For both tunnel shapes, the compressive axial forces generated on the pattern-1 are the smallest values comparing with other patterns; on the other hand, the maximum bending moments generated on the pattern-2 are the smallest values. The pattern-3 is obviously the least stable pattern in the view of both section forces. From the results of the tensile strain, the strains generated at the lining surface of the pattern-2 are the smallest values especially at the inner surface of tunnel crown and the outer surface of spring line for both tunnel shapes.

Since the concrete structure can withstand a small level of tensile stress comparing with compressive stress. The tensile strain, generated on the surface of tunnel lining, expresses the tensile stress level in the same way. The local crack at the surface of concrete lining, generated by the tensile stress, can cause the sever damage to the overall structure. Thus the most stable thickness pattern should be the thickness pattern, which causes the minimum tensile strain. From the above results, in the case of elliptic tunnel,

the pattern-2, which is thicker at the spring line but thinner at the crown and the invert, comparing with the uniform-thickness pattern, is considered to be the most stable pattern.

4 RESULTS AND DISCUSSIONS OF THE HORSESHOE-SHAPED TUNNEL

In this part, the test results of the horseshoe-shaped tunnel, as shown in Figure 3, are demonstrated and discussed. In this study, the horseshoe shape is written from two ellipses that have the same width. Four thickness patterns are used in this test. All of them have the same inner-sectional shape and lining area. The variation of the thickness pattern is designed to increase the thickness at the spring line. Since the total area of the lining must be constant, the thickness at the crown or the invert is reduced. The tests are carried out at the cover to width ratio (H_s/B) equals 0.5, 1.0 and 1.5.

The results of the section forces and the strain are shown in Figure 8(a)~(d). The pattern-1 (uniform thickness) causes the minimum axial forces, however at the invert, the tensile axial force is generated. For the maximum bending moment, the pattern-1 shows the smallest value. However in the view of the tensile strain, the pattern-2 and 4 cause the minimum strains at the lining surface for both inner and outer sides. From the plots of the maximum tensile strain and the cover to width ratio (H_s/B) of all patterns in Figure 9, the results show that there is no remarkable difference between pattern-2 and 4 until $H_s/B = 1.0$, however, when $H_s/B \geq 1.5$, the pattern-4 becomes superior. These results agree with the optimum lining shape analyzed by the authors (2001) with finite element method. In Figure 9, the alphabets above the bars mean the position where the maximum tensile strain took place: I = Invert, S = Spring line. They show that, when the tunnel depth is shallow, the invert is the severest point, but, when the tunnel depth becomes deep, the spring line becomes critical point instead.

Thus from the view of tensile strain, the pattern-4, which is thicker at the spring line, but thinner at the crown comparing with the uniform-thickness pattern, is the most stable thickness pattern.

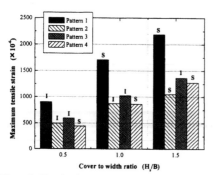

Figure 9. The plots between max strain and cover to width ratio.

5 CONCLUSIONS

From this study, it can be concluded that the stability of the noncircular tunnel, the elliptic and the horseshoe-shaped tunnel, can be improved by varying the thickness of lining without increasing the overall lining area. The thickness pattern that increasing the lining thickness at the spring line but decreasing at the crown, comparing with the conventional uniform-thickness lining, causes the minimum tensile strain on the lining surface. This study also shows that the variation of the lining thickness should be taken into consideration in the tunnel design procedure.

REFERENCES

Cau, D., Konda, T. & Nishimura, K. 1993. The basic study on the behavior of large-section with flat tunnel lining: *Proc. of the 48th annual conf. of the Japan society of civil engineers*, pp. 252-253 (in Japanese).
Takamoto, H., Kobayashi, M. & Koizumi A. 1996. Model experiment on the design method of the elliptical and rectangular cross sectioned shield tunnel: *Proc. of the 51st annual conf. of the Japan society of civil engineers*, pp. 306-307 (in Japanese).
Yingyongrattanakul N., Adachi T. & Tateyama K. 2001. Shape optimization of tunnel lining with the finite element method: *Proc. of the 56th annual conf. of the Japan society of civil engineers* (in Japanese).Hashimoto, H., Hanakusa, K., Kobayashi, M. & Koizumi A. 1996. An experimental study on the arbitrary-shaped shield tunnel: *Proc. of the 51st annual conf. of the Japan society of civil engineers*, pp. 306-307 (in Japanese).

2. Field measurements, monitoring, geophysics and site characterization

Modern Tunneling Science and Technology, Adachi et al (eds), © 2001 Swets & Zeitlinger, ISBN 90 2651 860 9

Seismic Prospecting in Underground Excavation to Investigate the Behavior of Surrounding Bedrock

H. SASAO
Tekken Corp., Tokyo, Japan

A. HIRATA
Sojo Univ., Kumamoto Japan

Y. OBARA
Kumamoto Univ., Kumamoto Japan

K. KANEKO
Hokkaido Univ., Sapporo, Japan

M. YAMAZOE
Recotech Corp., Kumamoto, Japan

N. NAKAMURA
Nittetsu Mining Co., Ltd. ,Tokyo, Japan

ABSTRACT: Excavation by blasting is a technique most widely used in tunneling and other underground construction sites. Dynamic pressures produced by blasting may damage an area around the tunnel. It is therefore critical to obtain accurate data of damages, both for safety management of excavation and evaluation of long-term mechanical stability of the underground structure. This paper discusses a compact vertical seismic prospecting system developed, aiming at creation of a system that facilitates measurement tasks, and that offers useful measurement data. The application results of the system are also presented, demonstrating its effectiveness in the management of bedrock excavation.

1 INTRODUCTION

Excavation by blasting is the technique most widely used in construction sites of underground structures including tunnels. It is thought that an area around the tunnel is damaged due to the force of dynamic pressures produced by blasting. Starting immediately after digging, the ground pressures and ground water infiltration accelerate the damage with the passage of time. The mechanical stability of a structure in bedrock is maintained by the interrelationship between three elements. These elements are hollow space, damaged zone and intact zone. It is therefore important to accurately grasp the extent of the damaged zone and the level of damage, both for safety management of excavation work and evaluation of long-term mechanical stability of the underground structure. So, a variety of measurement methods have been developed and put into practical use.

This paper discusses a compact vertical seismic prospecting system using a super-elastic alloy. In the development, we aimed at creation of a system that is reusable without embedding the device in boreholes, facilitating measurement tasks, and that offers useful measurement data. This paper also presents the application results of the system, and demonstrates that it is easily applied to the management of bedrock excavation.

2 SEISMIC PROSPECTING SYSTEM

The seismic prospecting (also called elastic wave exploration) is a measurement technique to evaluate the ground properties such as density and rigidity. This is done by detecting a wave transmitted through the ground by means of a probe incorporating vibration sensors inserted into a borehole (logging hole). When this technique is applied to determination of velocities of elastic waves (P wave and S wave), it is called "velocity logging". However, velocity logging is sometimes distinguished from "PS logging", the former determines the P wave velocity alone, whereas the latter considers the velocity of both P and S waves. From the measurements of the velocity of body waves, i.e., P and S waves, the elastic constant can be obtained. Since measurement of P wave velocity is easier than that of S wave velocity, the ground state is sometimes evaluated by the absolute value of P wave velocity alone.

Rock bolts are driven into almost all over the tunnel surroundings, which serve to reinforce the bedrock zone damaged by excavation. If boreholes for rock bolts are used for logging, the measurement range can be selected as desired. It is indeed possible to measure the entire region of tunnel wall. Another merit of the use of boreholes is that it is not necessary to drill holes dedicated for measurement. Consequently, the workload is reduced, resulting in easier daily management of work. However, since rock bolt holes are usually bored with a percussion type drill, defects of holes may occur, such as bent holes, roughened hole surfaces, and irregular hole diameters. These defects significantly affect the installation state of the vibration sensors. It is an adverse environment for ensuring effective elastic wave measurement.

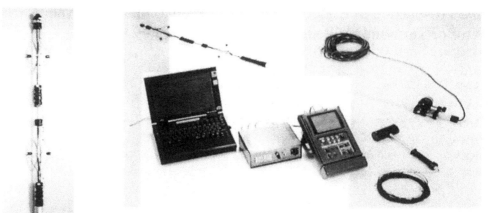

Figure 1.   Prospecting probe.   Figure 2.   Compact vertical seismic prospecting system.

Figure 1 shows the appearance of the probe of the compact vertical prospecting system we developed. This probe is provided with a spring for pressing the vibrating sensors against the hole wall. The spring is made of a super-elastic pipe (titanium-nickel alloy). This alloy has a property that, once a certain initial deformation is given, then a constant spring force is generated when deformed. The spring gives a stable pressing force to the vibration sensors even in a hole whose wall is not smooth. As a result, the vibration sensors can be secured at a desired position in a rock bolt hole. Moreover, this system does not require a complicated device to force the vibration sensors tight against the hole wall, that is used in conventional seismic prospecting systems. This facilitates the displacement and recovery of the vibration sensors in the logging hole, ensuring very easy measurement operations.

The measurement system is composed of, as shown in Figure 2, probe (signal detector), telescopic rod for inserting the probe, extension rod, extension cable, trigger hammer, signal conditioner, and waveform recorder. These devices are connected with each other by cable, forming a measurement system. On the middle portion of the probe, a super-elastic alloy tube divided in four segments is mounted, which keeps the probe on the center line of the logging hole. A compact accelerometer is installed on the back of the two super-elastic alloy segments facing each other. The expansion device on the grip of the telescopic rod expands the super-elastic alloy springs to secure the accelerometer against the logging hole wall. The applicable logging hole diameter ranges from 50 to 100 mm. Though the standard logging depth is 6 m, a larger logging depth is available by the use of an additional extension rod. The maximum depth already validated by experiments is 20 m. The trigger hammer is provided with a piezometric element on the side opposite to the impacting face. This hammer generates vibrations, and a trigger

signal is emitted on the hammering moment to the signal conditioner. The wave transmitted through the bedrock is detected by the accelerometer in the probe. After the signal conditioner processes the signal for noise cutting and amplification, the signal is digitally stored by the waveform recorder. The measured waveform is written on the memory card incorporated in the waveform recorder and transmitted to the personal computer by software specifically designed for this purpose. The functions of this software include downloading the waveform, reading the travel time, displaying the travel curve, displaying multiple waveforms on the same screen, and printing.

3   SEISMIC PROSPECTING DURING TUNNEL EXCAVATION

3.1 Site conditions and measurement method

Seismic prospecting was conducted for the purpose of investigating the behavior of the surrounding bedrock during tunneling. The excavated ground is tuffaceous sedimentary rock. The excavation method was top heading with blasting. The advance per blast was 1.2 m.

The measurement section is illustrated in Figure 3.   Immediately after the excavation of the upper section was completed, three logging holes were bored by a crawler drill, at the position 1.2 m behind the tunnel face, in the direction vertical to the tunnel wall. The holes were 65 mm in diameter, 4 m long. Using these holes, the variation of P wave velocity in the measurement section along with face advance was measured.

The relationship between measurement section and face advance is shown in Figure 4. The first measurement was done just after the drilling of logging holes. Since the hole wall surfaces were rough, only one set of prospecting probe was

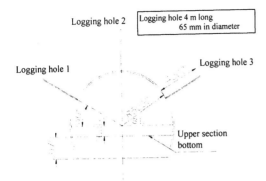

Figure 3.   Measurement section.

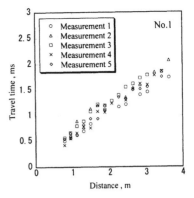

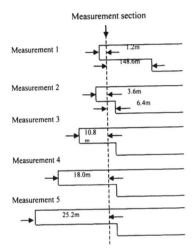

Figure 4.   Tunneling advance and measurement position.

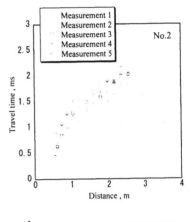

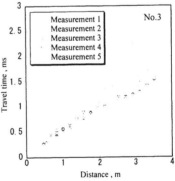

Figure 5.   Travel time vs. transmission distance.

employed. The probe was inserted into the hole, and gradually relocated at intervals of 20 cm. With this setting, the waveform through left and right channels was recorded. Then, the second measurement was conducted at the moment the lower section was excavated to 6.5 m from the measurement section and the upper section was further excavated by 2.4 m. After that, the measurement was done once a day, that is, once at a rate of 7.2 m face advance. We conducted the measurements five times in total, because we judged that further variation would not occur after the fifth measurement.

3.2 *Measurement results and considerations*

Figure 5 shows the relationship between the travel time of the P wave and the linear distance between hammering point and vibration sensors. The travel time was determined from the measured waveforms of the P wave.

The graphs in this figure are the results of logging holes Nos. 1 to 3 (left to right) respectively.

Though there are differences in minute points, the results of the left and right logging holes coincide with each other in travel time at 4 m distance, i.e., about 2.0 ms. The travel time in the case of logging hole No.2, with which the measurement was conducted upward from the tunnel crown, is greater, i.e., 2.5 ms at a depth of 4 m.

In a range from 1 to 2 m, there is a folding point of the relationship between travel time and transmission distance. It can be said that this point corresponds to the loosening area near the wall, caused by excavation. In the case of logging hole No.2 in the tunnel crown, a remarkable fold is found

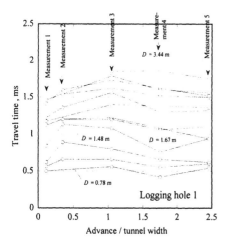

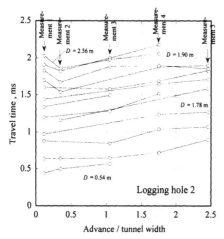

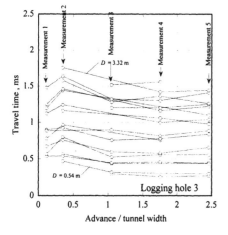

Figure 6. Variation of travel time along with tunneling advance.

near 1 m. This suggests that the loosening in the crown is more significant than that of lateral wall area on both sides. The inclination of the relationship is the reciprocal of P wave velocity. Hence, the greater is the inclination, the lower is the P wave velocity. The average velocity of the measurement points shown in Figure 5 is 1.5 km/s for walls on both sides and 0.8 km/s for the crown. In the zone deeper than the loosened area, the P wave velocity is 2.0 to 2.7 km/s.

Looking at the measurement results in the same depth, we know that the travel time in general tends to increase with the tunneling advance, though there is some variance. The graphs in Figure 6 represent the relationship between travel time and advance of excavation, with measurement section locations given as reference points. The advance is normalized by tunnel width. The graphs are the results of logging holes Nos. 1 to 3 from left to right. The circular symbols in the graphs are the results of measurements 1 to 5 from left to right.

In the walls on both sides, the travel time abruptly increases between measurements 1 and 2. In this lot, the face of the lower section came to 6.5 m behind the measurement section, and then the upper section advanced 2.4 m. The sudden increase of travel time can be explained by these excavations. The excavation of the lower section may exert a significant influence. In the right wall zone, the travel time gradually decreases after Measurement Point 2, converging to a certain value. In contrast, in the left wall area, the travel time tends to increase further, especially at greater depths in the bedrock. This tendency continues till the excavation advances about 10 m. Since, after excavation, the tunnel is maintained by rock bolts, shotcrete and steel supports, further loosening of bedrock near the tunnel wall is restrained, while the deeper bedrock continues to loosen.

In the crown zone, the bedrock tends to loosen continuously, and loosening in the bedrock near the tunnel wall is more noticeable.

4 INVESTIGATION OF THE ZONE DAMAGED BY BENCH BLASTING

4.1 Site conditions and measurement method

Seismic prospecting was conducted in the horizontal direction for investigating the loosening in the wall of a limestone mine excavated by blasting. The bedrock is composed of limestone and slate. The typical measurement section is shown in Figure 7. The measurement was done in the wall after excavation on each side, A and B. The Side A wall is purely composed of limestone.

On Side B, slate is covered with limestone.

210

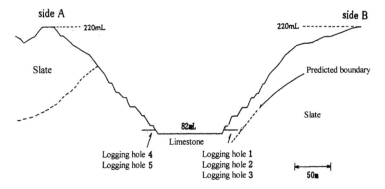

Figure 7. Typical measurement section.

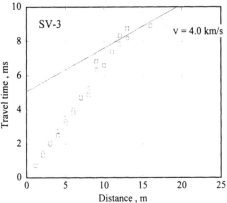
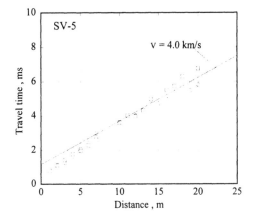

Figure 8. Travel time vs. transmission distance.

Logging holes were made horizontally about 1.5 m above the 82 m level by the use of a rotary drill.

Two logging holes 20 m long, 75 mm in diameter were made on Side A and three on Side B. With these logging holes, P wave prospecting was implemented. Only one set of probe was used in the measurement. It was relocated at 50-cm intervals in logging holes Nos. 1 and 2, and at one-meter intervals in logging holes Nos. 3, 4, and 5. The waveform data through two channels, right and left, were recorded.

4.2 Measurement results and considerations

The travel time of the P wave was determined from the measured waveforms. Figure 8 shows the relationship between typical travel time and linear distance from hammering point to vibration sensors with A and Side B logging holes. The marks O and □ in the graphs represent the different sensor channels. As is obvious in the graphs, the travel time with logging hole 3 is by far larger than logging hole 5. Average travel times in deep bedrock with elastic wave transmission speed of 4.0 km/s are regressed to

the straight line in Figure 8. Since the straight line well approximates the measured values, we can suppose that the P wave velocity in deep bedrock is around 4.0 km/s. The average P wave velocity was 2.3 km/s with logging hole 1, 1.4 km/s with logging hole 2, 1.5 km/s with logging hole 3, 2.3 km/s with logging hole 4, and 2.4 km/s with logging hole 5.

The folding point in the relationship between travel time and transmission distance, which may correspond to the bedrock zone damaged by blasting, is found at 12 m with logging hole 3, and 8 m with logging hole 5. With logging hole 3 on Side B where covering rock exists, among others, the P wave velocity in the damage zone is small, 1.5 km/s, whereas it is 2.4 km/s on Side A. For this reason, we can suppose that the damage is significantly larger on Side B than Side A.

5 CONCLUSIONS

A seismic prospecting system was developed to readily evaluate the damaged state of bedrock. This paper discussed the element techniques of the system, reporting application examples of site experiments.

The probe, which is the most essential feature of the system, was inserted into holes whose walls were rough, bored with a percussion drill. It was easy to make the vibration sensors incorporated in the probe closely contact the logging hole wall. We confirmed that the extent of the loosened zone in bedrock due to blasting and the P wave velocity in deep bedrock can be readily, yet precisely determined by the seismic prospecting system. With 20 measuring points, the prospecting was completed in about 30 minutes.

It was predicted that the elastic wave transmission in bedrock would vary along with the tunneling advance. The seismic prospecting during tunnel excavation demonstrated that the P wave travel time did vary. As a general rule, the bedrock around the tunnel tends to deform toward the interior of the tunnel. So, a weak surface and loosened zone may occur along the tunnel wall. The seismic prospecting with the measurement line orthogonal to the wall face is therefore effective for investigating weak and loosened zones.

Also in the investigation of the damage due to bench blasting, we used 20 m long logging holes, to confirm the travel time of the P wave in bedrock deeper than the damaged zone and to determine the distribution of P wave velocity. In this way we were able to calculate the average P wave velocity both in damaged zone and intact zone. Furthermore, we estimated the damaged zone width from the difference in P wave velocity distribution.

In the study on the relationship between travel time and transmission distance, we assumed that the P wave distribution is uniform in each zone to calculate the average P wave velocity. However, more detailed analysis of travel time distribution indicates that it is not always uniform. In such cases, by considering the measured velocity as velocity in a specific lot, it is possible to evaluate the inner state of bedrock. Moreover, since the measured waveforms contain information other than travel time, we can obtain further useful data by analyzing them. It can be concluded that seismic prospecting for loosened zones of bedrock around tunnels is an effective technique offering new information, which has not been supplied by conventional logging methods.

We reported measurement examples in application to the work management of new tunnel construction and control of remaining walls after bench blasting. The seismic prospecting system is equally capable of grasping the existence of weak surfaces and loosened zones in the bedrock around existing tunnels, by the same method as the application to new tunnels discussed here.

REFERENCES

Hirata, A., Inaba, T., Baba, S. and Kaneko, K. 1997. Compact VSP probe for inspection of rock mass quality. Proceedings of Environmental and Safety Concerns in Underground Construction. In Lee, Yang and Chung(eds.): 789-792. Rotterdam: Balkema.

Jaeger, C. 1972. Rock mechanics and engineering. Cambridge University press.

Kaneko, K., Nakamura, N., Hirata, A. and Ohmi, M. 1989. Technique for estimation of seismic Q by means of first arrival pulse. Butsuri-Tansa. 42, 4:235-244.

Nishida, M., Kaneko, K., Inaba, T., Hirata, A. and Yamauchi, K. 1989. Static rock breaker using TiNi shape memory alloy. Proc. of Int. Conf. on Marternsitic Transformations: 123-128. Rotterdam: Balkema.

Hirata, A., Yoshinaga, T., Horiba, N., Matsunaga, H. and Kaneko, K. 1999. Inspectin of damaged zone in tunnel walls by a compact vertical seismic prospecting system. The International Journal for Blasting and Fragmentation 3(1): 79-92

Hirata, A., Sasao, H., Yamazoe, M., Obara, Y. and Kaneko, K. 2000 Compact Vertical Seismic Profiling System and its Application in Underground Excavation. Proc. of GeoEng2000,
CD-Rom¥GeoEng2000¥papers¥G¥G0736.pdf

Modern Tunneling Science and Technology, Adachi et al (eds), © 2001 Swets & Zeitlinger, ISBN 90 2651 860 9

Evaluation of the geological condition ahead of the tunnel face by geostatistical techniques using TBM driving data

T. Yamamoto & S. Shirasagi
Department of Civil Engineering, Kajima Technical Research Institute, Japan

S. Yamamoto, Y. Mito & K. Aoki
Graduate school of Engineering, Kyoto University, Japan

ABSTRACT: In tunnel excavation by TBM, the ground condition ahead of and near the tunnel face is difficult to grasp because the face cannot be observed during tunnel driving. Because of this fact, it has been impossible to fully achieve the high-speed excavation capability of the TBM in ground having complex conditions. Therefore, the authors developed TBM Excavation Control System, the purpose of which is to grasp quickly and precisely, and in real time, the geological condition ahead of and surrounding the tunnel face. The special feature of this system is that the geological condition ahead of the TBM can be predicted with good precision by analyzing by geostatistical techniques both drill logging data from pilot boring and TBM driving data obtained during excavation.

1 INTRODUCTION

In tunnel excavation by TBM, the ground conditions ahead of and near the tunnel face are difficult to grasp because the face cannot be observed during tunnel excavation. This has made it impossible to fully manifest the high-speed excavation capability of the TBM in ground having complex conditions.

Therefore, the authors developed TBM Excavation Control System to understand quickly and precisely, and in real time, the geological conditions ahead of and surrounding the tunnel face. This system is designed to not only to manage normal measurements and construction, but also to provide information on conditions ahead of the face such as preliminary data on geological conditions and survey results (Drill logging); information during excavation such as TBM mechanical data (thrust load, cutter torque, penetration, etc.) to determine Rock mass strength (σ c(F)), Excavating energy (Qv), Gripper repellant force coefficient; and information after excavation such as ground data (state and weight of muck). This information is comprehensively analyzed and evaluated and used as feedback for the next excavation. The results of onsite applications and various indices have been found to be satisfactory for evaluating geological conditions.

However, there has been almost no systematic investigation of Drill energy coefficient for the area ahead of the face derived from Drill logging data, and its relationship with Rock mass strength, Excavating energy at the excavation site that have been derived from TBM machine data. Therefore, this paper will attempt to establish such relationships and effectively utilize them in evaluating geological conditions ahead of the face.

To improve the accuracy of the evaluation, geostatistical techniques were incorporated into the analysis and, using actual in situ data, we predicted these geological conditions at each stage of TBM excavation. These data were compared with actual measured TBM machine data, estimated unconfined compressive strength obtained by Rock Schmidt Hammer after excavation, and the results of tunnel wall observations. As a result, we were able to get a highly accurate understanding of the three-dimensional spatial distribution of the geological conditions. Moreover, as the excavation progressed we found that the accuracy improved.

2 OVERVIEW OF DRILL LOGGING AND TBM MACHINE DATA

In Drill logging system developed by the authors, data obtained while boring through the rock with a hydraulic drill (such as drilling speed, impact pressure, etc.) are measured and used to evaluate the ground conditions at various depths. Eq. 1, which is used to calculate the amount of energy required to bore a unit volume of rock (shown in Table 1 as "Drill energy coefficient"), confirmed that the geological conditions could be accurately evaluated.

In addition, TBM machine data consist of data on thrust load, cutter torque, penetration, cutter revolutions per minute, etc., which are successively recorded during excavation. In TBM Excavation Control System, these data are used to perform real-time calculations of Rock mass strength required for the ground evaluations (Eq. 2) and the Excavating energy (Eq. 3).

Table 1. Equations.

Eq. 1	$Ev=(ExN)/(VxA)$	Ev: Drill energy, E: drill energy per strike, N: striking times per second, V: drilling speed, A: area of cross section of drill
Eq. 2	$\sigma c(F)=F/(C_1xPe)$	σcF: Rock mass strength, F: thrust load, Pe: penetration, C_1: constant
Eq. 3	$Qv=F/A+2\pi NxTr/(AxV)$	Qv: Excavating energy, F: thrust load, A: area of cross section of TBM, N: cutter revolutions per minute, Tr: cutter torque, V: excavating speed

3 OVERVIEW OF GEOLOGICAL EVAUATIONS AHEAD OF THE TUNNEL FACE USING GEOSTATISTICS

"Geostatistics" refers to a statistical method that mainly deals with phenomena related to spatial variation and uses small sample sized to obtain highly accurate estimates of the entire spatial distribution. The impetus for developing this method came from the need to know ore grade and improve the accuracy of ore reserve. Geostatistics are presently used for numerous tasks in earth sciences.

Evaluations using geostatistics start out by making models of spatial trends and variancies that spatially characterize the physical quantity of the object being evaluated. For example, in the general spatial distribution of a physical value, there usually exist locations that show much higher or much lower values than surrounding areas, and there are quite a few such areas that show certain tendencies of mean value. Further, when comparing physical values of two points in a certain space, there is almost no correlation between the physical values of these points when the distance separating them is large. However, when they are close together, the physical

values are often similar. Such spatial structures are made into a model using the mean function and the covariance function, which includes location data.

Next, this model provides the basis for interpolating values for points that have no observed values. In this interpolation, changes in weight coefficients resulting from the distance between two points are used to make estimates of a linear sum of observed values. One of the most popular methods is "Kriging", a linear interpolation method.

This study marks the first time that the geostatistical techniques has been used to evaluate geological conditions ahead of a tunnel face. The procedure was as follows (see also Fig 1)

(a) Drill energy coefficient derived from Drill logging at each measuring location ahead of the face was converted into Excavating energy and Rock mass strength based on a predetermined correlation coefficient.

(b) The spatial distribution of these converted values was modeled using the mean function and the covariance function of the spatial structure.

(c) Kriging method was used to estimate (interpolate) the distributions of Rock mass strength and Excavating energy of the area ahead of the face (see Step 1 in Fig. 1).

At each step of TBM excavation, the derived Rock mass strength and Excavating energy were input as actual values, and the same type of method was used to estimate the distribution of physical values ahead of the face. In other words, as the excavation progressed the actual values derived therefrom were immediately fed back to the

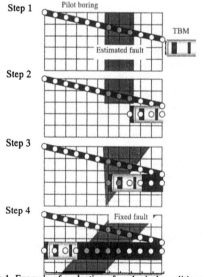

Figure 1. Example of evaluation of geological condition ahead of tunnel face by geostatistical techniques.

operation, vastly increasing the accuracy of the geological evaluation (Steps 2, 3 and 4 in Fig. 1).

By incorporating this method into the process, TBM machine data, which had previously been used only for geological evaluations at excavation points, could now be used to estimate geological conditions ahead of a tunnel face. Before that, gradually deteriorating geological conditions, cyclic changes in surrounding rock masses, etc., were qualitatively estimated by operator intuition, but the new method allows for quantitative estimates of conditions ahead of the face.

4 RESULTS OF GEOLOGICAL EVALUATIONS AHEAD OF THE FACE USING GEOSTATISTICS

Using geostatistical techniques, the tunnel in the present study, which extends a total of 3000 meters in an E-W direction, was subjected to data analyses and investigations. According to preliminary geological surveys, the geology within 2,000 meters on the eastern side is granite, while the western one within 1,000 meters is Tertiary sandstone and mudstone. Also, there is a major fault at the boundary between them. The rock on the western side is softer than the one on the eastern side. In addition, the section near the fault in the Tertiary is being metamorphosed by heat into hornfels due to the intrusion of granite.

Because Drill logging covered the entire length of this tunnel, data for Drill energy coefficient and TBM machine data also cover the entire length of this tunnel. In addition, unconfined compressive strength estimated by Rock Schmidt Hammer and wall observations covered nearly whole line, so these data were used for comparisons and investigations.

4.1 Correlation between Drill energy coefficient on the one hand and Rock mass strength and Excavating energy on the other

When the spatial characteristics of Drill energy coefficient derived from Drill logging are combined with the spatial characteristics of Rock mass strength and/or Excavating energy, we can find the correlation coefficient between the two. This is the first step in using geostatistical techniques.

Drill energy coefficient, Rock mass strength and Excavating energy at points spaced one meter apart were averages (one-meter moving average) of each datum within roughly 0.5m zones before and after the points, and the correlations were derived for each (Fig. 2). It should be noted that data were obtained for every 2 cm that the excavation advanced.

Looking at each correlation coefficient, we found that the correlation coefficients needed for the application of geostatistical techniques were satisfactory and that it was possible to establish a relationship for all data to predict geological conditions.

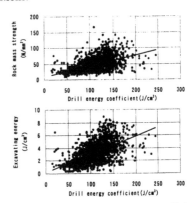

Figure 2. Correlation between Drill energy coefficient on the one hand and Rock mass strength and Excavating energy on the other.

4.2 Comparison with estimated and measured values

Using geostatistical techniques, we estimated the geological conditions on the TBM excavation line from Drill energy coefficient derived from Drill logging (Fig. 3). Figures 4-1 to 4-4 show unconfined compressive strength from the Rock Schmidt Hammer measurements as the actual ground strength, estimated Rock mass strength calculated from TBM machine data, Drill energy coefficient derived from Drill logging, and Drill energy coefficient on the TBM excavation line which were spatially estimated from Drill energy coefficient by geostatistical technique, respectively.

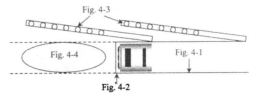

Figure 3. Estimated values and actual measurements.

In Figures 4-1 and 4-2, the absolute values are almost the same. This means that Rock mass strength predicted by the equation in Table 1 derived from TBM machine data can reflect the geological conditions at the excavation point with considerable accuracy.

Next, as shown in Figures 4-3 and 4-4, we can see that Figure 4-4 is less scattering than in Figure 4-3.

The geological structure of the site has orthogonal strike to the tunnel and has vertical dip dominantly. Therefore, even if the boring points are a little off, the trends and absolute values remain about the same. However, a detailed comparison with actual ground strength in Figures 4-1 and 4-4 shows a slight improvement in prediction accuracy in Figure 4-4 over Figure 4-3.

This means that even Drill energy coefficient derived from Drill logging that was conducted upward from the TBM-excavated line can be spatially interpolated using geostatistical techniques, and that geological conditions ahead of the TBM can be evaluated with greater accuracy.

Thus we can conclude that there are strong correlations between Figure 4-1 and Figure 4-2 and between Figures 4-1 and Figures 4-4, and that this method is very practical for geological evaluations.

4.3 Prediction results near the fault

As mentioned in the sections 4.1, 4.2, geostatistical techniques can be used to predict the geological conditions ahead of a tunnel face with a high degree of accuracy. This section will use examples of construction data of local geologically defective zone to investigate the applicability of the method in more detail.

The major fault that separates the Tertiary layers from the granite that was estimated from the preliminary geological survey was one of the biggest problems that remained in the TBM construction. At a point before the fault was expected to appear during actual construction, Drill loggings, which are usually conducted at one place, were conducted at three places from the face. As a result, the location of the fault was estimated to be at TD900 to 910 meters, so the surrounding ground was improved by cement injection.

This measure made TBM penetrated the fault zone without any big problems. However, since the locations of Drill logging were away from the excavation line, and since 3 tests were rigorously conducted, it would require a considerable amount of time and specialized knowledge to use data from each line to do a beforehand evaluation of the three-dimensional location of the fault plane.

Here, the fault position estimated by geostatistics using the results from the three Drill loggings and TBM machine data that were obtained as the work progressed were combined together in order to produce much more accurate results.

The upper part of Figure 5 shows the results of observations of muck taken during Drill logging and of the wall observations after excavation, while the lower part is the estimated geological map of the

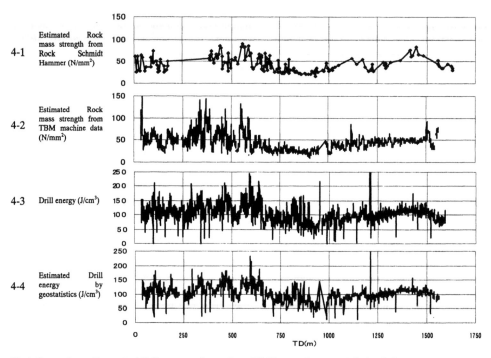

Fig 4. Comparison with an actual Drill energy and an estimated Drill energy by geostatistical techniques.

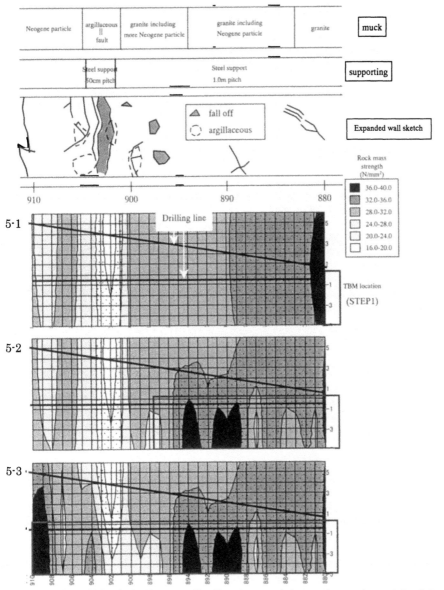

Fig 5. Comparison with longitudinal images predicted by using geostatistics at each step and expanded wall sketch of this tunnel.

longitudinal section that was derived from the geostatistical analysis of Drill energy coefficient and Rock mass strength. Figure 5-1 is a predicted diagram compiled with just the data from the three Drill loggings (Step 1), Figure 5-2 is the same type of diagram compiled from data from both the three Drill loggings and Rock mass strength to TD897.7m (Step 13), and Figure 5-3 is the same as Figure 5-2 except that Rock mass strength was measured to TD909.8m (Step 21).

Looking at Figure 5-1, we found that there was a zone with small Rock mass strength (σ c(F)=16.0 to 20.0 N/mm^2) near TD902m and Rock mass strength in the other sections was much greater. Examination of Figures 5-2 and 5-3, which were analyzed by adding Rock mass strength derived from TBM machine data at each step, confirmed the existence of hard zones (σ c(F)=36.0 to 40.0 N/mm^2) at TD889m to TD895m and TD908 to 910m that could not be confirmed in Figure 5-1. This and the other findings provided a more detailed distribution of Rock mass strength. Results of the wall observations

217

also confirmed that there were few fissures in the vicinity, and that the rock was relatively hard. In addition, Rock mass strength started to decrease at TD894m, where the amount of fine particles originating from the Tertiary mudstone began to increase, and was estimated to be lowest in the zone from TD901 to 905m, predicting the area which was believed to be the fault and which had undergone intensive fissuring, argillization, and weathering. Actually, the fault at that location was found to have completely changed into clay, so steel supports were installed every 50 centimeters.

The successive addition of TBM machine data to Drill logging results made the estimates more detailed and accurate at each step. Because the geological conditions at the present study area include the nearly vertical dip and layers whose dominant strike is orthogonal to the tunnel axis, no major unexpected temporal or spatial changes were seen in the area ahead of the face. However, near a boundary with faults and/or other geological structures that have gentle dip, we could expect there to be some major changes.

5 CONCLUSION

The use of geostatistical techniques enabled us to use one-dimensional data (Drill logging results, TBM machine data) to get a more accurate and rational understanding of the three-dimensional temporal and spatial distributions of geological conditions. We also confirmed that accuracy improved as the work progressed.

The present study marks the first time that geostatistical techniques have been used in the analysis of Drill logging and TBM machine data. Our results confirmed that this method can predict geological conditions ahead of a TBM face with a high degree of accuracy.

The incorporation of geostatistical methods in TBM Excavation Control System will likely enable geological evaluations of TBM face areas to be done more intelligently. Furthermore, the use of geostatistically-derived three-dimensional geological profiles for tunnels constructed using the TBM pilot boring method should enable more rational and swift selection of supports during enlargement.

REFERENCES

Aoki, K. et al. 1989. Application of geostatistics to obtain the geological profile by seismic prospecting (in Japanese). *Proceedings of the 21st Symposium on Rock Mechanics*: pp.136-140.
Fukui, K. et al. 1996. Estimation of rock strength with TBM cutting force and site investigation at Niken-goya tunnel (in Japanese). *Resources and Materials*: Vol.112, No.5, pp. 303-308.
Shirasagi, S. et al. 1999. Examination of the application of Drill Logging to predict ahead of the tunnel face (in Japanese). *Proceedings of the 54th annual conference of the Japan society of civil engineers*: VI-206, pp. 412-413.
Shirasagi, S. et al. 2000. Development of TBM navigator introducing geophysical and geological data, and application (in Japanese). *Proceedings of the 30th Symposium of Rock Mechanics*: pp. 298-302.
Yamamoto, Y. et al. 2001. Evaluation of the geological condition ahead of the tunnel face by geostatistical techniques using TBM driving data (in Japanese). *Proceedings of the 31st Symposium of Rock Mechanics*: pp. 186-190.

Modern Tunneling Science and Technology, Adachi et al (eds), © 2001 Swets & Zeitlinger, ISBN 90 2651 860 9

Application of CSAMT monitoring method and observational construction management to mountain tunnel

T. Fukui & T. Hirai
Toll Road Corporation, Osaka Prefectural Government, Japan

T. Morioka & A. Suga
Minoh Tunnel Construction Office, North Section,
Kajima, Taisei, Toa, Mitsui and Aoki Joint Venture, Japan

T. Matsui
Department of Civil Engineering, Osaka University, Suita, Japan

ABSTRACT: The CSAMT (controlled source audio magneto-telluric) method was applied to estimate the condition of the ground with fracture zones and heavily water-bearing zones for tunneling. The analyzed distributions of porosity and water content by volume resulted in good agreement with the results of actual ground condition through construction. Thus, the applicability of the method was confirmed. For the design of a T-shaped tunnel, three-dimensional finite element analysis was applied using the monitoring data of the ground. The tunnel was constructed by an observational method based on the data measured during construction. As a result, the validity of the design was confirmed by grasping the ground behavior at the intersection of the tunnels.

1 INTRODUCTION

The Minoh toll road tunnel has been constructed to extend northward the Shin-midosuji route running about 15 km to the north of the city center of Osaka, as an alternative to national road route 423. It is a road tunnel of a length of about 5.6 km that penetrates the Hokusetsu mountains of 200 to 600 m hight in the north of Minoh City.

In the north work section, a 3073m tunnel with a 346m working tunnel has been driven toward Osaka in the north of Minoh City. In the present paper, at around the intersection of the working and main tunnels, the applicability of CSAMT monitoring method for the ground investigation is examined. Then, the design method for the T-shaped tunnel is discussed based on 3-D finite element analysis.

2 PROJECT OUTLINE

Project name: Construction of the Minoh toll road
mountain tunnel (north work section)
Project site: Shimo-todoromi, Minoh City, Osaka
Prefecture
Main tunnel: 3037 m
Evacuation tunnel: 3113 m
Underground ventilation station: One station with a
set of equipments
Precipitator room: Two rooms
Shaft: 330 m
Working tunnel: 346 m
Temporary facilities: A set

3 GEOLOGICAL OUTLINE

Around the north portal of the tunnel, the Osaka groups mainly consisting of sandy gravels are distributed in a wide area, which are affected by the large-scale Satsukiyama fault system. Behind the distance of 200 m from the portal, a weak fractured zone mainly consisting of sandstone continues for several hundred meters. Behind the distance of about 700 m from the portal, both seismic velocity (more than 5 km/sec) and resistivity of the ground are high. This zone is the bedrock mainly composed of sandstones that was formed in the Triassic through Jurassic periods of the Mesozoic era.

4 GEOLOGICAL INVESTIGATION, DESIGN AND CONSTRUCTION FOR INTERSECTION OF WORKING AND MAIN TUNNELS

4.1 Objective of geological investigation

A working tunnel is connected to the main tunnel at right angle at the distance of 266 m from the north portal of the main tunnel. The intersection is located within a large-scale fractured fault zone. The ground condition around the intersection, however, can not be identified because no detailed data are available. Therefore, additional investigations are carried out. The CSAMT method is applied to investigate the ground condition and heavily water-bearing zones. The applicability of the method is examined through comparison between analytical results and actually observed results.

4.2 Procedure for investigation and analysis by CSAMT monitoring

The applied procedure for investigation and analysis is described below.

(i) To measure each apparent resistivity of ground for kinds of frequencies at 30 measuring points

(ii) To draw resistivity logs, and a resistivity profile of ground

(iii) To identify correlations of geology and groundwater to resistivity through calibration at existing boring locations

(iv) To estimate the geological structure of ground based on the resistivity structure

(v) To calculate the porosity and the degree of saturation of ground based on both available data of seismic velocity and resistivity through a physical property conversion mentioned below, and to estimate classification of ground and heavily water-bearing zone.

4.3 Conversion of physical properties

Both the resistivity and seismic velocity of ground can be expressed by Equations (1) and (2) respectively, using porosity and degree of saturation as variables.

Resistivity: Archie's formula

$$\left(\frac{1}{\rho}\right) = \left(\frac{1}{F} \cdot \rho_W\right) + \left(\frac{1}{\rho_c}\right) + \left(\frac{1}{\rho_0}\right) \quad \cdots \cdots (1)$$

$$F = a \cdot \phi^{-m} \cdot s^{-n}$$

where,

ρ : resistivity of rock mass obtained through analysis of CSAMT monitoring (Ω.m)

ρ_w: resistivity of pore water (Ω.m)

ρ_c: resistivity of conductive particle in rock (Ω.m)

ρ_0: resistivity of rock (Ω.m)

F : formation factor of resistivity

ϕ : porosity, S: degree of saturation

a,m,n: coefficients

Seismic velocity: Wyllie's formula

$$\left(\frac{1}{V_P}\right) = \left(\frac{1-\phi}{V_m}\right) + \left(\frac{\phi \cdot S}{V_f}\right) + \left(\frac{\phi \cdot (1-S)}{V_a}\right) \quad \cdots \cdots (2)$$

where,

V_p: velocity of P-wave in rock mass (km/s)

V_m: velocity of seismic wave in test piece (km/s)

V_f: velocity of seismic wave in pore water (km/s)

V_a: velocity of seismic wave in pore air (km/s)

ϕ : porosity, S: degree of saturation

Based on these equations, both the CSAMT and seismic wave monitoring results are converted to obtain the porosity and the degree of saturation. Then, two-dimensional profiles of porosity and

water content by volume (porosity multiplied by the degree of saturation) can be drawn, and the classification of ground and groundwater condition can be identified.

4.4 Comparison between profiling results and actual data in tunneling

(1) Comparison in working tunnel

Figure 1 shows the distribution of porosities and classifications of ground in the working tunnel. Figure 2 shows the distribution of water content by volume and the variation of amount of spring water at the tunnel face.

(i) Comparison in porosity and classification of ground (Fig. 1)

Analytical results: Up to the distance of 20m(the distance of 250m from the portal), the porosity was ranged between 5.5 to 8.5%, and the ground was classified by DI. From the distance of 20m through 110m, porosity was ranged between 8.5 to 16%, and the ground changed to class DII.

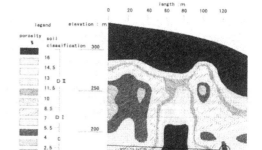

Figure 1. Porosities and classifications of ground in the working tunnel.

Actual results: At around the distance of 20m and the beyond, many irregular joints were developed and the weathering was outstanding. At the distance of 30m, AGF (All Ground Fasten) method was adopted as an auxiliary method. The geological condition became worse at the distance of 40m and the beyond so much as to enable easy excavation with a backhoe.

(ii) Comparison in water content by volume and amount of spring water (Fig. 2)

Analytical results: The zone between the distance of 50m and 90m was estimated as heavily water-bearing zone, because the water content by volume was ranged over 5.5%.

Actual results: Three 50m-long boreholes were drilled for dewatering in the above-mentioned zone. Spring water of about 1000 liters per minute was observed in all of the boreholes. Since the amount of spring water decreased with time, it must be under groundwater stored in a part of the ground.

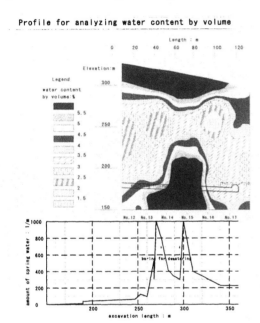

Figure 2. Water content by volume and amount of spring water at the face.

(2) Comparison in main tunnel

Figure 3 shows the distribution of porosities and classifications of ground in the main tunnel. Figure 4 shows the distribution of water content by volume and the variation of amount of spring at the tunnel face.

(i) Comparison in porosity and classification of ground (Fig. 3)

Analytical results: Within the 200-m zone from the intersection, the porosity fluctuated greatly from 3 to 8%. The ground was classified by DI to C. Behind the zone, the ground became to class C.

Actual results: In the above-mentioned zone, the ground was classified by DI to DII. Within the distance of 70 to 90 m from the intersection, AGF was adopted, while at the distance of 110 m, the excavation by breaker was replaced by blasting, in which the porosity was the lower of 2.5 to 4%. The rock type at the tunnel face generally varied so that soft and hard rocks alternated daily. In the zone

between 210 and 230 m from the intersection, the ground was classified by CII followed by DI subsequently.

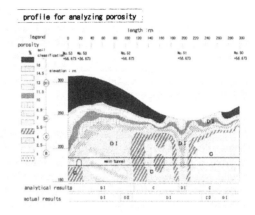

Figure 3. Porosities and classifications of ground in the main tunnel.

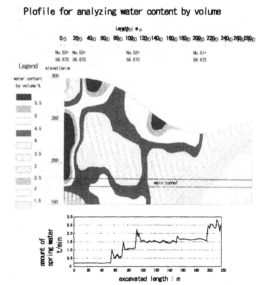

Figure 4. Water content by volume and amount of spring water at the face.

(ii) Comparison in water content by volume and amount of spring water (Fig. 4)

Analytical results: At the distance of about 100 m from the intersection, the water content by volume exceeded 4%, and a large amount of spring water was estimated to occur. Beyond the zone, the water content by volume might gradually decreased, reaching about 2% at the distance of 250 m from the intersection.

Actual results: Generally, the amount of spring water increased with the advance of excavation. At the distance of 100 m from the intersection, it reached a peak value of 2000 liters per minute. This location corresponded to the estimation of heavily water-bearing zone. Beyond the distance of 100 m from the intersection, spring water of more than 1500 liters per minute came out, followed by reaching 2500 liters per minute beyond the distance of 200 m from the intersection.

(3) Consideration

The actual work carried out in the working tunnel shows that considerable fracturing of ground occurred at porosity of 10% or higher, in which the ground was so weak that excavation by a backhoe could be possible. While, the amount of spring water increased greatly at water content by volume of 5% or higher, demanding precaution.

In the main tunnel, both the porosity and water content by volume were lower than those in the working tunnel. The ground was mostly classified by DI. The amount of spring water tended to increase with the advance of excavation. The difference may be because measuring points of resistivity profiling were set at intervals of 20 m, and because the ground was a complicated alteration of fracture zones and hard rocks. The unique topography beyond the distance of 200 m from the intersection may have affected the amount of spring water since the excavation at the depth of 50m was carried out right under and parallel to a stream running in a narrow valley.

Water content by volume is expected to provide a relatively accurate estimation stored underground water that decreases with time as in the working tunnel. On the other hand, in case where the total amount of spring water increases with the advance of excavation as in the main tunnel, water flowing in through cracks significantly affects it, and thus, the estimation based on water content by volume is not likely to be effective.

4.5 Design of intersection of tunnels

(1) Design method

Based on the result of CSAMT monitoring, the ground at the intersection of tunnels was assumed to be fractured as in the working tunnel. For designing the intersection, therefore, three-dimensional elastic finite element analysis was applied to check the stresses of tunnels in each excavation step. The modulus of deformation, which has the largest effect on analytical results, was obtained by back analysis

on crown settlement in the working tunnel followed by setting at E=500 MPa.

(2) Analytical result

Since the excavation of an evacuation tunnel was planned in parallel to the main tunnel 30 m away therefore, the following two methods were studied for connecting the working tunnel to the main and evacuation tunnels.

Case 1:The working tunnel is connected to the main and evacuation tunnels at right angle to each other. The intersection is cross-shaped one, as shown Fig.5.

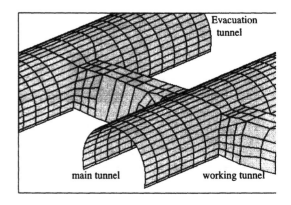

Figure 5. Cross-shaped analysis model.

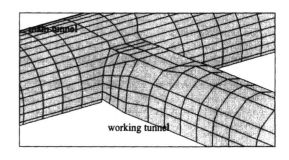

Figure 6. T-shaped analysis model.

Case 2:The working tunnel is connected only to the main tunnel at a T-shaped intersection, as shown Fig.6. The working tunnel is connected to the evacuation tunnel at a location in stable ground.

As a result of analysis, it was found that the stresses of shot Crete and steel supports at corners of the intersection exceeded the allowable limits greatly in Case 1. In Case 2 of a T-shaped intersection, all stresses were held below the allowable limit, where the shotcrete had a thickness of 250 mm, and steel supports of H200×200×8×12mm at an interval of 1000mm and rock bolts TD25×6000mm at an interval of 1000mm were used. These specifications were actually adopted.

4.6 Observational construction at intersection of tunnels

(1) Instrumentation plan

To measure the behavior of the ground at the intersection and examine the validity of the design, extensometers, rock bolt axial force meters, shot Crete stress meters, and steel support stress meters were installed as shown in Figure 7.

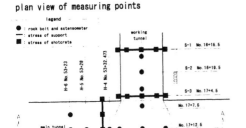

plan view of measuring points

Figure 7. Location of measuring equipment.

(2) Measured results

(i) Ground displacement

The displacement of the tunnel crown at the intersection almost converged after the excavation length of 20 m was preceded on one side (toward south). Subsequently, the excavation on the other side (toward north) hardly caused any displacement. Figure 8 shows distributions of ground displacement in respective cross section before the excavation toward north side. Ground displacement reached rather greater positions of 8 to 10 m above the tunnel crown. At around positions of 0 to 3 m above the tunnel crown, it becomes the maximum and then decreases linearly as increasing the distance. If the zone to the position of the maximum displacement is regarded as the loose zone, the measurements are almost identical to the analytical results.

(ii) Axial force of rock bolt

Axial force reached the maximum value after excavation length of 10 m on one side . Tensile force was 50 to 100 kN, increasing additional 30 kN during the excavation on the other side.

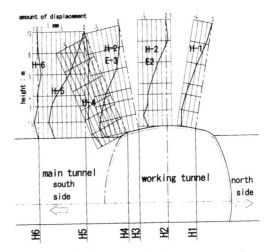

Figure 8. Distribution of ground displacement (results of measurement) in A-A cross section.

(iii) Axial force, Bending moment and stress of Steel support

Axial force increased slightly even at points 3D (D=tunnel diameter) or further away from the face. Bending moment converged near the point of 3D. Northward excavation had a slight effect on axial force, which increased about 10%, while bending stress remained unchanged. The maximum stress was 200 N/mm^2.

(iv) Stress of shotcrete

Stress of shotcrete converged after the excavation length of 30 m. The stress exhibited a similar tendency as for steel supports. The maximum stress was 12 N/mm^2.

(v) Crown settlement

Analytical and measured values of crown settlement at the intersection are shown in Figure 9 for comparison. Actual settlement converged at 35 mm, 10 mm smaller than the analytical value. As for the ratio of settlements in the initial southward excavation reached 90% of total settlement. While, measured settlement in the southward excavation reached 70% of total settlement and that in the northward excavation 30%.

(vi) Convergence

The analytical convergence was about 40 mm, while the actual convergence exceeded 60 mm. In the southward excavation, convergence reached about 90% of total convergence.

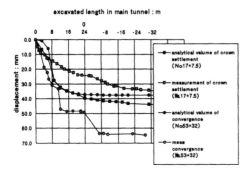

excavated length in main tunnel : m

Figure 9. Comparison of analytical and measured values of crown settlement and convergence.

(3) Consideration

In the analysis, full-face excavation was assumed, while in the actual construction, a top heading and short bench method was adopted. The execution was carried out carefully with driving forepoling and injecting silica resin, because of the excavation in fracture zones.

As a result of analysis, both stress and displacement reached 90% of the total values at the T-shaped intersection by the initial southward excavation. Various measurements also indicate a similar tendency.

There were slight differences between analytical and measured crown settlement and convergence. This may be because the ground in the analysis was assumed as a homogeneous elastic body, although the actual ground condition was fractured and in a complicated state. The ratio between measured crown settlement and convergence was 1:2. That is, the convergence was double the settlement. While the ratio in the analysis was approximately 1:1.3. This is attributable to a Poisson's Ratio v of 0.35 adopted for the analysis. Re-computation with a Poisson's Ratio of 0.40 resulted in a ratio of 1:1.7, closer to the actual one.

5 CONCLUSIONS

Porosity and water content by volume of the ground at the project site were obtained from the resistivity measured by the CSAMT monitoring method and the existing seismic velocity. Thereby, the ground with fracture zones and heavily groundwater-bearing zones were identified. As a result, the applicability of the method for the monitoring behind the tunnel face was confirmed.

The ground behavior at the intersection of tunnels was grasped in each construction phase by three-dimensional elastic finite element analysis, and observational construction was carried out based on various measurements. The validity of design was verified through the grasp of qualitative ground behavior. Stresses and displacement converged appropriately, and the intersection of tunnels maintained stability. Thus, both design and construction proved valid.

REFERENCES

Sawai.K,Kozato.T,Nishikata.U&Ohtomo.Y1996. Estimation of classification of ground based on resistivity and seismic velocity (in Japanese) *Proceedings of the 51 st annual conference of the Japan society of civil engineers:794-795*

Archie,G.E. 1983. The electrical resistivity log as an aid in determining some reservoir characteristics. *Trans, A.I.H.E.:146,54-67*

Wylie,M.R.,Gregory,A.R.&Gardner,L.W.1966.Elastic wave velocities in heterogeneous and porous media. *Geophysics 21: 41-70*

Modern Tunneling Science and Technology, Adachi et al (eds), © 2001 Swets & Zeitlinger, ISBN 90 2651 860 9

Nondestructive survey of shallow underground by electromagnetic method

T.Katayama , K.Ozaki
Kanden-kogyo Co., Inc., Osaka, Japan

Y.Yoshizu , E.Kurata
Kansai Electric Power Co., Inc., Osaka, Japan

T.Kozato
Construction Project Consultants Co., Inc., Osaka, Japan

Y.Ashida
Kyoto University , Kyoto, Japan

ABSTRACT: The electro-magnetic sounding system for shallow sub-surface prospecting was developed. This system is composed of the sounding instrument with the transmitter of 16 frequencies, the receiver and the interpretation software. In the present paper, the outline of the system and the inversion result by the field data are described. As a result, the high efficiency of this system was confirmed.

1 INTRODUCTION

Among the prospecting methods applied in the civil engineering field, resistivity method is being increasingly used; especially, high density electric prospecting called HDEP, that is two dimensional resistivity prospecting method, seems to be widely used. Among the several factors taken for this reason, the following factors may be considered.

· The relatively high accuracy of the prospecting
· The visual display of the results
· Easy estimation of the ground and geological interpretation of structure

On these backgrounds, utilization of the HDEP method is especially notable in civil engineering fields where the depth of investigation is almost less than 50 meters.

Of course, other electric/electromagnetic methods such as CSAMT and TDEM, for example, are being employed. These methods developed for investigating underground resources and being suitable for deep penetration depth, does not satisfy the resolving power for the shallow depth investigation less than 50 meters. On the other hand, the ground penetrating radar method, whose subjective scope is within few meters from the ground surface, has a limited application to the civil engineering field.

CSAMT method requires the measurement of both the electric field and the magnetic field. In the case of HDEP system, the hammering of many electrodes into the ground, is so difficult in some ground condition that the abandonment of investigation occurs occasionally for lack of alternative sounding method.

The electromagnetic (EM) method covers the depth of investigation in civil engineering field and obtains the resistivity only from the measurement of magnetic field without the electric field.

On these backgrounds, we developed the compact EM system based on frequency domain equipped with 16 channels frequencies at maximum, and then verified its validity by the comparison with the conventional resistivity prospecting method.

2 THE DEVELOPED SYSTEM

2.1 *Outline of the system*

In the first stage of developing the system, we manufactured 8-channel system for the prototype equipment. After repeating the test prospecting and the improvement using this system, we developed 16-channel system with much higher accuracy. The system is composed of the rectangular transmitter coil, the transmission unit with 12 volts battery as electric source, the magnetic field sensor unit inserted into protection box and the main body of receiver containing the receiver unit, A/D converter and the optical transmission unit.

Digitized EM data are sent to a personal computer through the optical transmission cable; hereupon data are processed and analyzed to display on the screen with real time.

2.2 *Main specification of the analogue unit*

(1) Transmitter unit(16-channel)

Transmission frequencies are allotted to each channel as follows, 364, 256, 182, 128, 91, 64, 45.5, 32, 22.8, 16, 11.4, 8, 5.7, 4, 2.8, 2 kHz. The transmission circuit has been developed by the resonance circuit so as to gain high output even by 12 volts buttery source.

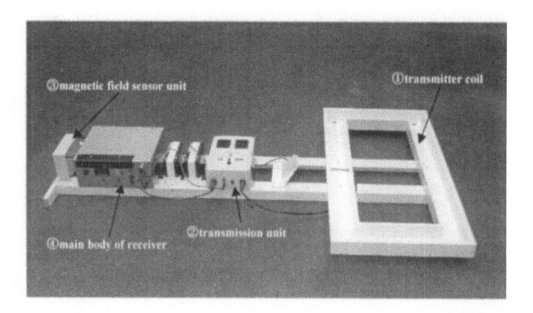

Figure labels: ③magnetic field sensor unit, ①transmitter coil, ②transmission unit, ④main body of receiver

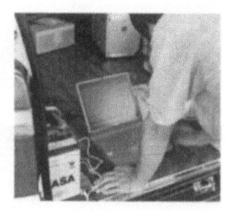

Figue1. Layout of EM inspection system.

Transmission is wholly practiced automatically for 16 channels to be transmitted in one second.

This system designed originally for car-born prospecting and therefore planned to operate by the car battery, is installed at the rear of automobiles to perform the running operation.

Transmitter coils: High-coil (above 32 kHz) and Low-coil (under 32 kHz), are separately looped onto the rectangular flame to let the transmission current flow in this coil by use of the highly stable constant current circuit.

(2) Receiver unit

In EM system, the apparent resistivity values are obtained by the following procedure.

First, the intensity of the primary magnetic field is measured. Next, the component imaginary of the secondary magnetic field derived from the primary magnetic response is measured. Multiplying the ratio of the primary and the secondary magnetic field as the calibration coefficient, the apparent resistivity against each frequency or each channel is obtained.

As the component imaginary becomes weak according to the decrement of frequency, the rise of detection accuracy is required. For this treatment, the high accuracy phase detection circuit was developed. For the present time, the accuracy of detection is 0.0001 volt.

The analog voltage signals are simultaneously digitized by A/D converter. Together with this, other observation data are sent to the personal computer.

As for the receiver sensor, coils are looped on the ferrite cores and set precisely on the same plane as the transmitter coil.

(3) Soft ware

For processing of measured values, the following are mainly prepared.

· High speed computation program of apparent resistivity
· Graphical display of apparent resistivity section
· Stacking process in the case of incorporation of noise
· One dimensional inversion program of EM system
· Two dimensional mapping program of the inversion result
· Car navigation program for recordings of its speed and position in the running operation (under development)

2.3 Theoretical formula of EM measurement

As for the setting of transmitter/ receiver coils in EM system, there are two ways to place the central magnetic field of coils vertically or horizontally to the ground shown in Figure2. In usual cases, the measuring using the vertical dipole is taken. The ratio of the intensity of magnetic field of the secondary magnetic field to the primary magnetic field is given by the following formula from (McNeil, 1980).

Vertical Dipole :
$$\left(\frac{H_s}{H_p}\right)_V = \frac{2}{(\gamma s)^2}\{9 - [9 + 9\gamma s + 4(\gamma s)^2 + (\gamma s)^3]e^{-\gamma s}\} \quad (1)$$

Horizontal Dipole :
$$\left(\frac{H_s}{H_p}\right)_H = 2\left[1 - \frac{3}{(\gamma s)^2} + [3 + 3\gamma s + (\gamma s)^2]\frac{e^{-\gamma s}}{(\gamma s)^2}\right] \quad (2)$$

where H_s = secondary magnetic field at the receiver coil
H_p = primary magnetic field at the receiver coil

$\gamma = \sqrt{i\omega\mu_0\sigma}$
$\omega = 2\pi f$
f = frequency (Hz)
μ_0 = permeability of free space
σ = ground conductivity (mho/m)
s = intercoil spacing (m)
$i = \sqrt{-1}$

VERTICAL DIPOLE   HORIZONTAL DIPOLE

Figure2. Setting of vertical and horizontal dipole.

2.4 One dimensional inversion of EM data

The theoretical formula (Fullagar, et. al, 1984) derived on the basis of Frehet-kernels integration formula due to the vertical dipole on the model ground assumed the multi-layered structure is adopted.

Also, as the numerical analysis of response of magnetic field requires the computation with high speed, the linear digital filter theoretical formula (Koefoed, et. al, 1971) was adopted. Modification of the model was practiced by matrix computation based on general nonlinear least square method.

2.5 Calibration test of measured values

In order to examine whether the conductivity, that is, reciprocal of resistivity, measured by this EM system indicates correct values or not, the field test was performed in the middle of Lake Biwa where the depth of water is more than 80 meters.

Validity of equipment of this EM system was confirmed by the fact that the conductivity of the instrument shows the agreement with the previously investigated value of the lake water against each channel.

2.6 Depth of investigation

The formula of skin depth of MT method is identically applied to this EM system. The skin depth, mainly for the surface layer, is defined as the depth within where the electromagnetic field in the case of high frequency influence.

As for the transmission frequency in the case of 16-channel, the lower limit is 2 kHz and the upper limit is 364 kHz. For instance, when the resistivity of ground is 50 Ω m, the corresponding lower limit of skin depth is 80 meters, while the corresponding upper limit of skin depth is 6 meters.

Accordingly, although skin depth depends on the resistivity of ground, investigation up to the depth of 50 meters at least is feasible by adopting appropriate channel of the system. In fact, the results of analysis of several other fields show that this has been achieved.

3 VERIFICATION TEST

3.1 Comparison with high density electric prospecting: HDEP

Among several resistivity methods, resistivity tomography or HDEP is considered to have the highest accuracy at the present time. Considering the same prospecting method from the ground surface, the comparison of EM system with HDEP is tried.

As a field investigated by HDEP, Takatsuka-yama (Mt.Takatsuka) fault located in the west of Kobe City shown in Figure3 is selected. According to the previous studies at the surroundings of the Takatsuka-yama, it is reported that Kobe group formation are located to the east and Osaka group formation to the west from the fault (Nakagawa, et. al, 1997).

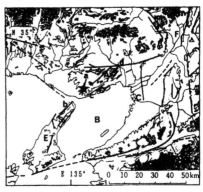

Map showing the surveyed site and the active faults in the central Kinki
(★: Surveyed site, A: Mt.Rokko, B: Osaka Bay,
C: Osaka, D: Kobe, E: Awaji Is. F: Kyoto,
a: Takatsukayama fault, b: Nojima fault, c: Suwayama fault, d: Arima-Takatsuki tectoniczone, e: Median tectonicline)
Figure3.Location of Takatsuka-yama fault.

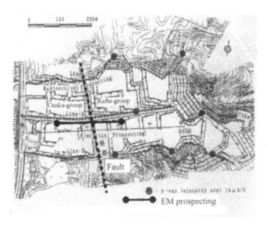

Figure4. Survey line of EM test investigation.

As shown in Figure4, EM prospecting using the 8-channel unit was performed on the almost same survey line as HDEP. The survey points were set with 2 meters interval on the line crossing the fault in its middle.

By using the result of one-dimensional inversion at each survey point, the resistivity section by the two-dimensional mapping was generated.

By the comparison of the section with that by HDEP shown in Figure5, it is recognized that both sections exhibit reasonably good agreement.

Judging from the result of EM system, the resistivity structure of Takatsuka-yama fault is clearly detected and the contrast between Kobe group and Osaka group is also enough distinguished.

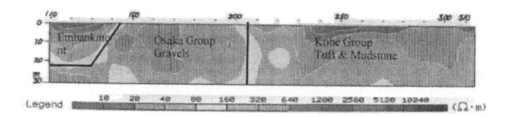

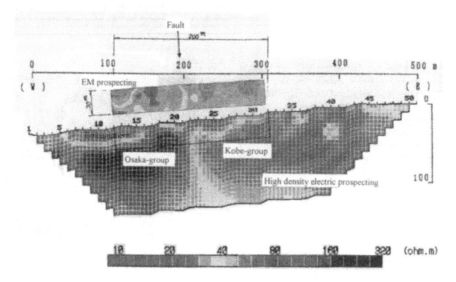

Figur5.Result of EM test at Takatsuka-yama fault.

228

3.2 *Test prospecting for shield pipe*

The field operation was carried out at the work site of setting the shield pipes to detect the presence of pipes itself and the loosened zone by setting work of pipe around the pipe in lengthwise.

The site for test is located at the suburbs of Ichinomiya City in Aichi Prefecture where shield pipe of the outer diameter of 2 meters was set at the depth of 10 meters. Field operation using 8-channel system was performed on the ground just over the pipeline with 1 meter interval on the survey line of 100 As understood from the geologic column in this figure, the ground is composed of downward from the surface, thin surface soil embankment, sand mixed with silt to the depth of 6 meters and gravel bed lying horizontally beneath.

The resistivity section exhibits that resistivity contrast varies around the depth of 6 meters and the low resistivity zone lies horizontally beneath the depth of 10 meters.

This pattern of contrast is explained as follows. The original resistivity of the ground might be $50\,\Omega\,m$, while the detected resistivity of the same part after setting of shield pipe is $40\,\Omega\,m$. This indicates the combined effect of the disturbance of the soil and the presence of pipes of reinforced concrete due to the high sensitivity of magnetic field to ferrite in EM system.

4 SUMMARY AND THE THEMES HEREAFTER

The EM system in frequency domain being capable of transmitting 16 channels of frequency at maximum was developed. If the more transmission channels are provided, the more prospecting data are obtained upon the same depth and the much more resolving power of investigation is improved, although the development of multi-frequency instrument requires the considerable cost and the efforts.

Now, the 16-channel EM system has already been completed and several verification test surveys have been finished. As a matter of course, the acquired results support higher accuracy than ever have been obtained.

The following themes hereafter should be fixed.

1) Rise of the transmission current to increase the prospecting depth and to obtain the higher data quality.

2) Development of higher speed transmission and display of the investigated image with real time in order to realize the car-born EM prospecting system.

3) Development of more small-sized instruments for the anywhere easy operation.

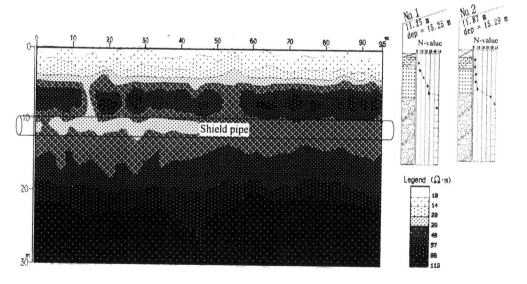

Figure6.Results of EM test on post-construction shield pipe

REFERENCES

McNeill, J.D. 1980. Electromagnetic terrain conductivity measurement at low induction numbers, Geonics Ltd. Technical Note TN-6

Fullagar, R.K. & Oldenburg, D.W. 1984. Inversion of horizontal loop electromagnetic frequency sound-ings, Geophysics, Vol. 49, No. 2

Koefoed, O., Ghosh, D.P. & Polman, G.J. 1971. Computation of type curves for electromagnetic sounding with a horizontal transmitting coil by means of a digital linear filter, Geophysical Prospecting, Vol. 20

Nakagawa, Y. & Shimizu, K. 1997. Era of activity of Takatsukayama Fault in Kinki — Active faults in Kinki(Part 1), Science Engineering Department Doshisha University Research Report 38th Vol. No.1 Separate Volume

Modern Tunneling Science and Technology, Adachi et al (eds), © 2001 Swets & Zeitlinger, ISBN 90 2651 860 9

A study on the evaluation of in-situ rock conditions by the tomographic method in tunnel geological investigations

S.Miki
Kiso-Jiban Consultants Co. Ltd., Japan

H.Shiroma
Research Institute, The Japan Highway Public Corporation, Japan

K.Inoue & K.Nakagawa
Department of Civil Engineering, Yamaguchi University, Japan

ABSTRACT: We have introduced the tomographic method to the seismic exploration in tunnel investigations, and have presented a way to evaluate the performances of the tomographic method based on geological observations. Geological observation records at the face have been utilized to measure the performances of the tomographic method and the reciprocal method. And these seismic methods have been examined in 35 tunnel sites. The performances of the tomographic method varied with tunnel sites. However, the tomographic method showed same or good performances compared with the reciprocal method in many tunnel sites. This leads to the conclusion that the tomographic method as well as the reciprocal method can be applied in tunnel investigations.

1 INTRODUCTION

It is important to obtain accurate geological conditions in a tunnel design. A seismic exploration is generally utilized as a method for the investigations, and support patterns for tunneling are determined with seismic wave velocities. In many cases, the reciprocal method such as Higiwara's method has been adopted as an interpretation method for the seismic explorations. Recently, the tomographic method has been developed with the advances of electronics and computer technologies, and it sometimes has been utilized as an interpretation method. However, its applications to tunnel investigations are limited, and the efficiencies of the tomographic method are scarcely evaluated in many geological conditions. There are few works on the quantitative comparison between rock mass conditions estimated from the tomograohic method and geological observations at the faces in tunneling.

Geological conditions in Japan are complex and the seismic velocity distributions of rock mass don't always show layer structures, however the reciprocal method has an assumption that the seismic velocity distributions should be layer structures. Therefore, we expect that the accuracy of a seismic exploration will be improved by using the tomographic method in addition to the reciprocal method.

Nakagawa et al. (2000) have studied the quantitative relations between the geological observation scores at the tunnel faces and seismic velocities surrounding rock mass. According to their criteria, we can estimate the scores from the seismic velocities.

Therefore, geological observation records at the tunnel face have been utilized to measure the performances of the tomographic method and the reciprocal method, and these seismic methods are examined. In order to evaluate the efficiency of the tomographic method, the comparison about the performances of both seismic methods has been made by using the geological observation scores.

2 OUTLINE OF THE TOMOGRAPHIC METHOD

The tomographic method is a suitable technique to solve complicated velocity structures. In this method, rock mass is divided into numerical cells as shown in Figure 1a, and the slowness, which is a reciprocal of the seismic wave velocity, of each cell is computed. Procedures of the tomographic method are composed of two computation parts. One is the computation of travel times and ray-tracing, which are forward parts. The other is the estimation of the slowness of the cells, which is inversion part. In this work, finite-difference extrapolation presented by Vidale (1988) is adopted as the computation technique in the forward part because this scheme can treated head waves properly. The method is outlined as follows.

The model is covered with finite-difference grid and divided into numerical cells, and each cell has the slowness. First step of the method is to compute first arrival travel times and expanding wavefront of all nodes by a finite-difference solution to the eikonal equation. Second step is the ray-tracing from a source point to receivers. In this work, ray-tracing starts from the receiver along a direction of the maximum gradient of the arrival times, and this technique can save computation times. Final step is the update of slowness of the cells for minimizing errors between the observation travel times and the computed travel times, and we have adopted SIRT(Simultaneous Image Reconstruction Technique) for minimizing the errors.

The tomographic method has the advantage of imaging complex velocity structures and velocity structures change continuously. The scheme also could easily treat travel time data measured in the boreholes and drifts, but the reciprocal method only treats travel times data measured on the surface.

3 DATA PROCESSING FOR EVALUATIONS

3.1 *Tomographic method*

The data provided for the tomographic analyses were obtained from the seismic explorations, which were carried out for the reciprocal interpretation method at the design stage in a tunnel project. Consequently, arranges of receivers and source points were not always suitable for the tomographic method. In this works, the tomographic method has the same disadvantages in seismic refraction measurements that the reciprocal method has, because travel times measured on the surface were only used for the tomographic analyses. The grid span in the analysis was 5m. The initial velocities of the rock mass were assumed to be uniform, and these values were equal to the average values which correspond to the rock mass grade C_{II} in the rock mass classification of the tunnel established by The Japan Highway Public Corporation(1992). Iterative computation in the analysis was ended when the RMS error value was less than 5% or the RMS error value did not reduced significantly for more iterative computations. When no seismic ray passed through the tunnel formation, the seismic velocity tomogram was not provided for the further examinations. It is due to the velocities of the numerical cells are not updated and remain initial values when no seismic ray passes the cells.

3.2 *Quantitative evaluation based on the geological observation records*

Support patterns are determined in accordance with the rock mass grade and the other factors. Therefore, it is not always suitable to evaluate the performance of seismic explorations with the support patterns. Nakagawa et al.(2000) have considered that geological observation records at tunnel face represent surrounding rock mass conditions. The geological observations have been done for almost all tunnels by using the unified record sheet. The description parameters in the record sheet are shown in Table 1. They have presented that categories C and D in the sheet have close relations to the seismic velocities, and compiled the relation between the scores of C+D and the seismic velocities as shown in Table 2. In this work, according to Table 2, the C+D scores were estimated from the seismic velocities

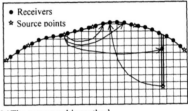

(a)The tomographic method

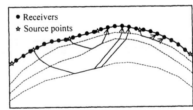

(b)The reciprocal method

Figure 1. Analysis models of the tomographic and the reciprocal method.

Table 2. Relations between the C+D scores and seismic wave velocities(Nakagawa et al. 2000).

Type	Velocity (km/s)	Scores			Rock mass grade
		C	D	C+D	
I	5.0~				
II	4.9~	1~2	1~2	2~4	A
III	3.5~				
I	3.8~5.0				
II	3.6~4.9	2~3	2	4~5	B
III	3.0~3.5				
I	3.3~3.8				C I
II	3.1~3.6	2~3	2~3	4~6	C II
III	2.0~3.0				
I	2.6~3.3				D I
II	2.6~3.1	3	3~4	6~7	
III	1.5~2.0				
I	~2.6				D II
II	~2.6	3~4	4	7~8	E
III	~1.5				

Table 1. Description parameters of geological observation sheet.

Categories	Classification parameters
A	Condition of face
B	Condition of excavation
C	Compressive strength
D	Weathering and deterioration
E	Spacing of joints
F	Condition of joints
G	Pattern of jointing
H	Condition of water seepage
I	Inferior by ground water

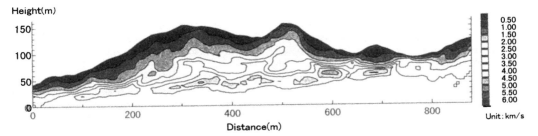

Figure 2. Seismic velocity tomogram in G-1 tunnel.

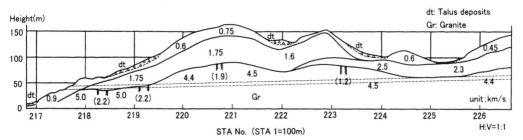

Figure 3. Seismic velocity section by the reciprocal method in G-1 tunnel.

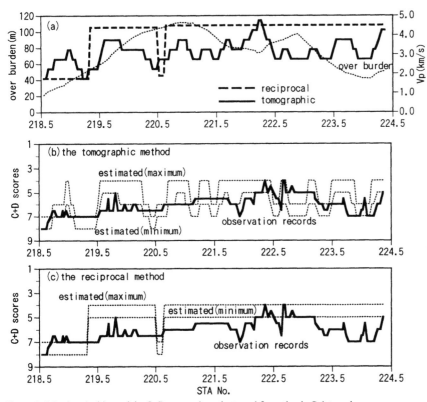

Figure 4. Seismic velocities and the C+D scores along the tunnel formation in G-1 tunnel.

233

along the tunnel formations, where the seismic velocities were interpreted by the reciprocal and tomographic methods. The validity of the tomographic method was evaluated by means of comparisons between the estimated C+D scores and the recorded C+D scores. If the consistency of the C+D scores estimated from the tomographic method has been almost equal to or has exceed that from reciprocal method, the validity of the tomographic method would be verified for rock mass characterizations. The consistency of C+D scores is defined as the ratio of the number of the observation records, which C+D scores are consistent with C+D scores estimated from a seismic exploration, to the total record in a tunnel.

3.3 Examined tunnels

Seismic exploration data taken at 35 tunnel sites were supplied for the tomographic analysis, and seismic tomograms for 24 tunnel sites were evaluated. These tunnels are located in West Japan and have been serviced. A length of the tunnels is 166~1582m, and an over burden of the tunnels is 20~157m. Rock types of the tunnel sites are granite(G), Paleozoic shale(S), Mesozoic alternations of shale and sandstone(A), schist(Sh), Mesozoic rhyolitic tuff(Tf) and so on. Granite and Mesozoic rhyolitic tuff belong to the type II in Table 2, and the other rock type belongs to the type I.

4 EVALUATIONS OF THE TOMOGRAPHIC METHOD

4.1 Results of the tomographic method

The results analyzed by the tomographic method and the reciprocal method for the same tunnel are shown in Figure 2 and Figure 3 as an example respectively.

Fig.4 illustrates the C+D scores obtained by the observation records and both seismic methods. Figure 4a shows the seismic velocities and over burdens along the tunnel formation, Figure 4b and Figure 4c show the C+D scores estimated from the tomographic method and the reciprocal method respectively. In these figures, the broken lines indicate the maximum and minimum estimated C+D scores and the solid line indicates the recorded C+D scores. When the solid line is between the maximum and minimum broken lines, it implies that the rock mass conditions estimated from the seismic method are appropriate.

4.2 Evaluations based on the consistencies of the C+D scores

Figure 5 illustrates the consistencies of the C+D scores estimated from the tomogrphic method and the reciprocal method for the examined tunnels. The consistency varies with tunnel sites, however it appears that the consistency for the tomographic method is higher than that for the reciprocal method for many tunnels. The average consistency for the tomographic method is 69%, and the average consistency for the reciprocal method is 56%.

In the case of the tunnels in granite, the consistency is more than 40% for the tomographic method, and that for the reciprocal method is about 40%. As a whole, the consistency for the tomographic method is higher than that for reciprocal method in excess of about 10%. In the case of the tunnels in Paleozoic shale, the consistency varies in both seismic methods. However, it appears that the consistencies for the tomographic method are roughly higher than that for reciprocal method. In S-11 tunnel, the consistency for the reciprocal method shows high value of 97%. However, the consistency for the tomographic

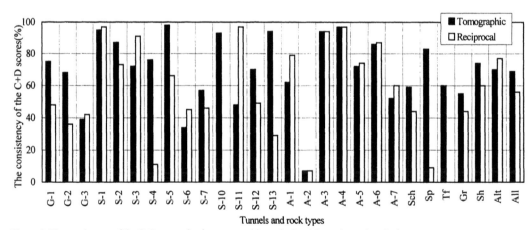

Figure 5. The consistency of the C+D scores for the tomographic method and the reciprocal method

method shows relatively low value of 48% against the reciprocal method. Contrarily, the consistency for the tomographic method in S-10 tunnel is 93%, however, that for the reciprocal method is 0%. In the two tunnels, the consistency for the both methods shows high values in permitting a tolerance of ±1 for the C+D scores. As a whole, the consistencies for the tomographic method are higher than that for the reciprocal method in excess of about 15%. When rock type is Mesozoic alternations of shale and sandstone in the tunnel site, the consistency for the tomographic methods is nearly equal to that for the reciprocal method. The consistency is more than 50% with the exception of A-2 tunnel. In permitting a tolerance of ±1 for the C+D scores, the consistencies in A-2 tunnel show high values for both methods. The consistency in Mesozoic alternations tends to be relatively high value compared with that in granite and Paleozoic shale.

The relations between the consistency and over burdens for the tunnels in granite, Paleozoic shale and Mesozoic alternation are illustrated in Fig.6. When rock type is granite, the consistencies for the tomographic method are slightly lower than that for the reciprocal method at 30~40m of over burden. However, the consistency for the tomographic method at most of over burden is higher than that for the reciprocal method. Variations in the consistency for the tomographic method to over burdens are less than that for the reciprocal method. In the case of the tunnels in Paleozoic shale, the consistency for the tomographic method is slightly lower than that for the reciprocal method at 60~80m of over burden. It

appears that the consistency for the tomographic method does not show significant variations for the over burden. Contrarily, the consistency for the reciprocal method tends to increase with increasing over burden. In case of the tunnels in Mesozoic alternations, it appears that the consistency for both methods is roughly equal. In all examined tunnels, the consistency for the tomographic method at the most of over burden is higher than that for the reciprocal method.

5 CONCLUSIONS

We have introduced the tomographic method to the seismic exploration in tunnel investigations, and have presented a way to evaluate the performances of the tomographic method based on geological observations at tunnel face.

The consistency of C+D scores varies with tunnel sites. However, the consistency for the tomographic method shows nearly equal or high values compared with that for the reciprocal method in many tunnel sites. The consistency for the tomographic method also shows nearly equal or high values at various over burdens. These facts lead to the conclusion that the tomographic method as well as the reciprocal method can be applied to the tunnel investigations.

A seismic exploration is important method in tunnel investigation, however disagreements between rock mass conditions estimated from a seismic exploration and actual conditions in construction stage sometimes cause problems (Suzuki et al. 1991).

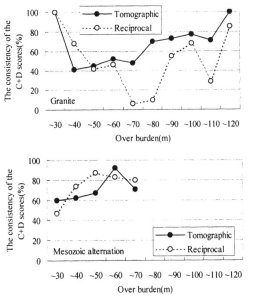
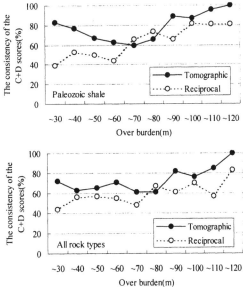

Figure 6. The consistency of the C+D scores for over burdens

235

This work is one of the approaches to improve reliabilities in tunnel investigation, and it will be contributed in rationalization of the preliminary tunnel designs and constructions.

ACKNOWLEDGEMENT

The authors would like to thank the organizations permit us to use the investigation data and geological observation records, and to Mr. Hiroyuki Iwamoto for the discussions on the subject.

REFERENCES

Nakagawa, K., T.Yasuoka, H.Kitamura, S.Miki, M.Fujimoto & N.Kimura 2000. Geological conditions at preliminary tunnel design stage and construction stage. *Journal of Construction Management and Engineering* 685/VI-48:33-43.
Suzuki, S., K.Furukawa, H.Inoue & K.Nakagawa 1991. A study on the evaluation for the preliminary design of supports based on the past records of tunnel construction by NATM. *Journal of Construction Management and Engineering* 427/VI-14:261-270.
The Japan Highway Public Corporation 1992. Classification of Rock and Soil for Tunnel and Slope. Rock mass classification in Japan. *Engineering Geology, Special Issue*:8-12.
Vidal, J. 1988. Finite-difference calculation of travel times. *Bulletin of the Seismological Society of America.* 78(6):2062-2076.

Modern Tunneling Science and Technology, Adachi et al (eds), © 2001 Swets & Zeitlinger, ISBN 90 2651 860 9

Systems for forward prediction of geological condition ahead of the tunnel face

T. Yamamoto, S. Shirasagi & M. Inou
Department of Civil Engineering, Kajima Technical Research Institute, Japan

K. Aoki
Graduate school of Engineering, Kyoto University, Japan

ABSTRACT: For safe and efficient tunnel excavation, it is important to accurately predict the geological conditions ahead of the tunnel face. The authors have proposed Comprehensive System for forward Prediction of Geological Conditions by combining the following methods: "TSP (Tunnel Seismic Prediction) system", "Reflection tomography", "Drill logging", "Velocity logging", and "Face image processing system".
This paper presents the features of the proposed system. The authors carried out performance tests by applying this system to an actual tunnel, and verified that this system is useful for practical investigation and evaluation of geological condition.

1 INTRODUCTION

In the construction of tunnels, precise prediction of the geological conditions ahead of the tunnel face is important for the economy, safety and efficiency of the project. In particular, sections of poor ground, such as faults and fracture zones, are likely to require changes in construction methods as well as tunnel support patterns, and could thus impede the progress of construction. Therefore, it is desirable to obtain data on sections of poor ground and properties of rock masses in advance.

In the past, pilot core boring from the tunnel face has often been performed out of necessity. However, this is relatively costly, and the excavation work often must be suspended while the boring is done. As a result, demands have risen for the development of new prediction system.

The authors believe that in order to obtain data on geological conditions and structure ahead of the tunnel face that can be applied to the tunnel support design work with minimal effect on the construction, it is necessary to be able to perform measurement and evaluation in a relatively short time, and to have better quantitative information on the properties of rock masses in sections of poor ground. Therefore, they have developed a prediction system by combining simple prediction systems consisting of "drilling logging" and "velocity logging", as well as systems using reflected elastic waves ("TSP measurement", and "Reflection tomography").

This paper will describe a summary of each prediction system, especially Reflection Tomography,

one of the Comprehensive system for forward prediction of geological condition. In addition, We can tell that an application of Reflection tomography to an actual tunnel for performance tests is fully suitable for measurement and evaluation.

2 DESCRIPTION OF THE SYSTEM

2.1 *TSP system*

TSP (Tunnel Seismic Prediction) system is used to obtain a position and inclination of a fracture zone and geological boundary 100 meters (150 meters in the case of hard rock mass) ahead of the tunnel face. This is based on reflected elastic waves (P waves) that can be detected from a discontinuities ahead of the tunnel face, through the analysis of oscillation waveforms generated by blasting on the tunnel wall.

2.2 *Reflection tomography*

The TRT© method uses a three-dimensional array of receivers placed in an Omni-directional Tube Array (OTA) installed around the tunnel at some distance behind the tunnel face. The OTA placement of receivers is predetermined by the tunnel geometry and forms the basis of the TRT© data acquisition system methodology. Up to 10 pre-amplified accelerometers are placed in a predetermined pattern on the rock surface (Fig. 1). The signal produced by hydraulic drills, breakers, blasting, and excavation by TBM (Tunnel Boring Machine) is collected in a standard 24-channel seismograph during the normal

24-channel seismograph during the normal excavating cycle.

Once collected, the data are transferred to the surface by CD and processed on a high speed Pentium III laptop. A 3-D reflection tomographic image of conditions ahead of the tunnel face is produced within a few hours. As the tunnel excavation advances, the array is reinstalled, and data acquired during the active excavating cycle. The entire data collection process takes one to two hours and conducted every 80 to 100m of tunnel advance to maintain a complete coverage.

Rock breaking activities associated with a tunneling operation can generate seismic waves at a number of source points at and adjacent to the tunnel face, and the signals are ideal for reflection requirements. The direct waves, as well as the waves reflected from anomalies ahead of and around the tunnel, are detected by the receivers in the array and are used to build a velocity model and to image ground anomalies.

The velocity model is defined within a rectangular block selected to include zones of interest. Normally, each rectangular block is oriented parallel to the tunnel axis and to the vertical direction. The velocity model forms a base for the TRT© to generate an image of anomalies of weaker and stronger rock ahead of and adjacent to the surveyed tunnel.

The principal concept of using seismic reflections for imaging ground conditions in three-dimensional space is shown in Fig. 2.

In a homogeneous media, for each source and receiver of known location, the locus of all possible reflector positions with a given two-way travel time defines an ellipsoid in three-dimensional apace. For a sufficient number of sources and receivers forming a three-dimensional array, each boundary reflecting seismic waves can be identified as an area where a majority of ellipsoids for pairs of sources and receivers intersect. Thus, each grid point in the volume examined may theoretically give rise to a reflection or scattering event.

A discrete image of reflecting or scattering anomalies is calculated for each point of a three-dimensional grid within a selected block of the rock space that includes all sources and receivers. The image is then smoothed by interpolation. A discrete value for the image at each point in the grid is calculated by stacking all seismic waveforms with each waveform shifted proportionally to the total distance from the source via this grid point to the receiver. The shift is calculated using the velocity model defined for the volume of the survey block. Using this approach, the resulting image is similar to a holographic reconstruction. The polarity of the value is positive for reflections from a weal to strong rock transition, and is negative for reflections from a strong to weak rock transition.

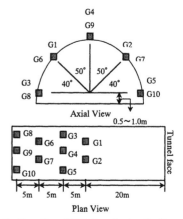

Figure 1. Configuration of OTA for seismic reflection survey.

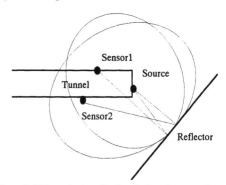

Figure 2. Using seismic reflection for imaging ground conditions in three-dimensional space.

The spacing of grid points sets the range for the desired resolution of the image. The shortest wavelengths of recorded seismic waves determine the available level of resolution. The dimensions of the block, which control the extent of the image, are inversely proportional to the desired resolution of the image. This limitation stems from constraints that control the time of computations required for the TRT© to generate an image within a preset block.

Typically, the initial velocity model is built by extrapolating the velocity tomogram obtained from direct waves and using other available data (velocity measurements, geological data, known voids). As the tunnel excavation advances, the velocity model should be continuously updated and improved based on comparison between tomographic predictions and the ground truth of geological mapping and observed rock conditions.

Several generalities can then be recognized from the technical concepts outlined above.

1) The resolution of ground images is inversely proportional to the distance from the array and directly proportional to the shortest detectable wavelength. In general, the resolution was the highest (1m or better) for the

first few meters from the tunnel face and deteriorated with distance.

2) An absolute error of distance measurement is determined by the precision of the velocity model can be improved by using geological and mechanical properties of the rock detected as the tunnel advances as feedback for velocity calibration.

3) The velocity models are different for blasting and rock breaking/splitting action. Seismic energy generated by blasting is dominated by P-waves, whereas drilling and mechanically breaking the rock typically radiate significant levels of S-waves. Also, the images generated from blasting sources appear to miss smaller anomalies associated with fractures in the rock while being more effective in detection of contacts between rock types.

4) Reasonably precise location of receivers and sources is required to ensure accuracy of the images.

2.3 Drill logging

The Drill logging system is used to evaluate the ground conditions at every depth of boring, through measurement and analysis of boring data, such as drilling speed and percussion pressure, obtained from a hydraulic drill used for tunnel excavation. It has been confirmed that a high correlation with the geological condition can be obtained when a Drill energy coefficient, as shown in the upper part of Fig. 3, is used in the evaluation. The Drill energy coefficient represents the energy required for breaking a unit volume of rock mass. It is large when rock mass

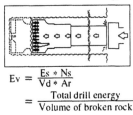

$$Ev = \frac{Es * Ns}{Vd * Ar}$$

$$= \frac{Total\ drill\ energy}{Volume\ of\ broken\ rock}$$

Ev : Drill energy   Es : Drill energy per 1 strike
Vd : Drilling speed   Ns : Strike times per a second
Ar : Area of cross section of drill

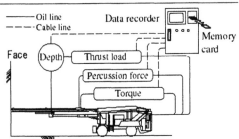

Figure 3. Summary of Drill logging.

at the point of boring is hard, and small when the rock mass is soft.

This system consists of two parts: the measurement system which is used to record boring data at the work site, and the data processing system which is used to analyze and evaluate the data at a site office. (See the lower part of Fig. 3). The measurement system uses pressure gauges to measure thrust load, percussion pressure, and torque as well as flow meters to measure drilling depth. Data measured by these sensors are recorded continuously on a memory card, which then is taken back from the work site to the site office where the data are analyzed using a personal computer.

2.4 Velocity logging

The Velocity logging system measures the elastic wave velocity of the ground, utilizing percussion drilling holes made by the above-mentioned Drill logging. (See Fig. 4)

Sensors are inserted into a percussion drilling hole to record oscillation waveforms at the time of vibration generated by blasting or hammering on the rock mass at the tunnel face. Then, after drawing the sensor out two meters, vibration is again generated and waveforms are recorded. These operations are repeated to the opening of the hole to complete the measurements. As data on measured waveforms are recorded simultaneously on a personal computer, analysis can be performed on the computer using special software. The results are displayed as the velocity distribution of elastic waves (P waves) according to the distance from the tunnel face.

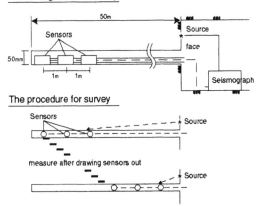

Figure 4. Summary of Velocity logging.

2.5 Face image processing system

Face image processing system is used not only to evaluate the geological conditions at the tunnel face but also to predict the geological conditions several

longitudinal geological map

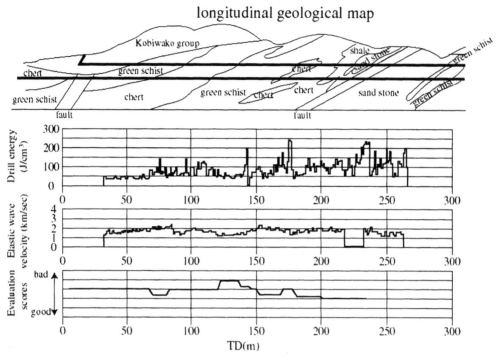

Figure 5. Comparison with geological condition on the one hand and Drill energy and elastic wave velocity on the other.

meters ahead of it. It is performed based on a geological map, which can be drown using pictures of the tunnel face taken by a digital still camera. This system enables observation of the tunnel face to be recorded in a short time.

3 RESULTS OF APPLICATION

3.1 An example of tunnel construction using Drill logging and Velocity logging in combination

Fig. 5 shows an example of tunnel construction using Drill logging and Velocity logging in combination. The figure also shows the results of these loggings, geological profiles, and evaluation scores representing geological conditions of the tunnel face (e.g. deterioration by percolating water). These evaluation scores were used for the ground evaluation during the observation at the tunnel face. In order to determine what geological information can be obtained based on Drill energy coefficient and the elastic wave velocity obtained from these loggings, the authors used a statistical method to compare the type of rock and the geological conditions at the tunnel face with each item (See Fig. 6 and Fig. 7). The results of the comparison show that face were determined quantitatively and accurately as follows:

i) The ground where Drill energy coefficient was 50 J/cm^3 or below was soft, consisting of argillaceous Paleozoic and Mesozoic, the "Kobi-

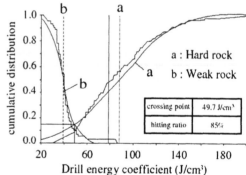

a : Hard rock
b : Weak rock

| crossing point | 49.7 J/cm^3 |
| hitting ratio | 85% |

Figure 6. Distinction of geology by using Drill energy.

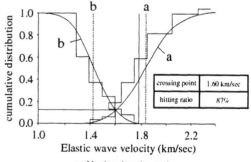

a : No deteriorative rock
b : Deteriorative rock

| crossing point | 1.60 km/sec |
| hitting ratio | 87% |

Figure 7. Distinction of deterioration by percolating water by using elastic wave velocity.

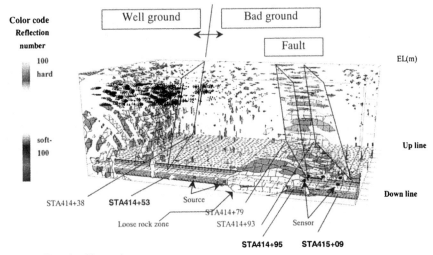

Figure 8. 3-dimensional isometric tomogram.

wako group", while the ground where Drill energy coefficient exceeded 50 J/cm^3 consisted of relatively hard chert or green schist. The hitting ratio was 85%.

ii) The ground where the elastic wave velocity was 1.6 km/sec of below had softened, and collapse/outflow was sometimes observed since the rock has been degraded by water. In contrast, the ground where the elastic wave velocity exceeded 1.6 km/sec hadn't been observed the deterioration. The hitting ratio was 87%.

3.2 Example of Reflection tomography

An application test of Reflection tomography was performed in a tunnel where the ground consisted of soft rock composed of mudstone and sandstone from the Tertiary. This system was initially developed by NSA of the United States for the prediction of mineral deposits. The authors modified this system for the prediction of geological conditions ahead of the tunnel face.

In the tunnel, a geological weak zone and a geological boundary were found during the excavation of the down line that was excavated already. The purpose of this test was to accurately predict the position of the weak zone and the ground boundary along the up line that is to be excavated in parallel with the down line at intervals of 50 meters.

Fig. 8 shows the results of analysis on Reflection tomography. It is obvious that the geographical boundary along the down line is located at STA414+53, while that along the up line is located at STA414+38, showing a difference of 15 meters.

Along the down line, a fault is located around STA414+95 to STA415+09, whereas along the up line, it is located around STA414+79 to STA414+93, showing a difference of 16 meters.

Moreover, a large loose rock zone is detected around STA414+80 to STA414+90.

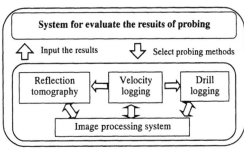

Figure 9. Contents of System for forward prediction of geological condition ahead of the tunnel face.

4 PROPOSAL FOR AN EFFICIENT SYSTEM FOR PREDICTING GEOLOGICAL CONDITIONS AHEAD OF A TUNNEL FACE

The authors have proposed the Comprehensive System for Forward Prediction of Geological Conditions consisting of four subsystems, as shown in Fig. 9. The results of application tests show that prediction can be made more accurate by combining the results of each subsystem, than evaluating geological conditions independently. Subsystems are operable with the same OS, and the results can be output using the same file format. This makes the comparison with the results of construction speedier, and the feedback to the construction site easier.

Fig. 10 shows the flow of the prediction of geological conditions ahead of the tunnel face.

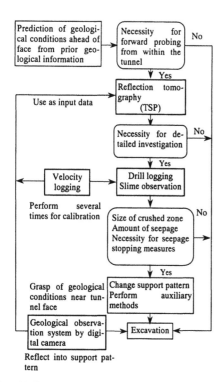

Figure 10. Flow chart of prediction of geological condition.

Miyajima, Y. et al. 1998. Results of prediction of geological conditions ahead of the tunnel face using Drill logging and Velocity logging in combination (in Japanese). Proceedings of the 53rd annual conference of the Japan Society of Civil Engineering: VI-103, pp. 206-207.

Shirasagi, S. et al. 2000. Three –dimensional Reflection Tomography applied to image ground conditions (in Japanese). Proceedings of the 55th annual conference of the Japan Society of Civil Engineering: III-B51.

Yamamoto, T. et al. 1998. Discussion on Applicability of Prediction System of Geological Conditions ahead of the Tunnel Face (in Japanese). Proceedings of the 53rd annual conference of the Japan Society of Civil Engineering: III-B117, pp. 234-235.

Yamamoto, T. et al. 2000. Prediction system of geological conditions ahead of the tunnel face (in Japanese). Proceedings of the Symposium on prediction methods of geological conditions ahead of the tunnel face: pp. 43-48.

5 CONCLUSIONS

This paper has described the Comprehensive System for Forward Prediction of Geological Conditions developed by the authors in recent years, as well as the results of applying this system to actual construction. Though the objectives of prediction may vary at each work site due to differences in geological and construction conditions, a common problem is that poor geological conditions, such as faults and fracture zones, cannot be predicted well, causing trouble at the time of construction. With the current prediction technology, it is difficult to provide all information efficiently through a single system. Therefore, it is necessary to examine the application of the combined prediction system while making the most of the characteristics of each system. Future study will focus on data acquisition at work sites under different geological and construction conditions, and the sophistication of prediction system based on these data.

REFERENCES

Aoki, K. et al. 1990. Evaluation of geological conditions by "Drilling logging system" (in Japanese). Proceedings of the 8th Japan Symposium on Rock Mechanics: pp. 67-72.

Modern Tunneling Science and Technology, Adachi et al (eds), © 2001 Swets & Zeitlinger, ISBN 90 2651 860 9

Deep excavation in Bangkok- characterization, measurement and prediction

S.Shibuya
Graduate School of Engineering, Hokkaido University, Japan

S.B.Tamrakar
Graduate School of Engineering, Hokkaido University, Japan

T. Mitachi
Graduate School of Engineering, Hokkaido University, Japan

ABSTRACT: Geotechnical engineering problems associated with construction of Bangkok Metro are highlighted how to predict short-term ground deformation in deep excavations of soft Bangkok clay. A concrete box at each station site was constructed by means of top-down construction method with diaphragm walls pre-installed 35 m deep in the dense sand layer through soft as well as stiff clay layers of 22 m thick. Prior to excavation work, the engineering properties of Bangkok clay, including the variation of clay stiffness with depth and with strain were thoroughly examined. In this paper, the predictions of soft clay ground behavior by FE analysis incorporating the results of site investigation are discussed in relation to the field behavior.

1 INTRODUCTION

A fundamental difficulty involved in predicting the deformation behavior of soft clay ground associated with excavation works, is attributed to the complexity of deformation characteristics of soft clay, which vary consecutively with location and with time. The elastic stiffness in the ground, for example, varies with the levels of stresses and strains. In a rigorous prediction, the variations of elastic moduli with stress and with strain ought to be properly accounted for in the form of non-linear soil model. However, in a practical prediction using a packaged FE programme incorporating the property of isotropic linear elasticity, the variation of elastic stiffness with depth, hence with stress, may be conveniently accounted for by dividing a single stratum into multiple layers, each having a fixed value of elastic modulus. The strain-level dependency of the elastic moduli may also be considered by choosing a single value of the elastic modulus in match with the induced ground strain level. However, the case history as such is rare in the literature (e.g. Anderbrooke et al., 1997).

It is a common practice in South East Asia to determine the elastic stiffness on the basis of the profile of undrained shear strength, s_u with depth using an empirical relation of $E_u = \alpha s_u$ (s_u: undrained shear strength and E_u: undrained elastic Young's modulus) where the modulus factor, α, for the excavation work of BKK clay, showing a variation of 200~500 (Bowels, 1988), 280~350 (soft) and 1200~1600 (stiff) (Hock, 1997), etc. Recently, Simpson (2000) has suggested to employ the specific value of E'

equal to around half of Young's modulus from seismic survey, E'_{max}, noting that the E'_{max} stands for Young's modulus at strains of about 0.0001 %.

In this paper, we describe a case study in which the effects of elastic stiffness in predicting ground deformation subjected to deep excavation were examined by means of FE analysis using elastic and elasto-plastic constitutive soil models. In this case study performed in Bangkok (BKK), the properties of Bangkok clay including the strain-dependency of stiffness were in detail investigated prior to the excavation work. The analysis focused on a pin-point subject whether $E'_{max}/2$ is applicable to predicting the deformation as well as pore pressure behavior of soft clay ground subjected to deep excavation.

2 GEOTECHNICAL SITE INVESTIGATION

A comprehensive site investigation was performed at Sutthisan site in November 1997 by conducting both field and lab tests, some results of which were employed for the excavation analysis of R-station. Both R-station and Sutthisan site belong to northern section of Metropolitan Rapid Transit Authority (MRTA) project of BKK. R-station is located 3 km north of Sutthisan site, both in central BKK area. It should be mentioned here that the soil profiles as well as properties are similar over the central BKK area.

The field tests included field vane shear (FVS), seismic cone (SCPT) and piezocone tests whereas the laboratory tests included oedometer test, undrained triaxial compression (MTX)) test and bender element (BET) test. Laboratory tests were performed on high-quality samples retrieved by using a fixed-piston thin wall sampler. The consolidation parameters from oedometer test and the stiffness variation with strain over a wide strain range from undrained triaxial test were manifested in the laboratory tests, while the small-strain shear modulus from SCPT, ground water profile from piezocone and undrained shear strength from FVS tests were obtained from the in-situ tests.

2.1 Basic Sub-soil Properties

The basic properties of BKK clay are shown in Figure 1, in which the variations of unit weight, γ_t, water content, w_n, stresses and pore pressure are examined against depth. The weathered crust extends down to 4 m depth and soft clay layer extends from 4 m to 15 m depths. About 7 m thick stiff clay lies below this soft clay layer. Below 22 m depth, there lies a dense sand layer. As seen in Figure 1, w_n in the upper soft clay (i.e., from 5 m to 10 m depth) was close to liquid limit, w_L, whereas it decreases beyond 10 m depth approaching towards plastic limit, w_P. The sensitivity ratio of this soft clay ranged from 3 to 6.

The ground water level measured on November 1997, was 1 m below the ground surface. Due to excessive pumping of underground water from aquifers, a non-hydrostatic distribution of pore water pressure was seen below 7 m depth. The pore pressure reached close to atmospheric pressure at 22 m depth, i.e., the interface between stiff clay and sand layer underneath. The overconsolidation ratio (OCR) value at the weathered crust was relatively high due to seasonal fluctuation of ground water level. In soft clay layer, it exhibits the value close to unity, suggesting the state of normal consolidation. On the other hand, the OCR value in the stiff clay layer (i.e., from 16 m to 22 m) was close to two (Shibuya & Tamrakar, 1998).

The concrete box at R-station was constructed by connecting the cast-in-situ reinforced concrete panels to form a diaphragm wall (DW), then by following the top-down excavation and bottom-up column procedure. The installation of two DWs, each 1 m thick and 25 m apart from one to the other started in September 1998. The bottom-end of DW was placed at 39 m depth from the original ground surface. The length of the station box was 226.8 m. The base slab concreting was completed in October 1999. The final depth of the excavation from the ground surface was 22.1 m.

The details of field instrumentation employed at R-station are shown in Figure 2. A total of five settlement markers were placed in order to monitor the profile of ground surface settlement behind the DW. In order to monitor deflection of DW as well as horizontal ground deformation, two inclinometers

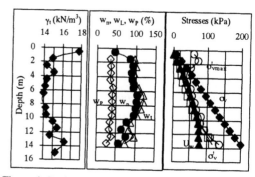

Figure 1. Basic properties.

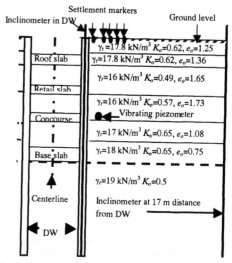

Figure 2. Instrumentations and structures.

were installed in DW and also in the soft clay at the distance of 4.5 m away from the DW. A vibrating wire piezometer was also installed prior to excavation with which rapid change in excess pore pressure could be detected. Total pore water pressure was measured at 18.3 m distance from the centerline of excavation behind the DW and at 11.6 m depth from the ground surface.

2.2 Soil parameters from in-situ and lab tests

The mechanical properties of subsoil, employed for the FE analysis are shown in Figure 3. The compression index, λ (=0.434C_c), in soft clay layer shows a tendency to increase gradually with depth. On the other hand, the swell index, κ (=0.434C_s) remains more or less constant throughout depth. The λ value is smaller in the stiff clay with the value of about 0.1.

The s_u value from FVS test with Bjerrum's correction, $s_{uFVS(cor)}$, is slightly lower than s_u from MTX test, s_{uMTX}. Similarly, the G_{max} value from MTX test ($G_{max}=E_{max}/3$) is lower than the comparable value

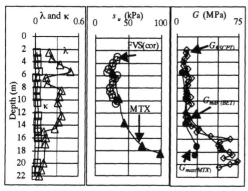

Figure 3. Soil parameters for Sutthisan.

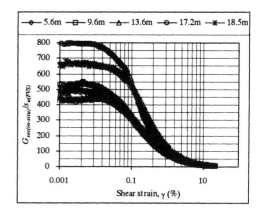

Figure 5. Rigidity index in relation with FVS.

from SCPT test, G_{SCPT}. Although the difference in the shear modulus between G_{max} and G_{SCPT} is relatively insignificant in the soft clay layer, it is more obvious in the stiff clay. This difference may be due to sample disturbance and also due to bedding error in the triaxial testing. Note that the G_{SCPT} value is considered free of any such disturbances. Hence it was employed in the FE analysis as a basic shear modulus at very small strains. The drained Young's modulus is estimated using $E'=2G(1+v')$, where v' stands for drained Poisson's ratio, which is taken as 0.25.

Figure 4 shows the results of undrained MTX test performed using the soft and stiff clay samples, each recompressed to the in-situ effective overburden pressure under K_o condition. Note that the stress-strain non-linearity is significant over a region of small strains. The stiffness values associated with two types of stiffness parameters; *i.e.*, $E'=E'_{max}/2$ and $E'=60s_{uFVS(cor)}$, employed in the FE analysis are indicated on the stiffness decay curve of a sample at 13.6 m, bearing in mind that the E'_{max} value is directly measured from SCPT test. It should be mentioned that the strain levels corresponding to $E'=E'_{max}/2$ and $E'=60s_{uFVS(cor)}$ are approximately at 0.1% and 1.0%, respectively.

Figure 5 shows the rigidity index, $G/s_{uFVS(cor)}$,

with strain. Note that the $G/s_{uFVS(cor)}$ value ranges from 425 to 800 at strain of 0.001% and from 300 to 525 at strain of 0.1%.

3 FE MESH AND FE ANALYSIS PERFORMED

3.1 *Finite Element Analysis*

A commercially available FE programme, SAGE-CRISP (e.g., Indraratna et al., 1992) was employed as it has often been used as a design-aid tool in geotechnical engineering for deformation analysis in many projects of South East Asia.

Figure 6 shows the FE mesh, together with boundary conditions considered in the analysis. The outer vertical boundary was taken at 150 m distance from the centerline of the excavation, which was roughly 6.8 times the excavation depth (thumb rule). The depth of bottom boundary was considered at 80 m depth.

Symmetrical condition was prevailed in the excavation box. Thus, the centerline of the width of excavation was assumed the axis of symmetry. Plain strain condition was assumed since the length of the DW box is much longer than the width.

Short-term, i.e., undrained FE analysis, was carried out since the analysis dealt with the behavior

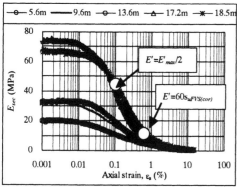

Figure 4. Stiffness decay curve.

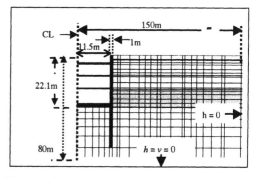

Figure 6. FE mesh and boundaries considered.

245

over a short period of about 13 months. Recommendation made by Ou et al. (2000) for the excavation work with DW in Taipei has to be mentioned here where the consideration of short-term analysis was mentioned.

Since soil properties such as compressibility, stiffness and undrained shear strength considerably vary with depth, the sub-soil was conveniently divided into 7 layers. Some of the properties considered for each sub-layer are also shown in Figure 2. It should be stressed that the soil properties of each sub-layer were obtained directly as the average values of relevant tests performed.

Soil layers, DW and four concrete slabs were represented by linear strain quadrilateral (LSQ) elements whereas temporary strut was represented by a line element (2 node beam element). Quadrilateral elements consist of only two unknown degrees of freedom; i.e., horizontal and vertical displacements. Two-node beam element selected for temporary beam has both displacement and rotational degrees of freedom at each node, which makes it capable of transmitting moments as well as axial forces. Three types of constitutive models, elastic, original Cam clay model (CCM) and modified Cam clay model (MCC) were applied to the clay layers, whereas elastic model was considered for sand layer and other structural elements such as DW and concrete slabs.

Horizontal displacement, h, along the vertical boundaries was fixed to zero, whereas horizontal and vertical displacements, h and v, along the bottom boundary was assumed to be zero. In modeling temporary strut, rotational freedom towards the DW side was fixed to zero and towards the centerline, horizontal displacement was retained to zero. Similarly, for the concrete slab elements, horizontal displacement of the edge towards the centerline of excavation was assumed zero.

Cam clay and elastic parameters; λ, κ, M, e_{cs}, p'_c and G or v' selected for each layer are shown in Table 1. Table 2 shows the in-situ stress parameters such as σ'_v, σ'_h and u_w adopted in the analysis. Table

Table 1. Soil parameters for excavation analysis.

Layer	$^{*}e_{cs}$	$^{**}e_{cs}$	λ	κ	M	$^{1)}K_w$	$^{2)}E'$	G_{SCPT}
m~m						MPa	MPa	MPa
0~1.8	2.07	2.06	0.21	0.03	1.38	72	1.78	9.9
1.8~4	2.47	2.46	0.29	0.03	1.38	184	1.78	9.9
4~9.2	3.13	3.11	0.38	0.03	1.38	373	1.50	12.7
9.2~15	3.17	3.16	0.32	0.03	1.38	710	1.73	14.6
15~18.5	1.80	1.80	0.14	0.04	1.05	683	$4.5^{3)}$	59.8
18.5~22	1.25	1.25	0.09	0.02	1.05	2598	$4.5^{3)}$	70.8
22~80	(elastic, undrained)							1125

$^{1)}K_w$= Bulk modulus of water = $100*K'$;

$^{2)}60*s_{uFVS(cor)}$: *MCC; **CCM; $K'=(1+e)p'/\kappa$

$^{3)}s_u$ for stiff layers are taken from MTX test

Table 2. In-situ condition prior to excavation.

Depth	σ_v	u_w	σ'_v	σ'_h	$p'_{c(MCC)}$	$p'_{c(CCM)}$
(m)	(kPa)	(kPa)	(kPa)	(kPa)	(kPa)	(kPa)
0.0	0.0	0.0	0.0	0.0	119.6	161.1
1.8	32.0	8.0	24.0	14.9	119.6	161.1
4.0	71.1	30.0	41.1	25.5	76.4	102.9
9.2	154.3	70.1	84.2	40.9	105.2	141.6
15.0	247.1	89.4	157.8	89.6	249.0	333.9
18.5	306.6	83.5	223.1	145.0	523.5	701.8
22.0	370.0	0.0	369.6	240.3	523.5	701.8
80.0	580.0	1472.0	891.6	445.8	N.A.	N.A.

Table 3. Properties of strut, wall and slabs.

	Thickness	E	v	γ_t
	(m)	(MPa)		(kN/m^3)
Diaphragm wall	1.0	2.80×10^4	0.2	24
Temporary strut		9.70×10^4	0.2	N.A.
Roof slab	0.9	2.30×10^4	0.2	24
Retail slab	0.7	2.3×10^4	0.2	24
Concourse slab	0.7	2.63×10^4	0.2	24
Base slab	1.8	1.97×0^4	0.2	24

3 shows the properties employed for DW, concrete slabs and temporary strut. The p'_c value was computed from the preconsolidation pressure, σ'_m, (obtained from oedometer test), K_0 and the equation of the yield locus given by Cam clay models. The p'_c value is necessary only for those mesh/material zones, where Cam clay model was assumed. CRISP uses in-situ stresses to calculate the initial values of void ratio. The coefficient of earth pressure at rest, K_o, was estimated using empirical equations given by Jaky (1944) and Wroth (1975), respectively as follows:

$$K_{NC}= 1-sin\phi'_{TC} \quad (1)$$

$$K_o=(OCR\, K_{NC})-[\{v'/(1-v')\}(OCR-1)] \quad (2)$$

where ϕ'_{TC} is the angle of shearing resistance from MTX test.

In case of elastic analysis, the elastic stiffness, E and G are required ($v'=0.25$ assumed). However, in case of CCM and MCC analyses, either G or v' has to be input along with other Cam clay parameters (λ, κ, e_{cs} and p'_c). As stated earlier, two types of stiffness parameters, $E'=60s_{uFVS(cor)}$ and $E'=E'_{max}/2$, were employed. It should be noted that neither of these values corresponds to the true elastic Young's modulus (refer to Figure 4).

Since top-down bottom-up excavation method was adopted, the top concrete slab was casted first and then the soil below this concrete slab was removed. Further excavations continued until the base slab was casted. Afterwards, the concrete columns

were built up upwards. It should be mentioned that the construction sequence of excavation up to the base slab concreting stage was simulated in the analysis. The sequence of excavation made in the analysis is described in the following:
1) Surcharge loading,
2) DW panel construction,
3) Surcharge unloading,
4) 1st stage excavation (excavate up to 1.8 m depth),
5) Temporary strut installation (H-beam with 30 cm × 30 cm),
6) 2nd stage excavation (excavate up to 4 m depth)
7) Roof slab concreting (0.9 m thickness)
8) Removal of temporary slab,
9) 3rd stage excavation (excavate up to 9.2 m depth)
10) Retail slab concreting (0.7 m thickness),
11) 4th stage excavation (excavate up to 14.4 m),
12) Concourse slab concreting (0.7 m thickness),
13) 5th stage excavation (excavate up to 22.1 m depth) and
14) Base slab concreting (1.75 m thick).

In the analysis here, each step of excavation was considered as one block. However, each block was divided into 100 increments. In the analysis, the initial loading due to traffic and set-up of construction equipments was assumed around 10 kPa. However, the load was removed once the DW was installed (as mentioned in the stages 1 and 3).

4 RESULTS OF FE ANLAYSIS

In each set of analysis results, the comparisons between prediction and measurement were made in terms of ground settlement, horizontal displacements of DW and retained soil (GS) and total pore water pressure change with time. In this paper, the examinations of ground settlement and horizontal displacements at 4th stage of excavation, i.e., before the base slab was casted, are shown.

Figures 7 through 10 show the comparison made with different constitutive models; MCC, CCM and elastic, using two stiffness values; $E'=E'_{max}/2$ and $E'=60 s_{uFVS(cor)}$. Note that in undrained analysis, the bulk modulus of water, K_w, was assumed varying with depth, i.e., 100 times that of effective bulk modulus of soil, K'.

Figure 7 shows the comparison of predicted and measured ground settlement profiles behind the DW. It should be mentioned that the measured data was corrected by considering the rate of ground subsidence, 1 cm/year, typically seen in central BKK. Note that there is no significant difference in the comparative results using different soil models. However, the effect of E' was significant; i.e. the maximum settlement observed was around 0.5 cm when $E'=E'_{max}/2$ is employed, whereas it varied from 3.3 cm to 4.4 cm with $E'=60 s_{uFVS(cor)}$. The maximum settlement of the ground was observed around 25 m

distance from the centerline of excavation. The effect of initial surcharge could be seen behind the DW, which allowed the ground to settle down even adjacent to the outer boundary. It seems that the predicted settlement with $E'=E'_{max}/2$ was closer to the measured profile.

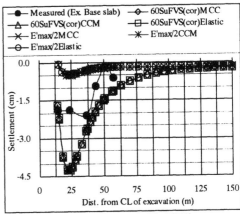

Figure 7. Ground settlement behind DW.

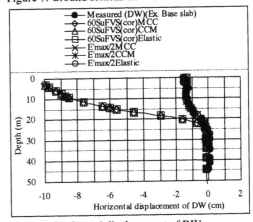

Figure 8. Horizontal displacement of DW.

Figures 8 and 9 show the comparison of horizontal displacements of the DW and of the ground behind the DW. Again, no difference is seen among the predictions by using different constitutive models. The predicted deformation was far larger than the measurement in case using $E'=60 s_{uFVS(cor)}$, whereas the small amount of deformation was successfully predicted with $E'=E'_{max}/2$ (Figure 8). Similar trend could be seen in Figure 9, in which the comparison of horizontal deformation in the ground is shown. Thus, it could be said that the prediction with $E'=E'_{max}/2$ may be better than $E'=60 s_{uFVS(cor)}$ in predicting the deformation of soft clay ground.

Figure 10 shows similar comparison for the behavior of total pore water pressure. The prediction of total pore water pressure was also not affected by the

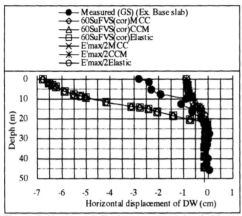

Figure 9. Horizontal displacement (retained side).

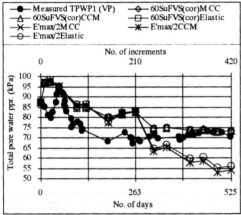

Figure 10. Pore water pressure comparison.

constitutive models. The predicted total pore water pressure at first increased to 97 kPa by using both the stiffness (almost equal up to 260 days) but it reduced with the advancement of excavation. Faster reduction in the total pore pressure was seen in case of using the high stiffness value down to 54 kPa than in using low stiffness value. It should be mentioned that pore pressure response was not a critical factor to influence the ground deformation.

On the basis of the predictions, we could say that the elastic analysis may be sufficient to predict the deformation behavior of soft clay ground subjected to deep excavation with rigid DW. This is because the ground strains induced refer to those below the yield point, for example, in Cam clay model incorporated in CRISP.

5 CONCLUSIONS

1. The elastic shear modulus from seismic cone test, $G_{f(SCPT)}$, was greater in value as compared to the results by MTX and BET tests. The G_{max} from laboratory tests being smaller than $G_{f(SCPT)}$ could be due to the effects of sample disturbance and bedding error in the laboratory. Accordingly, the elastic Young's modulus, E'_{field}, may be estimated from $G_{f(SCPT)}$ by assuming ν' value of 0.25, for example.

2. Rigidity index; $G_{sec(in-situ)}/S_{uFVS(cor)}$ at shear stains of about 0.1%, typical of averaged ground strains in excavation with rigid wall, ranged between 300 and 525 for both soft and stiff clays, implying that the use of $E'=60s_{uFVS(cor)}$ is inappropriate in predicting ground deformation in excavation work.

3. It is suggested that the FE analysis by using the linear-elastic soil model with $E'=E'_{max}/2$ may be the simplest and good enough for predicting relatively small ground deformation associated with excavation work with stiff retaining structure such as diaphragm wall.

6 REFERENCES

Anderbrooke, T.I., D.M. Potts & A.M. Puzrin 1997. The influence of pre-failure soil stiffness on numerical analysis of tunnel. *Geotechnique* 47(3):693-712.

Bowels, J.E. 1988. *Foundation analysis and design.* 4th Edition:McGraw-Hill.

Hock, G.C. 1997. Review and analysis of ground movements of braced excavation in Bangkok subsoil using diaphragm walls. *M. Engg. Thesis.* Bangkok, Thailand.

Indraratna, B., A.S. Balasubramaniam & S. Balachandra 1992. Performance of test embankment constructed to failure on soft marine clay. *Journal Geotechnical Engineering, ASCE* 118 (1):12-33.

Jaky, J. 1944. The coefficient of earth pressure at rest. *Magyar Mrenok Epitesz Kozloney.*

Ou, C.-Y., J.-T. Liao & W.-L. Cheng 2000. Building response and ground movements induced by a deep excavation. *Geotechnique* 30 (3):209-220.

Shibuya, S. & S.B. Tamrakar 1998. In-situ and laboratory investigations into engineering properties of Bangkok clay. In Tsuchida & Nakase (eds), *Characterization of Soft Marine Clays*: 107-132. Rotterdam:Balkema.

Simpson, B. et al. 200. Engineering needs. *Preprints of 2nd International Symposium on Pre-failure Deformations Characteristics of Geomaterials Keynote and Theme Lectures*, IS-Torino 99:42-157.

Wroth, C.P. 1975. In-situ measurement of initial stresses and deformation characteristics. *Proc. of the specialty Conference in In-situ Measurement of Properties, ASCE*, Rayleigh, North Carolina:181-230.

Modern Tunneling Science and Technology, Adachi et al (eds), © 2001 Swets & Zeitlinger, ISBN 90 2651 860 9

Rock response during construction of cavern station at Bagatza (Metro Bilbao)

J. Madinaveitia
Technical Manager of IMEBISA, Bilbao (Spain)

ABSTRACT: The Bilbao Metro's rock-bored subterranean stations are unique hollowed-out spaces. Located in a heavily built up area, Bagatza station had to be bored below buildings, requiring controls in real time of rock responses during the construction process.

1 LA LINEA 2

Inaugurated at the end of 1995, Line 1 of the Bilbao Metro runs for 28 kilometres, providing a service for the city of Bilbao and the towns lining the right bank of the river as it flows north to the sea.

The first six-kilometre phase of Line 2, currently under construction, runs through highly populated towns on the left bank of the river, joining Line 1 to provide a shared service in the city centre. (The 'Y')

This first phase involved one cut & cover station, another in viaduct and three rock-bored cavern stations, the one at Bagatza being the most challenging because of the proximity and importance of the buildings above the tunnel.

2 GEOLOGY

The Bilbao area is set in the Basque arch of Palaeozoic massifs, with Mesozoic and Tertiary series, that comprise the far western end of the Pyrenean chain.

The work zone is located on the left bank of the River Nervión, which has the typical morphology of an estuary close to the coast, i.e., gentle hills cut into by the lower reaches of rivers.

Tunnels were bored in a substratum of lutaceous rock with sandstone layers. Marls and bluish grey marly limestone were also present. Rock mass strength ranges between 30 and 60 Mpa.

This rock is crossed randomly by fairly insubstantial tabular layers (just a few centimetres thick) of igneous rock (diabases). However, simple rock mass strength can exceed 120 Mpa.

One of the most characteristic features of the rock in the works zone is its massive nature. This, together with the relatively scarce discontinuities, makes it highly impermeable. The rock is also very tenacious, being extremely difficult to break up during the boring process.

However, the rock becomes highly alterable in the presence of water and air, which means it has to be protected as fast as possible once the tunnel has been bored. Igneous rock in permanent contact with moving water throughout its geological history can turn into runny, almost liquid clays, which brings about the main danger of collapse when work is going forward.

3 FUNCTIONAL DESIGN OF THE CAVERN STATION

In line with the ideas of English architect Norman Foster, the station cavern was conceived and designed as the heart of the transport system. (Fig. 1) A single volume is excavated to take all the passenger service features and the trains themselves.

Figure 1. Station in operation.

Accesses from street-level are direct to the station's hollowed-out volume in the rock, thus eliminating corridors, bends and dead space. From the moment passengers decide to access the Metro, the route to follow, including escalators, is linear and direct, bringing them in seconds to the mezzanine distributor above the actual tracks.

The mezzanine itself is light in design and hangs from the cavern ceiling, accentuating the sensation of spaciousness in the subterranean area.

All stations have lifts for people with problems of mobility, together with permanent ventilation and emergency systems.

Just three materials were used in the construction of the station:

- Concrete, for the prefabricated panels used to make the resistant cavern skin. From the entrances, the horizontal and transversal joints of the panels modulate the space in which the passenger moves.
- Stainless steel, for the signposting, furniture and, in particular, the structure of the hanging distributor mezzanine and the stairways leading down to the platforms.
- Glass, for lift booths and the balustrades protecting the mezzanine walkway perimeters, giving them their particular feel of lightness.

4 PROCESS OF CONSTRUCTION

The rock in Bilbao is perfect for boring with roadheaders, which transmit few vibrations to the surrounding soil and do not create problems for people living in the buildings nearby.

Roadheader work in tunnels goes ahead in two phases: the first involves a heading, in sections of about 40 square metres, which are then completed in bench up to the 64 square metres of the total section.

In the station cavern, the section slightly exceeds 200 square metres and the crown is around six metres closer to the surface, in some cases reaching levels very close to the garages underneath the blocks of flats above.

At Bagatza, there is a twelve-storey building with three underground garages, which lies over one of the cavern headwalls.

All kinds of precautions had to be taken when constructing the cavern here, with a whole range of measuring equipment being used to control rock response at all times.

The cavern itself was bored in five phases. First to be bored was a pilot gallery measuring 37.93 square metres, the crown of which coincided with and slightly exceeded the definitive station crown. Excavation volume in this phase came to 20.24 % of the total.

Work on this phase was designed to provide the best safety conditions for the remaining excavation

work by reducing the section. This objective was favoured by the slightly oval form involved. It also helped by enabling us to take a series of real data concerning the nature of the rock being bored, the points of contact with water and, in particular, of potential meteorised (liquid clay) zones.

Shotcrete was used to support the pilot gallery. Fibre glass bolts were also employed at points where there was a risk of wedges being formed.

Rock response during the pilot gallery boring process was controlled at all times by external and internal measuring points as described below.

Figure 2. Internal Extensometers.

The second phase of boring involved the excavation of enlargements on either side of the pilot gallery to complete the general section of the upper part of the cavern, measuring some 97.35 square metres in all. Besides shotcrete, systematically placed steel bolts and metal ribs were used to support this phase of the excavation.

Rock response during the boring of the second phase was of course controlled by internal and external measuring equipment (Fig. 2). This was in fact the most complicated phase from the point of view of potential damage to the buildings standing over the tunnel.

Third phase boring consisted of the excavation of a deep central ditch. This was the simplest phase and also the one with least impact on rock stability, as some lateral zones remained unexcavated, thus ensuring the transmission of stresses.

Phase four involved the elimination of these lateral zones, which act as the shoulders of the cavern. They marked the complete width of the excavation and required redistribution of the stresses on the cavern perimeter rock. Support is similar to previous phases, with shotcrete, systematic metal bolts and metal ribs prolonging the ones set into place further above.

In this phase, geometric control of land deformations and rock stress measurements highlighted the need for reinforcement at isolated points to aid general rock mass equilibrium after the cavern had been excavated.

The fifth and final phase involved the excavation of an invert arch until the definitive section was achieved. The rock occupying the part of the future invert arch was extremely important in ensuring cavern stability, as it closed off the perimeter of the supported zone. Excavation here was carried out in several phases.

With boring finished and the new stress equilibrium in the rock assured (which involved further regular measurements over several weeks), preparations for executing the lining were begun.

Figure 3. Convergence measures.

The lining itself consisted of a ring of concrete, which in the zones open to the public was pumped behind the prefabricated concrete panels forming the architectural finish.

5 CONTROL STRATEGY

The experience acquired during the construction of Line 1 in the mid-nineteen nineties highlighted the importance of measurements taken externally before the heading reached the control section.

For this reason the control sections, located in principle every 25 metres in station zones, were reinforced on the surface with extensometers anchored at three different depths in competent rock, together with levelling points on the ground close to buildings nearby.

Extensometers were also set up on the interior, but initial measurements were not possible until after boring had passed through each section, which, as demonstrated during the work on Line 1, meant the loss of more than 50% of the absolute values that had to be controlled.

Pins were also placed to measure crown and shoulder convergence, in each of the excavation phases mentioned above. (Fig. 3)

The general control strategy was based on regular readings from all measuring equipment made at least once a day during the boring processes to create the pilot gallery and the enlargements. Data were immediately transferred, in accordance with preset files, to the officer in charge of monitoring the soundings. This monitoring provided information on rock responses in real time and enabled us to take any measures concerning reinforcement (increase in shotcrete thickness or in the number or density of bolts) while boring continued.

A range of equipment, for which a number of different people were responsible, was used for on-site measurements, which were concentrated into computer models giving the tendency curves for parameter measurements.

Works managers at each site and the Bilbao Metro construction manager had permanent access to this information, which enabled them to take the necessary measures at the right time.

The following equipment was used:

- Levelling points (Fig. 4), Leica model no. A 55.466, with accuracy levels to half a millimetre. The points are anchored into the ground from the bottom of a chest that goes down to the healthy rock and are protected by closed covers.

Figure 4. Levelling control point.

- Convergences between points anchored to the rock, or to the shotcrete, in sections of five (5) pins, one in the crown, two in the enlargement sidewalls and two in the bench sidewalls. Soil Instruments Ltd's MK II was the measuring equipment used.

- Rod extensometers sited externally (Fig. 5) or in the interior. Basically these are screwable rods anchored at different depths by means of a probe and which provide information on the common rock response at differing depths. IIC extensometers provided by Ingeniería de Instrumentación y Control, S.A. were used. (Automatic readings via a potentiometer and centralised connection box.)

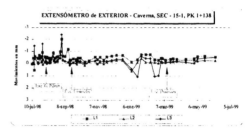

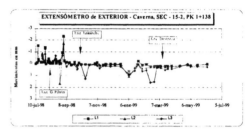

Figure 6. Extensometer diagram.

Figure 5. External extensometer.

- Extensometric bolts, consisting of a combination of conventional anchoring and multiple extensometer, set on the cavern interior as boring advanced. The measurement cords are set towards the anchor head at optional points on the inside of the probe bar. They are similar in form to support bolts, which means they can be set in place with the same equipment. As with the rod extensometers, IIC Ingeniería de Instrumentación y Control, S.A. equipment was used, in this case the company's MA 20 model.

Daily measurements were taken as long as the slopes of the deformation-time curves increased or remained constant; two or three measurements a week were taken when the slope showed a stable decrease; and, once stabilised, a monthly reading was sufficient until the definitive lining was put in place.

6 RESULTS

The Bagatza cavern proved to have the poorest rock yet encountered during boring. Approximately 47% of the rock bored came between types III (55> RMR > 45) and IV (45 > RMR > 20) in Bieniawski's classification.

As noted above, external extensometers provided the most accurate information. Measurements varied between 1 mm and 2.5 mm, very low rock stress values. (Fig. 6)

The most difficult measurements to complete were the convergences, basically because the control points became increasingly difficult to access as the phases of excavation went forward or were moved by the shotcrete. The controlled results came to around 0.75 mm, which were not significant.

Levelling points provided the most significant documentation, with values reaching 12 mm in the cavern crown vertical, in a section where the soil was of poorer quality.

7 CONCLUSIONS

Boring caverns in rock for the Bilbao Metro is fairly problem-free even when buildings are located over the cavern vertical.

Knowledge of rock stress re-equilibrium during boring is a real measure of operational safety. When the tendency curves in extensometer measurements and convergences do not tend towards stabilisation, support needs to be reinforced with greater shotcrete thicknesses.

Using shotcrete usually gives better results than putting more bolts into place to close the original systematic mesh.

Although facilitating the support configuration, the use of metal ribs does not actually lead to substantial increases in safety.

Modern Tunneling Science and Technology, Adachi et al (eds), © 2001 Swets & Zeitlinger, ISBN 90 2651 860 9

Excavation monitoring of a large cross section tunnel underpassing an existing railway

R. SASAKI, T. TAKAYAMA
Kobe Public works office, Hyogo Prefecture, JAPAN

M. TSUKADA
Nishimatsu Construction Co., LTD., JAPAN

M. KIMURA, S. TORII & H. NAKAGAKI
OYO Corporation, JAPAN

ABSTRACT: The Shin-Minatogawa tunnel with a diameter of 14m was excavated in loose hydrated sediments of the Osaka Group in Kobe City in a western part of Japan. The main concern of this construction was that the tunnel had to be excavated under an operating railroad with an overburden of only 13m. Due to the strict guideline on railroad subsidence, a 3-D real-time displacement observation system had to be established. Pore water pressure in sediments was also measured by the single borehole type multi-level piezometer (MP system) to estimate the stability of the cutting face. Use of the measurement results of the pore water pressure and displacement in 3-D, enabled identification of ground behavior achieved by prompt alteration of construction sequence or support measures at each construction stage. It was also possible to determine locations of drainage hole and necessity of rock bolts with no delay. The real-time and continuous monitoring scheme enabled the effective construction method in each step, which lead to safe construction.

1 INTRODUCTION

With the introduction of modern NATM in urban areas, the safer construction is becoming possible for a severer condition, with the selection of appropriate method of auxiliary support structures. However, the employment of those methods of construction does not guarantee a perfect control of the ground, and it has to be always backed up by performing carefully planned monitoring exercises.

The Minatogawa tunnel (683 m in length) had to be excavated with a thin overburden of 13m (slightly less than the tunnel diameter), while the excavation proceeded under a running railroad with a very strict subsidence control guideline. Therefore, a real-time 3-D displacement monitoring scheme had to be employed in order to get hold of ground behaviors at every construction stage, so that prompt responses would be possible when needed.

In this paper, the observed ground behavior is briefly introduced, and some counter measures and effectiveness of construction procedures were discussed, hoping that experiences from this tunnel would be useful references for future projects of similar characteristics.

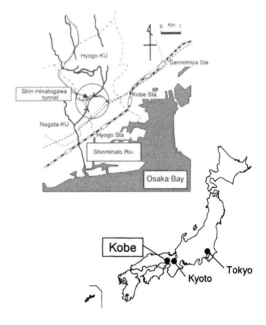

Figure 1. Location of the Shin-Minatogawa tunnel.

2 OVERVIEW OF THE SHIN-MINATOGAWA TUNNEL

The Shin-Minatogawa river originating from Mt. Rokko pours into the Osaka Bay through the urbanized Kobe City in the southeast Hyogo prefecture and has flooded several times in past. In order to increase the water intake capacity, the previous Shin-Minatogawa tunnel, damaged by the South Hyogo Earthquake, was decided to be remodeled by broadening its cross section making the tunnel diameter to 14 m.

The Quaternary sediments of Osaka Group consist of alternation of clay, sand and gravel, and the alluvium is distributed around the tunnel (see Figure 1). The Egeyama fault intersects the tunnel axis. The natural underground water level is 6 to 28m in depth from the ground level, while in some gravel layers the underground water had high water head.

The Kobe Electric Railway with the double tracks links the central area of Kobe City to a suburban residential area carrying about 70,000 passengers a day. The Shin-Minatogawa tunnel underpasses the railway with a minimum overburden of 13 m.

⑥ Existence of the Egeyama fault in a geologically complex zone.
It is of primary importance to monitor ground movement and restrain subsidence due to excavation and change of underground water level so that the effect of construction on the railroad transportation be absolute minimum.

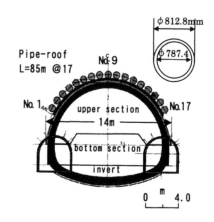

Figure 2. The tunnel standard section in the railroad crossing part.

Table 1. Geological composition.

Geology		Symbol	Elastic modulus	Note
Bank		B	E=20(MPa)	Rail way
Deposit talus		Dt	—	—
Alluvium		Asc	E=14(MPa)	—
Terrace deposit	Gravel	Dg	E=210(MPa)	—
	Cray	Dc	E=42(MPa)	—
Osaka group	Cray	Oc	E=80(MPa)	—
	Sand&Gravel	Osg	E=130(MPa)	Aquifer
	Sand	Osc	—	Loose

		Side drift	Main tunnel
Width	(m)	4.0	14.0
Cross-section area of excavation	(m²)	12.4	144
Shotcrete	(cm)	10	25
Steel support		H·125×125 (pitch 1m)	H·250×250 (pitch 1m)
Forepoling (Length)	(m)	2×5pipes	—
Pipe roof (Length)	(m)	—	85×17pipes
Lining	(cm)	—	60
Lining (invert)	(cm)	—	90

3 TECHNICAL PROBLEMS AND PROVISONS

3.1 Technical problems

Technical problems for excavation in the railroad crossing part are shown below.
① Large cross section of 144 m² in a loose ground with high water content.
② Thin overburden of about 1D for the first 85 m from the entrance and the crossing of the tunnel under the operated railroad.
③ High level of underground water and occasional spots with fairly high water head.
④ Loose ground materials with insufficient stiffness or strength indicated by the borehole lateral loading test and the plate loading test.
⑤Loose sand layers with small uniformity coefficient prone to a quicksand phenomenon.

3.2 Provisions

①Employment of the side drift method to make up for lack of ground strength, together with visual confirmation of the geological feature and the drainage.
②Use of the pipe roof method (ϕ813 mm×17 pipes), for the first 85m, to keep stability of the tunnel cutting face and the crown, and to restrain subsidence of the railroad (see Figure 2).
  To monitor the ground behavior, displacements in three orthogonal directions were measured in the four main sections.

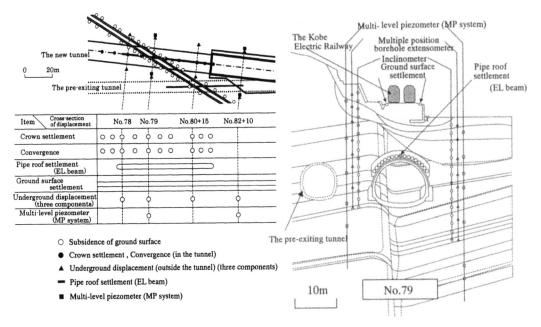

Item	Cross-section of displacement	No.78	No.79	No.80+15	No.82+10
Crown settlement		o o o	o	o o	o o o
Convergence		o o o	o	o o	o o o
Pipe roof settlement (EL.beam)					
Ground surface settlement					
Underground displacement (three components)		o	o	o	o
Multi-level piezometer (MP system)			o		o

○ Subsidence of ground surface

● Crown settlement , Convergence (in the tunnel)

▲ Underground displacement (outside the tunnel) (three components)

▬ Pipe roof settlement (EL beam)

■ Multi-level piezometer (MP system)

Figure 3. Measurement system in the railroad crossing part.

The automatic measurement system was composed of 42 settlement gauges (transducers of water pipe type) along the railroad orbit and a pipe roof settlement gauge (EL beam). On the other hand, single hole type multi-level piezometers (MP system) in 4 boreholes of two main sections were set to measure the water pressure of many gravel layers at the same time.

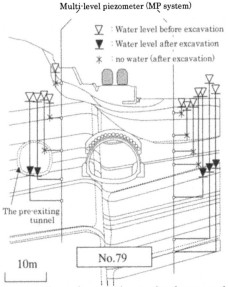

Figure 4. The underground water level measured by the MP system.

4 MEASUREMENT RESULTS

4.1 Ground water

The underground water level measured by the MP system steeply decreased with the deep well operation at No.82+10, but ground water level at No.79 went down only to the tunnel S.L. level (see Figure 4).

Therefore, the drainage by the deep well from the neighboring river tunnel was conducted in addition. As a result, the underground water level dropped down to the tunnel base level. During the side drift excavation, however, the local ground water trapped between the faults flew with sand several times. In upper section excavation, such a phenomenon did not occur and the water level drop achieved by the side drift method was remarkable.

4.2 Displacement

1) During installation of pipe roof

The pipe roof geometry and subsidence of the ground surface during excavation are shown in Figure 5. Because the diameter of each pipe is much smaller than that of side drift or upper sections of the main tunnel, it was expected that no significant displacements occur.

However, considerable deformation occurred as the installation of pipes progressed from the crown to shoulder part of the tunnel, yielding finally a subsidence of 12mm. This was probably due to the fact that the diameter of drill bit was approximately

40mm (a=20mm×2) larger than that of a pile. Also, this could have been attributed in part to the possible formation of a virtual tunnel with an equivalent diameter which is defined as the distance between two pipes installed in a symmetric position across the tunnel center line.

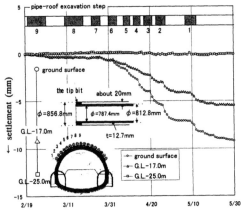

Figure 5. Ground subsidence during the pipe roof construction.

2) During side drift excavation

As for the side drift with 4m diameter, excavation was conducted approximately at the same time on both sides. The results of displacement measurements shown in Figure 6 revealed that the ground movement was greater during installation of forepoling of 2m length, than during excavation of main tunnel. It was concluded that this unexpected deformation could have been caused by vibration of drilling for rock bolts and use of water. Hence, it was decided not to use rock bolts in the shoulder part in the subsequent excavation stage of upper section.

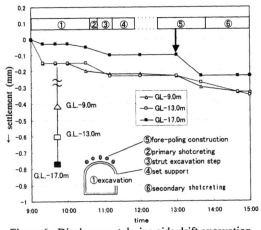

Figure 6. Displacement during side drift excavation.

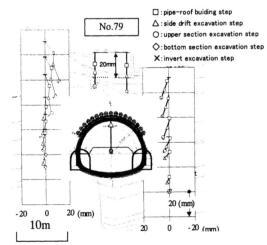

Figure 7. Monitoring results in the railroad crossing part.

3) During upper section excavation

Subsidence during the excavation of upper section was predicted to be about 30mm, using properties obtained from back analysis using measurement data taken during the drift excavation. This necessitated an emergency back up system to adjust railroad level.

However, the measurement picked up the subsidence of only 15mm, as shown in Figure 8. No discrete jump in deformation was observed for the pipe roof or for the ground surface at each excavation stage, which was more or less of the order predicted beforehand (see Figure 9).

Therefore, the excavation of the upper section of the tunnel was completed safely while the routine check on rails were performed twice. Further deformation which occurred during excavation of the lower section and installation of the invert concrete, was less than 2mm.

After all, the approximate ratios of deformation which occurred during the installation of pipe roof, the drift excavation and the upper section excavation were 35, 25 and 40 %, respectively.

4. 3 Comparison of displacements before arrivals of a tunnel face

Figure 10 shows ratios of displacement in relation to the final values, with respect to the positions of tunnel cutting face. It is noticed that the displacements emerged at a faster rate during drift excavation than during the upper section excavation. It has been a general observation that deformation rate is roughly proportional to the dimension of a tunnel. However, in this particular case a faster deformation rate was observed in the drift excavation stage since the two

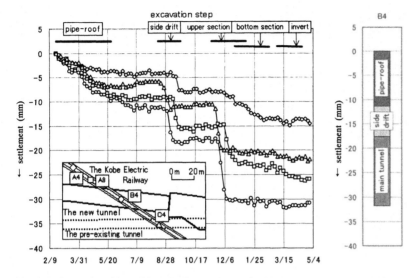

Figure 8. Ground surface settlement due to excavation.

drifts acted as if a tunnel having an equivalent diameter of the distance between the two drifts were being excavated.

The considerable deformation during drift excavation was because of the fact that it was the virgin excavation inducing natural development of deformation, whereas the deformation during the upper section excavation was confined because of the support structures resting firmly on the side wall constructed beforehand.

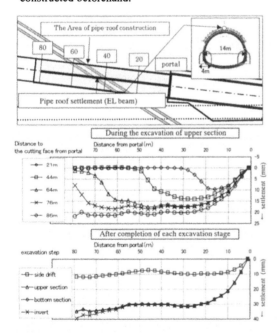

Figure 9. Monitoring results of pipe roof settlement.

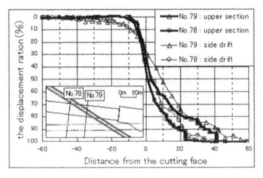

Figure 10. Comparison between the side drift and the upper section.

4.4 Construction method and displacement

The convergence measurement results are shown in Figure 11. A trend of displacement increasing is seen as monitoring section becomes farther from the railroad crossing section and the stiffness of supporting structures gets smaller. This is again an indication of the effectiveness of the supporting structure design employed at the railroad crossing section.

4.5 Excavation across the Egeyama fault

The Egeyama fault was believed to be troublesome in terms of tunnel stability during construction. However, only a minor set of discontinuities with strike of NW-SE was encountered. Ground materials around such discontinuities did not contain much

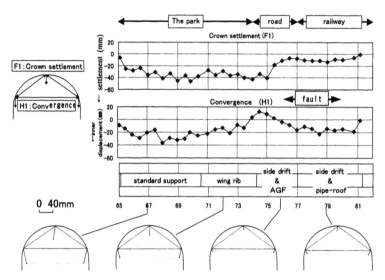

Figure 11. Monitoring results of the crown settlement and the convergence.

water nor they exhibited large strain. This fact suggests that faulting in a relatively soft ground has a tendency of scattering induced straining instead of producing one major discontinuity. This view is also supported by the fact that displacements in the fault zone are of the similar order as in other parts. Deformation moduli for the Osaka Group in this region were in the range of 50 to 100 MPa, which was of similar order to those in the fault zone.

5 CONCLUSIONS

The use of the real-time 3-D displacement monitoring and systematic observation procedure enabled safe construction of a large cross section tunnel underpassing the operated railroad. Decision to use pipe roof as the auxiliary reinforcement method in urban tunnel construction is usually based on the assumption that deformation behavior would be effectively controlled. However, the practice at the Shin-Minatogawa tunnel revealed that fairly large displacement could occur during installation of pipe roof probably due to the formation of a virtual tunnel having an equivalent diameter as wide as two pipes being installed at any given time. This would be a lesson to remember for future application of similar tunneling projects.

REFERENCES

M.Tsukada, M.Kimura & H.Nakagaki: The measurement of the single hole type multiple depth water pressure in the Osaka layer group, *Proceedings of the 2000 Symp. for the Society for Applied Geology*, 11-14. (in Japanese)

M.Kimura, S.Torii & H.Nakagaki: The pipe roof subsidence measurement example using EL beam sensor, *Proceedings of the technology forum 2000*, 161-162. (in Japanese)

R.Sasaki, T.Takayama, M.Tsukada, M.Kimura & S.Torii: Monitoring tunnel excavation under railroads, *Proceedings of Tunnel Engineering, JSCE, 2000*, 131-136. (in Japanese)

Modern Tunneling Science and Technology, Adachi et al (eds), © 2001 Swets & Zeitlinger, ISBN 90 2651 860 9

Evaluation of the load on a shield tunnel lining in gravel

H.Mashimo & T.Ishimura
Public Works Research Institute, Independent Administrative Institution, Tsukuba, Japan

ABSTRACT: It is important to accurately evaluate the load acting on a shield tunnel lining for the economical and rational design of shield tunnel lining. In Japan, the load calculated by Terzaghi's formula or overburden load is generally adopted for the design of tunnel segments. But some previous field measurements have shown that the actual load acting on the shield tunnel lining could be much smaller than that adopted for the design in the case of good ground conditions.

In this study, field measurements at two shield tunnel construction sites in gravel were carried out, and the load acting on the shield tunnel lining was evaluated by analyzing the field data to establish the rational design of the shield tunnel segments. The way of treating the self-weight of segments in the design was also investigated.

1 INTRODUCTION

It is well known that the segments production cost accounts for a large part of the total shield tunnel construction cost and one of the effective methods to reduce the shield tunnel construction cost is to design the segments more efficiently. The common design method for shield segments is to determine the load acting on the tunnel lining first, then determine the material and the cross sectional dimensions of the segments by structural calculations. It is, therefore, very important to evaluate the load accurately. Up to now, overburden earth pressure or loosening earth pressure calculated by Terzaghi's formula has generally been adopted as the vertical load acting on the tunnel lining for the segment design on the basis of previous field measurement data. But some of the field measurement results have recently shown that the load acting on the tunnel lining adopted in the design might be greater than the actual load, particularly in case of good ground conditions (Koyama et al. 1995). Also the effect of the ground reaction, especially to the self-weight of the segments has not been fully resolved (JSCE 1996). Therefore, it is necessary to carry out measurements and analyses of the earth pressure acting on the shield tunnel lining in order to establish a rational design method for shield tunnel lining.

In this study, field measurements of earth pressure, water pressure and strain in the reinforcing steel bars of the segments were carried out at two shield tunnel construction sites in gravel. The meas-

ured earth pressure and water pressure were compared with the value adopted in the segment design. The measured bending moment occurring in the segments was also compared with the calculated results using frame analysis to evaluate the load acting on the shield tunnel lining, and to determine the influence of ground reaction on the bending moment in the segments due to their self-weight.

2 OUTLINE OF FIELD MEASUREMENT

Field measurements were conducted at two shield tunnel construction sites as shown in Figure 1. The overburden height H at the measurement section of Tunnel A with a diameter D of 6.2m is 9.6m, giving an overburden height to diameter ratio of approximately 1.5. The overburden height at the measurement section of Tunnel B with a diameter of 4.75m is 12.1m, giving an overburden height to diameter ratio of approximately 2.5. The ring of both tunnels was composed of six reinforced concrete segments, with a thickness of 27.5cm and a width of 100cm at Tunnel A, and a thickness of 22.5cm and a width of 100cm at Tunnel B. The ground at the tunnel site and its vicinity, where the measurements were carried out, appeared to be composed of permeable gravels with the standard penetration test value (N value) greater than 50. Both tunnels were excavated by the Earth Pressure Balanced Shield method.

Table 1 shows the items of measurements carried out at Tunnel A and Tunnel B.

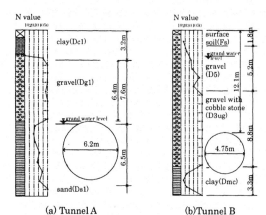

(a) Tunnel A   (b)Tunnel B

Figure 1. Description of the measured tunnels.

Earth pressure and water pressure acting on the tunnel lining were measured at one ring, as well as the strains in the reinforcing steel bars of the segments at two rings of each tunnel. The measured earth pressure is considered to be the total ground earth pressure, including pore water pressure. In order to measure the earth pressure and the strains in the reinforcing steel bars, earth pressure cells of 16cm in diameter at Tunnel A, and of 65cm x 32cm in length and breadth at Tunnel B, and strain gauges, were installed in the segments as shown in Figures 2 and 3 at the time of fabrication. Pore water pressure gauges were mounted in grouting holes, drilled through the segments as shown in Figure 4 after the backfill grouting materials were injected.

Table 1. Items of measurements.

Items	Number of Instrument	
	Tunnel A	Tunnel B
pore water pressure	4／RING×1RING	4／RING×1RING
earth pressure	8／RING×1RING	8／RING×1RING
strains in reinforcing steel bars	11／RING×2RING	11／RING×2RING

Figure 2. Detail of the earth pressure cell(Tunnel B).

Figure 3. Detail of the strain gauge attached to reinforcing steel bar (Tunnel A).

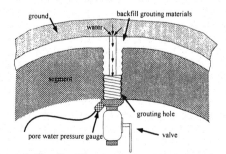

Figure 4. Detail of the pore water pressure gauge.

3 MEASUREMENT RESULTS BY EARTH PRESSURE CELLS AND PORE WATER PRESSURE GAUGES

The earth pressure measured by the earth pressure cells and the water pressure measured by the pore water pressure gauges are shown in Figures 5 and 6. The earth pressure and water pressure shown in the figures are those measured at the stable state about three months after a segment ring was assembled, and the measured earth pressure includes the water pressure.

It can be seen that the earth pressure acting on the segment reaches approximately 70kN/m^2 at the tunnel crown at Tunnel A. The ground water level around the tunnel is estimated to be at the tunnel crown level according to the ground water level in a borehole in the vicinity of the tunnel, and the measured water pressure corresponds to the theoretical hydrostatic pressure. Consequently the effective overburden earth pressure Pv=γH (γ：submerged unit weight) adopted as the design load reaches about 170kN/m^2 and the water pressure is nearly zero at the tunnel crown level. The measurement data, therefore, indicate that the load acting on the tunnel lining at the tunnel crown accounts for approximately 40 to 50% of the total amount of the effective

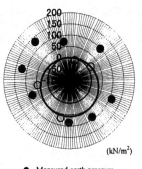

Figure 5. Earth pressure and water pressure distribution (Tunnel A).

260

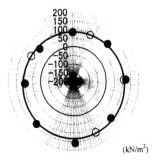

Measured earth pressure
Measured water pressure
Theoretical hydrostatic pressure

Figure 6. Earth pressure and water pressure distribution (Tunnel B).

overburden earth pressure and water pressure.

Also it can be seen that the earth pressure measured by the earth pressure cell reaches about 80 to 170kN/m² and is almost equal to the measured water pressure at Tunnel B. The ground water level around the tunnel estimated from the ground water level in the borehole is about 9m higher than the tunnel crown, and the measured water pressure corresponds to the theoretical hydrostatic pressure. Therefore it is presumed that only the hydrostatic pressure acts on the tunnel lining.

4 EVALUATION OF LOAD ON LINING FROM MEASURED BENDING MOMENTS

Frame analyses using the beam-spring model were carried out to estimate the earth pressure acting on the shield tunnel lining by comparing the calculated bending moments in the segments with measured ones.

4.1 Calculation Method

The beam-spring model adopted for the calculation is shown in Figure 7. Two parallel segment rings are modelled, where each ring consists of beam elements representing the segments. Rotational spring elements with a rotational stiffness coefficient k_θ represent the segment joints which connect the segments in the circumferential direction, and shearing spring elements with a shearing stiffness coefficient k_s represent the ring joints which connect the segment ring. The support of the ground that surrounds the tunnel is modelled by a continuous spring support with a coefficient of ground reaction k_r in a normal direction, which has no stiffness on the tension side. The ground conditions at the calculated section are shown in Figure 8.

Table 2 shows the values of the parameters used in the calculation. In the calculation, the effective

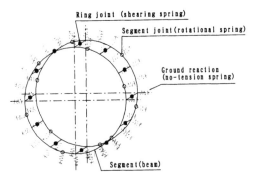

Figure 7. Beam-spring model.

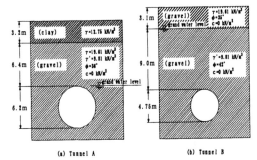

Figure 8. Ground conditions at the calculated section.

earth pressure and hydrostatic pressure, the models of which are shown in Figures 9 and 10, were adopted as vertical loads acting on the tunnel lining. To estimate the actual vertical load acting on the lining, calculations were carried out using three kinds of combination of vertical load, i.e. effective overburden earth pressure together with hydrostatic pressure, effective loosening earth pressure obtained from the following Terzaghi's formula (1) together with hydrostatic pressure, and only hydrostatic pressure.

$$P_v = \frac{B_1(\gamma - c/B_1)}{K_0 \tan\varphi} \cdot (1 - \exp(-K_0 \tan\varphi \cdot H/B_1)) \quad (1)$$

$$B_1 = R_c \cdot \cot((\pi/4 + \varphi/2)/2)$$

where P_v= Terzaghi's effective loosing earth pressure ; K_0= lateral earth pressure coefficient ; c,φ=cohesion and angle of internal friction of soil ; γ= submerged unit weight ; and R_c= tunnel radius.

The self-weight of the segments was not taken into account in the calculation at Tunnel A, as the data collection of strain in the reinforcing steel bars started after the assembly of a tunnel ring, while the self-weight was taken into account at Tunnel B. The stiffness coefficients of segment joints and ring joints were determined by laboratory tests using the actual joints or theoretical calculations, and the lateral earth pressure coefficient and the coefficient of ground reaction were determined from previous case studies in similar ground conditions.

261

Table 2. Parameters for calculation.

	Tunnel A	Tunnel B
Tunnel radius Rc (m)	5.925	4.525
Thickness of segment h (m)	0.275	0.225
Width of segment w (m)	1.000	1.200
Moment of inertia of segment I (m^4)	0.001733	0.001139
Elastic modulus of segment E_c (KN/mm^2)	31.44	32.36
Rotational stiffness coefficient of segment joint k_θ (MN·m/rad)	32.8~65.7	36.3~127.5
Shearing stiffness coefficient of ring joint k_s (MN/m)	1.96	1.96
Load	Effective overburden earth pressure and hydrostatic pressure Effective Tergaghi's loosening earth pressure and hydrostatic pressure Hydrostatic pressure	
	—	self-weight of segments
Lateral earth pressure coefficient λ	0.45	0.45
Coefficient of ground reaction k_r (MN/m^3)	50(After tail leaving)	50(After tail leaving) 1,10,100(Before tail leaving)

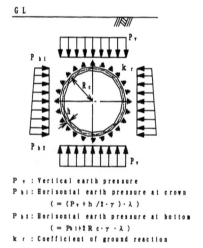

P_v : Vertical earth pressure
P_{h1} : Horizontal earth pressure at crown
  $(= (P_v + h / 2 \cdot \gamma) \cdot \lambda)$
P_{h2} : Horizontal earth pressure at bottom
  $(= P_{h1} + 2 R c \cdot \gamma \cdot \lambda)$
k_r : Coefficient of ground reaction

Figure 9. Earth pressure model.

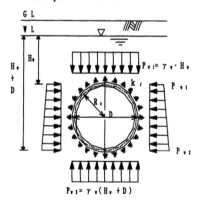

$P_{v1} = \gamma_v (H_v + D)$

Figure 10. Ground water pressure model.

4.2 Calculation results and consideration

4.2.1 Evaluation of load acting on tunnel lining

Figure 11 shows the comparison between the measured bending moments and the calculated bending moment at Tunnel A. The measured bending moments were obtained by using the measured strains in the reinforcing steel bars of the segments three months after assembly of a tunnel ring. To calculate the bending moments, it is assumed that the elastic modulus of reinforcing steel bars and concrete are Es=206kN/mm^2 and Ec=31kN/mm^2 respectively, ignoring the effects of the stress of the concrete on the tension side. The calculated results show the value for different value of vertical load. It can be seen that the calculated results using the effective loosing earth pressure give the closest agreement with the measured values. The effective loosing earth pressure based on the bending moments accounts for approximately 60% of the effective overburden earth pressure at the tunnel crown. This result is compatible with the effective earth pressure obtained from the earth pressure cell measurements which was approximately 40% to 50% of the effective overburden earth pressure.

Figure 12 shows the comparison between the measured bending moments and the calculated bending moments at Tunnel B. The measured bending moments were obtained in the same way as Tunnel A, assuming an elastic modulus of concrete Ec of 32kN/mm^2 and an elastic modulus of reinforcing steel bars Es of 206kN/mm^2. The calculated results show the value for different values of vertical load. It can be seen that the calculated results using the theoretical hydrostatic pressure give the closest agreement with the measured values. Therefore it is

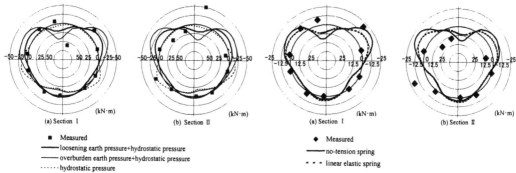

(a) Section I   (b) Section II

- ■ Measured
- —— loosening earth pressure+hydrostatic pressure
- —— overburden earth pressure+hydrostatic pressure
- ····· hydrostatic pressure

Figure 11. Bending moments distribution (TunnelA).

- ◆ Measured
- —— no-tension spring
- - - - linear elastic spring

Figure 13. Influence of the ground reaction model.

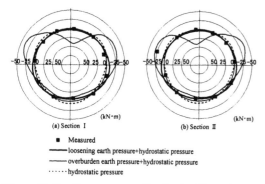

(a) Section I   (b) Section II

- ■ Measured
- —— loosening earth pressure+hydrostatic pressure
- —— overburden earth pressure+hydrostatic pressure
- ····· hydrostatic pressure

Figure 12. Bending moments distribution(Tunnel B).

presumed that only the hydrostatic pressure acts on the tunnel lining. This result is also compatible with the measurements obtained from the earth pressure cells and pore pressure gauges.

4.2.2 Influence of ground reaction model

In the calculation as above mentioned, ground reaction was basically modelled by the no-tension springs which do not resist on the tension side. But it is thought that the evaluation of ground reaction plays a very important role in the segment design. Figure 13 shows the influence of the ground reaction model on the calculated bending moments when the ground reaction was modelled by linear elastic springs. It can be seen that the maximum bending moments at both the positive and negative side obtained from the calculation using linear elastic springs account for approximately 60% of those using no-tension springs. This indicates that the calculated results using the linear elastic springs tend to give lower bending moments than those calculated using no-tension springs. Therefore more attention should be paid to determining the value of the lateral earth pressure coefficient and coefficient of ground reaction if the ground reaction is modelled by linear elastic springs.

4.2.3 Evaluation of ground reaction to self-weight of segments

Up to the present, it has been thought that there was no support around a shield tunnel at the stage of assembly of a tunnel ring in the shield machine, and that the stress occurring in the segments during assembly of a ring remained. Therefore in the conventional design method of shield segments, the bending moments occurring due to the self-weight have been calculated without taking account of the ground reaction. However, the bending moments due to the earth and hydrostatic pressure have been calculated taking account of the ground reaction. The sum of these calculated bending moments has been used to determine the material and the cross sectional dimensions of the segments. But according to recent experiences, it appears that little bending moment due to the self-weight of the segments occurred at the stage of assembling a tunnel ring, because of the improvements of backfill grouting technology, employment of the circle retainer and correct control of the jack thrust.

The measured bending moments that occurred due to the self-weight of segments before the tunnel ring left the tail of shield machine at Tunnel B is shown in Figure 14. To study the influence of the ground reaction, the calculated bending moments due to the self-weight of the segments with various coefficients of ground reaction k_r of 1, 10, 100MN/m³ are also plotted in the figure. It can be seen that the bending moments that occurred due to the self-weight before the tunnel ring left the tail of shield machine are very small and the calculated bending moments taking account of the ground reaction with the coefficient k_r of more than 100 MN/m³ are compatible with the measured values.

Furthermore, to investigate the influence of the ground reaction due to the self-weight of the segments after a ring leaves the tail of shield machine and is loaded by the soil, two kinds of ground reaction model were adopted in the calculation (see Figure 15). The standard model was the same model

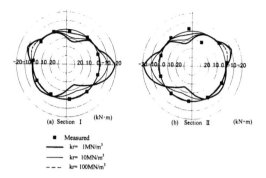

(a) Section I (kN·m)　(b) Section II (kN·m)

■ Measured
—— kr= 1MN/m³
—— kr= 10MN/m³
- - - kr= 100MN/m³

Figure 14. Bending moments distribution due to the self-weight of segments.

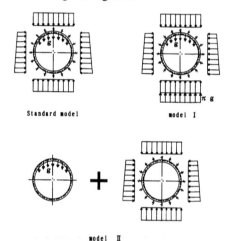

Standard model　　model I

model II

Figure 15. Model of the ground reaction to the self-weight of segments.

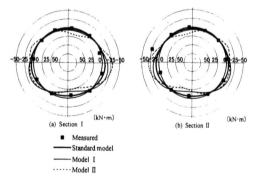

(a) Section I (kN·m)　(b) Section II (kN·m)

■ Measured
——Standard model
——Model I
······Model II

Figure 16. Influence of the ground reaction to the self-weight of segments.

mentioned in section 4.1, where the ground reaction to the self-weight of segments was taken into account as the reaction of spring support. In model I, the ground reaction to the self-weight of segments was taken into account as the fixed ground reaction acting on the invert equal to the weight of segments.

In model II, the bending moment that occurred due to self-weight was calculated without ground reaction. The calculation results for Tunnel B are shown in Figure 16. It can be seen that the value of the calculated bending moments using model I is approximately two to three times larger than those measured. The calculated bending moments using model II are approximately three to four times larger than those measured. From these results, it will be desirable to take account of the ground reaction to the self-weight of the segments based on the reaction of a spring support.

5 CONCLUSION

In this study, field measurements and frame analysis were carried out to evaluate the load acting on shield tunnel in gavel. The main results obtained from the study are as follows.
1) The water pressure acting on the shield tunnel lining is almost equivalent to the theoretical hydrostatic pressure.
2) In the case of an overburden height H to the tunnel diameter D ratio of nearly 1.5, the vertical earth pressure acting on the shield lining is equivalent to the effective earth pressure calculated by Terzaghi's formula, and in case of an overburden height to the tunnel diameter ratio of 2.5, only the hydrostatic pressure acts on the shield tunnel lining based on the field measurements conducted by the authors.
3) At the stage of assembling shield segments in the shield machine, small bending moments in the shield segments occurred due to their self-weight. The ground reaction to the self-weight of the segments could be taken into account by considering the reaction of a spring support.

REFERENCES

Koyama,Y. et al. 1995. In-situ measurement and consideration on shield tunnel in diluvium deposit. *Proceedings of tunnel Engineering, JSCE, Vol.7*: 385-390. (in Japanese)
Japanese Society of Civil Engineers (JSCE) 1996. *Japanese standard for shield tunneling, The third edition.*

Modern Tunneling Science and Technology, Adachi et al (eds), © 2001 Swets & Zeitlinger, ISBN 90 2651 860 9

Tunnel Convergence Measurement using Vision Metrology

S.Miura
Kajima Technical Research Institute, Tokyo, Japan

S.Hattori
Fukuyama University, Fukuyama, Japan

K.Akimoto
Shikoku Polytechnic College, Marugame, Japan

Y.Ohnishi
Kyoto University, Kyoto, Japan

ABSTRACT: In recent times, the vision metrology techniques using digital cameras have been widely used in precise inspection of industrial materials. The present paper discusses the application of this method to tunnel convergence measurement. While the vision metrology is an established technology in principle, many practical problems need to be solved before it can be applied to civil engineering problems. The experiment was carried out by applying the convergence measurement to the 15 m length of tunnel of 7m diameter. The results were compared with those measured with the conventional Total Station technique. The standard deviation of the difference in the measurements with the above two techniques was about 0.58 mm. Thus, the accuracy of the vision metrology was verified for practical use. We also discuss problems to be solved in putting the system to practical use.

1 INTRODUCTION

Until now, the convergence measure and the Total Station (TS) methods have generally been employed for tunnel convergence measurement during construction. However, the former requires the setting of pins and tensioning of the measuring tape. These operations not only interfere with tunneling operations, but also cause a safety problem because they involve the work in high places within the tunnel. The latter takes time for setting up and moving the instrument, and it is only accurate within several mm. Furthermore, with both methods, only the points where pins or targets are set, can be measured, and it is difficult to determine the superficial configuration of the tunnel walls and displacement distribution. However, there have been rapid changes in recent years in the environment surrounding tunnel construction in rock. Increased urbanization and a rise in demand for tunnels of large cross section have highlighted the importance of quickly obtaining measured data of high reliability and feeding it back into the construction.

The vision metrology (VM) is now being used in Europe and the United States to measure the configuration of relatively large objects such as automobiles, rockets and ships. It has been reported that VM systems have delivered accuracies generally about 1:50000 of the principal dimension of the object. With the appearance of high resolution digital cameras and high speed, large capacity personal computers, it has become relatively simple and in-expensive to use.

With this in mind, an investigation was made into the application of VM to tunnel measurement.

The present paper describes the results of an experiment in a tunnel and discusses problems to be solved in putting the system to practical use.

2 OUTLINE OF TUNNEL MEASUREMENT USING VISION METROLOGY

Figure 1.outlines the VM system for tunnel convergence measurement.

Retro-reflection targets were placed on the tunnel wall. Photographs of the targets were taken with a digital camera in such a way that multiple photographs could be taken from various positions and at various angles. Photographs of reference bars of known length are taken, together with reflection targets as measurement targets. The photographs are processed and analyzed by close range photogrammetry theory with personal computers, and the three-dimensional coordinates of the reflection target are calculated. By placing many reflection targets, information is obtained on the tunnel profile like the example of analytic results as shown in Figure 1. Furthermore, the tunnel displacement can be measured by periodically carrying out these measurements and calculating the difference between them.

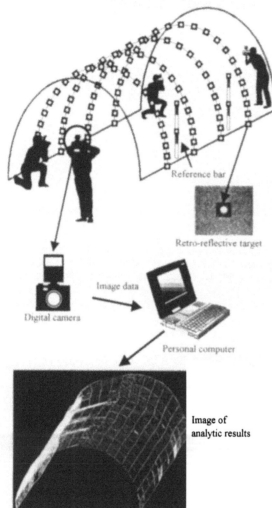

Reference bar

Retro-reflective target

Image data

Digital camera

Personal computer

Image of
analytic results

Figure 1. Overview of tunnel measurement system.

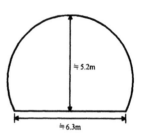

≒ 5.2m

≒ 6.3m

Figure 2. Tunnel cross section for the experiment.

Table 1. Equipment used in experiment.

Machinery and Materials	Manufacturer Model	Specifications
Digital camera	Kodak DCS660	Numbers of pixels: 3040×2008
Lens	Nikon	20 mm
Retro- target		Diameter 20mm
Reference bar		Length 900mm
Total Station	SOKKIA	Angle accuracy 2" Distance accuracy ±(1mm+2ppm)

Figure 3. Work condition using the working vehicle for setting targets.

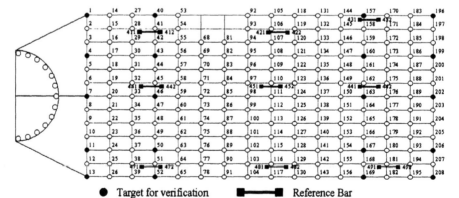

● Target for verification   ■■■■ Reference Bar

Figure 4. Target arrangement.

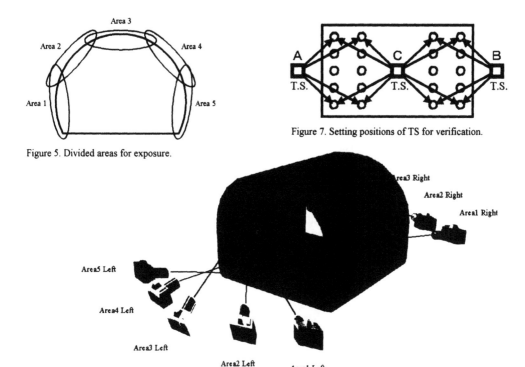

Figure 5. Divided areas for exposure.

Figure 7. Setting positions of TS for verification.

Figure 6. Camera station configuration.

3 FIELD EXPERIMENTS

3.1 Experiment outline

3.1.1 Tunnel and equipment used in experiment
Figure 2 shows a schematic cross-section of the tunnel used in the current experiment. The reflection targets were placed over a total tunnel length of 15 m. The configuration in the tunnel was then measured. Table 1 shows the main items of equipment used in the experiment

3.1.2 Target installation
Figures 3 and 4 show the views of placement work and the target arrangement. In Figure 4, the tunnel wall is rolled out and is viewed from the top. The target spacing is one meter and about 200 targets are placed over 15 m length. Furthermore, a total of nine reference bars were placed with the interval of three bars every five meters.

3.1.3 Photographic conditions
It is necessary to deal with where and how the photographs of the targets were taken. Unlike with conventional measurement targets, because a tunnel is long and narrow, photographs can be taken over a limited range. In the current experiment, therefore, the range of 15 m was divided into three zones. Each zone was further sub-divided into five blocks

Table 2. Comparison between VM and TS measurements.

Target No	X(mm)	Y(mm)	Z(mm).
1	−0.838	0.272	−0.268
4	0.092	0.418	−0.502
7	−0.369	−0.348	−0.288
11	−1.256	0.102	0.312
13	−1.222	−0.225	0.394
40	0.042	−0.039	−0.727
43	−0.131	0.583	0.425
46	−0.628	−0.611	−0.402
50	−0.433	0.219	0.323
52	−1.017	−0.750	0.493
157	−0.387	−0.112	−0.076
160	0.727	0.199	−0.124
163	1.352	−0.178	0.198
167	0.293	0.069	0.674
169	0.853	0.015	1.260
196	0.807	−0.159	−0.861
199	0.204	0.238	−0.087
202	−0.002	−0.312	−0.913
206	0.699	0.197	−0.129
208	1.214	0.422	0.297

*X,Y,Z Values: Subtraction TS form VM

Table 3. Difference from TS (standard deviation).

X(mm)	Y(mm)	Z(mm).	3-D coordinates(mm)
0.7789	0.3434	0.5462	0.5839

(Figure 5). Photographs were taken with a digital camera from the position shown in Figure 6.

3.1.4 Measurement to verify accuracy

To verify the measurement results of the VM, three-dimensional coordinates of several targets were measured. Twenty points were selected from the targets shown in Figure 4, and were measured by TS from three locations as shown in Figure 7.

3.2 Measurements results

The results of these 20 measurements were compared with those measured by VM (Table 2).

Table 3 also shows the standard deviations of the differences in the X, Y and Z coordinates and in the three-dimensional coordinates obtained by synthesizing them.

3.3 Summary of experimental results

In this experiment, 200 targets were placed for measurement over a 15 m length of tunnel and over 60 photographs were taken. It took about two days to place the targets and take the photographs, and one day to analyze the data.

Comparison with the TS showed a maximum difference of 1.352 mm. Furthermore, the standard deviation of the three-dimensional coordinates was 0.584 mm. Since the TS measurement itself has its own error of about 0.5 mm, it was found that VM was at least as accurate as TS.

4 CONCLUDING REMARKS

This experiment confirmed that the measurement accuracy was sufficient for the practical use. Advantages and disadvantages of VM were also assessed.

4.1 Advantages of vision metrology

a. Once the targets are placed, the field work is limited to taking photographs with a camera, even where many measurement points are required. Accordingly, measurement time in the field can be shortened. Furthermore, the necessary equipment can be made small and light, thus facilitating portability. For example, it may become possible to carry out measurements without needing to stop traffic in streets and railway tunnels.

b. The three-dimensional positions of many points can be remotely measured under no-contact conditions. Furthermore, it may become possible to carry out displacement measurements in dangerous areas, such as for surveillance of dropping rocks.

c. No special skills are required, as there are in survey equipment handling. Furthermore, measurements can be carried out even where survey equipment cannot be placed. For example, it may be applied to measurement of wall surface geometry of vertical wells.

d. It is certain that the performance development and price reduction of the equipment will be advanced in the future.

4.2 Disadvantages of vision metrology

a. Since the enhancement of photography network affects accuracy even for a small number of measurement points, targets need to be placed in areas other than the measurement points. This lowers efficiency. Although very big advantage exists in many points to be measured, no advantage prospects in some cases of discrete and a small number of measurement points.

b. Targets must be placed for measurement, and scaffolding needs to be secured (as with TS convergence measurement).

c. Currently, outdoor measurement is impossible in well-lighted places, since the targets need to be made luminous with a stroboscope.

4.3 Future projection

a. Application to the measurement of geometry and displacement may be expanded not only for the tunnels, but also for various civil structures. However, further studies about photogrammetric network design optimization are necessary to determine the relationship between conditions such as the number of targets placed and photographs taken, photo-taking methods, etc., and accuracy. It would be desirable to normalize the measuring work to make it compatible with the measurement targets in future.

b. Although this report has not explained in detail, the data processing and analysis are manned operations. Thus, the work time increases with the number of targets. Currently, studies are advancing on automated data processing and analysis.

c. It would be desirable to develop civil-measurement-purpose targets that have high durability and high visibility from afar. It is also necessary to develop methods for placing targets easily and enabling their long use.

REFERENCES

Akimoto, K., Hattori, S., Ohnishi, Y. & Miura, S. 2000. Application of Vision Metrology to Tunnel Profile Measurement. *Journal of the Japan Society of Civil Engineering* (under contribution)

American Society of Photogrammetry 1980. *Manual of Photogrammetry* (Forth Edition)

Atkinson, K.B.(ed.) 1996. *Close Range Photogrammetry and Machine Vision. Whittles Publishing*

Bickel, J.O., Kuesel, T.R. & King, E.H.(ed.) 1996. *Tunnel Engineering Handbook. Chapman & Hall*

Hattori, S., Akimoto, K., Okamoto, A., Hasegawa, H. & Imoto, H. 1999. CCD Camera Calibration without Control Points by Multi-Photography of a Target Field. *The Transactions of the Institute of Electronics, Information and Communication Engineers* Vol-J82-D-2: 1391-1400

Modern Tunneling Science and Technology, Adachi et al (eds), © 2001 Swets & Zeitlinger, ISBN 90 2651 860 9

Sequential application of several survey systems in tunnelling for ground classifications

M.Okamura & T.Hara
Toda Corporation, Tokyo, Japan

T.Kimura, K.Ishiyama & T.Hirano
Nishimatsu Construction Co., Ltd., Kanagawa, Japan

ABSTRACT: The effectiveness of the combination of three methods (TDEM , TSP , and DRISS) by means of collecting the precise information on the ground ahead of the tunnel face, was estimated through application to tunnel construction projects.   As a result, it was found that these methods, when used in combination, can provide higher accuracy than used separately, and can also enable safe and effective tunnel construction.

1. INTRODUCTION

Locating the presence of geological anomalies (such as faults, groundwater, and changes in rock type) ahead of the faces with a high accuracy is very important in tunnel construction methods from the view point of the safety and the efficiency.   In many cases, however, sufficient information cannot be obtained from conventional methods alone (such as field geological survey, seismic exploration, and boring survey), and unexpected heavy inflow and geological discontinuity are often encountered during excavation.

In order to obtain more accurate geological information than that by conventional survey methods, we have been carrying out studies to establish a new high-accuracy exploring system which consist of three methods, i.e. TDEM (Time Domain Electro Magnetic) method, TSP (Tunnel Seismic Prediction) method and DRISS (Drilling Survey System), and which can comprehensively estimate the differences in the ground properties ahead of the faces (e.g. resistivity value, presence of reflector, strength) both before and during excavation.

We verified characteristics of these three methods and their accuracies depending on the ground types, and applied Tunnel High-accuracy Exploring system to a tunnel construction site to establish as a tool for comprehensively estimating the various ground properties. As a result, the system was found to be effective for tunnel constructions.

2. TUNNEL HIGH-ACCURACY EXPLORING SYSTEM

The flowchart for applying Tunnel High-accuracy Exploring system to tunnel construction is shown in Figure 1, and an example of the system application in Figure 2. Although the three methods shown in Figure 1, in general, are separately carried out according to the conditions of the construction site, Tunnel High-accuracy Exploring system in the present paper performs two or all of these methods in combination depending on the geological conditions to enable cost- and time-effective tunnel construction.

The following sections outline each of the three methods.

2.1 *TDEM Method*

TDEM method (Hara & Saito 1995), which was developed and improved overseas as a means for underground resource exploration, is an electromagnetic prospecting method to survey the resistivity values of the ground. As shown in Figure 3, this method measures the underground resistivity by converting the transient phenomenon in the secondary magnetic field generated by the intermittent current passing through the transmitting sources installed on the ground surface into a time function.

2.2 *TSP*

As shown in Figure 4, TSP (Akashi et al. 2000) is

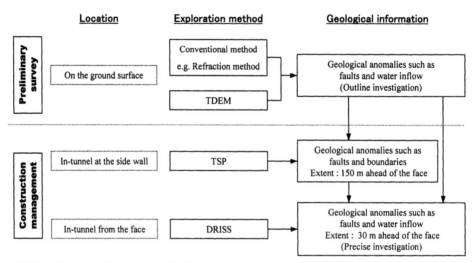

Figure 1. A flowchart of Tunnel High-accuracy Exploring system.

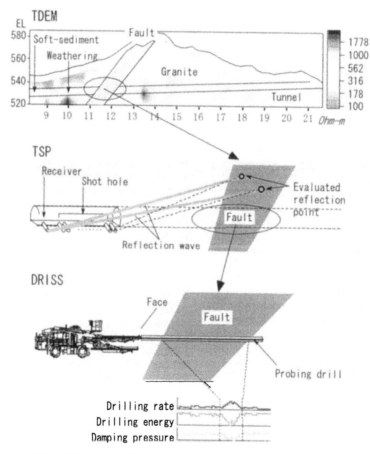

Figure 2. The strategy of Tunnel High-accuracy Exploring system.

based on the reflection method (one of the seismic prospecting method) to detect discontinuity in acoustical impedance located ahead of the face. This method uses unique devices and theories which enable short cycle times and cost-effective survey compared to other prospecting methods.

2.3 *DRISS*

DRISS (Hikima et al. 2000) is based on a similar principle to that used in conventional methods for exploring drilling. However, as shown in Figure 5, this system can automatically measure the hydraulic drilling data and determine the ground estimate parameters (e.g. drilling rate, drilling energy, and damping pressure) using such data. Furthermore, by taking into consideration the observed information, this system can estimate the geological conditions ahead of the face quickly and quantitatively.

3. EXAMPLE OF APPLICATION TO TUNNEL CONSTRUCTION

In order to precisely detect the location and extent of the hydrothermal alteration zones distributed in extremely hard hornfels with Vp value of 5.7 km/s

by the seismic refraction method, Tunnel High-accuracy Exploring system was applied in a tunnel construction site.

The resistivity value was first measured from the ground surface by TDEM method before tunnel excavation to ascertain the presence and approximate location of hydrothermal alteration zones and fracture zones. DRISS was then conducted when the face reached near the section with the geological anomaly detected by TDEM method, enabling drainage of groundwater as well as precise estimation of the geological conditions ahead of the face.

In addition, TSP was carried out when the face reached about 50 m from the section with the most geological anomaly across the tunnel route. DRISS was then conducted for the purpose of both ground exploration and groundwater drainage when the face neared the reflector (the anomaly's location) detected by TSP. Figure 6 compares the results by each of the three exploration methods and the types of ground encountered during tunnel excavation. As shown in this figure, TSP can detect the location and extent of hydrothermal alteration zones with better accuracy than seismic refraction method conducted on the ground surface. Furthermore, DRISS, carried out when the face reached near the predicted geological anomaly, found the accurate location and extent (7 m) of the geological anomaly and revealed that it is fractured and contains clay.

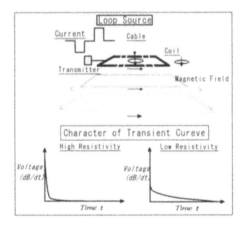

Figure 3. The principle of TDEM.

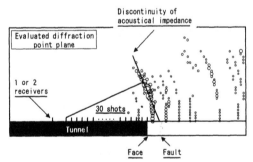

Figure 4. The principle of TSP.

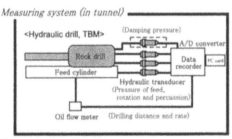

Figure 5. Outline of DRISS.

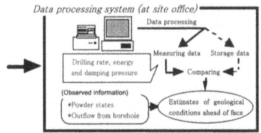

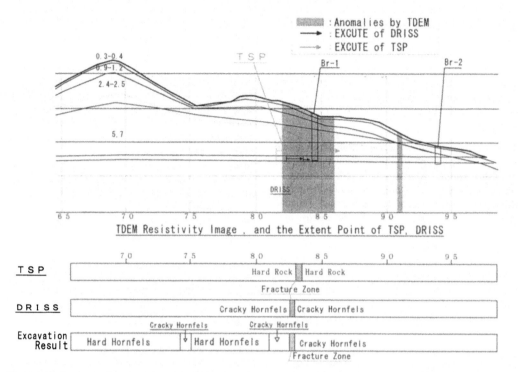

TDEM Resistivity Image , and the Extent Point of TSP, DRISS

Figure 6. Exploration result, and Excavation Result.

Applying Tunnel High-accuracy Exploring system enabled supports to be reinforced before the face reached near geological anomalies and the tunnel to be constructed in safety.

4. CONCLUSION

The effective system consisting of three methods was suggested. We conducted this system for hydrothermal alteration zones in the hard rock. As a result, the geological anomalies were prospected accurately by reasonable consideration.

In future, we will try this system on many tunnel projects with various ground and construction conditions and conduct studies on the appropriate combinations of exploration methods to establish a ground exploring system capable of effectively and precisely collecting the information on the ground ahead of faces.

REFERENCES

Hara, T. & Saito, A. 1995. Application of time domain EM survey in civil engineering. *Proc. ISRM 8th Congress*, Vol.1,: 85-88.
Akashi, T., Inaba, T. & Kimura, T. 2000. A potential of geological inference ahead of tunnel face by using seismic survey. *Proc. 10th Road Engineering Association of Asia and Australasia Conf.* No.216.
Hikima, R., Ishiyama, K., Tsukada, J. & Ishii, Y. 2000. The comparison of geological survey system using hydraulic drilling to face and core boring in soft rock. *Proc. 30th Symp. of Rock Mech.*: 238-242.(in Japanese)

Modern Tunneling Science and Technology, Adachi et al (eds), © 2001 Swets & Zeitlinger, ISBN 90 2651 860 9

Application of Several Electric Resistivities of Rock Masses for Tunnel Supporting Design

M. Nakamura
NEWJEC Inc., Osaka, Japan

H. Kusumi
Kansai University, Osaka, Japan

E. Kondo
The Kansai Electric Power Co. Inc., Hyogo, Japan

ABSTRACT: In this paper, the relations among electric resistivity, porosity and degree of saturation of various rock specimens investigated by laboratory tests, are compared with the field measurement data obtained by electric prospecting. In addition, it is tried that these results obtained laboratory test are applied on tunnel supporting design, and that using this relations, the rock masses around existing tunnel were estimated. As the results, it is recognized that this assessment method can be used on tunnel supporting design.

1 INTRODUCTION

Before the underground structures such as tunnels and large caverns will be constructed, seismic prospecting and electrical prospecting for geological surveys are often executed. However, it is few that the velocity of elastic wave and the electric resistivity of the rock masses around these construction sites obtained from these prospecting methods are directly used to the design.

In this paper, the relations among electric resistivity, porosity and degree of saturation of various rock specimens investigated by laboratory tests are compared with the field measurement data obtained by electric prospecting. In addition, it is tried that these results obtained laboratory test are applied on tunnel supporting design, and that using this relations, the rock masses around existing tunnel were estimated. As the results, it is recognized that this assessment method can be used on tunnel supporting design.

2 RESISTIVITY CHARACTERISTICS OF INTACT ROCK

2.1 *Outline of measurement*

The rock specimens using this study are three kinds of hard rock shown in Table 1. Each effective porosity is shown in the Table 1. Each rock specimen belongs to the class of hard rock in the column 50mm in the diameter and 100mm in height. Figure 1 shows the GS method used for measuring resistivity. As shown in Figure 1, the GS method uses size 80 copper mesh for both the current electrodes and voltage electrodes, utilizing 0.1N CuSO$_4$ as the solution for the electrodes and for the filter paper

Table 1. The used intact rock specimens.

Rock Type	The Number	Porosity
Rhyolite	8	0.008~0.030
Granite(fine)	4	0.009~0.012
Granite(medium)	9	0.009~0.014

sandwiched in the specimens. In the experiments, rock specimens of the same type were all placed in water and then simultaneously subjected to vacuum degassing for 96 hours, making them thoroughly wet (bringing the saturation level to 1.0), before being left in atmospheric conditions and then air-dried. Resistivity values were measured at the different water content conditions resulting. The resistivity value was obtained from the following formula (1).

$$\rho = \frac{V}{I} \cdot \frac{A}{L} \qquad (1)$$

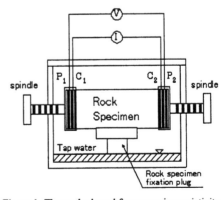

Figure 1. The method used for measuring resistivity.

Where, ρ : resistivity, V : voltage, I : current, A : cross-sectional area of specimen, L : electric current electrode spacing.

2.2 Characteristic of resistivity in non-saturation condition

Generally, the relationships between resistivity value and saturation can be expressed as exponential, as in the following formula (2).

$$\rho = \rho_s \cdot Sr^{-a} \tag{2}$$

Where, ρ : resistivity value, ρ_s: resistivity value when $Sr=1.0$, Sr : saturation, a : index.

Figure 2 shows the relationship for three kinds of hard rock specimens between effective porosity ϕ and resistivity ρ_s in formula (2). From this figure, it can be seen that for each of the rock specimens, the value of ρ_s decreases as effective porosity increases, and the decrease in the value of ρ_s is particularly striking for small values of effective porosity. As effective porosity increases, the rate of decrease in the value of ρ_s tends to decrease. It can be recognized that despite being deferent rock specimens, samples of the same rock types appear to approximate the relationship between ρ_s and effective porosity in the following formula (3).

$$\rho_s = m \cdot \phi^{-n} \tag{3}$$

Where, ϕ : effective porosity, m : coefficient, n : index.

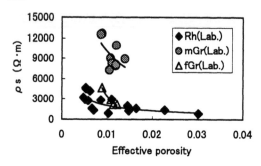

Figure 2. Relationship between effective porosity and ρ s.

3 RELATIONSHIP BETWEEN RESISTIVITIES AND POROSITIES OF INTACT ROCK AND FIELD ROCK MASS

In order to investigate the application to an in-situ rock mass, the investigation was made of the relationship between the rock class and the results of simple resistivity measurements at the rock sampling site where there was water inflow at the face, and where the rock was thought to be close to saturation.

The site used examination is an existing tunnel, with height 2.2m, width 2.0m and overall length of approximately 2400m, with the surrounding geology being mainly rhyolite and granites formed in the Mesozoic Cretaceous period. The granite consists of medium and fine-granite. A geological profile of the tunnel is shown in Figure 3. The water inflow is measured at the top of the tunnel as the tunneling face moved forward during its excavation.

The resistivity at the tunnel site was measured using simple resistivity measurement equipment with a four-electrode Wenner array, positioned with interelectrode spacing of 20cm. The state of the rock was evaluated using CRIEPI's classifications and a modification of the Q-system assessment method. Water inflow was a scale that distinguishes levels 2 (dripping and seeping only), 3 (inflow), 4 (small amounts of continuous inflow) and 5 (concentrated inflow from specific fissures).

Table 2 shows the relationship between porosity and the state of the rock mass. This information was used to infer the effective porosity from the classifications made at the site and from the state of the rock mass. Table 3 shows the resistivity values measured by simple resistivity measurement of the tunnel wall surface and the assessment of the rock mass inside the tunnel. Figure 4 shows the results of laboratory experiments for rhyolite and granite (medium & fine) with the simple resistivity measurement equipment. From this figure, it can be clarified that there is a good match between the both results and the approximation curve calculated from the effective porosity ϕ and resistivity value ρ_s, for the intact rock specimens obtained in the laboratory experiments and simple resistivity measurement.

As described here, three types of rock were investigated, and the simple resistivity measurement

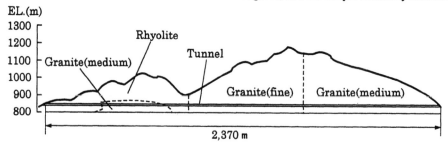

Figure 3. Established tunnel geology longitudinal section.

Table 2. Division contrast of the natural.

Assessment of The Rock Mass	Porosity
0~20	0.14~
20~40	0.07~0.14
40~60	0.035~0.07
60~80	~0.035
80~100	

Table 3. Assessment of the rock mass and simple resistivity measurement.

Rock Type	TD. (m)	Assessment of The Rock Mass	Water Inflow	ρ min (Ω·m)
Rhyolite	44.4	35	2	288
	46.2	43	2	731
	48.9	37	2	748
	232.4	58	4	429
	253.5	60	4	356
	325.0	25	3	225
	347.0	18	2	58
	372.5	20	2	20
Granite (medium)	424.4	18	2	94
	452.0	27	2	194
	546.0	52	3	179
	569.9	36	3	67
	596.8	76	2	756
	624.0	67	3	720
	653.0	57	2	217
Granite (fine)	966.2	46	2	334
	987.8	49	2	96
	1030.0	67	5	1642
	1186.6	55	3	985
	1265.8	73	2	708
	1295.8	45	2	436
	1392.6	71	3	804
	1445.6	41	2	307
	1498.6	72	3	700
	1581.6	49	5	71
Granite (medium)	1710.5	61	2	53
	1814.6	63	2	392
	1854.9	64	5	56

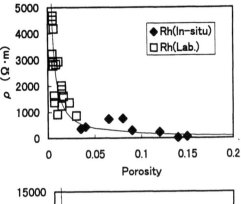

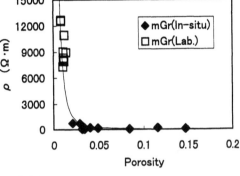

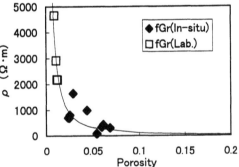

Figure 4. The comparison of laboratory resistivity result and in-situ resistivity result.

results corresponding to the effective porosity inferred from the state of the rock showed a good match with the approximation curve from the laboratory experiments. Consequently, this showed that by measuring the apparent resistivity value for in-situ rock it was possible to assess the physical characteristics of a rock mass from the formula (3) results obtained in the laboratory for the different rock types.

4 APPLICATION FOR SUPPORT DESIGN

In support design, mainly the classification of the rock mass is used. However, the decision of the support pattern is greatly influenced by the presence of inflow water. So, it is effective in the design to be able to take information on inflow water beforehand.

Then, the volume water content, which was the product of the porosity and degree of saturation, was attempted to design support pattern. Volume water content was calculated by formula (4) from the simple resistivity measurement result. The parameter in each rock kind is shown in Table 4.

$$\rho = c \cdot (Sr \cdot \phi)^{-d}$$

$$Sr \cdot \phi = (\rho/c)^{-1/d} \qquad (4)$$

Where, c and d : index.

Table 4. The parameter used on formula (4).

Rock Type	c	d
Granite(medium)	124	0.99
Granite(fine)	0.25	2.07
Rhyolite	111	0.70

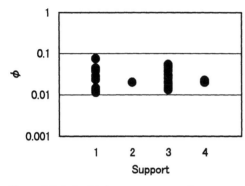

Figure 5. Relationship between support and ϕ.

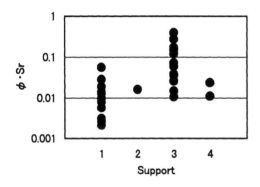

Figure 6. Relationship between support and $\phi \cdot Sr$.

Table 5. The specification of the support pattern.

Pattern	Specification
1	Shotcrete, Rock bolt
2	Shotcrete, Rock bolt, Channel steel
3	Shotcrete, Rock bolt, H-type steel
4	Concrete, H-type steel

Figure 5 shows the comparison with the support pattern results and the porosity ϕ obtained from assessment of the rock mass shown in Table 2. Similarly, Figure 6 shows the comparison with the support pattern results and the volume water content $\phi \cdot Sr$. The specification of the support pattern in these figures is shown in Table 5. It has been understood that the evaluation which uses volume water content is somewhat good the correlation with the support pattern as a result of the comparison. However, it is an examination problem for the difference in the same support kind.

5 CONCLUSION

In this paper, the results of simple resistivity measurement were found out to be compatible with the proposed formulas of effective porosity and resistivity using for the intact rock specimens, and the empirical formulas were demonstrated to be capable of assessing the physical characteristics of any rock masses.

In addition, the effectiveness of the evaluation by volume water content could be shown in the application for supporting design.

REFERENCES

Kusumi, H., Hatakenaka, Y., Nishikata, U., & Nakamura, M. (2000). Characterization of electrical resistivity for intact and jointed specimens and its application for in suit rock mass assessment. An International Conference on Geotechnical & Geological Engineering. Melbourne. CD-ROM.

Modern Tunneling Science and Technology, Adachi et al (eds), © 2001 Swets & Zeitlinger, ISBN 90 2651 860 9

Development and Application of Seismic Reflection Survey Looking Ahead of Tunnel Face using Hydraulic Impactor

T.Kato & H.Murayama
FUJITA CORPORATION, Technology Development Division, Kanagawa, Japan

T.Yanai
FUJITA CORPORATION, Tokyo Civil Works Branch, Tokyo, Japan

M.Murayama
Saitama Pref., Minano-Yorii Bypass Construction Office, Saitama, Japan

N.Shimizu
JGI Inc., Development Division, Tokyo, Japan

ABSTRACT: Recently, Tunnel Seismic Prediction (TSP) and in-tunnel Horizontal Seismic Profiling (HSP) have been applied to the geological estimation ahead of the tunnel face at several tunnel construction sites. However, since these methods require the use of explosives, permission to use explosives is required even for construction using a partial cutting machine or Tunnel Boring Machine (TBM), for which such permission is not required. So, we developed the seismic reflection method of predicting the geological condition ahead of the tunnel face using a non-explosive hydraulic impactor. The validity of this method was evaluated by comparing the estimated result of this method with the actual geological situation through field experiments.

1 INTRODUCTION

For the tunnel technician it is very important in control of the process, execution, and safety to predict the geological conditions ahead of the tunnel face accurately. However, because the geological condition is complex in Japan, the proper prediction may be difficult even though the preliminary survey was done. So, Tunnel Seismic Prediction (TSP) and in-tunnel Horizontal Seismic Profiling (HSP) have been applied to geological estimation ahead of the tunnel face at several construction sites in order to speed up the tunnel construction safely and economically (Inazaki & Chida 1993, Nishino et al. 1995). However, since these methods require the use of explosives, permission to use explosives is required even for construction using a partial cutting machine or Tunnel Boring Machine (TBM), for which such permission is not required. Therefore, the seismic reflection survey method using a non-explosive source is more applicable than conventional methods.

We developed the seismic reflection survey method to look ahead of the tunnel face using a hydraulic impactor , and verified the validity of this method through field experiments.

2 THE OUTLINE OF SHALLOW SEISMIC REFLECTION METHOD FOR TUNNEL

We developed Shallow Seismic Reflection Method for Tunnel (it is called as S²R-T) to predict the geological condition ahead of the tunnel face under construction. This method has the following charac-

teristics though a fundamental principle is the same as TSP, HSP.

1) The non-explosive hydraulic impactor is used as a seismic source.
2) The measurement is done using many receivers and many shot points.
3) The analysis has flexibility.

The outline of the measurement and the analysis are shown in the following sections.

2.1 *Measurement method*

Receivers are planted at constant intervals in the bottom of the tunnel, and shots using a hydraulic impactor are done one after another near the receivers as shown in Figure 1. The vertical component geophones along a tunnel axis and some three component geophones cross the axis are used so as to be able to acquire reflected energy from every direction.

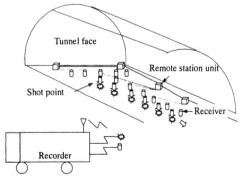

Figure 1. Measurement arrangement.

The hydraulic impactor will be shot just after the radio signal from a recording truck has been received. The data are temporarily stored in Remote Station Unit with the A/D converter, and are transmitted to the recording truck through the cable. Because drilling is unnecessary, the measurement work is so easy that it can be carried out in about a half day.

Hydraulic impactor makes use of energy which the piston accelerated by the gas pressure strikes the baseplate on the ground. The characteristics of the hydraulic impactor are as follows,

1) It is a nondestructive seismic source, and do no damage the ground.
2) High mobility by self-propulsion.
3) Applicable to both P and S wave exploration.

The hydraulic impactor is shown in Figure 2. And, as for this method, Vibroseis (S.E.G.J. 1999) and the explosive can be used as the seismic source instead of the hydraulic impactor.

Figure 2. Hydraulic impactor.

2.2 Analysis method

The analysis process of S^2R-T is summarized in the following sections. The horizontal structure just under the survey line is estimated according to the procedure described in 2.2.1, and the vertical structure of the ahead and the back of the survey line is estimated according to both procedures described in 2.2.2 and 2.2.3.

2.2.1 The analysis for horizontal structure just under the tunnel

This method uses common-midpoint (CMP) stack method for the surface seismic reflection data. The field records are processed in the flow a) shown in Figure 3, and the time section that represents the structure just under the survey line is made.

2.2.2 The structure analysis ahead and back of the tunnel face by enhancing the reflected wave from the front and the back

If the reflector is at almost right angles to the tunnel axis, the travel time of all the reflected waves are equal in the CMP ensemble. Therefore, stacking the-

se waves without NMO correction directly enhances the reflected wave. This is different from the previous flow in 2.2.1. Therefore the reflected waves from bottom is weakened, only the reflected waves from the ahead and the back of the tunnel face are enhanced.

2.2.3 The analysis according to the VSP processing

This analysis is basically the same as vertical seismic profiling (VSP) processing such as in the flow c) shown in Figure 3. The reflected waves from the ahead and the back of the tunnel face are enhanced after eliminating the outgoing waves like direct waves which will be aligned by time shifts. The reflector is supposed to be at mostly right angles to the tunnel axis in the same way as 2.2.2.

All the analysis results of 2.2.1-2.2.3 are indicated in the time section. The reflection image based on the flow of 2.2.3 and the estimated reflector distribution ahead of the tunnel face are shown in Figure 4. The white trough of the trace shows the reflected wave from the boundary with negative change of acoustic impedance, and the black peak of the trace shows the reflected wave from the boundary with positive change of acoustic impedance. The appearance position of reflectors ahead of the tunnel face is calculated using the travel time of the reflected wave and the average velocity of the direct wave.

S^2R-T can estimate the geological conditions ahead and back of the tunnel face and the section which has already been excavated. This leads to good prediction of the geological information ahead of the tunnel face by referring to the reflection image

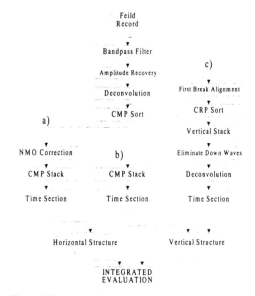

Figure 3. Processing Sequence.

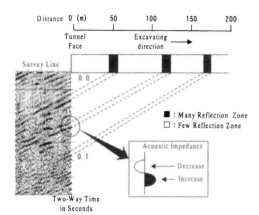

Figure 4. Result of the analysis according to VSP processing and the evaluation result.

of the section which has al-ready been excavated. In TSP, when the overburden is shallow, it is difficult to distinguish the reflected wave from the ahead of the tunnel face from that from the ground surface. Therefore, the geological interpretation for the shallow overburden section may be difficult (Koshino et al. 1996). However, since there often occurs the trouble during construction in the shallow overburden section and the portal the estimation of geological conditions for that area is more important. Because it is possible to specify the reflected wave from the ground surface in the analysis process in S²R-T, the exploration depth and precision are hardly affected by the overburden.

3 FIELD EXPERIMENTS

3.1 The outline of the exploration tunnel

The exploration experiment by S²R-T was done in the Kimo section of Minoyama tunnel that is located at Minano-cho, Saitama Prefecture Japan. This tunnel is composed of green schist (Gs), black schist

(Bs) and the alternation of strata which belong to Sanbagawa metamorphic rocks. A fold axis (a syncline axis) was presumed around the middle of the tunnel by seismic refraction method on the ground surface. There are some low-velocity zones having the width of several 10m. The largest low-velocity zone having the width of about 200m was presumed around a syncline axis. So, this method was tried to use in the field where the fracture zone has been predicted in this tunnel, and the applicability of this method was verified.

3.2 Exploration method

The experiments were carried out at the Minoyama tunnel at different places three times. In these experiments S²R-T and TSP have been tried. The explosive and the hydraulic impactor were used as a seismic source. The measurement arrangement of the 3rd experiment is shown in Figure 5 as an example. The receivers were planted at 1.5m or 3.0m intervals in the center of the bottom along the tunnel axis. The shot points were arranged in parallel about 1.0m apart from receivers at 3.0m or 6.0m intervals. The shots for a hydraulic impactor are done in the same place several times, these data are stacked in order to improve the exploration precision and the exploration depth.

3.3 Comparison with the experimental result and the excavation result

There are two ways of evaluating fracture zones in TSP. In the first way each reflector is considered as a boundary with geological change. In the second way the reflection image is classified into two groups such as "many reflection zone" and "few reflection zone"(Akashi&Inaba.1997). In the latter case, "many reflection zone" is generally considered as the zone with rapid geological change, in other words, fracture zone. On the other hand "few reflection zone" is considered as the zone with the stable geological

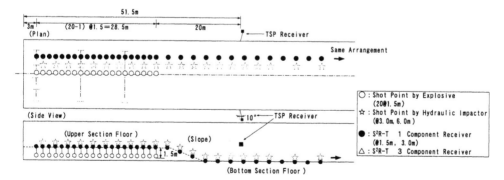

Figure 5. The measurement arrangement of 3rd exploration.

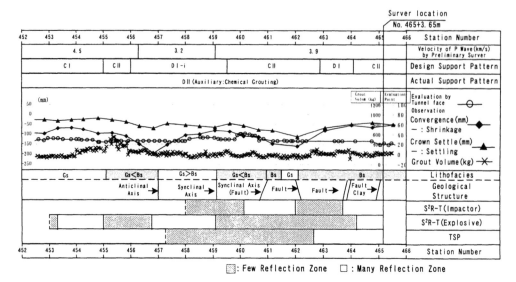

452	453	454	455	456	457	458	459	460	461	462	463	464	465	466	Station Number
		4.5				3.2					3.9				Velocity of P Wave(km/s) by Preliminary Survey

Figure 6. Comparison with the experimental result and the excavation result.

condition. S²R-T use the evaluation method based on the distribution of the reflections.

The experimental result of the 2nd is shown with the result of the preliminary survey, the control measure, the excavation and the observation and so on in Figure 6. The reflection image is classified into "many reflection zone" and "few reflection zone".

Each result mostly matches the spatial distribution of the fracture zone, the small fault, the fold axis, and the lithofacies classification, although the evaluation for S²R-T using hydraulic impactor is partly different from that for S²R-T using explosive. In Figure 6, it is considered that S²R-T using hydraulic impactor is highly precious in the range less than about 100m ahead of the tunnel face, and that S²R-T using explosive is highly precious to in the range of more than 100m ahead of the tunnel face.

In the section near the tunnel face, the results of each method almost coincide with each other. However, in the section far from the tunnel face, the result of each method doesn't coincide so much. This may be caused by the estimation error of P wave velocity which is used to calculate the appearance position of the reflector, and by the method of extracting the reflector. Therefore we will have to examine the analysis and evaluation method to improve the exploration precision of S²R-T and TSP.

4 CONCLUSION

We developed the new seismic reflection method of predicting the geological condition ahead of the tunnel face under construction. The validity of this method was evaluated by comparing the estimated

result of this method with the actual geological situation through field experiments.

Because S²R-T requires many hardware at this moment, the hardware in S²R-T must be simplified by the improvement of the measurement method. And, we will need to build up the analysis system which can evaluate quantitatively the distribution of the reflected energy. The present TSP and S²R-T use only P wave data. By making use of S wave data which can be emitted by hydraulic impactor, S²R-T will become more useful method to evaluate the geological characteristics of tunnel in detail.

REFERENCE

Inazaki, T. & Chida, K. 1993. Prediction ahead of the Tunnel Face using HSP Method. *Proceedings of the 25th Symposium on Rock Mechanics, J.S.C.E.*: 271-275.

Nishino, H., Yamamoto, M., Nakamura H., Hieda, H. and Nakamura, Y. 1995. Applicability of TSP (Tunnel Seismic Prediction) - System for Predicting ahead of the face under various geological condition. *Proceedings of Tunnel Engineering, J.S.C.E. Vol.5*: 347-352.

The Society of Exploration Geophysicists of Japan 1999. Handbook of geophysical exploration. Tokyo.

Koshida, Y., Yamamoto, H., Kasa, H., Utsuki, S. Kudou, S. 1996. An application of Tunnel Seismic Prediction in the shallow ground. *Proceedings of the 31st Japan National Conference on Geotechnical Engineering, J.G.S.*: 2251-2252.

Akashi, T. & Inaba, T. 1997. An Investigation on Seismic Survey Accuracy ahead Tunnel Face. *Proceedings of Tunnel Engineering, J.S.C.E. Vol.7*: 141-146.

Modern Tunneling Science and Technology, Adachi et al (eds), © 2001 Swets & Zeitlinger, ISBN 90 2651 860 9

Application of CCD Photogrammetry System to Measurement of Tunnel Wall Movement due to Parallel Tunnel Excavation

Y. Ohnishi, H.Ohtsu & S. Nishiyama
Dept. of Civil Eng.,Kyoto University

N.Okada & T.Seya
Ikeda Construction Office, Japan Highway Public Corporation

Y.Yoshida
Takamatsu Engineering Office, Japan Highway Public Corporation

T. Nakai & M. Ryu
Earth-tech Toyo Corporation

ABSTRACT: When the second tunnel parallel to the first one is excavated afterward, the first tunnel may receive some damage such as lining crack and wall deformation. When the tunnels are crossing faults, the damage may be very severe. In order to avoid the damage, an adequate measurement should be done for the prediction of tunnel movement for the first tunnel. Now we propose to use a new method for tunnel surveying system using CCD camera and photogrammetry. We shows an example in which 300 points were installed on the wall of the first tunnel and measured during construction of the second tunnel carefully. The results indicated that our new system is capable of measuring many points in the tunnel quickly, accurately, and safely,

1 INTRODUCTION

Monitoring is a key element for design and construction of tunnels. Monitoring can help determine the depth of the loosened zone, movements on discontinuities, in-situ modulus, and other items that may be dealt with in design. During construction, deformations are important indicators of the degree of stability. After construction, monitoring could be also an important tool for obtaining useful information to control the life of tunnels. However, conventional monitoring systems attach to the rock mass at point locations, so that an instrument may be measuring exaggerated or useless deformations because a joint or rock block is unfavorably located. Rock deformations of tunnels are monitored mostly by using tape extensometers and traditional surveying instruments. When monitoring of the behavior of the rock mass at many points, during and after construction, is planned to provide a verification of design and safety considerations, conventional monitoring systems are troublesome and time consuming. In fact, incorrect monitoring can lead to serious errors in decision making in the construction process.

The Houou Tunnel is being constructed as a secondary road tunnel near the existing road tunnel that was constructed in 1990. Because the two tunnels cross the Median Tectonic Line that is the largest active fault in Japan, monitoring the actual behavior of the existing tunnel is needed to evaluate various potential hazards. However, conventional monitoring systems are difficult to install monitoring equipment and measure without blocking the existing tunnel.

Recently, photogrammetry has been increasingly applied as a precise measuring tool in industrial works. Photogrammetry has the advantage of measuring deformation of an object by some photos with easy measurements. Traditional photogrammetric techniques and instruments are inappropriate for monitoring the behavior of rock mass because of disadvantages in cost performance, but the high resolution digital cameras of the present day have become available with low prices. Digital photogrammetry can permit easy installation along with a capability of real-time measurement and remove human skill in measurement work. This paper presents the results of digital photogrammetry that was applied to monitoring the actual behavior of the existing tunnel during construction of Houou Tunnel.

2 PRICIPLE OF PHOTOGRAMMETRY

The fundamental mathematical model of digital photogrammetry is an optical triangulation that describes the perspective transformation from two dimensional image coordinates into three dimensional object space coordinates. The ultimate extension of the principles is to adjust many photogrammetric measurements to ground control values in a single solution known as a bundle adjustment. The analytical process is so named because of the many lights rays that pass through each lens position constituting a bundle of rays. Any object point can be determined as the intersection of the corresponding rays from each of many images.

All parameters describing the perspective transformation process can be also determined without prior knowledge of camera positions and calibration parameters. The computational model of the bundle adjustment is based on the well-known collinearity equations:

$$x = \Delta x - c\frac{a_{11}(X-X_0)+a_{12}(Y-Y_0)+a_{13}(Z-Z_0)}{a_{31}(X-X_0)+a_{32}(Y-Y_0)+a_{33}(Z-Z_0)}$$

$$(1)$$

$$y = \Delta y - c\frac{a_{21}(X-X_0)+a_{22}(Y-Y_0)+a_{23}(Z-Z_0)}{a_{31}(X-X_0)+a_{32}(Y-Y_0)+a_{33}(Z-Z_0)}$$

where x and y are the observed image coordinates; X_0,Y_0,Z_0 and X,Y,Z are the object space coordinates of the camera positions and object points, respectively; the a's are functions of three rotation angles of each image; $\Delta x, \Delta y$ and c are the interior orientation elements; Δx and Δy are perturbation parameters which describe departures from collinearity due to lens distortion and in-plane and out-of-plane image distortion, c being the focal length. This model which includes the interior orientation elements is known as a self-calibrating bundle adjustment. This method has the advantage of obtaining high photogrammetric accuracies with cameras having unknown calibration parameters.

The collinearity equations can be recast into the following observation equations by linearization:

$$v = A_1X_1 + A_2X_2 + \Delta \qquad (2)$$

where A_1, A_2 are the design matrices of the unknown exterior orientation elements, vector X_1, and object point coordinates, X_2; Δ is the discrepancy vector, v is the vector of image coordinate residuals. Solution for X_1 and X_2 is according to the method of least-squares.

3 EXPERIMENTAL APPLICATIONS

The procedure for digital photogrammetry is generally illustrated in Figure 1. In order to discuss the applicability of photogrammetric techniques for the monitoring of structures in rock, a pilot study was conducted. In our laboratory, discrete object points were signalized with retro reflective targets. The geometric design of these targets were arranged on the wall and ceiling of the laboratory like rock wall of a tunnel. 10 targets were mounted on micrometers. The assessment on the performance of digital photogrammetric system was conducted by measuring the displacements induced by the micrometers.

The constructed tunnel model and data acquisition were the following conditions;

size of model:

3m width x 3m length, 3m above the floor

photographing distance: 15m

number of digital images: 20
number of object points: 90

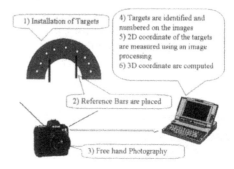

Figure 1. Procedure of Digital Photogrammetry.

A digital camera equipped with 18mm lens, which had 2.54 million pixels, was used for image acquisition and the data processing was performed on a laptop computer. An example of the images is shown in Figure 2. The image coordinates of the target points were measured using typical semi-automatic software.

Figure 2. An example of digital image

Since independent movements of micrometers had been carried out in three coordinate axes, it was possible to evaluate the discrepancy between true and measured values of displacements in each coordinate.

Consider simplified linear equation shown in Equation (2).

V = AX L

E(V)=0 ; E(L)=AX   (3)

where A is the design matrix with dimensions n × u (n > u) and rank R(A) ≤ u ; L is an n × 1 stochastic vector of observations; X is the non-stochastic vector of unknown parameters of dimension u; V is a stochastic n-vector of residuals; E is the expectation operators. A necessary and sufficient condition for the $E(L)=AX$ to be solved is as follows;

$$AA^+E(L)=E(L) \qquad (4)$$

where A^+ is a chosen member of generalized inverse. This equation can be viewed as being a consistency condition and if the equation (3) is consistent, then the rank of A determine whether X is unique, or whether many solutions exist.

If the design matrix A has no rank defect, a unique estimate exists. This is the least-squares estimate that can be obtained for x;

$$x=N^{-1}A^tPL \qquad (5)$$

where $N= A^tPA$ and P is the weight matrix.

In the full rank case $N^{-1}=N^+$, thus satisfying the consistency condition and also yielding best linear unbiased estimations while minimizing V^TPV. The variance-covariance matrix, Σx, of the parameter estimates, x, is given by

$$\Sigma x = \sigma^2 N^{-1} \qquad (6)$$

where σ^2 is the a posteriori variance factor. In order to ensure a unique solution for the parameter estimates, the coordinates of the control points are treated as fixed parameters in the least-squares adjustment because information required to overcome the network defect need to be imposed. This involves three translations, three rotations and a scale factor.

Because it was difficult to control the movements of micrometers along each coordinate correctly, the absolute values of displacements of X,Y,Z coordinates may not be evaluated accurately. The result is shown in Table 1.

The principal factors governing the precision of recovery of the parameters are the number and intersection geometry of the imaging rays. A very coarse accuracy indicator is as follows;

$$\sigma_c =qS\sigma \qquad (7)$$

where σ_c is the mean standard error of XYZ coordinate standard errors, the image scale is 1:S, σ the image coordinate standard error, q an empirical factor with a value ranging from 0.5 to 0.9 for network geometries.

In digital photogrammetry, it is necessary to design the photogrammetric network geometry to optimize the operation in terms of accuracy and economy, and to ensure that a certain design level of measurement precision and reliability will be attained. At data acquisition stage, a number of questions such as the number of camera stations required, the number of targets photographed in each digital image, and the adopted image geometry affording a strong camera self-calibration must be solved. The simulator we have developed has provided the means by which photogrammetric network planning can be carried out. This simulator has been created for use by non-photogrammetrist.

Table 1 indicated the accuracy specification for the measurement called for mean standard errors of XYZ coordinates were 0.18,0.07 and 0.23, respectively. Of primary concern in the pilot study was the accurate measurement of Z displacement that was the optical axis of a camera. For the network geometry adopted, X and Y coordinates were in plane coordinates and Z coordinate was out of plane ones. In convergent camera networks, in plane accuracies will be better than out of plane accuracies. A mean standard error of XYZ coordinates was 0.17mm.

Table 1. Result of photogrammetric measurements.

target No.	True values (direction)	meaured values		
1	1.15 (Y)	0.251	1.152	0.058
2	1.50 (Z)	0.057	0.077	1.533
3	1.00 (Y)	0.097	0.926	0.084
4	1.00 (X)	1.004	0.017	0.257
5	1.30 (Z)	0.061	0.150	1.249
6	3.15 (Y)	0.073	3.151	0.175
7	2.00 (Z)	0.023	0.093	2.118
8	3.00 (Y)	0.228	2.849	0.040
9	3.50 (X)	3.580	0.077	0.240
10	1.00 (Z)	0.188	0.037	0.785

Unit (mm)

4 FIELD EXPERIMENTS

In order to discuss the applicability of photogrammetry for dimensional measurements in a tunnel, an example will be briefly summarized. The example was in-situ monitoring measurements of the convergence at Houou tunnel. This tunnel is being constructed in parallel with the existing railway tunnel. The two tunnels crosses the largest active fault that is called Median Tectonic Line, and the distance between them is about 30m. Because the existing tunnel is being served for daily traffic, it is necessary that the monitoring system should be: reliable, easily installed, easily real-time read, and inexpensive. These requirements can be met through the digital photogrammetry

Targets were placed on the wall of the first line over 30 m length. Construction of the first line revealed that the fault zone crossed this site. Images by a digital camera with 2.54 million pixels were constructed under the following conditions;

focal length of lens : 18mm
number of images : 32
number of targets : 300

Target arrangement is shown in Figure 3. Targets appear as white circles in the figure.

Two independent measurements were carried out before and after the second line passed through the fault zone. The fault could be stimulated to move by artificial cause that was the excavating of the second line. Deformation of the first line was monitored as shown in Figure 4.

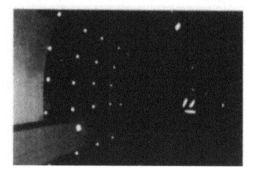

Figure 3.   Target arrangement.

The mean standard error of XYZ coordinates was 0.36mm. Arrows in this figure show the vectors of displacements of targets. The magnitudes of arrows mean those of displacements relatively. Fault displacements could be monitored using the digital photogrammetry. Thus, the hazards of movements on the fault zone crossing the site were evaluated.

The amount of uncertainties still remaining at the design stage of a tunnel is often so large that serious risks may be encountered during the actual construction operations. Monitoring the behavior of a tunnel under construction by the digital photogrammetry can provide a timely and corrective action for verification of design and safety considerations. The three dimensional measurements of deformations using the digital photogrammetry can permit a realistic evaluation of the design because sufficient data on their behavior are available at the construction stage. The larger the amount of uncertainties, the more useful will be the digital photogrammetry as the monitoring tool.

5 Conclusion

This paper has presented a digital photogrammetry for measuring three dimensional coordinates of targets situated on the wall of a tunnel. During construction, deformations are important indicators of the degree of stability. Because measurements must be made repetitively, the systems must be reliable, easily installed, and inexpensive. However, conventional systems that measure deformations at point locations require time consuming and highly-trained operators.

This paper demonstrates that a digital photogrammetry offers a practical and economical measurement tool that would yield less than 1mm positioning accuracies over large areas.

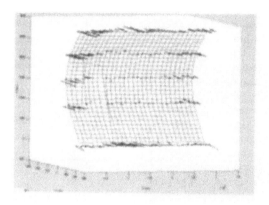

Figure 4. Results of monitoring the behavior of the first line when the second line passed through the fault zone.

Potential future uses of the digital photogrammetry lie in deformation monitoring of the real behavior of a tunnel at construction stage.

REFERENCES

Wolf P.R. & Dewitt. B.A. 2000. Elements of Photogrammetry with Applications in GIS, third edition, McGraw-Hill.

Ohnishi, Y., Zhang, C., Nishiyama, S., Hayashi, K.& Okamoto, A. 1999. A. Precise Close Range Photogrammetry System in Rock Displacement Measurement. '99 Japan-Koreas Joint Symposium on Rock Engineering, Fukuoka, Japan: 239-246

Ohnishi, Y., Zhang, C., Nishiyama, S., Hayashi, K.& Nakai T. 1999. Application of Photogrammetry on Rock Displacement Measurement. The 8th International Underground Space Conference,Xi'an: 239-246

Fraser,C.S.1982. Optimization of Precision in Close-Range Photogrammetry, Photogrammetric Engineering and Remote Sensing, Vol48, No.4 : 561-570

Fraser, C.S. 1984. Network Design Considerations for Non-Topographic Photogrammetry, Photogrammetric Engineering and Remote Sensing, Vol50, No.8 : 1115-1126

Modern Tunneling Science and Technology, Adachi et al (eds), © 2001 Swets & Zeitlinger, ISBN 90 2651 860 9

Semi-automated tunnel measurement by vision metrology using coded-targets

S.Hattori
Fukuyama University, Fukuyama, Japan

K.Akimoto
Shikoku Polytechnic College, Marugame, Japan

Y.Ohnishi
Kyoto University, Kyoto, Japan

S.Miura
Kajima Technical Research Institute, Tokyo, Japan

ABSTRACT: In application of vision metrology to tunnel deformation measurement, the number of target points amount to the order of a few of thousands. It is, therefore, impractical to identify all target points only by manual involvement. This paper discusses a further design of coded targets, which can be automatically recognized and identified. And it also reports on the semi-automatic procedures, which include estimation of approximations of orientation parameters of images, a new scheme of semi-automatic image triangulation, identification of other non-coded targets, as well as object coordinate determination by bundle adjustment. An experiment shows this procedures reduced the processing time down to 1/5 compared with manual processing.

1 INTRODUCTION

Vision metrology has been applied to industry. And it is now going to be used for tunnel deformation monitoring (Akimoto et al. 2001). With the method many points can be measured in short time of a range and measurement works less hinder the traffics of vehicles, in which retro-targets are placed on a tunnel profile and are multiple-imaged by a digital camera with an electric strobe. Object space coordinates of targets are determined by bundle adjustment together with interior and exterior orientation parameters of the images.

It is, however, not easy to identify (or to label) target images which are featureless white dots in under-exposed conditions. The number of targets, e.g. in an experiment in section 5, is about 200, the number of images is over 60 and the number of target image dots amounts to as many as 3000. Therefore it is apparently impractical to identify these points only by manual operation. The adjustment calculation may easily diverge, if only

two points are faultily labeled to each other. Moreover approximations of all the parameters are required at the start of adjustment, since observation equations are highly nonlinear. It is not productive to input them by manual involvement.

The authors have developed a semi-automatic measurement procedure by use of coded targets (CTs) (Hattori et. al.2001) . This paper discusses a design of CTs, the identification of image dots and the efficiency of measurement.

Since it hurts workability to give codes to all the targets, relatively small number of targets are coded, with which images are triangulated. Then other non-coded measurement targets (MTs) are labeled by multiple ray intersection and their approximate object coordinates are also evaluated.

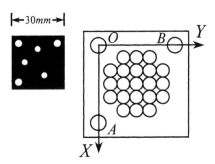

Figure 1. Concentric coded target.

Figure 2. Design of coded target.

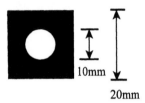

Figure 3. Measurement target.

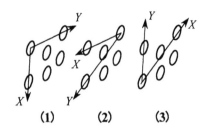

(1)   **(2)**   **(3)**

Figure 4. Possible combinations of axes.

2 DESIGN OF CODED TARGETS AND THEIR IDENTIFICATION

The idea of CTs is not new (Fraser 1997). Up until now two kinds of CTs have been released; a concentric type and a dot distribution type. The former (see Figure 1) is easy to identify but the number of possible code patterns is limited, while the latter goes reverse. The authors have employed the latter.

As shown in Figure 2, on a plywood plate six retro circles are placed as dots. Three of them (*AOB*) are to define coordinate axes, while the other three construct a code. These dots are allocated to three positions out of 21 circles such that they are spaced out at least one dot apart, which yields 420 kinds of code patterns.

The size of coded targets should depend on the image scale. The CT in Figure 2 is imaged to order of 20-30 pixels so that one dot is imaged about to three pixels.

Figure 3 shows a code-less measurement target (MT), which is designed to be imaged to 5-10 pixels, which show the best precision according to the authors the authors experiments.

CT images are identified in the following way. (1) The images are binarized at a suitable threshold to recognize target dots. By use of the well-known dilation algorithm (Gonzalez, 1992), dots of a CT are extracted when six dots are merged to one. Though error-causing noises little occur in a tunnel, fluorescent lights or reflections from fixtures or vehicles are possible sources. Hence eye check is mandatory in the process.

(2) A domain including a CT is clipped off and the centroid of six dots is calculated over the domain.

(3) The three dots remotest from the centroid are assumed *A*, *O* and *B* in Figure 2.

(4) As shown in Figure 4 there are some of possible combinations of *X*, *Y* and *Z* axes. If the projections of three dots for coding to the presumed *X*, *Y* axes come all within the line segments *AO* and *BO*, this combination is judged correct.

3 TRIANGULATION OF IMAGES

Images are connected sequentially to form a unified model by the help of CTs through following steps. The initial step is a definition of the origin of the object space coordinate axes and their directions. Among several options adaptable (Hattori 2001), the authors define the coordinate system by giving approximations of object coordinates of 4-6 CTs explicitly. Seven rank defects (corresponding to rotations, translations and scale) are compensated by fixing seven coordinates of three points (minimal constraints).

The first three images taken to include these initial CTs are bundle-adjusted to update starting rough parameter values.

The later images are connected to the previously established model. If three or more CTs, the object coordinates of which are already known are included, the closed form of similarity transformation can be conveniently used (See section 4). Otherwise the values must be manually input. Then all the parameters are updated by bundle adjustment of the images in hand til then. And again together with CTs which become newly available, bundle adjustment is executed. This iterative procedure is repeated until all the images are triangulated.

4 ESTIMATION OF ORIENTATION PARAMETERS BY CLOSED FORM OF SIMILARITY TRANSFORMATION

Let three or more CTs, the object coordinate of which are known be included in an image. And then let the image coordinates and object coordinates of the CTs be respectively denoted by

$$\boldsymbol{x} = \begin{bmatrix} x_i & y_i & z_i \end{bmatrix}^T \text{ and } \boldsymbol{X}_i = \begin{bmatrix} X_i & Y_i & Z_i \end{bmatrix}^T, (i=1,2l,,n;n \geq 3)$$

The collinearity equation is expressed by

$$\boldsymbol{x}_i = scale_i \ M(\boldsymbol{X}_i - \boldsymbol{X}_0) \quad (1)$$

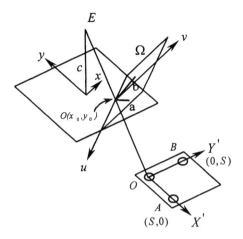

Figure 5. Estimation of the scale of a CT.

where M is a rotation matrix, $X_0 = [X_0 \ Y_0 \ Z_0]^T$ are camera station coordinates and $scale_i$ is the scale of an image of the i'th CT. $scale_i$ is estimated by comparing the size of the image and its design value. In this case the parallel projection must be assumed instead of the conventional central projection, since the scale of an image can not be defined with the central projection. As shown in Figure 5 OA and OB (design scale is S) of a CT are imaged onto the plane Ω perpendicular to the ray as line segments $\overline{oa}$ and $\overline{ob}$ respectively. $scale$ is evaluated by

$$scale = S/\sqrt{\overline{oa}^2 + \overline{ob}^2}. \quad (2)$$

And this holds for any objects imaged in small size.
Equation 1 divided by $scale_i$ leads to

$$u_i = x_i/scale_i = M(X_i - X_0). \quad (3)$$

Let the centroids of X_i's and u_i's be denoted by X_G and u_g respectively, and then

$$u_i - u_g = M(X_i - X_G) \quad (4) \quad \text{or} \quad \hat{u}_i = M\hat{X}_i \quad (4')$$

follows. The best rotation matrix M is defined as one minimizing the square of sum of discrepancies in Equation 4, which is obtained by singular value decomposition of the correlation matrix C;

$$C = \sum_{i=1}^{n} (\hat{u}_i \ \hat{X}_i^T) \quad (5)$$

to

$$C = V^T \Lambda W \quad (6)$$

where V, W are orthogonal matrices and Λ is a

diagonal matrix with positive singular values as diagonal elements. M is given by (Kanatani, 1993)

$$M = V^T W. \quad (7)$$

The camera station X_0 is also derived from Equation 3.

5 IDENTIFICATION OF MT'S

After all the images are connected and oriented in the unified coordinate system, code-less MTs are identified by intersection of multiple images. For expediting the identification process only possible combinations of images are processed as one group, by searching for the images which include common CTs.

The intersection point M of two rays is defined as the center of a line segment L_{min}, which is perpendicular to both the rays. The threshold of intersection can be derived in principle by the law of error propagation, but simply is given by

$$|X_1 - X_2| \le s\sigma_S, \quad (8)$$

where X_1 and X_2 are the position vectors of end points of the line segment L_{min} and σ_S is a root mean square of variances of camera station coordinates that is evaluated from the variance covariance matrix of parameters. The appropriate values of the factor σ_S ranges usually in 1.5-2.5 according to preliminary experiments.

6 EXPERIMENT AND CONSIDERATIONS

The usefulness of CTs was verified by an experiment at a test site of a dead railway tunnel. The test site is 7m in diameter times 15m in longitude. For more details see Miura et al.(2001). The figures referred to hereafter by the number followed by (Miura), like Figure 1(Miura), are all cited from his paper. As shown in Figure 4 (Miura), 16 profiles were taken with 1.0m spacing. Along a perimeter on every profile, 13 MTs as shown in Figure 3 were placed with their faces parallel to the wall. The total number of these MTs was 208. Further as shown in Figure 4 (Miura), nine scale bars of one meter long were placed on the ceiling and on the both sides of the tunnel. The scale bars had been precisely calibrated up to 5μm in advance. To reinforce the geometry, 11 back-to-back targets were placed on the floor and at the height of 0.4m reciprocally with 0.5m spacing as shown in Figure 6. At the photographing work this set of augmented MTs were shifted into longitude to cover all the area, overlapping to each other, as camera stations moved

287

into that direction. As the result practically 33 MTs were augmented (In the Miura's report this auxiliary placement of targets was not employed). Eventually the total number of MTs amounted to 272.

56 CTs were placed over the area. As shown in Figure 6 (Miura) the total area was parted to five zones with 5m wide in longitude. Images were taken such that each zone was convergently imaged and these zones were connected by additional exposures. The number of images was 66 in tatality, the number of CT images was 595 and the number of recognized MT image dots was 2936.

From the 66 images dots of CTs and Mts were recognized, and then the codes of CTs were automatically identified. No trouble occurred in code reading with the procedures discussed above. After checking the connectivity of images, they are connected in the order of images including the largest number of common Cts to ever connected images. Though the estimates of exterior orientation parameters obtained thus were not in good quality, the adjustment successfully converged due to wide convergence range of orientation parameters.

After all the images were connected, bundle adjustment with all the CTs was executed. The estimated measurement precision of target image coordinates was 3.9µm(1/2.3 pixels), which is apparently degraded compared to expected precision of 0.5µm in vision metrology. This was caused probably because dots of CTs were imaged to as small as about three pixels.

The estimated precision of camera station coordinates was 3.86mm. Hence the threshold of intersection is s =5.5mm in Equation 8. The identification process with this value gave rise to labeling 2426 out of 2936 dots(82.6%). The dots not labeled were all dark and lay in corners of images. From the result of the bundle adjustment with these 2426 dots, five dots were rejected successfully as blunders. As to blunder detection the simple 3σ method was used. The blunders were corrected and the remaining dot images were identified manually. The estimated precision of image coordinates was 1.0µ (about 1/10 pixel). The total processing time was about 60 minutes (50 minutes spent on manual work).

The maximum identification rate, 89.1% was obtained for s = 2.5. For excessively large threshold, e.g. s = 8, 92.7% of points were labeled, but many dots of different CTs were faultily labeled as the same CT (more than 20 label faults occurred). Though the bundle adjustment diverged, additional manual label fixing was needed. The 3σ method failed to find out blunders, which themselves showed small intersection errors but blurred the entire precision.

To compare the processing time to manual involvement, the above works were conducted by connecting images one-by-one and dots are

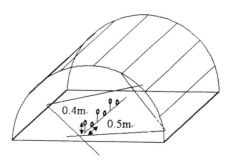

Figure 6. Back-to-back targets placed on the floor.

identified manually with the assistance of local intersection and resection techniques. Total manual processing time was about 5 hours, which means the automatic method could reduce the processing time to 1/5.

7 CONCLUSIONS

This paper discussed the semi-automatic measurement by use of specially designed coded targets. The total processing time from identification of target images through bundle adjustment was reduced to about 1/5 compared to manual process.

REFERENCES

Akimoto, K., S.Hattori, S.Miura & Y. Ohnishi 2001. Development of vision metrology combined with auxiliary observations for tunnel profile measurement. *IS-Kyoto2000. Modern Tunneling, Science and Technology. Nov.1.* Kyoto.

Fraser, C.S. 1997. Innovations in automation for vision metrology systems. *PE&RS.*15(90): 901-911.

Hattori,S., K.Akimoto.& H.Imoto 2001. Automated procedures with coded targets in industrial vision metrology. *IEICE, D-II (in Japanese)*, to appear.

Gonzalez, R.C. & R.E.Woods 1992. *Digital image processing*: Addison Wesley: 518.

Kanatani,K. 1993. *Geometric computation for Machine vision.* Oxford engineering science series 37: 105-117.

Miura,S., S.Hattori, K.Akimoto & Y.Ohnishi 2001. Tunnel convergence measurement using vision metrology. *IS-Kyoto2001. Modern Tunneling, Science and Technology. Nov.1.* Kyoto.

Modern Tunneling Science and Technology, Adachi et al (eds), © 2001 Swets & Zeitlinger, ISBN 90 2651 860 9

Development of vision metrology combined with auxiliary observations for tunnel profile measurement

K. Akimoto
Shikoku Polytechnic College, Marugame, Japan

S. Hattori
Fukuyama University, Fukuyama, Japan

S. Miura
Kajima Technical Research Institute, Tokyo, Japan

Y. Ohnishi
Kyoto University, Kyoto, Japan

ABSTRACT: This paper discusses an application of vision metrology to tunnel profile measurement. In general vision metrology is less reliable unless a strong observation network is available. In tunnel profile measurement, weak camera configurations are often inevitable because of its geometrical constraint. To cope with this difficulty, the authors have extended the basic strategy of vision metrology with a single camera to one reinforced by auxiliary observations with a total station. The augmented observation equations are solved by inner constraints, which are required of some modifications due to reduction of the rank of the null space of observation equations. The usefulness of the method is illustratively shown for a measurement experiment on a tunnel of 10m in diameter. The adjusted object space coordinates of targets are compared those independently observed by a total station for check. The root mean square of differences of the coordinates is reduced from 1.43 mm with the basic observation method to 0.95mm with the improved method.

1 INTRODUCTION

Vision metrology (VM) using a digital camera is applied to tunnel profile measurement (Miura et. al. 2001). A major key to a success is how to obtain good imaging configurations.

It is easy in dimensional inspection in industrial applications to configure camera positions to scope entire targets from outside of the objects (convergence photography), while in tunnel measurement the available range of camera stations is limited and images must be taken from inside out. This leads to limited coverage of each image and therefore many images are required to cover the total object space, especially in the longitude direction. This constraint tends to render the measurement network weak and cause unexpected distortions into the reconstructed shape.

To overcome this difficulty the authors employs combined system with camera observations and auxiliary observations with a total station (TS) in order to strengthen networks. A set of observation equations based on collinearity equations are reformulated to include TS observations. The augmented observation equations are solved by the least squares method with inner constraints. A well-known formula for inner constraints in VM is modified to be applicable to the observation equations involving reduction of rank defects in a design matrix.

2 PRINCIPLE OF VM AND ITS SOLUTION

The principle of VM is illustrated in Figure 1. Retroreflective targets are placed on a tunnel surface and are imaged with a flash light at various camera stations in a convergent way. A target P, its image p and the lens center O must be in a straight line. This relation is called collinearity equations (ASP 1980) and are written as;

$$
\begin{aligned}
x + \Delta x + c\, \frac{m_{11}(X - X_0) + m_{12}(Y - Y_0) + m_{13}(Z - Z_0)}{m_{31}(X - X_0) + m_{32}(Y - Y_0) + m_{33}(Z - Z_0)} = 0 \\
y + \Delta y + c\, \frac{m_{21}(X - X_0) + m_{22}(Y - Y_0) + m_{23}(Z - Z_0)}{m_{31}(X - X_0) + m_{32}(Y - Y_0) + m_{33}(Z - Z_0)} = 0
\end{aligned}
\tag{1}
$$

(x, y) : target image coordinates in image plane

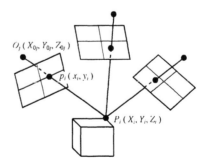

Figure 1. Principle of vision metrology.

$(\Delta x, \Delta y)$: lens distortions
c : principal distance
(X_0, Y_0, Z_0) : coordinates of camera station
(X, Y, Z) : target coordinates in the object space
(M_{ij}) : rotation matrix of the camera coordinate system associated with the object space coordinate system.

As for lens distortions, the authors use a conventional polynomial model. Because equations (1) is nonlinear about unknown parameters, they must be linearized around initial values and we obtain in a matrix form;

$$v + A_1 x_1 + A_2 x_2 + A_3 x_3 = e \qquad (2)$$

or

$$v + A x = e \qquad (3)$$

v : observation errors
x_1 : camera station coordinates and rotation angles (exterior orientation parameters)
x_2 : camera intrinsic parameters (interior orientation parameters)
x_3 : object space coordinates of targets
A_1, A_2, A_3 : respective design matrices of design matrices
e : discrepancies
$x = [\, x_1{}^T \; x_2{}^T \; x_3{}^T \,]^T$
$A = [\, A_1 \, A_2 \, A_3 \,]$.

This set of equations are solved by the iterative least squares to update approximations. If there are no absolute control points in the object space, which is not exceptional in industrial measurement, a least squares solution can not be obtained in a usual process because of rank defects in a design matrix. A solution vector x has the seven degree of freedom in the object space, i.e. three in translation, three in rotation and one in scale. To conquer this situation, additional constraints must be posed on equation (3) as;

$$B^T x = o \qquad (4)$$

with B : constraint matrix with 7 columns.

The least square solution with constraint by equation (4) is called the method of inner constraints. The matrix B is arbitrary as long as it belongs to the complementary space of that spanned by matrix A^T (Teunissen 1985). A representative choice of B is to form seven linearly independent vectors out of the basis vectors of the null space of A. Hereafter this matrix is written as G. Imposing the constraints

$$G^T x = o, \quad AG = O \qquad (5)$$

on equation (3) yields the minimum value of mean

variance of the estimated parameters x (Akimoto & Hattori, 2001). An analytical formula of G is already found out, by exploiting the special features of a design matrix of VM (Granshaw 1980). The inner constraint method is theoretically well-established and is featured by less cpu power compared with other methods like the generalized inverse matrix-based method, which uses singular value decomposition.

3 COMBINED MODEL OF VISION METROLOGY AND TOTAL STATION

In order to strengthen measurement networks in VM, some augmented observations with a total station (TS) are sometimes indispensable. Each set of observations for a target include horizontal angle α, vertical angle β and distance s (see Figure 2 and 3 for the definition of coordinate axes and angles). The observation equation for horizontal angle measurement is then written as;

$$\alpha + \alpha_0 = \tan^{-1} \frac{Z - Z_a}{X - X_a} \qquad (6)$$

α_0 : reference horizontal angle of a graduation disk of the TS
(X_a, Y_a, Z_a) : station coordinates of the TS.

The observation equation for a vertical angle measurement is expressed as (see Figure 3)

$$\frac{\pi}{2} - \beta = \tan^{-1} \frac{Y - Y_a}{\sqrt{(X - X_a)^2 + (Z - Z_a)^2}} \qquad (7)$$

And the observation equation for a distance measurement is given by

$$s = \sqrt{(X - X_a)^2 + (Y - Y_a)^2 + (Z - Z_a)^2} \qquad (8)$$

These equations must also be linearized around initial values. Thus we obtain a set of observation equations in a matrix form for observations with a TS as

$$v_a + H_3 x_3 + H_4 x_4 = e_a \qquad (9)$$

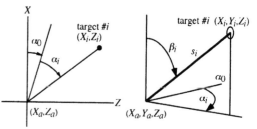

Figure 2. Horizontal angle.　　　Figure 3. Vertical angle.

v_a : observation errors
x_3 : target coordinates
x_4 : α_0 and TS position
e_a : discrepancies
H_3, H_4 : design matrices.

Incorporating equation (9) with equation (2) we obtain an extended set of equations for observations with a camera and a TS together as

$$\begin{bmatrix} v \\ v_a \end{bmatrix} + \begin{bmatrix} A_1 & A_2 & A_3 & O \\ O & O & H_3 & H_4 \end{bmatrix} \begin{bmatrix} x_1 \\ x_2 \\ x_3 \\ x_4 \end{bmatrix} = \begin{bmatrix} e \\ e_a \end{bmatrix} \quad (10)$$

The rank defects of the design matrix of equation (10) decrease from 7 to 4 because rotation around the X axis and Z axis and the scale of the space were fixed by TS observations. Least squares solutions of equation (10) can be calculated by the inner constraint the same way as in usual VM with a constraint matrix G, which is orthogonal to the design matrix of equation (10). This is obtained by removing three columns from equation (5) corresponding to the fixed axes and the scale and adding elements correspond to the matrix H_4 (Akimoto & Hattori, 2001).

4 EXPERIMENTAL RESULTS

An experiment for combined measurement system with a camera and a TS was carried out at a tunneling site, 10m in diameter and 20m in longitude. 63 retroreflective targets and nine back-to-back targets were placed on the lattice of the tunnel surface as shown in Figure 4. Back-to-back targets (see Figure 5) are ones observable both from front and rear. These targets were observed by both a camera and a TS. With a TS all the nine back-to-back targets were observed at three stations so that the targets were sighted at least from two stations. Besides 26 retro targets were observed with a TS for absolute accuracy check.

(a) Data processing
Target images are centralized to measure image coordinates. Figure 6 shows a target image sample. The centralization is executed in this way: An image area including a target is clipped out and divided into bright and dark subareas with an appropriate threshold. The centroid of the bright area is calculated as an image coordinate of a target image.

Images were connected to each other to form a unified model starting by giving approximations of object coordinates of six points in the first image. And approximations of all the object coordinates of targets and other parameters including camera station coordinate, reference horizontal angles of the graduation disk of a TS are yielded. Then iterative least squares adjustment is executed to determine precise parameter values.

Figure 5. Back-to-back target.

Figure 6. Sample of target image.

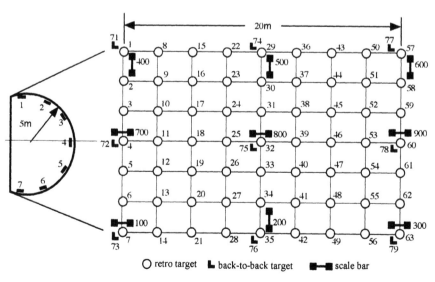

Figure 4. Target arrangement.

291

(b) Adjustment result with no scale bars included

Adjustment results are summarized in Table 1. Precision σ in the table means the standard deviation of the most probable values of unknown parameters x. The mean value of the precision of X, Y and Z coordinates of targets was 0.5141 mm. Table 2 shows the results of absolute accuracy check with the direct observations with a TS. In the table σ stands for the root mean square of differences between the adjusted coordinates and the corresponding values with a TS. The mean of three RMS differences in X, Y and Z was 0.9450 mm. Figure 7 shows the distributions of the differences.

(c) Adjustment result with only scale bars

Adjustment is executed with image coordinates and distance observations for nine scale bars, excluding TS observations. This is to verify the effectiveness of the VM-TS measurement system. The absolute accuracy comparison was also shown in Table 2. The mean RMS difference in VM with only scale bars was 1.4274mm, which is significantly larger than the value in the combined system.

(d) Discussions

According the experiment the VM-TS integration system has improved the accuracy significantly. At the experimental site a ventilation pipe was fixed on a ceiling of the tunnel, which highly hindered the photographing operation. And images covering the ceiling are inclined to be separated into two groups, each

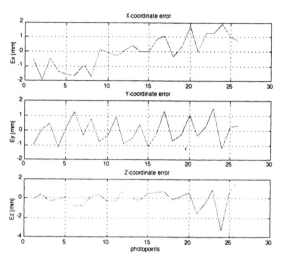

Figure 7. Distribution of the difference beween the target coordinates obtained from combined model and that from total station.

of which shared a half of the ceiling and poorly connected to each other group. Consequently the photogrammetric network became weak. As the result the shape of the tunnel surface reconstructed only by VM plus short scales was estimated to contain nontrivial distortions.

5 CONCLUSION

A VM system combined with a total station was developed for high accuracy tunnel surface measurement. And it was verified by an experiment that this system is very useful to strengthen measurement networks. This flexible system can be applied to construction sites where many factors degrading the imaging circumstances exist.

Table 1. Calculation condition and results.

Number of images	183
Number of target images	2403
Number of unknowns	1370
Iteration times	6
CPU time for 1 iteration	26s (450MHz PC)
Standard dev. in image plane	0.00138mm
Precision σ_X	0.6709mm
σ_Y	0.4403mm
σ_Z	0.3860mm
σ_{XYZ}	0.5141mm

Table 2. RMS difference of object coordinates of targets between by the combined system and the VM with scale bars.

		[mm]
	Combined system	VM with scale bars
s_X	1.078	1.962
s_Y	0.822	1.123
s_Z	0.945	1.001
s_{XYZ}	0.945	1.427

REFERENCES

Akimoto, K. & S. Hattori 2001. Revisit to zero order design in industrial vision metrology. *Proceedings of ASPRS Annual Conference, St.Louis*, Apr. 2001.

American Society of Photogrammetry 1980. *Manual of photogrammetry, fourth edition*. American Society of Photogrammetry.

Granshaw, S.I. 1980. Bundle adjustment methods in engineering photogrammetry. *Photogrammetric Record*, 10(56) : 181-207.

Miura, S., S. Hattori, K. Akimoto & Y. Ohnishi 2001. Tunnel convergence measurement using vision metrology. *IS-Kyoto 2001*.

Teunissen, P. 1985. Zero order design. In E.W.Grafarend & F.Sanso (eds), *Optimization and design of geodetic networks* :11-55. Springer.

Modern Tunneling Science and Technology, Adachi et al (eds), © 2001 Swets & Zeitlinger, ISBN 90 2651 860 9

Assessment of grouting effect in rockmass using borehole jack tests

G. Zhang , J. Chen & S. Wang
Wuhan Institute of Rock and Soil Mechanics, the Chinese Academy of Sciences, Wuhan 430071, China

Y. Li
Tianjin Investigation, Design and Research Institute of Water Resources & Hydropower, Ministry of Water Resources, Tianjin 300222, China

ABSTRACT: The effect of consolidation grouting is checked and evaluated on the high pressure tunnel of the underground pumping station of a water diversion project in China using the borehole elastic modulus test (BEMT), the results have shown that the grouting art used can not only have functions to increase the deformation parameter index of the surrounding rockmass but also effectively improve the anisotropic behavior of the rockmass. In addition, the checking result provides scientific basis for giving priority to choose a consolidation grouting art. The comparison of BEMT with the acoustic sounding method indicates that as another measure for choice, BEMT is effective for checking and evaluating the effect of consolidation grouting and gives the elastic modulus of rockmass that is of paramount importance and can not be obtained from the acoustic wave method. BEMT technique is the one worthy to be furthered.

1 INTRODUCTION

The secondary pumping station in the transmission line of a water diversion project in China is large underground works. The designed delivery head of the station measures 142 meters high and the outlet distributary and the arterial pipe are all subjected to a high compression. The high pressure tunnel will be lined with reinforced concrete, so its surrounding rockmass calls for a high capacity in both impermeability and strength. For this reason, the consolidation grouting is carried out. In order to prove the feasibility of cement grouting measure in technology that increases the strength and impermeability of the rockmass surrounding the high pressure tunnel lined with reinforced concrete and to select rational technological parameters, the consolidation grouting test has been carried out. The grouting effect has been evaluated quantitatively using such methods as hydraulic pressure test, geophysical sonic sounding and borehole elastic modulus test, thus providing scientific basis for both design and construction of grouting works.

The borehole elastic modulus testing (BEMT) has not been widely used in China, especially, seldom introduced to check the grouting effect on the surrounding rockmass. To counter this situation, the present paper lays emphasis on the description of BEMT results and makes a comparison of BEMT with the geophysical sonic sounding so as to popularize this advanced testing method in China.

2 CONSOLIDATION GROUTING TESTING

The consolidation grouting test has been made in the testing pumping station in the general transmission line. The testing tunnel is 24 m long and round in cross sectional shape, having an excavation diameter of 4.0 m and a final diameter of 3.0 m after lining. The tunnel axis is parallel to the outlet pressure tunnel.

The stratum outcrops in the testing tunnel are the third($\in_3 f^3$) and the second ($\in_3 f^2$) segment of Fengshan Group belonging to ante series of Cambrian system and the exposed stratum in boreholes is the first underlying segment ($\in_3 f^1$) of Fengshan Group. The characters of the strata are as below:

$\in_3 f^3$:limestone; dark grey marlite, having a structure of microlite.

$\in_3 f^2$:grey limestone; dark grey marlite (microlite in structure).

$\in_3 f^1$: in the upward: medium-thick layers of limestone and in the downward: thin layers of limestone sandwiched with marlite; grey limestone; marlite dark-grey.

The testing tunnel for grouting is divided into Regions A and B (Fig.1), measuring 8 m and 6 m long respectively. The grouting boreholes are arranged in rings, each ring having 8 holes at an equal distance and two neighboring holes meeting each other with 45°. Five rings are arranged in Region A, having a distance of 2 meters between them and five in Region B having distance of 1.5 meters.The alternate angle between the corresponding holes at two neighboring rings is 22.5 ° The technological parameters of the grouting tests in the two Regions are the same with the exception of different ring to ring distances, the detailed description is omitted here.

3 BEMT MEASUREMENTS

3.1 Principle and equipment

The principle of BEMT is that the piston, steel wedge or small flexible pillow in a borehole jack are used to apply a pair of uniform distributed forces on the borehole wall and the corresponding deformation of the borehole diameter is recorded simultaneously and then the deformation or elastic modulus of the rockmass is calculated in the light of the force -deformation relation(Goodman et al 1968):

$$E = A \cdot H \cdot T^*(\gamma, \beta) \cdot d \cdot \frac{\Delta p}{\Delta d} \qquad (3.1)$$

in which

E — deformation or elastic modulus of rockmass

A — modified coefficient relating the length diameter ratio of borehole jack(Davydora 1968)

H — modified coefficient for pressures that is dependent upon the way of measuring pressures

$T^*(\gamma, \beta)$ — modified coefficient relating Poisson's ratio and half contact angle β (Hustrulid 1976)

2β — contact angle, i.e. centering angle corresponding to the pressed arc segment of the borehole wall

d — borehole diameter

Δp — increment of pressure on force bearing surface

Δd -increment of borehole diameter d corresponding to Δp

The typical equipment used for BEMT is Goodman Jack(Goodman et al 1968) that is recommended by International Society of Rock Mechanics(ISRM). Unfortunately, the defects in Goodman Jack's structure lead to a measured elastic modulus of rock mass by far lower than the true value, so modification is needed(Heuze et al 1976). To counter this situation, Wuhan Institute of Rock and Soil Mechanics, the Chinese Academy of Sciences has developed BJ series of borehole elastic modulus apparatus(Li et al 1991). This new type of apparatus can directly determine the true elastic modulus of rock mass without further modification. The calibration test and the in-situ test show an error of only 5～10%(Li et al 1991,Zhang et al 2000).

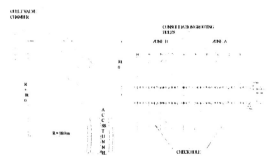

Figure 1. Layout of test tunnel.

The equipment used for our tests is BJ-130 borehole jack in BJ series.BJ-130 has the parameters ofA=0.915(Davydora,1968), H=1.0, $T^*(\gamma, \beta)$= $T^*(0.25, 22.5°)=2.141$(Hustrulid, 1976),d=130mm, making a reduced eq(1):

$$E = 254.7 \frac{\Delta p}{\Delta d} \qquad (3.2)$$

3.2 Layout of testing

Altogether six boreholes have been arranged before and after grouting, having total 90 measuring points to conduct BEMT measurement. Of these six boreholes, two (T1 and T2) were arranged in Region A before grouting and two (TB1 and TB2) were in Region B. 10 points were arranged in each of borehole of T1, TA1 and TB1 that only contain one stratum of $\in_3 f^3$, five being parallel to the stratum plane and the other perpendicular; whereas in each of boreholes of T2, TA2 and TB2 that contain two strata of $\in_3 f^2$ and $\in_3 f^1$, 20 points were arranged, each stratum having 10 points and similarly, half of the 20 parallel and the rest perpendicular to the stratum plane.

3.3 Loading way

A stepwise single cyclic load-unload way was used with pressure-deformations curve shown in Fig.2.

The loading curve from the final cyclic is used to calculate the elastic modulus of the rock mass. All in situ measured curves from each point are characterized by gentle in the downward and steeple in the upward. This is due to gradual compaction of fissures in the early loading stage and to gradual complete contact between the bearing platens of the borehole jack and the borehole wall. When the bearing platen bears a compression force greater than 10Mpa, the deformation curve becomes approximately linear with slope constant basically. In making data processing and calculating elastic modulus, the initial compressive interval in the

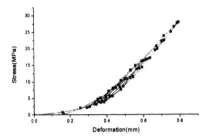

Figure 2. Relationship between borehole deformation and applied   pressure.

in situ measured curve that is not obvious is neglected and only the compressive segment that is over 10 Mpa   is considered.

4   TESTING RESULTS AND ANALYSES

Listed in Tables 1 and 2 are the *in situ* measured results of BEMT measurements before and after consolidation grouting in the testing tunnel of the secondary pumping station. It can been seen in the

Table 1. Measured elastic modulus before grouting.

Borehole	strata		elastic modulus		(Gpa)		
T1	$\epsilon_3 f^3$	depth (m)	4.0	5.5	6.5	8.0	9.0
		parallel	18.37	26.61	24.78	23.4	23.41
		Vertical	19.58	20.76	18.50	17.82	19.14
	$\epsilon_3 f^2$	depth (m)	4.0	5.0	6.0	7.0	8.0
		parallel	22.52	24.23	23.77	20.01	17.20
T2		vertical	19.42	20.51	17.21	18.17	19.38
	$\epsilon_3 f^1$	depth (m)	8.8	9.2	9.6	10.0	10.4
		parallel	22.71	19.69	25.91	22.06	13.83
		vertical	23.00	14.75	24.33	16.07	13.85

Table 2. Measured elastic modulus after grouting.

Borehole	strata		elastic modulus (Gpa)				
TA1	$\epsilon_3 f^3$	depth (m)	2.0	3.0	4.0	5.0	6.0
		parallel	20.36	19.49	25.49	29.57	21.86
		vertical	19.00	15.82	23.25	17.47	19.85
TA2	$\epsilon_3 f^2$	depth(m)	3.0	4.2	6.0	8.0	12.0
		parallel	25.64	21.90	19.98	23.58	18.21
		vertical	24.5	32.19	18.55	21.65	16.03
TA1	$\epsilon_3 f^1$	depth(m)	13.6	14.1	14.6	15.1	15.5
		parallel	19.14	28.40	22.81	17.08	21.50
		vertical	20.35	19.32	21.42	21.94	17.12
TB1	$\epsilon_3 f^3$	depth (m)	2.0	4.0	6.0	8.2	9.5
		parallel	23.07	23.21	27.14	23.16	23.49
		vertical	22.90	24.55	27.43	24.50	25.96
TB1	$\epsilon_3 f^2$	depth(m)	2.0	4.0	7.5	9.0	12.0
		parallel	24.95	29.81	20.15	25.09	22.71
		vertical	22.84	32.10	26.89	25.45	18.43
TB1	$\epsilon_3 f^1$	depth(m)	13.92	14.3	14.7	15.1	15.5
		parallel	21.63	25.74	25.85	22.59	19.35
		vertical	22.39	24.32	22.78	24.21	19.73

whole that the measured values exhibit a certain discrete character. For the sake of convenience in analyzing, the statistical analyses have been performed on the measured values, which gives the mean value and the mean quadratic error of the measured values as tabulated in Table 3. Table 3 also lists the inicreasing percentage η in the elastic modulus of each stratum due to grouting.

It can be seen from Table 3 that the consolidation grouting has a certain effects on increasing the elastic modulus of the surrounding rockmass. With the exception of isolated cases,the elastic modulus of each stratum has been increased to a certain extent due to grouting, and relatively, the increasing effect of the moduli of strata $\epsilon_3 f^2$ and $\epsilon_3 f^1$ is more remarkable than that in $\epsilon_3 f^3$. The comparison analysis on the elastic modulus of each stratum in Regions A and B makes the conclusion that as concern the technological arts in the two Regions, the one introduced in Region B is considerably better in increasing the elastic modulus of the rockmass than that in Region A. Generally speaking, Compared with the elastic module in grouting in Region A, those with grouting has increments of parallel and vertical directions of the rockmass without 2.29% and 9.B respectively and in Region B, the corresponding increments are 9.30% and 20.05% respectively.

Table 3.Elastic modulus in statistics before and after grouting.

Stratun		before grouting	After grouting			
			Region A		Region B	
		M±σ (Gpa)	M±σ (Gpa)	η(%)	M±σ (Gpa)	η(%)
$\epsilon_3 f^3$	parallel	23.11±2.94	23.35±3.97	1.04	24.01±1.67	3.89
	Vertical	19.16±1.06	19.07±2.67	-0.47	25.07±1.63	30.89
$\epsilon_3 f^2$	parallel	21.55±2.97	21.86±2.78	1.44	24.54±3.40	13.87
	vertical	18.93±1.21	22.54±5.94	19.07	25.14±4.81	32.81
$\epsilon_3 f^{11}$ 10.51	parallel	20.84±4.29	21.78±4.10	4.51	23.03±2.65	
	vertical	18.40±4.66	20.03±1.82	8.86	22.69±1.77	23.32
general	parallel	21.83	22.33	2.29	23.86	9.30
	vertical	18.83	20.55	9.13	24.30	29.05

The testing results also show that the consolidation grouting has a better effect in improving the anisotropy of the rockmass. Before grouting, the elastic modulus in each direction parallel to the stratum is higher than that in each vertical direction, in general these two has a relative difference of 15.93%, showing an obvious anisotropic character. After grouting however, either in Region A or in Region B, the elastic modulus of each stratum along the direction parallel to the stratum is increased much more greatly than that in the direction perpendicular to the stratum.   difference between the two moduli becoming smaller. Generally speaking, the relative difference of the two after grouting in Region A is 8.66% and 1.81% in Region B. Grouting improves the   anisotropic   situation   and   the   surrounding rockmass can be considered to be isotropic.

5 RESULTS FROM ACOUSTIC SOUNDING

In order to evaluate the effect of consolidation grouting comprehensively, acoustic sounding has been carried out on the rocks surrounding the testing tunnel before and after grouting using TYPE CE 9201 engineering checking device. The testing results are listed in Table 4.

Table 4. Sonic wave velocities in rockmass surrounding the testing tunnel before and after grouting.

Strata	number of Statistical Points	average sonic wave velocity before grouting (m/s)	number of statistical points	average sonic wave velocity after grouting (m/s)	increment percentage (%)
$\in_3 f^3$	1044	5310	866	5610	5.65
$\in_3 f^2$	403	5160	537	5470	6.01
$\in_3 f^1$	226	5070	184	5630	11.05

It can be known from Table 4 that the average wave velocities measured after grouting in strata of $\in_3 f^3$, $\in_3 f^2$ and $\in_3 f^1$ are increased by 5.65%, 6.01% and 11.05% respectively compared with those before grouting. The fact wave velocity is increased due to grouting indicates the improvement of rockmass quality. Consequently, the acoustic sounding results qualitatively lead to the conclusion that consolidation grouting does improve the surrounding rocks quality, this is the same as the conclusion from BEMT measurements. All this shows that the BEMT like acoustic sounding is also effective while used to check and evaluate the consolidation grouting effect and meanwhile gives the deformation parameter that is of paramount importance to the stability analysis on the surrounding rockmass, but the acoustic method fails to do so.

6 CONCLUSIONS

BEMT (Borehole Elastic Modulus Testing) has been used to check the effect of consolidation grouting that has been carried out in the testing tunnel of the secondary pumping station of a diversion project in China, which draws the following conclusions:

1. Consolidation grouting has a certain function to increasing the elastic modulus of country rockmass of the pumping station, which is more remarkable in strata of $\in_3 f^2$ and $\in_3 f^1$ than in $\in_3 f^3$.

2. The elastic modulus in the direction parallel to the stratum plane and that perpendicular to the plane after grouting in Region A are increased, after grouting, by 2.29% and 9.13% respectively and in Region B the corresponding increments are 9.30% and 29.05% respectively, showing that the effect of the grouting art performed in Region B is considerably better than that in Region A. It is suggested that the grouting art be used.

3. The comparison between BEMT and acoustic sounding test has shown that BEMT is an effective choice for checking and evaluating the effect of consolidation grouting and is worthy to be furthered. In the meantime, BEMT can not only increase the capability of the rockmass against deformation but also effectively improve the anisotropic behavior of the rockmass. It offers important parameters for design but the acoustic sounding test can not.

REFERENCES

Davydora, N.A. 1968. Interpretation of pressiometric tests results with consideration of loaded section final length *Proc. Int. Symp. Rock Mechanics*, Madrid

Goodman, R.E., Van, T.K. & Heuze, F.W 1968. The measurement of rock deformability in boreholes. *Proc. 0th Symp. on Rock Mechanics*. University of Texas, Austin

Heuze, F.E. & Salem, A 1976. Plate bearing and Borehole jack tests in rock — a finite element analysis, *Proc. 17th Symp. on Rock Mechanics*. Snowbird, Utah

Hustrulid, W.A. 1976. An analysis of the goodman jack. proc. *17th Symp. Rock Mechanics*. Snowbird, Utah

Li Guangyu & Zhou Baihei. 1991. A modified borehole jack [J]. *Chinese Journal of Geotechnical Engineering*, Vol.13(4):11~23

Zhang Guang, Wang Shuilin & Zhu Aimin. 2000. Measure-ment of rockmass deformation parameters for yellow river diversion project [J]. *Chinese Journal of Rock Mechanics and Engineering*, Vol.19(4):539~542

Modern Tunneling Science and Technology, Adachi et al (eds), © 2001 Swets & Zeitlinger, ISBN 90 2651 860 9

Rock stress measurement using the compact conical-ended borehole overcoring (CCBO) technique

Y. Ishiguro
Chubu Electric Power Co., Inc., Nagoya, Japan

Y. Obara & K. Sugawara
Kumamoto University, Kumamoto, Japan

ABSTRACT: In this paper, the results of rock stress recently measured in Japan are discussed. Firstly, the CCBO developed by the authors is described in brief. Then, the case examples of stress measurement for rock slope and underground cavern by the CCBO are demonstrated, then the state of stress in rock slope and the design of underground powerhouse are discussed from the rock engineering viewpoint, based on the measured stress.

1 INTRODUCTION

Rock stress is of prime importance for construction of rock structures, such as rock slope, underground opening and so on. Because that the mechanical behavior of rock mass is dominated by not only mechanical property of rock mass but also rock stress, and that the stability of rock structures strongly depends on the state of rock stress. Therefore, knowledge of rock stress is one of keys for design and stability estimation of rock structure, and it is desirable that rock stress is accurately measured.

The compact conical-ended borehole overcoring (CCBO) technique is promising measurement method of rock stress, which is developed by Sugawara & Obara (1999). This method is widely used to measure rock stress in the sites of mine, rock slope, tunnel and cavern in Japan, because that the three dimensional stress tensor can be determined from a measurement in a single borehole and that the method has an advantage to reduce the time, effort and cost for a series of rock stress measurements. Recently, this method is applied to clarify the state of initial stress and induced stress in several sites of underground powerhouse cavern, then the design and its validity are discussed, based on the measured rock stress.

In this paper, the results of rock stress recently measured by the CCBO in Japan are discussed. Firstly, the CCBO developed by authors is described in brief. Then, the case examples of rock stress measurement for rock slope and underground cavern by the CCBO are demonstrated, then the state of stress in rock slope and the design of underground powerhouse are discussed from the rock engineering viewpoint, based on the measured stress.

2 COMPACT CONICAL-ENDED BOREHOLE OVERCORING TECHNIQUE

2.1 *Field measurement system*

The field measurement procedure is illustrated in Fig.1. Firstly, a pilot borehole having a diameter of 76mm is drilled to the stress measurement station. Then, the bottom of the pilot borehole is formed into a conical shape with a bortz crown bit (Fig.2(a)), the surface of which is ground smooth with an impregnated diamond bit (Fig.2(b)). After bottom cleaning with water and acetone, the existence of cracks on the borehole surface is confirmed by a borehole camera. In the case that there is no crack on the surface, the 24 or 16 element conical strain cell shown in Fig.2(c) is directly bonded to the conical borehole bottom surface with glue. Finally, the stress around the bottom of the pilot borehole is relieved by the compact overcoring, that is a thin-walled core boring having a diameter of 76mm which coincides with that of the pilot borehole as shown by broken line in Fig.1. During this operation, the changes in strain are continuously measured and recorded by strain meter and computer. For this purpose, the cable is linked to the strain meter through the boring rods and the water swivel.

2.2 *Theory*

For estimation of the initial stress from the measured strains, the cylindrical coordinates (r,θ,z) and the spherical coordinates (ρ,θ,ϕ) are defined as well as the Cartesian coordinates (x,y,z) with the z-axis coincident with the borehole axis, as illustrated in Fig.3. The initial stress tensor $\{\sigma\}$ can be expressed as follows:

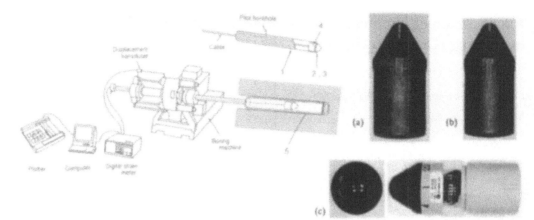

Figure 1. Field measurement system of the CCBO; (1) drilling a pilot borehole; (2) creating a conical borehole socket; (3) socket cleaning; (4) gluing the strain cell into the socket; (5) compact overcoring.

Figure 2. Special bit and strain cell; (a) bortz crown bit; (b) impregnated diamond bit; (c) 24 element strain cell.

Table 1. Strain coefficients in the isotropic case.

Poisson's ratio	A_{11}	A_{12}	A_{21}	A_{22}	A_{31}	A_{32}	A_{33}
0.10	1.002	-1.762	0.109	0.343	0.562	-0.724	-0.802
0.20	1.000	-1.752	0.022	0.365	0.519	-0.707	-0.818
0.25	0.999	-1.733	-0.021	0.373	0.496	-0.693	-0.821
0.30	0.997	-1.704	-0.065	0.380	0.474	-0.679	-0.822
0.40	0.989	-1.611	-0.154	0.386	0.426	-0.625	-0.823

Poisson's ratio	C_{11}	C_{21}	C_{31}	D_{11}	D_{21}	D_{31}	D_{32}
0.10	-0.155	0.655	0.246	0.082	1.542	0.802	-1.725
0.20	-0.263	0.641	0.185	0.095	1.627	0.860	-1.860
0.25	-0.317	0.636	0.155	0.101	1.673	0.886	-1.923
0.30	-0.371	0.632	0.126	0.108	1.716	0.911	-1.983
0.40	-0.481	0.630	0.071	0.123	1.787	0.953	-2.091

$$\{\sigma\} = \{\sigma_x, \sigma_y, \sigma_z, \tau_{yz}, \tau_{zx}, \tau_{xy}\}^T \quad (1)$$

where σ_x, σ_y, σ_z, τ_{yz}, τ_{zx} and τ_{xy} are the stress components in the Cartesian coordinates.

The strains are required to be measured at eight specified points on the conical borehole socket of radius 38mm, as shown in Fig.3. The strain measuring points are axisymmetrically arranged along a measuring circle of radius 19mm, by rotating 45° at a step. The specification of strain measuring points has been optimized through theory and experiment. In the 16 element method, the tangential strain ε_θ and the radial strain ε_ρ are measured at each strain measuring point, using a 16 element conical strain cell. The 24 element method requires the additional strain at each point, that is, the oblique strain ε_φ, as shown in Fig.3(b). Thus, the strains measured on a conical borehole socket can be denoted by

$$\{\beta\} = \{\beta_1, \beta_2, \ldots\ldots, \beta_n\}^T \quad (2)$$

where n is the number of strains; i.e. $n=16$ for the 16 element method; $n=24$ for the 24 element method.

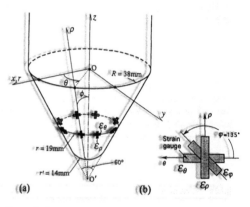

Figure 3. Arrangement of strain to be measured; (1) 16 element method; (b) strain of 24 element method.

The strains $\{\varepsilon_\theta, \varepsilon_\rho, \varepsilon_\varphi\}$ at a strain measuring point of a tangential angle θ are given, in the isotropic case, as follows:

$$\begin{Bmatrix} \varepsilon_\theta \\ \varepsilon_\rho \\ \varepsilon_\varphi \end{Bmatrix} = \begin{bmatrix} A_{11} + A_{12}\cos 2\theta, & A_{11} - A_{12}\cos 2\theta, & C_{11}, \\ A_{21} + A_{22}\cos 2\theta, & A_{21} - A_{22}\cos 2\theta, & C_{21}, \\ A_{31} + A_{32}\cos 2\theta, & A_{31} - A_{32}\cos 2\theta, & C_{31}, \end{bmatrix}$$

$$\begin{matrix} D_{11}\sin\theta, & D_{11}\cos\theta, \\ D_{21}\sin\theta, & D_{21}\cos\theta, \\ D_{31}\sin\theta - D_{32}\cos\theta, & D_{31}\cos\theta + D_{32}\sin\theta, \end{matrix}$$

$$\begin{matrix} 2A_{12}\sin 2\theta \\ 2A_{22}\sin 2\theta \\ 2A_{32}\sin 2\theta - 2A_{33}\cos 2\theta \end{matrix} \Bigg] \cdot \frac{\{\sigma\}}{E} \quad (3)$$

where E is the Young's modulus of rock and A_{11}, A_{12}, ..., D_{32} are the strain coefficients.

The values of the strain coefficients are dependent upon Poisson's ratio of the rock. They have to be

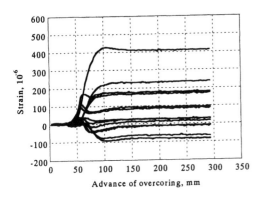

Figure 4. Changes in strain during overcoring.

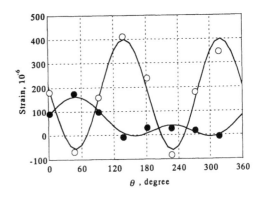

Figure 5. Comparing of measured strain with theoretical value.

evaluated by numerical analysis, since there is no analytical solution. The strain coefficients of the isotropic case computed by the BEM analysis are summarized in Tab.1.

Observation equation of the initial stress tensor $\{\sigma\}$ is expressed by the following matrix equation:

$$[A]\{\sigma\} = E\{\beta\} \qquad (4)$$

where $[A]$ is an n by 6 elastic compliance matrix. The elements of $[A]$ are computed by substituting the tangential angle θ of each strain measuring point in equation (3).

The most probable values of the initial stress components are determined by the least square method, providing the normalized expression of equation (4) as follows:

$$[B]\{\sigma\} = E\{\beta^{*}\} \qquad (5)$$

where $[B] = [A]^{T}[A]$ and $\{\beta^{*}\} = [A]^{T}\{\beta\}$. The most probable values of the initial stress $\{\sigma^{*}\}$ can be expressed as:

$$\{\sigma^{*}\} = E[C]\{\beta^{*}\} \qquad (6)$$

where $[C]$ is the inverse matrix of $[B]$.

2.3 Strain change during overcoring

Fig.4 shows a series of changes in strain on the conical bottom surface monitored by the 16 element conical strain cell. The lateral axis is the advance of overcoring. The changes in strain are rapid in all cases after the overcoring passed through the section of the strain measuring circle (about 50mm). From the convergence of strains, the elastic strains on the bottom surface which existed prior to the overcoring are estimated, and the most probable stress tensor is calculated from them according to the mathematical procedure as mentioned above.

The theoretical distributions of strain on the measuring circle, corresponding to the stress tensor

determined, are shown in Fig.5 by solid curves. These curves are agreement well with the measured strains plotted. It is concluded that the measurement is performed successfully.

3 CASE EXAMPLES

3.1 Rock slope

Mt. Torigata lies in the western part of Kochi prefecture in Japan as shown in Fig.6. It is consisted of limestone body, namely Torigara limestone deposit. The Torigata Limestone Mine is an open pit mine located at the top of Mt. Torigata at 1459m above sea level. The excavation in the Torigata Limestone Mine was started in 1971. The excavation level is 1220-1235m at present, and the scale of the open pit is 2.5km in the east-west direction and 1.0km in the north-south direction.

The rock stress measurement by the CCBO method was performed at the gallery that is excavated collected at 960m above sea level, from 250m depths beneath the excavation level at the location No.1 and 2, shown in Fig.6 (Nakamura et al. 1990; Obara et al. 2000).

The rock stress measurements were carried out in two boreholes drilled from the gallery excavated at 960m above sea level. The borehole at the location No.1 is drilled within limestone and borehole at

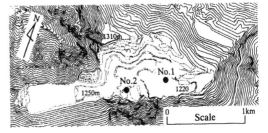

Figure 6. Plane view of Torigata limestone mine and location of rock stress measurement.

the location No.2 is within crystalline limestone. The thickness of the overburden is about 240-255m. Therefore, the overburden pressure is estimated to be 6.3 to 6.7MPa, assuming the unit weight of rock is 26.2kN/m^3.

The measurement results are summarized in Tab.2. The results for each borehole and average values are described in the table, and the principal directions are plotted on the lower hemispherical stereographic projection shown in Fig.7. The Young's modulus and the Poisson's ratio are 63GPa and 0.28 at location No.1, and 50GPa and 0.28 at location No.2.

The vertical stress component σ_z at location No.1 is larger than the estimated overburden pressure, while that at location No.2 is smaller than the estimated overburden pressure. This means that the rock near location No.2 is crystalline limestone and

that the Young's modulus is smaller than that at location No.1. However, the average value at location No.1 and 2 is in reasonable agreement with the estimated overburden pressure.

The horizontal normal stress components σ_x and σ_y are relatively large and 80 to 90 % the value of σ_z. Since the rock mass near location No.2 consists of crystalline limestone, the state of stress is considered to become hydrostatic. However, it is clear that large horizontal stress exists in the mountain.

The maximum principal direction of both results are 30 degrees from north to east with a 40 ~ 45 degree inclination. It is considered that the maximum principal direction is dominant because the directions from both results coincide each other. The intermediate and the minimum principal directions exist in a plane, although the directions of these results are different.

Table 2. Rock stress measured by the CCBO.

Location		No.1	No.2	Average
	σ_x	5.0	5.1	5.0
Stress component	σ_y	7.2	4.7	6.0
in MPa	σ_z	7.7	5.2	6.4
	τ_{yz}	-1.8	-2.5	-2.2
	τ_{zx}	-1.0	-1.0	-1.0
	τ_{xy}	2.0	0.5	1.3
Principal stress in MPa and principal direction (azimuth/dip) in degrees	σ_1	10.1 30/40	7.7 29/45	9.0 30/34
	σ_2	6.0 212/50	4.7 288/11	4.5 263/33
	σ_3	3.8 121/1	2.4 187/43	3.9 152/29

Remarks: The x-axis is the direction of East,
the y-axis is North and the z-axis is vertical.

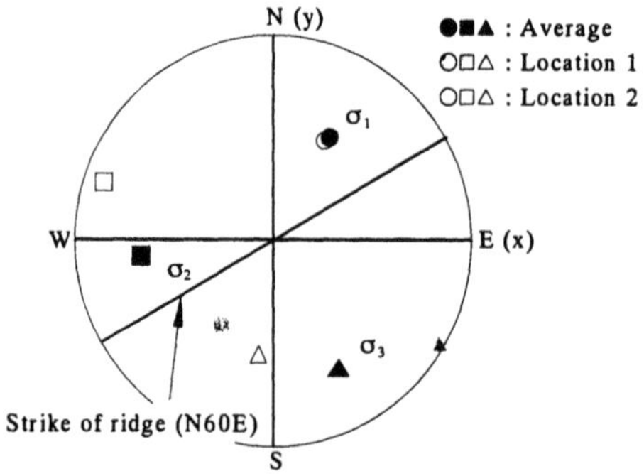

Figure 7. Principal directions at location I and II, and ridge of the mountain.

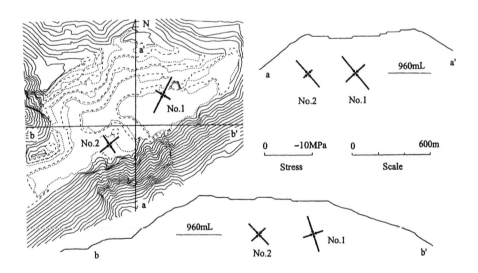

Figure 8. Magnitude and direction of principal stress on the horizontal and two vertical planes trending NS and EW.

The principal magnitudes and directions are plotted in the horizontal and the vertical cross sections shown in Fig.8. The location of the cross sections is shown in plan view. Here, the maximum principal directions are oriented NE-SW and are in agreement with the orientation of the mountain ridge at N60°W. The maximum principal stress generally exists along the ground surface and the minimum principal stress is normal near the surface of the mountain (Goodman, 1980). According to the vertical sections, however, the maximum principal direction of No.2 is oriented normal to the free surface in this field. Furthermore, the maximum principal direction at location No1 where is at the center of the mountain is steep. Therefore, it is concluded that the state of stress in this field is influenced by the regional stress due to tectonic activities.

3.2 Underground powerhouse

A pure pumped storage power plant with a maximum capacity of 1300MW is planned in an underground powerhouse. The plant consists of four units of reversible pump-turbine with a capacity 325MW and two units of main transformer with a capacity 740MVA. Therefore, the dimension of the power-

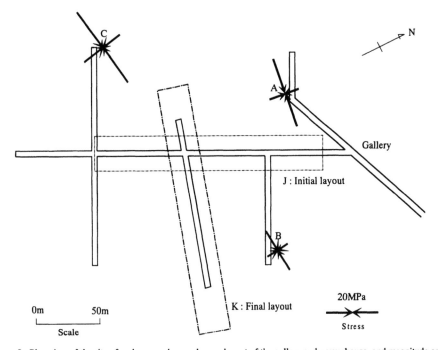

Figure 9. Plan view of the site of underground powerhouse, layout of the gallery and powerhouse, and magnitude and direction of principal stresses.

Table 3. Rock stress measure by the CCBO.

Location		A	B	C
	σ_x	25.5	16.5	32.5
Stress component	σ_y	11.9	9.3	17.6
in MPa	σ_z	8.1	14.0	10.1
	τ_{yz}	-1.0	-1.0	4.1
	τ_{zx}	-6.4	-6.7	-7.1
	τ_{xy}	-1.3	1.1	3.5
Principal stress	σ_1	27.6	22.3	34.8
in MPa and		94/18	82/40	82/15
principal direction	σ_2	12.1	9.1	19.4
(azimuth/dip)		359/13	350/2	179/24
in degrees	σ_3	5.7	8.4	6.0
		235/68	258/50	324/61

Remarks: The x-axis is the direction of East,
the y-axis is North and the z-axis is vertical.

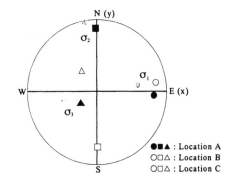

Figure 10. Principal directions on the lower hemispherical stereographic projection.

house becomes 182m in length, 24m in width and 47m in height.

The geology consists of granite, which is of the Cretaceous period of the Mesozonic era. A sound and intact granite is widely distributed with a low crack density and wide spacing. The powerhouse is located at a depth of about 570m below the surface.

The gallery for investigation is excavated at 21m over the powerhouse parallel and normal to the longitudinal axis of it. The initial layout of the powerhouse is drawn as the dotted line J in Fig.9, with consideration of the efficiency of a whole conduit system such as penstock and draft tunnel. In order to confirm the initial design for the direction of the longitudinal axis of the powerhouse (N26^{0}E), the rock stress measurements by the CCBO method were performed in three boreholes drilled from the gallery as shown in the figure.

The measurement results are summarized in Tab.3. The results for each borehole are described in the table, and the principal directions are plotted on the lower hemispherical stereographic projection shown in Fig.10. These results are almost the same measured by another methods (Ishiguro et al. 1997). The vertical stress component σ_z at all locations is smaller than the estimated overburden pressure 15MPa. The horizontal stress component σ_x is largest in the stress components at each location. It is noted that the large stress exists in the direction of East-West in the horizontal plane.

The principal magnitudes and directions are plotted as the arrows facing each other in the horizontal plane as shown in Fig.9. The state of horizontal stress at each location has a same tendency. That is, the direction of maximum horizontal is East-West, though their magnitudes are different. This fact means that there is no geological discontinuity to change the direction of maximum horizontal stress in this field. It is concluded that the state of stress in this field is uniform.

The maximum horizontal stress is acted from the direction of about 60 degrees against the side wall of the powerhouse. Therefore, the maximum depth of loosening region within the rock mass at tailrace side and penstock side were estimated 19m and 13m respectively by an elasto-plastic 2D FEM analysis. Considering this result and the direction of predominant joint set, the longitudinal axis of the powerhouse is changed from N26^{0}E to N105^{0}E as the dotted line K in Fig.9 (Ishiguro et al. 1999). As the result, the maximum horizontal stress is acted almost parallel to the longitudinal axis of the powerhouse, and the depth of loosening region can be reduced from 19m to 6m at tailrace side and 13m to 4m at penstock side, by the 2D FEM analysis.

At present, this powerhouse is in a stage of completion of the design. Since the construction of the powerhouse will be started in near future, the systematic monitoring for the behavior of the surrounding rock mass will be performed to verify the validity of its design.

4 CONCLUSIONS

Two case examples of rock stress measurement for rock slope and underground cavern by the CCBO, which is developed by the authors, were demonstrated, then the states of stress are discussed from the rock engineering viewpoint, based on the measured stress in this paper.

After describing the fundamentals of the CCBO in brief, the CCBO was firstly applied to the rock slope. Based on the results of rock stress measurement, it was clarified that the horizontal stress components exist, nevertheless the rock slope is located near the top of mountain and that the state of stress in this filed is influenced by the regional stress due to tectonic activities.

The rock stress measurement for the design of the underground powerhouse was performed by CCBO. It is concluded that the state of stress in this field is uniform. Furthermore, it is made clear that the loosening region at both side walls of the powerhouse is estimated to be large by an 2D-FEM elasto-plastic analysis under the boundary condition of the measured initial stress, in the case of the initial layout of it. Consequently, the longitudinal axis of the powerhouse was changed and determined as the maximum horizontal stress is acted almost parallel to it, considering initial rock stress and the direction of predominant joint set.

REFERENCES

Goodman, R.E. 1980. Introduction to Rock Mechanics. New York, John Willy & Sons.
Ishiguro, Y., H. Nishimura, K. Nishino & K. Sugawara 1997. Rock stress measurement for design of underground powerhouse and considerations. *Proc. Int. Conf. on Rock Stress, Kumamoto* : 491-498. Rotterdam, Balkema.
Ishiguro, Y., K. Nishino, A. Murakami, K. Sugawara & T. Kawamoto 1999. In-situ initial rock stress measurement and design of deep underground powerhouse cavern. *Proc. 9th Int. Cong. on Rock Mech., Paris* : 1155-1158. Rotterdam, Balkema.
Nakamura, N., R. Ohkubo, Y. Obara, S.S. Kang, K. Sugawara, K. Kaneko. 1999. Rock stress measurement for limestone open pit mine. *Proc. of 5th Int. Symp. on Field Measurements in Geomech.* : 375-380. Rotterdam, Balkema.
Obara, Y., N. Nakamura, S.S. Kang & K. Kaneko 2000. Measurement of local stress and estimation of regional stress associated with stability assessment of an open-pit rock slope. In: *Int. J. Rock Mech. Min. Sci.* 37: 1211-1221.
Sugawara, K. & Y. Obara 1999. Draft ISRM suggested method for in situ stress measurement using the compact conical-ended borehole overcoring (CCBO) technique. In: *Int. J. Rock Mech. Min. Sci.* 36: 307-322.

Modern Tunneling Science and Technology, Adachi et al (eds), © 2001 Swets & Zeitlinger, ISBN 90 2651 860 9

Blast Vibration Monitoring and Control of Twin Tunnels with Small Spacing

Wu Congshi & Liu Houxiang

Changsha Communications University, Changsha, Hunan 410076 P.R.. China

ABSTRACT: In an excavating project of highway twin tunnels which were separated by a space of 13.5 meters, the blasting vibration field in the existing tunnel was studied. The vibration was induced by the adjacent tunnel blasting. The numerical simulation of the vibration field was conducted. The Peak Particle Velocity (PPV) and the acceleration of blasting vibration were measured in the existing tunnel. The results of simulation and measurement showed that the maximum vibration level appeared at the wall and the arch fronting towards explosion source and that the minimum level was at the other side wall. Based on the data of site tests, the relation between the PPV and the initiation sequence used in cut blast holes was discussed and the PPV attenuation rules were proposed, which were used as a reference material for guiding the tunnel blasting. During the construction, the PPV of blasting vibration caused by adjacent tunneling blasting was monitored in the existing tunnel and was not permitted to exceed 6cm/s. Practice showed that these measures were successful in ensuring safety of the existing tunnel.

KEY WORDS: blast vibration, twin tunnels, blasting safety

1 INTRODUCTION

In some excavating projects of twin tunnels, the drill-and-blast method has to be used, owing to the rock strength and some other factors. The vibration effect of the tunnel blasting on the neighboring tunnel is an important problem that the tunnellers must pay close attention to.

This paper presents an excavation project of highway twin mountain tunnels which is situated in Guangdong Province, in the south of China .The project was divided into two stages. In stage 1, the north tunnel of the twins was constructed 8 years ago and was running. In stage 2, the south tunnel was to be constructed, which was parallel to the north one and had a 2437.5 m long. The two parallel tunnels, each with 11.3m wide, were driven in weathered granite with a 25.5m distance between centers. The thickness of rock between the two tunnels was about 13.5m. In the north one, there was a reinforced concrete ceiling (see Fig.1). The segment above the ceiling was an exhaust duct.

During the construction of the south tunnel, the north one must continue to be opened for traffic. So it is important how to minimize the vibration level induced by adjacent tunnel blasting to ensure safety in the running tunnel.

2 GENERAL SITUATION OF BLASTING

The south tunnel was excavated by full-face round blast method in feebly weathering intensity granite and by heading and horizontal bench blast method in severely weathering intensity granite. A parallel-hole cut and a v-type cut were employed respectively. The millisecond delays detonators that had 50 ms time interval between successive detonations were used. In the early days of excavation, the normal delay detonators were adopted and the time interval had a variation of 15ms to 150ms. The maximum charge per delay was between 7.8kg and 38.0kg. The depth of the round was about 3m .

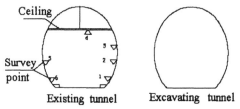

Figure 1. location of sensors in existing tunnel.

3 NUMERICAL SIMULATION AND TRIAL BLASTING

The numerical simulation in two dimensions of the blast vibration field in north tunnel was conduced with ADINA software. In general, the cut holes would produce higher ground vibration than other holes in tunneling blasting, therefore the cut hole explosion loading was used as the dynamic source in the simulation, which had a simplified triangular pressure history. The elastic modulus of the surround rock in simulation was 25 GPa, the density was 2650 kg/m^3 and Poisson ratio was 0.25. The distribution of the Peak Particle Velocity (PPV) at the perimeter of the north tunnel was that the maximum value occurred at the wall of concrete liner close to the blasting tunnel, the minimum value occurred at the other side wall far from the blasting tunnel, the radial quantity of PPV was large than the tangential one (see Fig 2). The result of simulation proved that the blast vibration would not affect the stability of the north tunnel. But it was difficult to estimate the behavior of concrete liner and ceiling in the existing tunnel owing to close blasting.

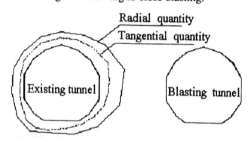

Figure 2. Distribution of PPV in the existing tunnel based on simulation.

In the initial 50m of excavating tunnel, a series of trial blasts were carried out. At the moment of blasts, vibration monitoring in the existing tunnel was taken, by positioning a few geophones and accelerometers at the surface of concrete liner and ceiling (see Fig 1). In the existing tunnel, the cross section of installing the sensors would be placed side by side with the advancement face of the driving tunnel. The macroscopic observation at the concrete liner and ceiling was conducted before and after blasting to determine whether new cracks had been produced by blasting vibration.

The results of monitoring indicated that the PPV at the concrete liner wall close to the blasting tunnel varied from 1.82cm/s to 7.61cm/s (radial quantity) and from 1.05cm/s to 4.26cm/s (tangential quantity). The maximum value of PPV occurred at the survey point 4 of the concrete ceiling (see Fig. 1), and the following value was at the survey point 3 of concrete liner. The minimum value was at the survey point 6.The acceleration had a similar phenomenon

to the PPV. The value of PPV and acceleration at the arch top was not measured due to the separation of the reinforced concrete ceiling, but it could not exceed the value of the concrete ceiling. No new cracks were found at the concrete liner and the ceiling after blasting. All of these survey phenomena were in agreement with those of the numerical simulation. No problems were experienced in the north tunnel when the south tunnel was blasting.

4 LIMIT OF BLAST VIBRATION

After a certain number of trial blasts, it had been ascertained that no new cracks had occurred and the old cracks had not expanded, and then a vibration velocity limit of 6cm/s was drawn up for the concrete liner of the existing tunnel. The limit seemed to be conservative, but the risk of the liner damage always existed during excavation. In drawing the limit, following factors had to be considered:

- The north tunnel had worked for eight years and the concrete liner and ceiling were corroded by waste gases from motor vehicle and by underground water to some extent .
- There were some micro-cracks in the concrete liner, and underground water had strained through the liner cracks.
- During blast operation, the traffic in the existing tunnel could not stop.
- The limit number of effective delays in the ranges of available detonators, the opportunity to increase charge mass per delay and to increase vibration level.

5 THE VARIATION LAW OF PPV WITH SCALED CHARGE

During succeeding blasting, in order to obtain the variation law of the PPV as a function of the scaled charge and to keep the vibration level of the blasting within control, vibration were measured in the tunnel lining along the longitudinal axis at various distances from the excavation face, every cross section of measurement had one survey point where the vibration value was the maximum in the concrete lining surrey points (see Fig 2) according to the trial blasts.

These data, added to that of trial blasts, allowed the extrapolation of the law:

$$V = 176.46 * \left(\frac{\sqrt[3]{Q}}{R} \right)^{1.76} \quad \text{(radial quantity)} \quad (1)$$

and

$$V = 78.00 * \left(\frac{\sqrt[3]{Q}}{R} \right)^{1.50} \quad \text{(tangential quantity)} \quad (2)$$

where V=vibration velocity, in cm/s; Q=weight of the maximum charge per delay, in kg; R=distance between the detonation point and survey point, in m. The correlation coefficient of equations r=0.8995 and r=0.9388,respectively. The scaled charge $(Q^{1/3}/R)$ varied from 0.0221 to 0.1573.

This relationship was used to calculate and design a charge that would induce vibration with PPV of the limit.

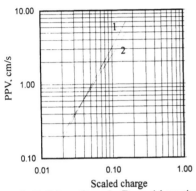

1—Radial quantity   2—Tangential quantity

Figure 3.The relationship between PPV and scaled charge.

6 THE EFFECTS OF CUT TYPES AND DELAY TIMES ON VIBRATION

When blasting full-face rounds and top heading, v-type cut and parallel-hole cut were employed. According to the recording of vibration, there were no obvious differences of vibration intensity between the v-type cut and the parallel-hole cut. But the delay times in cut blast holes became considerable for vibration intensity.

The vibration recording showed that, in great majority of the blasts, the values of vibration in cut holes were much higher than the values in the other holes. In cuts, the earliest firing blast holes had very poor free faces and, therefore, produced high vibration for a given amount of liberated energy. In trial blasts, the cut blast holes were fired by using the low numbers normal detonators, with15ms—25

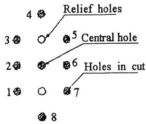

Figure 4.   Parallel-hole cut.

ms time interval. The vibration velocity often exceeded the 6cm/s and the depth of the round was 2m or less. After trial blasts, a kind of special detonators were adopted, which had 50ms time interval between successive delay numbers with a small scatter.

The improved delay allocation for parallel-hole cut was shown in Fig 4. Two relief holes were drilled in the cut, which could reduce the "tightness" of the cut, when the radial expanding strain wave reached a relief hole, wave reflection promoted fragmentation of the rock between the charge and relief hole, causing decrease in ground vibration. There were three forms of initiation sequence for the parallel-hole cut (see Fig 4). The form A was that central hole initiated first, after 50ms, the holes 1-4 and the holes 5-8 were fired with the No.2 and the No.3 millisecond-period delay, respectively, with 50ms interval between the two delay numbers. The form B was that, after initiating the central hole, the holes 1-8 were initiated with the same delay number No.2. The time interval between the central hole and the holes 1-8 was 50ms. The form C was that the all holes in cut were fired simultaneously with the No.1 delay number detonator.

The PPV was recorded in blasting .The distance between the geophone and the central hole was 19m.The results (see Table 1) indicated that the vibration induced by form A was lower than that by form B and C. In form A, the first initiated central hole would have the beneficial effect of reducing the degree of choking of the faces to which later firing holes shot. But in form C, the blast holes on the same delay would compete for a void volume sufficient for only one of the hole and prevent progressive relief of burden and, therefore, would produce high ground vibrations.

Table 1. The PPV from different initiation sequence forms in cut.

Initiation Sequence form		N*	H**	A***	PPV (cm/s)
Form A	Central hole	No.1	1	3.2	1.58
	Other hole in cut	No.2 and No.3	4	12.8	1.97
Form B	Central hole	No.1	1	2.6	2.29
	Other hole in cut	No.2	8	19.2	3.03
Form C		No.1	9	21.6	4.15

N*   ——Number of delay detonator;
H**   —— Holes per delay;
A***—— Amount of charge per delay (kg).

7 CONCLUSIONS

In driving the south tunnel, the existing tunnel was in safety, although blast operations were carried out at neighboring tunnel every day.

The limit of 6cm/s for the concrete lining was reasonable, which both ensured the existing tunnel against vibration danger and kept a high production potential in excavating tunnel.

The key to the success of driving the south tunnel was the vibration monitoring which tracked the advancement face in the parallel tunnel. The tunnellers could obtain the monitoring datum timely and modify the blast design (specially, the amount of charge per delay). Except the values of trial blasts, The PPV was kept within $\leqslant$6cm/s, no damage occurred at the surface of concrete lining during excavating tunnel.

According to the monitoring data at the surface of concrete lining, the variation law of PPV as a function of the scaled charge along the longitudinal axis of the existing tunnels was showed in formula(1) and formula(2).

The monitoring results indicated that the initiation sequence and delay time in cut played an important part in blasting vibration. The initiation sequence and delays needed to be allocated to meet the competing demands of maximum progressive relief of burden and minimum peak ground vibration. It is unsuitable that all the cut holes were initiated simultaneously. Except the central hole, a blast-hole mass per delay in the parallel-hole cut should not exceed four holes. The time interval between successive detonations should be about 50ms or greater.

REFERENCES

Dowding.C.H. 1985, Blast Vibration Monitoring and Control, Englewood Cliffs, N. J: Prentice--Hall.

Wu Congshi and Wu Qisu, 1990, A Preliminary Approach to Simulating Blast Vibration. In:Proceedings 3rd. International RockFragmentation by Blasting, Brisbane, Australia.

A.K.Chakraborty, A.K.Raina, M.Ramulu and J.L.Jethwa, 1998, Lake Tap at Koyna, World Tunnelling, Vol.11, No.9.

Berta.G, 1990, Explosives: An Engineering Tool, Italesplosivi Milano.

Modern Tunneling Science and Technology, Adachi et al (eds), © 2001 Swets & Zeitlinger, ISBN 90 2651 860 9

The monitoring result and consideration of vertical closed tunnels

K.Umeda , S.Koyama , T.Omichi
The Kansai Electric Power Co.,Inc. , Osaka, Japan

H.Sakamoto , K.Okaichi
Kumagai-Tobishima-Tekken-Takenakadoboku JV, Osaka, Japan

ABSTRACT: This paper reports the monitoring result and conclusions regarding the effects that is caused by new tunnel against the existing tunnel, when it is constructed nearby. The following results have been confirmed from the monitoring data.

Total thrust of shield machine, slurry pressure, and pressure of tail grout during excavation does not effect so much to the existing tunnel. The most of the increasing strain is caused by soil pressure relieve after passing shield machine. The effect caused by soil pressure relieve can be assumed by elastic analysis.

1 INTRODUCTION

The construction of new shield tunnel positioned near the existing structure are necessary to make good usage of underground space, there has been few data which monitored the effects on adjacent structures. But also analysis method has not been developed.

We constructed vertical closed tunnels which are positioned in diluvium sand layer. It was concerned about the influence for the first tunnel. Therefor we monitored the strain of segments and gap of segments to get the influence.

1.1 *The condition of new tunnel and existing tunnel*

The second tunnel (outer diameter: 7,600 mm) has been constructed close to the first tunnel (outer diameter: 5,600 mm) and the distance between two tunnels is only 1m.The second tunnel positioned just above the first tunnel. And both tunnels connected to each other to install 500kV cables.

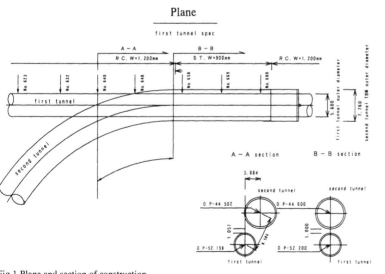

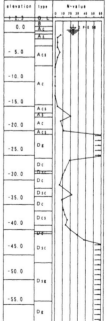

Fig.1.Plane and section of construction.

The second tunnel approaches to the first tunnel from a right angle then curves on 50m radius, and goes along 30m. The position of both tunnels is shown in Fig.1. The monitoring data for second tunnel excavation are shown in Table.1. The first tunnel has been constructed 7 months before the second tunnel excavation. Both tunnels were excavated by slurry shield method.

1.2 *Soil condition*

The bore hole data is shown in Fig.1. Both tunnels are positioned in diluvium sand layer which has high ground water pressure. The soil parameter of it is shown in Table.2.

Table 1. Monitoring data for second tunnel excavation.

Property	Curve section	Straight section
Heading water pressure	430 kpa	
Heading slurry pressure	510 kpa	
Backfill grouting pressure	610 kpa	
Propulsion force	33,000 kN	40,000 kN
Jacking speed	25 mm/min	22 mm/min
Over cut	100mm(right side)	0 mm

Table 2. Properties of diluvium sand.

Property	Value
N-value	over 60
Coefficient of uniform	4.54
Coefficient of permeability	$1.3*10^{-3}$ cm/sec
D_{50}	0.3 mm

2 MONITORING METHOD

To monitor the stress in segments and movement of segment joints, several instruments were installed on the surface of segment at 10m intervals, which is shown in Table.3. And also, the position of the instruments is shown in Fig.2. All the monitoring has been carried out by automatic devices except vertical movements of tunnel which is monitored by leveling instrument. The monitoring data have been transferred by telephone line to the control room and used for next excavation control value.

Table 3. Measuring list.

Measuring object	Measuring method	Number of instrument
Strain of segments	Strain gauge	7section*3piece*3point
Gap of segments	Cantilever gap gauge	4section*8point
Vertical movement	Level survey	7section

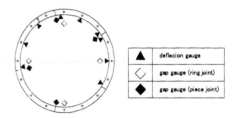

Fig.2.Position of measuring instruments.

Unit: μ +:Tension   -:Compression

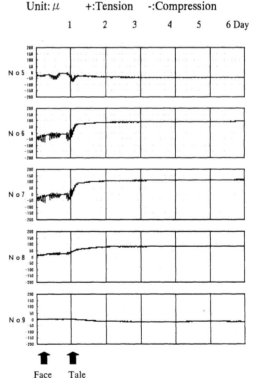

Fig.3.Strain vs. time graph of ring No.648 of first tunnel.

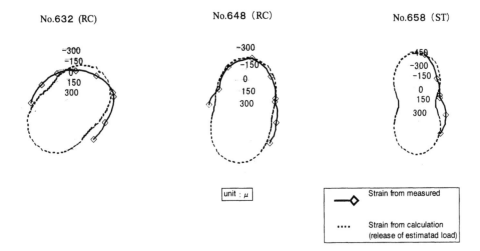

No.632 (RC)   No.648 (RC)   No.658 (ST)

unit : μ

—◇— Strain from measured

.... Strain from calculation
(release of estimatad load)

Fig.4.Strain caused by closed construction.

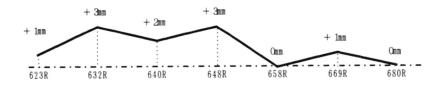

+ 1mm   + 3mm   + 2mm   + 3mm   0mm   + 1mm   0mm

623R   632R   640R   648R   658R   669R   680R

Fig.5.Vertical movement of the first tunnel.

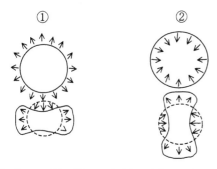

① ②

Fig.6.Image of the effects against the existing structure.

3 MONITORING RESULTS

3.1 *Stress in segment*

Strain vs time of ring No.648 of first tunnel is shown in Fig.3. We could say from this graph that most of the strain has been experienced after tail passed and very little of strain has been experienced during shield machine passing. The increase of strain has ceased within 3days after tail passed. Monitoring data of other locations of tunnel also showed similar tendency.

But the increasing strain curve by second shield is not the same as this result. We could say from Fig.4 that the first tunnel moved toward the second tunnel.

3.2 *Vertical movement of tunnel*

The first tunnel uplifted 2 to 3mm after the second tunnel constructed. The vertical movement of tunnel is shown in Fig.5.

3.3 *The gap of segments*

It has been observed that no movement of segment gaps has been experienced and also no leakage from the segment joint.

4 DISCUSSION

The effects against the existing structure by shield machine would be considered by the two items as follows:
①The first tunnel is pushed by thrust and grouting pressure of the second tunnel.

②By excavation of the second tunnel, the first tunnel moved toward the second tunnel due to release of existing soil pressure.

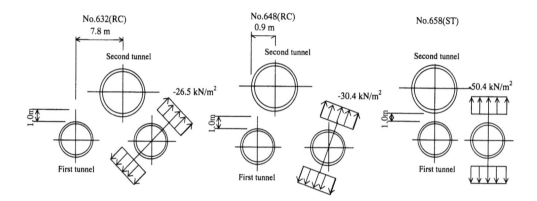

Fig.7.Analysis model of each section.

The monitoring results show that most of the movement is caused by ②, and ① is negligible. The reason of these results is that the second tunnel position is above the first tunnel, and soils around the both tunnels are dense sand. Shortage of tail grout volume might have caused the movement of the first tunnel. But it has been observed that enough tail grout was there later. The elastic analysis result which used 10% of stress release is very similar with the actual value. This result is shown in Fig.4. The analysis model of each section of Fig.4 are shown in Fig.7.

5 CONCLUSIONS

The most effect of adjacent tunnel construction is pressure release caused by the other tunnel. And it will be more after passing shield machine that during passing shield machine like this case in which the second shield machine passes above the existing tunnel, and having very dense soil condition. Increased moment and shear caused by ground pressure release can be assumed by elastic analysis. The additional reinforcement required for the existing structure also can be designed by initial value plus increased value, which can be assumed by elastic analysis.

Modern Tunneling Science and Technology, Adachi et al (eds), © 2001 Swets & Zeitlinger, ISBN 90 2651 860 9

Study on the tunnel load generation mechanism of twin tunnel

M. Yoshinaga
Sikoku Regional Bureau of Japan Highway Public Corporation, Kagawa, Japan

H. Kasamatsu & K. Ogawa
Sikoku Regional Bureau of Japan Highway Public Corporation, Kochi, Japan

S. Ito
Hinokio Tunnel Site Office, Kochi, Japan

S. Azetaka & H. Tezuka
KUMAGAI GUMI CO.,LTD, Tokyo, Japan

ABSTRACT: For expansion to four lanes of the Kochi Highway, the Hinokio Tunnel Project was drawn up to construct a new tunnel (Phase II tunnel) parallel to an existing tunnel (Phase I tunnel). The zones of weak ground that had experienced significant displacements in Phase I construction were prone to fail under the influence of the new tunneling. For monitoring the influence exerted by Phase II excavation upon the existing tunnel, this tunnel side was instrumented in advance. By the monitor scheme, safety of work was achieved. The measurement results are helpful for grasping the tunnel load generation mechanism.

1 INTRODUCTION

When a new tunnel is constructed in close proximity to an exiting tunnel, in a zone with adverse ground conditions where a significant displacement occurred, unfavorable effects are exerted by the new tunnel construction. Such effects cause a deformation of the existing tunnel, pulled toward the new tunnel, and an enlarged plastic zone of the existing tunnel. As a result additional loads are applied upon the existing tunnel (Phase I), thereby degrading the stability of the tunnel.

The project discussed here is construction of a new tunnel (phase II) parallel to and near an existing tunnel (phase I) in service, for expansion of a highway to four lanes. For the purpose of ensuring the stability of the existing tunnel, measuring instruments were installed in advance to know the behavior of the surrounding ground in response to the approaching new tunnel and the significance of its influence.

First, this paper shows, referring to the measurement results, the characteristics of deformation behavior of the surrounding ground influenced by the new tunnel excavation in progress. Then, it discusses the results of finite element analysis, and compares them with the measurements. Through these works, the authors investigate the mechanism of tunnel load generation in the twin tunnel.

2 OVERVIEW

2.1 Overview of the project

The Hinokio tunnel (expansion to four lanes) is located Otoyo-cho, Nagaoka-gun, Kochi prefecture, nearly at the middle of the Shikoku Mountains, on the Trans-Shikoku Highway. It is a road tunnel 2,541 m long, about 80 m^2 in excavation section.

This tunnel, running in the nearly north-south direction, is constructed in the west of the existing tunnel in service. The center-to-center distance between two tunnels is 3D (D = excavation diameter). This project passes through the Mikabu tectonic line that is one of representative weak grounds in Japan.

2.2 Overview of topography and geology

(1) Topography

The surroundings of the tunnel site are mountains 350 to 520 m above sea level. At STA27+40, the earth covering is the largest, 190 m. The overburden is in general from 100 m to 190 m except for the tunnel entrance. At the point of fault F5' reported here, the overburden is about 150 m.

(2) Geology

The geology of the site is divided by Mikabu tectonic line running in the east-west direction, into north and south zones. The north zone over 350 m is Sambagawa belt composed of crystalline schist. The south zone is Mikabu belt composed of Mikabu greenstones that were made under the effect of regional metamorphism from basic ultrabasic igneous rocks such as diabase, and basalt, as well as ultrabasic igneous rocks such as olivine.

(3) F5' fault

The results of geological survey by horizontal boring suggest that the F5' fractured zone was greatly influenced by Mikabu tectonic line. We found that the group of many faults is generally crushed. The Mikabu greenstones contain much chrolite which is an expandable clay mineral, in the F5' fault and prone to slaking. So creep property of this ground was anticipated. In addition, the strength of the ground is low. It was therefore estimated that the ground is prone to plasticization.

3 OVERVIEW OF THE CONSTRUCTION

3.1 Construction status of the Phase I line (record)

According to the work record, the area near the F5' fault is generally crushed, where numerous crushed zones different in size gather on both sides of the heavily crashed zone about 40 m wide. The tunneling in this area used D3-1 support pattern as shown in Figure 1. In this area chlorite green shist is crushed, becoming clayey. Chlorite, one of expandable clay minerals, was found in a wide range there.

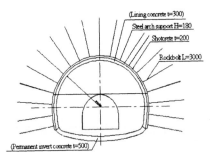

(Lining concrete t=300)
Steel arch support H=180
Shotcrete t=200
Rockbolt L=3000

(Permanent invert concrete t=500)

Steel support	Mu-29 Interval 1.0m
Rockbolt	L=4.0m n=19 Interval 1.2 m
Wire mesh	Upper/lower sections
Shotcrete	20 cm
Lining concrete	30 cm

Figure 1. Support pattern (Phase I tunnel).

Table 1. Maximum convergence, crown settlement (Phase I).

Observation point	Upper section convergence	Crown settlement
STA30+84	76 mm	44 mm

Properties of the ground by construction record and survey

q_u : unconfined compressive strength 50 to 100 kgf/cm²
γ : unit weight of soil about 2.9 to 3.0 kg/cm³
C : cohesion 0 to 0.1 kgf/cm³
ϕ' : angle of shear resistance 20 degrees
E : reduced elastic modulus by back analysis 6,000 kgf/cm²
λ : coefficient of lateral pressure by back analysis 1.05

The velocity of elastic wave was 3.0 km/sec. The tunnel was driven by the bottom heading upper half method, but it was difficult to maintain the face sta-

bility. It is in this area that the maximum displacement in the Phase I construction occurred. Table 1 shows the maximum convergence and crown settlement. In D3-1 lot, since some cracks occurred in the shotcrete was cracked, additional bolts were installed and invert concrete was placed to limit the displacement. Furthermore, concentrated water inflow took place, of 18 to 40 liters/min in the boring survey and 380 to 420 liters/min during drift construction.

3.2 Construction status of the Phase II line

(1)   Support pattern

Referring to the work record of Phase I line and the results of horizontal boring conducted in advance, we estimated that, in the F5' fractured zone (STA31+64 to STA32+30), a high risk was anticipated of remarkable loosening and displacement, causing falling down of the face and collapse of the crown during excavation, and broken rock bolts and cracked shotcrete after completion of excavation. Therefore, for excavation in this area, basic support pattern DII (S1) with an increased stiffness was adopted from the start of work. Furthermore, we set up a selection flow of support/ auxiliary methods and management reference values, which allowed us to reflect the measurement results upon the support pattern. The tunneling was implemented, while choosing suitable supports.

Figure 2. shows the basic support pattern.

(2)   Construction method

The full face cutting with auxiliary bench was in principle used. Early invert placement was done with shotcreting, up to the position 7 m from the face, once per two-cycle advance of the face. In addition, permanent concrete invert was placed once per week over an interval of 7 to 17 m from the face. Though 18 bolts 6 m long were initially planned, nine bolts were added to the pattern, which were cyclically driven from the first stage. In places where the face is especially unstable, one excavation cycle of 80 cm with DII (S2) pattern was adopted. For preventing crown collapse, steel pipe expansion bolts (Swellex forepiling) were driven in a range of 120 degrees at intervals of 60 cm. For stabilizing the face, shotcrete was applied to the face (50 mm thick or more). Table 2 shows the maximum convergence and crown settlement in the Phase II construction work. The monthly advance in the area with DII(S2) support pattern was 40 m.

4 MEASUREMENTS

4.1 Overview of the measurements

To obtain data of the influence exerted by the approaching new tunnel face upon the exiting tunnel,

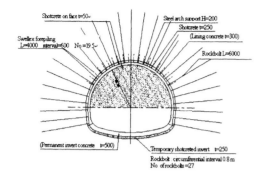

Figure 2. Support pattern (Phase II tunnel).

Steel support	Mu-200 Interval 1.0m
Rockbolt	L=6.0 n=27 Interval 0.8 m
Wire mesh	Upper/lower sections
Shotcrete	25 cm
Lining concrete	30 cm

Table 2. Maximum convergence, crown settlement (Phase II).

Observation point	Upper section convergence	Crown settlement
STA31+89.5	62 mm	37 mm

measuring instruments were installed in advance in the existing tunnel. The installation position was the fractured zone where the ground condition is adverse and remarkable displacements would occur, so the influence of twin tunnel would be especially important. The key measurement items were ground displacement (extensometer) for grasping the behavior of the ground around the tunnel, and the stress in the lining for knowing the influence upon the lining of the exiting tunnel. Other measurement items included crack displacement, convergence and crown settlement. The measurement positions are illustrated in Figure 3 and the instrument arrangement in the cross section in Figure 4.

4.2 Measurement results

The results of the measurements by the extensometers and lining stress gauges mentioned above accurately express the influences in the twin tunnel.

(1)   Extensometer

The measurements by the extensometers are shown in Figure 5. When the new tunnel face came at a distance about 2D, the part near the exiting tunnel wall began to move toward the existing tunnel. Then, as the face came closer, the portion in the vicinity of the new tunnel face significantly moved toward the new tunnel when the face was at about 0.5D. At 9 m position, the ground once began to move toward the Phase I line, but after the face passed away by about 1D, the ground moved toward Phase II line. Looking at the final ground displace-

ment distribution, we know that an inflection point of displacement exists between 6 and 9 m from the existing tunnel wall. When the face passed away by about 5D, the displacement nearly converged.

(2)   Lining stress meter

Figure 6 shows the measurement results of stress in the lining of the existing tunnel. When the distance from the new tunnel face became about 5D, compressive stress began to increase, and after the distance became about 2D, the compressive stress abruptly increased. After the face passed away by about 2D, the stress began to converge, and at about 5D, it converged. The increase in compressive stress was 40 kgf/cm^2 to 60 kgf/cm^2

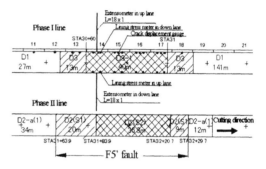

Figure 3. Measuring position.

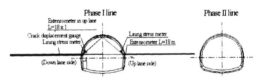

┌ Extensometer : ground displacement in the horizontal direction :
│   +elongation −contraction
│ Lining stress meter : strain in circumferential direction :
│   +compression −tension
└ Crack displacement gauge : crack width : +opening −closing

Figure 4. Arrangement of extensometers (cross direction).

(3)   Convergence and crown settlement

The convergence and crown settlement were 0 to 2 mm. Taking into consideration measurement errors, it can be said that the tunnel lining of the existing tunnel did not move at the measuring point.

Looking at the behavior of the twin tunnel on the basis of these measurement data, the mechanism of twin tunnel was revealed as follows. The ground around the preceding tunnel had been in the well balanced state. When its equilibrium was disturbed under the influence of the succeeding tunnel, tunnel loads were applied upon the preceding tunnel.

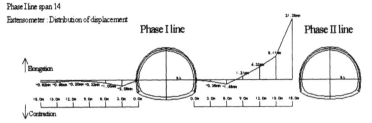

Phase I line span 14
Extensometer : Distribution of displacement

Phase I line

Phase II line

↑Elongation

↓Contraction

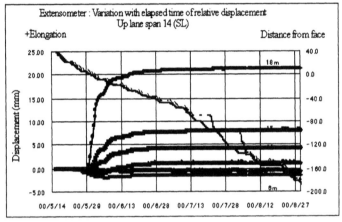

Extensometer : Variation with elapsed time of relative displacement
Up lane span 14 (SL)

+Elongation

Distance from face

Figure 5. Results of measurement by extensometers.

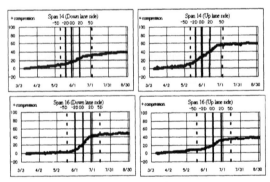

Figure 6. Results of measurement by lining stress meter.

5 LOAD GENERATION MECHANISM OF TWIN TUNNEL AND DISCUSSION

The influence of twin tunnel has been discussed referring to the variation in stress distribution around the tunnel, on the basis of twin-hole circle problem of the theory of elasticity and numerical analysis such as FEM. The results of conventional analysis by twin-hole circle provide larger stresses in between two tunnels, and with some ground properties, failure occurs there. In the case of numerical analysis such as FEM, by which the excavation of the succeeding tunnel (Phase II tunnel

with the present case), the preceding tunnel is pulled toward the succeeding tunnel side, and the displacement around the preceding tunnel is larger on the succeeding tunnel side. As a result, the increase of stress is greater in between two tunnels. This is the same result as the twin-hole circle approach. Neither of these results precisely represent the behavior of the ground around Phase I tunnel or stress in the lining concrete. As mentioned above, the excavation of the succeeding tunnel disturbs the stress equilibrium in the ground around the preceding tunnel, generating additional tunnel loads upon the preceding tunnel. In the present study, the distribution of the magnitude of such loads is inferred, on the basis of the measurement results, from the stress distribution in the lining or displacement distribution in the ground. Estimating from the increase in stress in the lining on both sides of Phase I tunnel, the loads are almost the same on both sides. Therefore, loads are likely to work uniformly.

The load generating position is estimated at about 6 m from the wall, referring to the extensometers.

The actual behavior of the twin tunnel was verified by numerical analysis as follows.

5.1 *Numerical analysis*

Finite element analysis was conducted for representing the measurement results, and quantitatively

314

investigating the mechanism of load generation in the twin tunnel. The analytical model is shown in Figure 7.

The analytical results are summarized in Figure 8. In the conventional excavation simulation, Phase II tunnel excavation is done separately and after Phase I excavation, which provides the phenomenon that Phase I tunnel is pulled toward Phase II tunnel during excavation of Phase II tunnel. With this approach, it is difficult to simulate the measured loads working upon Phase II tunnel. In consequence, it turned out that the behavior of extensometer with an inflexion point can be expressed by applying a magnitude of load to Phase I tunnel at the same time of Phase II tunnel excavation. The diagram shown in Figure 8 expresses the application of a load of 60 tf/m^2 from tunnel outside in addition to the external forces due to excavation. This result presents an inflexion point in ground displacement. The stress increase in the lining of Phase I is 81 kgf/cm^2 (measurement = about 40 kgf/cm^2) near the spring line on the left, and 67 kgf/cm^2 (measurement = 60 kgf/cm^2) near the spring line on the right.

As demonstrated by these results, the analytical results agree in general with the measurements. When designing a twin tunnel, it is therefore necessary to apply load α in addition to the external forces of excavation.

5.2 Load generation mechanism of twin tunnel

The stress equilibrium is disturbed in the ground around the preceding tunnel by succeeding tunnel excavation, resulting in increased stresses in the lining concrete of the preceding tunnel and displacements of the ground between two tunnels. The magnitude of the load due to such disturbance is supposed to be α. By numerical analysis, the magnitude of α was obtained, that agreed well with the measurement data. However, for more realistic approach, it is necessary to simulate the mechanism by which increase of stresses induces additional tunnel loads. A technique for achieving this approach is an elasto-plastic analysis by which surplus stresses work as external forces. The authors are currently studying the possibility of expressing the behavior of twin tunnel by such approach. The ground properties relating to plasticization are usually represented by cohesion C and internal friction angle ϕ. However, with these parameters alone, it is impossible to express the sensitive response of the ground indicated by measurement results. The authors propose a technique involving a "sensitivity coefficient θ" which is applied to the failure zone in Phase I tunnel to simulate the mechanism mentioned above. On the other hand, as for the load generation of the side heading method conventionally used in adverse ground conditions, a schematic diagram of a sup-

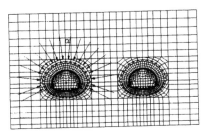

Figure 7. FEM analysis model.

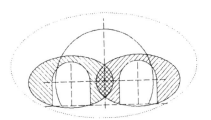

Figure 9. Mechanism of load generation in side drift method.

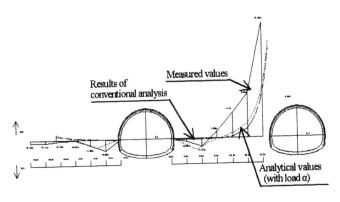

Figure 8. Analytical results.

posed mechanism is shown in Figure 9. In this diagram, loads in a range enveloping the influence area of two side drifts are applied on the preceding tunnel. The authors will continue to investigate the load generation mechanism of twin tunnel, resorting to these two concepts.

6 CONCLUSIONS

Though many twin tunnels were constructed, its behavior has not been sufficiently studied. There are just a few papers of investigation based on measured ground displacements and stresses, especially in a ground with an large earth covering, that is, with a high potential energy.

In the project discussed here, since some influence is anticipated, of Phase II tunnel excavation upon Phase I tunnel, the stress in lining of Phase I tunnel was measured, and extensometers were installed from Phase I tunnel side, to study the behavior of the ground around the tunnel. The measurement results revealed the fact that Phase II tunneling exerted influence upon the ground around Phase I tunnel, and the stress distribution in the lining of Phase I tunnel varied. Moreover, the elapsed variation change variation of the ground strain was obtained.

By numerical analysis which applies load α to the preceding tunnel, in addition to the external forces due to excavation, it was known that the behavior can be expressed satisfactorily. The authors will determine specific magnitudes of load α corresponding to different ground conditions and tunnel geometries, to establish a more reliable design technique of twin tunnel.

Modern Tunneling Science and Technology, Adachi et al (eds), © 2001 Swets & Zeitlinger, ISBN 90 2651 860 9

Determination of the horizontal subgrade reaction coefficient for the backside of shoring systems in clayey soil

H. Nakamura
Pacific Consultants Co.,Ltd, Director

H. Suzuki
Pacific Consultants Co.,Ltd, General Manager,Eng.Dept.

ABSTRACT: Deformation of shoring walls used in cut-and-cover excavations in clayey soil is significantly reduced when the walls are constructed in improved ground and supported by preloaded braces. At the same time, earth pressure on the backside of the walls is greater than it would be if the above methods were not employed. In such cases, when developing a design method, it is no longer valid to assume that the earth pressure on the backside is a known value independent of wall displacement. The use of a calculation method that assumes a ground spring on the backside may be considered. This paper proposes a method to determine the horizontal subgrade reaction coefficient based on a numerical calculation using the finite element method. It also discusses the results of applying this proposition to two sites: one with ground improved on the excavation side and another where preloaded braces were used.

1 INTRODUCTION

Two-dimensional elastoplastic analysis of successive excavations (hereinafter called the "elastoplastic analysis") is used in many cases for the shoring system design of open-cut tunneling in urban areas. In this analysis, shoring walls, strut supports, and ground resistance are modeled as elastic beam, elastic support, and elastoplastic distribution spring, respectively[1]. When shoring walls are constructed in improved ground or preloaded braces, however, deformation is significantly reduced, and the resultant backside earth pressre is greater than usual. It is therefore not appropriate to adopt a design method in which the backside earth pressure is given as a known amount regardless of wall displacement. It is rational to use elastoplastic analysis assuming a ground spring on the backside if wall displacement is significantly smaller than usual. However, due to the lack of an established method to determine such inputs as the horizontal subgrade reaction coefficient, it is difficult to use this method to estimate the behavior of shoring walls.

This paper proposes a method of determining the horizontal subgrade reaction coefficient for the backside on the basis of a numerical calculation using the finite element method. It also discusses a review of the adequacy and applicability of this proposed method of determining the above-described coefficient for two cases: one with improved ground on the excavation side and another where preloaded braces were used.

2 SETTING THE HORIZONTAL SUBGRADE REACTION COEFFICIENT FOR THE BACKSIDE

Generally, shoring walls in clayey soil behave differently above and below the excavated subgrade; namely, the horizontal displacement of the shoring wall is small above the lowest internal bracing tier while the walls are displaced inward below this location.

Therefore, it is considered rational that horizontal subgrade reaction coefficients for the backside be set for above and below the excavated subgrade respectively.

2.1 Method to determine the horizontal subgrade reaction coefficient below the excavated subgrade

(1) *Study method*
From studies of the author et al[2,3] concerning the horizontal subgrade reaction coefficient on the excavation side, it may be presumed that the equation to determine this coefficient for the backside below the excavated subgrade must contain such factors as the ground deformation coefficient, Poisson's ratio, and thickness of the subgrade concerned. Considering that the backside subgrade is characteristically backed up with the support, the effect of the provision of support (represented by the excavation stage) has to be included in these factors. Therefore, the calculation model shown in Figure 1 and Table 1 was used and an analysis simulating the excavation process was conducted of the horizontal subgrade reaction coefficient on the backside below the excavated

317

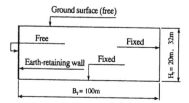

Figure 1. Calculation model used in the study on the horizontal subgrade reaction coefficient below the excavation bottom.

Table 1. Calculation conditions for the finite element method (below the excavated subgrade).

Case	H_T	ν	K_G
LF-1	20	0.480	9.8×10^5
LF-2	20	0.444	9.8×10^5
LF-3	32	0.480	9.8×10^5
LF-4	32	0.444	9.8×10^5
LS-1	20	0.480	9.8×10^2
LS-2	20	0.444	9.8×10^2
LS-3	32	0.480	9.8×10^2
LS-4	32	0.444	9.8×10^2

Note H_T: Model height ν : Poisson's ratio of ground
K_G: Shear spring constant of wall and ground (kN/m³)

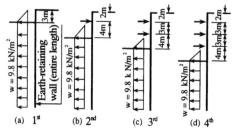

Figure 2. Example of loads and support locations using the finite element method (first to fourth).

subgrade. This coefficient was determined from the relationship between the wall displacement and modulus of deformation. Figure 2 shows the load during excavation process and the location of support.

(2) Horizontal ground displacement and horizontal Subgrade reaction coefficient

Figure 3 shows the result of a calculation using the finite element method, particularly, an example of horizontal ground displacement at the shoring wall position below the excavated subgrade in each excavation stage. This figure indicates that horizontal displacement of the shoring wall is curved. In addition, such displacement decreases along with a decrease in the thickness (called the "thickness of the subgrade") from the excavated subgrade to the top (the lower end boundary in this model) of hard ground below the excavated subgrade. There is also a slight difference in the tendency of the horizontal displacement during the first excavation from others. Moreover, it is evident that the effects of Poisson's ratio can be ignored and the distribution condition of horizontal displacement due to the effects of shear resistance of the shoring wall surface is similar. Finally, the horizontal ground displacement is slightly greater when shear resistance of the wall surface does exist than when it does not.

From the above review, it was found out that factors influencing horizontal ground displacement are the existence or non-existence of the support (that is, whether the support is provided during the first excavation or the second and subsequent stages), the thickness of the subgrade, the existence or non-existence of shear resistance on the shoring wall surface, the magnitude of the load, and the modulus of deformation. Therefore, the horizontal

displacement equation containing these factors was introduced.

Horizontal displacement of the ground was defined with a line graph for the sake of convenience of handling. With the break point located in the mid-point of ground thickness, the constant value representing the calculation using the finite element method was assumed within the range above the break point (from the break point to the excavated subgrade). On the other hand, below the break point (from the break point to the top surface of the lower hard ground), the straight line was assumed to approximate the calculation using the finite element method. At the break point, the value for the above range was assumed. The equation for this straight line was determined as follows according to the least square method (See Figure 4).

Upper displacement $\quad \delta_U = \dfrac{wH_0}{\alpha_U E}$ (1a)

Lower displacement $\quad \delta_L = \dfrac{wH_0}{\alpha_L E}$ (1b)

wherein, δ_U, δ_L =Horizontal displacement at the top and bottom of ground(m),w=Horizontal load acting on the ground (kN/m²), H_0=Thickness [from the excavated subgrade to the top of lower hard ground] of the subgrade (m), α_U, α_L=Coefficients (See Table 2), E=modulus of deformation(kN/m²)

Now, the horizontal displacement of the backside ground when the uniform load was applied could be calculated. Therefore, the horizontal subgrade reaction coefficient for the backside can be calculated as indicated below by dividing the uniform load by the displacement (Figure 4):

Upper horizontal subgrade reaction coefficient

$$k_{HU} = \alpha_U \frac{E}{H_0}$$ (2a)

Lower horizontal subgrade reaction coefficient

$$k_{HL} = \alpha_L \frac{E}{H_0}$$ (2b)

2.2 Method to determine the horizontal subgrade reaction coefficient above the excavated subgrade

(1) Study method
As the ground above the excavated subgrade is backed up

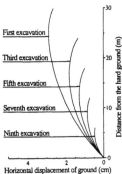

Figure 3. Typical horizontal displacement of ground (Case [LF-3]).

Table 2. Values for coefficients α_U and α_L.

Excavation stage	Coefficient	Shear resistance of earth-retaining wall surface	
		With resistance	Without resistance
1st	α_U	1.08	0.83
	α_L	20.00	7.70
2nd and after	α_U	1.39	1.25
	α_L	14.30	7.10

with the support, it develops no or extremely small, if any, horizontal displacement except when the preload is applied. In this condition, the shoring wall behaves independently from the horizontal subgrade reaction coefficient. Accordingly, this coefficient for the area above the excavated subgrade is determined only for cases where the preload is acting on it.

The review uses, as calculation models, an elastic finite element method model and a beam model on an elastic body shown in Figure 5.

It was decided to use, for the calculation using the finite element method, the model shown in Case [LF-1] for H_T=20 m in Figure 1 and Table 1 that was used in the study on the horizontal subgrade reaction coefficient for the area below the excavated subgrade. Namely, given bending and axial rigidities of the shoring wall, the horizontal displacement of the wall at a preload acting point was calculated.

It was assumed that the calculation value with the finite element method model is equal to that with the beam model on the elastic body while paying attention to the wall displacement depending on the preload acting point. Then, the relation expression was introduced between the horizontal subgrade reaction coefficient with the beam model on the elastic body and the ground deformation coefficient with the finite element method. An equation to determine the horizontal subgrade reaction coefficient used in the design is proposed on the basis of this equation. The numerical values used in the calculation are shown in Table 3.

(2) Horizontal subgrade reaction coefficient
The horizontal displacement of a shoring wall in the pre-

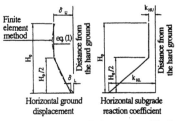

Figure 4. Horizontal ground displacement and horizontal sub-grade reaction coefficient.

Table 3. Values used in calculation.

Item	Setting method
(Common)	
- Excavation stages	1st (D=2m), 2nd (D=5m), 3rd (D=8m)
- Bending rigidity of earth-retaining wall	EI=15580 (Steel pipe sheet pile, Type III), 228400 (Column-type diaphragm wall), 17777kNm² (Diaphragm wall)
- Preload	P_0=98kN
(Finite element method)	
- Modulus of ground deformation	E_1=1960, E_2=9800, E_3=49000kN/m²
- Poisson's ratio	ν =0.480
(Beam on elastic body)	
- Horizontal subgrade reaction coefficient	k_{H1}=98, k_{H2}=980, k_{H3}=9800kN/m³

load acting position determined according to the finite element method is shown in Table 5. An example of the relationship between the horizontal displacement of the wall and the ground deformation coefficient is shown in Figure 6. As indicated, the displacement of the wall decreases with the increase in the deformation coefficient and the bending rigidity. It is also evident that displacement during the first excavation is greater than during subsequent excavation and that displacement is approximately equal for the second and third excavations.

Similarly, the horizontal displacement of the shoring wall in the preload acting position determined from the calculation of the beam on the elastic body is shown in Table 4. Relationship between the wall horizontal displacement and the horizontal subgrade reaction coefficient is shown in Figure 7. The trend similar to the calculation result using the finite element method can be observed.

Among the calculation results using the beam model on the elastic body , the wall horizontal displacement in the preload position was represented by the equation shown below. On the basis of the three calculation results obtained from the differing horizontal subgrade reaction coefficient, the three coefficients of C_1, C_2 and C_3 were determined for each excavation stage of each wall type. Note that Equation (3) was selected from trial calculation results using various functions because of its satisfactory matching with the result of the finite element method.

$$\delta = \frac{C_1}{\left(\log k_H\right)^{C_2} + C_3} P_0 \qquad (3)$$

Wherein δ =Horizontal displacement of shoring wall at the preload acting position (Beam model on the elastic body) (m), P_0=Preload(kN), k_H=Horizontal subgrade reaction coefficient (kN/m³), C_1, C_2 and C_3=Coefficients

Table 4. Horizontal displacement (cm) of shoring wall at the preload acting position.

Wall type	Excavation stage	Finite element method			Beam on elastic floor		
		E_1	E_2	E_3	k_{H1}	k_{H2}	k_{H3}
Steel pipe sheet pile	1st	11.26	2.46	0.57	8.87	2.27	0.33
	2nd	3.21	1.03	0.30	5.96	1.59	0.31
	3rd	2.26	0.92	0.28	3.37	1.31	0.31
Column-type diaphragm wall	1st	10.32	2.18	0.47	15.33	1.82	0.22
	2nd	0.94	0.43	0.15	1.30	0.49	0.14
	3rd	0.34	0.24	0.12	0.45	0.26	0.11
Diaphragm wall	1st	9.82	2.04	0.43	14.70	1.55	0.19
	2nd	0.19	0.14	0.07	0.58	0.15	0.06
	3rd	0.05	0.05	0.03	0.09	0.05	0.03

Note: For E_i and k_{Hi} refer to Table 3.

Table 5. Values for coefficient α_k in each calculation case.

Excavation stage	Earth-retaining wall	Value of coefficient α_k			
		E=1960	E=9800	E=49000	Average
1st (D=2)	Steel pipe sheet pile	0.102	0.092	0.095	
	Column-type wall	0.082	0.082	0.111	0.091
	Diaphragm wall	0.090	0.078	0.073	
2nd (D=5)	Steel pipe sheet pile	0.188	0.178	0.218	
	Column-type wall	0.142	0.127	0.163	0.178
	Diaphragm wall	0.327	0.116	0.108	
3rd (D=8)	Steel pipe sheet pile	0.186	0.176	0.243	
	Column-type wall	0.203	0.129	0.151	0.190
	Diaphragm wall	0.688	0.194	0.132	

Note: E (Modulus of ground deformation) unit: kN/m^2

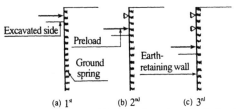

(a) 1st   (b) 2nd   (c) 3rd

Figure 5. Calculation model used in the study on the horizontal subgrade reaction coefficient above the excavation bottom (beam on the elastic body).

The coefficient of Equation (3) is determined from the calculation result of beam on the elastic body. This in turn enabled calculation of the horizontal displacement of the wall in the preload acting position as a function of an arbitrary horizontal subgrade reaction coefficient k_H for the excavation stage and wall bending rigidity shown in Table 3. Now, back calculation using this equation can be made to determine the horizontal subgrade reaction coefficient with the beam model on the elastic body, that provides a displacement equivalent to that of the shoring wall at the preload acting position determined according to the finite element method under the conditions shown in Table 3.

Then, the horizontal subgrade reaction coefficient thus determined by back calculation was related to the deformation coefficient of the ground according to the finite element method. The coefficient, α_k, calculated from this relationship is shown in Table 5. Note that the average shown in this table was determined by adding weight linearly proportional to the horizontal displacement of the finite element method.

$$k_H = \alpha_k E \qquad (4)$$

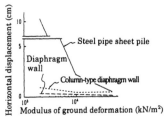

Figure 6. Relationship between the wall horizontal displacement and modulus of deformation (Finite element method, column-type wall).

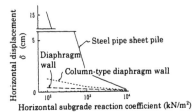

Figure 7. Relationship between the wall horizontal displacement and the horizontal subgrade reaction coefficient (Beam model on elastic body, column-type wall).

Wherein, k_H=Horizontal subgrade reaction coefficient (kN/m^3), α_k=Coefficient (m^{-1}), E=modulus of deformation according to the finite element method (kN/m^2).

As is evident from Table 5, the value of coefficient α_k was approximately equal in all excavations, except for a case with E=1960 kN/m^2 during second and third excavations. There is no definite relationship between the modulus of deformation and change in the bending rigidity of the shoring wall. It is also evident that the value for the first excavation is different from that for second and third excavations.

On the basis of the above considerations, it is proposed for the sake of practical convenience to handle the coefficient α_k as shown in the following equation:

First excavation $\alpha_k = 0.090$ (5a)
Second and subsequent excavations
     $\alpha_k = 0.180$ (5b)

3 COMPARISON WITH MEASURED VALUES

3.1 Comparison with an excavation site with improved ground

(1) Outline of the site to be compared

Site $A^{4)}$ is a subway construction site where soft ground is excavated to a depth of GL-19.6 m. The ground consists of an alluvial layer to a depth of about 20 m, under which a gravel (diluvial) layer with an N value of 30 or more exists. The excavated side was improved with chemico-piles (square in 1.5 m pitch) for a 10 m range from a GL-7.68 m point to the final excavation depth. The earth retaining structure was a ø450 column-type diaphragm wall

320

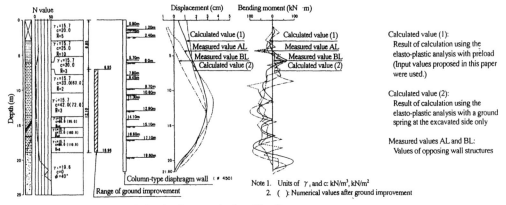

Figure 8. Comparison between calculated and measured values (Site A).

(spacing of soldier piles, 0.9 m), with an embedment made to a depth of 2 m into hard ground.

(2) Input value determination method

Inputs related to the excavated side ground, including the passive earth pressure, equilibrium earth pressure, and horizontal subgrade reaction coefficient, are calculated according to propositions [2), 5), 6)] of existing reports of the author et al.

Considering the soft clayey soil ground, the coefficient of the earth pressure at rest should be $K_0=0.8$. When the excavated side is to be improved, however, it is necessary to take into account the increase in the horizontal pressure due to implementation of ground improvement. As regards Site A, it was reported that displacement of the shoring wall was controlled by providing pressure-relief holes near the wall during driving of chemico-piles. In consequence, the earth pressure at rest before excavation was set to $K_0=0.8$.

The ground deformation coefficient used for calculation of the horizontal subgrade reaction coefficient was set to $E = 480c$ for the clayey soil layer and $E=2800N$ for the sandy layer. "E" is the modulus of deformation (kN/m^2) while "c" is the cohesion of subgrade (c' is used for the improved or sandy ground). "N" is the N value of the standard penetration test.

The effect of ground improvement with chemico-piles is represented by the result of $c'= 2c_0$, $c'= c_0+50$, and $c'= 150$, whichever is smallest. c' is the apparent cohesion (kN/m^2) after improvement, and c_0 is the cohesion (kN/m^2) of natural ground.

(3) Result of comparison

Ground improvement on the excavated side produces expansion pressure, causing the shoring wall to be displaced toward the backside. The measured value at both sites to be compared is based on the assumption that the condition immediately before excavation after ground improvement is the initial condition. Therefore, calculation for comparison is made only on the behavior after ground improvement.

The calculation result and site measurements during final excavation are compared in Figure 8.

It is known from this data that the calculated value for the shoring wall deformation is approximately the same as the measured value as a whole, except for the difference between them in the upper portion of the wall.

3.2 Comparison at the excavation site with preloading of internal bracing

(1) Outline of the site to be compared

Site B[7)] is a site where excavation was made to a depth of 13.95 m in the alluvial lowland along the basin of the Tsurumi River. The ground consists of a silt layer with an N value of 0 to 1 to a depth of GL-24 m. The upper portion of the sandy layer below the silt layer contains a mixture of silt. The lower portion contains gravel. Alternating strata of mudstone and fine sand exist below a depth of GL-30 m. The shoring wall is a diaphragm wall with t = 80 cm, and with embedment provided in the hard alternating strata of mudstone and fine sand. For the support, a concentrated steel bracing framing of three tiers was used.

(2) Input determination method

Similar to the case of 3.1 above, inputs related to backside ground were determined and the coefficient of earth pressure at rest was set to $K_0=0.8$. Inputs related to the excavated subgrade were calculated on the basis of propositions [2), 5), 6)] of existing reports of the author et al. that contain factors related to the displacement of shoring walls. Methods to determine soil constants, such as the ground deformation coefficient, etc. were similar to 3.1(1) above. Preloading values were introduced while establishing steps for each excavation stage.

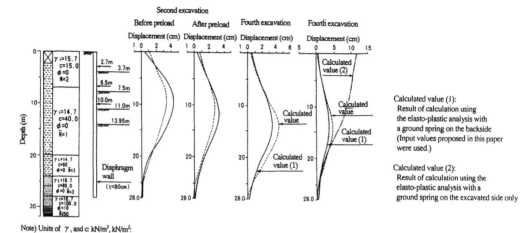

Note) Units of γ_t and c: kN/m³, kN/m²:

Figure 9. Comparison between calculated and measured values (Site B).

(3) Result of comparison

Calculated and measured values related to deformation of the shoring wall were compared as shown in Figure 9. The calculated value (1) in the figure is the result of calculation according to the proposed method. It is known that the calculated value is approximately the same as the measured value.

When preload is to be introduced, the behavior of the wall can be estimated accurately by entering the input value determined for the backside according to the propose method, into the elastoplastic analysis with the ground spring on the backside .

4 CONCLUSION

This paper discusses an in-depth study made on clayey ground by using the finite element method model. It introduces equations for the horizontal subgrade reaction coefficient as a function of the modulus of deformation, the depth from the excavated subgrade to the lower hard ground, and the shear resistance of the shoring wall surface. The equation also gives the minimum earth pressure as a function of the unit weight of soil, cohesion, and the bending rigidity of the shoring wall. These calculated values were compared with the measured values at the site. It was made clear that this method to set the horizontal subgrade reaction coefficient on the backside and minimum earth pressure enables highly accurate prediction.

This paper proposes a method based on dynamic considerations of the horizontal subgrade reaction coefficient on the backside and minimum earth pressure. Because of its rationality and substantial improvement in estimate accuracy, we concluded that this method should prove to be extremely useful for the design of shoring systems.

REFERENCES

1) H. Nakamura & K.Hirashima : Study on the horizontal subgrade reaction coefficient of embedment of shoring walls in clayey ground, Collected papers of the Japan Society of Civil Engineers, No.534／Ⅵ-30, 1996.3.

2) H. Nakamura & K. Hirashima : Study on the method to set the horizontal subgrade reaction coefficient for shoring system in consideration of the effect of lower hard ground, Collected papers of the Japan Society of Civil Engineers, No.595／Ⅵ-39, 1998.6.

3) H. Nakamura, H. Suzuki & K Hirashima : Method to set the horizontal subgrade reaction coefficient and minimum earth pressure on the backside of shoring system in clayey ground, Collected papers of the Japan Society of Civil Engineers, No.665／Ⅵ-49, 2000.12.

4) Y. Masuda & H. Irie, Y. Watanabe : Site measurement and design method for shoring walls in soft ground (Tatsumi Parking Area of No.8 Wangan Line), 41st lecture meeting of the Japan Society of Civil Engineers, pp. 451 ～ 452,1986.

5) H. Nakamura & K. Hirashima : Study on earth pressure on the backside of shoring walls in clayey ground, Collected papers of the Japan Society of Civil Engineers, No.504／Ⅵ-25, 1994.12.

6) H. Nakamura & K.Hirashima: Study on passive earth pressure of embedment of shoring walls in clayey ground, Collected papers of the Japan Society of Civil Engineers,No.528／Ⅵ-29, 1995.12.

7) M. Hebikawa, O. Nakamura, J. Mase & I. Sawada : Design and construction of shoring system during reconstruction of the Tokyo Stock Exchange (Phase 1: Construction of the new building), the 23rd presentation meeting of the Japanese Society of Soil Mechanics and Foundation Engineering, pp. 1563～1564, 1988

3. Maintenance of tunnel structures

Modern Tunneling Science and Technology, Adachi et al (eds), © 2001 Swets & Zeitlinger, ISBN 90 2651 860 9

Detection of Cracks on Tunnel Concrete Lining with Electric Conductible Paint

T. OKADA
Railway Technical Research Institute , Tokyo, Japan

S. KONISHI
Railway Technical Research Institute , Tokyo, Japan

T. MOHRI
Horyo Sangyo Co.Ltd , Osaka , Japan

K.TATEYAMA
Kyoto University , Kyoto , Japan

ABSTRACT: Cracks, which grow on concrete linings in tunnel, have been recognized to be one of serious problems for tunnel maintenance because it is part of potential greater failure. Development of convenient methods is requested to detect cracks on the concrete lining, precisely in early stages. Electric conductible paint contains carbon particles to achieve high electric conductivity. We employed this paint as a sensor for detecting cracks on the tunnel concrete lining. In this system, the paint is given a coat on the surface of concrete linings to be a line of 2-3cm width, and the electric conductivity of the paintshould be monitored due to measuring electric resistance of it. When a concrete lining cracks, the line of paint will be cut off at that place, and then the electric resistance will becomes infinity immediately. Therefore, occurrence of cracks can be detected precisely by monitoring the electric conductivity in control offices distant from the tunnel. We have made some experiments with the concrete samples tostudy : 1.Durability of the paint under various conditions. 2. Relationship between the size of concrete cracks and cutting of the paint. The experiment made it clear that the paint shows the good performances in the durability. It has been also made clear that the sensibility of the change in electric conductivity depends on the thickness of the paint. It means that various size of the cracks can be detected

1 INTRODUCTION

It is important for safety and durability of tunnels to detect cracks on tunnel linings in early stages. Even if a crack occurs individually, it is possible that the crack grows, merge with other cracks and lead to spallings. Observation and hammering are generally applied to detect deformations. But those methods have problems, which are shown below. 1. Vary of classification due to vary of skill in inspection. 2. Lack of time for inspection in railway tunnels and road tunnels. 3. Discontinuous observation. 4. High price.

Then some novel detection methods, such as an inspection car and a observation system with optical fiber, are developed and applied recently. A detection method with electric conductible paint, which are described in the paper, is one of the most available methods. Merits of the method are shown below. 1. Simple and precise detective method. 2. Continuous monitoring. 3. Cost effective. We have many experimental tests for confirming that electric conductible paint is applicable in practice. The paper describes concept of detection with the paint and results of the various tests.

2 CONCEPT OF DETECTIVE SYSTEM

Electric conductible paint contains carbons as its ingredient, and thus it shows high electric conductivity. We employed this paint as a sensor for detecting cracks that come into existence on the tunnel concrete lining. In this system, a line is drown on the surface of concrete linings by the electric conductible paint with 2cm to 3cm width, and the electric conductivity of the line is monitored due to measuring electric resistance of it. Figure 1 shows a concept of the detective system. When cracks are induced on the linings, the line of paint will be cut off at that place, and then the electric resistance will increase immediately. Therefore, occurrence of cracks can be detected precisely by monitoring the electric conductivity in a control office distance from the tunnel.

If many circuits with a pattern are made in a tunnel, we will be able to identify a rough position on which deformation of tunnel lining occurs

If chronological change is recorded automatically, we will be able to know when a deformation happens on tunnel lining.

Figure 2 shows a concept of the continuous observation system.

3 FIELD TEST

3.1 *Electric conductible paint*

Electric conductible paint is in general used for preventing static electricity, such as paints for floors. We used it on the market for the tests. It contains carbons and it is black. The specific gravity is from 0.2 to 1.2. Electrical resistance is 10 $^1 \Omega \cdot$ cm and below. The electric conductible paint is diluted with thinner at painting properly.

3.2 *Durability of paint*

Material of the electrical conductible paint is good weather proof. To confirm durability of it after painting on the concrete structure, we are exposing a line painted on a model lining in the field and observing it (Figure 3). The model lining is one thirds size model. The line is painted with a pattern on round direction. It is about 2 cm in width and 28 m in length. Although 6 months have passed after

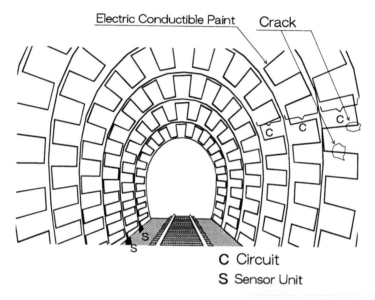

Electric Conductible Paint   Crack

C Circuit
S Sensor Unit

Figure 1. Concept of the detection method with an electric conductible paint.

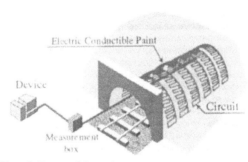

Figure 2. Concept of the continuous system.

Figure 3. D durability test with scale model lining.

painting, aging, such as peeling off and crumbling, can not be found and electric resistance is the same as initial value (350 kΩ). It is still on the half of a long term test. But, we consider that the electric conductible paint has satisfactory durability, because change of the environmental condition in tunnel due to variation of temperature and humidity is less than it in the field.

3.3 *Time for drying*

The electric conductible paint dries, as solvent evaporates after painting. It is thought generally that drying of the paint require a few week. We checked time for drying in experiments. Electric conductible paint was painted on glass plates (Figure 4). Plates were weighed at regular intervals with a chemical balance in order to investigate relations between time and weight of paint. Three cases with different thickness of the paint were carried out. Results are shown in Figure 5. It was known that weight levels off in few hours after painting. Time for drying is affected by temperature and humidity. But, it was known that the electric conductible paint will dries out for one day, if it is cured after painting.

3.4 *Detectable*

We studied a crack width of blocking conductivity due to bending test of mortal bars. The bars are 40mm in width, 40mm in height and 160mm in length. A line of the electric conductible paint was drawn with a brush on the bar. The line is 10mm in width. The bars were loaded with an unconfined compression test unit and cracks were made in bars.

A specimen is shown in Figure 6 and specimens afte

As load increases, a crack gradually widens. When a width of the crack become some size, a line by the electric conductible paint also cuts off. A crack width was measured with a π shaped gauge continuously and electric resistance of the painted line was also measured at the same time. Results are shown in Figure 9.

Values of crack width, at which electric resistance increase suddenly, were 0.25mm and 0.1mm in width. It is considered that difference between two results due to difference in thickness of the paint. Figure 9 shows parts in which values of resistance increase slowly in proportion as cracks open. The reason is considered that paints extend and sectional areas of those decrease gradually as cracks open. And, when width of cracks became 0.25mm and 0.1mm, paints cut off and electric resistance became infinity in each test. Because a width of the harmful crack, which affect safety and durability of tunnel lining, is considered 0.3mm to 0.5mm in generally, results of tests must suggest that the detective method with electric conductible paint is available.

Figure 4. Drying test.

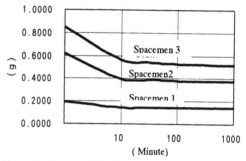

Figure 5. Chronological change of paint material weight

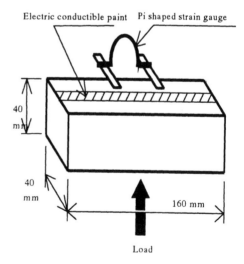

Figure 6. Spacemen.

327

Figure 7. Crack detection test with loading.

Figure 8. Spacemen after testing.

NO. 2-1

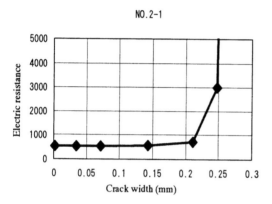

NO. 2-2

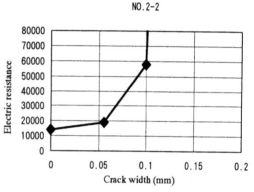

Figure 9. Relations between crack width and electric resistance.

4 CONCLUSIONS

CoNCLUTIONS We have the following remarks as some conclusions of the presentpaper.

(1) Although 6 months have passed after painting, aging of paint, such as peeling off, crumbling and change of conductivity, can not be found in the field test.   The electric conductible paint has satisfactory durability for detection system.

(2) Electric conductible paint becomes dry sufficiently, if it is cured for one day after painting.

(3) When width of cracks became 0.25mm and 0.1mm, electric conductible paint cut off and electric resistance became infinity in each test. Because a width of the harmful crack, which affect safety and durability of tunnel lining, is considered 0.3mm to 0.5mm generally, the

detection method with the electric conductible paint is available.

To establish the crack detection system, weare   planning the following.

1) Due to increase of bending tests and study on dtailed relation between thickness of paint and crack width when paint cut off.
2) Improvement of an electric conductible material.
3) Improvement of a painting method.
4) Confirmingtest in site.

Modern Tunneling Science and Technology, Adachi et al (eds), © 2001 Swets & Zeitlinger, ISBN 90 2651 860 9

Soundness investigation of tunnel concrete by core sampling

H. Saito & T. Uebayashi
Transportation Bureau, City of Osaka, Japan
N. Tasoko
Chuo Fukken Consultants Co., Ltd. , Japan

ABSTRACT: We report the results of research concerning 163 concrete cores sampled in order to clarify the condition within tunnel concrete and to grasp the mechanism about the tunnel concrete fall in Osaka Municipal Subway. Formally, there are two methods, observation and sounding, to judge whether tunnel concrete parts fall. However, these two methods don't clarify condition within tunnel concrete. Therefore, we adopted core sampling of tunnel concrete. We obtained the directions and depths of cracks and cold joints (joints separated by delayed concrete placing), the depth and size of honeycomb and void, the neutralization depths, alkali-aggregate reaction, and amount of salinity.

1. INTRODUCTION

The City of Osaka has major problems how to avoid the troubles the tunnel concrete parts fall to trains. There are two methods to judge whether tunnel concrete parts fall.

1) To observe cracks, leaks, honeycombs, and cold joints which appear on the tunnel surface;
2) To sound concrete parts by a hammer and ascertain whether the sound is clear or dull.

However two methods mentioned above don't clarify condition within tunnel concrete. The condition is important by reason of the two factors mentioned below.

1) Tunnel concrete parts have a tendency to fall if cracks and cold joints oblique within tunnel concrete parts (Figure 1.);
2) Tunnel concrete parts have a tendency to fall if honeycomb and void exit (Figure 1.).

To clarify the condition, we adopted core sampling of tunnel concrete. We report the results of research concerning 163 concrete cores as follows.

2. METHODS

The positions of sampled cores are shown in Figure 2.

1) On cracks and cold joints;(5 cm in diameter)
2) 15 cm above cold joints;(5 cm in diameter)
3) 30 cm below cold joints.(10 cm in diameter)

We sampled 5 cm diameter cores except for the position 30 cm below cold joints, where sampled 10 cm diameter cores

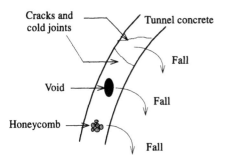

Figure 1. Cracks, cold joints, void and honeycomb within tunnel concrete.

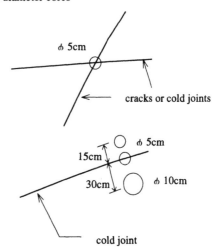

Figure 2. Positions of core sampling.

We sampled 30 cm long cores from the cut and cover tunnel, and 25 cm long cores from the secondary concrete lining of shield tunnel. (Figure 3.) We traced the direction and depth of cracks and cold joints (Figure 4.), and investigated the depth and size of honeycomb and void. (Figure 5.)

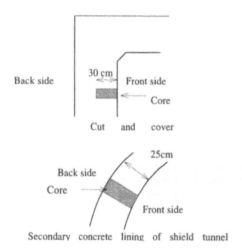

Cut and cover

Secondary concrete lining of shield tunnel

Figure 3. Core length.

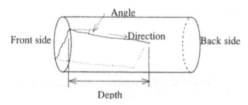

Figure 4. Tracing of cracks and cold joints.

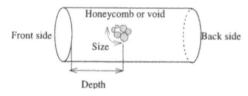

Figure 5. Investigation of honeycomb and void.

Table 1. Details of the cores.

	Cut and cover	Shield	Total
Crack	94	53	147
Cold joint	0	7	7
Others*	2	7	9
Total	96	67	163

* Above or below cracks and cold joints.

Eventually, we clarified the conditions of 163 cores. We state details of the cores in Table 1.

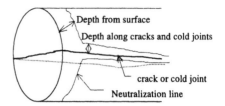

Figure 6. Neutralization depth of cores.

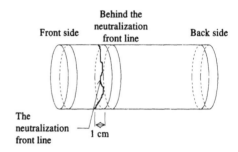

Figure 7. Measurement positions of salinity.

We measured the neutralization depths of the cores from the surface and the depths along cracks and cold joints. (Figure 6.) We judged the crack age occurred from the neutralization depth.

We observed the existence of alkali reactive aggregates through three-day curing, because alkali reaction arises from moisture.

We measured amount of salinity in front side, back side, and behind the neutralization front line. (Figure 7.)

3. RESULTS

We classified the cores on cracks and cold joints into three groups. (Figure 8.)
1) Horizontal;
2) Vertical;
3) Horizontal + vertical (bi-directional).
We show details of the cores on cracks and cold joints in Table 2.

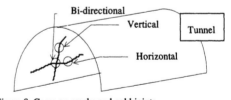

Figure 8. Cores on cracks and cold joints.

Table 2. Details of the cores on cracks and cold joints.

	Cut and cover	Shield	Total
Horizontal	18	21	39
Vertical	65	11	76
Bi-directional	11	28	39
Total	94	60	154

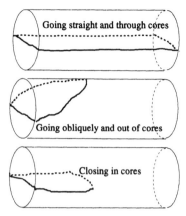

Figure 9. The direction and depth of cracks and cold joints.

We classified the directions and depths of cracks and cold joints into three groups. (Figure 9.)
1) Going straight and through cores;
2) Going obliquely and out of cores;
3) Closing in cores.

We show the classification of the directions and depths of cracks and cold joints in Table 3.

Table 3. The classification of the directions and depths of cracks and cold joints

	Cut and cover	Shield	Total
Going straight and through	62	48	110
Going obliquely and out of	18	22	40
Closing in	23	20	43
Total	103	90	193

We sampled 9 cores, which were above or below cracks and cold joints. We ascertained two of them included honeycombs and a crack respectively.

We show the neutralization depths from the surface in Figure 10.The neutralization depths are various, therefore we obtained the averages for passage time. The average depth for ten years is 1 cm, which nearly equals the depth of our other investigations.

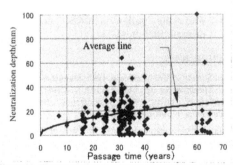

Figure 10. The neutralization depths from surface.

Table 4. The comparison with neutralization depths.

Dc* /Ds**	%
0 to 1/4	12
1/4 to 1/2	16
1/2 to 1	71
Over 1	1
Total	100

* Dc: The depth along the crack.
** Ds: The depth from the surface.

Table 5. The salinity measurements (%).

kg /m³	Front	Behind neutralization	Back
0 to 0.3	60	38	62
0.3 to 1.2	36	46	32
1.2 to 2.5	4	12	2
Over 2.5	0	4	4
Average	0.33 kg/m³	0.87	0.49

We compared with two neutralization depths from the surface and along the crack. We show the results in Table 4.

The neutralization depth is proportional to the square root of passage time. We suppose most cracks occurred at the same time with the construction stage or several years later.

We obtained two cores included alkali reactive aggregates. However no cracks occurs around the aggregates, because alkali reaction is small in degree. We show the salinity measurements in Table 5.

In Japan, 1.2 kg/m³ is given as the critical point that reinforcing bar corrodes. About 95 % of the salinity measurements are less than 1.2 kg/m³ in front and back sides. However 16 % behind the neutralization front line are more than 1.2 kg/m³, and the average behind the neutralization is the highest of the three averages.

4. DISCUSSION
4.1 Cracks and cold joints

We sampled 21(= 7 × 3) cores on ,above ,below cold joints. All cold joints go straight or down slightly and through cores. We ensured no honeycomb and void existed above cold joints, and a group of honeycomb existed below the core. The honeycomb is 8 to 15 cm in thickness, in the inner side of the core. We identified the honeycomb according to the sounding by a hammer.

38%(= 71/186) of the cracks are horizontal, and 62%(= 115/186) are vertical. 55 % (= 103/186) of the cracks go straight and through cores, and 22% (= 40/186) go obliquely and out of cores, and 23% (= 43/186) close in cores.

Thermal and drying shrinkage cracks have a tendency to go vertically, straight and through members. (Figure 11.) We suppose more than half of the cracks are thermal and drying shrinkage ones.

We cannot judge whether cracks are oblique in tunnel concrete so far as we observe the tunnel surface. We are trying whether sounding around cracks by a hammer is useful to judge the oblique crack. (Figure 12.)

We sampled 0.1 to 2.0 mm wide cracks. (Figure13.) We predicted crack width was proportional to crack depth. However there were cracks less than 0.5 mm in width to go through cores, and more than 1.0 mm to close in cores. We suppose vertical cracks have a tendency to go through because of thermal and drying shrinkage tension on a crack plane.

We found the honeycombs along some of the cracks which closed in cores. (Figure 14.) The cracks newer than 30 years old contain no honeycomb, because the control technology of ready mixed concrete has developed highly.

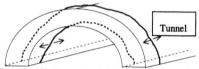

Figure 11. The thermal and drying shrinkage crack

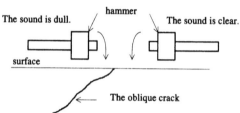

Figure 12. Sounding by a hammer

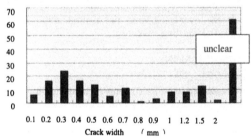

Figure 13. The range of the crack width.

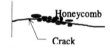

Figure 14. The honeycomb along the crack.

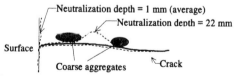

Figure 15. The neutralization front line between coarse aggregates.

4.2 *Neutralization depths*

The neutralization depth is proportional to the square root of passage time. The neutralization depth can be determined with Equation (1), that is Kishitani's equation most popular in Japan.

$$\left. \begin{array}{ll} t = \dfrac{0.3(1.15 + 3x)}{R^2(x - 0.25)^2}C^2 & (x \geqq 0.6) \\[3mm] t = \dfrac{7.2}{R^2(4.6x - 1.76)^2}C^2 & (x \leqq 0.6) \end{array} \right\} \quad (1)$$

where t = passage time(years); C = neutralization depth(cm); x = water-cement ratio; R = neutralization ratio.

The water-cement ratio is 54 % for t = 10 and C = 1. We suppose the tunnel concrete is dense and durable by this ratio.

We found one core which had 22 mm depth along the crack and 1 mm depth from the surface. (Figure 15.) Because the most progressive neutralization was between coarse aggregates, we suppose the concrete between coarse aggregates is sparse, and it is a rare case.

4.3 *Alkali-aggregate reaction*
We suppose the degree of alkali-aggregate reaction is quite small in Osaka Subway's tunnel.

4.4 *Amount of salinity*
Cores including amount more than 1.2 kg/m^3 of salinity are sampled near the sea. The salinity of seawater is supposed to be brought to tunnel concrete.

5. CONCLUSIONS

1) Most cracks and cold joints go straight and few go down slightly, but every crack or cold joint has no tendency to fall.
2) The neutralization of concrete is progressing gradually and the depth is various, but the neutralization does not reach the designed position of reinforcing bars.
3) Alkali-aggregate reaction is quite small in degree.
4) The salinity of concrete is almost low level, though cores near the sea show high salinity. And the salinity moves behind the neutralization front line.

REFERENCES

Japan Society of Civil Engineers 1995. Maintenance guideline (proposition) for concrete structures. *Concrete Library 81.*

Kishitani, K., et al. 1986. Neutralization. *The series of durability of concrete structures.*

Public works research institute, Ministry of Construction 1998.Soundness diagnosis manual (proposition) for concrete structures. *Joint research reports in light of the development of soundness diagnosis engineering for concrete structures.*

Modern Tunneling Science and Technology, Adachi et al (eds), © 2001 Swets & Zeitlinger, ISBN 90 2651 860 9

Comprehensive Safety Inspections of Sanyo Shinkansen Tunnels

T.Kondo
Track and Structures Department, West Japan Railway Company

M.Ichida
Track and Structures Department, West Japan Railway Company

ABSTRACT: In 1999, a series of accidents occurred in which pieces of concrete lining in the Fukuoka Tunnel and the Kita-Kyushu Tunnel on the Sanyo Shinkansen Line came loose and fell. These accidents greatly affected the trust had in the Sanyo Shinkansen Line. Therefore, West Japan Railway Company (hereafter, "JR West") conducted comprehensive safety inspections of the Sanyo Shinkansen tunnels more securely and thoroughly ever had(hereafter, "comprehensive inspections") to verify safety of the Sanyo Shinkansen Line. JR West has taken the completion of comprehensive inspections as a new starting point and began implementing various new safety measures. In this paper, we present an outline of comprehensive inspections, the results of the inspection, and technological development efforts related to tunnel maintenance that have been undertaken since the implementation of comprehensive inspections.

1 INTRODUCTION

In June 1999, a piece of concrete falling from the lower part of a cold-joint section hit a Shinkansen train running in the Fukuoka Tunnel on the Sanyo Shinkansen Line. In October 1999, a piece of concrete fell from a filling port on a sidewall in the Kita-Kyushu Tunnel. These two accidents severely damaged safety reputation of the Sanyo Shinkansen line. Although two accidents occurred at weak points, at a cold-joint section and at a concrete placement structure, JR West conducted comprehensive inspections to verify safety of the Sanyo Shinkansen Line more thoroughly and securely ever had , since previous inspections had failed to find indications of the accidents.

2 INSPECTION OVERVIEW

An outline of comprehensive inspections is shown in Table 1.

All concrete surfaces of the 142 tunnels on the Sanyo Shinkansen line, a total surface area of 590 million square meters and a total length of 280.5 km, were examined during comprehensive inspections. Comprehensive inspections consisted of the followings:

1. Hammering tests for all concrete surfaces.

2. Careful hammer testing at important structural areas or weak areas such as arch crowns and construction joint sections.

3. Implementing necessary measures indicated by inspection results, such as the removal of concrete fragments.

4. Verification of the installation conditions of tunnel fixtures.

Table 1. Inspection Overview.

object	Sanyo Shinkansen 142 tunnels total length 280.495km total surface area 590 million m²
term	25th, October 1999 ~ 15th, December 1999 (52days)
total amount of work	The total number of persons : 69,000 The total number of machines: 9800
remarks	The extension of the inspection hours :The operation cancelled between Hiroshima st. and Hakata st. from 11th, November till 15th, December

3 INSPECTION RESULTS

Comprehensive inspections results confirmed that 99.8% of the concrete surfaces were in good condi-

Table2. The results of Hammering tests

contents of measures		areas of measures	the number of a place	areas (m²)
hammering		all concrete surfaces		5. 9million
repair etc.	hammering off	honeycombs	1164	445. 7
		floating concrete	10, 205	1856. 9
	repair such as steel boards	honeycombs	178	195. 3
		dull-sounding sections around the important crack	283	212. 2
supervision etc.	individual supervision	dull-sounding sections around serious cracks without water leakage and a single crack with water leakage	982	339. 8
	confirmation of safety by the sampling examination	light-sounding sections of no being hammered off	23, 801	5261. 9
		dull-sounding sections without a serious crack	4525	1378. 4
no measures		normal-sounding sections		5. 89 million (99. 8%)

tion. 9690.2 square meters of concrete surface (approximately 0.2% of the total surface area) were shown to have some defects, and concrete fragments in an area of 2302.6 square meters were hammered off. However, concrete fragments within the remaining 7387.6 square meters could not be removed by hammering. (Table 2)

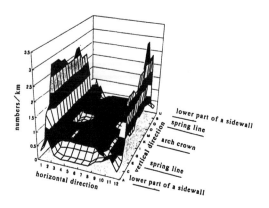

Figure1. The distribution of honeycombs, light-sounding sections and dull-sounding sections.

4 INSPECTION FINDINGS

Figure 1 shows the distribution of honeycombs, light-sounding sections and dull-sounding sections. the parallel direction to a rail, defects were found primarily around the construction joint sections, which are weak points caused by the construction process. At right angles to a rail, faults were seen comparatively more on arched areas than on sidewall areas.

5 APPLICATION OF THE INSPECTION DATA TO FUTURE MAINTENANCE WORK

The findings of comprehensive inspections were reported to the Tunnels Safety Issue Committee (Chairman: Professor Toshihisa Adachi, Kyoto University), which was established by the Japanese Ministry of Transport. The findings were also reflected in the contents of "Tunnel Maintenance Manual" published by Ministry of Transport in February 2000. In the manual, inspection systems were revised and important points were reconsidered.(Fig.2)

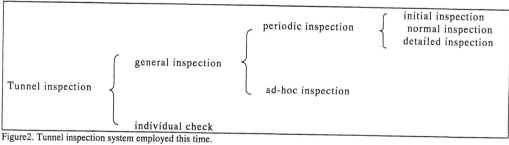

Figure2. Tunnel inspection system employed this time.

The following improvements in maintenance operations were also implemented:

5.1 Securing Mobility of Maintenance Machines

Various measures have been implemented to ensure the maneuverability of the inspection and to improve the precision of inspections for arched areas. On the Shinkansen lines, JR West developed an inspection car with booms which make it easier for an inspector to reach arch crowns. The car also has a platform for inspection of wide arched areas and has strong force of traction to pull railway wagons.

For conventional lines, JR West has also developed an inspection car that can run on and off tracks and that has similar booms and platforms.

5.2 Database of the Inspection Results

The collected data was compiled into a CAD system database to provide fast and easy methods for searching, updating and modifying the data. The constructed database system uses GIS.

Also, in order to fully ascertain and preserve the initial condition of the tunnels as a whole after comprehensive inspections, and to visually complement the diagrams and drawings of tunnel defects, continuous and color-imaged data of the surfaces of the tunnels were fully prepared using slit cameras.

5.3 Improvements of skills

A guidebook, which is a revised version of "Tunnel Maintenance Manual" was published by JR West. The guidebook includes knowledge gained from Comprehensive inspections, and details and explanation of tunnel inspections. All employees of JR West engaging tunnel inspections were trained using the guidebook. During the training, the point of maintenance methods and changes made to the conventional methods were emphasized.

Also, Sample honeycombs, cold joints and construction joint sections with derivative cracks were made for training and practicing hammering test in order to improve the skill of the inspectors. (Fig.3)

6 DEVELOPMENT OF NEW TECHNOLOGIES

A tunnel inspection method is required to ascertain the state of covering materials of the tunnels accurately, uniformly, and effectively. Then, based on the results of the analysis of inspection data, monitoring and/or repair measures must be devised. However, current inspection methods and the evaluation of inspection results rely, as always, on human implementation. It is necessary to improve precision of the inspection through the increased use of inspection instruments and systemization as well as through more uniform and speedy methods.

As for defects below concrete lining of the tunnels, careful observation and repairs are necessary in case of need. However, an inspection method available at the moment is an only core boring method etc. Therefore new technologies need to be developed.

As such, JR West has undertaken the following projects.

6.1 Development of an automatic inspection system that utilizes photographs of concrete surfaces

Currently, the primary method for inspecting surfaces of tunnels is visual inspection by human eyes. Therefore, it is difficult to conduct such inspections in a uniform, continuous manner, and it is also difficult to accurately record the findings gained from such inspections. Other problems include the possibility of evaluation of inspection results being affected by the capability of the inspectors. To resolve such problems and further improve the precision of inspections, JR West has examined tunnel concrete surfaces imaging systems that will be used to collect data and information which will be applied to the development of a future fully-automatic inspection system.

This project has led to the current development of a tunnel concrete surfaces imaging vehicle (Fig.4) that uses a laser detection system, is capable of operating at a maximum speed of approximately 9 km/h (when inspecting half of the tunnel cross-sectional area), and is precise enough to detect cracks as small as 0.5 mm wide. Also under development is a detailed analysis system capable of automatically detecting and analyzing cracks.

Figure3. Sample honeycombs, cold joints and so on.

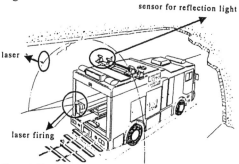

Figure4. The tunnel concrete surfaces imaging vehicle.

6.2 Development of Non-Destructive Methods for Inspecting Concrete Lining

The typical method used to inspect the inside of a concrete lining is hammering inspection. The strong points of this method are that it is simple and efficient, and faulty concrete can also be hammered off at the same time as the inspection. The method also has shortcomings, however -- inspection results can not be evaluated numerically and evaluation of concrete condition becomes more difficult as the depth of the covering increases.

To compensate for these shortcomings, JR West is developing a method for probing the inside of covering material using acoustic elastic waves (Fig.5). With this method, a magnetic-strain sensor is pressed against the concrete, the frequency is modulated over time as low-frequency sound waves are sent into the concrete. A reception sensor detects the dominant frequency and reflection level, attenuation, etc., from which the system attempts to evaluate the presence and depth of cracks, honeycombs, etc.

7 FUTURE UNDERTAKINGS

The projects presented in this paper are expected to make dramatic improvements in the precision of tunnel inspection methods compared with conventional inspection methods.

In regard to repairs for abnormalities discovered by inspections, there is a particular problem with the Shinkansen lines. Usually, within 1 to 2 hours after a repair has been made on a Shinkansen line, a train passes through the tunnel, generating negative pressure and wind pressure particular to tunnel conditions. As such, a repair must be adequately strong and have sufficient adhesive force to withstand these pressures, so current repair methods involve primarily the use of steel materials.

However, the deterioration of the bolts holding the steel materials in place could cause the steel material itself to fall, so it is very important that a substitute repair method is developed.

It is also important that operation efficiency be improved, such as means of efficiently transporting work materials and personnel and methods for easily erecting work site equipment such as scaffolding.

REFERENCES

Japanese Ministry of Transport 2000. Tunnel Maintenance Manual(in Japan)
Railway Technical Research Institute 2000. A draft of Tunnel Maintenance Manual(in Japan)
Track and Structures Department, West Japan Railway Company 2000. A guidebook for the Japanese Ministry of Transport's Tunnel Maintenance Manual(in Japan)

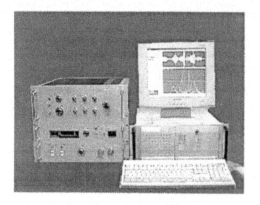

Figure5. Non-Destructive Methods for Inspecting Concrete Lining.

Modern Tunneling Science and Technology, Adachi et al (eds), © 2001 Swets & Zeitlinger, ISBN 90 2651 860 9

A New Non-Destructive Testing Method for Crack and Its Application to Tunnel Structure

T. Nakamura , N. Kawamura & Y. Hattori
Tokyo Electric Power Company, Tokyo, Japan

K. Egawa & J. Wu
Central Giken Co., Ltd., Tokyo, Japan

ABSTRACT: It is well understood that cracks in the concrete lining of a tunnel (particularly those that are progressing) can be a severe safety hazard to the tunnel structures. While information on crack depth is essential, the performance of existing crack detection methods is far from satisfactory. This paper presents a new testing technique for monitoring the cracking depth and progress. Field experiments have demonstrated that this new method yields results superior to the existing methods.

1 FORWARDS

When the crack depth in concrete structures can be properly monitored, it will improve the precision in predicting load-bearing capacity of structure component. It is well known that the existence of cracking in concrete lining, especially its progress, has profound influence on the safety of rock-embedded tunnel, which validate the extreme importance of monitoring the cracking status. Unfortunately, no satisfactory non-destructive method is currently available to detect the cracking depth, which is one of the most important parameters of concrete crack.

In this paper, the authors report a new high-precision non-destructive testing method. Unlike the existing ultrasonic-based non-destructive methods, this new method relies on the surface-generated impulse elastic wave to predict the crack depth, through measuring Rayleigh wave component and its energy propagation. This method is applicable to the cases in which the cracking surfaces are closed, or the cracks are filled, such as by water, powder or calcium lime.

The authors believe that precise measurement of concrete crack status (length, width and depth) as well as careful monitoring of cracking progress will lead to greatly improved precision in safety evaluation of civil structures, including tunnels.

2 EXISTING METHODS

2.1 Classification

The numerous existing crack depth measurement methods can be classified into three major categories.

Category 1: Measure the wave propagation time and distance, then estimate the crack depth according to geometry relationships. This type of method is the most popular one, and some representative methods include T_c-T_0 method (L-L formula), T method, BS method, Delta method (see Figure 1), S-S method, R-S method, etc(JSNDI,1994).

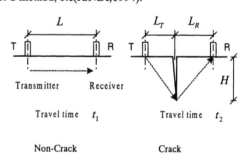

Figure 1. The Concept of Delta Method.

By Delta method, the depth of crack H is given by

$$H = \sqrt{\left\{\frac{L_T^{\,2} - L_R^{\,2} + (V \cdot t_2)^2}{2V \cdot t_2}\right\}^2 - L_T^{\,2}} \qquad (1)$$

where $V = L/t_1$ is the propagation velocity.

Category 2: Utilize the measurement of changes in phase angle of first arrival signals.

Category 3: Utilize the frequency analysis (predominant frequency of received signal).

2.2 Common characteristics and problems

Almost without exception, the methods mentioned above all utilize the ultrasonic wave. Theoretically, these methods can also be applied to hammer generated impulse elastic wave. However, they are rarely used in practice due to difficulties in determining the starting time. In addition, these methods have the following common characteristics.

1) Focus on the initial component of received wave propagation signal, such as first arrival time or phase angle, then use geometric relationships to predict the crack depth.
2) Utilize ultrasonic wave as the source and receiving media.
3) Emphasis on using the P-wave signal.

Because of their operation simplicity, these methods have been widely adopted in the metal-defect detection field. However, their applicability in the civil engineering industry is severely compromised by following issues:

1) Different structure scale: While the dimensions of most metal parts are in the range of several millimeters to centimeters, typical concrete structures can be as large as tens of centimeters or even larger such as for tunnels or dams.
2) Different material properties: Unlike metals, which are nearly homogeneous materials, concrete is largely heterogeneous in which the wave scattering and attenuation are much larger than those in metals.
3) Different crack conditions: The crack in metal material is normally free of filling such as water or granular, while in concrete cracks the filling is very common, especially when the crack surfaces are in contact. Furthermore, in reinforced concrete structures, steel bars frequently present in the cracks. Figure 2 shows us that the first arrival time t_2 becomes earlier due to the contact of crack surface.

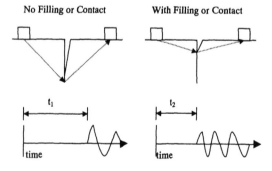

No Filling or Contact   **With Filling or Contact**

Figure 2. The effect of filling and contact.

With the awareness of these differences, it is easy to understand the following problems when the existing methods are applied to concrete structures directly. First, ultrasonic source wave is weak in energy, and the wave signal reflected at crack fronts will be gradually weakened by large scale concrete crack, leading to a much reduced S/N (signal/noise) ratio. Second, because of compression characteristics of P-wave, the water, filling or contact make P-wave penetrate through the crack surface easily. While the transmitted wave energy is close to or even exceed the reflected wave energy, the depth of crack will be greatly underestimated. Obviously, the introduced error is on the unsafe side and it grows even larger with deeper crack and consequently weaker reflected wave energy.

3 NEW METHOD BASED ON RAYLEIGH WAVE

3.1 Properties of Rayleigh wave

Rayleigh wave is one type of elastic waves that exists only in the adjacent of media surface, which are commonly referred as surface waves.

Some of the major characteristics of Rayleigh wave are (F.E.Richart etc., 1970):

1) Among the elastic waves generated by surface excitation, Rayleigh wave carries the most energy due to two reasons. First, Rayleigh wave is the dominant component of surface generated elastic waves. In the perfect-elastic domain, the total wave energy generated by a point source on the surface of semi-infinite medium is distributed as following: P-wave 7 percent, S-wave 26 percent and Rayleigh wave 67 percent. Second, the attenuation (mainly geometric attenuation) of Rayleigh wave is much smaller than that of P-wave and S-wave (or body waves). Some researches indicated that the Rayleigh wave amplitude attenuates by $r^{-0.5}$ on the surface of semi-infinite medium, where r is the source distance, while the body waves attenuates by r^{-2} (and by r^{-1} if within the medium). Therefore, the Rayleigh wave carries the major portion of wave energy propagating along the surface and is the easiest one to detect.
2) Rayleigh wave is strongly characterized by the shearing properties of the media in which it propagates. As demonstrated in Figure 3, the direction of the media particle movement is nearly perpendicular to the direction where wave propagates toward.
3) Rayleigh wave amplitude drops rapidly with increasing depth from the surface, consequently, the majority of Rayleigh wave energy exists within the immediate adjacent of surface of the semi-infinite medium with depth no more than one wavelength.

It is based on these characteristics of the Rayleigh wave that we developed this new non-destructive crack detection technique reported herein.

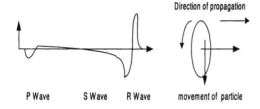

P Wave   S Wave   R Wave   movement of particle

Figure 3. Characteristics of Rayleigh wave.

3.2 Principles

As described before, the wave energy density of Rayleigh wave is gradually reduced by geometric attenuation, meanwhile the wave energy is dissipated through material attenuation due to concrete viscosity and other reasons. It is therefore possible to maintain the wave amplitude (or wave energy) constant through appropriate corrections that compensate these attenuations.

In addition to the geometric and material attenuation, Rayleigh wave energy can also be dissipated by the existence of cracks. The changing of wave energy before and after crack is more prominent with increasing crack depth or shorter wavelength, and it can be used to estimate the existence and depth of cracks. A schematic illustration of the principle is provided in Figure 4.

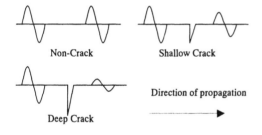

Non-Crack   Shallow Crack

Deep Crack   Direction of propagation

Figure 4. Principle of new method by Rayleigh wave.

The rigorous theoretic solution of correlation between crack depth H and ratio of wave amplitude change x (x = amplitude after crack / amplitude before crack) is nearly unobtainable. Therefore, we assume the following log-normal relationship (Eq.2), as shown in Figure 5, where C and C_0 are constants.

$$H = C \ln(x) + C_0 \tag{2}$$

When no crack exists, $H = 0$ and $x = 1$, which leads to $C_0 = 0$. C can be determined by laboratory testing.

A series of laboratory tests were performed on concrete block specimens with length of 869-884mm, width of 445-460mm and height of 452-520mm. The depth of crack in those specimens is 0, 72, 178, 325, and 503mm, respectively. Figure 6 shows the laboratory testing results, from which the correlation between C and Rayleigh wavelength can

be determined using the minimum square method as following:

$$C = -0.7429\lambda \tag{3}$$

where λ is the wave length and the crack depth H is given by:

$$H = -0.7429\lambda \ln(x) \tag{4}$$

Figure 5. Assumed relationship between H/λ and x.

Figure 7 shows the comparison between the actual crack depth and the evaluation value, from which it is clear that this method yields satisfactory results. Additional tests show that the filling in the crack, such as water or sand, has negligible effects on laboratory testing results, as expected.

This method is named "Surface Wave Method" since the Rayleigh wave is one type of surface wave.

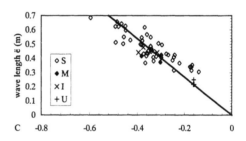

Figure 6. Relationship between C and λ, where symbol S, M, I, U illustrates the exciting source, small, mini, medium hammer and ultrasonic transmission.

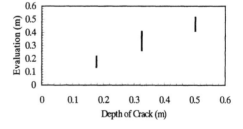

Figure 7. Evaluation to crack depth of blocks.

3.3 Measurement and analysis system overview

The measurement and analysis system consists of hardware sub-system and software sub-system. Depending on the subject being examined and the precision requirement, the hardware system is either based on portable PC (Rapid inspection system, as shown in Figure 8) or desktop PC (Specified inspection system, as shown in Figure 9).

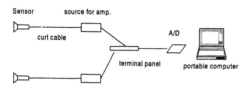

Figure 8. Rapid inspection system basing on portable computer.

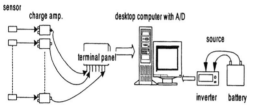

Figure 9. Specified inspection system basing on desktop computer.

The wave signal collected by hardware system is processed through an analysis system to calculate the crack depth and opening depth. The major steps in the evaluation process are:

1) Wave signal integration and baseline correction. The "Surface Wave Method" is based on the change of wave energy that relies on the media particle velocity. However, the direct measurement of particle velocity can be very difficult, so we use the acceleration sensor to measure the particle acceleration, and then compute the velocity through numerical integration, and finally perform the baseline correction.

2) Geometry attenuation correction according to the location of sensors and wave source.

3) Material attenuation correction, which takes into account the attenuation effect of different materials.

4) Thickness correction. Under certain circumstances, the concrete wall or lining thickness variation can affect the wave amplitude ratio, and this effect should be corrected.

3.4 Measurement error reduction techniques

Table 1 summarizes the principal types of measurement error, error source and reduction methods. Due to the poor uniformity of concrete material and adverse surface condition of tunnel lining, the measurement error of the "Surface Wave Method" can become excessive when compared to the methods based on initial signal arrival time or phase angle. Therefore, the error reduction technique is critical

for improving measurement precision, especially when the ratio of wave amplitude change must be examined.

Table 1. Principal error types, source and reduction methods.

Error Type	Random Error	Systematic Error
Error Characteristics	Irregular	Regular & large effect
Error Sources	Environmental Noise & Electrical Noise	System Drifting & Sensor Fixing Error
Error Reduction Methods	Repeated Measurement (theoretically, the S/N increases with $N^{0.5}$)	Calibration & Adopt "Two-way Excitation" Method

In order to reduce the systematic error, the authors developed the "Two-way excitation" method, whose principle is shown in the schematic illustration Figure 10. The source signal amplitude is denoted as S_0, while Ch1 and Ch2 receiving signal is denoted as S_1 and S_2, respectively. The excitation source is first placed at the left side of Ch1 and the measurement is taken both Ch1 and Ch2, then the configuration is reversed. The source is now at the right side of Ch2 and measurement is taken.

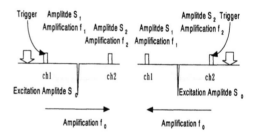

Figure 10. Principle of the "Two-way Excitation" method.

Equation 5 can be used to compute the amplitude ratio η_1 when the source is at the left side of Ch1, in which f_0 is the crack amplification factor ($f_0 < 1$) while f_1 and f_2 are the amplification factor for Ch1 and Ch2.

$$\eta_1 = S_2 / S_1 = f_2 \cdot f_0 / f_1 \qquad (5)$$

Similarly, when the source is at the Ch2, the amplitude ratio η_2 can be computed using Equation 6.

$$\eta_2 = S_1 / S_2 = f_1 \cdot f_0 / f_2 \qquad (6)$$

By averaging the two amplitude ratio, the crack amplification factor f_0 can be easily computed using the following equation:

$$f_0 = \sqrt{\eta_1 \cdot \eta_2} \qquad (7)$$

As indicated in Equation 6, the amplification factor for Ch1 (f_1) and Ch2 (f_2) have been cancelled by averaging these two measurements, thus this method effectively reduces the system error of Ch1 and Ch2 as well as the effects of sensor fixing irregularity.

3.5 Crack detection results

The performance of the "Surface Wave Method" was evaluated by comparing its predictions with measured results from a number of core boring tests on actual civil structures. In addition, the authors developed another crack depth detection method called "Velocity Method" that is similar to the Delta method mentioned before, but with impulse elastic wave instead of ultrasonic wave. Since the "Velocity Method" is capable of recording the complete initial waveform, it is very precise in measuring origin time.

Figures 11 and 12 show the crack detection results (in solid mark) and Core-boring measurement results on various civil structures, such as tunnel, wall and facing using "Velocity Method" and "Surface Wave Method". A comparison between the two figures shows that "Surface Wave Method" yields satisfactory estimations of crack depth, while the prediction results of "Velocity Method" come much lower than actual observation values.

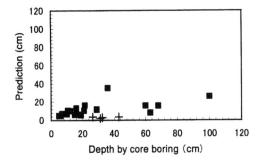

Figure 11. Prediction by "Velocity Method".

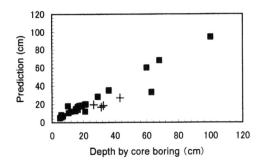

Figure 12. Prediction by "Surface Wave Method".

The authors performed several tests on the pressurized waterway tunnels, and the testing results are shown in "+" marks. It is clear that those prediction values are substantially lower than actual depth. A possible reason of this abnormality is the stress redistribution introduced by inner pressure release.

It is believed that the estimation by "velocity method" is close to the crack opening depth, and the "Surface Wave Method" should be used to estimate the crack depth (See Figure 13).

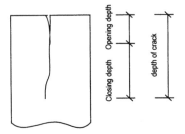

Figure 13. Definition of cracks depth.

3.6 Application in the tunnel stability investigation

This method has been applied in a complete investigation of a waterway tunnel, using mainly the rapid inspection system. The specified inspection system was used in some occasions. The tunnel has a length of 10 kilometers, a width of 3.03 meters and a height of 3.03meters. The testing program took three days and over a hundred cracks in longitudinal direction (parallel with the axis of tunnel) were investigated, which demonstrated the high efficiency of this new technique.

A summarization of the major testing results is presented as following:
1) The relationship between width and depth of crack: in this tunnel, the crack may be deeper with the increasing of width.
2) 2)The relationship between location and depth of crack: in this tunnel, the testing results show that the depth of cracks located on arch section is usually shallower than those located on sidewall. This observation is easy to understand since the crack on sidewall occurs mainly due to bending, while the crack on arch section is likely due to compression deviation.
3) Although the lining is made of plain concrete, the authors believe the occurrence of cracking does not necessarily indicate that the crack penetrates the lining.

While these testing results agree well with common engineering judgment, some exceptions do exist due to different cracking mechanism. Therefore, it is not appropriate to predict the depth of crack simply according to its width and location.

3.7 Summarization and characteristics of the "Surface Wave Method"

The proposed testing procedures for the "Surface Wave Method" are shown in Figure 14, where the explanation of corrections is described in 3.3.

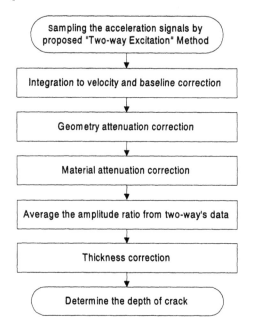

Figure 14. Flow chart for the "Surface Wave Method" testing.

As described before, the "Surface Wave Method" has following important characteristics:

1) The hammer-generated impulse elastic wave carries more energy than ultrasonic wave does, which is beneficial for deep crack detection.

2) Rayleigh wave depends on the material's shearing properties. Both theoretical analysis and field-testing results show that the crack filling should have small effect on the detection result due to its small shearing stiffness (e.g., water, granular and calcium lime). Similarly, the effect of crack surface contact on the testing results is also small if little compression force is applied on the crack surface.

3) Since this method is based on the changing wave energy during propagation, it is not sensitive to the existence of steel bar in the crack due to the relative small cross section area of such bars.

4) The detection resolution and coverage depths of this method are mainly controlled by the wavelength of Rayleigh wave. Because most Rayleigh wave energy is confined within a depth about one wavelength, longer wavelength is necessary for deep cracks. However, for shallow cracks, shorter wavelength is preferred for better detection resolution.

5) However, it should be pointed out that the crack shearing stiffness may increase as compression force is applied on the crack surface, which cause the prediction depth by "Surface Wave Method" to be lower than actual value. The correction technique is waiting for being developed.

4 CONCLUSIONS

While many different methods have been developed for the purpose of measuring concrete crack depth, they are mostly based on a single principle, which relies on the characteristics (travel time or phase angle) of the earliest arriving body wave. Consequently, performance of these methods is strongly affected by the contact and filling material between the crack faces, especially when the crack is deep. In contrast, the proposed "Surface Wave Method" is based on the principle of energy density loss of Rayleigh wave, which is insensitive to the contact or filling between the crack faces.

Experimental evidence shows that this proposed technique provides more accurate and reliable estimation of crack dimensions than the existing methods. When proper wavelength is selected, this method is capable of detecting cracks deeper than 1 meter from the surface.

This new technique has important applications in the maintenance and safety evaluation of existing tunnel structures, such as cracking development monitoring. Besides, the predicted depth of crack will also be an important parameter for evaluation of residual strength of concrete lining.

The authors are in the process of obtaining a patent for the "Surface Wave Method".

5 REFERENCES

Japanese Society for Non-Destructive Inspection. 1994. *Non-Destructive Testing Method for concrete structures*. Yokendo: Tokyo. (in Japanese)

F.E.Richart, Jr., J.R.Hall, Jr. & R.D.Woods, 1970. *Vibrations of Soils and Foundations*. Prentice-Hall, Inc., New Jersey.

Railway Technical Research Institute, 2000. *Railway Tunnel Maintenance Manual (Draft)*. (in Japanese)

Modern Tunneling Science and Technology, Adachi et al (eds), © 2001 Swets & Zeitlinger, ISBN 90 2651 860 9

Development of new grouting material for tunnel rehabilitation

T. Asakura
Dept. of Earth Resources Engr., Kyoto University

S. Kohno
Technology Development Dept., Civil Engr. Div., Shimizu Corporation

T. Kiuchi
Technology Dept II., Civil Engr. Div., Shimizu Corporation

ABSTRACT: It is common that the void is formed behind the tunnel lining on the crown portion of the aged tunnel due to the geological reasons. It is strongly recommended that those void is filled with the grouting material to improve the stability of the tunnel.Authors newly developed the grouting method called "Aqua Grout Method"using the new grouting material consisting of cement, bentonite, super absorbent polymer, set accelerating agent and water. Through its applications into the actual aged tunnels, the Aqua grout method with new material is found to be very effective as the back-fill method for the voids behind the lining of the aged tunnel.

1. INTRODUCTION

In the tunnel constructed prior to the introduction of the New Austrian Tunnelling Method (NATM), the void was generally formed behind the tunnel lining mainly around the crown portion due to the concrete pumping technologies at that time and geological reasons. When the horizontal earth pressure acts on the tunnel lining with such void under the occurrence of the earthquake or uneven ground stress, the tunnel lining could vertically tend to deform leading to the reduction of the stability of the tunnel. Consequently, it is strongly recommended that the void is filled with the adequate grouting material when its presence is identified.

Conventionally, the air mortar or air milk is widely used as the grouting material for the void due to its low cost. However, since fluidity of such material is very high, the grounting material could easily flow out into the bottom portion of the tunnel and subsequently, down to the lower portion along the tunnel axis. While it flows, it could also flow into the small cracks in the ground and lining, and drains. As a result, the void on the crown portion could not be sufficiently filled. Also, the segregation of the grouted material may occur through grouting when the ground water is present around the void. In pumping the material from the mixing plant in the yard and conveying distance is large, the pumping pressure may crush the trapped air in the material and several cares are needed in keeping the quality.

The authors have examined the corresponding problems embeded in the conventional grouting materials and newly developed the grouting method called "Aqua Grout Method" using the new grouting material(Tachibana 2000).

2. OUTLINE OF NEW GROUTING MATERIAL

2.1 Required Conditions

For the grouting material specially desiged to the void on the crown portion of the tunnel, following qualities may be needed:

(1) The flow of the material into the portion other than the void on the crown is limitted and the material may be grouted only to the planned void (ability of the restricted grouting);

(2) Resistant ability to the segregation of the grouting material against the ground water or high pressure through pumping is very high;

(3) Uniaxial compressive strength of the material for 28days should be at least $1N/mm^2$;

(4) Once the material is grouted into the void on the crown portion, it may rapidly harden and never drop through the cracks of the lining;

(5) The weight of the material is small so that it could reduce the burden for the aged lining;

(6) Construction and its management for the grouting work is simple.

2.2. *Design and Proportion of New Material*

Through the investigations upon the side effects of the air in the conventional grouting material on the stability of the material, the super absorbent polymer in stead of air has been determined to be used in the new grouting material. The super absorbent polymer absorbs the water and the particle may expand more than 50 times in volume and strongly resist to the separation of water even under high pressure.

However, the super absorbent polymer itself absorbing the water is soft and unstable. Consequently, in order to add the strength to the polymer with water, the cement and set accelerating agent are chosen as components. Also, for improving the viscousity of the material, bentonite is used. Bentonite plays an important role in generating the thixotropical aspect in which the high fluidity is guaranteed under the pressure or vibration while viscousity becomes to be high when the pressure is released. Due to such aspect, the material easily flows through pumping pipe and it steadily stands inside the void once it is released from the pipe.

As a result, the components of the new material are cement, bentonite, super absorbent polymer, set accelerating agent and water. For the reference, the example of the propotion of the material is shown in Table 1. The proportion for the actual construction may be designed based upon that prior to the start of the construction with the actual equipments. The actual proportion could be modified correspondin: to the conditions given in each case such as th‹ mixing ability of the mixer, quality and temperature of water.

Table 1. Example of proportion.

Components for 1m³			
Cement	bentonite	Polymer + SAA	water
350kg	285kg	8.4kg	774kg

SAA stands for set accelerating agent

Since all components are prepared as powder, the procedure in manufacturing the grouting material is very simple by just mixing all the components with water in the suitable mixer. The required area of the yard for the mixing plant could be reduced more than 30% than that for the conventional grouting material such as air mortar.

3. VERIFICATIONS OF DESIGNED ABILITIES

Figure 1 shows the table flow test in which material after being extracted from the cylinderical case on the table receives the 15times cyclic shocks through the table. It is confirmed that the material becomes

to be fluid to expand and stop corresponding to each shock. It denotes the thixotropical aspect of the material and it is highly recommended that such tests are implemented prior to the start of the construction.

Figure 1. Table flow test.

In order to confirm the resistant ability of material to segregation by water and filling ability into the limitted void, the material is pumped into the plastic case resembling the void inside the water. The objective of the test is to examine whether the case is filled without any material segregation in the water. Through the test, it is confirmed that the water keeps to be clear even while pumping and case is being filled. It is also observed that the material is being easily pumped into the case with low pressure. (see figure 2). While grouting, ph value in the water is observed to be neutral.

Figure 2. Test of resistant ability to segregation.

In modelling the void with 30cm depth behind the lining on the crown, clear plastic board is installed keeping 30cm interval from the ground. Actually, since grouted material into the void on the crown may flow and expand into all sides, the model is designed so that all sides are free. The grouting pipe is positioned in the center and its end is fixed 5cm apart from the plastic board. The purpose of the test is to examine that grouting material could touch the plastic board and fill the 30cm interval in stead of flowing into all sides. Figure 3 shows that the 30cm interval has been filled with the grouted material without the excessive flows into all sides and it is confirmed that the material owns the very high ability of the restricted grouting.

Figure 3. Test of ability of the restricted grouting.

It is very important that grouted material should not drop through the cracks on the lining particularly for the case of keeping the one lane open to the traffic during grouting works. Conventionally, prior to the construction using the air mortar or air milk, the additional care such as crack sealing to prevent the dropping through the cracks on the lining is needed. The figure 4 is the test to examine the relationship between the crack thickness of the lining and amount of leakage of the material. The 50cm thick lining is modelled using the styrol form and three penetrating cracks which are 5mm, 10mm and 20mm thick from the left in the photo are prepared.

Figure 4. Test of leakage through cracks.

The form with cracks is positioned inside the clear plastic box. The grouting material is directly poured from the top of the clear box into the form and the degree of material going through the cracks is observed. The results of the test demonstrate that the material went through the 5mm and 10mm cracks by only 55mm and 120mm depths from the top of the form, respectively. For the crack of 20mm, the material went through to the bottom and subsequently stopped dropping. Practically, it is easily understood that the crack whose thickness is over 10mm penetrating the lining may not actually exist in the tunnel in common use and this material does not go through the cracks of the tunnel lining in actual grouting without the additional care such as crack sealing.

4. ACTUAL APPLICATIONS

(1) Case of old railroad tunnel
The Aqua grout method has been firstly introduced into the backfill works on the crown of the actual railroad tunnels made of the bricks which are over 90 years old (Tachibana 1998),. (Nakura 1998), (Kohno 1998), (Miyase 1998). All the components of the material are poured into the mixers in the plant and mixed grouting material is pumped into the grouting point about 150m away. The grouting pipe is installed through the drilled hole and fixed at the lining so that the end of the pipe is positioned inside the void 5cm lower than the ground surface. The grouting should be stopped when the pumping pressure at the grouting pipe exceeds $2kg/cm^2$ or grouting material starts dropping through the neiboring pipe denoting that the void around the pipe is filled. Through the deteriorated joints, grouted material has never been leaked out while grouting. Also, the water originally dropping through the joints is still clear even while material is being grouted and it is easily observed that the segregation does not occur in the water.

(2) Case of long railroad tunnel
For grouting in the long railroad tunnel during the limitted time between the last train and first train, the grout work should be efficiently proceeded manufacturing a large amount material during the short time. The continuous and automatic grouting system has been developed and used in the actual long tunnel(Asakura 1999). The system consists of the sub-system located on the ground and that positioned on the cars for grouting work. In the sub-system on the ground, all the component as powder such as cement are stored in the silos. Prior to the start of construction work, the each component except for water is automatically mixed in powder and air-pumped into the silos in the sub-system on the cars. The sub-system on the cars are pulled to the grouting point by the motored cars. Upon construction, premixed powder is automatically mixed with water corresponding to the designed proportion and grouting material is continuously generated. Through the actual use, it is confirmed that the production rate and quality of the grouting material are very stable due to the continuouis mixing. Also, working conditions are very good inside the tunnel with the low noise and without any dusts due to the perfect ly closed system.

(3) Case of long road tunnel
For the long road tunnels with the heavy traffic, it is not practical for the grouting material to be pumped from the mixing plant positioned outside the tunnel

to the grouting point. Also, it is important the size of the equipments for grouting inside the tunnel is small so that high visibility of the drivers of the passing vehicles may be guaranteed leading to the reduction of the possibility of accident. In the application of the Aqua grout method to this situation, the mixing plant may be located in the yard outside of the tunnel. The mixed material is pumped into the agitator track waiting in the yard and conveyed into the grouting point. At the grouting point, only the small pump and mobil lift are prepared and pump receives the grouting material from the agitator track using agitator and poures the material into the void behind the lining. In this case, total length of the equipments at the grout position is small such as less than 30m. Figure 5 shows the equipments at the grout position during construction.

Figure 5. Equipments at grout position.

After the construction using the Aqua grout method, the core borings are performed along the tunnel axis to confirm the filling of the voids. Figure 6 shows the example of the recovered core indicating that the void is perfectly filled with the grouted material. As shown in this example, it is commonly observed that the wooded support is used for excavation and void is divided in the upper and lower portions of the support. Due to the high ability of filling of the material, both portions are perfectly filled.

5. CONCLUSIONS

The Aqua grout method newly developed by the authors is found to be very effective as the back-fill method for the voids behind the lining of the aged tunnel. Through the proper maintenance of the aged tunnel with the suitable technologies including the Aqua grout method, the life of the tunnel may be largely prolonged leading to the safe and cost-saving tunnel rehabilitation. Furthermore, authors will continue the research of the method and propose the more practical grouting system.

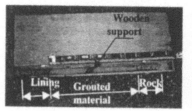

Figure 6. Example of recovered core.

REFERENCES

Asakura, T., C. Sano, T. Omyo, K. Nishikawa & S. Kohno 1999. New Grouting Method using Continuous Grouting System. *JSCE Annual Conference*(VI):434-435(in Japanese).

Kohno, S., T. Asakura, M. Kawashima, K. Yasui & S. Kikuchi 1998. Application of New Grouting Material and its Practice. *JSCE Annual Conference*(VI):226-227(in Japanese).

Miyase, H., T. Asakura, K. Kuribayashi, K. Yasui & K. Nakura 1998. Monitoring upon Application of New Grouting Method. *JSCE Annual Conference*(VI):224-225(in Japanese).

Nakura, K., T. Takahashi, D. Tachibana, M. Soranishi & Y. Masuda 1998. Experimental Study on New Grouting Method. *JSCE Annual Conference*(VI):222-223(in Japanese).

Tachibana, D., S. Konishi, S. Kohno, M. Soranishi & Y. Masuda 1998. Development of New Grouting Material for Existing Tunnel. *JSCE Annual Conference*(VI):220-221(in Japanese).

Tachibana, D., S. Kohno & T. Asakura 2000. Development of a New Backfilling Grouting Material Applied for Rehabilitation of Aged Tunnels. *Vol.11 No. 3 Concrete Research and Technology*:129-138(in Japanese).

Modern Tunneling Science and Technology, Adachi et al (eds), © 2001 Swets & Zeitlinger, ISBN 90 2651 860 9

A fundamental study on dynamic response of tunnel reinforcement structure for high speed railway

T. Kameda
Institute of Engineering Mechanics and Systems, University of Tsukuba
T. Nishioka
Institute of Engineering Mechanics and Systems, University of Tsukuba

ABSTRACT: Recently, there have been some reports of deteriorated concrete linings in railroad tunnels. In order to protect train tracks from falling debris, the installation of reinforcement structure is proposed, which makes possible the installation work at night without suspending train services. Therefore, this proposal should assume the situations where the reinforcement structure would be left unfinished by some reason. In this study, a numerical analysis was carried out on the safety of trains passing through the tunnel without backfilling being conducted, using the measurement data of pneumatic pressure fluctuation and vibration. The analysis results confirmed that the proposed reinforcement structure maintains sufficient safety even in an unfinished condition and also suggested that it is desirable to install displacement-restricting spacers on the ends of the structure to prevent vibration.

1 INTRODUCTION

In 1999, some pieces of tunnel lining from a tunnel in Kitakyushu fell off the inner wall and hit the top part of a train that was passing through the tunnel at the time. In the same year, a similar type of accident occurred in Hokkaido suspending the railway line service for a certain period of time. Considering the nature of these accidents, one can easily predict the possibility of larger debris falling from tunnel walls at any time soon. For reasons of safety it is therefore urgent to take the necessary steps to reinforce deteriorated tunnel linings. As part of such efforts, in the following section, a certain method for installing the reinforcement structure over the inner surface of a tunnel to protect the train track and trains against the fall of lining debris has been proposed. This idea has two major features:

- The proposed installation is based on the ring-by-ring process, which allows workers to finish each process in a short time. Thus, the installation work itself may be conducted without stopping the train service.

- Each ring is divided into segments, which are attached onto the tunnel wall. This does not require a large space for installation and therefore requires no removal of existing catenaries in the tunnel.

Despite these advantages, there is still the possibility that the reinforcement structure may be left unfinished due to problems during the installation work, and the first train of the day passes through the tunnel. Such situations must be taken into account when considering the proposed reinforcement system. In particular, the situation considered the most dangerous is when the reinforcement structure has been assembled to form an arch but no backfilling has been done because of lack of installation time. Possible external forces acting on the reinforcement structure during passage of trains in the daytime include fluctuation of pneumatic pressure in the tunnel and vibration by trains. Data on each of these is necessary to make a numerical analysis of the stability of the reinforcement structure against those forces. A large number of research projects were conducted on pneumatic pressure fluctuation that occurs as trains pass through the tunnel, many of which are related to the Shinkansen (Tanaka et al. 1999; Kage et al. 1993). However, most of them were intended to study the impacts of compressive waves, which are generated by the entry of a train into the tunnel and turn into shock waves at the exit of the tunnel, on the surrounding environment (Kashimura et al. 1994a; Kashimura et al. 1994b; Sasoh et al. 1994). Almost none were made from the viewpoint of pressure given by the pneumatic pressure fluctuation to the wall surface of the tunnel. Particularly, for the existing railway lines, it is very difficult to find the research, except for the subway (Shiratori et al. 1992). This study intended to check the safety of the reinforcement structure without backfiller against the passage of trains in the daytime. For this purpose, pneumatic pressure fluctuation and vibration acceleration were measured in existing railroad tunnels. Laboratory wind tunnel testing was also conducted to learn the effects of wind on the ends of the reinforcement structure caused by train passage so as to acquire data about external forces to be considered in the numerical analysis. In addition, dynamic finite element analysis was conducted based on that measurement data in order to identify the safety of the reinforcement structure and promote fundamental study for the application of the proposed reinforcement structure to high-speed railways.

2 PROPOSED TUNNEL REINFORCEMENT STRUCTURE

A tunnel reinforcement structure, as shown in the Figures 1–2, was proposed with the aim of installing the structure to the tunnel wall during train intermissions at night, without suspending the train service. As indicated by the figures, the structure consists of independent rings each measuring 1 m in the direction of the tunnel axis. This composition is designed to allow workers to finish each ring even in a short period of time. Each ring is composed of four segments. These segments are assembled as they are attached to the inner wall of the tunnel. No removal of catenaries is necessary. The structure is made of stainless steel. Each segment is designed to sufficiently sustain a static load of 1 tf/m. The segment is 16 mm thick.

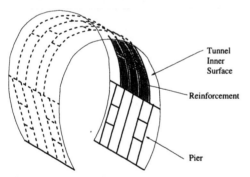

Figure 1. Concept of reinforcement structure.

3 MEASUREMENT OF PNEUMATIC PRESSURE FLUCTUATION AND TRAIN WIND

As shown in the Figure 3, a pressure meter and an anemometer were attached at three locations on the wall surface of the tunnel to measure pressure fluc-

tuation and train wind caused by passing trains. The pressure meter used is PD-80H0005 manufactured by the ST Research Institute; it can measure up to 1,013 ± 50 hPa. The anemometer used is Model 1550 made by Japan Kanomax and is capable of measuring from 0 to 50 m/s. The sampling interval of either measurement is 1/4000 sec. The measurement results show no difference depending on the height of the measurement; the data was almost the same for all heights. The Figure 4 and 5 show the pressure and wind speed data measured at Point C in Figure 3 when a train passed through a single-track tunnel at 58.2 km/h respectively. The anemometers used in this study measure absolute values only. Considering the past studies on train wind, it was assumed that wind at Section A-B in Figure 5 blows in the direction opposite to the train's traveling direction and wind at Section B-C in Figure 5 blows in the same direction as the train's traveling direction. These assumptions were used for the analysis of the effects of train wind in the following discussion.

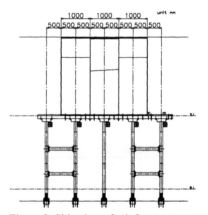

Figure 2. Side view of reinforcement structure.

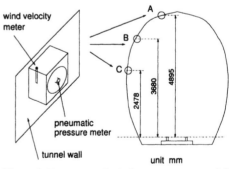

Figure 3. Pressure and wind measurement position.

4 EFFECTS OF TRAIN WIND ON THE ENDS OF THE REINFORCEMENT STRUCTURE

As it is presumed that force derived from train wind acts on the ends of the reinforcement structure, wind tunnel tests were conducted to check this. The test setting includes a 1.8 m long plate placed in front of the simulated tunnel wall; the distance from the plate to the simulated wall is changed in the range of 20 mm to 100 mm. Four conditions were simulated between the plate and the wall: (1) no filler; (2) filler only on the windward side; (3) filler only on the leeward side; and (4) filler entirely in-between. Both uniform flow and turbulent flow were created in all cases for comparison. The test results indicate that due to the action of turbulent flow, a large difference of pressure occurred on the ends and the center only when the filler was applied to the leeward side between the tunnel and the reinforcement structure shown in Figure 6. No significant difference was confirmed depending on distance from the wall ranging from 20 mm to 100 mm. The measurement results with the wall-to-plate distance of 20 mm are shown in the Figure 7. The length scale is normalized by its length. The result suggests that the area up to 0.4 m from the windward edge would be affected by wind (wind pressure coefficient of 4 to 2). Appropriate correction was thus made to pneumatic pressure fluctuation in the area 0.4 m from the end of the windward side when numerical analysis was conducted.

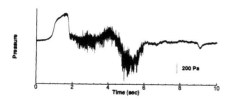

Figure 4. Pneumatic pressure.

Figure 5. Train wind.

5 VIBRATION BY TRAIN PASSAGE

In order to consider the effect of the vibration by train passage in numerical analysis, the acceleration time series is required as the external excitation source. To

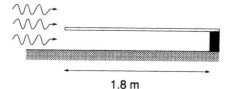

Figure 6. Wind tunnel experimnet.

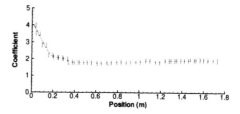

Figure 7. Wind pressure coefficient.

measure the acceleration of vibration caused by the passage of a train, acceleration meters were installed immediately under the track and at the bottom part of the tunnel wall. The acceleration meter used is piezo-electric type PV44A manufactured by Rion Co., Ltd. The sampling interval was 1/12,000 sec. Comparison between these two measurements shows an attenuation of about 1/10 at the peak for the bottom part of the wall. As the input of the numerical analysis, the acceleration recorded at the bottom of the tunnel wall is used. Acceleration waveforms recorded at the bottom of the wall are shown in the Figure 8. The train passage speed was 58.2 km/h. The figure shows 12 peaks, which were caused by wheels passing the measurement point. The magnitude of acceleration is within 1 m/s^2 except the peaks.

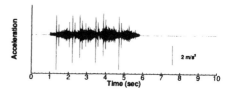

Figure 8. Acceleration.

6 METHOD AND RESULTS OF NUMERICAL ANALYSIS

A steel reinforcement structure without backfiller was simulated by dynamic finite element analysis to analyze the behavior of the reinforcement structure during the passage of a train through the tunnel. Using the LS-DYNA, hexahedral elastic elements, having 8 integration points per element, were used in the analysis. Although the behavior was expected to change due to the condition of engagement of the segments, an elastic constant of 1/10, which is that of the base material, was used in this study. Considering the number of rings installable during the night, the length of the analysis model in the tunnel axial direction was set at 3 m (3 rings). The proposed reinforcement structure has been designed to ensure its maximum integration to the tunnel, by installing spacers between the structure and the tunnel even when the backfilling is insufficient. Thus, rigid elements that restrict displacement of the reinforcement structure only in the direction toward the tunnel wall surface were placed at the positions corresponding to the spacers, in order to create an analysis model as close as possible to that of the proposed reinforcement structure. The model used in the analysis had 703 elements and 1,520 nodes. External forces considered in the analysis were the measurement results of pneumatic pressure fluctuation with the corrected effects by train wind on the ends and the measurement results of the acceleration. An representitive numerical analysis result of effective stress is shown in the Figure 9. The analysis results indicate that the maximum value of effective stress was about 13 MPa momentarily (point A in the Figure 9) and that the stress values remained in the elastic range in the entire reinforcement structure. The results also confirmed that vibration of about 0.5 mm in amplitude, with its fulcrum at the engagement point (C in the Figure 9), occurred in the area where no spacers were installed at the ends (region B in the Figure 9).

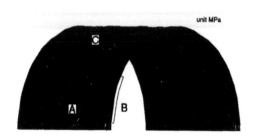

Figure 9. Example of numerical analysis result.

7 CONCLUSION

1. The proposed reinforcement structure was proven able to maintain its minimum required safety against the passage of a train in the daytime even with insufficient backfilling.

2. In the case of no spacers on the ends of the reinforcement structure, vibration, with its fulcrum at the engagement point, occurred. It is considered desirable to install spacers also on the ends

of the reinforcement structure in future installation work for tunnels in service.

8 ACKNOLEDGEMENT

This study was supported by the Program for Promoting Fundamental Transport Technology Research from the Corporation for Advanced Transport & Technology (CATT).

REFERENCES

Kage, K., H. Miyake, & S. Kawagoe 1993. Numerical study of compression waves produced by high-speed trains entering a tunnel (2nd report, effects of shape of hood). *Journal of Japan Society of Mechanical Engineering* 59(560), 159–165. In Japanese.

Kashimura, H., T. Yasunobu, T. Aoki, & K. Matsuo 1994a. Emission of a propagating compression wave from an open end of a tube (2nd report, relation between incident compression wave and impulsive wave. *Journal of Japan Society of Mechanical Engineering* 60(569), 71–77. In Japanese.

Kashimura, H., T. Yasunobu, T. Aoki, & K. Matsuo 1994b. Emission of a propagating compression wave from an open end of a tube (3rd report, experiment on magnitude of emitted impulsive wave. *Journal of Japan Society of Mechanical Engineering* 60(575), 57–63. In Japanese.

Sasoh, A., O. Onodera, K. Takayama, R. Kaneko, & Y. Matsui 1994. Experimental study of shock wave generation by high speed train entry into a tunnel. *Journal of Japan Society of Mechanical Engineering* 60(575), 71–78. In Japanese.

Shiratori, T., Y. Matsudaira, K. Konishi, & C. Sakurai 1992. Airflow characteristics in underground railway station (1st report, airflow pattern and reduction of train wind effect). *Journal of Japan Society of Mechanical Engineering* 58(546), 182–187. In Japanese.

Tanaka, Y., K. Kikuchi, & R. Takahashi 1999. Pressure wave radiated from a portal when a train enters or exits from a tunnel. In *Proc. of 31st Fruid Dynamics Conference*, pp. 281–284. In Japanese.

Modern Tunneling Science and Technology, Adachi et al (eds), © 2001 Swets & Zeitlinger, ISBN 90 2651 860 9

Damage to mountain tunnels by earthquake and its mechanism

T. Asakura
Department of Earth Resources Engineering, Kyoto University, Kyoto, Japan

S. Matsuoka
Tekken Corporation, Tokyo, Japan

K. Yashiro
Railway Technical Research Institute, Tokyo, Japan

Y. Shiba & T. Ōya
Taisei Corporation, Tokyo, Japan

ABSTRACT: It is generally said that mountain tunnels are little damaged by earthquake. However, recent case studies of the damage to mountain tunnels caused by earthquakes also show that they are likely to be damaged when 1) the scale of the earthquake is large, 2) there are earthquake faults near the tunnel or 3) there are special conditions. We collected information on the tunnels which suffered damage from earthquake to study the damage mechanism of mountain tunnels. We analyzed collected data, classified damage patterns and preformed simulation analyses and model tests. Based on the study results, we concluded that we were able to classify damage patterns of mountain tunnels by earthquake into following three patterns; 1) damage to tunnel entrance and portal, 2) damage to tunnels at a fractured zone, 3) damage to tunnels by sliding of fault. We were able to prove these mechanisms by the results of simulation analyses and model tests.

1 INTRODUCTION

As tunnels are surrounded by the ground, they have good earthquake-resistance if the ground is stable during an earthquake. Therefore, it is generally said that the earthquake-resistance is not necessarily required for tunnels in the stable ground. At the 1995 Hyogoken-nanbu Earthquake, however, 10 tunnels among at least one hundred mountain tunnels in service in and around the disaster area had serious damage to need repair and reinforcement. We reconfirmed that if an earthquake was larger, even mountain tunnels would have been damaged. There are few studies on the earthquake damage mechanism of mountain tunnels. On the other hand, more and more tunnels have been constructed recently in the ground of low strength and low earth covering, to require the establishment of a method to design mountain tunnels in consideration of the effect of earthquake. We performed this study to acquire the basic knowledge of the earthquake damage mechanism of mountain tunnels.

2 DAMAGE TO MOUNTAIN TUNNELS BY EARTHQUAKE

2.1 Past damage to mountain tunnels by earthquake

Mountain tunnels and other structures have suffered damage from large earthquakes. Table 1 summarizes the damage to mountain tunnels in Japan caused by large-scale earthquakes, from the 1923 Kanto earthquake to the 1995 Hyogoken-nanbu earthquake.

2.2 Damage patterns of mountain tunnels by earthquake

We analyzed above-mentioned data and concluded that we were able to classify damage patterns of mountain tunnels by earthquake into the following three patterns.

1) Damage to tunnel entrance and portal
2) Damage to tunnels at a fractured zone
3) Damage to tunnels by sliding of fault

Detailed damage patterns are shown below.

2.2.1 Damage to tunnel entrance and portal

Tunnel entrances and portals are likely to suffer from earthquake because they often exist in the loose ground where earthquake motion is amplified and the ground deforms to a large extent. The damage to tunnel entrances and portals include the inclinations of and cracks in portal walls and cracks in the lining near tunnel entrances. This damage pattern is the most popular except for damage caused by landslides around tunnel entrances.

Typical examples are the damages to the Imaihama Tunnel at the 1978 Izu-oshima-kinkai earthquake and the Higashiyama Tunnel at the 1995 Hyogoken-nanbu earthquake (Figure 1).

The Higashiyama tunnel was constructed with concrete blocks in 1928. The geology is a relatively loose quaternary formation. Its earth covering is less than 10m. In this tunnel, existing cracks expanded; new cracks occurred in the portal wall and some cracks with spalling occurred at the shoulder of the arch along the tunnel axis at the earthquake.

Table 1. Past damage of mountain tunnels by earthquake in Japan.

Year, Name	Magnitude	Epicenter	Tunnel Performance
1923 Kanto	7.9	Sagami Bay (Depth: unknown)	Extensive, severest damage to more than 100 tunnels in southern Kanto area
1927 Kita-Tango	7.3	7km WNW of Miyazu, Kyoto (Depth: 0km)	Very slight damage to 2 railway tunnels in the epicentral region
1930 Kita-Izu	7.3	7km west of Atami, Shizuoka (Depth: 0km)	Very severe damage to one railway tunnel due to earthquake fault crossing
1948 Fukui	7.1	12km north of Fukui City (Depth: 0km)	Severe damage to 2 railway tunnels within 8km from the earthquake fault
1952 Tokachi-oki	8.2	Pacific Ocean, 73km ESE off the Cape Erimo (Depth: 0km)	Slight damage to 10 railway tunnels in Hokkaido
1961 Kita-Mino	7.0	Near the border between Fukui and Gifu Prefectures (Depth: 0km)	Cracking damage to a couple of aqueduct tunnels
1964 Niigata	7.5	Japan Sea, 50km NNE of Niigata City (Depth: 40km)	Extensive damage to about 20 railway tunnels and one road tunnel
1968 Toikchi-oki	7.9	Pacific Ocean, 140km SSE off the Cape Erimo (Depth: 0km)	Slight damage to 23 railway tunnels in Hokkaido
1978 Izu-Oshima-kinkai	7.0	In the sea between Oshima Isl. and Inatori, Shizuoka (depth: 0km)	Very severe damage to 9 railway and 4 road tunnels in a limited area
1978 Miyagiken-oki	7.4	Pacific Ocean, 112km east of Sendai City, Miyagi (Depth: 40km)	Slight damage to 6 railway tunnels mainly existing in Miyagi Prefecture
1982 Urakawa-oki	7.1	Pacific Ocean, 18km SW of Urakawa, Hokkaido (Depth: 40km)	Slight damage to 6 railway tunnels near Urakawa
1983 Nihonkai-cyubu	7.7	Japan Sea, 90km west of Noshiro City, Akita (Depth: 14km)	Slight damage to 8 railway tunnels in Akita, etc.
1984 Naganoken-seibu	6.8	9km SE of Mt. Ontake, Nagano (Depth: 2km)	Cracking damage to one hydraulic power tunnel
1987 Chibaken-toho-oki	6.7	Pacific Ocean, 8km east off Ichinomiya Town, Chiba (Depth: 58km)	Damage to the wall of one railway tunnel at Kanagawa-Yamanashi border
1993 Notohanto-oki	6.6	Japan Sea, 24km north of Suzu City, Ishikawa (Depth: 25km)	Severe damage to one road tunnel
1993 Hokkaido-nansei-oki	7.8	Japan Sea, 86km west of Suttsu, Hokkaido (Depth: 35km)	Severe damage to one road tunnel due to a direct hit of falling rock
1995 Hyogoken-nanbu	7.2	Akashi strait (Depth: 18km)	Damage to over 20 tunnels, about 10 tunnels required repair and reinforce

Figure 1   Cracks in the portal wall of the Higashiyama Tunnel.

2.2.2 *Damage to tunnels at a fractured zone*

If a tunnel is constructed in the ground whose earth covering is large, it generally has good earthquake-resistance due to the stiff rock mass. However, a few tunnels with large earth covering have also suffered damage from earthquake.

A typical example is the damage to the Rokko Tunnel at the 1995 Hyogoken-nanbu earthquake.

The Rokko Tunnel is a long railway tunnel in Mesozoic granite with a length of over 16km. There were troubles when the tunnel was constructed because of a number of fractured zones at high pressure water.

Tunnel are as follows.

1) Compressive and shear failure with spalling at the arch
2) Compressive failure with spalling at joints between the arch and side wall
3) Spalling of the lining at longitudinal construction joints

These damage patterns are superimposed in Figure 2.

Figure 3 shows the longitudinal profile of and earthquake damage locations in the Rokko Tunnel. Almost all damage locations coincide with fractured zones.

Figure 2.   Damage patterns observed in the Rokko Tunnel.

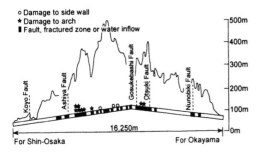

Figure 3. Longitudinal profile and damage locations of the Rokko tunnel.

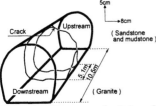

Figure 4. Damage to the lining of the Shioyadanigawa River Diversion Tunnel (Schematic illustration).

Figure 5. Damage to the lining of the Shioyadanigawa River Diversion Tunnel.

2.2.3 Damage to tunnels by sliding of a fault

Earthquake fault crossing sometimes causes severe damage to tunnels by sliding of the fault at earthquake. There are some examples of damage such as those of the Tanna Tunnel at the 1930 Kita-Izu earthquake, Inatori Tunnel at the 1978 Izu-oshima-kinkai earthquake, hydraulic power tunnel at the 1984 Naganoken-seibu earthquake and Shioyadanigawa River Diversion Tunnel (Figure 4, Figure 5).

The downstream portal of the Shioyadanigawa River Diversion Tunnel is near the Suma fault and the middle of the tunnel crosses the Yokoo-yama Fault. Near the Suma Fault, ring cracks occurred at the earthquake. At the Yokoo-yama Fault, the upstream side of the tunnel moved 8 cm to the right and 5 cm upward relatively to the downstream, and a number of cracks occurred in the arch, side wall and invert concrete. The ground of the downstream side of the Yokoo-yama Fault consists of granite, whereas the upstream side consists of sandstone and mudstone of the Miocene (Figure 4, Figure 5).

In the following Chapters, we show the results of numerical analyses and model tests to reproduce the above-mentioned three earthquake damage patterns and discuss the earthquake damage mechanism of tunnels.

3 THE DAMAGE MECHANISM OF TUNNEL ENTRANCES AND PORTALS

We performed numerical analyses to reproduce earthquake damage patterns of tunnel entrances and portals.

3.1 Analysis model

The dynamic behavior during an earthquake of tunnel entrances and portals, which often exist in the loose ground with low earth covering, is greatly influenced by the dynamic behavior of the ground because mountain tunnels have a smaller unit weight than that of the surrounding ground. Therefore, we used the seismic deformation method.

Analyses were performed by frame analysis where a tunnel was modeled by beams and the ground by springs. Figure 6 shows the analysis model of a standard single track railway tunnel. Table 2 shows analysis conditions in the normal state.

We supposed three earth pressure modes in the normal state shown in Table 3 because it was impossible to estimate the real earth pressure acting on the tunnel lining.

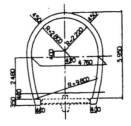

Figure 6. Analysis model.

Table 2. Analysis conditions (in the normal state).

Earth covering	5m
Location of foundation bed	Lower end of tunnel side wall
Thickness of tunnel lining	45cm
Surrounding ground	Sandy soil
Young's modulus of ground	$E=0.75kN/mm^2$
Modulus of subgrade spring	$k=25.7N/m^3$ ($k=1/2 \times 1 \times E^{-3/4}$)
Strength of concrete	$f_{ck}=18N/mm^2$
Young's modulus of concrete	$E_c=24kN/mm^2$
Unit weight of concrete	$\gamma=23.5N/mm^3$

Table 3. Earth pressure modes in the normal state.

Model	Vertical pressure	Horizontal pressure	Assumption for load setting
A	None	None	Assuming that ground is stable.
B	Loosened bedrock	K_0=0.5	Assuming that ϕ is about 30°.
C	None	K_0=0.2	(As a load between A and B)

Table 4. Analysis conditions (during earthquake).

Modulus of subgrade spring	k=51.4N/m³ (k=1/2 × 2 × E⁻³⁄⁴)
Seismic intensity	0.2
Velocity of shear elastic wave	109.5m/s
Natural period of subsurface layers	0.41s
Ground level displacement	1.23cm

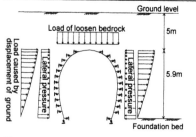

Figure 7. Analysis model (during an earthquake).

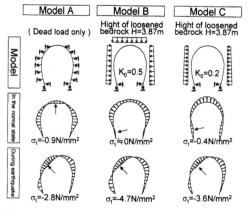

Figure 8. Bending moment diagram and the maximum tensile stresses.

Table 4 shows analysis conditions during an earthquake. In this analysis, we assumed a horizontal seismic intensity of about 0.2 to study mainly an qualitative damage mechanism of tunnel entrance and portal. We calculated the ground level displacement from that intensity and converted it into nodal forces to use for the analysis model (Figure 7).

3.2 Result of the analyses

Figure 8 shows the bending moment diagrams and the maximum tensile stresses in the normal state and during an earthquake. In the normal state, the deformation

modes were different from each other but the maximum tensile stresses at the lining were small and no cracks occurred on the lining. On the other hand, during an earthquake, all the bending moment diagrams roughly looked alike regardless of the mode of earth pressure. The maximum tensile stresses of the tunnel lining occurred almost at the same location on the shoulder of the arch, and the maximum tensile stresses exceeded the tensile strength. (When the compressive strength is 18N/mm², the tensile strength is 1.6N/mm².) This result roughly agreed with the actual locations of cracks caused by an earthquake at tunnel entrances and portals, and proved that cracks could occur on the shoulder of the arch at tunnel entrances and portals at an earthquake of the horizontal seismic intensity of about 0.2.

4. DAMAGE MECHANISM OF TUNNELS AT A FRACTURED ZONE

4.1 The analysis model

We performed numerical analyses to reproduce damage patterns of tunnels at a fractured zone at an earthquake.

At the Rokko tunnel, cracks which were slanted to the lining occurred at the earthquake. Therefore, we used an FEM code which were able to simulate generation and growth of cracks. Figure 9 shows the analysis model. We mainly focused on grasping the crack mechanism of the lining qualitatively and simplified the modes of shear deformation of the ground during an earthquake into the following two cases; a) when the angle of incidence of shear wave is vertical, and b) when the angle of incidence of shear wave is 45° against the vertical line.

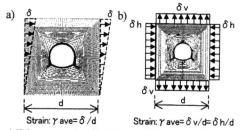

a) When an angle of incidence of shear wave is vertical
b) When an angle of incidence of shear wave is 45° against the vertical line.
Figure 9. Analysis models.

Table 5. Analysis conditions.

Young's modulus of ground	500N/mm²
Poisson's ratio of ground	ν =0.3
Young's modulus of concrete	E_c=26kN/mm²
Compressive strength of concrete	f'$_{ck}$=18N/mm²
Tensile strength of concrete	f$_{td}$=1.9N/mm²
Poisson's ratio of concrete	ν =0.167

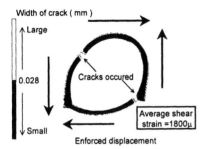

a) When the angle of incidence of shear wave is vertical.

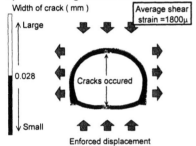

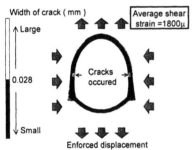

b) When the angle of incidence of shear wave is 45°
against the vertical line.

Figure 10. Distribution of width of crack.

4.2 *Effect of shear wave*

Figure 10 shows the distribution of crack width when the average shear strain of the ground caused by shear wave is $1.8*10^{-3}$. The crack width is calculated by the following equation on the assumption that one crack occurs on one element.

Width of crack = Plastic strain of the element
$\times$ equivalent length of the element

4.2.1 *When the angle of incidence of shear wave is vertical.*

As the shear strain of the ground increased and reached $6.3*10^{-4}$, cracks occurred at the joint between the invert and side wall and at the shoulder of the arch at the same time. As the shear strain became large, the width of crack became wider the crack penetrated the lining when the shear strain reached $1.8*10^{-3}$.

4.2.2 *When the angle of incidence of shear wave is 45° against the vertical line.*

Cracks occurred at the arch crown and the center of the invert when the ground is vertically compressed. On the other hand, cracks occurred at the springline of both sides when the ground is horizontally compressed. We performed parameter studies by changing intensity of shear deformation, we found that the intensity affected only the width and depth of cracks but did not the location of cracks. From these results, we found that tunnels tend to have cracks and compressive failures at the crown of the arch and the springline at a fractured zone by a large bending moment caused by earthquake. However, the direction of crack was only vertical to the lining and we were not able to reproduce cracks slanted to the lining like those observed at the Rokko Tunnel.

4.3 *Effect of initial load acting on the lining.*

If the strength of the ground around the tunnel is low as in a fractured zone, an initial load like the squeezing earth pressure may sometimes act on the lining. We analyzed other cases with some initial load enforced on the lining. The direction of the earth pressure was assumed to be horizontal. An equally distributed load ($80kN/m^2$) was applied at the boundary first and then a shear strain corresponding to the deformation caused by earthquake was added.

Figure 11 shows the distribution of width of crack when the average shear strain of the ground caused by a shear wave is $1.8*10^{-3}$, where the angle of incidence of shear wave is 45° against the vertical line. As the shear strain became larger, cracks occurred not only inside the springline and at the joint between the invert and side wall but also at the shoulder of the arch. We performed parameter studies by changing the intensity of initial earth pressure, and found that the intensity affected only the width and depth of cracks but not the location of cracks. From these results, we found that when mountain tunnel suffered damage from the shear deformation caused by an earthquake, cracks slanted to the lining like those observed at the Rokko Tunnel occur depending on the mode of initial earth pressure acting on the lining.

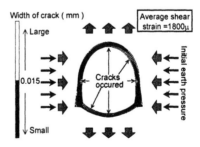

Figure 11. Distribution of width of crack.

5. DAMAGE MECHANISM OF TUNNELS BY SLIDING OF A FAULT

5.1 Testing model

We performed model tests to reproduce damage patterns of tunnels by sliding of a fault.

We used a 1/30 scale tunnel lining test unit that models the Shinkansen standard tunnel (Figure 12). The test unit consists of loading bolts, cylindrical rubbers and a reaction frame. After a lining made of mortar was set in the test unit, step loading was carried out with displacement controlled. At all points except the loading point, the cylindrical springs made of hard rubber simulating the ground were set to induce subgrade reaction proportional to deformation. Figure 13 shows the modeling of sliding of fault, and Table 6 the properties of the model.

To simulate faults, we changed the type of rubber at the center of the model. The rubber on one side was hard (supposing Diluvium) and that on the other side was soft (supposing alluvium). The tunnel section was closed by a strut made of steel.

Load were applied by enforcing displacements on the side wall of Alluvium side to simulate right-lateral slipping of a fault.

5.2 Result of the test

Figure 14 shows the crack chart of the lining obtained from the model test. Cracks followed the processes of, 1) occurrence near the fault from the left side wall to the

Figure 12. Tunnel lining test unit.

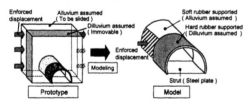

Figure 13. Modeling of sliding of fault.

Table 6. Property of model.

Lining	Material	Mortal
	Compressive strength	f_a=18N/mm²
	Young's modulus	E_c=26kN/mm²
	Thickness	t=2cm
	Width	w=60cm
Spring	Material	Cylindrical rubber
	Modulus of subgrade spring	K_1=1,900 (N/mm) (hard type)
	Modulus of subgrade spring	K_2=80 (N/mm) (soft type)

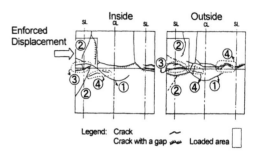

Figure 14. Crack chart (model test).

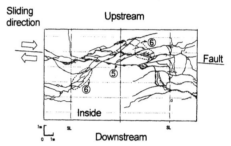

Figure 15. Crack chart (Shioyadanigawa River Diversion Tunnel).

arch, 2) extension to the tunnel axis, 3) occurrence of right-lateral shear cracks and 4) occurrence of a number of parallel cracks to final tunnel fracture.

Figure 15 shows the crack chart of the Shioyadanigawa River Diversion Tunnel at the part of crossing the Yokoo-yama Fault. Cracks followed the processes of , 5) occurrence of right-lateral shear cracks along the Yokoo-yama Fault, and 6) occurrence of a large number of diagonal cracks crossing the Yokoo-yama Fault. We found that the mode of cracks of the model test was similar to that observed at the Shioyadanigawa river Diversion Tunnel.

6. CONCLUSION

We conclude that we can classify damage patterns of mountain tunnels by earthquake into the following three patterns; 1) damage to tunnel entrances and portals, 2) damage to tunnels at a fractured zone, 3) damage to tunnels by sliding of a fault. We have proved these mechanisms based on the results of simulation analyses and model tests.

REFERENCES

Asakura, T., Shiba, Y., Sato, Y. & Iwatate, T. 1996. Mountain Tunnels Performance in the 1995 Hyogoken-Nanbu Earthquake, *Special Report of the 1995 Hyogoken-Nanbu Earthquake*: 265-276, Committee of Earthquake Eng.: JSCE

Modern Tunneling Science and Technology, Adachi et al (eds), © 2001 Swets & Zeitlinger, ISBN 90 2651 860 9

The tunnelling at the crush zone in the Sanbagawa Metamorphic Belt

Y. Yoshida
Takamatsu Technical Office, Japan Highway Public Corp.

Takamatsu, Kagawa, Japan

ABSTRACT: This paper is a summary report on the state of progress of tunnelling at the fault-crush zone of the Sasagamine Tunnel. The Sasagamine Tunnel, which is the longest mountain tunnel of Kochi Expressway, has two traffic lanes. This tunnel has been constructed in the area of the Sanbagawa Metamorphic Belt, a famous belt which is located between the Median Tectonic Line and the Mikabu Tectonic Line in Japan. The characteristic of this geology is that the belt is composed mainly of muddy crystalline schists and countless crush zones with fault clay. The result of the measurement at the crush zone, the maximum value of the initial inner cross section displacement of the first day after the excavation, was over 30mm. The final value of the inner cross section displacement was over 100mm. The temporal invert on the heading bench was effective in the control of the inner cross section.

1 INTRODUCTION

In terms of size, Shikoku Island is the fourth largest among Japan's islands and is composed of four prefectures. Within its total area of 19,000km², Shikoku has a population of 4,200,000.

The routes of the expressway in the Shikoku region, as shown in Figure 1, are planned to take a form like the infinity mark, with the crosspoint at the Kawanoe Junction in Kawanoe City, Ehime Prefecture.

The expressway network in the Shikoku region covers an aggregate distance of over 660km, of which approx. 390km are in operation at present. Major cities in Shikoku are interconnected by this network, which is connected to the expressway in Honshu (the main island of Japan) via three different routes of the Honshu-Shikoku Bridge.

2 TOPOGRAPHY AND GEOLOGY OF SHIKOKU ISLAND

Approx. 85% of Shikoku is steep mountainous areas. The peak is Mt. Ishizuchi which is 1,982m above sea level. The rest of the island is hills and plains.

The characteristic of the geological structure of Shikoku is, as shown in Figure 2, its division by three clear tectonic lines, extending east and west, creating four clear-cut geologic provinces. The Sanbagawa Metamorphic Belt lies along the south

Legend:
● Location of the Sasagamine Tunnel
▲ Honshu-Shikoku Bridge route
① Kobe-Naruto route
② Kojima-Sakaide route
③ Onomichi-Imabari route

Figure 1. Expressway network in the Shikoku region.

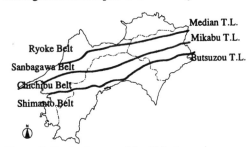

Figure 2. Geologic map of the Shikoku region.

side of the Median Tectonic Line which extends approx. 1,000km from the Kanto area in Honshu Island to Kyushu Island.

In the Shikoku region, the Sanbagawa Metamorphic Belt lies in the central part of the island, extending east and west over a distance of approx. 250km.

The Belt is at its widest, with a width of approx. 30km, in the central part of the island. Most of the rock mass is crystalline schist.

3 KOCHI EXPRESSWAY

Kochi Expressway, at present, extends from Kawanoe Junction toward the south up to Ino Interchange in Kochi Prefecture. The part of the expressway that covers a distance of approx. 70km, running through cities such as Nankoku City and Kochi City is already in operation.

The expressway crosses the Shikoku Mountains which lie between the Kawanoe Junction and the Nankoku Interchange, facing the Seto-Naikai Inland Sea and the Pacific Ocean, respectively. The crossing covers a straight line of approx. 50km. A typical mountainous route, this section of the expressway involves 19 tunnels including the Sasagamine Tunnel. The aggregate of those 19 tunnels is approx. 26km.

4 SASAGAMINE TUNNEL

The Sasagamine Tunnel, which is 4,307m, is the longest tunnel among the expressways in the Shikoku region. It passes through Mt. Sasagamine which is 1,016m above sea level and located on the border of Ehime and Kochi prefectures. The proposed highest point is 445m above sea level in the middle of the tunnel. The characteristic of the geology is that the rock mass is mainly composed of muddy crystalline schists. And sandy crystalline schists are partially visible. Countless fault-crush zones with fault clay also exist in them.

The longitudinal axial direction of this tunnel basically runs from NNW to SSE. The strike of the rock mass is approx. N50° W~EW, and its dip is approx. 70~80° S.

The tunnelling was begun with the upper half advancing excavation method by blasting from both its north and south portals. A lot of faults and crush zones were found at the tunnelling, and argillic alterations at the crush zones were observed clearly. Figure 3 shows the vertical section of the geological survey of the Sasagamine Tunnel on the Kochi prefecture side.

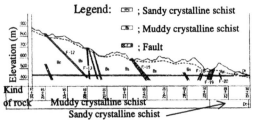

Figure 3. Geological profile of the Sasagamine Tunnel.

5 GEOLOGICAL CHARACTERISTIC OF THE CRUSH ZONE

This crush zone exists in an area between the 1,600m point and the 1,655m point from the south portal. Though the outcrop was found in the geological survey before the tunnelling, it was difficult to estimate where it would appear from the heading in the tunnel.

There are stratum of strongly-weathered muddy crystalline schists and stratum of clay in the area of this crush zone, The cohesion of the muddy crystalline schist was so weak that they were easily smashed by a tremor. And the clay stratum containing a soft-sediment quartz was greasy and weak. They formed the alternation of strata. The strike of the stratum was approx. EW, and its dip was approx. 75~85° S.

The changes in the condition of the heading are listed below:
at the 1,602m point
··· The condition of the heading begins to get unstable.
The rocks can be penetrated by a hammer.
at the 1,615m point
···A clay stratum is found under the right part of the heading.
The clay stratum begins to appear a lot thereafter.
at the 1,618m point
···A clay stratum is found all over the heading.
at the 1,624m point
···Fault clay is found all over the heading.
at the 1,652m point
···A muddy crystalline schist is found under the right part of the heading.
The muddy crystalline schist begins to appear a lot thereafter.
at the 1,655m point
··· There is no clay stratum found at the heading.
A grain-size analysis of the rock mass material at the 1,635m point, which has a relatively stable condition at the heading in this crush zone, was

attempted. The materials were analyzed on the basis of the grain-size analysis method of soils (JIS 1204).

The results from the analysis of the samples of the material under 2mm in size are as follows:

Sand (2mm-75 μ m) $\cdots$ 41%
Silt (75 μ m-5 μ m) $\cdots$ 35%
Clay (under5 μ m) $\cdots$ 24%

According to the rule of the Japan Unified Soil Classification System, they belong to the clayey loam. Moreover, the result of the clay mineral assay shows that there is no montmorillonite mineral in this fault clay.

Secondly, the unconfined compressive strength test was attempted for the gathering of rock mass produced by excavation at the heading in this crush zone. However, it was unsuccessful because of its fragility. That value of the compressive strength was estimated to be approx. 1.0N/mm^2 based on my experience and experienced judgment.

Based on this estimated value, the competence factor is estimated to be approx. 0.23 because the thickness of the overburden at this point is approx. 240m.

6 TUNNELLING AND MEASUREMENT OF THE ROCK MASS

The supporting at this crush zone, as shown in Figure 4, was constructed with the stiffest pattern among the standard supporting patterns designed for the tunnels of the expressway.

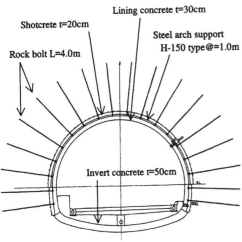

Figure 4. Supporting pattern design before the countermeasure.

Table 1 shows the measured results of the horizontal inner cross section displacement during the seven days after the excavation at the five points in the area of the crush zone mentioned in chapter 5.

Table 1. measured values of the horizontal inner cross section displacement.

measured point (from the south portal)	measured value (after the excavation) (unit : mm)				
	1st day	2nd day	3rd day	4th day	7th day
1,627m point	13	18	21	25	41
1,632m point	14	39	60	73	---
1,637m point	33	54	62	65	---
1,643m point	19	24	26	28	---
1,645m point	3	6	10	31	---

Legend: --- After a countermeasure

Table 2 shows the measured results of the roof settlement during the seven days after the excavation at the same five points.

Table 2. measured values of the roof settlement.

measured point (from the south portal)	measured value (after the excavation) (unit : mm)				
	1st day	2nd day	3rd day	4th day	7th day
1,627m point	3	4	5	7	11
1,632m point	3	6	11	13	---
1,637m point	7	9	10	10	---
1,643m point	7	8	9	10	---
1,645m point	6	10	13	44	---

Legend: --- After a countermeasure

Next, table 3 shows the measured result of the horizontal inner cross section displacement during the seven days after the excavation in the rest of the crush zone.

Table 3. measured values of the horizontal inner cross section displacement (in the rest of the crush zone).

measured point (from the south portal)	measured value (after the excavation) (unit : mm)				
	1st day	2nd day	3rd day	4th day	7th day
219m point	0.8	1.1	1.3	1.6	1.8
727m point	0.5	1.2	2.1	2.5	3.2
859m point	1.3	2.6	3.5	4.2	5.0
1,357m point	1.2	2.1	2.6	3.0	3.3
1,577m point	1.0	2.5	2.8	3.1	3.9
1,884m point	1.5	3.7	5.4	6.0	7.3

The characteristic of the horizontal inner cross section displacement in the crush zone is the remarkable increase of the displacement at the initial stage.

7 INSTABILITIES OF THE HEADING AND THEIR COUNTERMEASURES

With the remarkable increase of the displacement at the initial stage, the collapse of the heading and the damage to the constructed supporting gradually became worse and worse. The collapsing and damage were particularly noticeable both in the left half of the heading and under its right wall.

Some typical instabilities of the heading at the crush zone section and their countermeasures taken are as follows: heading collapse: application of the ring cut method, drive a fiber-glass reinforced plastic rock bolt into the face (length=3m, 10 ~ 25 pieces/cross section); roof collapse: forepoling (length=3m, 7 pieces/cross section); shotcrete damage: drive a reinforcing longer rock bolt into the heading (confer with Figure 5) (length=6m, type=TD24, 7 pieces/cross section × 6 spans & 11 pieces/cross section × 25 spans); rock bolt bearing plate damage: simultaneous systematic bolting and longer bolting reinforcement (some modifications of the standard supporting pattern design); unstable convergence of the inner cross section displacement: construction of a temporal invert on the heading bench.

The supporting pattern design after the countermeasure is shown in Figure 5.

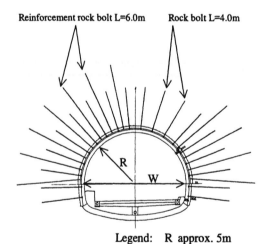

Legend: R approx. 5m
W approx. 10m

Figure 5. Supporting pattern design after the countermeasure.

The size of the reinforcing longer rock bolt was decided on the basis of the result of the axial force measurement. Table 4 shows the measured results of the axial force of the rock bolt in the area of the crush zone.

Table 4. measured values of the axial force of the rock bolt (at the crush zone).

measured point	axial force (maximum value)		
(from a wall)	left side	left shoulder	roof
0.0~0.5m	-0.4kN	+0.8kN	-0.6kN
0.5~1.5m	+0.9kN	+1.1kN	-0.2kN
1.5~2.5m	+1.8kN	+0.8kN	-0.5kN
2.5~3.5m	+1.1kN	+0.6kN	+0.6kN
3.5~4.0m	+0.3kN	+0.2kN	+0.8kN

Legend: + tension, - compression

Figure 6. Construction site of the south portal.

8 EVALUATION OF AN EFFECT OF THE COUNTERMEASURES

Table 5 shows the measured results of the horizontal inner cross section displacement after the reinforcement work provided the longer rock bolt at the 5 points in the area of the crush zone mentioned in chapter 5.

Table 5. measured values of the horizontal inner cross section displacement (after the reinforcement work provided the longer rock bolt).

measured point (from the south portal)	timing of the work (after the excavation)	measured value (unit : mm/day)
1,627m point	8 days	1.0~2.0
1,632m point	6 days	5.0~7.5
1,637m point	5 days	8.0~10.0
1,643m point	4 days	1.0~7.0
1,645m point	4 days	4.0~10.0

The remarkable increase of the displacement was temporarily weakened by this countermeasure. However, it was unfortunate that a clear convergence of the inner cross section displacement could not be recogniged here.

If the phenomenon is that an inner cross section displacement continues to occur, it will be difficult to get enough space for the tunnel sectional area.

So, it was decided to construct a temporal invert on the upper bench in order to stabilize the upper half of tunnel section.

Shotcrete, which strengthens in a short time, was used for the temporal invert construction to close the upper half of tunnel section immediately.

The thickness of the temporal invert is 15cm, and its design strength is $18N/mm^2$, which is the same as the one used in the side wall.

Table 6 shows the measured results of the horizontal inner cross section displacement after the temporal invert work at the 5 points in the area of the crush zone.

Table 6. measured values of the horizontal inner cross section displacement (after the temporal invert work).

measured point (from the south portal)	timing of the work (after the excavation)	measured value [unit : mm/day]
1,627m point	14 days	0.2~0.3
1,632m point	12 days	0.5~1.0
1,637m point	10 days	1.0~1.5
1,643m point	9 days	0.8~1.0
1,645m point	8 days	0.3~0.5

Table 7. final measured values of the horizontal inner cross section displacement at the crush zone and at the stable area of the rock mass.

measured point (from the south portal)	measured value (after the excavation) (unit : mm)	remarks (kind of a rock at the heading)
219m point	5.2	sandy crystalline schist
727m point	6.9	sandy crystalline schist
859m point	7.6	sandy crystalline schist
1,357m point	14.8	muddy crystalline schist
1,577m point	21.9	muddy crystalline schist
1,627m point	118	crush zone (fault clay)
1,632m point	136	crush zone (fault clay)
1,637m point	84.0	crush zone (fault clay)
1,884m point	27.2	muddy crystalline schist

As the result of the construction of the temporal invert, the convergence of the inner cross section displacement could be recogniged clearly.

Table 7 shows the measured results of the final value of the horizontal inner cross section displacement at the 5 points in the area of the crush zone mentioned in chapter 5 and at the stable area of the rock mass in the rest of the crush zone.

As a summary of this chapter, Figure 7 shows the typical daily transition of the inner cross section displacement at the crush zone of the Sasagamine Tunnel.

The ratio of the displacement arising from the excavation of the upper half of tunnel section and the one arising from the excavation of the lower half of tunnel section was approx. 1 to 1 respectively. However, there was a tendency for displacement arising from the excavation of the lower half of tunnel section to increase when the displacement arising from the excavation of the upper half of tunnel section was small.

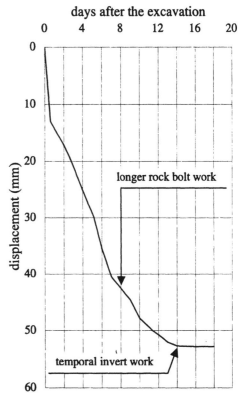

Figure 7. typical daily transition at the crush zone of the Sasagamine Tunnel (1,627m point from the south portal).

9 CONCLUSION

There are 5 important points in the construction at the crush zone of the Sasagamine Tunnel. They are listed below.

(1) Plastic flow earth pressure

The characteristic of the displacement in the crush zone is the remarkable squeezing in a short period of time at the initial stage after the excavation.

Judging from the competence factor, this phenomenon is considered to be caused by the plastic flow earth pressure (the action of the plastic collapse load).

(2) Loosened zone

According to the present theory, the area of the the loosened zone of the rock mass around a tunnel which has a diameter of approx. 10m is approx. 6m in case of high earth pressure.

However, judging from the measured results of the rock mass, the area of the loosened zone was over estimated by 6m in the Sasagamine Tunnel.

(3) Temporal invert

During tunnelling, adopting a supporting method with a high load-carrying capacity is thought to be the best countermeasure against plastic flow earth pressure.

The effect of temporal invert work can be evaluated by the change of the displacement after the temporal invert work on the upper bench.

(4) Temporal invert with shotcrete

Temporal invert work with shotcrete is considered a very effective method because it does not need much preparation and its design strength can rise in a short time.

(5) Reinforcing longer rock bolt

When the application of the reinforcing longer rock bolt is adopted as the countermeasure against plastic flow earth pressure at the tunnelling, it is hard to judge the proper time to begin work.

It will be less effective if the countermeasure is too late because the loosened zone expands in the rock mass around the tunnel with the passage of the time.

Lastly, as the final road plan, we are constructing a second tunnel in parallel as a one-way traffic tunnel with two lanes.

The experience in this summary report will be reflected in the construction of the new tunnel.

Modern Tunneling Science and Technology, Adachi et al (eds), © 2001 Swets & Zeitlinger, ISBN 90 2651 860 9

Maintenance of the Undersea Section of the Seikan Tunnel

M. Ikuma
Manager of the Sapporo Construction Office
Japan Railway Construction Public Corporation

ABSTRACT: This paper introduces the results of measurements made over a period of 12 years after the Seikan Tunnel was opened for service. Because this tunnel is a very long undersea tunnel, the necessity of clarifying the long-term behavior of the tunnel structure was recognized from the start. Various measurements are made to survey the condition of the ground and lining concrete and data thus obtained is used to assess the soundness of the tunnel structure. Judging from the data thus far obtained, the tunnel structure appears to remain in a good condition.

1 INTRODUCTION

The Seikan Tunnel is a large-scale, long undersea tunnel that connects Honshu, the central island of four main islands in Japan, to Hokkaido ,the northern one of Japanese main islands, by land. In the middle of the Tsugaru Straits, it runs at a depth of 240 m underwater (2.35 MPa of water pressure). The soil consists of Tertiary deposits and many faults formed in strata. In addition, igneous rocks are widely distributed in this area of Honshu. The Japan National Railway began inclined shaft excavation on the Hokkaido side in January 1964. After the Japan Railway Construction Public Corporation (hereinafter called the JRCC) was

established in March 1964, the JRCC proceeded with the construction of the Seikan Tunnel and the tunnel opened for service in March 1988. The railway through the Seikan Tunnel was named the Tsugaru Kaikyo Line and is operated by the Hokkaido Railway Company.

Because the JRCC has accumulated knowledge and expertise in the construction process and also owns the property, it is responsible for the maintenance of the 24.2-km long central section of the tunnel structure including the undersea section shown in Figure 1. The main tunnel is used for scheduled train service, and the inclined shaft, pilot tunnel, and service tunnel are used for ventilation, drainage, patrolling, and evacuation.

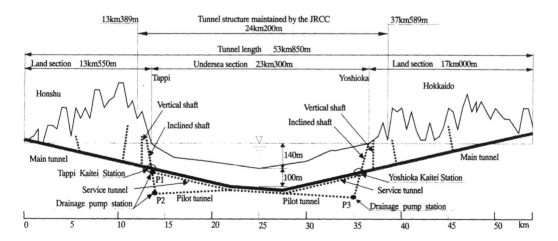

Figure 1. Longitudinal section of the Seikan Tunnel.

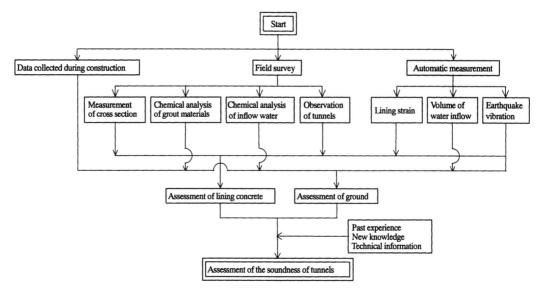

Figure 2. Flow of tunnel maintenance planning.

2 FLOW OF MAINTENANCE PLANNING

When performing maintenance, the soundness of the tunnel structure must be assessed. Specifically, the soundness is assessed based on data obtained for the condition of the ground and lining concrete. Figure 2 indicates the maintenance flow for the Seikan Tunnel and items to be surveyed.

3 ASSESSMENT OF GROUND CONDITIONS

The Seikan Tunnel is designed so that water pressure acts on the outer periphery of an injection zone in the ground established around lining concrete to allow inflow water to permeate through the injection zone and be routed to the backside of the tunnel lining, thus preventing water pressure from acting directly on lining concrete. It is important, therefore, to verify the soundness of the injection zone in ground that is exposed to water pressure.

3.1 *Change in the volume of inflow water*

With the maintenance of an undersea tunnel, the volume of water inflow is an important indicator used to identify external changes and must be continuously monitored over a long period of time. For the Seikan Tunnel, it is monitored by integrating data provided by flow meters installed at 27 points (16 points in the main tunnel, 5 points in the service tunnel, and 6 points in the pilot tunnel) and data provided by pumping discharge in three drainage pump stations.

When the tunnel opened for service in March 1988, the total volume of inflow water was about 32 m³ per minute. After that, it decreased. Although it decreased to 24 m³ per minute, it suddenly increased to 27.5 m³ (an increase of about 3.5 m³ per minute at that time) due to the Hokkaido Southwestern Offshore Earthquake that occurred in July 1993. Afterwards, it decreased again, reaching 22.8 m³ in March 2000. Immediately after the 1993 offshore earthquake occurred, there was a prominent increase in the volume of inflow water at the land section of the main tunnel on the Honshu side. In

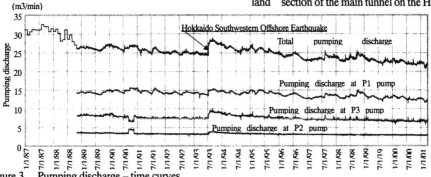

Figure 3. Pumping discharge – time curves.

the undersea section, the volume of inflow water increased in the fault which made a serious flooding at construction at the 16-km point and in the pyroclastic rock dikes at the 21-km point. This indicates that an earthquake increases inflow water going down through fissures in rocks.

In addition, the volume of water inflow in the land section is affected by precipitation.

3.2 Chemical analysis of inflow water and water inflow pressure

Inclusions in inflow water have remained relatively unchanged and stable since the tunnel opened for service. A change in water pressure in the injection zone in the ground and at the back of the injection zone is monitored in the service tunnel near the center of the undersea section (at three points 10 m, 15 m, and 20 m from the wall of the service tunnel). The water pressure is in a stable condition, as shown in Figure 4. Figure 5 depicts the relation between depths of boring holes, and water inflow pressure.

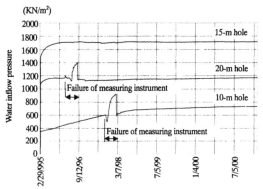

Figure 4. Inflow water pressure-time curves at the 26-km point.

3.3 Earthquake information

Seismic observations are made using accelerometers (at both tunnel ends, four points where the construction works were difficult, and two points on land for a total of 8 points) and seismographs for identifying the position of a hypocenter from preliminary tremors (at both tunnel ends, and at each entrance of the inclined shaft on the Honshu side and the Hokkaido side for a total of 4 points) that were installed on the undersea section of the Seikan Tunnel and on land.

Four large earthquakes occurred in 12 years after the Seikan Tunnel opened for service. When the Hokkaido Southwestern Offshore Earthquake occurred on July 12, 1993, 214 gals maximum was recorded on the ground surface at Kikonai in Hokkaido and 56 gals on the undersea section at the 33-km point. When the Hokkaido Eastern Offshore Earthquake occurred on October 4,

1994, 65 gals were recorded on the ground surface at Kanita in Honshu while no earthquake vibration was detected on the undersea section of the tunnel since the vibration did not reach 5 gals, which is the minimum acceleration needed to activate an accelerometer. The acceleration in the underground tunnel was found to be much smaller than that on the ground surface. The same result was obtained with other earthquakes observed during the tunnel construction.

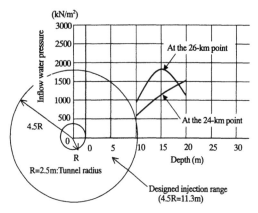

Figure 5. Relation between depths of boring holes and inflow water pressure.

3.4 Chemical analysis of for water stop grout materials

Chemical analysis is performed (on a nonscheduled basis) using the grout materials in the ground sampled by boring to verify whether the grout has deteriorated over the years or not. Water-glass and cement were used as grout materials for dewatering. Chemical analysis was performed in 1980, 1992, 1993, 1994, and 1999 to examine changes in the composition of grout materials and to verify the products of deterioration. As a result of this chemical analysis, neither the ettringite nor the Friedel salt, the products of deterioration generated through reactions between seawater and cement ingredients in the grout materials, were detected. Although no change was detected in the primary components (SiO_2 and Al_2O_3) in the grout materials, Na_2O and CaO tended to decrease while MgO tended to increase. Since the composition of the injected grout materials is not always uniform, it cannot be determined whether a change in the composition of the grout materials occurred or not.

Concerning a change in the composition of the grout materials that may occur in contact with seawater, surveys should be carried out to verify whether CaO and Na_2O continue to decrease and MgO to increase and what effects the changes in the composition may give to the properties of the grout materials.

4 ASSESSMENT OF LINING CONCRETE

If structural defects develop in the tunnel structure, the conditions of lining concrete will change. Measurement is made to detect a change in the conditions of lining concrete.

4.1 *Measurement of the cross section*

Deformation of cross sections is measured in the undersea section of the Seikan Tunnel. In 80 cross sections of the main tunnel and 86 cross sections of the pilot tunnel, the service tunnel, and the inclined shafts the measurement is performed at about 1-km intervals. Furthermore, intervals are shorter in the points where the construction work was difficult. Figure 6 indicates the measuring points of a cross section of the main tunnel.

All measuring lines in the cross sections have been measured 36 - 38 times since they were first measured in March 1988. In the present measurement system for the main tunnel, a temporary coordinate system is automatically determined when any two measuring points in a cross section are collimated using one electro-optical transit; using this data in the coordinate system, the temporary coordinate of each measuring point in the cross section is calculated and relevant data is automatically imported to a data card. Then each length of measuring lines is calculated with the data.

In the main tunnel a total of 480 measuring lines (6 measuring lines at each cross section × 80 cross sections) have been established. There are 146 measuring lines with fluctuation of ±1.0 mm or less from start of measurement in a total of 160 vertical measuring lines (each length of about 4 m). The maximum elongation in the measuring lines was 1.45 mm and the maximum contraction was 1.43 mm. There

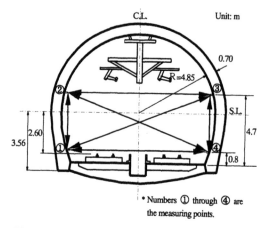

* Numbers ① through ④ are
the measuring points.

Figure 6 Measuring lines of the main tunnel cross section.

are 118 measuring lines with fluctuation of ±1.0 mm or less from start of measurement in a total of 160 horizontal measuring lines (each length of about 9 m). The maximum elongation in the measuring lines was 2.41 mm and the maximum contraction was 3.41 mm. There are 125 measuring lines with fluctuation of ±1.0 mm or less from start of measurement in a total of 160 diagonal measuring lines (each length of about 10 m). The maximum elongation in the measuring lines was 2.32 mm and the maximum contraction was 2.76 mm. As a result of this measurement, the fluctuation of measuring lines was small enough, i.e., about -3.4 mm at a maximum, and no abnormalities were found after the entire main tunnel had been visually inspected. Therefore, the tunnel remains in good condition.

Figure 7 shows the fluctuation of measuring lines over time at the point where maximum value was measured.

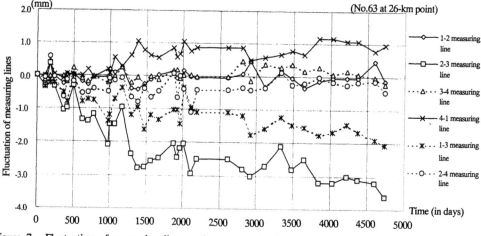

Figure 7. Fluctuation of measuring lines – time curves in the cross section of the main tunnel with the maximum contraction.

There was no large deformation in the cross sections of the pilot tunnel and the service tunnel as would require reinforcement. Therefore, all of the tunnels appear to remain in a good condition.

4.2 Observations of tunnels

Deterioration of repair materials used in the pilot and the service tunnels was noted. Although clogging caused by stripping of the equipment covering waterway that was provided on the lining of the main tunnel for leaking water or by the generation of bacteria was observed in the waterway, there was no change in the conditions of lining concrete.

4.3 Measurement of the strain of lining concrete

The strain of lining concrete of the main tunnel is measured using a high-sensitivity strain meter. A change in the strain during an earthquake has been detected. In addition, measurement is always performed to check the elongation and contraction of lining concrete. Strain meters are installed on the surface of lining concrete at four points in the undersea section where the construction work was difficult. Figure 8 indicates the layout for lining strain meter sensors. The strain of lining concrete changes like a wave in a one-year cycle. A great number of strain data observed at the beginning of measurement were contractive. As of March 1999, however, the ratio of the number of shrinkage strain to that of tensile strain was fifty-fifty. The strain changes according to the season; the strain tends to elongation from fall to spring and it tends to contraction from spring to fall. This strain cycle remained the same after strain measurement began. The strain change proceeds at 50 to 150×10^{-6} per cycle in terms of total amplitude. Figure 9 shows strain changes of the A-section over time at the 21-km point in pyroclastic rock dikes.

With regard to factors responsible for changing the strain, results indicated that temperature as well as the ebb and flow of tide affect the daily, monthly, and annual changes in the strain. Although temperature change, tide level, and wind pressure were considered, it is still difficult to provide providing precise accounts of the mechanism of changes in strain is still difficult. In addition, strain changes with air pressure produced when a train passes.

5 CONCLUSION

Data obtained from various measurements is stable and leads to the conclusion that the tunnel structure remains in a good condition. Based on the results of surveys and measurement:
(1) The undersea tunnel structure remains in a good condition.

(2) Although the volume of inflow water increases right after a large earthquake, it is gradually decreasing. The fluctuation of inflow water seems to be attributed to changes of the amount of inflow water passing through fissures in rocks.
(3) No change in the inclusions of inflow water was found.
(4) The impact (acceleration) of an earthquake on the undersea tunnel is much less than that on structures on the ground surface.
(5) Although the products of deterioration generated through reactions between seawater and cement ingredients in the grout materials were not found, monitoring must be continued to determine if there is a change in the composition of the grout materials.

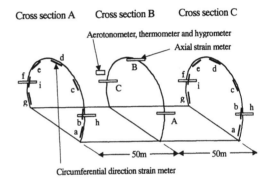

Figure 8. Layout of lining strain sensors.

(6) The change in tunnel cross sections was minimal and no abnormalities were detected in lining concrete.
(7) Although changes in the strain on lining concrete of the main tunnel occur in a cyclic manner, the strain has a minimal impact on the integrity of lining concrete.

A geological map, construction work record sheets, reports on injection conditions, and results of measurement made during construction and after the tunnel opened for service have been prepared for the section where the construction work was difficult in order to allow tunnel maintenance work. Future research will involve formulation of criteria for assessment of the soundness of the Seikan Tunnel based on these materials.

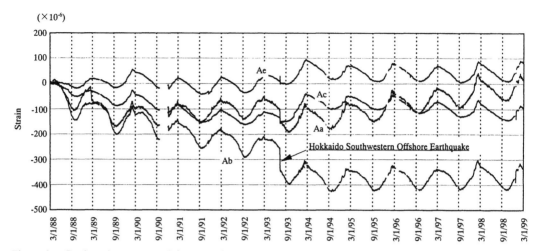

Figure 9.   Strain – time curves of the A-section at the 21-km point in pyroclastic rock dikes.

4. Design and construction of mountain tunnels

Modern Tunneling Science and Technology, Adachi et al (eds), © 2001 Swets & Zeitlinger, ISBN 90 2651 860 9

A proposal of new rock mass classification for tunneling

W. Akagi, T. Ito & H. Shiroma
Expressway Research Institute, Japan Highway Public Corporation

A. Sano & M. Shinji
OYO Corporation

T. Nishi
Shimizu Corporation

K. Nakagawa
Yamaguchi University

ABSTRACT:In the paper, the authors propose a new quantitative assessment method of rock condition based on observational classification obtained at the tunnel face, which will enable to know either strength or deformability of rock mass, further more appropriate amount of rock bolts, thickness of shotcrete and size and pitch of steel ribs just after the blasting procedure. The study is confirmed by the database of up to the 6,101 sections of tunnels constructed by Japan Highway Public Corporation.

1 INTRODUCTION

NATM is generally considered a tunnel construction method that utilizes field measurement results as a judgment criterion, and a vast amount of measurement data has been accumulated over the years. From the time NATM was introduced, JH has been collecting records of observation data, using a face observation data sheet devised by JH. In the field of tunneling, expectations are growing for even greater effort not only to ensure construction safety by making utmost use of accumulated measurement data, but also to build tunnels in a rational and economical way.

Since 1997, Japan Highway Public Corporation (JH) has been working to diversify its standard tunnel support patterns applicable to NATM tunnels by developing new patterns or modifying existing ones on the basis of performance data for the tunnel support patterns by JH for NATM applications. Table 1 shows the standard support patterns of JH. The objective of this diversification is to select a support pattern best-suited to the geological conditions, instead of applying a standard pattern more or less uniformly. Flexible selection of tunnel support patterns requires modifying, if necessary, the support pattern chosen at the preliminary study stage according to field measurements. JH introduced new framework for the condition of rock mass on tunnel faces in 1996, and compared the conventional support determination method based on the conventional face observation items and a new rock mass classification based on new face observation items (hereafter referred to as the "JH method")[1]. Figure 1 shows the tunnel face observation data sheets used in JH method. Preserved records of observation data obtained by this method has been accumulated since then, and the amount of data thus accumulated has become large enough for use in analyses.

In this paper, the authors proposed the degrees of contribution of different observation items to tunnel support determination are assessed as weights in order to evaluate observation results on tunnel face quantitatively for rational tunnel construction. Results thus obtained are then applied to past project data to verify the validity of the new evaluation system for the condition of rock mass on tunnel face during excavation.

Table 1. Standard tunnel support patterns

	Length of Excavation	Rock Bolt		Thickness of Shotcrete	Steel Rib		Stiffness ratio by Hokuriku Method
		Length	Radial Spacing		Upper Section	Lower Section	
B-a	2.0m	3.0m	1.5m*	5cm	-	-	1.0
CI-a	1.5m	3.0m	1.5m	10cm	-	-	1.99
CII-a	1.2m	3.0m	1.5m	10cm	-	-	2.06
CII-b	1.2m	3.0m	1.5m	10cm	125H	-	2.57
DI-a	1.0m	3.0m	1.2m	15cm	125H	125H	3.84
DI-b	1.0m	4.0m	1.2m	15cm	125H	125H	3.92

*Upper section only

Rock Mass Classification Data Sheet

Japan Highway Public Corporation

Name of tunnel:		Date:		
Station +	Distance from portal m	Section No.		Support pattern:
Over burden m	Type of rock:	Group of rock (1~5):		Code of rock:

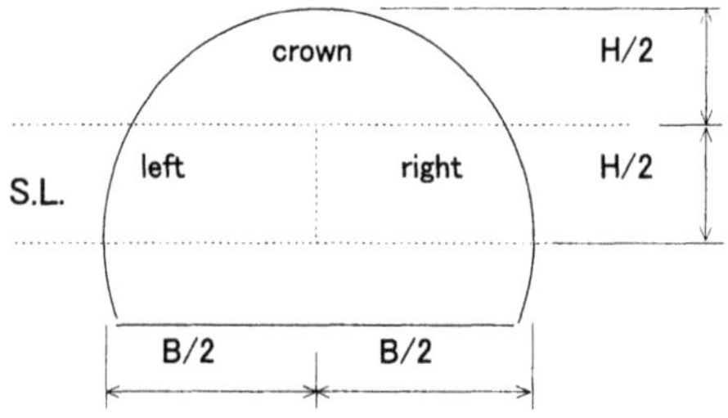

Special conditions _____

_____

Collapse _____

Inverted arch closed _____

	Parameter	Rating					
A. Unconfined strength (MPa)	Unconfined strength	>100	100~50	50~25	25~10	10~3	3<
	Point load	>40	40~20	20~10	10~4	4<	
	Hammering	A piece of rock placed on the ground cannot be broken by hammering	A piece of rock placed on the ground can be broken by hammering	A piece of rock hold by hand can be broken by hammering	Pieces of rock can be broken by hitting each other	A piece of rock can be broken by bare hands	A piece of rock can be crushed by a finger
	Rating	1	2	3	4	5	6
B. Weathering alteration	Weathering	Unweathered	Weathered only along joints		Weathered to inside of rock		Soil
	Alteration	No alteration	Clay in joints		Altered to inside of rock		Heavily altered
	Rating	1	2		3		4
C. Spacing of joints.	Spacing	d≧1m	1m> d≧50cm	50cm> d≧20cm	20cm> d≧5cm	5cm> d	
	RQD	>80	80~50	60~30	40~10	20	
	Rating	1	2	3	4	5	
D. Condition of joints	Joint separation	No separation	Partly separated (Width < 1mm)	Mostly separated (Width < 1mm)	Separated (Width1~5 mm)	Separated (Width >5 mm)	
	Joint fillings	None	None	None	Thin clay (< 5mm)	Thick clay (>5mm)	
	Joint roughness	Rough	Smooth	Partly slickensided	Slickedsided		
		1	2	3	4	5	
E. Joint orientations	Strike perpendicular to tunnel axis	1:Drive with dip Dip 45~90°	2: Drive with dip Dip 20~45°	3:Irrespective of strike Dip 0~20°	4: Drive against dip Dip 20~45°	5: Drive against dip Dip 45~90°	
	Strike parallel to tunnel axis			1: Dip 0~20°	2: Dip 20~45°	3: Dip 45~90°	

Ground water inflow per 10m tunnel.

Possibility of alteration by water shall be judged by both present and future conditions.

F. Ground water	Inflow	< 1l/min	1~20l/min	20~100l/min	> 100l/min
	Rating	1	2	3	4
G. Alteration by water	Alteration by water	None	Loosened	Softened	Flow
	Rating	1	2	3	4

Figure 1. Tunnel face observation data sheet.

2 JH METHOD

2.1 Grouping of rocks

It is quite possible that the degrees of contribution of different evaluation categories of each observation item vary depending on rock types. Having analyzed field observation data accumulated within JH and examined the relationship between support patterns and convergence for different rock types, Yagi et al. report that rocks can be classified into a number of groups according to deformation behavior[2]. For the purposes of this study, the authors divided general rock groups into four classes according to compressive strength in the fresh state and the modes of subsequent weathering and deterioration, and investigated the degrees of contribution of observation items in each class. For simplicity, the rock groups are hereafter referred to by the rock group names shown in Table 2.

2.2 Evaluation of concerning division on tunnel face

At an introduction of JH method, tunnel engineers require consideration of cases where two or more rock types appear at the tunnel face or where the degree of weathering or deterioration varies within the tunnel face. For the purposes of this study, therefore, it was decided to divide the tunnel face into three sections (crown, right shoulder and left shoulder) and evaluate them independently. Table 2 also shows the concept of division on tunnel face. The relationships between average, minimum, weighted average and other values for each of the three sections and support patterns for different tunnels were examined for the longitudinal direction. Many tunnel engineers, however, said that the crown area should be given greater weight, and it was found that the area ratio between the crown and shoulder areas of a typical two-lane highway tunnel was approximately 1.28:1. From these reasons, it was decided to use a weighted average evaluation method in which ratings for the crown area, which greatly affects face stability, are given greater weight (two times the "right/left shoulder" value for the crown area in the case of rating representing the tunnel face) than those for the shoulder areas. In cases where the

tunnel face is divided into the three sections, the weighted average of rating values is calculated from the follwing equation. The results of this equation are given in increments of 0.25. In cases where a rating value for a particular observation item is to represent the tunnel face, a weighed value is used.

$$[rating] = (left + 2*crown + right) / 4 \quad (1)$$

3 RELATIONSHIP FOR WEIGHTING AMONG CATEGORIES

3.1 Basic concept

In order to investigate the relationships between support patterns and the observation items, it is necessary to quantify support stiffness. Although various methods, both empirical and theoretical, for support stiffness evaluation have already been proposed, there is as yet no established method. To decide on how to obtain numerical values for support stiffness to be compared with face observation results, a questionnaire survey of tunnel engineers directly involved in tunnel face observation was conducted. The respondents were a total of 209 engineers employed by JH, technical consulting engineers and tunnel contructors. The intent was to find out what they thought of each of the methods and to use a method that yields results that are considered the closest to the tunnel engineers' impressions. The reason for this approach is that in view of the fact that adequate theoretical solutions to support stiffness calculations have not yet been obtained for all rock types ranging from hard to soft rock, the authors think that the use of criteria relying substantially on the judgment of engineers charged with regularly making rational judgments concerning constructibility is permissible.The authors adopt one of the empirical method named as "Hokuriku method". This method evaluate the limit support stiffness as sum of thickness of shotcrete, number of rock bolts and the size of steel ribs. This methods were applied to the standard support patterns used by JH and arranged the form of the stiffness ratio to pattern B-a (see Table 1).

Table 2. Grouping of rocks

[ ] means number of samples

Strength Weathering	Hard	Medium	Soft
	Group1 [Massive hard rocks]	**Group2** [Massive medium, soft rocks]	
Massive	Granite (68), Paleo&Mesozoic sandstone (192), Quartz porphyry (65), Granodiorite(695), etc Total 1650 [123] data	Tuff (1201), Tuff breccia(331), Phyolite (341), Andesite (479), Sandstone (67), Basalt(3), etc Total 2698 [673] data	
		Group3 [Planar medium rocks]	**Group4** [Planer soft rocks]
Planar	▬▬▬▬	Paleo&Mesozoic Shale(1279), etc Total 1279 [111]data	Green schist (218), Black schist (27), Tertiary mudstone (229), etc Total 474[105] data

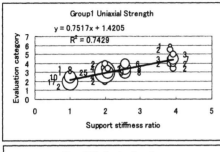
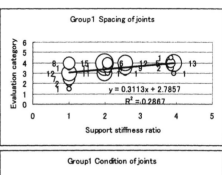
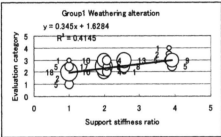
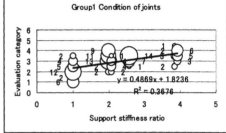

Figure 2. Evaluation results distribution for items concernig "Rock group 1".

3.2 Weighting among evaluation categories

3.2.1 Data used in analysis

The data used for the purpose of analysis was taken from data on tunnels from 1996 to 1998, during which tunnel face observations were conducted using the new face observation form. Since the new tunnel face observation method was introduced at tunnel construction sites throughout the country, almost all face observation data obtained at the sites of the tunnels constructed during the period mentioned above are included.

The data thus collected cover 53 tunnels, 66 work sections and 9,690 cross sections. Of these, 6,101 cross sections were built using one or more of the six standard support patterns, namely, B-a, CI-a, CII-a, CII-b, DI-a and DI-b (see Table 1). These data are used for analysis.Prior to the degree-of-contribution study, the data items shown in Table 2 were screened, according to the conditions described below, to determine the data items to be used in the analysis.

    (i) Screening by overburden conditions
    (ii) Screening by standard deviation
    (iii) Screening based on water inflow/deterioration evaluation

To eliminate the influence of water inflow/deterioration evaluation on support determination, only data with a water inflow/deterioration rating value of "1," that is, data deemed almost free from the influence of water inflow and deterioration, were extracted, and the other cross sections were excluded.

Thus, the number of data sets (cross sections) extracted under the conditions mentioned above became 1,012 for 49 tunnels. In Table 2, a number in parentheses indicates the number of cross sections adopted, which corresponds to about 17% of all standard support pattern data.

3.2.2 Determination of weights for evaluation categories concerning rock mass strength

The JH method is rock mass rating system for tunneling in which the four general observation items related to rock mass strength, namely, "compressive strength," "weathering," "spacing of joints" and "condition of joints" are rated on a combined scale of 100, and adjustment points for "water inflow," "deterioration due to water" and "strike and dip of discontinuities" are subtracted from the rating points measured on the 100-point scale. In this study, the maximum rating for each observation item is determined by quantitatively evaluating the degrees of contribution of the four observation items to support determination.

3.2.3 Weighting within rock groups

As an example, Figure 2 shows evaluation result distributions for different observation items for Rock Group 1, with support stiffness shown on the horizontal axis and evaluation category distribution on the vertical axis. Notice that since the horizontal axis shows support stiffness expressed by the Hokuriku method, as mentioned earlier, CI-a and CII-a or other patterns corresponding to them, and DI-a and DI-b or other patterns corresponding to them can be clearly distinguished. The size of a circle and an accompanying number indicate the number of points. Regression lines estimated from the distributions are also shown in Figure 3. As clearly shown in the figure, for all observation items in Rock Group 2, the lower the rating, the

Table 3. Rating of category of each rock groups.

		Inclination of regression line	Occupation ratio of each group	Rating of category	1	2	3	4	5	6
Group1	Uniaxial strength	0.752	39.68%	40	32	24	16	8	0	
	Weathering alteration	0.345	18.20%	18	12	6	0	-	-	
	Spacing of joints	0.311	16.42%	16	12	8	4	0	-	
	Condition of joints	0.487	25.71%	26	19	13	6	0	-	
	Total	1.895	100%	100						
Group2	Uniaxial strength	1.113	31.28%	31	25	19	13	6	0	
	Weathering alteration	0.777	21.84%	22	15	7	0	-	-	
	Spacing of joints	0.725	20.38%	20	15	10	5	0	-	
	Condition of joints	0.943	26.50%	27	20	13	7	0	-	
	Total	3.558	100%	100						
Group3	Uniaxial strength	1.427	32.37%	32	26	19	13	6	0	
	Weathering alteration	1.149	26.07%	26	17	9	0	-	-	
	Spacing of joints	0.603	13.68%	14	10	7	3	0	-	
	Condition of joints	1.229	27.88%	28	21	14	7	0	-	
	Total	4.408	100%	100						
Group4	Uniaxial strength	1.005	37.08%	37	30	22	15	7	0	
	Weathering alteration	0.414	15.28%	15	10	5	0	-	-	
	Spacing of joints	0.558	20.59%	21	15	10	5	0	-	
	Condition of joints	0.733	27.05%	27	20	14	7	0	-	
	Total	2.710	100%	100						

higher the stiffness of the supports used becomes. This means that a graph showing support pattern (light=>heavy) on the horizontal axis and evaluation (good=>poor) on the vertical axis rises to the right. However, there are observation items, such as crack spacing for Rock Group 1 shown in Figure 2 for which rating values remain more or less constant, regardless of changes in support stiffness. This is thought to mean that if a rating value increases substantially with sup-

port stiffness, the observation item concerned greatly affects support stiffness. Conversely, if a rating value does not change substantially as support stiffness changes, then the observation item concerned does not substantially affect support determination. In other words, the slope of a regression line indicates the degree of contribution of an observation item to support determination.

Attention was paid, therefore, to the slope of regression line, and the degree of contribution of each observation item was set so that greater slopes give higher rating values (maximum rating points). Thus, it was decided to evaluate tunnel face condition in terms of the sum of rating values for each observation item, and rating points were allocated so that the slope values for each observation item add up to 100%. The maximum rating values for each observation item thus determined are shown in Table 3. It shows rating point allocations in which evaluation categories for each rock group are spaced more or less equally.

As the tables clearly indicate, "compressive strength" shows a high degree of contribution in all rock groups, followed by "condition of joints". The degree of contribution of compressive strength tends to be higher and joint information (spacings, conditon) lower than in the RMR classification system[3], which is generally said to be suitable for application to hard rock. In the evaluation of the stability of hard rock ground, where the strength of ground and rock fragments is high, judgment as to falls of blocks of rock is important. Consequently, information of joints becomes necessary, thereby increasing these evaluation categories' degrees of contribution. In the case of medium-hard to soft rock ground, which is typical in Japan, strength of ground and rock fragments is very important for ground stability evaluation and tunnel support determination. This

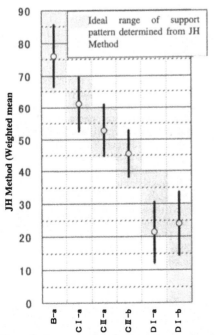

Figure 3. Average rating for tunnel support pattern.

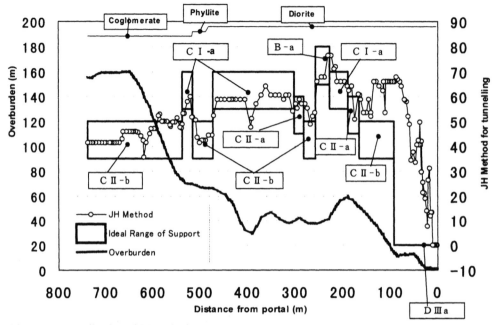

Figure 4. An application of JH method.

is probably why the degree of contribution of the unconfined compressive strength of rock fragments was given relatively large weight.

4 APPLICATION OF JH METHOD

Figure 3 shows the average rating values and their standard deviations for different support patterns based on Table 3. The shaded areas in the figure indicate approximate ranges of rating values for support determination, based on the results obtained. These ranges, each of which covers about 15 to 20 rating points, include the corresponding support patterns. Figure 4 shows ground ratings in the longitudinal direction of a two-lane highway tunnel based on the rating point ranges shown in Table 3. The tunnel to which the new rating method was applied was constructed after data were collected to build the tunnel face observation database. The data for the tunnel, therefore, had not been used as part of the database. In Figure 4, the tunnel support patterns that were actually used for the tunnel are shown with the corresponding rating points, using rating point ranges defined by reference to Figure 3: B-a: 65-80 points; CI-a: 55-70; CII-a: 45-60; CII-b: 35-50; DIa: 20-40; DI-b: 10-30; and DIII-a: 0. Overburden depths are also shown in Figure 4.

As clearly shown in Figure 4, the rating points not only fall mostly within a reasonable range from the support patterns actually used except in the portal zone, but also accurately express the good/poor conditions

of the ground. Considering the fact that support patterns at tunnel portals are determined by factors other than the ground conditions such as overburden depth and eccentric loads, noncorrespondence between the rating points and the support patterns is self-evident. JH method, therefore, can be considered a very useful method of ground evaluation.

5 CONCLUSION

The relationships between rating results obtained by the new rating method and tunnel support patterns have been investigated, and ways to weight evaluation categories to achieve good correspondence with support patterns chosen were considered. A satisfactory rating method has been developed, but, in applying this method, it is important to minimize individual differences between human observers and evaluate the tunnel face conditions as objectively as possible. Over the past several years, the authors have organized JH's effort to achieve these goals and conducted presentations of the new rating method and face observations at tunnel sites. The authors will continue these efforts and accumulate more data necessary for appropriate ground evaluation.

REFERENCES:
1)Nakata M., Mitani K., Yagi H., Nishi T., Nishimura K. and Nakagawa K.,1999, *A new proposal of rock mass classifica-*

tion for tunnel faces based on analyses of observation records, Challenges for the 21st Century, pp.113-119.

2) Yagi H, Otsu T., Mitani K.and Yoshizuka M.,1997,*Results of trial applications of new support patterns (in japanese)*, Highway Technology, pp.26-35.

3) Bieniawski, Z. T.,1989, *Engineering Rock Mass Classifications*, John Wiley & Sons.

Modern Tunneling Science and Technology, Adachi et al (eds), © 2001 Swets & Zeitlinger, ISBN 90 2651 860 9

Rock mass classifications study with the neural network theory

P.G.C. Lins
Universidade Federal da Bahia, Salvador, Brazil

T.B. Celestino
THEMAG Engenharia Ltd. & Universidade de São Paulo, São Carlos, Brazil

ABSTRACT: Rock mass classification systems and the neural networks have similar characteristics: both employ a data base for their development and weights are used in the processing. The main rock mass classification systems, *Q* and *RMR*, can be written as local neural representations. Distributed neural representations can increase the reliability in geomechanical classifications, but the most important contribution of the work is to open new possibilities of interpretation of the classification process. The paper shows local representations of both *Q* and *RMR* systems, and the further development of a neural network to identify wall and roof instability with good results.

1 INTRODUCTION

Conventional rock mass classification systems are an important tool in several stages of a tunneling project. Those systems and the neural networks have similar characteristics: both employ a data base for their development and weights are used for the processing. However the rock mass classification systems are not based on a consistent theoretical framework, like the neural networks.

Neural network are models, inspired on the biological nervous systems, which can "learn" by examples. They have the ability for generalization and the capacity to work with imprecision and uncertainty in the real world.

The geomechanical classification systems group the rock mass into classes with similar behavior from the engineering view point. The Rock Mass Rating System (*RMR*), proposed by Bieniawski (1973) and the *Q* System proposed by Barton et al. (1974) are the most used systems in the engineering practice.

Several neural network classification models have been developed. Cecil's (1970) database was used by Lee & Sterling (1992) in the construction of an autoassociative model, which was improved by Yang & Kim (1996). Yang & Zhang (1998) applied a neural network to the Rock Engineering System also using the database compiled by Cecil (1970). Feng (1995) proposed a neural classification for stability, blastability and drillability of rock mass. The classification of Yang & Zhang (1997) relates the *Q* system parameters to the behavior of the rock mass. Other new technologies have also been applied to improve classification schemes like multivariate discriminant

models (Yun-Mei & Xi-Cai 1991, Nakata et al. 1999).

Neural networks are also applied in several fields of rock mechanics research like constitutive modeling (Ghaboussi et al. 1991), identification of rock properties (Nie & Zhang 1994, Zhang et al. 1991), geophysics (Huang & Williamson 1996), quantification of joint roughness (Lessard & Hadjigeorgiou 1998).

The approach adopted in this work consists of an initial exploration of the classification theory using the neural network framework, followed by a discussion of the possibilities in representing the geomechanical classification systems as neural networks. Historical data are analyzed, and comments are presented. The results show that neural networks representations can improve the classifications or help in the development of new applications. An example of such a new application is presented, with respect to underground excavation wall and roof instability.

2 NEURAL NETWORK BASICS

There is a large number of neural network models. Rumelhart et al. (1986) identify eight major aspects of these models: 1) A set of processing units; 2) A state of activation; 3) An output function for each unit; 4) A pattern of connectivity among units; 5) A propagation rule for propagating patterns of activities through the connectivity network; 6) An activation rule for combining the inputs that come to a unit with the current state of that unit to produce a new level of activation for the unit; 7) A learning rule whereby patterns of connectivity are modified by experience; and 8) An environment within which the system must operate.

The major aspects of these systems are represented in Figure 1. The sets of processing units are indicated by circles in the diagram. For each time instant t, each unit u_i has an activation value, denoted by $a_i(t)$; this activation value is passed through a function f_i to produce an output value $o_i(t)$. This output value can be passed through a set of connections (indicated by lines or arrows in the diagram) to other units in the system. Associated with each connection there is a real number, usually called weight or strength of the connection, denoted by w_{ij}, which determines the effect that a first unit a_j has on a second unit a_i.

The input values in each unit are combined to produce a net input for the unit, usually in the form:

$$net_i = \sum_{j=1}^{n} w_{ij} x_i \qquad (1)$$

where x_i = the input value in the unit u_i; w_{ij} = weight of the connection from unit a_j to unit a_i. These calculations can be made by functions in the so called functional link networks (Pao 1989, Looney 1997).

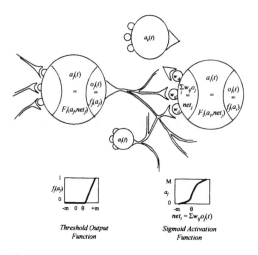

Figure 1. The basic components of a parallel distributed processing system (Rumelhart et al. 1986).

Given a network of simple computing elements and some entities to be represented, the most straightforward scheme is to use one computing element for each entity. This is called local representation. It is easy to understand and easy to implement because the network mirrors the structure of knowledge it contains (Rumelhart et al. 1986).

In the so called distributed representation, there is no direct relationship between units and their representation. The knowledge is distributed in the network. This approach has the capacity to produce more complex mapping between input and output of the net. The most used neural networks have this kind of representation.

3 PATTERN CLASSIFICATONS THEORY

The concept of classification involves the learning of likenesses and differences of patterns that are abstractions of instances of objects in a population of nonidentical objects. When it is determined that an object from a population P belongs to a known subpopulation, a pattern recognition is done. Classification is the process of grouping objects together into classes (subpopulations) according to their perceived likenesses or similarities. The subject area of pattern recognition includes both classification and recognition (Looney 1997).

Learning is done by a system when it records its experience into internal system changes that causes its behavior to be changed. Human beings learn from experiences by accumulating rules in various forms such as associations, tables, inequalities, equations, relationships, logical implications and others. Classification is a form of learning which induces from antecedent attributes the classes that are consequences. Reasoning is a process of applying general rules, equations, relationships, etc., to an initial collection of data, facts, etc., to deduce a resultant decision. Recognition is a form of reasoning, while classification is a form of learning.

Figure 2 represents subpopulations $S_1,...,S_4$ of a population P of nonidentical objects, along with the processing that recognizes a sample object. Attributes of an object are sensed or measured to yield a pattern vector that is transformed into a reduced set of features. The object is recognized from these features by the recognizer. Let $\{m_i: i = 1,...,P\}$ be variables for pattern measurements to be made on objects selected from P. A feature extractor T transforms the pattern vector $\mathbf{m} = \{m_1,...,m_P\}$ into a feature vector $\mathbf{x} = \{x_1,...,x_P\} = T(\mathbf{m})$. A pattern recognizer is a system to which a feature vector is given as input, and which operates on the feature vector to produce an output which is the unique identifier (name, number, codeword, vector, string, etc.) associated with the class to which the object belongs. Pattern recognition based on measurements of physical attributes are called concrete. In the world of ideas, patterns are based upon the attributes of concepts and mental models, which is abstract pattern recognition (Looney 1997).

Each object of a population can be represented by a point in the feature space. Class separation is made in the feature space. Figure 3 presents a population P which has two features, and can be grouped in two classes. In Figure 3a it is clear that a hyperplane can separate the classes. This is not the case of Figure 3b.

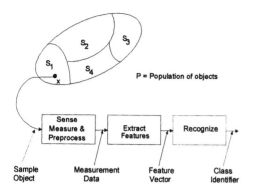

Figure 2. The recognition/classification process (Looney 1997).

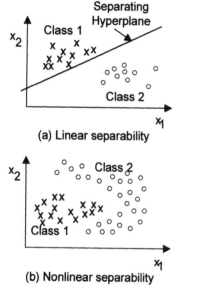

(a) Linear separability

(b) Nonlinear separability

Figure 3. Linear and nonlinear separability in feature space (Looney 1997).

3.1 *Classifications with neural network*

Neural networks can easily make the classification and recognition process. Take a N-dimensional feature space separable by a hyperplane H determined by a discriminant function in the form:

$$D(\mathbf{x}) = w_1 x_1 + w_2 x_2 + \ldots + w_{N+1} x_{N+1} = 0 \qquad (2)$$

For each feature vector ($\mathbf{x}$) the value of $D(\mathbf{x})$ is positive or negative or zero. Positive or negative values of $D(\mathbf{x})$ denote that the feature vector belongs to the right or left half-space respectively. A zero value for $D(\mathbf{x})$ denotes that the feature vector belongs to the hyperplane H. Note that, for convenience, the feature vector ($\mathbf{x}$) is taken with dimension ($N+1$), and the

($N+1$)th component of the vector is always equal to 1 (Looney 1997).

The single neural network model shown in Figure 4 can make the classification process. The neuron computes the discriminant function in the form:

$$D(\mathbf{x}) = \sum_{i=1}^{N+1} w_i x_i = 0 \qquad (3)$$

The value of $D(\mathbf{x})$ is passed through a threshold output function, which yields $y = +1$ if $D(\mathbf{x}) \geq 0$ or $y = -1$ if $D(\mathbf{x}) < 0$. The class identifier is y.

The model shown in Figure 4 can only be applied for linearly separable problems. Only networks with more than one layer can make nonlinear class separations.

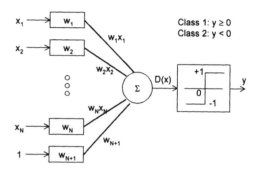

Figure 4. Linear discriminant neural network.

4 A LOCAL REPRESENTATION TO *RMR*

The *RMR* system can be represented as a modification of the model presented above: a single perceptron with 32 input units (Figure 5), processed in a neuron with a stepped output function (Figure 6). The weights can be the same used in conventional *RMR* classification (see Bieniawski 1989). The output of the model is an integer number which represents the rock mass class.

The input feature vector can be fragmented in six parts, each one related to the six input parameters of the *RMR* system (Figure 5). In each of the six parts of the feature vector there is only one unit value element, all other elements are zero. The non-zero element identifies the category of the parameter.

Conventional neural network training algorithms can be used to train this network and create or adapt classifications similar to conventional *RMR*. The developer must observe that conventional training algorithms do not have responsibility with the weight hierarchy (i.e. better field conditions have greater weight), and specific algorithms are needed.

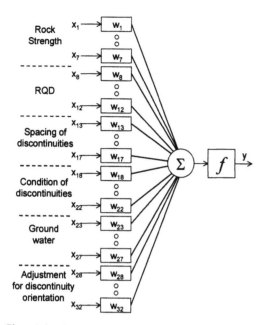

Figure 5. Local representation to *RMR* system.

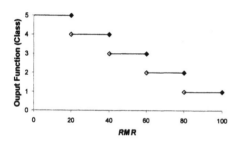

Figure 6. Output function for *RMR* network.

5 A FUNCTIONAL LINK REPRESENTATION FOR *Q*

The *Q* system needs particular attention to be written as a neural network. Figure 6 shows the proposed model. A Sigma-Pi processing unit is used. This unit produces a net input by multiplication (Rumelhart et al. 1986) in the form:

$$net_i = \sum_{i=1}^{n} w_i \prod_{i=1}^{n} x_i \qquad (4)$$

where x_i = the input in the output unit; w_i = weight of the connection from the hidden units J_j to the output unit. In a more general context this kind of processing unit is called functional link (Pao 1989, Looney 1997).

The proposed model shown in Figure 6 has a 60-dimensional input feature vector. The first element of the feature vector is the numeric value of RQD divided by 100. The other elements are grouped in relation to the other five parameters of the *Q* system. For each group there is only one unit value element, the other elements in the group are zero. There are also three individual elements to represent construction of portals or intersections and joint spacing greater than 3 meters. The weights in the input layer can be the same presented by Barton et al. (1974) for J_r and J_w, and their inverse for J_n, J_a, and *SRF*. The first weight of the input layer is taken as one. The sum of synaptic connections of the output layer is taken as one.

The network has a hidden layer with six processing elements that compute the values of RQD, J_r, J_w, and the inverse values of J_n, J_a, and *SRF*. The activation value of the output layer is numerically equal to the *Q* value. A stepped output function similar to the one presented in Figure 6 is used to make the model produce an integer number as output that represents the rock mass class. The activation in the output unit represents the numerical value of *Q*. The representation structure described above is only one of the existing possibilities. The network has the *Q* system as "teacher" in the model just described. Other "teacherless" classification methods are being studied, i.e., non-supervised classification. In the future it may be possible to compare similar responses to those provided by the *Q* system.

The model was coded and preliminary tests indicate that it work well. As expected by using the weights presented by Barton et al. (1974) the network made the classification correctly.

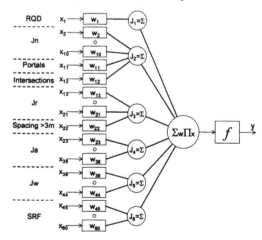

Figure 6. Functional link representation to *Q* system.

The backpropagation algorithm was adapted to train the proposed network. The Cecil (1970) database is about half of the database used by Barton et al. (1974) to develop the Q system. A training process with modification of only the *SRF* weight was performed with the database of Cecil (1970). The resultant weights are very similar to those presented by Barton et al. (1974). In training with fixed weights, the error is still propagated as conventional training (see Rumelhart et al. 1986).

This kind of training can be interesting in some situations when the objectives are adaptations in a classification process. It is interesting to remember that in the updating for the Q system presented by Barton & Grimstad (1994) several modifications were proposed for the system, but the main change was made exactly on the *SRF* weights.

6 A DISTRIBUTED MODEL

There are some important points in the design of a neural classification model: 1) representative data must be available; 2) the selection of a network architecture that gives a good answer to the problem; and 3) the input parameters should be available and follow a standard so that they can be applied to others projects A neural network was designed to identify wall and roof instabilities in each case record in the database (Figure 7). Cecil's (1970) database was used in the simulation. The network has 14 input parameters (6 Q systems parameters, D tunnel span, 6 parameters for joint orientation, and the Q value). A two-dimensional vector is the output of the network; the first element of the vector indicates wall instability with value 1 and zero for a safe state. The second output vector element similarly indicates roof instability.

Generalization ability is the capacity to given answers never presented to the system before. To test the capacity of generalization of the neural network the database was divided into two sets, one for training (90% of the patterns) and the other for testing (10% of the patterns). The network was trained with 83 cases (patterns) and learned the training set. After 743 training cycles the sum of square errors was found to be equal to 0.008933.

Table 1. Results for the neural network test.

Case	Actual		Neural net	
	Wall	Roof	Wall	Roof
84.	0.00	1.00	0.33	0.99
85.	1.00	1.00	0.33	0.99
86.	1.00	1.00	0.00	0.98
87.	0.00	0.00	0.00	0.18
88.	0.00	1.00	0.39	0.90
89.	0.00	1.00	0.00	0.99
90.	0.00	1.00	0.03	0.99
91.	0.00	0.00	0.01	0.01
92.	0.00	1.00	0.00	0.92

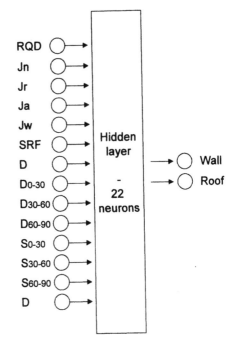

Figure 7. Neural network to identify wall and roof instabilities.

The test set, 9 cases (patterns), was submitted to the trained network, and the results are presented in Table 1. There are only two wrong classifications in 18 output values. It is interesting to see that the network failed only in predicting wall instability.

The presented model is only one from an extraordinary number of possible combinations that can be easily made with neural networks.

7 CONCLUSIONS

Neural networks and geomechanical classifications have several similarities. The neural network theory can help to improve the classifications or help in the development of new applications. The neural network algorithms can be used to create classifications similar to classifications used in engineering practice. It is also possible to adapt a conventional classification to a particular geological condition.

The main geomechanical classifications can be represented as neural networks. Distributed neural network classification can handle tasks that conventional classification cannot. The exploration of the distributed neural classification can help to obtain a better field characterization. An example of distributed neural network model was presented with good results for the prediction of instability during excavation.

REFERENCES

Barton, N. & Grimstad, E. 1994. Rock mass conditions dictate choice between NMT and NATM. *Tunnels & Tunnelling*, October: .39-42.

Barton, N., Lien, R. & Lunde, J. 1974. Engineering classification of rock masses for the design of tunnel support. *Rock Mechanics* 6(4): 189-239.

Bieniawski, Z.T. 1973. Engineering classifications of jointed rock masses. *Trans. S. Africa Instn. Civ. Engrs.* 15(12): 335-344.

Bieniawski, Z.T. 1989. *Engineering rock mass classifications*. New York: Wiley.

Cecil III, O.S. 1970. *Correlation of rock bolt-shotcrete support and rock quality parameters in Scandinavian tunnels.* Ph.D. Thesis, University of Illinois, Urbana. (Also Swedish Geotechnical Institute, Proceedings No. 27).

Feng, X. 1995. A neural network approach to comprehensive classification of rock stability, blastibility and drillability. *International Journal of Surface Mining, Reclamation and Environment* .9: 57-62.

Ghaboussi, J.; Garrett, J.K.Jr. & Wu, X. 1991. Knowledge-based modeling of material behaviour with neural networks. *Journal of Engineering Mechanics* 117(1): 132-153.

Huang, Z. & Williamson, M.A. 1996. Artificial neural network modelling as an aid to source rock characterization. *Marine and Petroleum Geology* 13(2): 277-290.

Lee, C. & Sterling, R. 1992. Identifying probable failure modes for underground openings using a neural network. *Int. J. Rock Mech. Min. Sci. & Geomech. Abstr.* 29(1): 49-67.

Lessard, J.-S. & Hadjigeorgiou, J. 1998. Quantifying joint roughness using artificial neural networks. *Int. J. Rock Mech. Min. Sci.* 35(4-5), Paper No.093.

Looney, C.G. 1997. *Pattern recognition using neural networks: theory and algorithms for engineers and scientists.* New York: Oxford University Press.

Nakata, M., Mitani, K., Yagi, H., Nishi, T., Nishimura, K., Nakagawa, K. 1999. A new proposal of rock mass classifycation for tunnel faces based on analyses of observacional records. In Alten et al. (eds), Challenges for the 21st Century: 113-119. Rotterdam: Balkema.

Nie, X.-Y.; Zhang, Q. 1994. Prediction of rock mechanical behavior by artificial neural network. A comparison with tradicional method. In *IV CSMR, Integral approach to applied rock mechanics*; Santiago Chile.

Pao, Y.-H. 1989. *Adaptative patter recognition and neural networks*. Reading, MA: Addison-Wesley.

Rumelhart, D.E. et al. (eds) 1986. *Parallel distributed processing: Explorations in the microstructure of cognition.* Massachusetts: MIT Press. 2v.

Yang, H.S. & Kim, N.-S. 1996. Determination of rock properties by accelerated neural network. In *North American Rock Mechanics Symposium; Proc., Quebec, June 1996.* Rotterdam: Balkema.

Yang, Y. & Zhang, Q. 1997. A hierarchical analysis for rock engineering using artificial neural networks. *Rock Mechanics and Rock Engineering* 30(4): 207-222.

Yang, Y. & Zhang, Q. 1998. The application of neural networks to rock engineering systems (RES). *Int. J. Rock Mech. Min. Sci. & Geomech. Abstr.* 35(6): 727-745.

Yun-Mei, L. & Xi-Cai, L. 1991. A multivariate discriminant model for classification of rock masses. In Beer et al. (eds), *Computer Methods and Advances in Geomechanics*; 1991 Rotterdam: Balkema.

Zhang, Q.; Jiarong, S. & Nie, X. 1991. Application of neural network models to rock mechanics and rock engineering. *Int. J. Rock Mech. Min. Sci. & Geomech. Abstr.* 28(6): 535-540.

Modern Tunneling Science and Technology, Adachi et al (eds), © 2001 Swets & Zeitlinger, ISBN 90 2651 860 9

Rock Support Design for Tunnelling in Nepal Himalayan region (With a Case study of Khimti I Hydropower Project, 60 MW)

Gyanendra Lal Shrestha
Acting Director
Butwal power company Ltd
Lalitpur, Nepal.

ABSTRACT: Nepal Himal consists of various types of rocks and geological structures. Thus the underground construction works through these domains varies in terms of excavation, support system requirement and stability. The critical issues with regard to the underground works are geotechnical. Hence investigations are necessary to clarify the type and magnitude of the risks and how they can be reduced or managed. Experiences have shown that same type and scale of ground investigation is not suitable to all the project sites. The Khimti 1 Hydropower Project (KHP) is the first project applying Q-system in Nepali Himalayan Rock. The method was found to be appropriate for drill and blast tunnels in jointed, fractured and sheared rock, which tend to overbreak. The main advantage of the NMT is that each stretch of tunnel is evaluated and it takes optimum advantage of the self-supporting capacity of the rock. Site specific modification would be necessary in determining support for the underground construction.

1 INTRODUCTION

The Mining Industry has not been developed in Nepal with the result that only a few kilometres of tunnel have been constructed to date. The history of systematic tunnel construction in Nepal started only 35 years ago with the construction of the 500 m long tunnel for Tinau Hydroelectric Project (1024 kW) in Butwal. The project also includes an underground powerhouse constructed at about 40 m below the riverbed level. Since then tunnel construction work has proceeded. A summary of the available data on projects in Nepal that have involved tunnelling is given in Table 1.

Table 1. Tunnelling projects in Nepal.

Name of Project	Size m2	Length
Doti Small Hydel	4	116
Surnaya Irrigation	4	800
Andhi Khola Hydel	6-7	2400
Tatopani Small Hydel	5	800
Test Adit for Karnali	4	550
Arun II Test Adit	4	1400
Jhimruk Hydel Project	8.5	1050
Andhi Khola Shaft	12.5	250
Powerhouse	60	36
Tatopani Small Hydel Desilting Basin	25	28
Arun III Test Caverns	22	45
Jhimruk Inclined tunnel a) 1:12 slope	9	140
b) 1:1 slope	4	270
Marsyangdi Hyd el Headrace Tunnel	6.4	7119
Pressure Tunnel	5	75
Khimti	11-14	12800
Modi		2298
Kali G	43	6300
Kulekhani I Hydel	5.6	6327
Kulekhani II Hydel	5.6	5848
Puwa		3678
Tinau Hydel	4.0	2500

Except for the 800 m long tunnel of Surnaya Gad Hill Irrigation Project, all the tunnels have been constructed for Hydropower projects.

Kali Gandaki Hydropower Project, Melamchi Diversion Scheme and Middle Marshyandi Hydropower Project are the presently on-going projects in Nepal that require tunnelling.

2 NECESSITY OF TUNNEL IN NEPAL

Nepal has got an extreme topography ranging from the 8848m Mt Everest summit to the Terai flat land with elevation 300 masl . The rivers in this topography offer a potential for hydropower, which could help Nepal's economic growth. Maximum power can be generated either by constructing tunnels or by building dams. But a high dam always submerges a large area of land upstream and imposes the risk of flooding the downstream area if the dam is breached. On the other hand tunnels have very small environmental impact and do not impose any risk to the downstream area. Thus water-carrying tunnels are the better alternative for the maximum power generation.

Similarly, cities and towns have been developed often on the opposite sides of the high hilly ranges. Tunnelling through these hills is one of the possible means of shortening and speeding journeys.

Large projects in Nepal usually need outside financial assistance. The outside assistance does not come through if the project is not justified as environmentally acceptable. Surface road and canal construction in hilly region often induce environmental degradation and, so are less attractive for the outside financial assistance. On the other hand, tunnel roadway and water carrying tunnel not only avoid surface instability problem but also do not cause environmental degradation. So these tunnel networks easily attract outside financial assistance and national development can go ahead.

With the rapid population growth in the Kathmandu Valley and in other cities, surface road networks are getting very congested. It is globally understood that the large cities need underground transport networks. Sooner or later, Nepal will also need to construct this type of network.

3 GEOLOGICAL RISKS

Tectonic movement along with all its geological elements has disturbed the geology in Himalayan region. It has not only caused the fragile geology but also resulted with the series of different types of rocks and geological structures. Thus the underground construction work through these domain varies in terms of excavation, support system requirements, stability and potential for rock burst and squeezing in the underground openings.

Wide disturbed zones are associated with the faults. Minor faults and shearing are common places. Even in good rocks, intense jointing can be observed. Weathering is expected to be deep and extensive, especially under the saddle and this is the area where the cover over the underground structures is low, and in areas where the rocks have been disturbed by faulting and shearing. In addition, major stress relief joint systems can be expected close to the deep valleys. Underground construction will be very sensitive to these adverse geological conditions.

Lining will be required to prevent leakage into highly permeable ground across the saddle areas where the water table is likely to be below tunnel elevation. Heavy support may be needed in highly stressed ground resulting from high cover or tectonic forces. Ground water inflows may be encountered along faults and highly permeable rocks and could prove troublesome where the head is high or pumping is required in downgrade headings. It can be considered that the risk of displacement along an active fault during the life of the project is low and less of a potential problem than the existing ground disturbance created by the major inactive faults that cross the alignment and which could cause difficulties in excavation.

Detailed geotechnical and geological investigations should be carried out before the design for the main project is started.

4 CHOOSING SITE INVESTIGATION METHODS

The hazards associated with ground may jeopardise a project, if these are not properly assessed or understood. With the properly supervised and interpreted geotechnical investigation, underground condition, hazards and other design and construction parameters which were beneath the site become known. Otherwise the costs of remedying wrongly designed works or mobilising alternative construction methods are usually far in excess of the cost of the original site investigation.

The critical issues with regard to these underground works are geotechnical. An understanding of the ground through which the excavation will be carried out and areas through which the portal will be developed will be essential. Hence investigations are necessary to clarify the type and magnitude of the risks and how they can be reduced or managed

The investigation includes geological mapping, geophysical survey, core drilling, test tunnelling, insitu testing and laboratory testing.

4.1 If good outcrops are available at project area geological mapping will be very useful. Geophysical test needs to be correlated with the core drilling, so that it provides inexpensive and quicker sub surface information. For the underground works at shallow depth and for a small budget geological data can be obtained by mapping and geophysical tests only.

4.2 The standard approach to underground investigation is to drill testholes. In the case of the good rock condition, recovery of the samples from the testholes will be good. For the area of simple geology, the data will be representative of a large area and worth drilling the test holes.

4.3 In the case of the crushed rock, recovery of the samples from the testholes will be minimal. For the area of the high ground cover, drillholes would be deep and hence expensive. Although they could provide good site-specific data at tunnel elevation, that might only be representative of short tunnel lengths because of the complexity of the geology. In this condition test tunnels is considered to be a valuable means of investigation because it provides a first-hand look at the ground, potentially at tunnel elevation, but also allows the testing of different techniques of excavation and support and permits pre-construction of tunnel access. The test adit is also driven, to investigate the rock conditions considered to be representative of some of the poorer ground along the tunnel route and to gain early access to a crucial section of the main tunnel line.

4.4 In tunnelling projects where data is difficult or expensive to obtain during the feasibility or design stages, it is now common to undertake investigations concurrently with excavation, for example by drilling ahead of the face to anticipate ground conditions or to use modern technique on the tunnel face. Thus it is not always essential that the critical issues are fully resolved prior to commencing construction, providing that they have been recognised, and can be dealt with, and that the requisite data can be obtained in a timely way to minimise the risks during construction and operation. However this approach must also be reflected in the way that the construction contracts are written to allow flexibility in design and a degree of shared risk between the owner and contractor.

5 ROCK SUPPORT DESIGN

In Himalayan region, rock quality changes in a short section of underground construction. Drill and blast method adapts better in such condition. Degree of mechanisation depends on the cross-section of the excavation and scale of the project.

The principal objective of the rock support is design of efficient and economical support for underground excavation. Rock strength around the excavated space is utilised to support itself. None of the available classification systems provides ability to design the correct support for underground construction for this region. Therefore site-specific modifications are usually required as compliments to the established support design methods. The applied support needs to be monitored specially in the critical areas. As per the observation additional support may be required.

Where there is a possibility of encountering water it is important to specify that a probe hole should be drilled ahead of the face. Typically, such a probe hole should extend 2-3 times tunnel diameter ahead of the face at all times. A geotechnical engineer should very carefully monitor the drilling. Penetration rate and the quantity and colour of the water return should be recorded on probe-hole log. Sudden change in the penetration rate will indicate the presence of the hard or soft zones, and deviation from normal water quantity and colour may indicate water-bearing fault or fracture zone. This probing is particularly important if a major fault zone that acts as a water barrier is to be traversed. This type of problem can only be solved satisfactorily if there has been sufficient advanced planning, by both the engineer and the contractor, to ensure that an agreed course of action has been mapped out and that appropriate equipment has been mobilised before the fault is exposed.

6 CASE STUDY: KHP I HYDROPOWER PROJECT

The Khimti Hydropower Project (KHP) site is located in the Dolakha and Ramechhap District, approximately 175 km due east of Kathmandu. The project is a "run of the river" hydroelectric power project designed for an installed generating capacity of 60 megawatts. The power plant will utilize a drop from 1272 to 586 m above sea level in Khimti Khola and has the highest head of 686 m in Nepal. A concrete diversion weir will divert up to 10.75 cumecs of water from the Khimti Khola into the 7.9 km long headrace tunnel and then through 913 m long, steel lined penstock, inclined at 45° and 100 m horizontal, to an underground powerhouse (70 m long, 11 m wide and 10 m high). The powerhouse is 420 m under the ground surface and 893m inside. It will contain five horizontal Pelton units. Total tunnel length in this project is 12.8 km with diameter in the range of 3 to 5 m including adits.

6.1 GEOLOGY OF THE PROJECT AREA

The project area lies in the Midland schuppen zones of the Melung Augen Gneiss. The rock of this zone is mainly grey, coarse to very coarse grained, porphyroblastic augen gneisses (63%), occasionally banded gneisses (12%), and granitic gneisses (7%), with bands of very weak green chlorite and bright grey talcose schists (18%) parallel to the foliation in 5 to 15 m intervals. Structurally, the zone is bounded by two major faults, the Midland Thrust and the Jiri Thrust to the south and north, respectively. The area is also influenced by several minor thrust faults characterised by very weak sheared schist with clay gouge running parallel to the foliation plane with 3 sets of clay filled prominent joints.

6.2 BASIS FOR ROCK SUPPORT DESIGN

The principal objective of the rock support is design of efficient and economical support for underground excavation. Rock strength around the excavated space was utilised to support itself.

In Khimti rock masses are divided into main five classes in order to ease and speed up the decision for correct tunnel support. The tables also suggested the procedure for excavation and application of required rock support depending on rock class. There is additional especial adopted support recommendation for sub horizontal layers in significantly different rock quality (alternate of competent gneiss and incompetent schist) because almost 90% of rock along headrace tunnel dips sub horizontally (dip <20°).

The methods are proceeded in 3 stages. First tunnel log is prepared and the six parameters for quantifying rock mass quality are collected after each round of blast before installation of temporary support. Required amount of rock support is designed using NMT, 10 Adapted Rock Support Design and Observation and Monitoring methods. The recommended support is installed immediately or afterward depending on the stand up time of rock.

The applied support is monitored and observed in critical areas where deformations are noticed. Additional support like rock bolts, shotcrete, reinforced ribs of shotcrete and concrete lining are recommended according to monitoring data and observations where necessary.

The following 3 steps were used for the underground excavation and support design in Khimti Project:

6.2.1 Q-system or Norwegian Tunnelling Method (NMT)

NMT or also called Design-as-you-drive, utilizes a quantitative rock mass classification by Q-system, appropriate use of temporary reinforcement such as bolting and wet fibre reinforced shotcrete, and supplementary reinforcement and support according to Q-based permanent support design. The main support is a combination of rock bolts in pattern and fibre reinforced shotcrete. Shotcrete and rock bolts offer great flexibility with respect to the amount of support provided. Shotcrete thickness, rock bolt spacing and the spacing and thickness of reinforced ribs of shotcrete on poor ground can be varied with the greatest ease to suit rock conditions. Reinforced ribs of shotcrete are mainly recommended in Extremely to Exceptionally poor rock (with squeezing) but concrete lining is also recommended in very small amount. The method is short time consuming,

has rapid advance rate, improves safety and environment and often provides the most economical tunnel support system. Cast concrete lining was also used occasionally in KHP, mainly in overbreak areas with high ground water discharge.

The Q-System uses simple equations to numerically account for the most important size parameters affecting rock mass stability.

The equation uses six different parameters,

$$Q = RQD/J_n \times J_r/J_a \times J_w/SRF \ldots\ldots\ldots\ldots \quad (1)$$

Where RQD   Rock Quality Designation
J_n Number of joint sets
J_r Joint roughness
J_a Joint alteration number
J_w Joint water leakage or pressure
SRF   Stress reduction factor

Equation 1 is made up of three terms, the first term (RQD/J_n) represents the block size, the second term (J_r/J_a) represents inter block friction angle and the last term (J_w/SRF) is a measure of the active stress.

6.2.2 Adopted rock support design:

None of the classification system provides ability to design the correct support for excavations in different type of rock masses. Therefore additional 10 design principles as compliments to the Q-system are introduced (Rock Committee, November 1998) in order to establish a proper and correct tunnel support design. The four important hazards connected to the geological conditions elucidated at the Khimti are main reasons to add 10 design principles, which are as follows:

Erosion and slaking of the tectonised or weathered parts of the rock mass

Squeezing ground at the poorest quality along the tunnel

Raveling ground mostly connected to the poorer parts of the tunnel (unstable rock in general).

Swelling ground where swelling clay is prevailing

6.2.3 Monitoring and observation method

Applied support in the tunnels is monitored by tape extensometer in critical area where deformation is noticed and continuous observations are made in changing condition of applied support. Deformations are noticed mainly in class III, IV and V rock containing very weak schist (Unconfined compressive strength between 1 to 15 MPa) bands greater than 20 cm thick. The rock squeezing process caused deformations. Additional final support is decided according to the monitoring data and field observation.

6.3 ROCK SQUEEZING

Squeezing is a stress problem and is normally considered to be the result of overloading of the rock mass. When the tangential elastic stress around an opening is of the same magnitude as the in-situ strength of the weak rock mass or the stress is greater than the strength of the weak rock mass, squeezing may occur.

The term weak rock is considered below Very poor rock (Q value is less than 1 defined by Bhasin and Grimstad, 1996) but in practical it is found difficult to define weak rock by classification system because rock mass is inhomogeneous in nature due to faulting, shearing, jointing, weathering, altering, alternation of competent and incompetent rock etc.

It is a long-term stability problem, which concentrates on time-dependent behavior of rock mass. If rock strength is less than field stress or overburden load in any direction the rock squeezing will occur.

Singh (1993) proposes that squeezing occur when the Q-value and the stress level satisfy the relation

$$H > 350*Q^{1/3} \qquad (2)$$

Where H = vertical rock cover (m)

Q = Rock mass index according to the Q-system.

Bhasin and Grimstad (1996) presented the condition of squeezing in weak rock (Q value is less than 1) by calculating ratio of tangential stress (σ_θ) and compressive strength (σ_{cm}). According to ratio, the following three conditions are established.

No squeezing if the ratio is less than 1.

Mild and moderate squeezing if the ratio is 1 to 5.

Heavy squeezing if the ratio is greater than 5.

6.3.1 Method of rock support in squeezing ground

The philosophy of rock support method in squeezing ground in Khimti was the application of flexible support (shotcrete and pattern rock bolting) with invert lining at first for providing room for deformation. The squeezing rate was monitored and observed in affected area in weekly or day to day basis depending on squeezing rate. Additional support was recommended after deformation monitoring. The amount of additional support depends on deformation and damage of applied support.

In Khimti, support was provided according to the Tables and the applied support was monitored in class III, IV and V rock (containing very weak schist) by visual at first and by tape extensometer if deformation was noticed. Specially class V rock containing very weak schist area steel fibre was increased from 50 kg to 70 kg per cubic meter to make shotcrete more flexible to bear load. Similarly concrete invert overlapping with shotcrete was provided in very weak schist area. According to monitored data after invert concreting, squeezing rates gradually/abruptly decreased. At some places applied shotcrete were badly cracked and cracks measured were upto 15 cm wide. Grouted rock bolts worked efficiently in squeezing area represented by twisting of end plates. Additional reinforced ribs of shotcrete were applied in critical area of squeezing with invert concrete whereas additional bolts with shotcrete were applied in minor squeezing area.

It was experienced that reinforced ribs of shotcrete with invert concrete provided more flexibility and worked effectively in the mild rock squeezing area

6.3.2 Monitoring and observation in squeezing area

The critical sections of the tunnel have been monitored by tape extensometer and observation for the applied rock supports were also recorded According to observation, deformations in different locations have been noticed by shotcrete cracking generally after 2 to 3 weeks of excavation. The deformed sections were monitored immediately after deformation was noticed. Monitoring and observation were carried out until deformation stopped. In some places additional support like reinforced ribs of shotcrete or rock bolts had to be added due to considerable deformation. Maximum deformation recorded is 40 cm (deformation as large as 5.5% of the tunnel diameter) in Adit 1 Downstream headrace tunnel from chainage 500 m to 600 m.

The effectiveness of the rock bolt strongly depends on the bearing capacity of the rock mass under the end plate and on the ability of the nut plate blocking system to undergo high plastic deformation without failure. Due to mild squeezing in Khimti only few bolts (about 1% of rock bolt end plates or nuts) were failed.

It was found that squeezing took place also in weak rock like weak schist and decomposed gneiss (even in Q value greater than 1) where strength of rock mass was exceeded by rock stresses. For example, squeezing was noticed only in a single band (10-20cm thick) of very weak schist intercalated within massive strong gneiss (Q value was 1 to 6). Therefore, this example clearly indicates that the strength of rock is more important than the term weak rock defined by classification system.

In Khimti mild-squeezing phenomena was noticed at different locations of the tunnels after 2 weeks or 20 m behind face mainly in weak schist and decomposed gneiss having Uniaxial Compressive Strength of 1 to 15 MPa with overburden of 80 m to 420 m deep.

The ratio of σ_θ and σ_{cm} in different squeezing locations of tunnels was calculated and plotted in Chart 1 to satisfy the above conditions. According to the graph, 31 locations satisfied the conditions and 5 were below squeezing limit but squeezing were noticed.

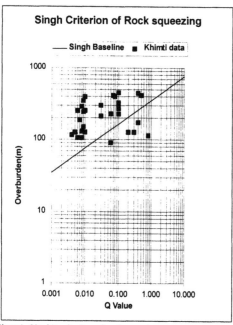

Singh Criterion of Rock squeezing

— Singh Baseline ■ Khimti data

Chart 1. Singh's criterion of rock sqeezing.

6.4 PROGRESS OF TUNNEL EXCAVATION

The progress of tunnel excavation depends on several factors like geological condition, equipment, manpower skill, method of excavation etc. At the beginning the progress rate of the project was slow because of poor geological condition than the expected from feasibility report, inefficient equipment, semi mechanized method of excavation etc. After one and half years, the system was completely changed into mechanized system using boomer for drilling, robot machine for shotcreting, scooptram, wheel loader and dump truck for mucking. The mechanisation speeded up the progress so as to complete the excavation work on time.

In Khimti the excavation record progress rates are 63 m/week, 46 m/week and 73 m/week (in both headings) in 25 m², 14 m² and 11 m² cross sectional area of tunnels respectively.

7 CONCLUSIONS AND RECOMMENDATIONS

Detailed geotechnical and geological investigations should be carried out before the design for the main project is started. Experiences have shown that same type and scale of ground investigation is not suitable to all the project sites.

Some site-specific modification would be necessary in determining support for the underground construction. There should be capabilities to provide support within a short stand up time so that the support will be most effective and prevent potential overbreaks. Monitoring needs to be done to observe the performance and adequacy of the support provided.

The NMT with some practical site specific modifications, such as the design principles used in the Khimti I Hydropower Project, would be an appropriate method of determining rock support in many of the future tunneling projects planned for Nepal.

The theory being used for Rockmass Classification and Support Design in Nepal were developed as per experiences and researches in other parts of the world. But the Nepal Himalayas are the youngest range, very active and geologically different from rocks in other parts of the world. So further adjustment is needed to make these theories applicable to excavating tunnels in Nepal Himalayas.

REFERENCES

Butwal power company Ltd 1996. Melamchi diversion scheme: Bankable feasibility study report.

Barton et al., 1974, Engineering classification of rock masses for the design of tunnel support Rock Mechanics, v 6(4) pp. 36-88.

Barton. N., and Grimstad. E., 1994, Rock mass conditions dictate choice between NMT and NATM., Tunnels and Tunnelling, UK.

Singh. B., 1993, Workshop on Norwegian Method of Tunnelling, CSMRS, New Delhi in "Bhasin. R., and Grimstad. E., 1996, The Use of Stress-Strength Relationships in the Assessment of Tunnel Stability., Tunnelling and Underground Space Technology, Vol. 11., No. 1, Great Britain, pp 93-98."

Grimstad E., and Barton N., 1993, Design and methods of rock support. Rec. Norwegian Tunneling Today., Norwegian Soil and Rock Engineering Association., Bup. 5., Tapir Publishers. pp. 59-64.

Grimstad F., and Barton N., 1993, Updating of the Q-system for NMT. Proc. Int. Symp. On Sprayed Concrete., Norwegian Geotechnical Institute, Norway, pp. 1-10.

Ishida T., and Ohta YM 1973, Geology of the Nepal Himalayas. Saikon Pub. Co. Ltd., Tokyo Japan, pp. 35-45.

Stille. H., Ivarsson. H., and Agaard. B., 1998, Third Site Visit Report from the Rock Committee's, Khimti I Hydropower Project, Nepal. pp 5-6.

Terzaghi. K., 1946 (revised 1968, reprinted 1977), Rock tunnelling with steel support, Commercial Shearing Inc.

Modern Tunneling Science and Technology, Adachi et al (eds), © 2001 Swets & Zeitlinger, ISBN 90 2651 860 9

Assessment of Susceptibility of Rock Bursting in Tunnelling in Hard Rocks

Ö. Aydan
Tokai University, Department of Marine Civil Eng., Shimizu, Japan

M. Geniş
Karaelmas University, Department of Mining Eng., Zonguldak, Turkey

T. Akagi
Toyota National College of Technology, Department of Civil Eng., Toyota, Japan

T. Kawamoto
Aichi Institute of Technology, Department of Civil Eng., Toyota, Japan

ABSTRACT: The overburden of tunnels and underground caverns has been increasing in recent years. As a result, rock-bursting phenomenon has become probably one of the major concerns for the stability of such structures. This study is concerned with the assessment of susceptibility of rock bursting problem in tunnelling. The authors present a method for such a purpose by extending an earlier method developed by the authors for tunnels in squeezing rocks. This method is compared with other methods proposed for the predictions of rock bursting in underground excavations and it is applied to some tunnels where rockburst problems encountered to check its validity. And then its several applications to some tunnels with great overburden and under constructions are given and discussed.

1 INTRODUCTION

Squeezing and rock bursting problems in underground excavations are often encountered and they are major modes of failure in both short-term and long-term. Squeezing problem was investigated by the first author and his group in detail and a method for predicting the squeezing potential and deformation of tunnels was put forward and its validity was confirmed through applications to a number of tunnelling projects (Aydan et al. 1993, 1995, 1996). While squeezing problem is observed in weak rocks, rock-bursting problem is commonly seen in underground excavations in hard rocks. Rockburst could be particularly a very severe problem during the excavation as it involves detachment of rock fragments with high velocity. Mont Blanc tunnel in France, Gotthard tunnel in Switzerland, Dai-Shimizu tunnel and Kanetsu tunnel in Japan are some of the well known examples of rock bursting in tunnelling. Rockburst problems are also one of the common instability modes in deep mining in hard rocks and numerous examples are reported from South Africa and Canada (Bosman & Malan, 2000; Kaiser et al. 1993, 1996).

Several methods are proposed to assess the susceptibility of rock bursting in underground excavations. Ortlepp and Stacey (1994) recently present a detailed review of these methods. These methods can be broadly classified as energy methods, elastic-brittle plastic method, extensional strain method. Nevertheless, none of these methods is validated for assessing the susceptibility of rock bursting in tun-

nelling and its intensity. In this article, the authors presents a method for the assessment of susceptibility of rock bursting in tunnelling in hard rocks and this method is essentially a slight extension of their method proposed for tunnels in squeezing rocks. After presenting the fundamentals of the method, it is compared with other methods to check its merits and de-merits. Furthermore, its several application to a tunnel with great overburden and under construction is given and discussed.

2 ROCKBURST PHENOMENON

2.1 *Physical characteristics of Rock Bursting*

Rock bursting is generally associated with the violent failure of brittle hard rocks such as igneous rocks, gneiss, quartzite and siliceous sandstone. It is well known phenomenon of instability in mining for long time. When hard rocks are tested under uniaxial loading conditions, the fragments of rocks can be thrown to a considerable distance once the peak strength of rock is exceeded. The failure surface is mostly associated with an extensional straining. Rock bursting in underground excavations is quite similar to that under laboratory conditions. When rock bursting occurs in underground openings, rock fragments detach from surrounding rock and are thrown into opening in a violent manner like bombshells. The less severe form of rock bursting is observed as spalling.

2.2 Mechanical characteristics of rock bursting

It is known that rock bursting is said to occur in hard rocks having high deformation modulus while squeezing is observed in weak rocks having a uni-axial strength less than 20-25 MPa. Figure 1 shows typical stress-strain responses for both bursting and squeezing rocks. Bursting rocks are characterized with their high strength, higher deformation modulus and brittle post-peak behaviour. On the other hand, squeezing rocks are characterized with low strength, smaller deformation modulus and ductile post-peak behaviour.

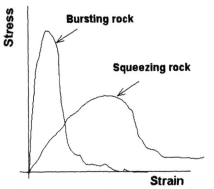

Figure 1. Typical stress-strain responses of squeezing and bursting rocks.

The violent detachment of rock fragments during rock bursting is associated with how the stored mechanical energy is dissipated during the entire deformation process. As shown in Figure 2, if the intrinsic stress-strain response of rocks could not be achieved through its surrounding system in laboratory tests or underground openings, a certain part of stored mechanical energy would be transformed to kinetic energy. This kinetic energy results in the detachment of rock fragments, which may be thrown into the opening with a certain velocity depending upon the overall stiffness of the surrounding system and deformation characteristics of the bursting material (Jaeger and Cook, 1979). The first author observed this phenomenon even in granular crushed quartz samples confined in acrylic cells and dry initially sheared Fuji clay. It is observed that wrapping samples with highly deformable rubber-like strings greatly reduced the violent detachment of fragments as seen in Figure 3. Although such materials could not delay or increase the overall confinement, they act as dampers to reduce the velocity and acceleration of detaching rock fragments, which may be one of very important observations in dealing with rockburst problem in underground excavations.

For rocks those exhibit bursting phenomenon, the following identity must hold:

$$E_S = E_K + E_T + E_P + E_O \quad (1)$$

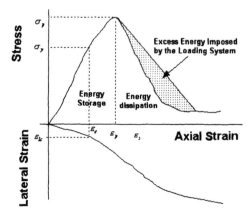

Figure 2. A simple illustration of the mechanical cause of rock bursting phenomenon.

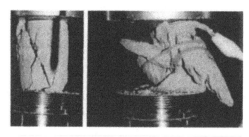

Figure 3. Post-failure views of dry initially sheared Fuji clay samples unwrapped and wrapped with rubber strings.

Where E_S, E_K, E_T, E_P and E_O stand for stored mechanical energy, kinetic energy, thermal energy, plastic work done and other energy forms, respectively. For a very simple case of one-dimensional loading of a block prone to bursting, one can easily derive the following equation for the velocity of the rock fragment if one assumes that the stored mechanical energy is totally transformed into kinetic energy (i.e. Arıoğlu et al. 1999, Kaiser et al. 1996):

$$v = \frac{\sigma_c}{\sqrt{\rho E}} \quad (2)$$

Where σ_c, ρ and E are uniaxial strength, density and elastic modulus of the detaching rock block. Furthermore, the maximum ejection distance d of the block for a given height h of the opening and horizontal ejection can be easily obtained from the physics as follows:

$$d = v\sqrt{\frac{h}{g}} \quad (3)$$

Where g is gravitational acceleration. Figure 4 shows the ejection velocity and throw distance of block as a function uniaxial strength of surrounding rock mass.

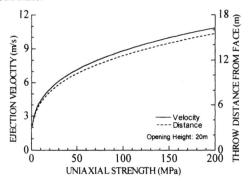

Figure 4. Ejection velocity and throw distance of rock fragment for an opening height of 20m as a function of uniaxial strength of surrounding rock.

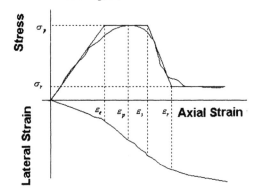

Figure 5. Illustration of strain limits for different states of rock under compressive tests.

3 A METHOD FOR PREDICTING ROCK BURSTING AND ITS INTENSITY

Bosman & Malan (2000) recently reported that the overall behaviour of hard rocks could be very similar to that of squeezing rocks. The fundamental difference between squeezing and bursting is probably the strain levels associated with different states as illustrated in Figure 5. As noted from Figure 1, the strain levels for bursting rocks are much smaller than those for squeezing rocks. Figure 6 shows a plot of normalized strain levels by the elastic strain limit defined in Figure 5 for a uniaxial compressive strength range between 1 and 100 MPa. This figure is an extension of the earlier plot for squeezing rocks together with new data. The horizontal axis of the figure is the uniaxial strength of surrounding rock. It should be noted that we do not differentiate intact

rock strength and rock mass strength on the basis of our own experiences and databases for rock masses and intact rocks. In other words, if the uniaxial strength values of intact rocks and rock masses are similar, their mechanical behaviour should be quite similar to each other.

The empirical relations shown in Figure 6 are those previously proposed for squeezing rocks by Aydan et al. (1993,1996) and they are also applicable to hard rocks with bursting potential. Furthermore, the empirical relations for other mechanical properties, which are required for analyses, can be applicable for rocks with bursting potential. For circular tunnels under hydrostatic initial stress state as shown in Figure 7, the strain levels and plastic zone radii can be obtained as follows (Aydan et al. 1993, 1996):

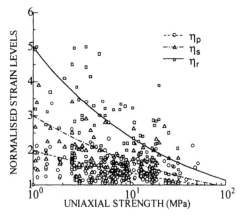

Figure 6. Comparison of normalized strain levels and empirical relations for squeezing and bursting rocks.

Elastic state

$$\xi = \frac{\varepsilon_\theta^a}{\varepsilon_\theta^e} = 2\left(\frac{1-\beta}{\alpha}\right) \leq 1 \qquad (4)$$

Elastic perfectly plastic state

$$\xi = \frac{\varepsilon_\theta^a}{\varepsilon_\theta^e} = \left\{\frac{2[(q-1)+\alpha]}{(1+q)[(q-1)\beta+\alpha]}\right\}^{\frac{f+1}{(q-1)}} \qquad (5)$$

$$\frac{R_{pp}}{a} = \left\{\frac{2[(q-1)+\alpha]}{(1+q)[(q-1)\beta+\alpha]}\right\}^{\frac{1}{(q-1)}} \qquad (6)$$

Elastic perfectly plastic & brittle plastic state

$$\xi = \frac{\varepsilon_\theta^a}{\varepsilon_\theta^e} = \eta_{sf}\left\{\frac{\frac{2[(q-1)+\alpha]}{(1+q)(q-1)}(\eta_{sf})^{\frac{(1-q)}{f+1}} - \frac{\alpha}{q-1} + \frac{\alpha^*}{q^*-1}}{\beta+\frac{\alpha^*}{q^*-1}}\right\}^{\frac{f+1}{(q^*-1)}} \qquad (7)$$

$$\frac{R_{pp}}{a} = \left\{ \frac{\frac{2\left[(q-1)+\alpha\right]}{(1+q)(q-1)}\left(\eta_{sf}\right)^{\frac{(1-q)}{f+1}} - \frac{\alpha}{q-1} + \frac{\alpha^*}{q^*-1}}{\beta + \frac{\alpha^*}{q^*-1}} \right\}^{\frac{1}{(q^*-1)}} \tag{8}$$

Where $\beta = \dfrac{p_i}{p_0}$; $\alpha = \dfrac{\sigma_c}{p_0}$; $\alpha^* = \dfrac{\sigma_c^*}{p_0}$; $q = \dfrac{1+\sin\phi}{1-\sin\phi}$;

$q^* = \dfrac{1+\sin\phi^*}{1-\sin\phi^*}$

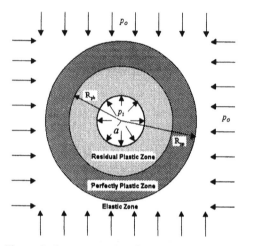

Figure 7. States around a circular tunnel and notations.

The approach presented above can also be extended to situations, which involve complex excavation geometry and initial stress states. However, the use of numerical methods will be necessary under such circumstances as described by the authors previously (Aydan et al. 1995).

4 COMPARISONS WITH OTHER METHODS AND APPLICATIONS

As mentioned in the introduction, energy methods, extensional strain method and elastic-brittle plastic method are used to assess the bursting susceptibility of hard rocks. These methods are briefly described herein and some equations are presented in order to make some comparisons with the method presented in the previous section.

4.1 Energy Method

Energy methods are used in mining for long time and it is based on the linear behaviour of materials. When the material behaviour becomes non-linear, it becomes difficult how to define the energy. The overstressed radius of rock around a circular tunnel

under hydrostatic stress state can be obtained with the use of Mohr-Coulomb yield criterion and elastic stress components as

$$\frac{R_p}{a} = \left\{ \frac{(q+1)(1-\beta)}{\left[(q-1)+\alpha\right]} \right\}^{\frac{1}{2}} \tag{9}$$

The total energy per unit area in the overstressed zone is then obtained as follows

$$w_{el} = \frac{1+\upsilon}{E}(p_o - p_i)^2 a\left[1 - \left(\frac{a}{R_p}\right)^2\right] \tag{10}$$

Figure 8 shows the relation between overburden and elastic energy and radius of overstressed zone different uniaxial compressive strength of rock. The potential of bursting is quite high if the strength of rock is low. Kaiser et al (1996) suggested the following values for assessing the intensity of rock bursting on the basis of in-situ observations and the capacity of support members as given in Table 1.

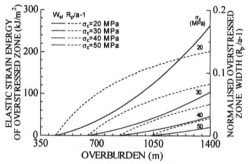

Figure 8. Relation between overburden and elastic energy and radius of overstressed zone.

Table 1. Relation between energy and intensity of bursting.

Intensity	Energy (KJ/m^2)	Velocity (m/s)
Low	<5	<1.5
Moderate	5 - 10	1.5 - 3
High	10 - 25	3 - 5
Very high	25 - 50	5 - 8
Extreme	>50	>8

4.2 Extensional Strain Method

Stacey (1981) proposed the extensional strain method for assessing the stability of underground openings in hard rocks. He stated that it was possible to estimate the spalling of underground cavities in hard rocks through the use of his extensional strain criterion. The extensional strain is defined as the deviation of the least principal strain from linear behaviour. This definition actually corresponds to the definition of initial yielding in the theory of plasticity. This initial yielding is generally observed, at the

40-60 percent of the deviatoric strength of materials. If this criterion is applied to circular tunnels under hydrostatic initial stress state, the radius R_o, at which the extensional strain is exceeded, can be shown to be

$$\frac{R_o}{a} = \left(\frac{\left[\frac{1+\upsilon}{E}(p_o - p_i) \right]}{\varepsilon_c} \right)^{\frac{1}{2}} \quad (11)$$

Since

$$\varepsilon_c = \upsilon \varepsilon_e \text{ and } \sigma_c = E \varepsilon_e, \quad (12)$$

one can easily obtains the following:

$$\frac{R_o}{a} = \left(\frac{1+\upsilon}{E} (\frac{1-\beta}{\alpha}) \right)^{\frac{1}{2}} \quad (13)$$

4.3 Elastic - Brittle Plastic Method

In elastic-brittle plastic method, the strength of rock is reduced from the peak strength to its residual value abruptly. If this concept is applied to circular tunnels under hydrostatic initial stress state, the radius R_p of plastic region can be obtained as follows:

$$\frac{R_p}{a} = \left\{ \frac{\frac{2-\alpha}{(1+q)} + \frac{\alpha^*}{q^*-1}}{\beta + \frac{\alpha^*}{q^*-1}} \right\}^{\frac{1}{(q^*-1)}} \quad (14)$$

4.4 Comparisons and Applications

The above methods are compared with the proposed method by considering a circular tunnel under hydrostatic initial stress state. The uniaxial strength of rock mass is assumed to be 20 MPa and the internal pressure was set to 0 MPa. The parameters required for analysis are obtained from empirical relations proposed by Aydan et al. (1993, 1996). In the computations, overburden is varied and the radius of plastic zone or overstressed zone is computed. Figure 9 compares the computed radius of plastic zone or overstressed zone. As expected from theoretical relation (13) of the extensional strain method, the overstressed zone must appear at shallower depths as compared with predictions of the other methods. The other three methods predict the yielding at the same depth. This difference is due to the value of yielding stress level associated with the extensional strain criterion. The radius of the plastic zone, estimated from the elastic-brittle plastic method, becomes quite large, and it even exceeds the one estimated

from extensional strain method. The estimations from the proposed method and the energy method are quite close to each other. They are also more reasonable as compared with estimations from other methods.

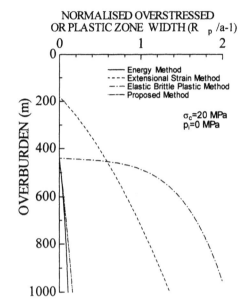

Figure 9. Variations of radius of plastic zone or overstressed zone with overburden, estimated from different methods.

The proposed method is applied to a tunnelling project, which is now under construction. The tunnelling project is associated with an expressway construction and passes beneath high mountains in the Central Japan. Rock mass properties for this tunnel, which is 10km long and 12m in diameter, are estimated from the empirical relations developed for RMR classification by Aydan and Kawamoto (2000). Figure 10 shows the variations of overburden RMR and estimated level of bursting or squeezing and tunnel wall deformation along the tunnel alignment. Only the 600m long section of this tunnel excavated at the time of computations. The preliminary deformation measurements are quite close to the estimations shown in Figure 10.

The final application is concerned with the comparison of the deformation behaviour of a circular tunnel in bursting and squeezing rock masses (Figure 11). In the computations, the value of the competency factor for both situations was chosen as 1. As expected from the behaviour of squeezing and bursting rocks shown in Figure 1, the tunnel wall strains becomes larger for tunnels in squeezing rock as compared with that in bursting rock mass. In addition, the radius of plastic zone in squeezing rock is larger than that in bursting rock.

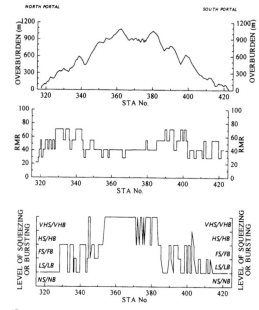

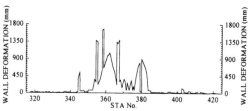

Figure 10. Predicted results for a 10km long expressway tunnel under constructions.

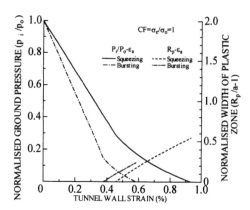

Figure 11. Comparison of computed tunnel wall strain and plastic zone of circular tunnels in squeezing rock and bursting rock.

5 CONCLUSIONS

The authors presented an extension of a method, which was previously proposed for tunnelling in squeezing rock, for tunnels in bursting hard rocks. It seems that many empirical relations proposed by the authors previously can also be used for rocks prone to bursting with some confidence. Since the proposed method is capable of handling the mechanical behaviour of rocks prone to bursting, the estimations should be more reasonable than those from the elastic-brittle plastic model. The energy method also yields similar results to those estimated from the proposed method. Although the proposed method is quite promising to predict the behaviour of tunnels under rock bursting conditions, it is felt that further studies are necessary. Especially the estimation of ejection velocity of rock fragments is quite important, as it is closely associated with the safety of workers during excavations.

REFERENCES

Arıoğlu, E., Arıoğlu, B. & Girgin, C. (1999): Stability problems in tunnels excavated in massif-brittle rocks (in Turkish). *Mühendislik Jeolojisi Bülteni*, 27, 69-78.

Aydan, T. Akagi & T. Kawamoto (1993). Squeezing potential of rocks around tunnels; theory and prediction. *Rock Mechanics and Rock Engineering*, 26(2), 137-163

Aydan, Ö., T. Akagi, T. Ito, J. Ito & J. Sato (1995). Prediction of deformation behaviour of a tunnel in squeezing rock with time-dependent chracteristics. *Numerical Models in Geomechanics NUMOG V*, 463-469

Aydan, Ö., T. Akagi & T. Kawamoto (1996). The squeezing potential of rock around tunnels: theory and prediction with examples taken from Japan. *Rock Mechanics and Rock Engineering*, 29(3), 125-143.

Aydan, Ö., & T. Kawamoto (2000). The assessment of mechanical properties of rock masses through RMR classification system. *GeoEng2000, Melbourne*.

Bosman, J.D. & Malan, D.F. (2000): Time dependent deformation of tunnels in deep hard rock mines. News J., ISRM, 6(2), 8-9.

Kaiser, P.K., Jesenak, P. & Brummer, R.K. (1993). Rockburst damage potential assessment. *Int. Smp. on Assessment and prevention of Failure Phenomena in Rock Engineering*, Istanbul, 591-596.

Kaiser, P.K., McCreath, D.F. & Tannant, D.D. (1996): Canadian rockburst support handbook. Geomechanics Research Centre, Laurentian University, Sudbury.

Jaeger, J.G. & Cook, N.G.W. (1979): *Fundamentals of Rock Mechanics*. 3rd Ed., Chapman and Hall, London.

Ortlepp, W.D. & Stacey, T.R. (1994): Rockburst mechanisms in tunnels and shafts. Tunnelling and Underground Space Technology, 9(1), 59-65.

Stacey, T.R. (1981): A simple extension strain criterion for fracture of brittle rock. *Int. J. Rock Mech. Min. Sci. & Geomech. Abstr.*, 18,469-474.

Modern Tunneling Science and Technology, Adachi et al (eds), © 2001 Swets & Zeitlinger, ISBN 90 2651 860 9

Design of large scale road tunnel based on the behavior of discontinuous rock mass -Ritto Tunnel-

Y.Sakayama
Second Construction Division, Kansai Bureau, Japan Highway Public Corporation, Japan

H.Niida
Otsu Construction Office, Kansai Bureau, Japan Highway Public Corporation, Japan

Y.Ohnishi
School of Civil Engineering, Kyoto University, Japan

Y.Tanaka
Nishimatsu/Shimizu/Okumura Joint Venture, Japan

S.Inosaka
Nishimatsu/Shimizu/Okumura Joint Venture, Japan

ABSTRACT: The geology, where Ritto Tunnel is located, mainly consists of discontinuous hard granite. Compared to ordinary tunnels, the tunnel cross section is very large ($180m^2$) and the tunnel width is very wide (18m). The aim of this study is to predict and prevent the falling or sliding of rock mass or blocks along discontinuities before excavation of tunnel. In this paper, we explain the application of key block analysis on site to investigate the stability of tunnel based on the behaviors of discontinuous rock mass not only during the construction of a tunnel but also after opening to the public.

1 INTRODUCTION

Over the past few years a large number of studies have been made on economical, rational and safe design and construction of mountain tunnels, and many useful results were acquired. However, in many cases, enormous cost and time was consumed to cope with the falling or sliding of rock mass or blocks, which couldn't be predicted because of the complexity of the rock mass structure in advance. It is difficult to grasp the dynamic properties and distributions of discontinuities in the rock mass because they exist on a variety of scales.

Key block theory, which can find the falling or sliding of rock mass along discontinuities, is one method to examine the stability of the tunnel in discontinuous rock mass. Key blocks, which are the rock masses surrounded by tunnel face and discontinuities, can be detected by the key block analysis.

The purpose of this study is to predict and prevent the falling or sliding of rock mass along discontinuities, before excavation of tunnel. In this paper, we explain the development of a computer simulation method with user friendly interfaces to apply the key block analysis to investigate the stability of tunnel face based on the behaviors of discontinuous rock mass and to design supplementary supports if detected key blocks are unstable.

2 CONSTRUCTION OUTLINE

Ritto Tunnel, a large scale flat section tunnel which is 3.8km long, is located in Shiga Prefecture of about 10km from the end of the south of Lake Biwa to the east-southeast, as shown in Figure 1. It passes through the mountainous zone called *Konan Alps*. The cross section is 18m width and 12m height. There are 3 lanes (each lane is 3.75m width) and the shoulder of a road is 2.5m width. The ratio of width to height is 0.65. The typical cross section of the tunnel is shown in Figure 2.

In the excavation of Ritto Tunnel, a pilot tunnel by Tunnel Boring Machine (TBM) was followed by enlargement by drill and blast of top heading and bench method in New Austrian Tunneling Method (NATM) to construct a large scale tunnel safely and quickly. Firstly, a working tunnel, which was 300m long, was excavated about 1 km to the west from the east portal because of the topographical condition and the temporary storage yard of muck waste. Secondly, a pilot tunnel by TBM was excavated to the west. Finally, a pilot tunnel is now under enlargement by NATM. Construction plan is shown in Figure 3.

Figure 1. Location of Ritto Tunnel.

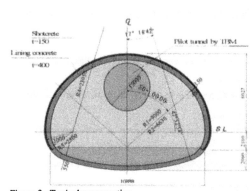

Figure 2. Typical cross section.

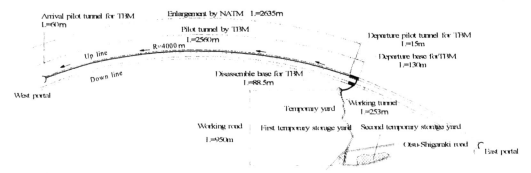

Figure 3. Construction plan.

3 GEOGRAPHICAL FEATURES AND GEOLOGY

A mountainous region around the tunnel is 300 to 600 meters above the sea level and a comparatively gentle slope is seen at the summit of the mountain. In addition, steep V shape valley develops along swamp and river around the tunnel. The geology of the tunnel mainly consists of biotite granite of the latter period of Cretaceous period called *Tanakami* granite, which is fresh and hard. Maximum unconfined compressive strength is 100 MPa and seismic velocity is more than 4.7 kilometers per second. However, a lot of small scale faults and fractures are distributed in this area. Longitudinal geological section is shown in Figure 4.

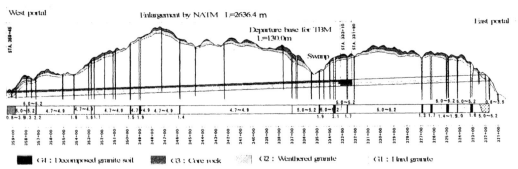

Figure 4. Longitudinal geological section.

4 INTRODUCTION OF THE KEY BLOCK ANALYSIS ON SITE

4.1 *Key block theory*

In discontinuous rock mass, blocks of rock mass which have a variety of shapes and sizes are formed geometrically along discontinuities. When the rock mass is excavated, the new shape of block appears on the excavated surface shown in Figure 5. The key block theory was proposed by Goodman and Shi (1985). All blocks formed in the rock mass are classified based on the directions of discontinuities by stereographic projection. Then, key blocks which have the possibility of falling or sliding are detected based on the direction of excavated surface.

4.2 *Technical investigation to introduce key block theory*

4.2.1 *Introduction of key block theory*
Ritto Tunnel is one of the tunnels of second *Tomei Meishin* - expressway and now under construction examining standard support system for large hard rock tunnel in Japan. Moreover, the tunnel has a possibility for rock masses to fall or slide along discontinuities not only because the rock mass have a lot of discontinuities but also because the cross section of the tunnel is very large and flat. Therefore, a key block analysis was introduced based on the behaviors of discontinuous rock mass during the construction of a tunnel as well as after opening to the public.

4.2.2 *Development of key block analysis program*
It is possible to detect key blocks all along the tunnel exactly by using the program developed in Ritto Tunnel. This computer simulation method with user friendly interfaces can calculate not only the stability of key blocks but also the design of supplementary supports if necessary. The items examined before developing a program are shown in the following.

1. It can cope with an optional cross section and alignment of tunnel to make it have a generality.
2. The shapes and locations of key blocks are detected precisely based on the geometrical information of discontinuities, which are absolute three dimensional coordinates, strike, dip, alignment for tunnel and so on.
3. The result of key block analysis can be acquired in a short time for the daily management.
4. It is possible to eliminate and/or revise the data of discontinuities easily.

4.2.3 *Investigation of strength of discontinuity*
Before applying a key block analysis to the site, the input information to the program is absolute three dimensional coordinates, strike and dip of discontinuities, strength of discontinuities (cohesion C and angle of internal friction ϕ), unit volume weight, and so on. Among them, strength of discontinuities is one of the most important information which strongly affects the stability of the key block. In the example of key block analysis applied before, same values of strength of discontinuities were used regardless of conditions of discontinuities. However, the strength of discontinuities is different depending upon the existence of clay in the discontinuities. Therefore, two cases are examined. As a result of the examination shown in Figure 6, cohesion C is 2.0 KPa and angle of internal friction ϕ is 30° in the case of existence of clay filling. On the other hand, cohesion C is 0 KPa and angle of internal friction ϕ is 37° in the case of no existence of clay filling. Further, we are now re-examining strength of discontinuities to get more accurate C and ϕ by simple shear test.

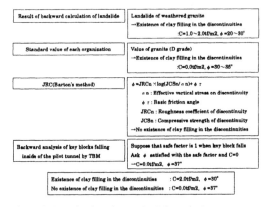

Figure 6. Investigation of strength of discontinuity.

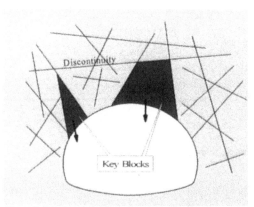

Figure 5. Key blocks in discontinuous rock mass.

5 APPLICATION TO THE TUNNEL

5.1 *Flow of the application to the tunnel*

In the excavation of Ritto Tunnel, a pilot tunnel by Tunnel Boring Machine (TBM) is followed by enlargement by drill and blast in New Austrian Tunneling Method (NATM). Key block analysis can be divided into two stages. The first stage is after completion of the pilot tunnel (ϕ 5m) and the second stage is under construction by NATM. Figure 7 shows the flowchart for the application of key block analysis in each stage.

At the first stage, based on the information of discontinuities acquired during the excavation of the pilot tunnel, detected key blocks are supported before enlargement by NATM. At the second stage, based on the observation of the excavated surface after enlargement, key block analysis is applied again in addition to the information of discontinuities which were not observed at the first stage. Furthermore, we re-examine the stability of key blocks detected at the first stage and supplementary supports are installed in case of unsafe ground condition.

5.2 *Detection of unstable key blocks*

At the first stage, 38 key blocks were detected in total all along the tunnel (2.6km). According to the flowchart shown in Figure 7, 7 key blocks were judged to be unstable because they could not be supported by standard supports.

Figure 8 shows positions of these 7 key blocks. Almost all of these 7 key blocks have slender wedge shape in the up-down direction because the discontinuities in the vertical direction are dominant in-situ.

5.3 *Supplementary support of unstable key blocks*

Supplementary supports were installed to these 7 key blocks from inside of the pilot tunnel before enlargement by NATM. An example of supplementary supports of key block No.4 is shown in Figure 9.

As for the material of support, we used skin friction type of rockbolts, i.e. Connectable Swellex, because they had large resistance for shearing along discontinuities and adhesive strength to rock mass was strong in comparison with ordinary rockbolts. In addition, the most suitable position, number and length of rockbolts were calculated in this method in order to anchor the rock mass outside of the key blocks and to maintain the stability.

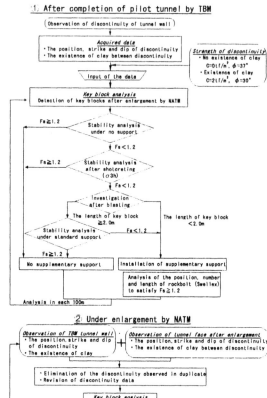

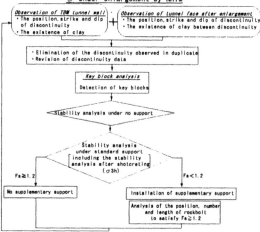

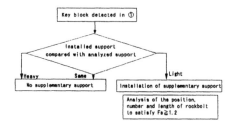

Figure 7. Flowchart of key block analysis.

38 key blocks were detected in total based on the information of discontinuities
7 key blocks were supported supplementary from inside of the pilot tunnel by TBM

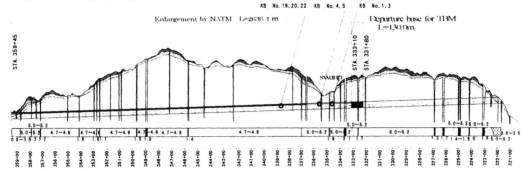

Figure 8. Position of supplementary support.

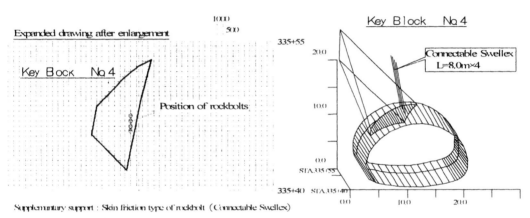

Supplementary support : Skin friction type of rockbolt (Connectable Swellex)

Figure 9. An example of supplementary support.

6 VERIFICATION OF THE EFFECTIVENESS OF SUPPLEMENTARY SUPPORT TO KEY BLOCKS

6.1 Three-dimensional joint displacement measurement

One discontinuity of key block No.19, which was judged to be unstable and was supported supplementary, is now under measurement. Three-dimensional joint displacement measuring instrument (measuring accuracy : 5/1000 mm) was installed crossing the detected discontinuity from inside of a pilot tunnel before enlargement by NATM directly below key block No.19. Behaviors of the discontinuity are measured and analyzed every thirty minutes automatically until the face pass through it .

6.2 Example of enlargement by NATM just under key block

The effectiveness of supplementary support to key block was verified by the falling of rock mass under enlargement just below key block No.3. Rock mass of 5m height along a discontinuity which is a composition of key block No.3. Rock mass above the tunnel section was supported and fixed by skin friction type of rockbolts. The outline of supplementary support is shown in Figure 10.

7 CONCLUSION

Key block analysis was presented in 1980s, but it was not put to a practical use until recetly. In Japan, key block analysis was applied to the excavation of large scale underground rock cavern for the first time in 1996. But this method was not applied to the long road tunnel running through the mountain with complex geological condition. Ritto Tunnel is the first site that this method was applied all along the tunnel based on real discontinuity information observed in-situ. Ritto Tunnel is now under enlargement by NATM applying key block analysis.

401

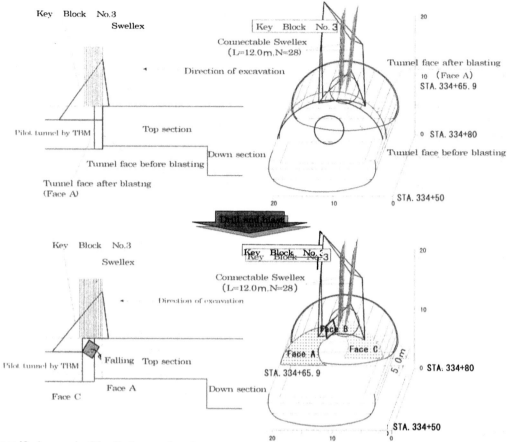

Figure 10. An example of the effectiveness of supplementary support.

However, the method to evaluate the strength of discontinuities is open to further discussion.

REFERENCE

Goodman, R. E. & Shi, Gen-hua 1985. Block theory and Its Application to Rock Engineering : Prentice Hall.

Modern Tunneling Science and Technology, Adachi et al (eds), © 2001 Swets & Zeitlinger, ISBN 90 2651 860 9

Behavior of mountain tunnel with large cross section in earth ground

N. Tomisawa
Manager, Civil Engineering Div. Konoike Construction Co., LTD

T. Matsui
Professor, Dept. of Civil Engineering, Osaka University

M. Hino
Graduate student, Dept. of Civil Engineering, Osaka University

ABSTRACT: Although tunnels with large cross section, such as tunnels in the Second Tomei and Meishin Highway and two-lane tunnels with sidewalk, have recently increased, their design and construction methods have not been established yet. There are many problems to design and construct such tunnels especially in earth ground. NATM with center diaphragm is one of measures to solve the problems. In this paper, behavior of a tunnel with large cross section in earth ground, which was constructed by applying NATM with center diaphragm, is firstly described, based on the measurement results. Then, a practical finite element analytical method is proposed, to predict behavior of large-scale mountain tunnel with center diaphragm in earth ground.

1 INTRODUCTION

NATM is generally applied to construct mountain tunnels in Japan. The method had made construction of tunnels easy and safe in case of rather bad geological condition and large cross sectional area. In the meantime, cross sections of highway tunnels have been enlarged because of increasing traffics and spaces for sidewalks. Such large-scale tunnels are generally designed individually because standard design technique for them have not established yet. And locations and geological conditions of newly planned tunnels have become more difficult. Under such conditions, it is important to predict deformation behavior of surrounding ground of large-scale tunnels especially in urban area to consider influences on environments such as ground settlements.

Pre-support methods are usually applied to secure stability of cutting faces of tunnels in earth ground. Besides, especially in cases of tunnels with large cross section, it is needed to increase number of divided heading sections to keep ground

deformation in very small level. In such cases, it is important to predict deformation behavior of surrounding ground in each construction stages.

Amanosan No.1 tunnel mentioned in this paper is a highway tunnel with three lanes mainly in earth ground of sands and gravels. To ensure stability of the tunnel and control ground settlements, the center diaphragm method was applied and many kinds of measurements were carried out during the construction.

In this paper, based on the measured data, behaviors of the large-scale tunnel with center diaphragm in earth ground and deformation of the surrounding ground are described, followed by discussing their finite element prediction method.

2 OUTLINE OF CONSTRUCTION

Amanosan No.1 tunnel is a 214m-long highway tunnel with three lanes as shown in Figure 1(Tomisawa and Matsui,2000). The shape of its excavated cross section is 16m wide and 11.5m high, and its area is 145m^2 as shown in Figure 2.

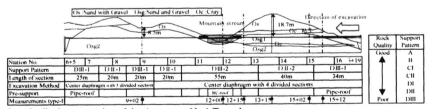

Figure 1. Longitudinal cross section of the Amanosan No.1 Tunnel.

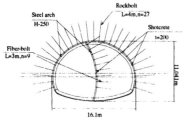

Figure 2. Example of support pattern (D II-2).

Table 1. Mechanical properties of ground.

Geology	Elastic Modulus N/mm²	N value
Os1	60.7-74.3	33-60
	–	–
Osg1	65.3	54
	38.0-142.0	20-60
Os	–	–
	157.0-177.0	60<
Osg2	123.0-125.0	60<
	115.0-234.0	60<

Upper column: investigation data before tunnel construction
Lower column: investigation data during tunnel construction

The geological condition of the ground mainly consists of unconsolidated sands and gravels of Pleistocene. Table 1 shows the mechanical properties of the ground. The maximum overburden is 18.7m high.

The center diaphragm method was applied to excavate the tunnel.

3 DEFORMATION BEHAVIOR OF GROUND BASED ON MEASURED DATA

Deformation behavior of ground is shown below, based on the measured data (Tomisawa and Matsui, 2000). Figure 3 shows an example of measured settlement changes in the excavation sequence

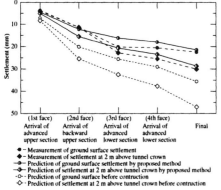

Figure 3. Settlement changes in the excavation sequence (No.15+02).

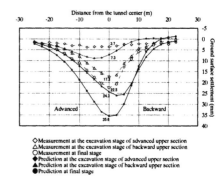

Figure 4. Distribution of ground surface settlement at each excavation stage (No.15+02).

(broken lines) with the predictions analyzed by using 2-D elastic FEM before the construction (dotted lines). Solid lines show the predictions by proposed method mentioned later on. Although predictions show increasing tendency from the third face to the fourth face, measurements hardly show such tendency. This might be because the actual interval may be different from the analytical conditions.

Figure 4 shows an example of measured and predicted distributions of ground settlements. Although measured values are different from predictions quantitatively, the shapes of both distributions are similar. It seems that deformation behavior of surround ground is almost like elastic.

4 OUTLINE OF PROPOSED ANALYTICAL METHOD

4.1 Back ground

In Japan, FEM is generally used to analyze behaviors of tunnels. And 2-D elastic analyses are supposed to be enough to understand ground deformations during tunnel constructions by following reasons.

(1) 3 dimensional treatments
3-D analyses are rarely applied because of problems on cost and time. Besides, quality and quantity of geological information obtained before tunnel constructions are not enough to apply 3-D analyses.

(2) Non-linear treatments
Geological materials are often treated as linear materials because enough geological information to express non-linear behavior is not obtained. Besides, non-linear analyses are often unstable on their convergence.

(3) Time-dependent treatments
Although properties of supports change with elapsed time, equivalent values are used because they are sufficient to predict deformation of ground but stress.

404

In most of tunnel analyses, predicted values are converged ones in each construction stage of subsequent cutting faces. Nevertheless, in actual construction controls, the predicted values have to be compared with measured data that do not always converge within each stage, because subsequent faces may influence them.

4.2 Outline of proposed method

Amanosan No.1 tunnel was constructed by applying center diaphragm method with 3 or 4 divided sections with the maximum overburden of approximately 1D. The results of measurements did not converge in each construction stage, because intervals between faces were very close. In this study, therefore, the analysis sequence shown in Figure 5 is considered to fit predictions to measurements.

(1) Setting of excavation stress release ratio

There are four cutting faces in a cross section. The ratio for the first face is calculated by using axially symmetrical 2-D analysis, considering actual overburden and cross-section area.

The second to fourth face excavations are considered as enlargement. Then, the ratios are calculated by using axially symmetrical 2-D analysis, considering actual overburden and enlarged cross-section areas.

(2) Setting of settlement characteristic curve caused by approach of subsequent faces

When applying center diaphragm method, deformation behaviors of surrounding ground are influenced by subsequent faces because interval of faces are often less than 2D. Then, settlement curves are calculated by using axially symmetrical 2-D analysis, for cases of several intervals between faces.

(3) Setting of settlement characteristic curve on distance from tunnel wall in cross section direction

Settlement ratios to converged values vary with distance from tunnel wall. Then, relationships between settlement ratios and distance from tunnel wall are calculated.

(4) Setting of deformation moduli of geological materials

Geological materials are treated elastically. Conditions of initial stresses are hardly measured. And deformation moduli obtained from laboratory and in-situ tests are often not suitable as the representative one of the ground as a whole. Then, it is assumed that deformation moduli can be decided based on a preceding ground settlement at a point close to the tunnel crown of the first face.

(5) Prediction of deformation behavior of surrounding ground

Using parameters obtained in (1) to (4) mentioned above, deformation behavior of the surround-

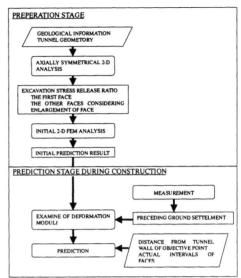

Figure 5. Sequence of proposed method.

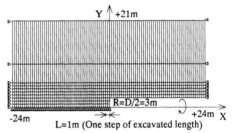

Figure 6. Example of meshes used in axially symmetrical analysis.

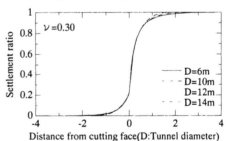

Figure 7 Effect of excavation diameter on settlement characteristic curve.

ground are calculated by 2-D FEM corresponding to actual interval between faces.

4.3 Decision of parameters and characteristic curves used in proposed method

Parameters and characteristic curves described in (1) to (3) in the previous section are decided by using axially symmetrical 2-D analyses corresponding to actual overburden and cross-section area. Figure 6 shows the FEM mesh for the analyses of the first

405

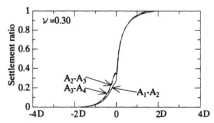

Figure 8 Effect of enlargement on settlement characteristic curve.

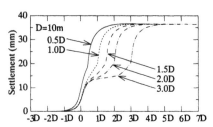

Figure 9. Effect of interval between neighboring faces on settlement characteristic curve.

face. Diameters on the mesh of each stage are calculated as the equivalent circle ones to total area of each cutting face. The Y axial area is decided as 21m considering the fourth face diameter of 14m (R=7m) and its overburden of 1D.

(1) Excavation stress release ratio

Figure 7 shows settlement characteristic curves at the points of tunnel crown changing excavation diameter in cases of D=6m (A1=28m^2), D=10m (A2=79m^2), D=12m (A3=113m^2) and D=14m (A4=154m^2). The ratio at the distance of one excavation length of 1m for the first face of D=6m is 0.586. Figure 8 shows settlement characteristic curves changing enlargement in three cases from A1 to A2 (enlargement ratio: 228%), from A2 to A3 (143%) and from A3 to A4 (136%). The ratios at the distance of one excavation length of 1m are 0.527, 0.533 and 0.501 respectively.

(2) Settlement characteristic curve caused by approach of subsequent faces

Figure 9 shows an example of settlement characteristic curves changing interval between faces in case of enlargement of D=10m. It is seen from this figure that the closer the distance from subsequent face, the larger the settlement ratio to the converged value.

(3) Settlement characteristic curve on distance from tunnel wall in cross section direction

Figure 10 shows settelments characteristic curves changing distance from tunnel crown in case of D=6m. The smaller the distance from tunnel crown, the wider the range of settlements before arriving faces and the later the converged period.

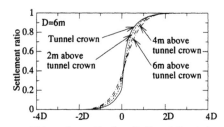

Figure 10. Effect of distance from tunnel crown on settlement characteristic curve.

5 APPLICABILITY OF PROPOSED METHOD

Based on the sequence and results described in the previous chapter, the authors tried to predict deformation behavior of the surrounding ground of the Amanosan No.1 tunnel, considering actual intervals of faces, followed by comparing with measured data.

The deformation moduli were examined, based on the measured value of the nearest point to the tunnel crown at the first face arrival, resulted in 2.33 times the values used in the initial analysis before construction.

The solid lines shown in Figure 3 are the predictions of ground settlements by the proposed method. They fit to the measurements much better than those before construction. Therefore, the applicability of the proposed method can be confirmed.

CONCLUSION

In this paper, the deformation behaviors of large-scale tunnel with center diaphragm in earth ground were described, followed by proposing its prediction method. The proposed method is based on 2 dimensional FEM, considering 3 dimensional effects such as multi-face excavations and distance from tunnel crown, and modifying deformation moduli based on measured data of preceding ground settlement. The applicability of the proposed method was confirmed by comparing the measured and predicted results.

REFERENCES

Tomisawa,N. & Matsui,T. 2000. Case history of a flat tunnel with large cross section in unconsolidated ground of gravel and sand. *Proc. AITES-ITA 2000 World Tunnel Congress*: 505-511.

Modern Tunneling Science and Technology, Adachi et al (eds), © 2001 Swets & Zeitlinger, ISBN 90 2651 860 9

Simplified behavior models of tunnel faces supported by shotcrete and bolts

L. Cosciotti & A. Lembo Fazio
University of "Roma III", Italy

D. Boldini & A. Graziani
University of Rome "La Sapienza", Italy

ABSTRACT: Ground-support interaction in the vicinity of a tunnel face is a typical three-dimensional problem, nevertheless the tunnel support design is usually based on simplified plane strain analyses. The results of such analyses strongly depend on the choice of ground stress release at the time of lining installation. The paper focuses on tunnels excavated in difficult ground conditions, where the application of shotcrete and face reinforcement by fiber-glass bolts is widely used as a primary measure to reduce ground deformation and ensure safe working conditions. The results of an exhaustive parametric study based on 3D axisymmetric models are presented, taking into account the effect of shotcrete hardening and density of reinforcement. An Italian case history of a tunnel driven in difficult conditions is also reviewed. Finally on the basis of the results, a strategy is proposed in order to enhance the capability of two-dimensional analyses to give a more realistic prediction of stress conditions in lining and reinforcement systems.

1 INTRODUCTION

Tunnels in difficult ground conditions (e.g. soft rocks under high overburden stress) can today be efficiently driven by full-face excavation, thanks to the early application of shotcrete and face reinforcement by fiber-glass bolts.

The green shotcrete layer forms a continuous and flexible ring near the tunnel face, which can tolerate large strains without undergoing structural damage. Steel arches can also be applied to create a stiffer primary lining.

Longitudinal fiber-glass bolts driven into the ground core to be excavated has proved effective in solving face stability problems and in assuring safer working conditions.

Even if this construction method is widely adopted (Kovari et al. 2000), the available design methods are not completely satisfactory so that empirical approaches are often preferred.

Of course 3D numerical models enable a realistic simulation of the construction process (Moussa &Wagner 1997), which is known to strongly affect the final load conditions of the lining and tunnel deformation. Nevertheless 2D plane strain analyses are more popular because 3D modeling is time-consuming and requires more experience in interpreting results.

Further critical issues in the design of tunnel support in difficult conditions are the prediction of ground stress-strain behavior at the full scale of the tunnel, the time-dependence of the support load due to progressive hardening of the shotcrete, and the possible decrease in the long-term strength of disturbed rock masses.

This paper focuses on the importance of modeling shotcrete hardening and face reinforcement to correctly predict the final load on the lining.

3D axisymmetric analyses have been performed (FLAC code, Itasca 1998) where the shotcrete behavior is represented by an elastic model with time-dependent stiffness and where the ground core reinforced by fiber-glass bolts is represented by a homogenized material model.

A simple strategy is finally proposed to enhance the capability of conventional plane strain models on the basis of 3D model results.

2 INFLUENCE OF SHOTCRETE HARDENING

The axisymmetric problem of a circular tunnel (radius a) driven through a rock mass with isotropic in situ stress S has been considered. The construction process has been modelled by applying the step-wise procedure illustrated in Figure 1.

The mechanical properties of the ground and the geometry data, assumed in the numerical analyses are summarized in Table 1.

An elastic constitutive law characterised by time-dependent stiffness has been used for the shotcrete.

Table 1. Geometry and mechanical properties.

tunnel	initial isotropic stress (S)	1.5 MPa
	tunnel radius (a)	5 m
	shotcrete lining thickness (t_s)	0.2 m
	Young modulus at 28days (E_s)	24 GPa
ground	uniaxial compr. strength (σ_c)	375, 750, 1500 kPa
	cohesion (c)	131, 262, 525 kPa
	friction (ϕ)	20
	stability ratio N_s	2,4,8
	Young's modulus (E)	90, 900, 9000 MPa
	Poisson's coefficient (ν)	0.3
	relative stiffness (β)	0.094, 0.94, 9.4

Increase in the elastic modulus as a function of time has been expressed by the empirical relationship (Chang 1994):

$$E_s(t) = c_1 \cdot E_{s,28} \cdot e^{c_2/t^{c_3}} \qquad (1)$$

where c_1, c_2, c_3 are material constants respectively equal to 1.062, -0.446 and 0.6; the Poisson coefficient is assumed to be time independent (ν_s = 0.3).

The aging time for each shotcrete ring has been calculated in accordance with the tunnel advance rate.

Three different rates of tunnel advance, v, have been considered, namely 2, 4 and 8 m/day. These low advance rates are typical of excavations in difficult ground conditions. Two limit cases, of unsupported tunnel (which is equivalent to assume v = ∞) and lined tunnel but with constant shotcrete stiffness $E_{s,28}$ (equivalent to v = 0), have also been analysed.

In difficult ground conditions, shotcrete is applied as close as possible to the tunnel face so that the unsupported tunnel length is reduced to the bare minimum. In the present analyses, two typical unsupported tunnel lengths l_u have been considered, respectively equal to 1 and 2 m.

Moreover the influence of the stability ratio (N_s=2·S/σ_c) and ground-support stiffness ratio (β=E·a/$E_{s,28}$/t_s) have been investigated.

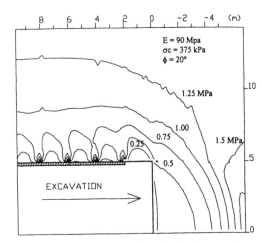

Figure 2. Typical radial stress distribution in the ground near the face.

Figure 2 represents a typical radial stress distribution in the ground close to the face. The formation of a bearing arch spanning from the face to the first shotcrete ring can be observed. The load within each shotcrete segment is far from being uniform, on the contrary, it is highest (up to 1.5 times the mean value) at the edge of the segment looking towards the face.

Figure 3 shows the convergence profile along the tunnel for different advance rates (elastic ground). When the tunnelling speed increases, shotcrete is loaded at an earlier age and hence at lower stiffness, its support action is therefore less effective, giving rise to larger deformation and lower final load on the lining.

The same trend is evidenced for an elasto-plastic ground, even if the magnitude of convergence is higher. In Figure 4, the final load (mean value within a segment) on the lining is represented as a function of the advance rate for various elastic moduli of the ground.

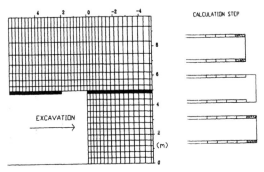

Figure 1. Grid details at the face and schematic of calculation steps.

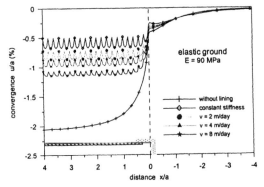

Figure 3. Normalized convergence for different advance rates

For the same advance rate, the load obviously increases as the ground becomes less stiff.

The reduction in load resulting from the fact of taking into account shotcrete hardening, appears particularly significant in poor soft rocks, where the stress conditions are generally more severe and an over prediction of load would lead to unnecessary conservatism in lining design. In spite of this, the reduction in final load, due to progressive hardening of the shotcrete, is proportionally less important for decreasing strength of the rock mass, as shown by Figure 5, where the final load q is scaled to the value obtained for a constant stiffness lining (v = 0).

3 SIMPLIFIED MODELS FOR SHOTCRETE SUPPORT DESIGN

As originally proposed by the Convergence-Confinement "CC" method (Lombardi 1973), the effect of the excavation advance can be simulated by a progressive reduction in the initial pressure S applied at the tunnel wall, controlled by the relaxation factor λ

$$q = (1 - \lambda) \cdot S \qquad (2)$$

The stress release λ at the time of lining installation (Figure 6) depends on the distance x from the face to the section where the lining is installed (Panet and Guenot, 1982), but also on the ground-lining relative stiffness β and on the plastic behavior of the ground (Kielbassa & Duddeck 1992, Bernaud & Rousset 1992).

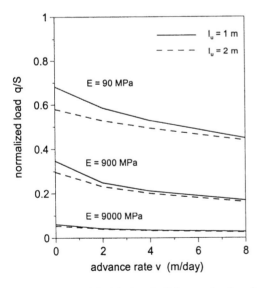

Figure 4. Normalized final load on the lining as a function of the advance rate.

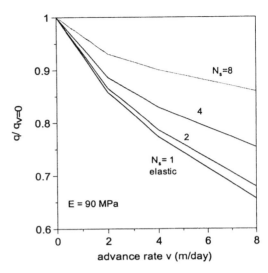

Figure 5. Influence of the stability ratio on shotcrete hardening effects.

Following the conventional CC Method the dependence of the λ factor on the relative stiffness β is usually disregarded, and λ is assumed to be only a function of the face distance x, according to the well known Panet & Guenot (1982) expression.

The results of 3D analyses can be used to improve the accuracy of the conventional CC method

The approach proposed herein consists in taking advantage of 3D results to back-calculate a λ factor which accounts for the influence of relative stiffness β, advance rate v, and stability ratio N_s of the tunnel. The back calculation is performed in the q, u plane, starting from the point q_{eq} (Figure 6) corresponding to the final load given by the 3D analysis and moving back along a lining curve characterized by a constant $E_{s,28}$ shotcrete modulus.

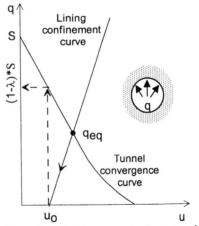

Figure 6. Schematic of the Convergence-Confinement method and back calculation of the λ factor.

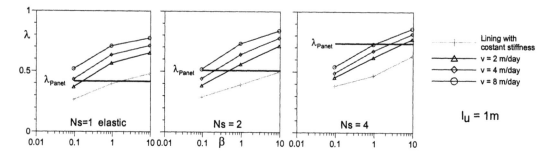

Figure 7. Release factor λ as a function of the stiffness ratio for different ground strength and advance rates.

Figure 7 shows the relief factor λ obtained through the outlined procedure, as a function of β for different N_s values and advance rate v. The unsupported tunnel span l_u near the face has been kept constant and equal to 1 m. The wide range of variation of such corrected λ values with respect to the λ factor predicted by the Panet formula underlines the strong influence of shotcrete stiffness on the final load.

It is worth observing that while in the case of elastic ground the conventional procedure generally leads to overestimated loads (stemming from underestimated λ), in the case of low-strength elasto-plastic ground (e.g. N_s=4), it can lead to largely underestimated loads.

It is deemed that the set of graphs in Figure 7 can also help the designer to properly estimate the λ factor to be used in 2D numerical models, which are usually applied in the design of support structures for non-circular tunnels.

4 FACE REINFORCEMENT BY FIBER-GLASS BOLTS

Fully grouted fiber–glass bolts installed at the face in the direction of the tunnel axis can be modeled by special beam-elements connected to the grid (Swoboda & Marence 1991), or by a homogenized material which represents the macroscopic behavior of the ground core with improved stiffness and strength (Graziani 2000).

The latter approach seems best suited to the axisymmetric model where the effect of the reinforcement is to be smeared around the axis anyhow.

Constitutive laws of the composite material are obtained on the basis of the following assumptions: elastic ideal-plastic behavior for both the ground and the reinforcement, purely tensile force within the bolts, and perfect bonding between grouted bolts and ground (Graziani 2000).

The homogenized material therefore exhibits an isotropic stiffness and strength properties with higher value in the direction of the bolts.

The smeared reinforcement model has been implemented in the numerical FLAC code (Itasca 1998) and a number of analyses have been carried out to investigate the influence of bolt density for different ground conditions. Typical properties assumed for fiber-bolts are: elastic modulus E_b, 20 GPa, tensile strength σ_b, 500 MPa, bolt section A_b, 0.0014 m^2. The construction process has been schematized as in the previous set of analyses, but now a bolt system of length l_b = 20m is also applied every 12 m of face advance. Shotcrete stiffness is assumed constant for the sake of simplicity.

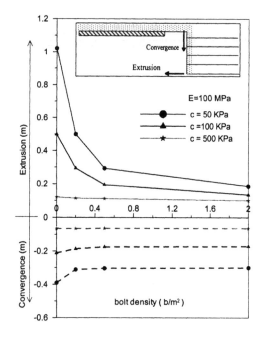

Figure 8. Influence of bolt density on extrusion and on convergence at the face.

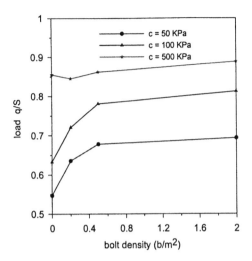

Figure 9. Influence of bolt density on the normalized load acting on the lining.

Figure 8 shows the effects of bolt density n_b on face deformation for various cohesion values of the rock mass. Face reinforcement turns out to be very effective in reducing extrusion and convergence at the face in very poor rock conditions, while a substantial amount of strain reduction is gained for a medium-low reinforcement ratio where a small reinforcement density is sufficient to halve extrusion. Increases in bolt density of over 0.5 b/m² do not further reduce the deformation in any of the analyzed cases.

Another point to be carefully considered in the design of the support system is the increase in the final load on the lining resulting from a reduction in stress relaxation of the ground core by means of bolt reinforcement (Figure 9).

5 APPLICATION TO THE S. VITALE TUNNEL

The proposed numerical model has been applied to the real case of the S. Vitale tunnel where shotcrete and face reinforcement have been extensively applied.

The S. Vitale railway tunnel (Poma et al. 1995) is located in the South of Italy inside a clay shale formation (Argille Multicolori), characterized by a close fissure network.

Instability phenomena were frequently met during tunneling. Thereafter it was decided to adopt full face excavation (section area 118 m²) with fiber glass bolts (length 18m) installed every 10 m of advance.

The support system consists of a shotcrete layer applied (1 m stretch) close to the face and a final concrete lining cast in place. The ground properties

and the parameters of the structural materials assumed in the numerical model are listed in Table 2.

Table 2. Parameters assumed in numerical analysis.

ground	shotcrete &concrete	fiber glass bolts
S= 1.9 MPa	$E_{s,28}$ = 22 GPa	A_b = 0.00196 m²
c = 30 kPa	t_s = 0.2 m	n_b = 1.1 b/m²
φ = 18°	$E_{c,28}$ = 30 GPa	E_b = 10 GPa
E = 200 MPa	t_c = 0.8 m	$σ_b$ = 400 MPa

In Figure 10, measured and calculated displacements along the tunnel axis (extrusion) are compared. Total extrusion after 10 m of advance predicted by the numerical model slightly underestimates the measured value by about 0.2 m.

Calculated displacement profiles at different excavation steps deviate from the measured profiles mainly at the face and at the end of the reinforcement zone where the homogenized model tends to overestimate bolt stress and therefore the confinement effect produced.

Some discrepancy between model prediction and reality could also stem from the axisymmetric modeling which cannot precisely represent the real construction steps where the invert is cast separately after the vault.

Taking the progressive hardening of both the shotcrete and concrete lining into account, the predicted mean hoop stress in the primary and final lining are respectively equal to 15 MPa and 6 MPa.

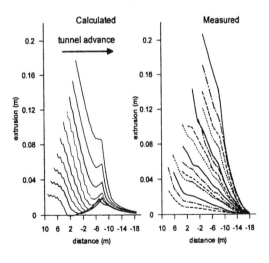

Figure 10. Comparison between measured and calculated extrusion.

6 CONCLUSIONS

The findings of a numerical analysis of a tunnel constructed by using the shotcrete and face reinforcement method, can be summarized in the following points:

• the reduction in final load due to the progressive hardening of the shotcrete, is proportionately less important as the strength of the rock mass decreases, but it is not negligible;

• the ground stress release factor to be assumed in the plane strain model increases with the ground-support stiffness ratio and with the advance rate;

• reinforcement of the face by fiber glass bolts reduces tunnel convergence but, at the same time, it increases the loads on the final lining;

• bolt effectiveness in reducing tunnel convergence is not linearly proportional to the number of bolts: a critical bolt density can be identified, above which the convergence will no longer decrease; moreover, the better the mechanical properties of the ground, the lower the bolt influence on the stress and strain conditions near the tunnel face.

REFERENCE

Bernaud D. & G. Rousset 1992. La "nouvelle méthode implicite" pour l'étude du dimensionnement des tunnels. *Rev. Franc. Geotech.*, 60, 5-26.

Chang Y. 1994. Tunnel support with shotcrete in weak rock – A rock mechanics study. *Ph.D. Thesis, Division of Soil and Rock Mechanics, Royal Institute of Technology*, Stockholm.

Graziani A. 2000. Evaluation of rock bolts effectvness in reducing tunnel convergence in squeezing rocks. *Rivista Italiana di Geotecnica*, 1, 64-72

Itasca 1998. FLAC, User's Manual, Version 3.40. *Itasca Consulting Group Inc., Minneapolis, Minnesota.*

Kielbassa, S & Duddeck, H 1992. Stress-strain field at the tunnelling face - Three-dimensional analysis for two-dimensional technical approach. *Rock Mech & Rock Eng.*, 24, 115-132.

Kovari K., F. Amberg & H. Ehrbar 2000. Mastering of squeezing rock in the Gottard Base Tunnel. *World Tunnelling*, 13, 5, 234-238.

Lombardi, G. 1973. Dimensioning of tunnel linings with regard to constructional procedure. *Tunnel & Tunnelling*, July, 340-351.

Moussa A. & H. Wagner 1997. Effect of construction speed on the behavior of NATM tunnel. *9th IACMAG*, Wuhan, Yuan ed., 1419-1424. Balkema: Rotterdam.

Panet M & A. Guenot 1982. Analysis of convergence behind the face of a tunnel. *Tunnelling 82*, 197-204, *Brighton.*

Poma, A., F., Grassi & P. Devin. 1995. Finite difference analysis of displacement measurements for optimising tunnel construction in swelling soils. *Field measurement in Geomechanics 4th Int. Sym.* ,225-236, Bergamo.

Swoboda G. & M Marence 1991. *FEM modelling of rock bolts. IACMAG 91, Cairus*, 1515-1520.

Modern Tunneling Science and Technology, Adachi et al (eds), © 2001 Swets & Zeitlinger, ISBN 90 2651 860 9

Simulation Analysis on Deformation of Soft Rockmass Due to Excavation of Tunnel

W.Zhu, S.Li, S.C.Li, Y.Zhang, S.Wang, W.Chen,
Institute of Rock and Soil Mechanics, The Chinese Academy of Sciences, Wuhan, China

ABSTRACT: This paper carries out a vast amount of FEM numerical simulation and prediction on the deformation and failure possibility of the rockmasses surrounding the tunnels under various excavating and supporting conditions in Taiwan. The rockmasses concerned belong to grade V and VI according to the Taiwan classification system of tunnel's surrounding rockmass. The part of the computing rigimes is carried out according to the method of large deformation analyses. Based upon the above work, a series of curves that predict the rockmass deformation have been drawn, which is of significance to guiding both designing and constructing of tunnels in future.

1 INTRODUCTION

As we know, the main factors relating to whether the design and construction of a tunnel is successful include the geological environment, parameters of rockmass, overburden depth, geostress field, cross-section shape of the tunnel, engineering size and other engineering factors etc in the region surrounding the tunnel. In terms of soft rockmass, rational selection of excavation and supporting method is a matter of paramount importance that has a bearing on the success or failure of the tunneling.

The engineering circle both at home and abroad mainly introduces the method of empirical analogue, for instances Q system or RMR of South Africa system, to carry out the supporting design for tunnels. In China, there exist a variety of classification methods that depend upon the professions such as railway, water conservancy, coal industry, war industry and the later national standard. Unfortunately, those classifying methods can not be of scientific quantization easily, so they fail to predict the interaction of surrounding rockmass and supporting system, thus giving no quantitative data about the developments of the supported work in dynamics, such as settlement of roof, convergence of inner space and forces on the supported structure.

In the context of the above reason, it is urged that a kind of predition system of tunnel deformations be established in the engineering scope, especially for a soft surrounding rockmass.

2 STUDYING METHODS

Previous researchers have conducted numerical simulations on the excavation and supporting of tunnels in soft rockmass (Moussa et al. 1997 and Sterpi et al. 1997). However, few persons have carried out analyses of large deformations and systematic studies concerned. The authors, by consulting the methods of previous researchers (Bathe et al, 1976) for large deformation analyses, proposes their own method.

Taking quantities of representative regimes as the object to be studied, this paper carries out FEM numerical simulations on the excavation and supporting of tunnels. Because of the adaptability of FEM, both natural and engineering factors can be considered comprehensively, such as problems of two and three dimensions and problems of small and large deformations of surrounding rocks.

This research topic is the co-operation project between Wuhan Institute of Rock and Soil Mechanics, CAS and Sinotech Engineering Consultants, LTD. With respect to the surrounding rocks of grades V and VI, considered are different buried depths, different rockmass mechanical indexes, different opening diameters and a variety of excavating and supporting methods. Then the numerical analysis is performed to obtain values of tunnel's surrounding rockmass deformation, size of plastic zones and force-bearing situation etc, whereby a library of data is established. A special computing programe has been developed for this purpose that can take into account large deformation of rockmass subject to deteriorative environments for analysis and computation.

3 PRINCIPLE AND METHOD OF NUMERICAL ANALYSES

The programes for FEM analysis are EP-3D that is used for elasto-plastic analyses of small deformations and EPL-3D that is used for analyses of large deformations. A brief description is given below to the principle and method of the latter.

3.1 Equilibrium equation

The column matrix of $\{\psi\}$ is employed to express the vector sum of the generalized internal and external node forces and the equilibrium equation system that is suitable for small displacements can be derived from the principle of virtual displacement:

$$\{\psi(\delta)\} = \int[\overline{B}]^T\{\sigma\}dv - \{P\} = 0 \tag{1}$$

In the case of large displacements, matrix $[\overline{B}]$ is correlated with matrix $\{\delta\}$:

$$[\overline{B}] = [B_0] + [B_L] \tag{2}$$

where the first term is linear one whereas $[B_L]$ is caused by the non-linear deformation being the function of $[\delta]$.

If Newton-Raphson method is used for solving eq.(1), the relation of $\{d\delta\}$ with $\{d\psi\}$ will be of

$$\{d\psi\} = \int[dB_L]^T\{\sigma\}dv + [\overline{K}]\{d\delta\}$$

which has the final form of

$$\{d\psi\} = [K_T]\{d\delta\} \tag{3}$$

where the tangential stiffness matrix of $[K_T]$ is

$$[K_T] = [K_O] + [K_\sigma] + [K_L] - [K_R] \tag{4}$$

$[K_O]$ being the nonlinear stiffness matrix of small deformation, $[K_\sigma]$ the matrix of initial stresses, $[K_L]$ the matrix of large displacement and $[K_R]$ the matrix of load correction.

3.2 Method for searching solutions

When Newton-Raphson Method is used to solve the problem of large displacements, we have the following iterative expressions of

$$\{\Delta\delta\}_n = -[K_T]^{-1}\{\psi\}_n \tag{5}$$

$$\{\delta\}_{n+1} = [\delta]_n + \{\Delta\delta\}_n \tag{6}$$

and the computing procedures are:
(a) find elasto-plastic solution as the first approximation value of $\{\delta\}_1$;
(b) compute $[\overline{B}]$, σ and $\{\psi\}$ respectively;
(c) compute $[K_T]$;
(d) compute $\{\Delta\delta\}_1$ and $\{\Delta\delta\}_2$ according to eq.s (5) and (6);
(e) go into step (6) and make repeating iteration until a $\{\psi\}_n$ small enough is obtained.

3.3 Method and character of programs

(a) capable of performing analyses of small or large elasto-plastic deformations using D-P or M-C yield criterion;
(b) reflect interaction of rock-bolt system using a bolting-column element and a bar element;
(c) capable of simulating process of construction of excavation and supporting;
(d) simulate effects of discontinuities, faults for example using weak-plane elements or joint elements.

Table 1. Mechanical Parameters of rockmss.

rock grade	burried depth	75m				150m				300m				500m			
		C^* (MPa)	ϕ^{**}	σ_{cm}^{***} (MPa)	E^{****} (MPa)	C (MPa)	ϕ	σ_{cm} (MPa)	E (MPa)	C (MPa)	ϕ	σ_{cm} (MPa)	E (MPa)	C (MPa)	ϕ	σ_{cm} (MPa)	E (MPa)
V	Hard rock (σ_{cm}=50MPa)	0.17	48	0.88	1000	0.28	42	1.27	1000	0.46	37	1.84	1000	0.66	32	2.41	1000
	medium hard rock (σ_{cm}=25MPa)	0.14	42	0.64	600	0.23	37	0.92	600	0.38	31	1.33	600	0.53	27	1.74	600
	soft rock (σ_{cm}=5MPa)	0.09	29	0.30	400	0.14	24	0.43	400	0.22	19	0.63	400	0.31	16	0.82	400
VI	fault material	0.05	25	0.16	200	0.10	23	0.30	200								

* : cohesion, **:friction angle, ***: uniaxial compression strength, ****: Young's modulus

4 NUMBER OF ANALYSIS CASES AND EXECUTION OF NUMERICAL SIMUL ATION

4.1 Mechanical indexes of rockmass

Shown in Table 1 are the mechanical indexes of the surrounding rockmass. In Table 2 and 3 we generalize the cases that are analyzed.

5 EXECUTION OF NUMERICAL SIMULATION

5.1 Domain of computation and simulation of excavation

The computation range measures 160 meters in the longitudinal axis X of the tunnel, 60 meters along the direction Y perpendicular with the tunnel (on the horizontal plane) and 130 meters in the vertical direction Z. The 3-D isotropic element with 8 nodes is introduced and altogether 7,200 nodes and 6,048 elements are meshed.

The mining or excavating is simulated using the method of "air elements". Two kinds of excavation order, i.e., standard and non-standard that are shown in Figure 1 and 2 are used for the excavating sequence of the tunnel.

5.2 Supporting method and simulation on blasting effects

A bolt-rock column element is used for bolts and the steel camber is simulated using the equivalent method whose stiffness converted onto the neibouring concrete.

The effect due to blasting is taken into consideration by reducing both elastic modulus and the strength indexes of the surrounding rocks. The blasting effect scope reaching one meter around

Table 2.   Ratio of squeezing degree of rock(σ_{cm}/p_0).

rock grade	burried depth	75m	150m	300m	500m
V	hard rock (σ_{cm}=50MPa)	0.470	0.339	0.245	0.193
V	medium hard rock (σ_{cm}=25MPa)	0.339	0.245	0.177	0.139*
V	soft rock (σ_{cm}=5MPa)	0.159	0.115*	0.084*	0.066*
VI	fault material	0.013*	0.024*		

* $\sigma_{cm}/p_0 \leq 0.15$ belong to high degree of squeezing, large deformation analysis will be used.

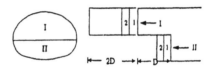

Figure 1. Standard excavating order.

Table 3.   Summarization of analyzed cases.

group	tunnel span (m)	buried depth (m)	rockmass type		lateral pressure coefficient	construction sequence	effect of blasting	number of cases	number of large deformation analyses
A	10	75,150, 300,500	V	hard and medium hard rock	K=1.0	standard construction	no	14	6
			VI	fault material					
B	10	75,150, 300,500	V	medium hard rock	K=1.0	non-standard construction	no	6	3
			VI	fault material					
C	15	75,150, 300,500	V	hard, medium hard and soft rock	K=1.0	standard construction	no	14	6
			VI	fault material					
D	15	75,150, 300,500	V	medium-hard rock	K=1.0	non-standard construction	no	6	3
			VI	fault material					
E	10	75,500,	V	soft rock	K=1.5	standard construction	no	3	2
			VI	fault material					
F	15	75,500,	V	medium-hard rock	K=1.5	standard construction	no	3	2
			VI	fault material					
G	10	75,150, 300,500	V	medium-hard rock	K=1.0	standard construction	unfavorable blasting	6	3
			VI	fault material					
H	15	75,150, 300,500	V	medium-hard rock	K=1.0	standard construction	unfavorable blasting	6	3
			VI	fault material					

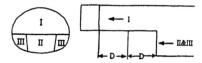

Figure 2. Non-standard excavating order.

tunnel walls is considered in the case of a tunnel of 10m in diameter and reaching 1.5m in the case of 1.5 m in diameter.

The deductions in rockmass strength are:

for the rockmass of grade V:

the values of the young's modulus E, cohesion C and friction angle ϕ are taken as 0.8 times the initial ones and for the rockmass of grad VI:

0.8 times initial modulus is taken for E and C and ϕ remains unvaried.

6 ANALYSES OF COMPUTATIONAL SIMULA -TION RESULTS

6.1 Settlement of tunnel roof

(a) The poorer the rock quality and the thicker the overburden and the wider the span, the greater the settlement. For regimes of A, B,C,D,G and H, the ratio of the roof settlement over the tunnel diameter is about 0.2~0.5%, when the overburden is about 100 meters thick where the rock belongs to medium hard character in lithology. But when the buried depth reaches much greater (for example) 500 meters this ratio can reach up to 1.5~2%; for soft rock, the ratio can be even to 3.5~5%(Figure 3).

(b) the settlement of the roof caused by the non-standard excavation is greater than that caused by standard excavation. In the case of regimes A and B where a 10-meter span tunnel is built, the settlement of the tunnel roof caused by the non-standard excavation is 20% greater than that caused by the standard excavation. If the span is 15 meters, the former is about 30% greater than the latter; and the highest difference of about 70% (nearly 10 centimeters) may result.

(c) In consideration of the unfavourate blasting effect, the settlement is more serious than the case of non-blasting, the roof settlement will have about 20% increments for regimes A and G and for a 15 meter span, the increment in settlement is about 30%.

(d) With respect to higher geostress, the roof

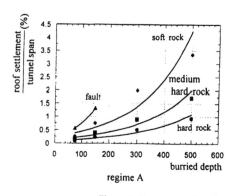

regime A

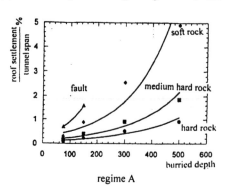

regime A

Figure 3. Curve group for relevant settlement to burried depth for regime A and C.

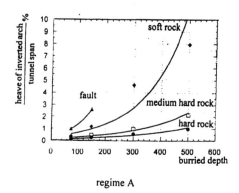

regime A

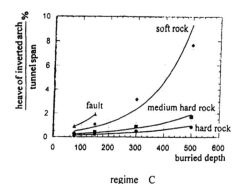

regime C

Figure 4. Curve group for relevant heave of inverted arch to burried depth.

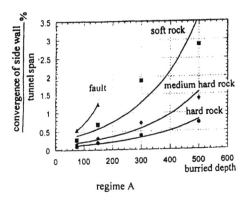
regime A

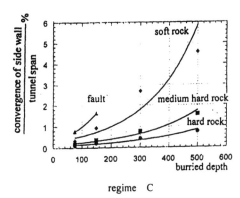
regime C

Figure 5. Curve group for relevant convergence of side wall to burried depth.

settlement becomes smaller for the most cases and if the environment is deteriorated, the result turns out contrary when instability occurs due to supporting failure. For example, the soft rock with a lateral pressure coefficient of K=1.5 in regimes A and E, the settlement of the roof is 60% greater than the rock with K=1.0

6.2 Heave of the inverted arch

The law of the inverted arch heave is similar with the roof settlement but the magnitude of the former is higher than the latter. Besides, various factors pay more considerable effect on the inverted arch heave than the roof settlement and it is the case especially for poorer lithological character and thicker overburden.

For the medium-hard rock, with respect to regimes A, B, C, D and G, the ratio of the inverted arch over the tunnel span increases from about 0.2% to 2~5% as the burried depth increases from 100 meters to 500 meters, and this ratio can be up to 8% also if the rock is soft (Figure 4).

6.3 Convergence of side wall and plastic zone of surrounding rocks

Various factors, except geostress, display the affecting law on the side wall convergence and on the plastic zone similar to that on the roof settlement. As concerns medium-hard rock, the side wall convergence's ratio to the tunnel span is similar to the situation of the roof settlement.

In the case of regimes A, B, C, D, the convergence of the side wall due to non-standard excavation is 30~50% greater than that caused by the effect of blasting, the side wall convergence in regimes A and G is about 30% greater than the situation without blasting effect, while the side wall convergence for a 15-meter span tunnel will be increased by about 46%.

For regimes of A and C, the maximum convergence of the side wall of a 15-meter span tunnel is about 50% greater than that of a 10-meter span tunnel (Figure 5).

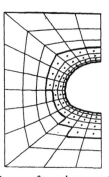

Figure 6. Plastic zone of tunnel cross section for case 38 (span of 15m).

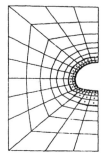

Figure 7. Plastic zone of tunnel cross section for case 18 (span of 10m).

In the case of regimes A and E, for soft rock a high lateral stress will result in 46% increment in the side wall convergence.

417

The plastic zone in the surrounding rockmass becomes larger as the above mentioned factors increase. If the tunnel is built in soft rockmass with a great burried depth, the thickness of the plastic zone for a 15-meter span is over two times as great as that for a 10-meter span, (Figure 6 and 7), the magnitude approaches the tunnel's radius. If the geostress is high, the plastic zone around the roof and inverted arch will increase whereas that around the side walls reduces. The unfavourate effect due to blasting is more serious on the increment in plastic zone around the floor.

6.4 Longitudinal deformation of tunnel and stable rate of plastic zone

(a) If the rocks are good in quality or the burried depth is not great, the development of longitudinal deformation or plastic zone becomes stable in short time, say, becomes unvaried on the whole within about three advancing turns.
(b) In the case of poor rocks or great overburden, it takes about 4~6 advancing turns for the above deformation or plastic zone development to become stable.

6.5 Failure distribution of concrete lining

For regime A, the tensile cracking of some cases has a large scope and mostly takes place in lower half of the tunnel; in some cases, the tensile cracking develops into the places around the roof. But for regime B, the situation is somewhat better.
For regime C where the tunnel has a large span of 15 meters, the tensile cracking becomes worse obviously. For some regimes, the whole lining enters into the tensile- cracking state. Compared with regime C, regime D is improved to a certern extent.
The above shows that the extent to which the lining bears tension under the non-standard excavating scheme is lower than that under the standard excavation scheme.

7 CONCLUSTION AND DISCUSTION

1. The settlement of the roof, heave of the inverted arch and the convergence of the side wall will increase when the burried depth becomes great, the rockmass quality becomes poor and the tunnel span get expanded. In most cases, the ratio of the roof settlement or floor heave or side wall convergence to the span will increase from 0.2% to about 2% as the overburden thickness increases from 100 to 500 meters; this ratio, especially for soft rocks, reaches the maximum value of up to 5~8%.

2. The non-standard excavation scheme will generally result in a higher convergence of the rockmass than the standard scheme does. In most cases, the former is 20% higher than the latter when the tunnel diameter is 10 meters and 30%~50% higher when the diameter is 15 meters. The maximum difference can reach up to 70%. However, the lining is subjected to lower tension under the non-standard excavation than under the standard excavation.

3. The surrounding rockmass when subject to unfavourate blasting effect bears displacements 20~30% greater than when subject to no blasting effect. If the tunnel span is 15 meters, this difference in displacement can be high up to 30~45%.

4. Generally, when the initial lateral geostress is high, the convergence of both roof and floor will decrease whereas the sidewall convergence will increase. However, when the overburden thickness is considerably great and the rockmass quality is very poor, either roof or floor convergence will increase abruptly. In the case of soft rock or of large opening span, the side wall may bear convergence increments as 50% as high.

5. The development of deformation or plastic zone in the rockmass at the place about three advancing turns away from the working face of the tunnel tend to be stable if medium grade rockmass is encountered and the burried depth is not high. However, if the rockmass is weak or the burried depth is high the above development will not be stable until four or six turns are completed.

Acknowlegements

This work is supported by SINOTECH Engineering Consultants, LTD. of Taiwan.

REFERENCES

Bathe K. J. and Ozdemir H. 1976. Elastic- Plastic Large Deformation Static and Dynamic Analysis. Compt, Struct., 6 (2): PP81~92.
Moussa A. and Wagner H., 1997. Effect of construction speed on the behavior of NATM tunnels. Proc. of 9th Int. Conf. on Comp. Meth. and Adv. in Geom. Wuhan, China, Yuan(ed.) PP1421-1423 Balkema, Rotterdam, ISBN 90 5410 904 1.
Sterpi D. and Cividini A., 1997. Numerical analysis of tunnels in strain softening soil. Proc. of 9th Int. Conf. on Comp. Meth. and Adv. in Geom. Wuhan, China. Yuan(ed.) PP1391-1394. Balkema, Rotterdam, ISBN 90 5410 904 1.

Modern Tunneling Science and Technology, Adachi et al (eds), © 2001 Swets & Zeitlinger, ISBN 90 2651 860 9

Behavior of tunnels built in sedimentary rocks with joints dominant in one direction

T. Koyama, S. Nanbu, Y. Suzuki & Y. Tasaka
Tokyo Electric Power Services Co., Ltd, Tokyo, Japan

K. Kudo
Tokyo Electric Power Company, Tokyo, Japan

ABSTRACT: The purpose of this study is to compare the results of tests conducted prior to tunnel excavation with the behavior of tunnels during excavation of rock with anisotropy in terms of strength and deformability; verify the causes for the difference between them; and establish an analysis method which is capable of predicting the actual behavior of tunnels.

1 INTRODUCTION

Sedimentary rock with well-developed cleavage planes and joints, which are considered to have been tectonically generated, is found in many places in Japan. Constructing tunnels with proper stability in such rock requires high technologies in various fields, from survey and design to construction. Particularly, technologies for evaluating the mechanical characteristics of rock, which are necessary for the design of tunnel supports and the prediction of their behavior, are an essential element. Methods to predict precisely the behavior of rock surrounding the tunnel during excavation are also important.

This paper outlines two methods: a method to evaluate the deformation characteristics of rock with well-developed cleavage and joints which are essential factors for precisely predicting the behavior of tunnels driven through such rock, and a newly developed method to predict the behavior of tunnels constructed in rock with cleavage surfaces or joints

dominant in one direction. In addition, the results derived from these methods are compared with measurement data obtained during excavation of a tunnel.

2 SURVEY LOCATION

An underground power plant in Kazunogawa Power Projoect, constructed by the Tokyo Electric Power Company (TEPCO) in Yamanashi Prefecture (Fig.1) and an approach tunnel leading to it were selected as survey locations.

The ground consists of alternating layers of sandstones and mudstones, both belonging to the Kobotoke Formation which was formed between the Upper Cretaceous of the Mesozoic Era and the Paleogene period of the Cenozoic era. The rock surrounding the underground plant has well-developed cleavage planes and joints. The joint has an east-west strike and a steep dip to the north (Fig. 2).

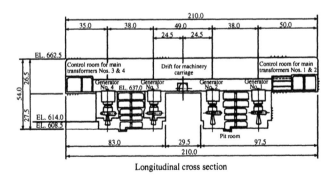

Longitudinal cross section

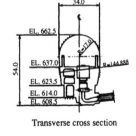

Transverse cross section

Figure 1. Structure of the underground power plant.

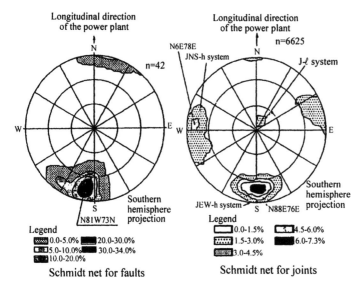

Schmidt net for faults   Schmidt net for joints

Figure 2. Schmidt net for faults and joints.

Anisotropic deformation was therefore expected, and this required that the supports are designed by taking into consideration its effects, as well as the locations of the caves for the plant and those of nearby tunnels.

3 EVALUATION OF ANISOTROPIC DEFORMATION BASED ON THE RESULTS OF IN-ADVANCE TESTS

The relationship of the results of rock tests, plate loading tests, and borehole loading tests with the angle between cleavage/joint and loading direction is shown in Figures 3-5.

The borehole loading test results did not show noticeable anisotropic deformation. On the other hand, although with some variation, anisotropic deformation as shown in Table 1 was observed in the results by the other two tests.

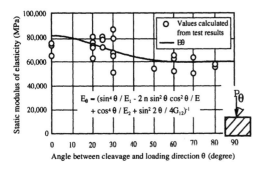

$$E_\theta = (\sin^4\theta / E_1 - 2 n \sin^2\theta \cos^2\theta / E + \cos^4\theta / E_2 + \sin^2 2\theta / 4G_{12})^{-1}$$

Figure 3. Rock test results.

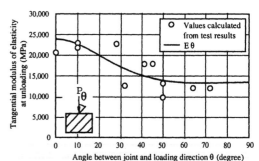

Figure 4. Results of plate loading test.

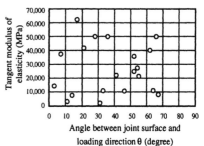

Figure 5. Borehole loading test results.

Table 1. The anisotropy of deformation obtained from test results.

Classification of tests	Modulus of elasticity (MPa)		E1/E2
	E1*	E2**	
Rock test	82,000	60,000	1.4
Plate loading test	24,000	13,500	1.8
Elastic wave exploration	***68,000	***52,300	1.3

* In the direction of joint surface
** In the direction perpendicular to joint surface
*** Dynamic modulus of elasticity

420

4 EVALUATION OF ROCK DEFORMATION CHARACTERISTICS BASED ON ROCK BEHAVIOR DURING TUNNEL EXCAVATION

Using the convergence monitored in the approach tunnel leading to the underground power plant (section with an overburden of 400 m) and the measurement results by the multiple-extensometers installed at the plant, the modulus of elasticity in agreement with actual displacements was calculated. Although with variation, the modulus of elasticity in the direc-

tion of joint surfaces (direction passing transversely through the underground plant) E1, which is used as the indicator for the extent of isotropic deformation, was on average three to four times as large as that in the direction perpendicular to joint surfaces (longitudinal direction of the tunnel cave) E2, larger than the values obtained from the results of rock tests and in-situ rock mass evaluation tests (Figs 6-7).

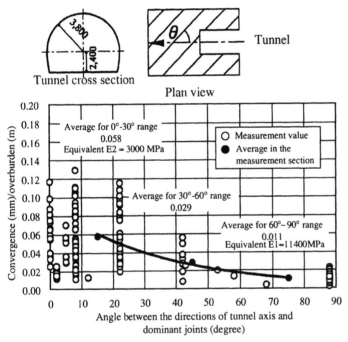

Figure 6. Convergence of approach tunnel.

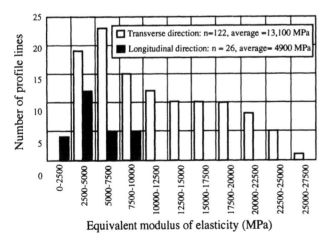

Figure 7. Equivalent modulus of elasticity of rock around the power plant.

421

5 DEVELOPMENT AND APPLICATION OF ANALYSIS METHOD CONSIDERING THE BEHAVIOR OF ROCK JOINTS

The major reason for the large difference between the deformation characteristics of rock derived from in-situ test and from the monitoring results of displacements during excavation of the tunnel and the underground power plant is that the deformation of rock mass with joints that loosened due to tunnel excavation was larger than expected.

5.1 *Development of an analysis method for strain-softening behavior considering the failure of joints*

The analysis method developed by the authors, which considers the effects of the failure of both discontinuity surfaces dominated in one direction and rock substrate (Fig 8), can evaluate strain-softening behavior after failure as well as the anisotropy in terms of rock mass strength and deformation.

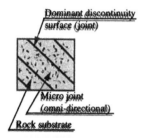

Figure 8. Model for rock with discontinuity.

5.2 *Evaluation of strength characteristics of joint surfaces*

The single shear test apparatus as shown in Figure 9 was developed to evaluate directly the strength characteristics of rock specimens with joints. The test results are shown in Figure 10.

5.3 *Example of application of the analysis method*

The deformation of the approach tunnel was analyzed using the non-linear FEM method which can consider the effects of the above-mentioned joint failure. The analysis conditions are shown in Table 2, and the analysis results in Table 3 and Figure 11. By considering the effects of joint failure, this method yielded results which are in better agreement with measurement results than the orthotropic elastic solutions calculated using test results.

6 CONCLUSION

The results of tests and analysis in terms of the deformation characteristics of underground tunnels excavated in sedimentary rock with well-developed cleavage planes and joints were compared in this report. It was found that for predicting the behavior of a tunnel excavated in rock which has joints and anisotropic deformation characteristics and which could collapse during excavation, non-linear analysis considering the effects of the collapse is effective.

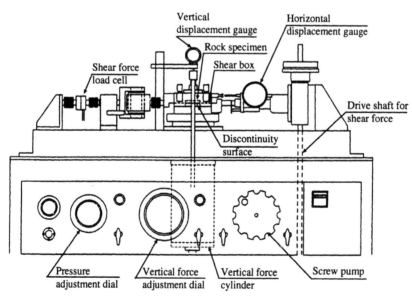

Figure 9. Single shear test apparatus.

Legend	Regression equation	Correlation coefficient	Number of data
· · · · :Without calcite	$\tau p = 0.8 + \sigma\,ntan\,30°$	0.73	16
—— :With calcite	$\tau p = 0.03 + \sigma\,ntan\,39°$	0.90	16

Legend	Regression equation	Correlation coefficient	Number of data
· · · · :Without calcite	$\tau p = 0.3 + \sigma\,ntan\,32°$	0.87	16
—— :With calcite	$\tau p = 0.03 + \sigma\,ntan\,36°$	0.93	16

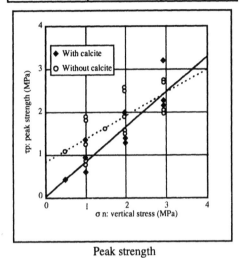

Peak strength

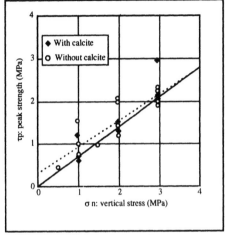

Residual strength

Figure 10. Results of single shear test.

Table 2. Analysis conditions.

Orthogonal anisotropy elastic analysis	Modulus of elasticity of rock	E1	10,000	MPa
		E2	5,000	MPa
Non-linear analysis considering failure of joints	Modulus of elasticity of rock	E	40,000	MPa
	Initial modulus of elasticity of joints	kso	10σ n	MPa /cm
	Failure strength of rock	c p	1.5	MPa
		φ p	58	Degree
	Residual stress of rock	c r	1.5	MPa
		φ r	58	Degree
	Failure strength of joint	c j	0	MPa
	Residual strength of joint	φ j	50	Degree

Table 3. Analysis results (equivalent to an overburden of 460 m).

Classification	Convergence (mm)			Joint density
	u1*	u2**	u2/u1	
Measurement data	5	27	5.4	
Anisotropy of elasticity	4.5	13.0	2.9	
Failure of joints (non-linear analysis)	5.2	18.2	3.5	5 nos./m
		23.5	4.5	10 nos./m
		36.9	7.1	20 nos./m

* Tunnel axis is perpendicular to joint surface.
** Tunnel axis is in the direction of joint surface.

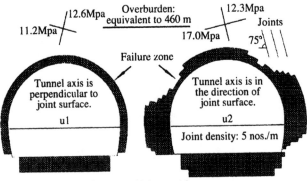

Figure 11. Failure zone for non-linear analysis.

REFERENCES

Tasaka, Y., Uno, H., Ohmori, T. & Kudoh, K. 2000. A Joint and Rock Failure Strain-Softening Model and Its Application to The Excavation simulation of Large-Scale Underground Caverns: *Proc . Jpn. Soc. Civil Engineers III-*51,652:73-90 (in Japanese).

Modern Tunneling Science and Technology, Adachi et al (eds), © 2001 Swets & Zeitlinger, ISBN 90 2651 860 9

Research of Observational Method on the Groundwater in the Tunnel Approach Crossing

Y. Ohnishi & H. Ohtsu
University of Kyoto, Kyoto, Japan

H. Ishihara & N. Okamoto
Hanshin Expressway Public Corp., Kyoto, Japan

T. Yasuda & K. Takahashi
Pacific Consultants Co., Ltd., Osaka, Japan

ABSTRACT: Groundwater issues related to mountain tunnels are discussed from the aspects of construction and environment. The former is related to groundwater control for safe construction, such as prevention of cutting face collapse due to large amount of inflow. The latter is related to impact of construction work on the natural environment including the groundwater system as well as surface water, which has been a key issue in recent years. This paper describes observational method executed in Inariyama Tunnel intended to carry out construction with higher reliability, including groundwater control. The methods of analysis and assessment employed are reported as well. Since the tunnel crosses an existing waterway tunnel at a separation of 27 m, the control of the impact of groundwater changes on the waterway tunnel was considered to be a crucial issue. The observational method for groundwater adopted in the project is a method for carrying out streamlined and reliable construction with high safety and cost efficiency. This is realized not only by a preliminary survey and numerical analysis for estimation but also by confirmation of the validity of construction methods by measurement and numerical analysis at each stage of excavation to grasp the impact on the waterway tunnel while feeding back the information to subsequent work.

1 INTRODUCTION

Inariyama Tunnel is a mountain road tunnel through alternating strata of Mesozoic/Paleozoic sandstone and slate. Its planned route involved critical areas in regard to groundwater, such as the approach to an existing waterway tunnel and passing directly under a river system designated as a protected environment. The observational method for groundwater employed in this project is carried out by making judgements supported by numerically analyzed predictions based on tracking of complicated field experience. In the case of Inariyama Tunnel, suitable construction methods were proposed by this method and adopted for each section, including the particularly important zone approaching an existing waterway tunnel.

The procedure was as follows: In the first phase, a large-scale 3-D groundwater prediction model is formulated based on information on the topography, geology, and groundwater obtained prior to the commencement of excavation, incorporating the approach to the other tunnel, topographical and geological conditions, and aquifer structures. In the second phase, the complicated groundwater behavior concomitant with the tunnel excavation is estimated beforehand by 3-D saturated/unsaturated seepage flow analysis using the prediction model. In addition, data obtained during execution are fed back to the prediction model to re-correct the model. In the third phase, estimation is made for the sections to be constructed while improving the estimation accuracy. The three-phase procedure was reiterated as the tunneling proceeded, with constant numerical analysis using conditions as close as possible to in-situ conditions. Based on the analysis, the optimum construction methods were selected, and measures to protect the groundwater environment of the area were proposed. Assessment was also conducted regarding the impact of the construction on the environment.

2 ENVIRONMENT

2.1 *Geological outline of the section near crossing*

The topography of the area where the tunnel would cross the existing waterway tunnel is a gentle slope in the mountains at a late stage of weathering erosion with hills and alluvial fans distributed in the piedmont area. The surrounding ground comprises the Tanba group (slate, shale, sandstone, and chert) of Paleozoic Medium Carboniferous period to Mesozoic Upper Jurassic period. Among the strata, slate and sandstone form alternations. The strike stretches in the NW-SE directions. The dip is

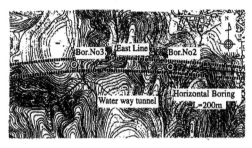

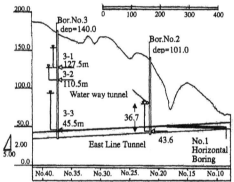

Figure1. Location of the crossings of adjacent tunnels Measurements Points No.2,No3:Vertical Boring No.1: horizontal Boring.

generally towards the south, with several fault zones penetrating in NE-SW directions. The tunnel excavation was planned to begin from the bottom towards the intermountain area as shown in Figure 1.

2.2 Hydrogeological characteristics

According to Lugeon test results during No. 2 boring, the permeability coefficients (K) of bedrock with few fissures and fault zones near the crossing are 3.0×10^{-5} to 1.0×10^{-4} cm/sec and as high as 1.0×10^{-3} to 1.0×10^{-2} cm/sec, respectively. Judging from drilled cores, fractured faults are found at 2 to 5 m intervals, forming a fracture zone as a whole. The groundwater level in the boreholes was recognized at around GL-60m, which was near the water level of the waterway tunnel. No.1 horizontal boring shown in Figure1 was carried out simultaneously with the beginning of tunnel excavation. During horizontal drilling, severe inflow of 500 to 700 l/min occurred in the fracture zones, while the inflow was not more than 100 l/min in other sections. The coefficient of permeability in the fracture zones determined from Lugeon test results was 0.8 to 2.0×10^{-2} cm/sec.

2.3 Tunnel approach conditions

Inariyama Tunnel has two cutting faces for the east line tunnel and west line tunnel. The calotte excavation and bench method was adopted for these faces with a cross-sectional area of 80 m². The existing waterway tunnel, having a constant flow rate of 12,000 to 13,000 m³/hour, is an important lifeline supplying drinking water. After 37 years of service, it might involve deterioration of lining concrete, but it was impossible to stop the water supply to repair or strengthen the waterway tunnel. According to the execution record of this waterway tunnel, severe inflow occurred during excavation of this tunnel at areas of concentrated fissures in the fault zones. Inariyama Tunnel was going to skew-cross under the waterway tunnel with a separation of 27 m at an angle of 57°.

3 OBSERVATIONAL METHOD FOR GROUNDWATER

3.1 Formulation of large-scale 3-D analysis model

Judging from the results of past geological surveys, the distribution of high-permeability fracture zones, and the state of crossing under a waterway tunnel, the most advantageous numerical analysis method is 3-D seepage analysis based on the finite element method, which can easily incorporate three-dimensional lithological distribution and construction conditions, such as excavation rate and processes. As the model for this method, a fracture zone model was considered most appropriate, as the fracture zones distributed in the sand-slate alternation would significantly affect the groundwater behavior. The fault zone distribution in the 3-D analysis model is shown in Figure.2. This Figure shows the 3-D models by preliminary analysis and those corrected in the process of excavation. Each model consists of 64,923 to 107,502 elements and 70,992 to 116,200 nodes. The data scale increased each time the model was corrected to satisfy the actual conditions newly acquired through excavation. As seen from the Figure, Inariyama Tunnel reaches the first fault zone, F5, as it proceeds. Fault zones F3 and F2 are encountered directly below the waterway tunnel. The high coefficient of permeability of 10^{-2} cm/sec determined by Lugeon testing was adopted for fault zones, for which high permeability had been suggested. The permeabilities of other layers were assumed as given in Table 1. The initial

Table 1. Hydraulic properties of 3-D model (Step 2).

Code	Geol-ogy	Coefficient of permeability X, Y, Z direction (cm/sec)	Water content (% by volume)	Specific storage (cm-1)
s11-s15	Alter-nation	$1.0 \times 10-5$	5	0.0001
F1-F7	Fault zone	$1.0 \times 10-2$	7	0.00001
Os-Oc	Osaka group	$1.0 \times 10-4$	10	0.0001

Table 2. Contents of observational method at each step.

Step	Work	Purpose	Judgement/Assessment
Step 1	Horizontal boring from portal Fault 5	Confirmation of the scale of fault Confirmation of inflow amount through Fault 5	Confirmation of scale of fault Ground properties around fault Confirmation of auxiliary methods for fault zone
Step 2 TD 0 m	Numerical analysis based on the results of Step 1 3-D seepage analysis	Inverse analysis of permeability coefficient Formulation of 3-D model	Inflow amount during crossing Impact on waterway tunnel Cut-off for Fault 5
Step 3 TD 200 m	Tunnel excavation	Monitoring of excavated ground Search boring (L=30 m)	Quantitative assessment of ground properties
Step 4	Numerical analysis based on the results of Step 3 3-D seepage analysis	Prediction of impact during crossing Necessity of pregrouting Prediction of effect of cut-off	Inflow during crossing Impact on waterway tunnel Necessity of cut-off at crossing zone
Step 5 TD 290 m	Passing directly below the waterway tunnel	Monitoring of ground directly below waterway tunnel Search boring (L=30 m) Confirmation of inflow during crossing	Quantitative assessment of ground properties directly below waterway tunnel Inflow amount during crossing
Step 6 TD 310 m	Completion of excavation of crossing zone	Confirmation of inflow after crossing	
Step 7	Numerical analysis based on the results of Step 6 3-D seepage analysis	Prediction during and after crossing and future prospect	Impact on waterway tunnel Necessity of postgrouting

groundwater level was assumed to be the water level in the boreholes.

3.2 Flow of observational method for groundwater

The flow of observational method carried out for Inariyama Tunnel is shown in Figure. 3. Detailed contents of each step are given in Table 2. Basically, numerical analysis was conducted based on the preliminary survey and 3-D saturated/unsaturated seepage flow analysis, and the impact on the waterway tunnel was estimated using monitoring data and water inflow into the tunnel under excavation as indices. In addition, measurement and

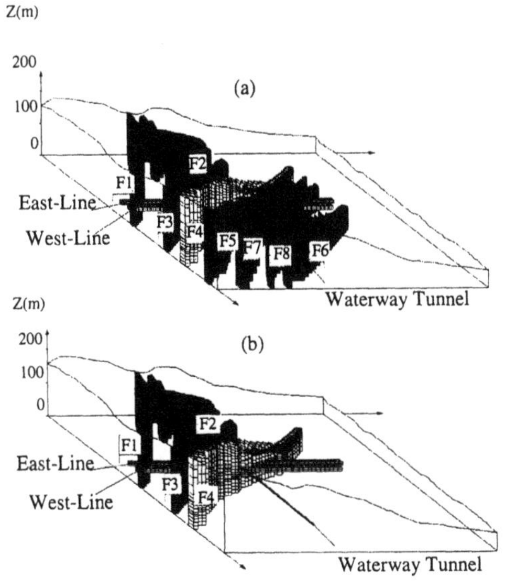

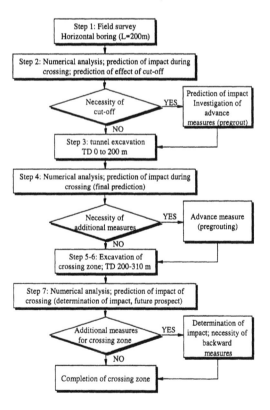

Figure 2. Distribution of fault zone in the 3-D geological profile   - (a) step2 model   (b) step7 model.

Figure 3. Flowchart of observational method.

427

Table 3. Measures to suppress impact on groundwater near crossing.

Process	Purpose	Type
Before excavation	1) Improvement of permeability of weak ground 2) Advance strengthening of weak ground	- Cut-off grouting - Advance fore piling
During excavation	1) Selection of excavation method that does not damage ground 2) Early support (Control ground loosening) 3) Cut-off from tunnel inside	- Introduction of mechanical excavation - Controlled blasting - Early closure, inside cut-off
After excavation	1) Control of ground loosening 2) Cut-off of leakage/inflow in tunnel	- Ground improvement around tunnel - Cut-off grouting from tunnel inside

Table 4. Hydraulic properties of 3-D model (Step 4: after excavation).

Code	Geology	Coefficient of permeability in each direction (cm/sec)			Water content (% by volume)	Specific storage (cm^{-1})
		X	Y	Z		
s11-s15	Alternation	1.0×10^{-5}	1.0×10^{-5}	1.0×10^{-5}	5	0.0001
s13	Alternation	1.0×10^{-5}	1.0×10^{-4}	1.0×10^{-5}	5	0.0001
F1-F7	Fault zone	1.0×10^{-3}	1.0×10^{-3}	1.0×10^{-2}	7	0.00001
Os-Oc	Osaka group	1.0×10^{-5}	1.0×10^{-5}	1.0×10^{-5}	10	0.0001

numerical analysis were repeated according to the phase of excavation to confirm the validity of the construction methods and grasp the impact on the waterway tunnel.

3.3 Establishment of permissible impact and proposal of optimum groundwater protection measures

The impact of excavating Inariyama Tunnel on the waterway tunnel includes the following: Firstly, the excavation would drain an enormous amount of water from the waterway tunnel, a life line supplying drinking water; and secondly, the large inflow can loosen the ground between the two tunnels. However, the impact on the waterway tunnel cannot be directly observed, as water is drawn off through several fault zones. For this reason, the impact on the waterway tunnel was evaluated in terms of the amount of inflow determined by numerical analysis. Since this required high reliability of the analysis model, it was essential to verify the analysis results by observing the actual inflow from excavation in process and monitoring of the surrounding ground.

On the other hand, the permissible impact on the waterway tunnel was determined in terms of the ratio of the change in the flow rate to normal flow rate, and the criterion was established to be not more than the seasonal variation, i.e., 4.0 to 7.6 m^3/min, which is 2 to 3% of the total flow rate. Based on this permissible impact, groundwater protection measures for suppressing the impact were examined and prepared for the stages before, during, and after the excavation of the area near the crossing under grade as given in Table 3.

3.4 Verification of estimation for near-crossing area

3.4.1 Results of estimation by numerical analysis at Step 2

In Step 2, the properties of the fault zones (presence/absence, width) were added and corrected based on the field data from horizontal boring. The horizontal boring was reproduced in the analytical

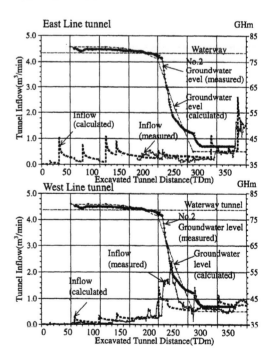

Figure 4. Comparison between calculated and measured the groundwater level and at the each tunnel inflow.

428

Table 5. Relationship between tunnel inflow and infiltration from waterway tunnel.

Model step	Inflow into Inariyama tunnel (m³/min)	Infiltration from waterway tunnel Q_0 (m³/min)	Impact ratio Q_0/Total (%)
Step 2	0.4-0.8	1.15	0.58
Step 4	0.4-1.9	2.23	1.11
Step 7	0.8-1.8	2.87	1.43

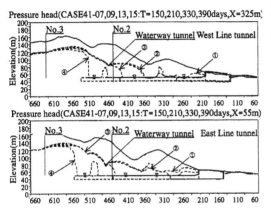

Pressure head(CASE41-07,09,13,15:T=150,210,330,390days,X=325m)

Pressure head(CASE41-07,09,13,15:T=150,210,330,390days,X=55m)

Figure 5. Predicted lowering of groundwater level in the longitudinal direction by 3-D analysis.
(①after 50days,②270 days,③330days,④450days) .

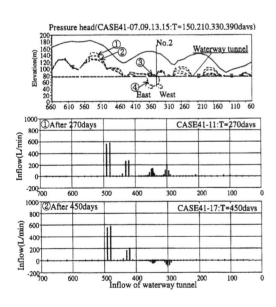

Pressure head(CASE41-07,09,13,15:T=150,210,330,390days)

①After 270days   CASE41-11:T=270days

②After 450days   CASE41-17:T=450days

Inflow of waterway tunnel

Figure 6. The lowering of groundwater levels in the waterway tunnel, Inflow water of waterway tunnel at each excavation.

model to verify the hydraulic constants. Using this analytical model, the groundwater behavior near the crossing concomitant with excavation of the main tunnel was estimated to assess the impact on the waterway tunnel. The result of this estimation was used for assessing the need for correction of the permissible impact values and protective measures for crossing excavation under grade.

3.4.2 Result of numerical analysis at Step 4

Step 4 is a numerical analysis incorporating the field experience up to the section near the crossing under grade. The field data indicated anisotropy of the permeability of the formation including the fault zones. Accordingly, the coefficients of permeability for the fault zones and alternation of sand and slate (s13) were corrected to an isotropic coefficients as given in Table 4.

Figure 4 compares the results of monitoring and numerical analysis for the area of crossing under grade. Comparisons are made between the estimation and actual values of groundwater level and water inflow in East and West tunnels at observation point No. 2 near the crossing. Both the estimated lowering of the groundwater level in borehole No. 2 and the tunnel inflow agree with the measurements.

Figure 5 shows the longitudinal groundwater level distribution as the excavation proceeds. Groundwater under the waterway tunnel is drained to Inariyama Tunnel, causing an unsaturated area between the tunnels.

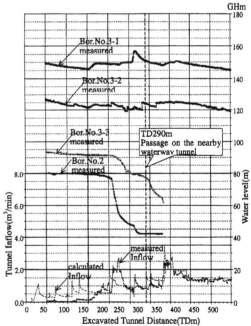

Figure 7. Comparison between calculated and measured by a long time (tunnel total inflow and the groundwater level).

Figure 6 shows groundwater level distribution along the waterway tunnel. The bar graph in the Figure indicates the water inflow into the waterway tunnel. Whereas the inflow is positive in all sections before the crossing of Inariyama Tunnel under grade, it turns negative near the crossing after the tunnel face passes through, suggesting lowering of the groundwater level. The behavior of groundwater drawn to Inariyama Tunnel is thus reproduced.

Figure 7 shows the 3-D water pressure distribution at Step 2. Table 5 tabulates the water inflow in Inariyama Tunnel and drought from the waterway tunnel when the tunnel face passes Fault 5 and Fault 3 near the crossing estimated by Step 4 to Step 7. When passing Fault 5, inflow of 650 to 720 l/min occurred both in East and West lines of Inariyama Tunnel, and the crossing of Inariyama Tunnel under grade was expected to cause groundwater drought of 1.8 m^3/min from the waterway tunnel. At Fault 3, an increase in the inflow and drought of approximately 2.87 m^3/min were expected. Figure 8 shows the changes in the inflow in Inariyama Tunnel and groundwater level at observation points up to Step 7.

3.4.3 *Impact assessment and proposal for execution*
As stated above, the behavior of groundwater near the crossing was precisely grasped, agreeing with monitored data, by reiterating correction of the analysis models and numerical analysis as construction proceeded. Thus the prediction of the impact on the waterway tunnel achieved high reliability. In actual construction, waterproofing by grouting, which had been investigated before excavation, was found unnecessary, judging from the tunnel inflow and the impact ratio as low as 1 to 2%. No particular measures were therefore taken to address the impact on groundwater.

4 SUMMARY

The observational procedure adopted for Inariyama Tunnel included formulation of a large-scale 3-D groundwater prediction model prior to the beginning of construction and repeated feedback of field data to improve the accuracy of the prediction analysis models, while repeating prediction analysis to select streamlined construction methods.

In the case of Inariyama Tunnel, horizontal boring was reproduced in the model based on the properties of fault zones obtained from the boring to verify the hydraulic constants and predict the behavior of groundwater near the crossing as the tunnel face proceeded, thereby assessing the impact on the existing waterway tunnel and investigating the necessity of protective measures. The accuracy of the model was improved by repeated correction to make it conform to the observation as the excavation proceeded. As a result, the groundwater behavior

near the crossing, which conformed to actual findings and monitoring data, was grasped beforehand providing reliable impact assessment results. In actual construction, no particular measures were taken in regard to the impact on groundwater, judging from the weak impact estimated on the waterway tunnel.

5 CONCLUSIONS

It is often difficult to predict the stability and permeability of the ground before constructing underground structures such as tunnels. The role of observational method, in which the behavior of tunnel inflow, groundwater, and surface water is observed and measured with the results being analyzed for feedback to subsequent construction, has therefore become increasingly important. It is particularly necessary to address groundwater and take protective measures quickly and adequately when the tunneling involves critical sections, such as passing other tunnels over or under grade as in the case of Inariyama Tunnel. In such a case, the impact on the neighboring environment or community should be grasped accordingly, and the information should be fed back to subsequent construction. Reliable and rational tunneling will be achieved by such efforts.

The achievement of observational method for groundwater includes precise prediction of the complicated behavior of groundwater with the proceeding of excavation in the area where Inariyama Tunnel crossed a waterway tunnel under grade and verification of the results. It also includes the establishment of permissible impact on the tunnel crossing over grade, proposal of optimum methods, and verification of the final impact assessment based on the actual data. Whereas numerical analysis of groundwater behavior has been widely applied to preliminary assessment, it was incorporated in the excavation processes in the Inariyama Tunnel project, enabling the operators to provide groundwater assessment useful for effective and cost-efficient construction with consideration to the environment. Also, the latest data quickly fed back to the prediction of subsequent conditions led to a control system that provides precise assessment in consideration of construction and environmental aspects. This method is a promising and flexible technique applicable to many other tunnels.

6 REFERENCES

Ohnishi, Y. & Tanaka, M. 1996. Evaluation of the effect of tunnel excavation for surrounding groundwater. *Proc. IGCS. Japan*.31: 2137-2138 (in Japanese).
Ohnishi, Y., Tanaka, M., Yasuda, T. & Takahashi, K. 1997. Evaluation of the effect of tunnel excavation for surround-

ing groundwater. *Proc. IGCS. Japan.* 32: 1997-1998 (in Japanese).

Ohnishi, Y., Ohtsu, H., Yasuda, T., & Takahashi, K., 1998. Assessment of influence in ground water surroundings at urban tunnels. *Proc. ITA.* World Tunnel Congress '98: 489-494.

Ohnishi, Y., Ohtsu, H., Yasuda, T., & Takahashi, K.,1999. Analysis of groundwater behavior around the crossing of adjacent tunnels. *Proc. ITA.* World Tunnel Congress '99: 147-154.

Modern Tunneling Science and Technology, Adachi et al (eds), © 2001 Swets & Zeitlinger, ISBN 90 2651 860 9

Main considerations on UDEC modeling of tunnel excavations and supports

S.G. Chen, H.L. Ong, K.H. Tan & C.E. Tan
ST Architects & Engineers, Singapore

ABSTRACT: Main considerations on the UDEC modeling of excavations and supports are investigated including modeling cycle, model size and joint geometry, and 3D effects. A case study is carried out to demonstrate the modeling methodology of the UDEC modeling. It indicates that the main considerations in UDEC modeling must be taken into account to obtain reliable modeling results.

1 INTRODUCTION

Natural rock mass differs from other geomaterials such as soil and concrete, in which rock joints dominate the deformation and failure of the rock mass. The discrete element code UDEC is specially designed to model jointed rock mass which treats the rock mass as a discontinuous medium containing of rock material and rock joints separately (Cundall 1980). Rock joints may be deformed and failed in both normal and shear directions. The UDEC is very physically suitable to the nature of the rock mass and thus widely used to simulate the jointed rock mass.

However, many difficulties are usually encountered in the practical modeling. First, the UDEC modeling can trace the construction procedure that may have multiple excavations and supports. The modeling at every stage must be converged in numerical to ensure no unbalanced force is transferred into the next stage. Otherwise, inaccurate modeling results may be obtained. Second, even having many rock joints from bore logging data and seismic information, only major rock joints are usually involved in the computational model depending on the model size. Using a small model size can involve more joints and thus hybrid DEM/BEM scheme is a good option because the model can be much small. Finally, the UDEC is a two dimension-based code and three-dimensional effect in simulating rock excavations and supports cannot be simply considered and requires special treatment according to the nature of the tunnels.

This paper is to investigate main considerations in the UDEC modeling of tunnel excavations and supports including modeling cycle, model size and joint geometry, and 3D effects. A case study is then carried out to demonstrate the modeling methodology of the UDEC modeling tunnel excavations and supports.

2 MODELING CYCLE

Similar to other programs, the UDEC modeling can follow the construction procedure such as the generation of in situ stress, excavations and supports of multiple tunnels, as well as other service loads after the completion of the tunnels as shown in Figure 1. Each stage in the modeling is actually a new individual modeling upon the numerical convergence at the previous stage and requires reaching force equilibrium to satisfy the requirement of numerical convergence. The 'save file' can be used to transfer data flow to next stage. Some values can also be modified, e.g., setting zero deformation after the generation of in situ stress.

The in situ stress distribution in jointed rock mass would not be uniform due to the presence of rock joints. Its generation can be performed in two ways. One is to adopt the traditional way which is widely used in other numerical modeling (Plaxis 2000). The equivalent in situ stress is applied to one horizontal side and one vertical side at the far field boundary while the another two sides of the model are fixed. As a special feature in the UDEC, the uniform in situ stress can be directly input to the model before any excavation and a consolidation stage is followed until reaching force equilibrium. The stress remains and will be transferred to next stage. The deformation is reset to zero since no deformation is considered before excavations.

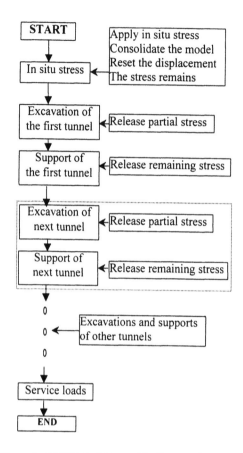

Figure 1. UDEC modeling multiple excavations.

Upon the generation of in situ stress, the excavations and supports of multiple tunnels as well as service loads can be modeled. Care must be taken that each modeling starts only when the previous modeling is completely converged. Otherwise, unbalanced force will be transferred to the next modeling to induce an inaccurate result.

The UDEC is based on dynamic analysis with applying 'static relaxing damping' to reach force equilibrium (numerical convergence) to solve static problems. Two methods can be used in the UDEC to check whether numerical convergence is reached. One is to print the unbalanced force-time. Only when the curve tends to be horizontal as well as the unbalanced force becomes very small, the modeling can be considered completely converged. The another one is to set several historical measurement points in the model to monitor the values changing with time. When the historical curves turns to be horizontal, the modeling can be considered completely converged.

3 MODEL SIZE AND JOINT GEOMETRY

The computational model would reflect the joint geometry but must be not so large that the modeling becomes impossible.

3.1 Model size and boundary condition

Tunnels are always located in an infinite or semi-infinite region while the computational model has a limited size with fixed far field condition. In continuity-based numerical modeling, it is common to take the model size at least five times the tunnel dimension to satisfy the modeling accuracy. For UDEC modeling, the model size would be even bigger due to the presence of rock joints. However, an bigger model implies more rock joints being introduced and longer running time is needed. In fact, sine the joint element (contacts) occupies more computer memory and thus the modeling with a bigger model becomes very difficult and even impossible.

A usually adopted solution is to reduce the amount of the rock joints in the model by simplifying the joint geometry based on an equivalent concept. E.g., it is often used to enlarge the joint spacing, say one joint instead of ten joints, to represent actual joint geometry by inputting equivalent joint parameters. Another way is to keep the actual joint geometry at closer area and simplify the joint geometry at farther distance. It sounds reasonable as joints at farther distance have less contribution but it makes the determination of joint parameters very difficult. This is because that the equivalent rock mass has the contributions not only from rock material but also from rock joints. The better way might be to employ the hybrid DEM/BEM scheme proposed by Brady et al. (1984 & 1987) that has been built in the UDEC. By employing the hybrid DEM/BEM scheme, the model can be much smaller to be able to involve more rock joints.

3.2 Hybrid DEM/BEM scheme

The hybrid DEM/BEM scheme is specially designed to model jointed rock mass in an infinite or semi-infinite region which has been built in the UDEC. The hybrid DEM/BEM adopts the manner similar to the FEM-based approach of the hybrid FEM/BEM scheme (Manolis and Beskos 1987). In the hybrid DEM/BEM scheme, the near area (DE region) is modeled by the DEM and farther zone (BE region) is modeled by the BEM as shown in Figure 2. The rock joints are introduced only in the DE region and the rock mass in BE region is treated as an equivalent continuous elastic medium.

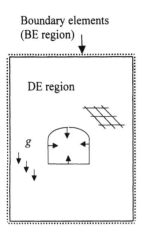

Figure 2. Hybrid DEM/BEM scheme model.

The scheme manipulates the direct boundary constraint equation to yield a stiffness matrix to find out the reactions at the interface of DE and BE regions. The reactions are then conversely applied to the interface nodes, which are emerged to the total force for each block in contact with the interface. The achieved interface nodal displacements and forces at the interface are then used to determine stresses and displacements at interior points in the DE domain. The capability of the hybrid DEM/BEM scheme was verified in modeling tunnel excavation in an infinite rock mass (Chen *et al.* 2001).

3.3 Joint geometry

Rock joints dominate the tunnel stability and should be reflected properly in the computational model. As stated by Hart (1993), major joints possibly affecting tunnel stability should be introduced in the computational model. As the rock joints at the crown have more risk to fail, extra care must be taken in introducing the joints at the crown. Natural rock masses are usually multiply jointed with several joint sets and some random joints. The introduction of the rock joints in the model would include three steps. First, the joint distribution can be traced from borehole logging data and joint mapping at current and adjacent constructions. Site seismic investigation may provide additional information for identifying weak zones and faults. Second, a statistical analysis on the joint strikes and dip angles with deviations would be conducted to classify the joint sets. Finally, a filtering should be carried out to identify the major joints most susceptible to rock tunnel stability. The joint amount involved depends much on the size of the computational model.

Figure 3. A practical joint geometry.

Many researchers prefer to use the commands JSET or VORONOI in the UDEC to get regular joints due to lack of geological information and the operation being much easier. However, using these commands hardly reflects the actual geological condition of joint geometry. Site mapping indicates that rock joints are usually neither regular nor continuous and some random joints are always observed. Although the deviation of joints in strike and dip angle can also be taken into account, using such commands often produces some blocks with very small size that makes the modeling uneconomical and even crashing. Therefore, in practice, manual generation of joint geometry by experienced geologist is strongly recommended to produce a reasonable and economical computational model. Figure 3 illustrates a practical computational model.

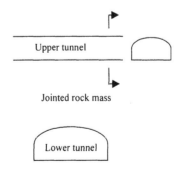

Figure 4. A tunnel locates above the another.

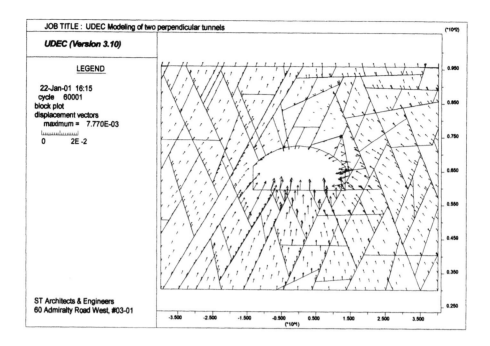

Figure 5. Displacement distribution in rock mass around the two tunnels.

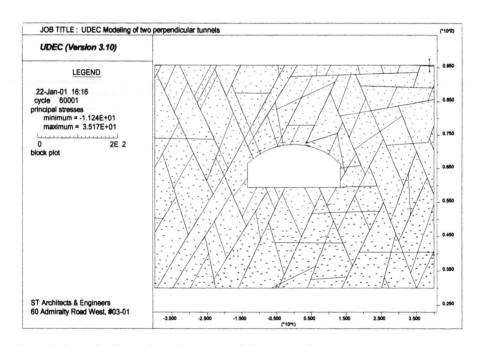

Figure 6. Stress distribution in rock mass around the two tunnels.

Care should be taken in some artificial lines including excavation outline, material interface and some extra lines for easier mesh generation. These lines are not actual rock joints but can only be treated as joint elements in the UDEC model. In practice, a joint type so called glue joint having very high stiffness and no failure is applied to present these artificial lines.

As stated previously, the blocks are subdivided into triangular finite difference elements to present the deformability of rock material. Practice found that the mesh size of the finite difference elements does not affect the modeling accuracy apparently comparing to continuity-based numerical methods as more deformation occurs at rock joints rather than rock material. This may be the reason that some discrete element methods treat blocks as rigid with no deformation (Cundall & Hart 1993; Chen 1999).

4 3-D EFFECT

The UDEC is a two-dimensional program and no 3-D space can be physically involved. In practice, the rock supports (rockbolt and shotcrete) are usually applied after the excavation of the section and some deformation has occurred before the installation of the rock supports. It was found that 25~75% of full deformation (stress release) would have been released depending on the distance of the support section away from the working face (Chen et al. 2001).

The 3-D effect for rock support can be achieved manually in the UDEC modeling by adjusting the stress release that can be executed with programming of the built in FISH language. The concept follows the construction procedure that partial stress is released first and the remaining stress is then released after the rock support is installed. Taking into account the possible creep deformation of hard rock, 50% and 75% of stress is often assumed to release before the installation of rock support for shotcrete and rockbolts, respectively.

On the other hand, in case of multiple tunnels in which some may be perpendicular to the others and could not be ignored as shown in Figure 4. The direct introduction of them in the model obviously does not reflect the actual case because such a model implies a very long tunnel in perpendicular direction with such a section (Brady & Brown 1993). An approximation can be made by applying velocity boundary condition to the excavation outline of such tunnels which includes two steps (Itasca 2000b). The first step is to find out the displacement by carrying out an individual modeling in which only perpendicular tunnels with their supports are introduced in the computational model. The displacement will be converted to velocity by

ensuring the final displacement be consistent. The second step is to involve the perpendicular tunnels in the computational model and apply the velocities to the excavation outline to simulate the effect of the excavating and supporting the perpendicular tunnels.

5 CASE STUDY

The project case has two perpendicular tunnels, one is perpendicularly above the another, located in heavily jointed rock mass as shown in Figure 4. The upper tunnel is 7 m high and 13 m wide. The lower tunnel is 12 m high and 25 m wide. The separation distance between the two tunnels is 20 m.

The construction procedure is to excavate and support the upper tunnel first. The lower tunnel is then excavated and supported. The stability of the separation distance becomes the main concern in the support design. The rock support is designed based on Q-system (Grimstad & Barton 1993). The upper tunnel is supported by rock bolts with 3.0 m in length and 2.2 m in spacing and shotcrete of 45 mm thick. The lower tunnel is supported by rock bolts of 5.0 m in length and 2.2 m in spacing and shotcrete of 75 mm thick.

The UDEC modeling is used to investigate the stability of the separation distance by employing the hybrid DEM/BEM scheme. The joint geometry is manually identified based on the borehole logging data and geological mapping during the construction nearby as shown in Figure 3. The rock material is assumed as an elastic medium and its properties are obtained from lab tests including uniaxial compression test and Brazil tension tests. The rock joints are assumed to obey the Barton-Bandis model and their properties are determined from site tests including joint roughness profile and Schmidt rebound tests.

The UDEC modeling involves two steps as stated earlier. First, only the excavation and support of the upper tunnel is modeled by carrying out an individual UDEC modeling to find out the displacements at the crown, bottom and on the side walls of the tunnel. The displacements are then converted to velocities based on the relation between displacement and velocity. The displacement is 2.1 mm at the bottom of the upper tunnel. As the running time is 3 second, the velocity is converted as 0.7 mm/sec. The excavation and support of the lower tunnel is then modeled by another UDEC modeling in which the upper tunnel is removed together with the rock mass above it. The velocity of 0.7 mm/sec is applied to the bottom of the upper tunnel.

For the 3-D effect on rock supports, it is assumed that 50% stress is released before applying rock bolts and shotcrete. The remaining 50% stress is then released afterwards.

Figures 5 & 6 illustrate the final displacement and stress distribution, respectively. It is shown that the displacement at the bottom of the upper is 2.1 mm which indicates that the velocity is correctly applied. The maximum displacement and stress at the separation are 3.0 mm and 11.2 MPa, respectively. The modeling results suggest that the separation between the two tunnels with the designed supports is stable.

6 REMARKS

The main considerations in UDEC modeling of tunnel excavations and supports are investigated including the model cycle, model size and joint geometry, and three-dimensional effects. A case study is then carried out to demonstrate the UDEC modeling methodology by taking into account of the main considerations. It can be remarked that the main considerations in UDEC modeling are very important and must be taken into account to obtain reliable modeling results.

REFERENCES

Brady B.H.G., Coulthard M.A. and Lemos J.V. (1984) A hybrid distinct element-boundary element method for semi-infinite and infinite body problems. *Proc. Computer Techniques and Applications Conference*, North-Holland Publishers, pp. 307-316.

Brady B.H.G. (1987) Boundary element and linked methods for underground excavation design. *Analytical and Computational Methods in Rock Mechanics* (edited by Brown, E.T.), Chapter 5.

Brady B.H.G. and Brown E.T. (1993) *Rock mechanics for underground mining*. 2nd edtion, Chapman & Hall, London.

Chen S.G. (1999) Discrete element modeling of jointed rock mass under dynamic loading. *PhD thesis*, Nanyang Technological University.

Chen S.G., Ong H.L., Tan K.H., Tan C.E. and Zhao J. (2001) UDEC modeling of rock tunnel excavations and supports. *2001 ISRM Symposium – 2nd Asian Rock Mechanics Symposium*, Beijing, October.

Cundall P.A. (1980) UDEC - A generalised distinct element program for modelling jointed rock. *Report PCAR-1-80*, Peter Cundall Associates, U.S. Army, European Research Office, London, Contract DAJA37-79-C-0548.

Cundall P.A. and Hart R.D. (1993) Numerical Modelling of Discontinua. *Comprehensive Rock Engineering* (Edited by Hudson, J.A.), Vol. 2, pp. 231-243.

Grimstad E. and Barton N. (1993) Updating of the Q-System for NMT. *Proceedings of the International Symposium on Sprayed Concrete – Modern Use of Wet Mix Sprayed Concrete for Underground support*, Norwegian Concrete Association, Oslo.

Hart R.D. (1993) An introduction to distinct element modelling for rock engineering. *Comprehensive Rock Engineering* (Edited by Hudson, J.A.), Vol. 2, pp. 245-261.

Itasca Consulting (2000a) *Personal communication on rock face effect in tunnel construction*, February.

Itasca Consulting (2000b) *Personal communication on displacement boundary conditions*, November.

Manolis G.D. and Beskos D.E. (1987) Boundary element methods in elastiodynamics, Unwin Hyman, pp. 282.

Plaxis consultant (2000) *PLAXIS user manual*, Ver. 7.1.

Modern Tunneling Science and Technology, Adachi et al (eds), © 2001 Swets & Zeitlinger, ISBN 90 2651 860 9

Analysis on Tunnel Lining Deformation and Effect of Countermeasures for Earth Pressure

T. Asakura
Dept. of Earth Resources Engineering Eng., Kyoto University, Kyoto, Japan

Y. Kojima & K. Yashiro
Railway Technical Research Institute, Tokyo, Japan

H. Shiroma
Japan Highway Public Corporation, Research Institute, Tokyo, Japan

K. Wakana
SHO-BOND Corporation, Tokyo, Japan

ABSTRACT: Filed measurements, model tests and numerical analyses are underway to establish a standard to evaluate soundness of deformed tunnel lining. The present paper reports on the results of (1) case studies of tunnels deformed by squeezing earth pressure; (2) 1/30 scale model lining tests; (3) crack propagation analysis, and (4) development of design method for countermeasures against deformed tunnels. Notably it refers to the effects of countermeasures against earth pressure. To sum up the contents; 1) Back-fill grouting can vastly improve the strength of defective tunnel lining with opening behind lining; 2) rock-bolting or inner reinforcement exhibit an ample effect of reinforcement, provided it is preliminarily treated with back-fill grouting; and 3) invert concrete excels in the effect of displacement suppression, but for the purpose of enhancing the structural strength, combination with other lining countermeasure is advocated.

1 INTRODUCTION

It is said that tunnels are durable when compared with any other structures. However, railway tunnels in Japan, mostly aged and constructed through mountains whose geographical and geological conditions are complex, are often deformed due to earth pressure and other causes after put into use, to require urgent countermeasures. Moreover, at the tunnels constructed by the basic method (wooden-pillar support method) before NATM was introduced, structural faults caused at their execution accelerate the deformation of tunnels.

Up to now, it has been thought to be difficult to design countermeasures against such deformed tunnels analytically. Then, an empirical design method, where similar examples are referred to, has been mainly applied. Accordingly, the authors have carried out a number of field measurements, experiments and analyses, and have proposed a new design method where both the standard design and an analytical method are applied.

In the present paper, the authors deal with the squeezing earth pressure, by referring to (1) case studies performed until the design method was suggested, (2) model lining tests and (3) knowledge concerned with the effect of countermeasures acquired by crack propagation analysis. Moreover, an outline of the new design method for countermeasures against deformed tunnels is shown.

2 CASE STUDIES OF DEFORMED TUNNELS

Here the discussions are focussed on deformation by squeezing earth pressure. In the following, two features are taken up to investigate the lining structure, deformation behavior and effects of countermeasures.

2.1 Case I (Tsukayama Tunnel)

(1) Outline of the tunnel: The tunnel is a double-track railway tunnels, as old as 30 years (opened to service in 1967), length of 1766m. It is located on the Japan Sea side where the sedimentary soft rocks belonging to "Green Tuff region" is widely distributed, specifically situated at the anticline wing partially constituting an active folding. Overburden of the deformed area is about 70m and the surrounding rock is a mud rock liable slaking (uniaxial compression strength 3~6 MPa, natural water content 25~40%, competence factor 2~4). The lining thickness is 50cm, and with no invert provided.

(2) Behavior and countermeasures: Soon after opened to service, the roadbed heaved and the side wall was squeezed, resulting in a bending compressive failure of the arch crown.

To counter this deformation, the sectional opening was closed with invert concrete in 1970. In consequence, as illustrated in Figure 1, which shows the result of convergence measurement, the maximum convergence rate 36mm/year before provision of invert concrete almost vanished after that.

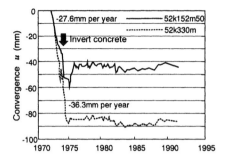

Figure 1. Convergence rate before/after invert concrete.

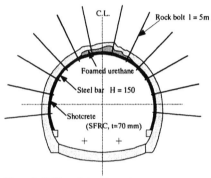

Figure 2. Outline of countermeasure.

In 1990, however, with lapse of 20 years, a relatively wide area of the crown which suffered the compressive failure dangled with a lot of shear cracks radiating from the vertical axis. With no time lost, an emergency step was taken to avert a threat-ening collapse of the crown. This was followed by additional countermeasures, as shown in Figure 2, including back-filling, rock bolting and inner lining (steel fiber reinforced concrete).

(3) Deformation mechanism: Figure 3 illustrates a series of deformation mechanisms which occured. First, a squeezing earth pressure was generated within the surrounding rock mass which had fallen into a secondary stress state when the tunnel was excavated. With lapse of time, the side wall bulged and as a result a strong negative bending moment developed in the crown until a compressive failure occured. When an invert concrete was provided, however, the lining deformation ceased to progress and an axial force built up in the lining steadily, which promoted the compressive failure of the crown. On the contrary, the loosening area of the rock mass gradually expanded with a resultant growth of the loosening vertical earth pressure acting on the arch. At the same time, radiating fissures originated from the failed region of the crown in the arch and ultimately the crown came to droop.

2.2 Relationship between convergence and effect of countermeasures

Figure 4 compares the effects of countermeasures against the deformation by squeezing earth pressure in terms of convergence velocity (annual reduction of inner lateral distance between side walls (mm/year)). Figure 4 shows that countermeasures have been chosen according to the convergence velocity and verifies that (1) invert concrete has a largest effect of suppressing the deformation and that

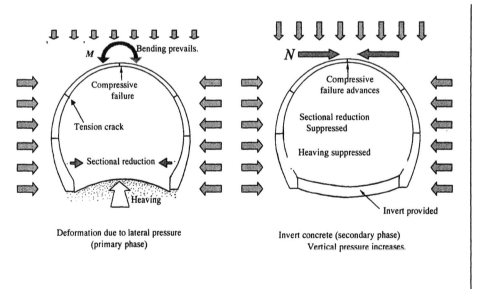

Figure 3. Deformation mechanism.

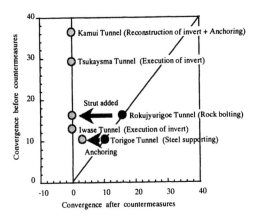

Figure 4. Comparison convergence before and after measure.

(2) rock bolting and inner lining is sufficiently effective when the deformation is small.

When the sectional opening is closed with invert concrete, however, an excessive axial force can build up in the lining, which promotes the compressive failure of the arch. Therefore, it is important to pay attention to the whole structural strength of lining.

2.3 Comments

From these case studies, the following things have been revealed about the behavior of tunnel linings deformed by squeezing earth pressure and the countermeasures to be taken:

①The lining structure and its defects (notably, presence of invert, configuration of side wall, lining thickness of the arch or opening behind the lining) have significant influence on the behavior of deformed lining. Therefore the first thing to be done as a countermeasure should be elimination of these defects or reinforcement to compensate them.

②When the deformation is nothing serious, back-fill grouting or rock bolting will be sufficiently effective. Countermeasures should be designed to match the of deformation behavior.

③Invert concrete has a large effect of suppressing the deformation. When the deformation is something serious, it will be important to execute additional reinforcement of the lining.

3 MODEL EXPERIMENT ON TUNNEL LINING

In this chapter the authors intend to discuss on the results of 1/30 scale model tests device they have developed, specifically referring to ①influence of structural defects, and ②effect of back-filling.

3.1 Test unit

As shown in Figures 5, a test unit for a 1/30 scale model of the Shinkansen standard tunnels (equivalent to 2-lane highway tunnels) capable of direct loading (Asakura et al., 1992, 1995) was prepared. The test unit consists mainly of loading/reaction members (consisting of a loading/reaction plate, a cylindrical spring made of hard rubber and double threaded screw bolts), a reaction frame, a bed plate, etc. Within the cross section, a total of 11 sets of the loading/reaction members were set in 11 rows along the tunnel axis to facilitate three-dimensional experiments. At loading points, a steel cylinder was set to directly cause displacement of the lining model. At all points except for the loading points, the hard rubber cylindrical spring was set to induce subgrade reaction.

3.2 Experimental procedure

Table 1 shows the materials and their properties used for the experiments. As for similarity relations, only geometric similarity (scale:α) was taken into consideration, assuming the same degree of strength for all materials. Consequently, before cracks occur, deformation and displacement correspond by $1/\alpha$ and stress corresponds by 1/1 under the same loading pressure. It is considered that, since the lining behavior will be affected by the crack behavior, the similarity will vary after cracks occur.

The experimental procedure is as follows. (1)The lining model on which strain gauges had been set were installed in the test unit together with load cells and displacement meters. (2)Step loading was carried out by displacement control. (3)The experiment was terminated either by (a)ultimate failure of the mode, (b)the stroke limitation of the bolts for load-

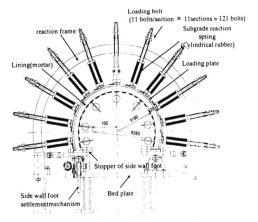

Figure 5. Outline of tunnel lining test unit.

Table 1. Experimental material and their properties.

Experimental material		Property
Lining	Mortar	$E = 1.5 \times 10^4$MPa $\sigma c = 30$MPa
Subgrade reaction Spring	Hard rubber	Spring constant 78N/mm
Foot reaction spring	Steel plate	It's a condition that there is invert.
Back-fill grouting	Rubber plate	$E = 3.0$MPa Thickness 0.15,0.30mm
Inner reinforcement	Phosphor bronze plate	$E = 1.2 \times 105$MPa
	Carbon fiber sheet	Fiber area weight 20g/m² $E = 2.4 \times 10^5$MPa

E: Young's modulus, σc: Unconfined compressive strength

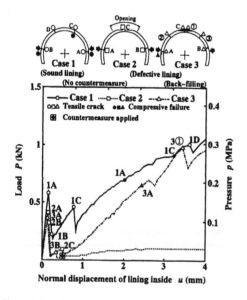

Figure 6. *P- u* curve (case 1 to 3).

ing, or (c)reaching the maximum loading level of the loading unit specified by the design.

3.3 Experimental results

(1)Effect of lining deficiency: Figure 6 shows the relationship between load and normal displacement of the lining at the loading point derived from the following two cases where lateral load was applied from both sides. Case 1 shows a sound lining without insufficient lining thickness and openings behind the lining. Case 2 represents a defective lining with openings behind the lining and insufficient lining thickness (1/2 of the regular design thickness) at its crown. As can be seen from the Figure, the sound lining model maintains its durability supported by arch action after cracks occurred. Whilst the

defective lining demonstrates brittle failure due to initial cracking.

(2)Effect of back-filling: Figure 6 also shows the effect of back-filling from a case of model experiment where back-filling was applied with soft rubber after cracking had been induced at both side walls and crown by lateral loading from both sides. As is obvious from the Figure, after the application of back-filling, the load increases as the progress of displacement. From this it can be concluded that the lining durability can be restored by the application of back-filling and that back-filling is an effective countermeasure.

4 NUMERICAL ANALYSES

In this chapter the authors intend to give an outline of the frame analysis they have developed to take account of the cracking behavior. Further they intend to report on the results of employing their technology in studying the lining strength and comparing the effects of practical countermeasures for deformed linings.

4.1 Crack propagation analysis method

In the crack propagation analysis, a structural model (a frame analysis model taking the ground as the spring, the lining as the beam and the crack as the plastic hinge) is formulated for each progress stage of cracking for the sake of calculation and the results of calculation are added up to express the total crack propagation. Thereby the lining element at each stage is assumed as a liner elastic body.

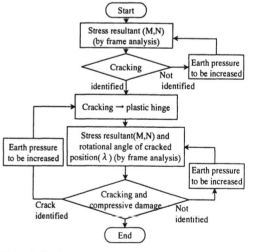

Figure 7. Crack propagation analysis procedure.

As illustrated in Figure 7, in the crack propagation analysis, the crack generation is evaluated form the stress resultant; the structural model is formulated for each stage of crack generation (analysis step); and then the calculation is repeated until the lining element hits the limit of the compressive strain. The crack is formulated by pin-connection to calculate it on the safe side. As for the propagation of cracking, the calculation is continued with the pin-connected model and the final result is expressed by piling up the stress resultant, the displacement, etc. in each stage.

4.2 Comparative analysis

(1) Modeling of tunnel and counter measures: An imaginary lining model of the standardized Shinkansen section was set. And opening behind the arch, soft rock as the ground condition and squeezing earth pressure (horizontally distributed load) were assumed. Figure 8 illustrates the concept of a tunnel lining analysis model. Meanwhile, Table 2 lists up the physical properties of lining and ground.

Five countermeasures which have been often applied for deformed tunnels are selected; back-fill grouting, rock bolting, inner lining, inner reinforcement(carbon fiber sheet) and invert concrete.

Table 2. Physical properties of lining and ground.

Lining	Thickness	70cm
	Young's modulus	2.1×10^4 MPa
	Unit volume	23.5 KN/m³
	Structural defect	Opening behind lining (arch 60°)
Ground	Young's modulus	5.0×10^2 MPa
	Earth pressure	Horizontal prevalent pressure

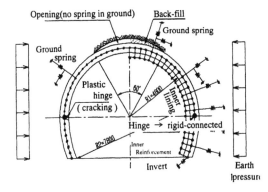

Figure 8. Concept of analysis model.

(2) Results of analysis: Figure 9 shows the relation between displacement and earth pressure by the type of countermeasures. From Figure 9, it is understood that the best effect of improving the structural strength belongs to back-fill grouting. Any other

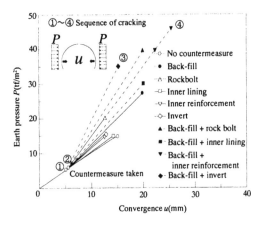

Figure 9. Convergence vs. earth pressure.

countermeasure alone cannot be expected to produce good effect. Better effect will come only when it is coupled with back-fill grouting. It is notable that combination of back-fill grouting plus inner reinforcement or rock bolting will bring about a great increase of structural strength. As for the rigidity improvement effect, the invert concrete, which increases the strength in the loaded direction, excels.

5 DESIGN METHOD OF COUNTERMEASURES AGAINST DEFORMED TUNNELS

The authors proposed a design method for countermeasures against deformed tunnels, on the basis of the results of case studies, model tests and numerical analyses mentioned above. Then, the "Design Manual of Countermeasures for Deformed Tunnels" (Railway Technical Research Institute, 1998.2) was adopted, established based on these results.

5.1 Designing system for countermeasures against deformed tunnels

A designing system to propose the application of the standard design and analytical procedures, is implemented in the design manual, including the traditional methods (application of past examples in similar conditions, and special cases where the application of the above procedures is disqualified). This system is shown in Figure 10.

The situation of deformation is classified into four reinforcing ranks (I–IV) first as shown in Figure 10. The rank IV is the most serious reinforcing rank among the four ranks. The reinforcing ranks are individually shown for each of three external causes, (1) squeezing earth pressure, (2) asymmetrical earth pressure and (3) loosening vertical earth pressure. Table 3 shows the reinforcing ranks for the deformation caused by the squeezing earth pressure. For the deformation caused by squeezing earth pressure, an appropriate rank is determined by the convergence velocity by using Table 3, which is corrected according to the response to structural faults, structure of lining and size of earth pressure, to finally determine the reinforcing rank.

5.2 Application of standard design

To apply the standard design, a reinforcing pattern (combination of reinforcing methods) is decided depending on the reinforcing rank (I–III). Then, the standard design is applied with reinforcing methods narrowed down to suit the field situation. Figure 11 shows an example of standard design for countermeasures against squeezing earth pressure.

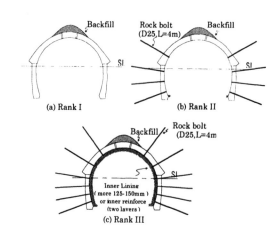

Figure 11. Examples of Standard design.

Table 3. Reinforcing rank (squeezing earth pressure).

Reinforcing rank	I	II	III	IV
Progressive rate ( ): Convergence velocity [mm/year] (by convergence measurement)	Exist (- 3)	Rather large (3 – 10)	Large (10 -)	Notably large

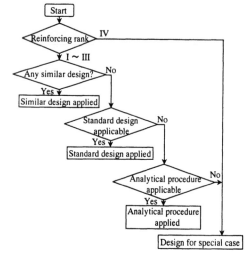

Figure 10. Design scheme for planning countermeasures for deformed tunnels.

6 CONCLUSIONS

Case studies, model testing and numerical analyses of deformed tunnel lining came up with such ample information as follows;

(1) The tunnel lining strength with structural defects can be vastly increased by back-fill grouting.
(2) Given mandatory execution of back-fill grouting, the reinforcing effect of rock bolting or inner reinforcement will be given full justice.
(3) The rigidity enhancing effect of invert concrete is admittedly large, but in order to increase the structural strength, execution of other measures such as rock bolting, inner reinforcement or inner lining will be necessary.

In addition, the authors have developed a design method for countermeasures against deformed tunnels, based on the above knowledge and published a "Design Manual of Countermeasures for Deformed tunnels."

REFERENCES

Asakura, T., et al., 1992. Analysis on the behavior of tunnel lining –Experiments on double track tunnel lining- : QR of RTRI, Vol.33, No.4

Asakura, T., et al., 1995. Countermeasure for deformed tunnel lining by tunnel reinforcement: 8th International Congress on Rock Mechanics, ISRM

Kojima Y., Asakura T., et al., 1998. Design Method of Countermeasures for Deformed Tunnel: QR of RTRI, Vol.39, No.1

RTRI, 1998. Design Mauual of Countermeasures for Deformed Tunnel (in Japanese)

Modern Tunneling Science and Technology, Adachi et al (eds), © 2001 Swets & Zeitlinger, ISBN 90 2651 860 9

Theory and practice of tunnel lining design

N.S.Bulychev & N.N.Fotieva
Tula State University, Tula, Russia

ABSTRACT: The objective of the paper presented is to pay attention of professors and students, engineers and designers to progressive modern theory and new effective methods of tunnel linings investigation and design. The basic principle in modern theory of underground structures is that of joint contact interaction of the rock mass and the lining taking into account the lining and the rock mass as elements of a common deformable "rock-support" system. All results of the theory are in a good agreement with experimental investigation including full-scale measurements moreover that agreement appears in not particular but in overall. The theory of underground structures is not only theory but also dozens of important underground structures that have been constructed using the methods based on the theory and exploited during 20 and more years.

1 INTRODUCTION

The advancement of sciences may be imagined as an orderly progression of forms from accumulation of scientific facts and their empirical generalisation to the creation of a theory as the top of development of any domains of science. That way has been gone by the science of the tunnel lining (generally, underground structure) design. In this respect one can gain greater insight into the progress of development of that theory called the Underground Structures Mechanics.

2 BRIEF HISTIRICAL REVIEW

The first attack the problem of the tunnel lining design was application of the ordinary design methods considering underground structure as ordinary one. The design method was discussed as consists of two steps: design (calculation, as such) and loads determination. It was proposed that the loads are conditioned only rock mass properties and not depend from the lining characteristics.

The two most widespread until quite recently design schemes of the tunnel linings are represented in Figure 1. The tunnel lining in Figure 1a is simulated by three times statically undetermined frame. The tunnel lining scheme shown in Figure 1b is a closed frame of the bedded ring beam somewhat more times statically undetermined according to the

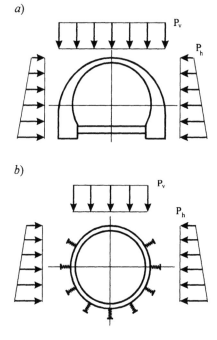

Figure 1. Schemes of the tunnel lining design:
a - upon fixed loads - support pressure;
b – upon fixed "active loads" taking into account a "passive loads" due to the reactions at the elastic supports.

number of the discrete elastic supports simulating elasticity (Winkler) base (rock mass). However the problem of design even in that case is not difficult.

It should be pointed out that the design scheme presented in Figure 1b (model of bedded ring beam) and the corresponding method of tunnel lining design was developed by Russian engineers-designers when the Moscow Underground construction was started (Bodrov & Matery 1936).

The problem was only determination of magnitude and distribution of the load. Thus the rock pressure problem came into existence. It was proposed that the rock pressure was conditioned only by rock mass properties being object of the Rock Mechanics.

A grate body of the rock press hypothesis was offered. The extensive literature on this problem is reviewed in works by Bulychev et al. (1974), Arioglu (1995), etc. One of the first widely spread in Russia notion about rock pressure mechanism by Protodyakonov is represented in Figure 2 (Protodyakonov 1930). The rock pressure value depends only on rock properties and sizes of a zone generating it.

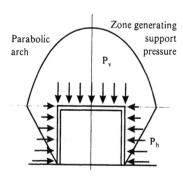

Figure 2. Support pressure due to the Protidyakonov's theory.

The formulae put forward by Terzaghi (Terzaghi 1943) based on the similar mechanism have been used extensively, particularly in North America.

At the present time the traditional calculation methods of both fixed and active loads (Fig. 1) as well as a century-long efforts to understand the rock pressure independent of the tunnel lining characteristics are not correct in general. That is proved today. In spite of that those methods are discussed and recommended for use at present time (IBSM 1995, ERTC 9 1997, WG2 ITA 2000).

Notions about tunnel lining function were sufficiently changed by R.Fenner (1938), A.Labasse (1949) and at last K.Ruppeneyt (1954). They have given us insight into the mechanism of rock mass and lining (support) interaction. The curve in Figure 3 illustrate response of lining to tunnel walls dis-

placements resulting in establishment of equilibrium. Owing to them a great step has been made towards the theory.

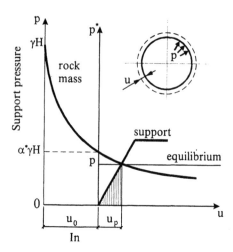

Figure 3. Scheme of interaction between rock mass and tunnel lining (u_0 is initial displacement before the lining is installed, u_p is joint displacement of rock mass and lining as result of their interaction).

Thanks to that investigations it has been found the principal peculiarity of underground structure, namely, the interaction between a structure and surrounding rock mass. It has been established that rock pressure or, more precisely, contact stresses on the interface between rock mass and support, is not fixed but depends from both technology (u_0) and support stiffness (u_p). Due to that peculiarity underground structures methods have been recognised as not having analogies among other engineering structures in respect to the design.

It is well known that generally recognised the New Austrian Tunnelling Method is based on the philosophy following relations represented at the Figure 3 (Muller 1990). The concept of the NATM is to construct a tunnel on the basis of scientifically established principles and ideas which have been proved in practice, the main idea being that by mobilising the bearing capacity of the rock mass, optimum safety and economy can be achieved. The load on supporting elements is reduced by permitting a controlled deformation of the rock surrounding the excavation.

It needs to be noticed that the picture (Fig. 3) was published by B.Matveev (Russia) and F.Mohr (Germany) simultaneously (1952) and independently of one another.

3 FUNDAMENTALS

The basic principle in modern theory of underground structures is that of joint contact interaction of the rock mass and the lining. In accordance with that principle the lining and the rock mass are elements of a common deformable "rock-support" system subjected to external loads and actions (Bulychev 1989, 1994a, Bulychev, Fotieva & Streltsov 1986).

There are also two postulates in the base of the mechanics of underground structures. The first of them is that the linear deformable medium (the elastic medium) can be taken as the rock mass model. It comes from the following. It is obviously that the u_0 initial displacements of an underground opening surface before the lining is installed must be excluded from the design scheme of the lining. According to one-dimensional analysis it means that the vertical axis p is to be moved to the new position p^* as shown in Figure 3.

The whole process of interaction between rock mass and lining takes place along the interval u_p of its joint displacements from the lining erection up to the equilibrium establishment. As the value of displacement u_p depending on the deformability of the lining is small the relation between stress and strain in the rock mass at that interval may be considered as linear (Bulychev 1994b).

This postulate is very important. The linear deformable model allows to divide the stress state of the rock mass surrounding an underground opening into the initial and the additional ones. The additional (removed) stress state is produced by extraction of rock during the formation of the opening and is the cause of rock deformation and rock pressure. Because of the first postulate the mathematical elasticity theory methods developed in the classical works by N.Muskhelishvily, D.Sherman, S.Lekhnitsky and others may be used to solve the problems of underground structures design and the mathematical models of underground structures stress-strain state based on rigorous analytical solutions of the problems can be developed. The analytical methods allows to obtain both direct and inverse solutions (back analysis) and optimal solution too.

All statements and results of the theory are being obtained mathematically from those positions due to analytical solutions of the problems. So that the mathematical apparatus is a means to deduction of all the recent knowledge and further development of the theory.

The second postulate is that excluding of the initial displacements is taken into account by introducing the α^* decreasing coefficient ($\alpha^* \leq 1$) to the components of the initial stress state (Fig. 3). As it follows from Figure 4 the initial displacement u_0 is connected with the influence of the tunnel face on the tunnel wall displacements.

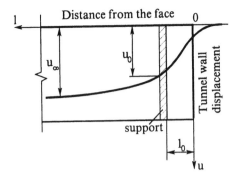

Figure 4. Tunnel wall displacement near the tunnel face.

The curve of displacements shows the lining loading during the face moves further and further. Excluding the initial displacements of the tunnel walls and mechanical meaning of the α^* coefficient may be presented as following

$$u_\infty - u_0 = u_\infty(1 - u_0 / u_\infty) = \alpha^* u_\infty \qquad (1)$$

The α^* coefficient depends from the l_0 distance between the lining installation place and the tunnel face. Analysis of the u_0 initial displacements the forms of the α^* coefficient are related to the Rock Mechanics and performed taking into account physically nonlinear behaviour of the rock mass. There are several theoretical and empirical formulae for that coefficient. For instance the empirical formula obtained from the data of numerical modelling is given (Bulychev 1994a):

$$\alpha^* = 0.64 \exp[-1.75(l_0/r)] \qquad (2)$$

In accordance with above principles and postulates the theory and methods of underground structures design are extended to all types of linings and supports and all kinds of loads and actions.

4 BRIEF RVIEW OF DESIGN SCHEMES

According to the basic principle in modern theory of underground structures of joint contact interaction of the rock mass and the lining the lining and the rock mass are elements of a common deformable "rock-support" system. It should be emphasised once more that according to the underground structures mechanics it is not necessary the support pressure to be prescribed as input data for the lining design. Normal and shear contact stresses on the rock-lining interface are determined as intermediate result during the process of a contact interaction problem solving.

447

Modern design scheme of underground structure is schematically shown at Figure 5. Thanks to that design scheme the unity of approaches of all kind of scientific research methods (analytical, numerical, all kind of experimental modelling, full-scale measurements etc.) has been reached.

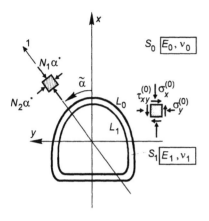

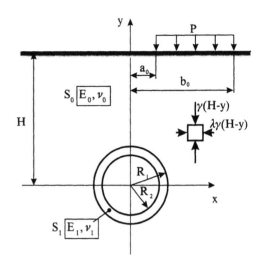

Figure 6. Design scheme of shallow tunnel lining.

Figure 5. Design scheme of an arbitrary cross-section tunnel lining interacting with the rock mass subjected to a tectonic intact stress state.

Note also that solution of elastic contact problems is much more difficult mathematically than those shown in Figures 1a, b. The solution of elastic contact problem design scheme was developed by N.N. Fotieva only in 1972 (Fotieva 1974).

The mathematical modelling is based on analytical solution of the elasticity theory plane contact problem (Fig. 5) for a ring S_1 of an arbitrary shape (with a single axis of symmetry) simulating the tunnel lining supporting an opening in an infinite linearly deformable medium simulating the rock mass. The mechanical properties of the S_0 infinite medium are characterised by the E_0 deformation modulus and the v_0 Poisson's ratio, the S_1 ring material is characterised by the E_1, v_1 deformation characteristics. The $\alpha^* N_1$ and $\alpha^* N_2$ principal initial stresses in the S_0 area simulate the action of tectonic forces.

In the case of shallow tunnel the lining is subjected to not only initial stress state of ground mass action but also surface loads P from buildings and vehicles as shown in Figure 6. By the way the loads from buildings may be applied both before and after the tunnel construction. Note that the undersea and under-river tunnels may be considered as particular cases of shallow tunnels (Fotieva & Bulychev 1997).

Linings of parallel closed tunnels are subjected besides all mentioned loads and actions to mutual influence (Fotieva & Antziferov 1988, Fotieva & Kozlov 1992).

Design method of a tunnel lining subjected to seismic actions, namely, action of longitudinal and transverse seismic waves, taking into account the external seismic stresses in the rock mass depending from degree of the Earthquake intensity and rock mass properties (Fotieva 1980).

As above pointed the analytical solutions of the elasticity theory contact problems have received practically for all types of lining and support including cast iron tubbings, steel-concrete multi-layer lining of circular openings (Bulychev et al. 1986), etc. on all kinds of loading and action, sprayed concrete and rock bolting supports including internal pressure, underground water pressure etc.

It is said there is nothing more practical than a good theory. Modern results of underground mechanics are in a good agreement with experience of experimental investigation including full-scale measurements moreover that agreement appears as not particular but in overall. Design schemes of contact interaction of rock-support have more high level of agreement with data of experience and experiments than previous design schemes of the fixed support pressure. Figure 7 illustrates the measured and calculated the radial normal contact stresses on the interface surface between the cast-iron tubbing tunnel lining and surrounding rock mass. The underground tunnel was constructed in Palaeozoic clays under Neva River in Leningrad. Deformation characteristics of the rock mass are the following: E_0 = 700 MPa, v_0 = 0.35. It is pertinent to note that the maximal normal pressure on the lining acts not from the top down as it was presumed traditionally but has horizontal direction.

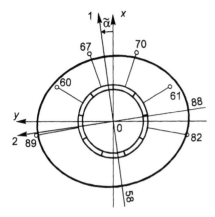

Figure 7. Full-scale measurement results of normal contact stress (rock pressure) on underground tunnel lining and the design curve of the stress (kPa).

5 CONCLUSION

Formation of the Underground Structure Mechanics (the theory of underground structures design) has been completed. That event implies raising the science to new level. The unity of approaches to design of all types of supports and linings on all kind of loads and actions has come about through the principle of interaction between support and rock mass. The notion about support and rock mass as elements of a common deformable system being in the contact interaction with one another has provided a requirement of mutual correspondence between modern methods of underground structures investigation such as analytical, numerical modelling, fotoelastic modelling, equivalent material modelling etc.

The underground structures mechanics is not only theory but also dozens of important underground structures such as the Baikal-Amur railway tunnels, the "Sportivnja" Station of S.-Petersburg Underground, hydraulic and pressure tunnels of the Rogun and the Irganay Hydroelectric Power Stations, irrigation tunnel of the Northern Kebir Water Storage in Syria, vertical shafts of deep mines of the Donetsk Coal Basin and the Norilsk Non-Ferrous Basin, utility and sewerage tunnels in the following cities: Cheboksary, Gomel, Kaspiysk, Kostroma, Krasnodar, N.Novgorod, Rjazan, Rostov-on-Don, Sochy, Tula, Voronege etc. that have been constructed using the methods based on the theory and exploited during 20 and more years. So that the validity of the theory has been tested by many years successful exploitation of dozens tunnels and other underground structures constructed.

The underground structures mechanics as undergraduate course has been included in curriculum and has been successfully teaching at mining and transport universities since 1976.

The purpose of this paper is to pay attention of professors and students, engineers and designers to progressive modern theory and analytical methods of tunnel linings investigations and analysis.

ACKNOWLEDGEMENT

The work has been carried out within the framework of project having received the grant supported by Council on Grants by President of Russian Federation and State Support of Leading Scientific School.

REFERENCES

Arioglu, E. 1995. Optimum support of development roadways. *Geomechanical Criteria for Underground Coal Mines Design*: 185-240. IBSM. Katowice: Central Mining Institute.

Bodrov, B.P. & B.F. Matery 1936. *Ring in elastic medium (design methods and instances). Bulltin 24, Vol II.* Metroproekt (in Russian).

Bulychev, N.S. 1989. *Mechanics of underground structures in instants and problems.* Moscow: Nedra (in Russian).

Bulychev, N.S. 1994a. *Mechanics of underground structures.* Moscow: Nedra (in Russian).

Bulychev, N.S. 1994b. Towards a Methodology for Mechanics of Underground Structures. In Z.Rakowski (ed.). *Proc. of the Int. Conf. Geomechanics '93*: 3-8. Balkema: Rotterdam.

Bulychev, N.S., B.Z. Amusin & A,G. Olovyanny 1974. *Design of permanent support of of mining working.* Moscow: Nedra (in Russian).

Bulychev, N.S. & N.N. Fotieva 1999. Once more about the modern theory of underground structures design. In T.Alten et al (eds). *Proc. of the World Tunnel Congress '99.* Vol. I: 327-333. Balkema: Rotterdam.

Bulychev, N.S., N.N. Fotieva & V.A. Bessolov 1992. Design of temporary support forming later on an element of permanent tunnel lining. *Proc. of the Int. Congr. Towards New Worlds in Tunnelling*: 257-261. Rotterdam: Balkema.

Bulychev, N.S., N.N. Fotieva, A.S. Sammal & I.I. Savin 2000. Theoretical aspects of monitoring and back analysis in tunnels. In T.R.Sacey et al. (eds). *AITES-ITA 2000 World Tunnel Congress.* Symp. Series S24: 73-78. The South African Inst. Of Mining and Metallurgy.

Bulychev, N.S., N.N. Fotieva & E.I.& Streltsov 1986. *Design and computation of permanent working supports.* Moscow: Nedra (in Russian).

Bulychev N.S., N.N. Fotieva, V.A. Bessolov, S.N. Vlasov, & K.P. Bezrodny 1997. Modern tunnelling technology and theory of lining design. *Proc. of the Int. Symp. on Rock Support - Applied Solution for Underground Structures /* ed. by E. Broch, A. Myrvang, & G. Stjern: 49-59. Lillenhammer, Norway.

Fenner, R. 1938. Untersuchungen zur Erkenntnis des Gebirgsdruckes. *Gluckauf:* 32:681-695, 33:705-715.

Fotieva, N.N. 1974. *Design of Non-Circular Cross-Section Tunnel Linings.* Moscow: Stroyizdat (in Russian).

Fotieva, N.N. 1980. *Design of support of underground structures in seismically active regions.* Moscow: Nedra (in Russian).

Fotieva, N.N. & N.S. Bulychev 1987. Inverse problem of design of temporary concrete lining with the aid of measured displacements. *Proc. of the 6th Int. Congr. of Rock Mech.* Montreal, 2:893-895.

Fotieva N.N. & N.S. Bulychev 1997. Design of undersea and under-river tunnel linings. *Proc. of the 1st Asian Rock Mech. Symp.: ARMS' 97-* Seoul, Korea 13-15 Oct. 1997. Hi-Keun Lee, Hyung-Sik Yang & So-Keul Chung (eds): 211-216.Rotterdam: Balkema

Fotieva, N.N. & S.V. Antziferov 1988. Design of multy-layer lining of complex of parallel circular cross-section tunnels on earthquake seismic actions. *Mechanics of underground structures:* 30-38. Tula State Technical University (in Russian).

Fotieva N.N. & A.N. Kozlov 1992. *Designing linings of parallel openings in seismic regions.* Moscow: Nedra (in Russian).

Fotieva N.N. & A.S. Sammal 1993. Design of linings of large cross-section tunnels undergoing construction in rock. *Proc. ISRM Int. Symp. Eurock'93*: 545-549. Rotterdam: Balkema.

Fotieva N.N., G.B. Kireeva & K.E. Zalessky 1996. Design of Multi-layer tunnel linings in transversely isotropic rock. *Proc. ISRM Int. Symp. EUROCK'96*: 909-914. Rotterdam: Balkema.

Geotechnical aspects of the design of shallow bored tunnels in soils and soft rock. W. Wittke (ed). 1997. *Recommendations of the ISSMFE Working Committee ERTS 9.* Berlin: Ernst & Sohn.

Labasse, H. 1949. Les presions de terrains autour des puits. *Revue Universelle des Mines*: 5(3).

Matveev, B.V. 1952. About graphical representation of the operation mechanism of mining support. *Ugol,* 11: 19-22 (in Russian).

Mohr, F. 1952. Gebirgsdruck und Ausbau. *Gluckauf.* Bd 88, 27/28: 675-683.

Muller, L. 1990. Removing misconceptions on the New Austrian Tunneiiing method. *Tunnel & Tunnelling, Special Issue*: 15-18.

Protodyakonov, M.M. 1930. *Rock pressure and mining support. Part one. Rock pressure.* Moscow: Gostechizdat (in Russian).

Ruppeneyt, K.V. 1954. *Some problems of rock mechanics.* Moscow: Ugletechizdat (in Russian).

Terzaghi, K. 1943. *Theoretical soil mechanics.* New York: John Willey.

WG 2, ITA 2000. Guidelines for the design of shield tunnel lining. *Tunnelling and Underground Space Technology.* Vol 15, 3:303-331.

Modern Tunneling Science and Technology, Adachi et al (eds), © 2001 Swets & Zeitlinger, ISBN 90 2651 860 9

Experimental study on static behavior of road tunnel lining

H.Mashimo & N.Isago
Public Works Research Institute, Independent Administrative Institution, Tsukuba, Ibaraki, Japan

H.Shiroma & K.Baba
Japan-Highway Research Institute, Japan-Highway Public Corporation, Machida, Tokyo, Japan

ABSTRACT: The proper design method for tunnel lining should be proposed considering mechanical characteristic for the purpose of reducing construction cost and improving safety of tunnel structure. The loading experiment and the analysis, which accounted for the development process of cracks on the basis of the experimental results, were carried out to grasp mechanical characteristic and to examine the structural strength of lining. It was shown that the structural strength was improved by adopting steel fiber reinforced concrete for the lining under the load condition that dominated the effect of bending moment, and that safety of tunnel structure was improved due to preventing the falling of concrete debris under every load condition.

1 INTRODUCTION

The number of construction of tunnel has been increasing due to the development of road traffic recently, and the section of tunnel also has been larger. The lining of tunnel is expected to have mechanical and useful functions. The design of thickness and the construction method of lining are mainly determined by past experience and same kind of examples of construction. Much proper design method for tunnel lining should be proposed considering mechanical characteristic. Effect of new material for lining should also be examined. It could lead to reducing construction cost and also improving safety of tunnel structure.

In this study the loading experiment was carried out to obtain the basic data to grasp mechanical characteristic of tunnel lining, which was made of plain concrete or concrete with steel fiber. The numerical analysis was carried out to examine the structural strength of lining considering the development process of cracks on the basis of the results.

2 METHOD OF RESEARCH

2.1 *Method of loading in experiment*

Concrete specimens simulated lining was shown in Figure 1. The shape was semicircle, the outer diameter was 9700mm, and the thickness was 300mm. The load was acted to the specimens in the radius direction by 34 jacks. There were two jacks per section every 10 degree in the experiment. Two jacks in one section were placed at both 30cm and 70cm

high from the bottom of specimen. The loading was continued till the collapse of specimen, or till the maximum load where the loading machine could act on the specimen. The loading board was inserted between the two jacks and specimen to be able to reproduce the real loading condition.

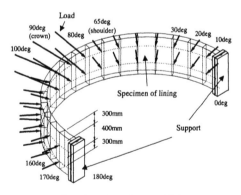

Figure 1. Outline of specimen for tunnel lining.

Two loading types were used in this experiment. The image of loading type (A) was shown in Figure 2. It dominated more effect of axial force than that of bending moment. The loading was done in the 17 sections every 10 degree in the coordinate that regulated anticlockwise starting from the right bottom of support as shown in Figure 1. Loading started in 17sections (34 jacks) up to 20kN/jacks. Loading was continued in 3 sections (6 jacks) from 80 degree to 100 degree after that. The rest of 14 sections (28 jacks) was held to keep the force, which were simulated as the spring considering ground reaction in the analysis.

Loading type (B) was also shown in Figure 2. It dominated more effect of bending moment in shoulder part, that is, about 65 and 115 degree of lining, than that of axial force. The loading was done in the 9 sections, which was from 10 degree to 40 degree, 90 degree, and from 140 degree to 170 degree, in the same coordinate as shown in Figure 1. Loading started in 9 sections (18 jacks) till the axial force was introduced. Loading was continued in 1 section at 90 degree after that. The rest of 8 sections (16 jacks) was held to keep the force like loading type (A).

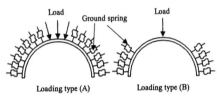

Loading type (A)　　Loading type (B)

Figure 2. Image of loading type.

The method of control in loading was load-control method in loading type (A), and displacement-control method in loading type (B). The modulus of reaction for each jack was about from 100 to 200 MN/m calculated by the reaction force and the stroke of jack in the experiment.

2.2 Specification of lining in test

The specification of each specimen and the result of material test for cylindrical specimen was shown in Table 1. Case I was the basic case, whose specification was based on technical standard for road tunnel (1989), which was 300mm-thick and plain concrete in material with loading type (A). Case II, which was used SFRC (Steel Fiber Reinforced Concrete) for its material, was tested as the same loading pattern to grasp the influence of steel fiber for lining.

Case III, which had the same specification as case I, was tested with loading type (B) to grasp the effect of loading condition. Furthermore case IV, which had the same specification as case II, was tested with loading type (B).

2.3 Method of analysis

The results of experiment were examined by analysis considering the occurrence and development of cracks in lining.

Nonlinear finite element method was used in this analysis. Modified Newton-Raphson method were adopted for solution of nonlinear equations. Plane-stress element for material and Kupfer's failure criterion were adopted. Compressive strength, modulus of elasticity, and Poisson's ratio was obtained by the

Table 1. Specification and result of material test for experiment.

Case Number	I	II	III	IV
Thickness(mm)	300	300	300	300
Outer radius(mm)	9700	9700	9700	9700
Material of lining	plain	SFRC	plain	SFRC
Loading type	(A)	(A)	(B)	(B)
Mixture rate of steel fiber [%/volume]	no	0.5	no	0.5
Diameter / length of steel fiber [mm]	no	0.8/60	no	0.8/60
Compressive strength(MPa)	26.3	19.9	26.7	21.2
Modulus of elasticity(GPa)	20.65	16.85	18.24	22.07
Poisson's ratio	0.20	0.16	0.20	0.16

value in material test shown in Table 1. Tensile strength was calculated on the basis of compressive strength as following equation (1).

$$f_t = 0.23 f'_c{}^{2/3} \qquad (1)$$

where f_t = tensile strength; and f'_c = compressive strength.

Characteristic of compression was simulated quadric curve till strain reached 0.002 and no stiffness decrease after that.

Characteristic of tensile was regulated by tensile-softening curve. Matsuoka et al. (1996) used allowable crack width in plain concrete as 0.02mm. In this analysis allowable crack width in plain concrete was set as the same value, and ratio of residual strength was 1.0. Allowable crack width in SFRC was calculated by material test, and ratio of residual strength was 0.55. The relation of tensile softening curve were considered as linear in both plain concrete and SFRC.

The analysis was done without considering self weight. Loading in analysis was continued as the same step in experiment and the load was acted in the point, not at the face. Isago et al. (2000) carried out the same kind of experiment and calculated the modulus of spring considering ground reaction as 50MN/m per section by frame analysis. In this study the modulus of spring was set as the same value. The springs were considered that tension was negligible and were acting in the section which held the force in the experiments.

3 RESULT OF EXPERIMENT

3.1 Results of displacement and strain

Load-displacement curve at 90 degree of each case was shown in Figure 3. Concerning loading type (A), the collapse of lining specimen occurred in both case I and II. Part of shoulder at 65 degree was collapsed when the load of 390kN/jack was acted at 80, 90, and 100 degree in case I. The collapse also occurred in case II at 65 degree when the load of 330kN/jack was acted at 90 degree. Displacement of case II was increasing to the load more slightly than that of case I in the load of from 200kN/jack to 300kN/jack from figure 3. However, it had no clear relation between the load and the displacement at crown whether the lining had steel fiber or not. It was concluded that steel fiber had little effects to improve the structural strength under the load type (A), such as the influence of axial force was prominent. While case I, whose material was plain concrete, had the falling of concrete debris from specimen around part of shoulder part, case II had no falling of debris. Then the effect of reinforcement by steel fiber was shown even under the load condition that axial force was dominant.

Concerning loading type (B), whose control method was displacement-control, case IV had more maximum load that was able to act than case III.

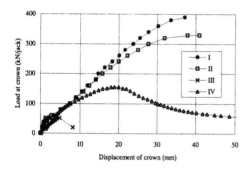

Figure 3. Load-displacement curve at crown in experiment.

Load-strain curve at 65 degree of each case was shown in Figure 4. The positive value of strain means tensile, and the negative value means compression. Stress of tensile was shown up to 0.0002 in figure, because the more value of strain was thought to be unreliable in experiment. Concerning case I and II, the collapses of lining were decided by the limit of compression at shoulder part of lining. The structural strength was regulated by the collapse with compression for loading type (A). In case III and IV, the strain of shoulder part did not reach the limit of compression.
Load-strain curve at both 87.5 and 45 degree of case III and IV was shown in Figure 5 and Figure 6. Concerning case III and IV, the collapse of lining was not decided by the reaching of compression limit of concrete, even in the sections that occurred crack. Although the compressive strength of specimen of case III was bigger than that of case IV, the maximum load of case IV was bigger than that of case III. The maximum load was not obtained by the effect of compression. It was found that the structural strength was regulated by the development of cracks, and that steel fiber influenced on the struc-

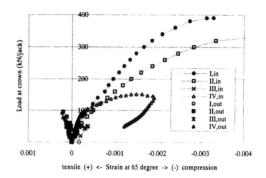

Figure 4. Load - strain curve at 65 degree in experiment.

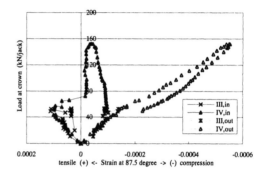

Figure 5. Load - strain curve at 87.5 degree in experiment.

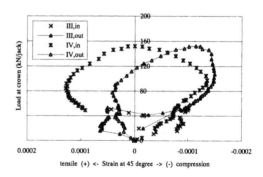

Figure 6. Load - strain curve at 45 degree in experiment.

tural strength under the condition of the load type (B) such as the bending moment was prominent. It was thought that the tunnel lining mixed with steel fiber could improve the structural strength under such loading condition.

3.2 Results of crack occurrence

The first crack occurred at the inner part of crown in the load of 80kN/jack, and the second cracks occurred at the outer part of 65 degree and 115 degree in 160kN/jack in case I. Furthermore in case II, the first crack occurred in the load of 60kN/jack, and the second crack occurred in 240kN/jack at the same position as case I. Although comparing the crack width between case I and case II was difficult because of the different strength of specimens, the crack in case II was relatively dispersed. The falling of the concrete debris was not found to occur in case II.

Concerning loading type (B), the first crack occurred at the inner part of crown in the load of 49.6kN/jack (in the displacement at crown of 0.26mm), the second crack occurred at the outer part of 65 degree in 53.9kN/jack (2.66mm), furthermore the third crack occurred at the inner part of 40 degree in 50.5kN/jack (4.48mm) in case III. In case IV, the first crack occurred in the load of 57.0kN/jack (in the displacement at crown of 3.8mm), the second crack occurred in 100.0kN/jack (7.60mm), and the third crack occurred in 149.8kN/jack (16.4mm), at the same position as case III. The disperse of cracks in case IV was also observed as well as case II.

4 RESULT OF ANALYSIS

Load-displacement curve of case I and II comparing the results of the experiment with the analysis at crown was shown in Figure 7. The loads that occurred cracks in both results from the figure were difficult to find, although the relations between the load and the displacement were almost similar in both cases. The line in Figure 7 that the mark changed means that the strain at the outer part of crown exceeded -0.0035, which was considered the limit of compression for concrete. It was thought that the swelling of concrete at the outer part near crown was constrained by the loading board in the experiment, and that the strain continued to be increasing without the collapse of specimens near crown.

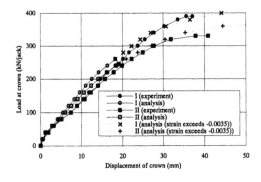

Figure 7. Load-displacement curve at crown (Case I and II).

Load-strain curves of case I and II comparing the results of the experiment with the analysis at 65 de-

gree were shown in Figure 8. The collapse with compression was clearly examined from the figure, for the strains of outer part of 65 degree was exceeding -0.0035, which was the limit of concrete compression. The crack at every case was occurred at the point, for the tensile stress exceeded almost 0.0002 from both the experiment and the analysis.

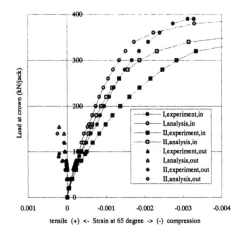

Figure 8. Load - strain curve at 65 degree (Case I and II).

Load-displacement curve of case III and IV comparing the results of the experiment with the analysis at crown was shown in Figure 9. The relations between the load and the displacement was almost similar till the occurrence of the second crack, which was occurred at the inner part of 65 degree, in case III, although the experiment was carried out in the deformation-control and reproducing the relation in every situation was difficult. Furthermore the relation in analysis was slightly different from the one in experiment in case IV at low load level. It was predicted that the strength of cylindrical specimen was a bit different from the strength of lining specimen in test. However, the tendency of their relations was

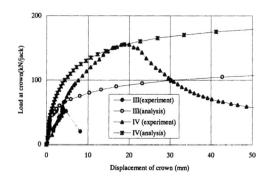

Figure 9. Load-displacement curve at crown (Case III and IV).

almost similar up to the maximum load in case IV. Concerning case III, the evaluation of load was almost possible, although the load-deformation curve changed without the drop of load after the occurrence of crack.

Load-strain curve of case III and IV comparing the results of the experiment with the analysis at 65 degree was shown in Figure 10. It was found that the maximum load was not decided by the collapse with compression from both the analysis and the experiment. The inner strain of case III in analysis exceeded -0.0035, but the structure had already been instable, and it was thought to be negligible.

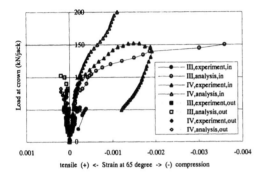

Figure 10. Load - strain curve at 65 degree (Case III and IV).

The situation of crack occurrence in case I and II was shown in Table 2. The load that the crack occurred in analysis was smaller than that in experiment in comparison with the analysis and the experiment. It was thought that the load of occurrence of crack in experiment was obtained after the crack was actually found in the experiment, and that finding the hair-crack in the experiment was difficult.

Table 2. Situation of crack occurrence (Case I and II).

Position	Case I				Case II			
	Experiment		Analysis		Experiment		Analysis	
	Load*	Width**	Load	Width	Load	Width	Load	Width
90 deg.			60	found			55	found
(outer)	80	found (0.15)	80	0.18	60	found	60	0.02
	160	0.50-1.5	160	0.51			100	0.10
							150	0.22
			200	0.76			200	0.43
			250	1.2			240	0.79
65 deg.			113	found			110	found
(inner)	160	found (0.10)	160	0.05			150	0.03
			200	0.10			200	0.03
			250	0.17	240	found	240	0.09

* Load: load at crown [kN/jack]
**Width: crack width [mm]

The order of crack width was almost the same comparing the analysis with the experiment, although to evaluate the width was difficult because it might vary every position even if the crack had same origin. The crack developed almost as the shape of line at first and occurred in the part of its vicinity as the load greatly increased in case I. However the occurrence and development in case II was observed to disperse. The same result was obtained from the analysis. The width of crack in case II was smaller because of their disperse, although the compressive strength of case I was bigger.

The situation of crack occurrence in case III and IV was shown in Table 3(a) and (b). The tendency of crack width to the displacement at crown was almost similar in case III, although to reproduce the drop of load after the occurrence of cracks was difficult. It was found that the crack penetrated at the shoulder part in the load of 60kN/jack in the analysis of case III, and that the cracks also occurred at the inner part of 40 degree in the same load.

Table 3(a). Situation of crack occurrence (Case III).

Position	Case III					
	Experiment			Analysis		
	Load	Disp	Width	Load	Disp	Width
90 deg.				45	1.38	found
(inner)	49.6	0.26	found			
	50.7	2.37	0.35-0.50			
	53.9	2.66	0.40-0.80			
				60	3.13	0.50
	42.7	3.45	0.50-1.0			
	50.5	4.48	2.0-6.0			
65 deg.	53.9	2.66	found			
(outer)				60	3.13	found (found and penetrated)
	42.7	3.45	0.20-1.2			
	50.5	4.48	0.25-6.0			
40 deg.	50.5	4.48	found			
(inner)				60	3.13	found

Table 3(b). Situation of crack occurrence (Case IV).

Position	Case IV					
	Experiment			Analysis		
	Load*	Disp**	Width***	Load	Disp	Width
90 deg.				37.5	0.70	found
(inner)	57.0	3.80	found			
				70	2.00	0.16
				97.5	4.11	0.48
	113.8	9.57	0.55-1.4			
	144.7	14.8	2.0-4.0			
				147.5	15.2	2.5
65 deg				70	2.00	found
(outer)				97.5	4.11	0.48
	100.0	7.60	found			
	113.8	9.57	0.10-0.20			
				147.5	15.2	0.63
	149.8	16.4	0.70-2.8			
40 deg.				97.5	4.11	found
(inner)	149.8	16.4	found			

* Load: load at crown [kN/jack]
** Disp: displacement at crown [mm]
***Width: crack width [mm]

455

The position of cracks in case IV was almost same as case III. However the second crack at 65 degree occurred dispersedly, not to penetrate the section in both the analysis and the experiment of case IV. The load of second crack in case IV was 70kN/jack while the third crack at 40 degree was 97.5kN/jack in analysis.

It was thought that the resistance for tensile was improved by the effect of using steel fiber, and that the lining with steel fiber was harder to occur the instability of structure by the development of cracks in three sections than the lining made of plain concrete. It led to improvement of structural strength.

5 CONCLUSION

In this study the loading experiment and the analysis of tunnel lining were carried out to obtain the basic data of the mechanical characteristic, to grasp the effect of new material like steel fiber for lining, and to examine the structural strength considering the influence of cracks. Results were shown as follows:

(1) The difference of the load between the occurrence of the first crack and the collapse of lining is large.

(2) The lining with steel fiber leads to the disperse of cracks and steel fiber has the effect of preventing the falling of concrete debris from the lining under every loading condition.

(3) The lining with steel fiber does not lead to increase of structural strength under the condition that the axial force is prominent.

(4) The lining with steel fiber leads to constraining the development of cracks and the lining with steel fiber is harder to occur the instability of structure than the lining made of plain concrete. Then it leads to the improvement of the structural strength.

The following points must be solved for the proposal of design method for the economical and durable tunnel lining.

(1) The experiment and analysis should be carried out under various load conditions, for the structural strength of lining would be varied by their conditions.

(2) The behavior of cracks should be evaluated properly, for it has influence on the whole behavior of lining.

(3) The relation among the maximum load, collapse and crack should be examined, especially in case that the third crack is occurred.

REFERENCE

Japan Road Association, 1989. *Technical standard for road tunnel (Part of structure) and the explanation.* (in Japanese)

Matsuoka, S. & Masuda, A. et al. 1996. A study on simulation of tunnel lining which involves crack. *Proceedings of Japan society civil engineers* No.554/III-37: 147-155 (in Japanese)

Isago, N. & Mashimo, H. et al. 2000. Basic study of mechanical characteristic for tunnel lining: JSCE Committee of tunnel eng., *Proc. of tunnel engineering, Tokyo, November 2000.* (in Japanese)

Modern Tunneling Science and Technology, Adachi et al (eds), © 2001 Swets & Zeitlinger, ISBN 90 2651 860 9

Applicability of steel fiber reinforced high-strength shotcrete to a squeezing tunnel

M. Hisatake
Kinki University, Osaka, Japan

T. Shibuya
Chizaki Kogyo, Co.Ltd., Tokyo, Japan

ABSTRACT: In order to clarify the reasonable conditions under which a steel fiber reinforced high-strength sprayed concrete lining (SFRS) can be applied to a squeezing tunnel with time dependency; a numerical analysis of non-linearity is conducted. The advance velocity of the tunnel face progress, the time dependency of the strength of SFRS, the elasto-plastic behavior of the steel supports, and the visco-elastic characteristics of the ground are all taken into account in the analysis. A comparison between analytical and field measurement results shows that the results obtained through this analysis have sufficient accuracy for practical applications. Consideration is given to the use of SFRS to a squeezing tunnel.

1 INTRODUCTION

Shotcrete is now recognized as one of the most powerful tunnel supports, and it has become the main support for mountain tunnels. Research into the improvement of the ability of shotcrete to support tunnels is being conducted, and steel fiber reinforced high-strength shotcrete (referred to as 'SFRS' from now on) has been developed.

The main features of the mechanical characteristics of SFRS are its high strength and its tenacity after its peak strength. The value of the strength of SFRS increases rapidly after its application, and it reaches almost double that of ordinary shotcrete. The tenacity of SFRS, which is achieved through the use of steel fiber, is effective for keeping the tunnels stable.

The behavior of a squeezing tunnel with SFRS is affected not only by the mechanical characteristics, such as the strength and the tenacity of SFRS, but also by the time dependency of the mechanical characteristics of the ground surrounding the tunnel and the conditions under which it is constructed.

The objective of this paper is to clarify the reasonable conditions under which SFRS can be applied to a squeezing tunnel with time dependency. A numerical analysis of non-linearity is carried out by taking such factors into account as the time dependency of both the mechanical characteristics of SFRS and the ground surrounding the tunnel, the strain softening behavior of SFRS, the elasto-plastic behavior of the steel supports, and conditions such as the advancing speed of the tunnel face progress.

Firstly, the mechanical characteristics of SFRS are expressed by non-linear stress-strain-time relationships based on experimental results. Secondly, an analytical approach to non-linearity is demonstrated. Thirdly, an analytical method is applied to the squeezing tunnel with time dependency, and the applicability of the method is demonstrated through a comparison of the field measurement results and the analytical results. Finally, some considerations are given through a parametric study.

2 EXPRESSION OF MECHANICAL PARAMETERS OF SFRS

The stress-strain relationship of SFRS, which contains 1% steel fiber, shows non-linearity and changes with time after its application. Table 1 shows the time-dependent values of the initial tangent modulus of elasticity(E_i), the uniaxial strength(σ_f), and a non-linear parameter(R_f) obtained through experiments.

The stress-strain relationship ($\sigma - \varepsilon$) of SFRS before the peak strength can be expressed by a hyperbolic curve (Kondner, R.L., 1963), namely,

$$\sigma = \frac{\varepsilon}{a + b\varepsilon},$$

(1)

where a and b are constants, but are functions of time. The tangent modulus of elasticity (E_c) at an arbitrary stress in the uniaxial compression tests can

Table 1. Time-dependent values of E_i, σ_f, and R_f.

(a)

Days	$E_i(MPa)$
$0 < t \leq 0.0241$	0.001
$0.0241 < t \leq 0.125$	$5975t - 144$
$0.125 < t \leq 0.25$	$35149t - 3787$
$0.25 < t \leq 28$	$467t - 5615$
$28 < t$	18702

(b)

Days	$\sigma_f (MPa)$	R_f
$0 < t \leq 0.125$	$12.7t$	$0.647t + 0.690$
$0.125 < t \leq 28$	$6.1(\ln t) + 16.2$	$-0.005t + 0.815$
$28 < t$	36.7	0.672

be derived from Equation (1) as a stress function, in other words,

$$E_c = \frac{\partial \sigma}{\partial \varepsilon} = \frac{(1 - b\sigma)^2}{a} . \qquad (2)$$

The theoretical peak strength, σ_{ult}, and the initial tangent modulus of elasticity, E_i, are related to a and b as follows:

$$\sigma_{ult} = \lim_{\varepsilon \to \infty} \frac{\varepsilon}{a + b\varepsilon} = \frac{1}{b} ,$$

$$E_i = \lim_{\varepsilon \to 0} \frac{\partial \sigma}{\partial \varepsilon} = \frac{1}{a} . \qquad (3)$$

The substitution of a and b in Equation (3) into Equations (1) and (2) yields the following equations:

$$\sigma = \frac{\varepsilon}{\dfrac{1}{E_i} + \dfrac{R_f}{\sigma_f} \varepsilon} , \qquad (4)$$

$$E_c = E_i \left(1 - \frac{\sigma}{\sigma_f} R_f \right)^2 , \qquad (5)$$

where, $R_f = \dfrac{\sigma_f}{\sigma_{ult}} .$

Equation (5) indicates that E_c is fixed by E_i, σ_f, R_f and the axial stress (σ).

SFRS does not easily decrease in strength after it reaches its peak strength, especially during the early stages of its application. Strain softening, however, does gradually occur one day after its application. A linear line can express the stress-strain relationship after the peak strength. Table 2 shows the tangent

modulus of elasticity E_c after the peak strength for which E_{28} is the value of E_c at t = 28 days.

By applying the experimental values in Tables 1 and 2 to Equation (4), the stress-strain relationships of time dependency are fixed. Figure 1 shows constitutive relationships for SFRS obtained from equation (4). It can be recognized that the non-linearity, the time-dependent strength, and the strain-softening characteristics are reasonably expressed.

Table 2. Tangent modulus of elasticity after the peak strength.

Days	$E_c(MPa)$
$t \leq 1$	0.0
$1 < t \leq 28$	$E_{28} \times (t - 1) / 27$
$28 < t$	$E_{28} = -817$

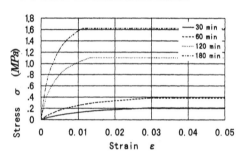

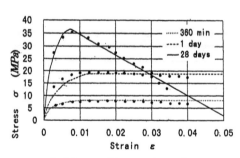

(• :Measured by experiments)

Figure 1. Stress-strain relationships of the SFRS of time dependency obtained by Equation (4).

3 MECHANICAL PARAMETERS OF THE GROUND

Figure 2 shows ground surface settlements caused by a shallow tunnel excavation. A back analysis of the results in Figure 2 indicates that the creep function of the ground can be expressed by a logarithmic function, as shown in the following equation (Ito, T. & M. Hisatake, 1981),

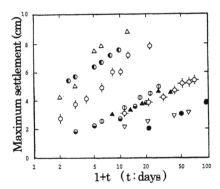

Figure 2. Ground surface settlements caused by a shallow tunnel excavation.

$$\phi(t) = \alpha + \beta \ln(1+t) \tag{6}$$

where, $\alpha = 1/G$, $\beta = 1/G_r$, and G and G_r are the shear modulus of elasticity and the retardation shear modulus of elasticity, respectively.

At this site, a circular tunnel was driven in the upper site before the main tunnel construction. When the creep function was back-analyzed, the same function as in Equation (6) was obtained and coefficients α and β showed the same value, namely, 0.125 (1/MPa).

4 ANALYTICAL PROCEDURE

A circular tunnel under a state of uniform initial stress is analyzed by considering the following conditions:
(1) visco-elastic behavior of the ground
(2) advancing velocity of the tunnel face progress
(3) effect of the three dimensional excavation, such as tunnel radial displacement in a profile through a tunnel axis
(4) non-linearity and strain-softening behavior of the stress-strain relationship of the shotcrete
(5) time dependency of the strength and the tangent modulus of elasticity of the shotcrete
(6) elasto-plastic behavior of the steel supports
(7) conditions at the time of application such as the interval distance between the steel supports and the distance between the tunnel face and the position of the lining construction.

The analytical procedure is as follows. The tunnel radial displacements of the steel supports (u^s) and SFRS (u^c), which are caused by earth pressure values p_s, p_c, respectively, can be expressed by Equation (7), for which the time scale is divided into intervals by the time values $t_k, (k = 1,2, \cdots (m+1))$ and where $t_1 = 0$, $t_{m+1} = t_i$, and t_i is the time after the steel supports and SFRS have been constructed at the same time.

$$u^s(t_{m+1}) = \frac{a_0^2 L}{A} \sum_{k=1}^{m} \frac{p_s(t_{k+1}) - p_s(t_k)}{E_s(t_k)}$$
$$u^c(t_{m+1}) = \frac{a_0^2}{h_0} \sum_{k=1}^{m} \frac{p_c(t_{k+1}) - p_c(t_k)}{E_c(t_k)} \tag{7}$$

where a_0 is the tunnel radius, A and E_s are the cross section area and the modulus of elasticity of one steel support, respectively, L is the interval distance between the steel supports, and h is the thickness of SFRS. Total pressure p_w, acting on the boundary of the tunnel, is

$$p_w(t_{m+1}) = p_s(t_{m+1}) + p_c(t_{m+1}) \tag{8}$$

It is possible to consider the steel supports and SFRS being constructed at the same time; therefore,

$$u^s(t_{m+1}) = u^c(t_{m+1}) \tag{9}$$

After solving equations (7) and (9) for $p_c(t_{m+1})$, the substitution of $p_c(t_{m+1})$ into Equation (8) yields the following equation:

$$p_s(t_{m+1}) = Z_1(t_m) p_w(t_{m+1}) + Z_2(t_m) \tag{10}$$

where Z_1, Z_2 are as follows:

$$Z_1(t_m) = \frac{1}{1 + h_0 L E_c(t_m)/(A E_s(t_m))}$$

$$Z_2(t_m) = Z_1(t_m) \left\{ E_c(t_m) \sum_{k=1}^{m-1} \frac{p_c(t_{k+1}) - p_c(t_k)}{E_c(t_k)} - p_c(t_m) \right\}$$
$$+ Z_1(t_m) \frac{L h_0}{A} E_c(t_m) \left\{ \frac{p_s(t_m)}{E_s(t_m)} - \sum_{k=1}^{m-1} \frac{p_s(t_{k+1}) - p_s(t_k)}{E_s(t_k)} \right\}$$

On the other hand, the boundary condition on the radial displacement of SFRS becomes

$$u^s(t_{m+1}) = \Delta u(t_{m+1})$$
$$- \frac{a_0}{2} \int_{l=0}^{t_{m+1}} \phi(t_{m+1} - \tau) \frac{\partial}{\partial \tau} p_w(\tau) d\tau \tag{11}$$

where Δu is the displacement on the tunnel boundary which would be brought about after time $t_0 = (2a_0 + W)/V$ if the lining were not constructed. V is the velocity of the tunnel face progress and W is the distance between the tunnel face and the lining nearest to the face. Δu is calculated by the equation

$$\Delta u(t_{m+1}) = \alpha_0 \int_0^{t_0 + t_{m+1}} \phi(t_0 + t_{m+1} - \tau) \frac{\partial}{\partial \tau} f(V\tau) d\tau$$
$$- \alpha_0 \int_0^{t_0} \phi(t_0 - \tau) \frac{\partial}{\partial \tau} f(V\tau) d\tau$$

where $\alpha_0 = a_0 p/2$ in which p is the uniform initial stress of the ground and f is a characteristic curve on the radial displacement of the tunnel boundary in a profile through a tunnel axis. After solving u^s from Equations (7) and (10), the substitution of u^s into Equation (11) leads to the following equation, namely,

$$p_w(t_{m+1}) = \frac{\Delta u(t_{m+1}) + M_1(t_m) + K_1(t_{m+1})}{M_2(t_m) + K_2(t_{m+1})} \quad (12)$$

where M_i, K_i are

$M_1(t_m) = -B_1(t_m)Z_2(t_m), \qquad M_2 = B_1(t_m)Z_1(t_m),$

$K_1(t_{m+1}) = -B_2(t_m) + 0.5a_0\phi(t_{m+1})p_w(0)$

$+\frac{1}{4}a_0\left[\begin{array}{l} \{\phi(0) - \phi(t_{m+1} - t_m)\}p_w(t_m) \\ + \sum_{k=1}^{m-1}\{p_w(t_{k+1}) + p_w(t_k)\}\{\phi(t_{m+1} - t_{k+1}) - \phi(t_{m+1} - t_k)\} \end{array} \right]$

$K_2(t_{m+1}) = 0.25a_0\{\phi(0) + \phi(t_{m+1} - t_m)\}, \quad B_1(t_m) = a_0^2 L/(AE_s(t_m)),$

$B_2(t_m) = a_0^2 L\left\{ \sum_{k=1}^{m-1}\frac{p_s(t_{k+1}) - p_s(t_k)}{E_s(t_k)} - \frac{p_s(t_m)}{E_s(t_m)} \right\}/A$

In the above equation, $p_w(t_{m+1})$ is determined successively in term of the values already obtained. p_s, p_c and $u^s(=u^c)$ are easily calculated with Equations (7), (8) and (10). Subsequently, circumferential axial stress values σ_{ns} and σ_{nc} of the steel supports and SFRS, respectively, are obtained by considering the equilibrium conditions between the earth pressure and the axial stress.

$$\sigma_{ns}(t_{m+1}) = \frac{a_0 L}{A}p_s(t_{m+1}), \sigma_{nc}(t_{m+1}) = \frac{a_0}{h_0}p_c(t_{m+1})$$

5 CONSIDERATIONS

In order to check the applicability of the numerical procedure to a squeezing tunnel, the results obtained through the analysis are compared with those from

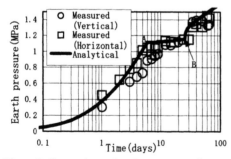

Figure.3. Comparison of earth pressure acting on the steel-pipe supports obtained by the analysis and the field measurements.

Table 3. Mechanical parameters and construction conditions.

h (cm)	25.0
p (MPa)	8.2
a (cm)	380.0
V (cm/day)	150.0
α (1/MPa)	0.0125
β (1/MPa)	0.0125
A (cm^2)	106
E_s before yield (MPa)	21400
E_s after yield (MPa)	1940
L (cm)	70.0
Yield strain of steel support	0.0013
W (cm)	1500.0

Table 4. Time-dependent values of E_i, σ_f and R_f of ordinary shotcrete.

(a)

Days	E_i (MPa)
$0 < t \le 0.0241$	0.00051
$0.0241 < t \le 0.125$	$2996t - 72$
$0.125 < t \le 0.25$	$17622t - 1899$
$0.25 < t \le 28$	$234t - 2815$
$28 < t$	9376

(b)

Days	σ_f (MPa)	R_f
$0 < t \le 0.125$	$6.4t$	$0.647t + 0.690$
$0.125 < t \le 28$	$3.1(\ln t) + 8.1$	$-0.005t + 0.815$
$28 < t$	18.4	0.672

Table 5. Tangent modulus of elasticity after the peak strength of the ordinary shotcrete.

Days	E_c (MPa)
$t \le 1$	0.0
$1 < t \le 28$	$E_{28} \times (t-1)/27$
$28 < t$	$E_{28} = -8210$

the field measurements. Figure 3 shows earth pressure acting on the steel-pipe supports into which cement mortar is injected, and the analytical and the field measurement results are compared. Values for the mechanical parameters and the conditions upon application are shown in Table 3. At this site, ordinary shotcrete is employed and input values of this ordinary shotcrete are shown in Tables 4 and 5.

Analytical points A and B in Figure 3 indicate the yield points of the steel-pipe supports and the shotcrete, respectively. The horizontal and the vertical earth pressure measurements are almost equal. This means that the analytical assumption, whereby the vertical and the horizontal initial stress levels are equal, is adequate. The earth pressure increases with time until it reaches point A, at which time the steel-pipe supports yield. Then, the increase in earth

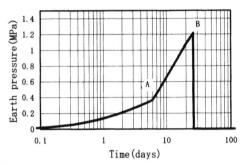

Figure 4. Time-dependent earth pressure of ordinary shotcrete.

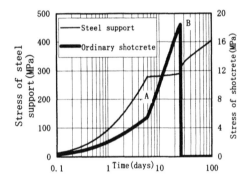

Figure 5. Time-dependent axial stresses of the steel-pipe supports and ordinary shotcrete.

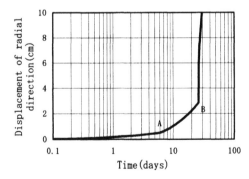

Figure 6. Analytical displacement of the tunnel in a radial direction.

pressure stops temporarily, for about twenty days, until the shotcrete yields at point B. Such analytical behavior is in good agreement with that of the field measurements. The increase, the stop, and the resumption of the increase in earth pressure can be explained as written below.

Figure 4 shows the time-dependent earth pressure of the shotcrete in the analysis. The stiffness of the steel-pipe supports decreases a great deal after its yield, and the increase in earth pressure of it is not recognized, as shown in Fig.3. On the other hand, the earth pressure of the shotcrete increases rapidly after point A, until the axial stress of the shotcrete reaches its peak strength (point B). Ordinary shotcrete does not have tenacity, so a decrease in strength occurs at point B and the earth pressure of the shotcrete is suddenly released. This means that the ground-supporting effect of the shotcrete can no longer be expected after it yields. Therefore, the earth pressure of the steel-pipe supports increases again after the release of the earth pressure of the shotcrete (Fig.3, point B).

The time-dependent axial stress levels of the steel-pipe supports and the shotcrete are shown in Figure 5 in order to make the ground-supporting characteristics of both types of supports clear. The yielding of one type of support has a great influence on the movement of the other type of support. Therefore, conditions such as the interval distance between steel supports, the type of steel supports employed, the thickness of the shotcrete, the advancing velocity of the tunnel face progress, and so on should be reasonably determined in order to keep a valance between the two types of support system.

An analytical displacement of a tunnel in a radial direction is shown in Figure 6. The displacement increases with time and the increment of the displacement increases somewhat at point A due to the yielding of the steel-pipe supports. The shotcrete does not yet yield at this stage; therefore, the increment of the displacement is not radical. After the yielding of the shotcrete at point B, however, the displacement increases drastically. These results show that the yielding of both supports must be avoided in order to maintain the stability of the tunnel.

The field measurements of the displacement play a very important role in this case. If the yielding of the steel-pipe supports is noticed through the field measurements of the displacement, it is possible to keep the tunnel stable by constructing other supports during the twenty days before the yielding of the shotcrete occurs.

The above considerations indicate that shotcrete does not easily yield if it has high strength, and that tunnel displacement does not increase drastically if it has tenacity after the yielding. The application of SFRS to this type of severe tunnel may be reasonable, since SFRS not only has higher strength than ordinary shotcrete, but it also has sufficient tenacity.

Some considerations are given in the following for a case in which SFRS is constructed at this site with steel-pipe supports.

Figure 7 shows the time-dependent earth pressure for the steel-pipe support and SFRS. The steel-pipe supports yield at a comparatively small displacement

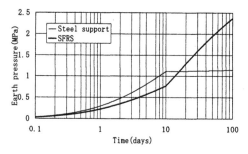

Figure 7. Time-dependent earth pressures of the steel-pipe supports and the SFRS.

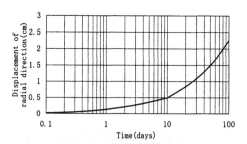

Figure 8. Displacement of the tunnel constructed with the steel-pipe supports and SFRS.

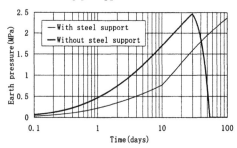

Figure 9. Earth pressure of SFRS with and without the steel-pipe supports.

in the same way as is shown in Figure 6. The strength of SFRS is sufficient high; therefore, it does not yield easily and it can support the earth pressure steadily.

The tunnel displacement for this case is shown in Fig.8. SFRS does not yield; thus, a drastic displace-

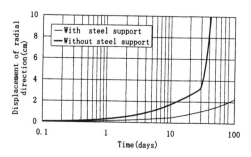

Figure 10. Displacement of SFRS with and without the steel-pipe supports.

ment cannot be recognized. From the results of Figures 7 and 8, it is understood that a primary lining which has both steel-pipe supports and SFRS is very effective for the construction of a tunnel in a squeezing ground.

In Figures 9 and 10, the earth pressure and the displacement of SFRS shown in Figures 7 and 8, respectively, are compared with those analyzed in a case where only SFRS is used in the same tunnel. SFRS is found to yield about thirty days after its application; thus, the construction of this tunnel using only SFRS would not be recommended. The movement of SFRS after its yielding, however, is much more moderate than the movement of ordinary shotcrete coupled with steel-pipe supports, as shown in Figures 4 and 6. From the considerations mentioned above, it can be said once again that the combination of SFRS and steel-pipe supports is adequate for a squeezing tunnel.

6 CONCLUSIONS

1. The movement of the primary lining of a tunnel constructed with both SFRS and steel supports is analytically clarified, taking nonlinear stress-strain-time relationships of SFRS into account.

2. Field measurement and analytical results show that it is very effective if both SFRS and steel supports are used for the primary lining in the construction of a tunnel in a squeezing ground.

REFERENCES

Kondner, R.L., 1963. Hyperbolic stress-strain response; cohesive soils, *J. Soil Mech. Fdns. Div. ASCE, 89, SM1,* pp.115-143.
Ito, T. & M. Hisatake, 1981. Ground pressure acting on arbitrary shaped tunnel lining in a visco-elastic ground, *Proc. Japan Society of Civil Engineers,* Vol.307, pp.51-57 (in Japanese).

Modern Tunneling Science and Technology, Adachi et al (eds), © 2001 Swets & Zeitlinger, ISBN 90 2651 860 9

Use of Electric Gradient to Increase Shotcrete Early Age Strength

D. A . Ferreira
Dept. of Geotechnical Engineering , São Carlos Engineering School, University of São Paulo. Brazil

T. B. Celestino
Professor, Dept. of Geotechnical Engineering , São Carlos Engineering School, University of São Paulo,
Civil Engineering Manager, Themag Engenharia Ltda., São Paulo
Brazil

ABSTRACT: The paper presents results of an experimental program adopting a physical method to increase the early-age strength of shotcrete. The method consists of applying electric gradients to the fresh shotcrete. The first results already published were obtained using non-conventional mixes and materials for shotcrete and electrodes. Strength increases of about 100% at 12-hour age have been reported. In the test program reported here, materials and mixes usual for shotcrete applications and electrodes made out of the same rebars used for lattice girders have been adopted. Scale effects have also been investigated, reproducing dimension ratios more similar to those of tunnel support . The influence of each different material employed is analyzed. The results obtained were even more significant than those of the previous program, with higher strength increase.

1 INTRODUCTION

The interest to develop a physical method to increase shotcrete early-age strength for tunnel support is due to negative side effects of some chemical accelerators. Previous papers by the authors (Ferreira 1997; Ferreira et al. 1997; Ferreira & Celestino 1999, 2000) have described a method based on the application of electric gradient to fresh concrete mixes. The first results have been encouraging, and strength increases at 12-hour age of the order of 100% have been reported. Strength decrease at 28-day age has been negligible, if any. The previous studies, however, were carried out using materials and mixes which were not usual for shotcrete. Laboratory specimens adopted, in addition to being orders of magnitude smaller, had aspect ratios not typical of shotcrete support shells.

Tests reported herein were carried out on specimens prepared with the same materials and adopting mixes similar to those employed for shotcrete used for tunnel support in the City of São Paulo. Galvanized wire mesh electrodes from the previous programs were also substituted by rebars. The reason is that if this method is to be applied in the field, the most feasible electrodes are lattice girders. Scale effects were also investigated by testing specimens larger than the previous ones, with the ratio between exposed surface area and thickness close to that of shotcrete shells. The results in terms of strength increase at early age were even better than those of previous studies.

2 PREVIOUS EXPERIMENTAL PROGRAM

The physical method used here for early-age strength increase consists of applying an electric gradient to the recently mixed concrete by means of electrodes placed in the fresh material. The electrodes are connected to a DC source, and an electric current is thus established. The electric gradient, which depends on the voltage and the distance between electrodes, is applied minutes after the addition of water to the mixture and kept constant for five-and-a-half hours.

The arrangement of the laboratory tests was presented by Ferreira & Celestino (1999). An ampmeter is used to measure the electric current. Electric terminals between the electrodes are connected to a voltmeter and are intended to measure the effective electric gradient. Thermo-pairs are installed to monitor temperature variation during the test.

In previous tests carried out by Ferreira (1997) concrete samples had dimensions 24 x 11 x 11 cm, subjected to four values of electric gradients: 0.44; 0.87; 1.30 and 1.70 V/cm. The cement used was CP

II F 32, a blended cement with maximum carbonate content of 5%.

Crushed basalt with 20-mm average dimension was used as coarse aggregate. For each test, two samples with the same mix were cast: one of them was subjected to the electric gradient, while the other one (the reference specimen) was kept under the same conditions, but without electric gradient application (electric gradient $i_E = 0$).

The most important results of the previous program were:

- The response in terms of electric current and temperature depended on the electric gradient and the water-cement ratio.
- The rates of strength increase (indirectly determined by means of a penetrometer) were positively affected by the electric gradient.
- Strength increase rates were higher for higher electric gradients, being in excess of 100% for a number of specimens.
- The loss of compressive strength at 28-day age for specimens subjected to the highest gradients was negligible.
- Temperature increased with both the electric gradient and the water-cement ratio. Typical temperatures recorded were in the range of 30 to 40 °C and the maximum was 60 °C.

The question of temperature increase was carefully analyzed in order to find out whether that was the only reason for strength increase. It has been widely known that high temperature during setting results in strength increase at early age, but compromises strength results at ages greater than 7 days. According to Verbeck & Helmuth (1966), the quick hydration at early age caused by high temperature results in a non-uniform distribution of hydrated products, decreasing long-term strength .

This point has been investigated by running tests in which specimens were artificially heated by means of a water bathe to the same temperature of the specimen subjected to electric gradient. The results showed that the effect of the electric current is stronger than that of heating only.

3 MATRIALS FOR THE NEW TEST PROGRAM

The experimental results obtained by Ferreira (1997) were very good, but there is the interest to extend testing conditions to those closer to shotcrete for tunnel support. There is also the interest of understanding the strength increase phenomena in the light of physical chemistry. The latter is not the scope of this paper.

Materials and mixes used in the previous program are not typical of shotcrete for underground support. So, cement, aggregate and mixes for the new program were changed.

Two types of cement were selected for analysis: CP II E 32 and CP III 32 RS. CP II E 32 is a slag–modified cement. CP III 32 RS is a sulfate resistant blast furnace slag cement. CP II E 32 has been used for tunnel support shotcrete in São Paulo, as well as for experimental programs carried out by Figueiredo (1993, 1997) and Silva (1993). CP III RS cement is not frequently used for underground works. The reason to adopt it in this experimental program was to investigate the behavior of sulfate resistant cement under the effects of electric gradient.

With regard to aggregates, materials selected met requirements of grain size distribution for shotcrete. Typical materials used for tunnels excavated in the City of São Paulo have been selected. Crushed and clean granitic rock from that region was used in this experimental program.

Electrodes previously used consisted of 1-mm diameter, 14 x 27 mm galvanized wire mesh. For underground applications, the most probable electrodes will be the lattice girders. For this reason, the same type of rebars used for that purpose was selected for the new program. Figure 1 shows the electrodes adopted, consisting of welded $\phi = \frac{1}{4}$" rebars.

Concrete mixes usual for shotcrete specifications were selected, with incorporated mortar rate α and aggregate rate m such that low rebound would result. Following Figueiredo's (1997) suggestions, $\alpha = 80$ and $m = 5$ were adopted, whereby the rate of incorporated aggregate $m' = 3$ was obtained. The concrete mixes used are shown in Table 1, one of which had been used in the previous program for comparison.

Table 1. Concrete mixes used.

Material	Mix 1 (Ferreira, 1997)	Mix 2 (Shotcrete Mix)
Cement (kg/m^3)	357	396
Fine aggregate, sand (kg/m^3)	1078	1345
Coarse aggregate, gravel (kg/m^3)	707	436
Water/cement ratio	0.66	0.55

4 EXPERIMENTAL PROGRAM DESCRIPTION

Specimens with 24 x 10 x 10 cm dimensions (Scale 1) and 35 x 35 x 5 cm (Scale 2) were used, exploring the effect of larger scale with respect to the previous

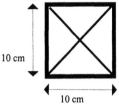
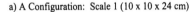
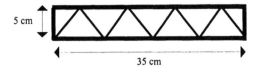

a) A Configuration: Scale 1 (10 x 10 x 24 cm)　　　b) B Configuration: Scale 2 (5 x 35 x 35 cm)

Figure 1. Electrode configuration.

program. The ratio between the exposed surface area and the thickness was also increased to 245 cm, closer to the range of 350 cm, usual for tunnels, and better than 24 cm from the previous program.

Electric gradient of 1.7 V/cm was adopted, based on the previous results (Ferreira, 1997). Strength was indirectly determined by means of a penetrometer. Two different penetrometers were used: one adapted and described by Ferreira (1997) and a Meynadier Penetrometer analyzed by Prudêncio (1993). A brief description of the first penetrometer was also presented by Ferreira & Celestino (2000). The reason to use two penetrometers was the difference in strength obtained with mixes 1 and 2. Each instrument was adequately sensitive for each strength range.

In order to investigate the separate influence of each material on the efficiency of the electric gradient method, the first set of specimens was cast. Mix 1 was adopted, changing one material in each test (cement, aggregate and electrode) and keeping the others as in the previous program. Specimens for these tests had 24 x 10 x 10 cm dimensions and the adapted penetrometer was used for strength evaluation.

In order to investigate the influence of the concrete mix, a second set of specimens was cast adopting the same dimensions of the first set (24 x 10 x 10 cm), all new materials and mix 2. As shown in section 5.1, these materials individually presented good results under the electric gradient method. In the second set of tests all the new materials were used in combination with A Configuration rebar electrodes. The third set of specimens had the same materials and mix of the second set, dimensions of Scale 2 (5 x 35 x 35 cm) and B Configuration electrodes. For tests on specimens with mix 2, the Meynadier penetrometers was used.

5 TEST RESULTS AND ANALYSIS

5.1 *Analysis of different materials*

The separate influence of each material on the performance of the electric gradient method was analyzed with the first set of specimens based on the results of electric current, temperature and penetration resistance. For comparison, results obtained by Ferreira (1997) were used.

Figure 2 shows the results of electric current and temperature obtained by substituting one material at a time. Results from the previous program are also shown for comparison.

Electric current was lower (0.25 to 0.35 A maximum), for concrete cast with CP II E 32 and CP III 32 RS cements when compared to CP II F 32 cement, for which the maximum current reached 0.40 A. Both granitic aggregate and rebar electrode resulted in increased electric current, in the range of 0.60 to 0.65 A.

With respect to temperature, the general trend was the same as that for the electric current, however less pronounced. Temperatures between 33 and 36 °C were obtained with the new cements as compared to the average 39 °C from the previous program. The use of granitic aggregate and rebar electrode resulted in increasing the temperature to the range of 43 to 45 °C.

Figure 3 presents the results of penetration resistance obtained with the adapted penetrometer. Even though electric current and temperature values were lower with the new cements, penetration resistance showed values similar to those obtained in the previous program. In comparison to the reference specimens (i_E = 0), the penetration resistance of concrete subjected to electric gradient reached levels about 5 times higher at 8-h age.

For specimens with granitic aggregate and rebar electrode, penetration resistance was much higher than that obtained by Ferreira (1997). This trend could be anticipated based on the indices of higher electric current and temperature, which also indicate efficiency.

5.2 *Concrete mix study*

The analysis consists of comparing the performance of mix number 2 in terms of electric current and temperature. It is not possible to compare penetration results directly to those obtained in the

465

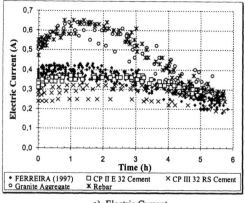

a) Electric Current

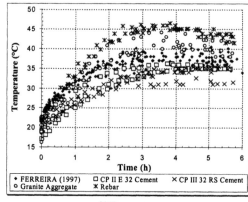

b) Temperature

Figure 2. Electric current and temperature results.

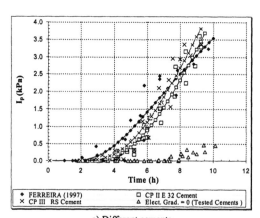

a) Different cements

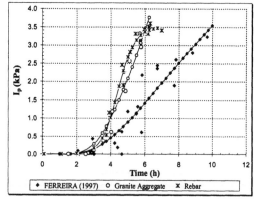

b) Granite aggregate and rebar electrodes

Figure 3. Penetration indices versus time. Concrete with new materials and Ferreira (1997).

previous studies because different penetrometers were used. Comparison will be made with correlated compressive strength values, as shown in Table 2.

Figure 4 shows electric current and temperature results. Comparison to similar results obtained by Ferreira (1997) shows that with the new mix, the electric current was only slightly lower, and there is no significant difference in results for the temperature.

Penetration resistance results are shown in Figure 5. Resistance for concrete subjected to electric gradient is about 200% higher than that for reference specimens with $i_E = 0$.

The uniaxial compressive strength of specimens from test sets 2 and 3 can be estimated by using an expression proposed by Prudêncio (1993) for the Meynadier penetrometer:

$$f_c = \frac{IP - 1.031}{66.24} \qquad \text{(MPa)} \qquad (1)$$

For mix 1 specimens the compressive strength can be obtained from the following expression obtained for the adapted penetrometer by Ferreira (1997):

$$f_c = 0.043 * I_p \qquad \text{(MPa)} \qquad (2)$$

Table 2. Compressive strength estimate based on penetration indices.

	Previous Materials (MPa)		New Materials (MPa)		
Age (h)	Refer.	Scale 1	Refer.	Scale 1	Scale 2
5	0.00	0.01	0.01	0.13	0.21
6	0.00	0.03	0.04	0.22	0.30
8	0.02	0.10	0.14	0.39	0.48
10	0.05	0.28	0.26	0.55	0.64
12	0.12	0.62	0.34	0.69	0.77

Table 2 presents results of compressive strength estimated by using Equations 1 and 2 with the values of Figures 5 and data from Ferreira (1997).

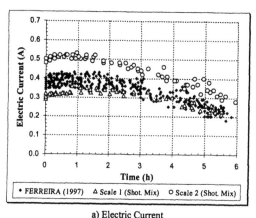
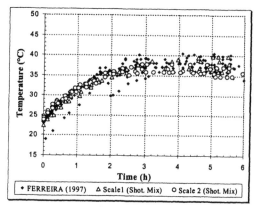

a) Electric Current   b) Temperature

Figure 4. Electric current and temperature: mix 2 and Ferreira (1997). Scale effects.

5.3 Scale influence study

Scale influence analysis was carried out by means of test sets number 2 and 3. Materials were CP II E 32 cement, granitic aggregate and rebar electrode with configurations shown in Figure 1 .

Electric current and temperature results are shown in Figure 4 . Absolute values of electric current are not to be directly compared because the areas of the specimens are different. They can be analyzed in specific terms by means of the resistivity. For tests on specimens with 10 x 10 x 24 cm dimensions, resistivity during the first hour was 4.3 Ω.m, and 3.5 Ω.m for specimens with 5 x 35 x 35 cm dimensions. The increase of electric resistivity with scale tends to decrease the specific electric current for larger dimensions. With respect to temperature, there is no practical influence of scale.

Figure 5 shows that the mechanical strength is higher for specimens with larger dimensions. Since the temperature increase was the same for both scales, there seems to be a stronger effect of electric current for larger specimens.

Table 3 shows ratios of temperature increase with the geometrical scale, in terms of volume and area of the exposed surface.

6 CONCLUSIONS

The method for concrete early age strength increase based on the application of electric gradients has shown good results when applied to a wide combination of materials, mixes and dimensions. These results have been used to improve the method and to obtain information for the explanation for the phenomena in the light of physical chemistry. So far, 3 types of cement, 2 aggregates, 3 different mixes and 2 types of electrodes have been tested. Scale effects have also been investigated. Strength increase through 10-hour age have been observed in all cases.

Aggregate type seems to be more important for test results than the type of cement. The type of electrode has also been shown to be very important.

Rebar used for lattice girders seems to be the most efficient material for electrode, which is important when thinking of possible field applications. Post-test conditions have also been favorable. No significant corrosion or electrolysis of the steel has been noticed in the time scale of days or months after the test. About concrete mix, higher water-cement ratios lead to higher electric current values. Results in terms of mechanical strength have been better with lower water-cement ratios, which shows that the method may be efficient for field shotcrete conditions. Strength results with mix 2 and rebar electrodes were 7 times higher for 6-hour age and 4 times higher at 8-hour age when compared to similar tests on mix 1.

Larger dimensions resulted in practically no change in temperature. Both electrical resistivity and mechanical strength increased.

Higher mechanical strength for larger scale is a very important fact. First, strength increase may be higher for field scale than for laboratory scale. Second, since there was no influence of scale on temperature, this is another strong argument that the reason for strength increase is not based on temperature increase.

Table 3. Temperature increase for specimens with different dimensions.

Dimensions (cm)	ΔTemp./Volume ($^\circ$C/cm^3)	ΔTemp./A$_{superf.}$ ($^\circ$C/cm^2)
10 x 10 x 24	5.93 x 10^{-3}	5.93 x 10^{-2}
5 x 35 x 35	1.95 x 10^{-3}	0.98 x 10^{-2}

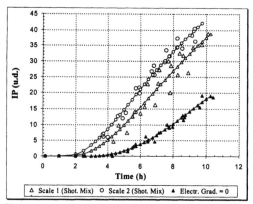

Figure 5. Penetration indices with the Meynadier Penetrometer: concrete mix 2.

7 ACKNOWLEDGMENTS

Funds for this research program have been provided by FAPESP -- the Research Funding Agency for the State of São Paulo, Brazil, -- under contract Nr. 98/02595-0. Prof. Nélio Gaioto from the University of São Paulo, São Carlos Campus, kindly lent the penetrometer to be adapted. Prof. Antonio D. de Figueiredo from the Polytechnic School, University of São Paulo, has offered important suggestions and lent the Meynadier penetrometer. Holdercim Brasil donated all the cement used in the experimental program.

8 REFERENCES

Ferreira, D. A. & Celestino, T. B. 1999. Strength increase in concrete by means of the application of electric gradients (In Portuguese). In: IBRACON 41, Seção 2.1 Concretos Especiais, código 2-1-02 (15 pg.). *Anais do 41° Congresso Brasileiro do Concreto*. Salvador (BA), Brasil (*em CR-ROM*).

Ferreira, D. A. & Celestino, T. B. 2000. Physical method to increase shotcrete early age-strength. *Proc. Conf. Shotcrete for Underground Support VIII*, ASCE, Campos do Jordão, pp. 94-107.

Ferreira, D. A. 1997. *Study about a physical method for the increase of early-age strength of concrete with the application of electric gradients for tunnel support.* (In Poetuguese). M.Sc. Dissertation, Escola de Engenharia de São Carlos – University of São Paulo

Ferreira, D. A., Celestino, T. B., Kupermann, S. C. & Bortolucci, A. A. 1997. Preliminary investigation about an alternative method for shotcrete early-age strength (in Portuguese).

Jornada Sul-Americana de Engenharia Estrutural XXVIII, Proceedings. São Carlos (SP), Brasil. Pp. 2139-2148

Figueiredo, A. D. 1993. *Shotcrete: Factors intervening in the process quality control.* (in Portuguese) M.Sc. Dissertation, Escola Politécnica, Univ. S.Paulo.

Figueiredo, A. D. 1997. *Control parameters and mix design of steel fiber reinforced shotcrete.* (in Portuguese) Doctoral Thesis, Escola Politécnica, Univ. S.Paulo.

Prudêncio Jr, L. R. 1993. *Contributions for shotcrete mix design.* (In Portuguese) Doctoral Thesis, Escola Politécnica, Univ. S.Paulo.

Silva, M. G. (1993). *Shotcrete with silica fume addition.* (In Portuguese) M.Sc. Dissertation, Escola Politécnica, Univ. S.Paulo.

Verbeck, G. J. & Helmuth, R. H. 1966. Structures and physical properties of cement paste. *International Congress on the Chemistry of Cement 5*, Proceedings. Tokyo, p. 1-32.

Modern Tunneling Science and Technology, Adachi et al (eds), © 2001 Swets & Zeitlinger, ISBN 90 2651 860 9

Influence of Invert Construction Procedure on the Deformation And Internal Force of Tunnel Lining

Zhu Hehua   Liu Xuezeng   Ye Bin
(Department of Geotechnical Engineering , Tongji University, Shanghai , 200092,China)

Liu Jun
(Guangzhou Science and Technology Committee, Guangzhou, China)

ABSTRACT: There are two kinds of invert construction procedure: one is constructing invert after initial lining and another is constructing invert after secondary lining. In this paper, taking the express way tunnel in Zhejiang Province as project background, according to the measured data, with FEM technique, the influence of different construction procedures of invert was evaluated on the convergent deformation and structural internal force, which provides guidance for rationalization construction of tunnel.

1 INTRODUCTION

Invert is an important part of lining structure of tunnel. It can not only make the mechanical system of lining structure which bears the loads of rock more rational and increase the degree of security of tunnel construction, but also extend the lifetime of lining structure and increase the reservation of tunnel construction. There are some problems in the design and construction of tunnel interiorly now. Because of the slow construction progress and unadvanced management, most inverts are constructed after the upper lining made up of arches and walls have been finished and stabilized, so, it is too late for invert to word as expected. There are few studies on the mechanics of invert construction. According to experiment of large-scale model and by using FEM, the stability of gross cavity, initial lining and secondary lining with and without invert is studied [1][2]. According to the model test and by using FE technique, the best occasion of invert construction and the longitudinal effect are studied.

In this paper, taking the express way tunnel as project background, the measured data in situ and the FEM technique are used. The influence of different construction procedure of invert is analyzed, constructing invert after initial lining and constructing invert after secondary lining, on the convergent deformation, structural internal force and interaction between surrounding rock and structure. Then the best occasion of invert excavation is confirmed.

2 ENGINEERING GEOLOGICAL CONDITION

Ren Hu Ling Tunnel is in the express way linking Shangyu and Sangmen. The left line is from FK85+200 to FK87+120 and right line is from K85+220 to K87+077. The total length of the two lines is 3777m. The gross opening which is surrounded by Class II rock is 11.77m wide and 9.99 high and it belongs to large and long tunnel. The maximal buried depth of the tunnel is 103m and the average buried depth is 60m. The covering is clay containing crushed stones, being $0-3.5$m thick. Under the overburden is weathered or micro-weathered felsite. Joints and crannies are developed and the rock is in the loose shape of berrica. The tunnel mostly lies in zone of joints and fracture, where the geological condition is complex and the surrounding rock is cracked and unstable. Ground water is mainly in the cranny of broken belt and bedrock and water content is little. However, because of the good water conductivity of cracked rock, it is possible that ground water leak into the tunnel when excavating. Due to many landslips taking place when excavating, invert should be constructed in time to form rational structural system to undertake loads. This measure is good to avoid the accident of landslip, decrease deformation of tunnel and structural force and increase the degree of security and the reservation of safety.

3 SIMULATION OF INVERT CONSTRUCTION WITH FEM

In order to analyze lining convergence and structural force after invert has been excavated, we simulate the whole procedures of tunnel excavation using GeoFBA2D software of Tongji University[3] [4]. There are two kinds of invert excavation: (1) Excavating invert and backfilling after initial lining have been finished; (2) Excavating invert and backfilling after secondary lining has been finished. Simulating mesh for FEM is as Fig.1.

The tunnel is excavated using two-step methods and the initial lining is composed of bolt and shotcrete, steel net and steel grid. The interval distance of steel grids is 1m. The length of the bolt is 3.5m. The thickness of steel grids is 15cm. The secondary lining is plain concrete of 40cm thick, which is simulated with beam element. The invert is plain concrete and simulated with filling element. The excavation depth of invert is 154cm.

The simulative procedure of the first kind of construction is:

excavating upper step → constructing initial lining → excavating lower step → constructing initial lining → excavating invert → backfilling → constructing secondary lining.

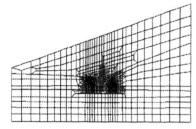

Figure. 1 FEM mesh

The simulative procedure of the second kind of construction is: excavating upper step → constructing initial lining → excavating lower step → constructing initial lining → constructing secondary lining → excavating invert → backfilling.

4 COMPARATIVE ANALYSES OF MEASURED DATA AND CALCULATED RESULT

4.1 Convergent deformation and settlement of the top of arch

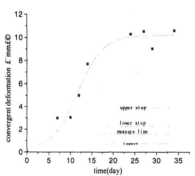

Figure 2   Convergence deformation and time （FK85+350）

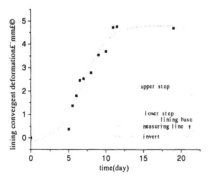

Figure 3.   Lining convergence deformation and time （K86+940）

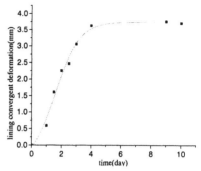

Figure 4. Lining convergence deformation and time（K86+945）

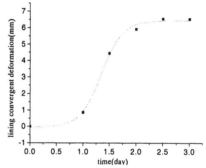

Figure 5. Lining convergence deformation and time （K86+952）

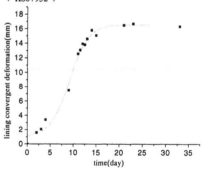

Figure 6. Lining convergence deformation and time(FK85+270)

In order to investigate the convergent deformation of the surrounding rock and the secondary lining after invert excavation embedded inserts are put into the lower step and the base of the secondary lining. Fig. 2 shows the relation between the convergent deformation of the surrounding rock and time in the case that invert is excavated after the construction of initial lining and before the construction of secondary lining. Fig.3-5 shows the relation between convergent deformation of the secondary lining and time (the embedded inserts are put into the base of the secondary lining, 0.5m away from the top of the tunnel) in the case that invert is excavated after the construction of secondary lining. The measured data indicate that the invert excavation will cause the deformation of the surrounding rock and the secondary lining. In the case of excavating invert after the

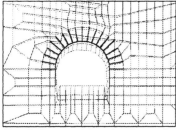

Figure 7. Convergent deformation of surrounding rock after invert construction

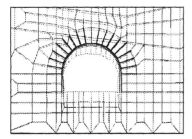

Figure 8. Lining convergent deformation after invert construction

initial lining has been finished convergent deformation of the surrounding rock reaches 11.28mm ; in the case of excavating invert after the secondary lining has been finished the deformation of the secondary lining varies according to different geologic condition and the deformation reaches 4.69mm, 3.71mm, 6.35mm in section K86+940, K86+945, K86+952 respectively. In the entrance of the left line (Section FK85270, Class II rock) the convergent deformation reaches 16.35mm as shown in Fig.6. Commonly, if surrounding rock is in normal condition, the deformation of the surrounding rock in the case of excavating invert before the construction of the secondary is larger than that of the secondary lining in the case of excavating invert after the construction of the secondary lining as a result of the resisting effect the secondary lining acts upon the surrounding rock. However, it is possible that bad construction quality (excessive excavation of rock under the base of secondary lining caused by blasting) or bad geological condition (geological fault) will cause too large deformation of secondary lining, decrease the degree of security and reduce the capacity to resist adverse factors. The simulative calculation using FEM shows that if invert is excavated after the construction of secondary lining, the convergent deformation reaches 15.2mm (Fig.7, calculation and measure dealing with the same position) ; if invert is excavated after initial lining has been finished, the deformation of the surrounding rock reaches 16.52mm (Fig. 8, calculation and measure dealing with the same position). The measured data indicate that backfilling can resist the deformation of the surrounding rock and the secondary lining efficiently. The deformation of the surrounding rock is under control about 5 days after backfilling of invert ; and to the deformation of the secondary lining, the time is shorten to 2—3 days.

The settlements of the secondary lining in different sections caused by the invert excavation after the construction of the

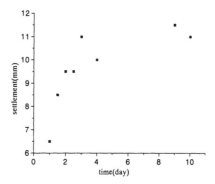

Figure 9 Lining settlement and time（ K86+945 ）

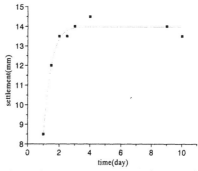

Figure 10. Lining settlement and time（ K86+952 ）

secondary lining are shown in Fig.9 and Fig. 10 . the settlements reach 11mm in section K86+945 and 13mm in section K86+952, The magnitude of simulative calculation being 12.54mm. The settlement is mainly caused by the convergent deformation arising from stress releasing in the spring of arch and the deformation of the surrounding rock (reserved settlement is neglected because of the quality of construction). During the actual invert excavation, due to the excessive excavation of the surrounding rock under the base of the invert the settlement of the secondary lining is rather large and the top of arch even separates from the surrounding rock. The settlement of the top of arch reaches 5.21mm if invert is excavated after the construction of the initial lining.

4.2 Analyses of Structural Force and Interaction between Surrounding Rock and Structure

Due to the different procedure of invert construction, the internal force of initial lining and secondary lining, the reservation of safety of secondary lining and the mechanical system undertaking loads vary after tunnel is finished. Finally, the lifetime of tunnel is influenced by all these factors.

By analyzing the two kinds of invert excavation have different influence upon structural force of lining and interaction between surrounding rock and structure. If invert is excavated after the construction of secondary lining, the maximal moment of secondary lining is 218.35KN/m, the maximal shearing force being 118.33KN/m, the maximal internal force of initial lining being 1005.80KN/m, the maximal axial force of bolts being 91.32KN/m, the interactive force between the surrounding rock and the initial lining being 5628.60KN/m ; if invert is excavated before the construction of the secondary lining, there is no internal force in the secondary lining, the maximal internal force of the initial lining being 1642.90KN/m, the maximal axial force of the bolts being 100.37KN/m, the maximal interactive force between the surrounding rock and the structure being 5632.70KN/m. The internal force of the initial lining of the former is 63% more than that of the latter and the interactive force between the surrounding rock and the structure is close to each other. It is found by comparative analysis that the secondary. Lining and the initial lining bear loads acted upon by surrounding rock at the same time if invert is excavated after the construction of the secondary lining ; however, if invert is excavated before

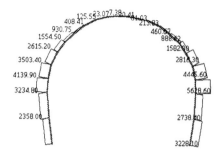

Figure 15. Interaction between structure and surrounding rock

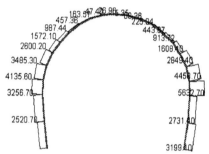

Figure 16. Interaction between structure and surrounding rock

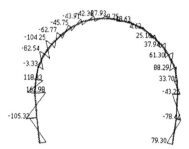

Figure 12. Lining shearing force

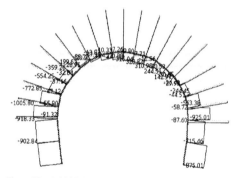

Figure 13. Initial lining and bolt internal force

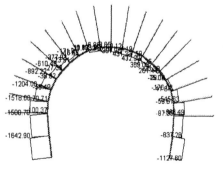

Figure 14. Initial lining and bolt internal force

the construction of the secondary lining, all the loads are undertaken by the initial lining solely. So, in view of structural force and interaction between the surrounding rock and the structure, excavating invert after the construction of the initial lining and backfilling in time is good to form the rational mechanical system. This also increases safety of the secondary lining and reduces the adverse factors during construction.

5 CONCLUSION

According to the measured data and by using FE technique, the following conclusions are obtained:

(1) Under the same geological condition of the surrounding rock the deformations of the surrounding rock and the secondary lining vary according to the different invert construction procedure. The deformation of the surrounding rock in the case of excavating invert after the construction of the initial lining is larger than that of the secondary lining in the case of excavating invert after the construction of the secondary lining.

(2) Structural force of initial lining in the case of excavating invert after the construction of initial lining is larger than that in the case of excavating invert after the construction of secondary lining. The interaction force between surrounding rock and initial lining of two kind of excavation are almost same.

(3) Invert excavation after the construction of initial lining is good to form rational mechanical system to bear loads, to reduce the influence on the secondary lining, and to increase the reservation of the secondary lining, so that the lifetime of the tunnel is extended.

REFERENCES

1 Wang Mingnian, Wong Hanmin. Mechanical study of tunnel invert. Chinese Journal of Geotechnical Engineering. 1996, Vol.18 (1):46~53
2 Chou Wenge. Experiment study on construction time and longitudinal effect of tunnel invert. Chinese Journal of southwest traffic University. 1993, No.6: 68~73
3 Zhu Hehua, Yang Linde & Hashimoto Tadashi. Back analysis of construction of deep excavation and deformation prediction. Chinese Journal of Geotechnical Engineering. 1998, Vol. 20 (4):30~35
4 Zhu Hehua, Ding Wenqi & Li Xiaojun. Handbook on Shuguang design and construction analysis software of Tongji University. Tongji University, Shanghai, 1997.

Modern Tunneling Science and Technology, Adachi et al (eds), © 2001 Swets & Zeitlinger, ISBN 90 2651 860 9

Efficient TBM Driving by Observational Construction Management

H.Niida
Japan Highway Public Corporation,Kansai Branch Office,Otu Construction Office, Otu Higashi Section,Shiga pref.,Japan

F.Kamada & T.Nishizono & H.Shigenaga & S.Mutaguti & Y.Miyajima & Y. Sawamura
Kajima Corporation, JDC Corporation, DAI NIPPON Construction JV,Shiga pref.,Japan

ABSTRACT: The Second East Meishin Expressway, parts of which are under construction work in different locations in Japan today, is expected to play the role of a trunk line of future transportation. The road is of such high specifications that it has six lanes (or three vehicle lanes on one side) and provided with a berm on both sides throughout the route. Therefore, tunnels need an extremely large cross section of 180 square meters for excavation, and a Tunnel Boring Machine (TBM) Pilot Drift Widening technique is adopted in the Rittou tunnel. Although the TBM method is featured by its high construction speed compared with the NATM technique, a number of delayed construction cases are reported in which TBM excavation is suspended due to geological trouble under complicated Japanese geological conditions. In this context, the authors developed a construction system before starting TBM construction. The system is based on observational construction management mainly forward prediction combined with auxiliary techniques to ensure prompt and continued excavation without suspending TBM operation even in areas of bad earth quality. As a result, we could take proper measures promptly when poor geological quality was encountered so that we could complete the excavation of approximately 2,600m at a satisfactory mean monthly progress of 338m without experiencing remarkable trouble, and thus the effectiveness of this technique has been confirmed.

1 INSTRUCTION

The Rittou tunnel of the Second Meishin Expressway has been designed and constructed so that the tunnel has as large an excavation cross section as about 180m² and aspect ratio of 0.65 or so. For the west section of the down line, the TBM technique will be used to excavate a pilot drift of 5m diameter and then the drift will be widened until the cross section of the main tunnel will be achieved.

The TBM excavation method is adopted in order to secure high-speed driving. However, in the construction of the up line executed prior to the down line TBM driving, some sections were identified to involve faults and poor ground qualities that may impede driving of the TBM.

Therefore, different sorts of potential geological trouble were assumed before starting TBM construction of the down line and "fore piling", "urethane void filling" and other auxiliary techniques were arranged beforehand. In addition, we applied a observational construction management consisting of forward prediction by drill logging. This system was built up so that collected geology information would be used for feedback and the most proper auxiliary technique would be used for construction. As a result, efficient TBM driving can be carried out without experiencing remarkable geological trouble. This paper introduces the outline of the system and the result of application.

2 LOCATION OF THE PROJECT AND BRIEF GEOLOGY

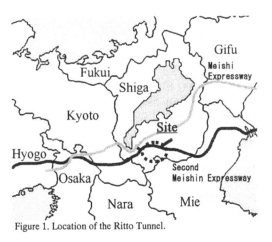

Figure 1. Location of the Ritto Tunnel.

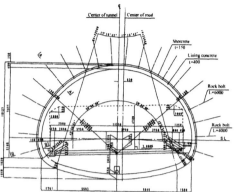

Figure 2. Normal Cross Section of Tunnel.

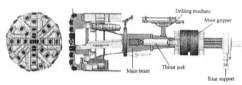

Figure 3. Schematic View of TBM Structure.

The location of the subject tunnel is shown in Fig. 1. The Rittou tunnel is located in a mountainous area of Kurita-gun (gun means a province) of Shiga prefecture, 10km east-southeast from the south end of Lake Biwa. It is surrounded by mountains of 300 to 600m elevation above sea level. Although relatively gentle slopes are seen in the crest area, steep cliffs are found in valleys and along rivers. The basement rock of the tunnel section consists of coarse grains of biotite granite, posing a fresh and uniform hard rock zone. There exists, however, complicated small fracture zones and shear zones, around which very fragile sections containing clay and decomposed granite soils are found.

3 OUTLINE OF CONSTRUCTION

Name of project: Down line west operation of the Rittou tunnel, the Second Meishin Expressway
Site for operation: Kurita-gun (county), Shiga prefecture to Ootsu City, Shiga prefecture, Japan
Construction period: From October 30, 1998 to March 12, 2002
Details of tunnel section (Operation No. 1)

(1)   Total length: 2,635m

(2)   TBM drift: 2,469.3m
(Diameter of 5.0m, excavated cross section of about 20m^2)

Table 1. Specifications of TBM.

Name	Major specifications
Type	Open
Diameter of bore	ϕ5.0m
Length of machine	15.4m
Cutter diameter/cutters	ϕ432mm/38 pieces
Motor HP/cutter torque	6 units of 180 kW/128.5/257tf-m
Thrust/thrusting stroke	Max. 932 tf/1,500mm
Gripper type	Horizontal facing
Gripper force	Max. 2,440 tf

(3)   Main tunnel widening: 1,838m
(NATM excavated cross section of about 180m^2)

(4)   Lining: 1,683m

(5)   Takeoff working pit: 94m (excavated cross section of about 32m^2)
The normal tunnel cross section is shown in Fig. 2.

4 OUTLINE OF TBM

The down-line west section continuously is driven by using the tunnel boring machine (or TBM) that was also used in the up-line west section constructed earlier. In the initial phase, the Rittou tunnel section was considered to have a uniform geology of 70% B-class natural ground. Therefore, a TBM of an open type is adopted, which fits the excavation of hard rock natural ground excavation. During the construction of the up line, however, a cave-in accident took place in a section of poor geological condition that we could not predict in the preliminary investigation. In the cave-in section, we had to suspend operation for three months in total while discussing remedial measures, making preparation, and executing the measures. For prevention of such possible suspension during the operation of the TBM also in the down line, modifications were provided to the machine such as extension of the side supports and reduction of the slit aperture ratio in order to improve the workability in the section of poor geological condition by controlling the fall of side wall and working face. In addition, a drilling machine (COP1838 by Atlas) was installed on the bench at the rear of the roof support. The drilling machine was used for auxiliary techniques such as fore piling and drill logging for the purpose of prediction of geological condition ahead of face. In addition to the above devices, a rotation mechanism was mounted on the boom to allow arch-shaped fore piling to the working face.

5 OBSERVATIONAL CONSTRUCTION MANAGEMENT

To perform TBM construction, a observational construction method was applied mainly consisting of a system (a TBM navigator) to synthetically control geological data, machine data and other information. The TBM navigator accumulates "drill logging data" as the information about forward geological condition and "muck belt scale information", "TBM machine data" as the working face information. These pieces of data are displayed in real time on the display of the personal computers in the operator cab and the field office. The following section describes drill logging and the muck belt scale.

5.1 *Drill logging*

To acquire geological information ahead of the face, we adopted a drill logging method taking into account the workability and accuracy. Table 2 shows the points in selecting the forward prediction techniques and the features of drill logging method. The drill logging system consists of a drilling machine that makes a hole in front of the working face to obtain geological data, enabling the staff members to determine the hardness of the rock ahead. Determination is made using the drilling energy, which means the "hydraulic drill energy needed to destroy a specific volume of rock" (Equation 1).

$$Ev = \frac{Es \times Ns}{Vd \times Ar}$$

$$= \frac{Total \cdot work \cdot of \cdot hydraulic \cdot drill}{Volume \cdot of \cdot destroyed \cdot rock} \qquad (1)$$

where Ev =drilling energy; Vd =driving velocity; Ar =the cross-sectional area of the drill; Es= work of hydraulic drill.

The greater the drilling energy is, the harder the rock is determined to be by the equation. To evaluate the information from drill logging, all data was accumulated for the initial section of about 150m and calibration was conducted by comparing the data

Figure 4. Drill Logging.

with the actual geological condition of made excavation. As a result, it was found that drilling energy of 150 (J/cm^3) less indicates such soft rocks that are liable to involve geological trouble. Further, if cracks are developing, or where the boring hole collapses in weak ground, wide variances are noticed in drilling energy because the drilling rod jams when boring is made. We determined such sections to be of poor geological condition. In addition the slime coming out of the hole while it is bored was sampled for observation in order to determine the degree of rock metamorphism. Moreover, a borehole camera was used to observe the inner wall of the hole for more detailed forward prediction. Drill logging was aimed at the point 30m ahead of the face taking into account the reliability of accuracy and workability. Drill logging was conducted every 30m driving of TBM throughout the tunnel length. TBM driving was suspended during drill logging, while the machine was serviced for maintenance and thus search did not affect the availability of the TBM driving.

5.2 *Muck belt scale*

If cave-in occurs at the working face, more muck has to be taken in by the volume of natural ground that has fallen into the cave than the original volume of muck that should be taken in by normal TBM

Table 2. Points for selection of forward prediction .

Item	Point for selection of techniques	Features of drill logging	
Necessary Information	Hardness of rocks	Evaluation by drilling energy	
	Metamorphism-affected region	Evaluation by slime	
Workability	Can be conducted during construction cycle	Work time:	3hours
		Search distance	30m
		To be conducted simultaneously with cutter replacement, TBM servicing, continuous belt conveyor extension	
Rapidity	Immediate evaluation of search result to supply feedback data to current operation	1 hour for analysis and evaluation time	

driving. Utilizing this theory, a belt scale was mounted on the muck belt conveyor to monitor the volume of taken-in muck, which allowed us to earlier find cave-in through comparison of the rate of muck weight per specific time with the TBM driving velocity.

5.3 *Flow of TBM operation*

To enhance observational construction method, we configured a system to use collected information rapidly and exactly. For the concept of the system, refer to Figure 5 showing the flow of observational construction management.

6 AUXILIARY TECHNIQUES

A menu of auxiliary techniques was created beforehand as shown below, taking the geological trouble that occurred in the up line for reference in order to help planning of remedial measures for poor geological condition and cave-in accident at the working face. If the working face falls down in real construc-

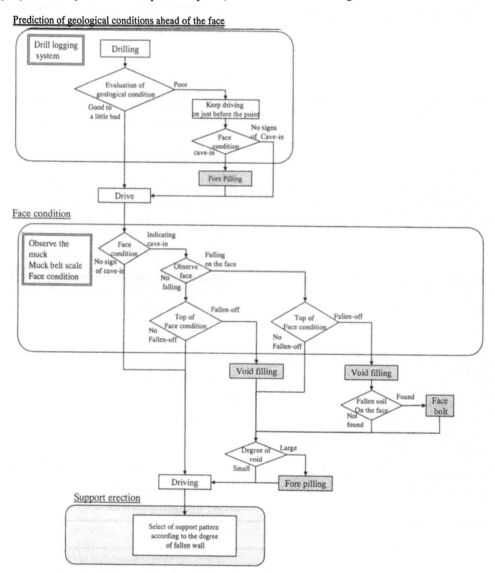

Figure 5. Flow of observational construction management.

tion, the cave-in would expand with time. Therefore, taking quick measures is essential if cave-in occurs.

6.1 Fore pilling (figure6)

This technique is used to provide prior support against poor geological condition. It is also used to reinforce top and front of face where a cave-in accident has been occurred. Operation is conducted by using the drilling machine mounted on the bench at the rear of roof support. The procedure of operation consists of: drilling holes by self-boring bolts; then consolidating the natural ground in vicinity by silica resin injection. Ten bolts of 10m length were used per section at every 3m TBM driving.

6.2 Face bolt (Figure7)

This technique is used to restrict the take-in of the muck that bears the improved body filled for reinforcement of the face top during TBM driving by consolidating the muck deposited in front of the TBM caused by cave-in. To do this, first drill (using a self-boring fiber bolt of 2.0m length) the muck in front of the face by leg drill through the clearance between cutters inside the chamber, and then consolidate the muck by silica resin injection.

6.3 Void filling (Figure8)

This technique fills the void created by falling with urethane in order to prevent the cave-in from expanding. To achieve void filling, first insert a 4m PVC pipe through cutter clearance, protect the face disc with Styrofoam etc. so that the cutter will not be constrained later by urethane and then fill the void with urethane. For injection, we used a 20-times foaming material suitable to void filling.

7 RESULT OF CONSTRUCTION

During our TBM construction, seven cave-in accidents including great and small were encountered, against which we could take proper and prompt action in faults and sections of poor geological condition owing to the observational construction management mainly consisting of drill logging. As a result, the work schedule was secured satisfactorily, minimizing time-out due to suspended TBM driving. Table 3 lists major data of the construction work.

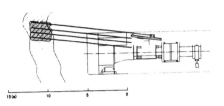

Figure 6. Schematic fore pilling.

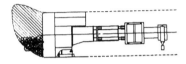

Figure 7. Schematic face bolt.

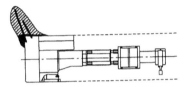

Figure 8. Schematic void filling.

7.1 Sixth cave-in accident (Fore piling applied to the section of poor geological condition predicted by drill logging; see Figure 9)

At the point approximately 1,530m from the start of TBM driving, relatively favorable geology continued and TBM driving advanced with ease. However, a small cave-in took place 2m ahead and 0.3m upward of the working face due to a locally poor geological condition. Drill logging had revealed beforehand that a section of poor geological condition existed at this point, and it would continue for about 11m. Therefore, we promptly decided to carry out three sets of fore piling to prevent the cave-in from expanding as well as for reinforcement of the front ground. As a result, the cave-in did not develop to the face top, although falling continued toward ahead of the face, allowing us to break through this challenging section. Total length of the poor geological section was 11m in total, and two days were spent for the operation.

8 CONCLUSION

Our TBM driving construction in the down line west of the Rittou tunnel achieved a successful record of mean monthly progress of 338.3m. The result may indicate that the geology was generally favorable

479

Table 3. Result of TBM operation.

Driving period	1999.11.18~2000.6.26	Maximum daily progress		36.5m
Excavated length	2469.3m	Maximum monthly progress		516.6m
Calendar months	7.3	Mean monthly progress		338.3m
Calendar days	222	Mean daily progress	Per calendar day	11.1m
Work days	191		Per work day	12.9m
Operation days	158		Per operation day	15.6m

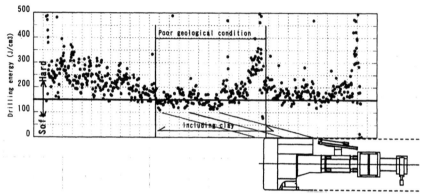

Figure 9. Fore pilling applied.

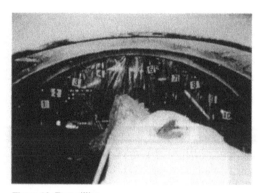

Figure 10. Fore pilling.

Figure 11. TBM tunnel.

throughout the total length, with D-class quality being only 10% or so. However, the actual geology

was more complicate than expectation so that hard and soft rocks were interwoven even within the TBM face of 5-m diameter. We experienced seven cave-in accidents including the cases introduced above. One fault was so big in the falling scale that it lasted 30m in total and twelve days were required to cope with it. However, we did not have any critical TBM driving suspension in these situations. We consider that this owes to the fact that prediction of poor geological conditions in high accuracy by drill logging enabled us to make prompt determination. In addition, the system of supposed application of auxiliary techniques allowed us to deal with different cases successfully. A rational and efficient construction procedure for TBM driving in natural hard-rock ground has been established during our tunnel project. We believe the concept of this system will make an effective and versatile solution in future TBM construction if we add small modifications according to the conditions of different sites.

REFERENCES

Y.Miyajima, T.Yamamoto, T.Kato, F.Kamada 1998.Result of Forward Prediction Using Combined Drill Logging and Velocity Logging, *Abstracts of the 53rd Annual Academic Lecture Convention of Japan Society of Civil Engineers(in Japanese)*, 206-207.

Modern Tunneling Science and Technology, Adachi et al (eds), © 2001 Swets & Zeitlinger, ISBN 90 2651 860 9

Study on the application of a large TBM to Hida Highway Tunnel

K. Miura, M. Kawakita, T. Yamada & N. Sano
Japan Highway Public Corporation, Tokyo & Nagoya, Japan

K. Ryoke
Taisei Corporation, Tokyo, Japan

ABSTRACT: This paper presents the results of study on the application of a large TBM (dia.12.84m) to Hida tunnel, which is the second longest highway tunnel (10.75km long) in Japan. The improved open type TBM, which combines advantages of open type TBMs' rapid excavation abilities in good rocks and shielded type TBMs' stable progress and safety work conditions in weak rocks, has been developed in order to suit the geological conditions of the tunnel. The new ventilation system, which uses the lower half section of the bored tunnel for ventilation ducts instead of the shaft, has also been developed for the first time in the world.

1 INTRODUCTION

Hida tunnel lies on the route of Tokai-Hokuriku Highway, connecting Nagoya (Pacific side in central Japan) and Toyama (Japan Sea side). It is the second longest highway tunnel in Japan, with its length approximately 10.75 km, and consists of a two-lane main tunnel and an emergency tunnel.

This study aims to analyze the suitability of TBM method applied to Hida tunnel project and to develop the appropriate TBM. It has consequently been decided to adopt the TBM method for both tunnels.

This paper presents the features of this project and the results of this study, including the specifications of a large TBM and its countermesures for unfavorable difficult geological conditions, and

describes the outline of newly developed ventilation system.

2 OUTLINE OF THE TUNNEL

2.1 Specifications of the tunnel

Tunnel Diameter: Main tunnel 12.84 m
                 Emergency tunnel 4.5 m
Total length: approx. 10,750 m
Design Speed: 80 km/h
Longitudinal gradient: 2% upwards from Northern portal (Toyama side) toward southern portal (Nagoya side)
Smallest radius of curvature: R=1,500m
Distance between main and emergency tunnel: 30 m center to center (See Figure 2)
Maximum overburden: approx. 1,000 m

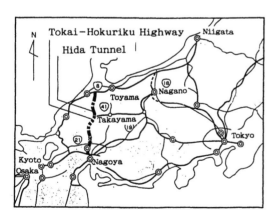

Figure 1. Location of the Hida Tunnel.

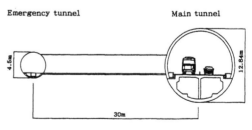

Figure 2. Distance between main tunnel and emergency tunnel.

2.2 Geologic conditions

Hida tunnel passes through mountainous region around Mt. Mominuka (1,744 m above sea level).

A geologic longitudinal section is shown on Figure 3. The estimated geological conditions consist of following the direction of tunnel driving from north to south, Shirakawa granite (Cretaceous Period) comprising 18%, Nohi rhyolites (Welded tuff of Cretaceous Period) 34%, and Hida metamorphic rocks (from Jurassic through Triassic Periods) 48%. In addition, intrusive rocks of Tertiary Period are deemed to be present.

The geological conditions of the region are generally good, with over 90% being estimated as medium hard rocks (of rock classification B to CII).

However, the following types and conditions of ground can be encountered as potential excavation problems due to peculiar geological features in the region:

- Large and high-pressure water inflow in rock class D and/or high overburden section.
- Squeezing in faulted and intrusive weak rocks.
- Collapse of fractured rocks.
- Rock burst at high overburden section.
- Settled discharge water volume at Toyama side portal is estimated approx. 22 m³/min, and hence heavy inflows during excavation are anticipated.
- Rock mass strength ranges from 160 to 360 MPa for Nohi rhyolites, and up to 85 MPa for Hida metamorphic rocks.
- Quartz contents of rocks, which would affect the excavation efficiency and cutter consumption, are approx. 50 % for Shirakawa granite, approx. 37 % for Nohi rhyolites and between 23 to 50% for Hida metamorphic rocks.

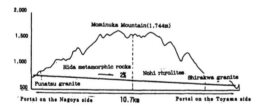

Figure 3. Geologic longitudinal section.

2.3 Climate

The regions around both portals are known for heavy snowfall (Maximum 3 m with a recurrence interval of 10 years) and low temperature (the lowest temperature between −7 to −10 degrees Celsius in January).

It is especially difficult to provide measures against heavy snow and avalanche, and to maintain access roads and portal area in the region near Nagoya side portal.

2.4 Landscape

Ogimachi Historic Village of Shirakawa-go, located near the Toyama side portal, has been registered as "World Heritage" in December 1995.

The alignment and elevation of the route of Tokai-Hokuriku Highway was decided by considering landscape preservation, and to be invisible from the historic village. The same consideration is also given to construction works. The temporary construction yard and the portal of work adit are lacated beyond a ridge behind the historic village.

3 OUTLINE OF THE MAIN TUNNEL

3.1 Selection of excavation method

If the tunnels are driven by the drilling+blasting method from both portals, the tunneling works from Nagoya side would be interrupted in winter because of heavy snowfall and avalanche, and be anticipated to encounter large water inflow problems with down grade 2 % excavation. This method, if chosen, will possess construction schedule problem.

The ventilation shaft will be required to construct in national park area. It causes to increase construction cost and induce environmental problems.

Driving the tunnels by TBMs in one direction from Toyama side has advantages and seems more suitable for the following reasons:

i) Expected geological conditions are fairly stable and suitable for TBM excavation;

ii) The emergency tunnel (driven by 4.5 m dia. TBM) will precede grasping the geological conditions of the whole tunnel route. It has also a role of drainage for the main tunnel excavation;

iii) Since the cross section of the main tunnel is circular, the lower half section could be effectively used as ventilation ducts, therefore eliminating the necessity of ventilation shaft; and

iv) Due to the situation mentioned above, the construction period driven by TBM would be shorter and the construction cost would be lower in comparison with the drilling+blasting method from both portals.

3.2 Selection of TBM type

Considering the geological and executive conditions, the types of the TBMs (Single shield, Double shield, Open and Improved Open types) were compared and studied in order to choose the appropriate TBM type.

The following factors were discussed:

i) Suitability for rapid excavation in good rocks.

482

ii)   Minimizing the delay in unfavorable and unstable rocks.
iii) Applicable economical support systems.
iv)   Compatibility to geological changes.
v)   Preventing TBM from trapping in squeezing rocks, fractured rocks, etc.

As a result, the Improved Open type TBM is chosen. It combines both features of open and single shield type TBM. Rapid excavation and application of economical support system (mainly combination of shotcrete and rock bolts) are possible in good rock conditions with driving by open type TBM mode (See Figure 4).

Stable advance and safety works are expected in unfavorable and unstable rock conditions with driving by single shield TBM mode with reinforced concrete liner installed at inside the shield tail. The shield length is designed as short as possible to prevent it from being trapped in squeezing rock conditions.

The auxiliary methods are applicable just behind cutterhead and shield tail with drilling equipments and etc.

3.3 *Specifications of TBM*

The specifications of the TBM are shown in Table 1.

The continuous conveying system will be used for mucking out. It contributes to keep rapid excavation of TBM and clean working circumstances in tunnel.

Table 1. Specifications of main tunnel TBM.

Item	Specifications
Machine Diameter	12.84 m
Diameter and Number of Cutters	19 inches × 91 pcs (including 3 over cutters)
Thrust	33,354 kN
Rotating Speed of Cutterhead	0 to 4 rpm
Cutterhead Torque	Low speed: 31,843 kN-m (at 1.275 rpm)
	High speed:11,772 kN-m (at 4 rpm)
Cutterhead Power	4,250 kW
Auxiliary Propel Force	55,427 kN
Main Gripper Force	44,145 kN × 2

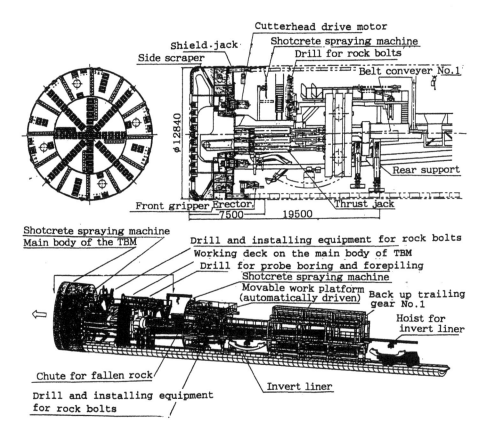

Figure 4. Improved Open type TBM.

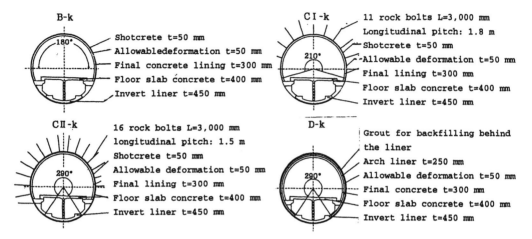

Figure 5. Support systems.

3.4 *Support system and excavation radius*

Support systems are shown in Figure 5. Reinforced concrete invert liner will be installed for the whole length of the tunnel. In rock class ranging from B to CII, shotcrete and rock bolts are used for initial support, executed just behind the shield tail, in front of main gripper. Reinforced concrete arch liner will be adopted for rock class D. In case that rock fall occurred in rock class B to CII and the fallen rocks piled up on the shield, the simplified steel liners are being considered.

The excavation radius was decided to be 12.84 m by considering the use of reinforced concrete liner.

3.5 *Countermeasures against unfavorable rocks*

The anticipated geological conditions that may cause excavation problems during tunnel construction and require countermeasures against them are summarized in Table 2.

It is planned that the excavation of the emergency tunnel would precede that of the main tunnel by 3 to 4 km. This situation offers geological information to the main tunnel in advance and enables to provide measures for problematic conditions, both from the main tunnel and emergency tunnel.

3.6 *Structures underneath the floor slab*

Since the main tunnel is of a circular cross-section, it is planned to utilize the lower half section as ventilation ducts. Precast invert liner constitutes a part of the structures. Another sidewalls, center wall and floor slab will be constructed by cast-in-place concrete. The re-bar works, form works, placing concrete and curing will be done underneath mobile decks.

These construction works are independent from the excavation progress, but they will be simultane

ously carried out to keep the moderate space behind the TBM back up system.

The completed floor slab is used for temporary roadway during construction.

3.7 *Power supply*

The commercial electric supply is limited up to 3,500 kW (2,500 kW for the emergency tunnel and 1,000 kW for the main tunnel).

The six diesel generators (Capacity of 1,500 kW each, including spare one) are provided to supply 7,500 kW electric power for the main tunnel execution.

3.8 *Assembly of the TBM*

The TBM was assembled in the main tunnel because of the following objections;

i) Portal area is too narrow to assemble the machine during continuing tunnel excavation.
ii) Adit with a large cross-section is necessary and the excavation of emergency tunnel has to be interrupted during installation of main tunnel TBM.

The assembling cavern (cross-section: approx. 458m^2, length: 41 m) was constructed in the main tunnel. The cross-section is shown in Figure 6.

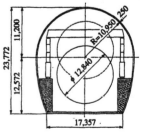

Figure 6. Cross-section of assembly cavern.

Table 2. Countermeasures against unfavorable rocks.

Significant Trouble	Countermeasures			Specification and Condition of Equipments for Countermeasures
	Ahead of Cutterhead	At Face	Behind of TBM	
Large and high-pressure water inflow	-Drilling exploratory and drainage holes ahead of cutterhead from a drilling pit while TBM can bore the tunnel continuously. -Drilling drainage hole(s) through cutterhead while TBM stops boring. -Grouting to reduce and stop water inflow -Drainage boring from a connection tunnel excavated from the emergency tunnel.	-Dewatering facility -Drilling drainage hole(s) behind the shield tail by hydraulic drifters. -Grouting into the water bearing stratum to reduce and stop water inflow.	-Adoption of tunnel liner for more endurable rock support. -Use of quick setting grout behind the liner.	-Rapid drilling/boring machines (Equipped in front of main gripper) (Use for exploratory boring, drilling drainage holes and grouting) -Dewatering facility -Injection pump and mixing plant for grouting, arranged in backups.
Rock fall and Collapse of fractured rocks and/or fault zone. Cave-in	-Exploratory boring which also has a role of drainage. -Grouting to improve ground cohesion -Forepiling with grouting to stabilize surround ground -Pre-improvement from a connecting tunnel at emergency tunnel.	-Grouting into loosened rock zone. -Forepoling with grouting to stabilize crown area.	-Use of tunnel liner with quick backfill grouting behind the liner.	-Drilling and grouting facilities for exploratory boring, forepiling and forepoling in front of main gripper.(Hydraukic drifters) -Grout injection facility, equipped in backup.
Trapping in squeezing rocks	-Exploratory boring. -Grouting to improve ground cohesion. -NATM excavation of crown part of main tunnel accessed from the emergency tunnel.	-Change to single Shield mode excavation method, driving with shield jacks and tunnel liner and enable to advance TBM continuously. -Enlargement of excavation diameter by over cutting system. -Injection of friction reducing material behind the shield skin.	-Use of tunnel liner with quick backfill grouting behind the liner.	-Over cutter (Overcutting:+100mm)
Machine subsidence in weak rocks	-Grouting to improve ground cohesion.	-Grouting to improve bearing capacity of the ground under the TBM shield. -Pitching control of upward deviation of the excavation axis.	-Grouting to increase the bearing capacity of the base ground under the invert liner. -Foot piles to resist for sinking of full circumference liner.	-Drilling equipment -Grouting facility
Unstable face	-Exploratory boring which also has a role of drainage. -Grouting to improve ground cohesion -Forepiling with grouting to stabilize surround ground. -Face bolts. -Pre-improvement from a connecting tunnel at emergency tunnel. -NATM excavation of crown part of main tunnel accessed from the emergency tunnel.	-Grouting into loosened rock zone. -Forepoling with grouting to stabilize crown area.	-Use of tunnel liner with quick backfill grouting behind the liner.	-Drilling equipment -Grouting facility

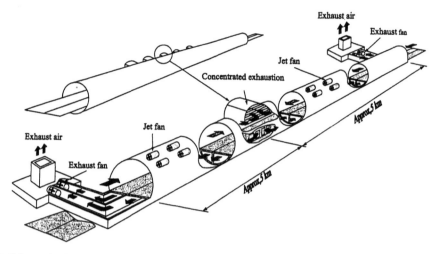

Figure 7. Schematic diagram of the ventilation system.

4 VENTILATION SYSTEM

The schematic diagram of the ventilation system is shown in Figure 7.

This system utilizes the lower half section of the main tunnel as ventilation ducts, and adopts longitudinal ventilation with the extracting method.

Fresh air is inhaled from both portals. Polluted air is extracted through the concentrated exhaustion inlet near the middle of the tunnel and exhausted through the ducts under the roadway to the portal section.

Generally in case of ventilation of the extraction method, natural draft and unbalance of traffic volume in a two-lane tunnel (Face-to-face traffic) result in the difference between the location of the highest air pollutant exhaust concentration of vehicles and exhaustion vents. The jet fans are used to adjust draft in the tunnel and cancel the difference between them.

Hida main tunnel has an advantage that it can freely select a location of exhaustion vent for canceling the difference, by means that the ventilation ducts are provided for the whole length of the tunnel.

In case of a fire, it is planned that air stream in the tunnel must be stopped first in order to secure the passengers and keep evacuation routes, and the smoke of fire is exhausted from the nearest vent to the ventilation ducts.

This ventilation system eliminates the construction of underground ventilation station, ventilation shaft and electric dust collector chambers.

As a result, it is expected that the construction period and cost and operation cost of ventilation facilities will be reduced.

5 CONCLUSIONS

Outline of the Hida tunnel project has been described in the above paragraphs. This project is characterized by many features, including: the use of TBMs for the emergency and main tunnels; the tunnel length exceeding 10 km; the large diameter of the TBM for the main tunnel being 12.84 m; and the ventilation system that makes use of sectional configuration of the main tunnel eliminating the ventilation shaft and underground facilities.

The length of the longest tunnel excavated by TBMs in Japan at present is less than 7 km, and the largest diameter is 8.3 m. Consequently, the Hida tunnel will be the longest and largest tunnel excavated by TBM in Japan.

The possibility to encounter large and high pressure water inflow, collapse in unstable rocks and trapping in squeezing rocks during excavation of the tunnel must be kept in mind because it is the first time to bore the tunnels with TBMs in mountains which have more than 1,000m overburden.

The Improved Open type TBM was chosen as the appropriate TBM for this tunnel conditions. The countermesures against unfavorable rocks are also studied beforehand. The emergency tunnel, which would precede the main tunnel, has an important role to provide geological conditions and to drain the ground water for the main tunnel excavation.

REFERENCES

Japan Highway Public Corporation. March 1998, Guidance for the TBM Design and Construction.

Modern Tunneling Science and Technology, Adachi et al (eds), © 2001 Swets & Zeitlinger, ISBN 90 2651 860 9

Constructing a Tunnel through the Embankment of National Road No. 4 Using the AGF Method

M. Miwa
Director, 3ʳᵈ Shinkansen Div. Shinkansen Dept.
Japan Railway Construction Public Corporation

M.Ogasawara
Deputy General Manager Technology Dept.No2
Civil Engineering Technology Division
Obayashi Corporation

ABSTRACT: The construction work for the Morioka to Hachinohe section of the Tohoku Shinkansen is proceeding smoothly with a completion time set at December 2002. In this section, there are twenty-one short and long tunnels. The second Itsukaichi tunnel is one of these tunnels. It is a relatively short tunnel of 1,175 m. The entrance to this tunnel is located directly under national road No. 4 where approx. 15,000 vehicles pass daily. The tunnel must be routed under a small earth covering of 2 m to 5 m thick.
The tunnel was excavated using the NATM method. In the portion where the tunnel intersects the road, an all ground fastening (AGF) method was used as a supplementary method.
As a result, the top end and the face were prevented from collapsing and the tunnel was excavated while keeping the sinking of the ground surface to a minimum without interfering with road traffic.

1 INTRODUCTION

The tunnel intersects national road No.4 built on an embankment at an angle of 30 degrees. The portion where the tunnel passes under the road is about 50-m long (see Figure 1). The smallest earth covering is 2 m at the toe of slope on the Aomori side and the lar-

gest earth covering is 5 m at the top of slope on the Morioka side. We had to excavate the tunnel by all means while keeping the sinking of the ground surface to a minimum without interfering with road traffic.

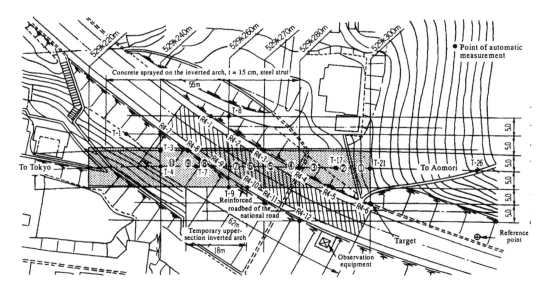

Figure1. Top view of where the tunnel intersects the road.

2 SUMMARY OF GEOLOGICAL FEATURES

Estimates were that the soil would be broadly classified into upper, middle, and lower layers according to the geological features confirmed by observing the face. The upper layer is soil fill used to form the ground for the road and mainly consists of sandy silt mixed with cobbles. It contains a great deal of decayed plant matter and the N value is 6 to 8. The middle layer consists of topsoil, fine silty sand and fine silty sand mixed with cobbles. The topsoil consists of organic soil and sandy silt; the soil thickness is about 1 m and the N value is 4. Fine silty sand and fine silty sand mixed with cobbles are unconsolidated. The lower the layer, the larger the degree of consolidation, and the N values are 11 to 22. The lower layer contains tuff breccia and the N value is more than 50. The degree of consolidation was low overall and the tuff breccia disintegrated and was locally assimilated into soil.

3 SELECTING A SUPPLEMENTARY METHOD

In excavating the tunnel using the NATM method, a supplementary method was used to solve the problems related to construction work.

3.1 *Construction conditions*

Conditions for the tunnel construction at the portion where the tunnel intersects the road are as follows:
Geological conditions are:
(1) A small earth covering in the portion right under the road, i.e., 2 to 5 m.
(2) Because the soil on which the temporary tunnel support is erected is fine silty sand, the soil has a small bearing capacity.
(3) The roadbed of the national road is an embankment that may disintegrate or collapse.
(4) Because the tunnel axis obliquely crosses the national road, uneven earth pressure may be applied to the face of slope at the entrance to the tunnel.
Environmental conditions are:
(1) Because national road No. 4 is a trunk road in the Tohoku region, the traffic cannot be controlled.
(2) Because there is unbought land at the entrance to the tunnel on the Morioka side, there is no space for work to be done at the tunnel entrance.
(3) The road curves and slopes and is constructed on the face of a mountain slope. Therefore, the conditions for road construction are bad.
(4) Optical cables are buried at a point 1.8 m under the sidewalk along the rise of the national road.

3.2 *Comparison of execution methods*

To select the most appropriate execution method that would enable us to prevent the ground surface from sinking or collapsing, to ensure the safety of road traffic, and to execute scheduled work processes smoothly and economically, four execution methods were compared after considering the conditions described:
(1) Tunnel excavation (road decking)
(2) Changing the route of the national road
(3) Pipe roof protection
(4) AGF method
These methods were compared by assessing the safety, the effects on the surrounding environment, the time necessary for completion, and the cost efficiency; the decision was made to use the NATM method in combination with the AGF method as a supplementary method.

3.3 *Designing the AGF method*

Because the collapse of the tunnel top must be prevented and the sinking of the ground surface must be minimized, the lap length of steel pipes was set to 6.65 m and they were designed so that steel pipes would be installed so as to enclose the tunnel in double layers. The steel pipes were designed to be installed over a wider range , i.e., 150 degrees, to ensure the reliability of their reinforcing capacity.

The steel pipe used was 114.3 mm in diameter, 6-mm thick and 12.65-m long. The lap length of this steel pipe was set to 6.65 m and the interval of steel pipe placement in a circumferential direction was set to 45 cm. The AGF method was used for tunnel sections that are 72.5-m long (11 shifts).

3.4 *Other supplementary methods*

(1) Measures to stabilize the face
If the layer of the road embankment appears in the face, this hampers maintenance of a solid face. The following measures were taken to keep the face solid and stable:
1) In the upper section, supplementary benches were installed at the position of SL+1.5 m.
2) Concrete was sprayed to a thickness of 5 cm to prevent the face from disintegrating.
3) Reinforcement bolts (glass fiber bolts 4-m long) were installed for each new face.
(2) Measures to prevent the ground surface from sinking
Because a load sustained by the AGF acts on the upper-section ground via the leg portion of installed temporary supports and the bearing capacity of the upper-section ground is small, there was concern over the sinking of the temporary supports. To prevent them from sinking, the following measures were taken:
1) To disperse loads that act on the upper-section ground, the H-200 temporary supports (pitch: 1 m, sprayed thickness: 25 cm) with wing ribs were used.

2) To increase the bearing capacity of the leg portion, reinforcement bolts 3m long were installed on the bottom of the temporary support and urethane was injected. The installation was executed by two times. First one was across under the bottom of the lower section of face, and second was right below supports after setting supports. The improvement area by the first injection was about 600mm across, and second was about 800mm.

(3) Measures to protect the face of slope

Because the tunnel axis obliquely intersects the national road, uneven earth pressure from the tunnel construction work being done near the face of slope along the national road may cause the face of the slope to collapse. To prevent this, embankment work was done using cement-improved earth.

(4) Reinforcing the road bed of the national road

To prevent the national road from sinking, the asphalt pavement in the section where it intersects the tunnel was changed to reinforced concrete pavement (30-cm thick).

4 EXECUTION OF WORK

4.1 *Work procedures*

As shown in Figure 2.

4.2 *Work done using the AGF method*

(1) Selecting the type of grout

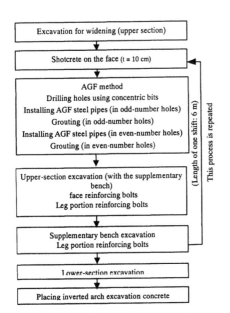

Figure 2. Excavation procedures.

Test grouting was done at the face 5 meters short of the point where the tunneling work would start.

Four different types of grout (silica resin, urethane, special fast-setting cement, and a combination of cement and bentonite) were used for testing, and grouting conditions (quantities of grout, grouting pressure, leak conditions, etc.) were recorded and

Procedures for tunnel excavation work

(1) Excavating upper section A1
(2) Shotcrete on upper section A1 the first time and on the face
(3) Excavating upper section A2
(4) Shotcrete on upper section A2 the first time and on the face
(5) Excavating upper section A3
(6) Shotcrete on upper section A3
(7) Installing temporary supports for the upper section
(8) Excavating the upper-section supplementary bench
(9) Shotcrete on the lower part of the upper section the first time
(10) Installing temporary supports for the lower part of the upper section
(11) Shotcrete on the upper section the second time and on the lower part of the upper section the second time
(12) Installing rock bolts on the lower part of the upper section, installing reinforcing bolts on the leg portion, and installing reinforcing bolts on the face
(13) Excavating the lower section (one side)
(14) Shotcrete on the lower section the first time
(15) Installing temporary supports for the lower section
(16) Shotcrete on the lower section the second time
(17) Installing rock bolts on the lower section
(18) Excavating the inverted arch
(19) Placing inverted arch concrete

This process is repeated | *In the portion directly under the road, the upper section is first excavated.*

compared. In addition, the face into which grout had been injected was excavated to check the status of improvement. As a result, urethane was selected because of its superior performance in terms of ground permeability, quick improvement, and cost efficiency.

After the tunneling work was completed, the condition of the ground was inspected and the objective of improving the ground had been attained. The results obtained with this check were as follows:

1) Fracture grouting was done for areas around AGF steel pipes, weak ground, and cracks as well as at points where inflow water comes out in such a way that injected grout ran like veins.

2) Fracture grouting was also done for areas around grouting bolts and for silty sand layers so that injected gout would form strata.

(2) Grouting method

To cut down on the time for grouting work, three hoses of different lengths (12 m, 8 m, and 4 m) were set inside the AGF steel pipe. The ends of these hoses were caulked and three injection pumps were activated simultaneously to inject grout, which is called a simultaneous grouting method.

(3) Hole drilling method

As the soil was unconsolidated, it was possible that the ground would disintegrate and drilled holes would bend due to water coming out of drilled holes.

To keep the deterioration and bending to a minimum, the double-pipe drilling method was used to drill holes.

(4) Construction machinery

In doing the tunneling work using the AGF method, a drill jumbo (130 kg class) used for ordinary rock bolt work was used. Machinery used for excavation included twin headers, breakers, and backhoes.

4.3 Excavation

(1) Upper-section excavation (Figure 3)

After the AGF work was completed, shotcrete on the face was broken with a breaker and the excavation work was done using a twin header or a breaker and a backhoe. The upper-section ground (A1, A2 and A3) except the supplementary bench (SL + 1.5 m, bench length: 3 m) was excavated first. Then, the supplementary bench along with upper ground was excavated at the same speed (1 m pitch). When the upper-section ground was excavated to the face where the AGF was to be executed for shift placement, the remaining 3 m of the supplementary bench was excavated.

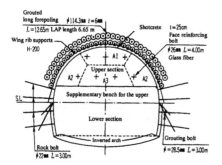

Figure 3. Cross section of the tunnel and installations.

(2) Lower-section excavation

When the upper-section excavation was completed for the entire section and displacement returned to normal, the lower-section ground was excavated using the same machinery as was used for the upper-section excavation.

4.4 Supplementary work methods and additional measures (Figure 4)

(1) Installing additional bolts and additional leg grouting bolts

If the sinking of the top end caused by excavation was anticipated and if it might take time for the sinking to stop, additional bolts and additional leg grouting bolts were installed before starting the excavation work.

(2) Placing the temporary upper-section inverted arch

Because the bearing capacity near the SL was insufficient, a temporary upper-section inverted arch was put in place if the sinking of the ground surface and the tunnel top end did not stop despite the installation of additional bolts and leg grouting bolts.

(3) Spraying concrete on the lower-section inverted arch and inverted arch supports

Although the sinking stopped after completion of the upper-section excavation, there was concern that the ground surface or the tunnel top end would sink again during lower-section excavation. In order to minimize the sinking caused by lower-section excavation, concrete was sprayed on the lower-section inverted arch to a thickness of 15 cm and inverted arch supports (H-150 set at 2-meter intervals) were installed.

4.5 Measurement

Outside the tunnel, the sinking of the ground surface was measured. Inside the tunnel, the displacement of the internal area was measured.

(1) Measuring the sinking of the ground surface

To measure the sinking of the ground surface, twenty target prisms were installed on the face and top of slope along the road and measurements were recorded automatically using observation equipment. Data was sent to the office via a telephone line. Measurements were taken every hour.

(2) Measuring the displacement of internal area

The displacement of internal area was measured at 3-meter intervals using a three-dimensional measuring instrument.

5 RESULTS OF THE TUNNELING WORK

5.1 Sinking of the ground surface

The sinking of the ground surface began to increase when the face reached the 529k280m point. The largest sinking, i.e., 63 mm, was observed around the 529k256m point.

5.2 Sinking of the tunnel top end

The sinking of the tunnel top end was similar to that of the ground surface. The largest sinking, i.e., 54 mm, was observed around the 529k256m point.

5.3 Comparison of the results of FEM analysis and actual measured values

Before the start of excavation work, FEM analysis (elasticity analysis using the finite element method) was performed. Based on the results of this analysis, additional measures were taken. According to the results of FEM analysis performed for the 529k266m

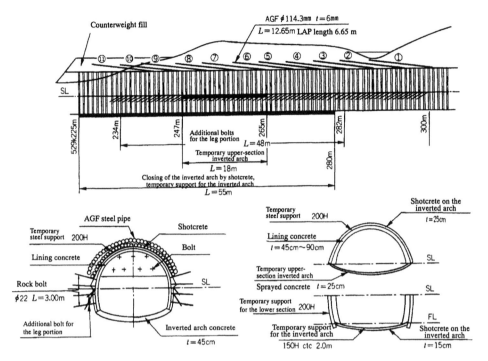

Figure 4. Supplementary work methods and additional measures.

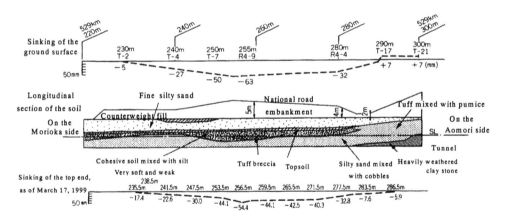

Figure 5. Sinking of the ground surface and tunnel top end.

point, the final sinking of the tunnel top end and that of the ground surface were estimated as 45 mm and 38 mm, respectively.

The actually measured value of the sinking of the tunnel top end was about 45 mm. This value, however, was measured after temporary supports were installed. Because the displacement prior to the installation of temporary supports was not included in this value, the actual sinking is believed to be larger than the actual measured value. Reasons estimated values were different from actual measured values

The reasons the actual measured values were larger than those obtained by FEM analysis are:
(1) Variations in the nature of the soil that forms the cross section of the heading
In the leg portion of temporary upper-section supports, topsoil was distributed. Below this topsoil, there were alternate layers of unconsolidated clay and silty fine sand. Therefore, the soil that forms the cross section of the face was softer than was initially estimated.

491

(2) Insufficient reinforcement

The bearing capacity of the leg portion of the temporary supports could not be increased to a satisfactory level despite the installation of leg grouting bolts. Therefore, the sinking of the leg portion led directly to the sinking of the tunnel top end.

6 CONCLUSION

In the present tunneling work, there was almost no displacement of the width of internal area while sinking of the tunnel top end, the ground surface, and the temporary upper-section supports was noted. This evidently resulted from the insufficient bearing capacity. Supplementary work methods based on the results of FEM analysis were developed and additional, effective measures implemented quickly. In executing the excavation work, the sinking of the ground surface and the displacement of the internal area were measured and sinking was kept to a minimum.

Reference:

Eijiro Tamura, et al. "Constructing a Tunnel through the Embankment of National Road No. 4 Using the AGF Method" ("Tunnel and Underground," volume 31st, 3rd issue)

Modern Tunneling Science and Technology, Adachi et al (eds), © 2001 Swets & Zeitlinger, ISBN 90 2651 860 9

Ground reinforcement for a tunnel in weathered soil layer beneath Han riverbed in Korea

Tae. Yang
Dept. of Environmental and Civil Eng., Kimpo College, Korea

J. Woo
Dept. of Civil Eng. Design, Kyungbok College, Korea

S. Lee
Dept. of Civil Eng., University of Seoul, Korea

ABSTRACT: When building tunnels beneath riverbeds where very large quantities of groundwater inflow exist due to high water head and the soil supporting conditions are very poor because the soil consists of sand and silt, it is necessary to get grouting and pipe roof installed in the region where ground reinforcement is in need to decrease permeability. According to this result of horizontal boring and laboratory soil testing, ground reinforcement was achieved by L.W grouting for range of 3.0 times the tunnel radius, to increase stability of the tunnel we used the ring cut method, 0.8m for one step excavation, shotcrete with 25cm thick, steel lib with H-125×125, and a temporary shotcrete invert 20cm thick was installed to prevent deformation of the tunnel.

1. PREFACE

This paper is on a single circular tunnel 90-meters long, a part of 1,288-meters totally long tunnel constructed in Han riverbed from Yoido to Mapo of Seoul city, Korea.

Soil layer under it consists of sand and silt, and the soil condition is too poor to support a tunnel. Also large quantities of groundwater inflow from the tunnel face according to the high water pressure from the mean water level of Han river.

Before excavating this tunnel, perimeter of this tunnel was reinforced by L.W grouting, and to prevent from collapsing of the tunnel crown, pipe reinforcement grouting and fore-poling were installed. Some part of this tunnel was reinforced by urethane and legging sheet.

To decrease opening of the tunnel excavation face we used the ring cut method for upper half, and partial excavation for lower half, and 0.8m for one step excavation, shotcrete with 25cm thick, steel lib with H-125×125. To diminish shaking and shock to the tunnel, the tunnel was excavated not by an explosion but by man and excavation machine, and to prevent deformation of the tunnel a temporary shotcrete invert 20cm was installed after excavating of upper half. The deformations of the tunnel were monitored by regular and main monitoring. A tunnel was successfully constructed by NATM after these steps were completed.

2. SOIL INVESTIGATION AND GROUND-WATER CONDITION

2.1 The geographical formation beneath the Han-riverbed

According to the vertical section of line 5 of Seoul subway, a tunnel beneath Han riverbed was planed to locate from 15.6m to 37m beneath a riverbed.

A fault zone was found existed ahead of tunnel when horizontal borings were performed to certify the condition of the foundation in front of tunnel. And this fault zone was consisted of sand and silt. The geological distribution is shown in "Figure 1. . Section of the geographical formation."

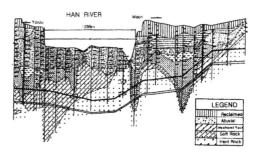

Figure 1. Section of the geographical formation of the tunnel perimeter.

2.2 Methods of soil investigation

The investigation of the geographical formation of the tunnel perimeter beneath riverbed was conducted. by loading in the bored holes. Groundwater leakage condition was tested for inflow from the tunnel face.

2.3 Results of the investigation of the geographical formation

Table 1. Results of horizontal borings.

division	boring length (m)	boring angle (o)	result	ground-water inflow (ℓ/min)	remark
A	70	5°	0~30 m : impossible to gain sample 30~42 m : check the slime 42~55.5m : sampling 55.5~70m : check the slime	300 400	
B	21.2	0°	sampling by triple core barrels		lab. soil test

Table 2. Results of loading tests.

test	depth of test (m)	av. radius of rubber	soil pressure Po (kN/m²)	yield pressure Py (kN/m²)	fail pressure Pf (kN/m²)	coefficient of reaction Km (kN/m²)	coefficient of elasticity Em (kN/m²)
1	0.7	4.178	269.7	1456.2	1858.4	8566.2	46527.9
2	1.7	4.313	192.2	1650.5	1842.7	8012.1	44922.5
3	2.7	4.034	313.8	1625.9	1885.8	9134.9	47903.8
max.	2.7	4.313	313.8	1650.5	1885.8	9134.9	47903.8
min.	0.7	4.034	192.2	1456.3	1842.7	8012.1	44922.5
Av.	1.7	4.175	258.9	1577.9	1842.7	8571.1	46451.4

Table 3. Results of laboratory soil tests.

sample	horizontal depth (m)	w_n (%)	atterberg LL	atterberg PI	unconf. comp. test q_u (kN/m²)	direct shear test C (kN/m²)	direct shear test Ø (°)	US CS
A	2.0~3.0	21.5	NP	NP	69.6	34.3	35	SM
B	8.5~9.0	22.7	NP	NP	52.9	23.5	32	SM
C	16.4~16.8	25.7	NP	NP	37.2	18.6	29	SM

*Gs is 2.66

Table 4. Summary of results.

division	cohesion (C) (kN/m²)	interfriction angle(ø) (°)	unit weight (γ_t) (kN/m³)	coefficient of deformation(E) (kN/m²)	US CS
range	18.6~34.3	29~35	19.4~22.6	44922~47903	SM
av.	25.5	32	19.7	46451.4	SM

The samples were obtained nearly same as natural state by triple core barrels and tested in the laboratory in Tables 1-3.

3. GROUND REINFORCEMENT SCHEME

When building tunnels beneath riverbeds where very large quantities of groundwater inflow exist due to high water head and the soil supporting conditions are very poor, following items are the most serious problems.

To excavate a tunnel successfully stopping up groundwater comes into the tunnel and ground reinforcement are the most important points.

· large groundwater incomes when a tunnel is excavated
· extremely short time to self-support of tunnel face
· boundless quantity of a source of groundwater located under the river
· large deformation of a tunnel and convergence delay of a tunnel deformation

3.1 L.W. injection design

After investigations were achieved about the fault zone, the tunnel was designed as follows. Injection pattern and grouting schemes can be found in "Figures 2-3."

· injection range : 3.0 × R(radius of tunnel)
· average injection ratio : 20.9%

$$S \times e = Gs \times Wn$$

$$porosity(\eta) = \frac{e}{1+e}$$

$$injection\ ratio(\lambda) = \eta \times \alpha \times (1+\beta)$$

here　S : degree of saturation(1.0)
　　　e : void ratio
　　　Gs : specific gravity (2.66)
　　　w_n : natural content of water (23.3%)
　　　α : filling ratio(50%)
　　　β : loss ratio(10%)
· injection pressure : 1176.8 ~1765.2 kN/m² (4~6 times of water head)
· injection rate : 25 ℓ/min
· injection : double lot

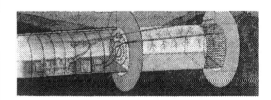

Figure 2. L.W. grouting scheme.

Table 5. Mixing ratio of L.W. grouting.

Mix No.	injection quantity (ℓ)	A-liquid(ℓ)		B-liquid		
		water	Water	cement (kg)	water (ℓ)	w/c(%)
I	1,000	250	250	95	470	500
II	1,000	250	250	115	463	400
III	1,000	250	250	150	452	300
gel-time(sec)		injection ratio(%)		remak		
92		80				
57		20				
45		extra inj. Quantity				

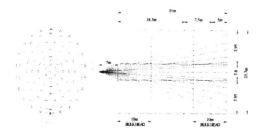

Figure 3. Injection pattern scheme.

3.2 Results of L.W. grouting

The results of L.W Grouting are summarized as follows

- The range of injection pressure is from $12 kg/cm^2$ to $18 kg/cm^2$, it is from 4 to 6 times water head.
- The injection quantities per hour were nearly uniform but as the injection time increase injection pressure increase linearly, this type of injection is same to the one that grouts were lost in the cracked rock.
- The shape of injection was as   fracture grouting. The grouts were filled first in the void and the loose part of the aimed injection region around.
- The range of L.W grouting injection ratio was from 19.8% to 23.4%, and average injection ratio was 20.9%.
- It was possible to do lugeon test for checking permeability before and after L.W. grouting. According to the results of the test, the coefficient of permeability is 10^{-4}-10^{-5}cm/sec before grouting but 10^{-5}-10^{-6} cm/sec after grouting. Such change of permeability was due to joint of rock. Permeability was reduced enough to excavate tunnel safely.

4. STEEL PIPE ROOF REINFORCEMENT INJECTION FOR TUNNEL CROWN

To reinforce shear of the rock, to prevent falling of the rock masses from the tunnel crown, to confirm safety of the excavation face of upper part, and to increase second effect of the L.W grouting, steel pipes were installed 120^0 of tunnel crown, and grouts were injected into the pipes in "Figure 4."

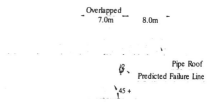

Figure 4. Pipe reinforcement injection for tunnel crown.

5. TUNNEL CUT AND COVER

Because the self-supporting condition was extremely poor after the tunnel was opened in the fault zone, the ring cut method was used to excavate the tunnel and to minimize opened tunnel face. And as soon as the small section excavation was over by man and machine, shotcretes were covered the full of the excavating face. We used 0.8m one step excavation, shotcrete with 25cm thick, steel lib with H-125×125. After upper half was excavated a temporary shotcrete invert 20cm thick was installed to prevent deformation of the side lower part of the tunnel shown in "Figure 5. Standard section."

6. MONITORING OF DEFORMATION OF TUNNEL

To observe the deformation of the tunnel and to check the suitability of the tunnel construction, regular and main monitorings were performed.

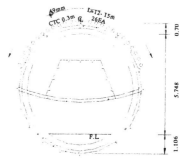

Figure 5. Standard section of tunnel beneath riverbed(horseshoe type).

Regularly internal deformation and settlement of the tunnel crown were monitored shown in Figure 6.

Figure 6.   Regular monitoring scheme.

Table 6. Results of regular monitoring.

division		deformation(mm)		
		internal deform. (upper half)	internal deform. (lower half)	crowm settlement
weathered soil	range	19.7~64.0	12.8~26.0	14.0~24.0
	av.	44.8	19.9	17.8
soft rock	range	4.8~7.4	1.6~2.2	3~10.0
	av.	6.0	1.8	6.7

Mainly ground displacements, rockbolt axial force, shotcrete stress were monitered for main monitoring shown in Figure 7.

Figure 7. Main monitoring scheme.

6.1 Analysis of monitoring

Working the process of regular monitoring, large deformations were noticed. Causes are that temporary shotcrete invert didn't work well and the groundwater continuously entered through bottom of the tunnel and made the shotcrete bottom soft and promoted deterioration. Study on how to prevent deformation of the bottom due to the softening and deterioration by groundwater is needed.

Ground reinforcement technique for tunnel in weathered soil beneath riverbed will be developed.

According to the results of main monitoring, the deformation of a tunnel constructed in a weathered soil region beneath a riverbed that is reinforced by L.W. grouting and pipe roof is similar to a tunnel constructed in soft rock with the same conditions.

Results of shotcrete stress is 500.1-2141.8 kN/m^2. Results of rockbolt stress is 31.4-422.7 t. Those of ground displacement is 0.61-2.58mm around the whole soil layer.

7. CONCLUSIONS

A tunnel was successfully constructed by NATM in the region where the soil condition was extremely poor such as a fault zone and the high water head was existed. The conclusions were as follows.

1) The range of L.W. grouting injection ratio was from 19.8% to 23.4%, and average injection ratio was 20.9%

2) According to the results of the lugeon test, the coefficient of permeability is 10^{-4}-10^{-5}cm/sec before grouting but 10^{-5}-10^{-6} cm/sec after grouting

3) According to the results of main monitoring, the deformation of a tunnel constructed in a weathered soil region beneath a riverbed that is reinforced by L.W. grouting and pipe roof is similar to a tunnel constructed in soft rock with the same conditions.

4) Working the process of regular monitoring, large deformations were noted. The causes are first, temporary shotcrete invert didn't work well, second, groundwater continuously entered through the bottom of the tunnel and made the shotcrete bottom soft and promoted deterioration. Further study on how to prevent deformation of the bottom due to the softening and deterioration caused by groundwater is needed and expected.

REFERENCES

Jongtae Woo & Gangchun Seo 1996. *Geological analysis and ground reinforcement of a tunnel beneath Han riverbed*. Office of Subway Construction , Seoul Metropolitan Government

Taeseon Yang & Jongtae Woo 1998. A case study for quality conformation and maintenance monitoring tunnel underpassing Han river. *J. of Korea Institute for Structural Maintenance Inspection*: 185-194

Korea Institute of Construction Technology 1994. *Multiple Grouting Technique as Reinforcement by Steel pipe*. Seoul

Modern Tunneling Science and Technology, Adachi et al (eds), © 2001 Swets & Zeitlinger, ISBN 90 2651 860 9

Stability and water leakage of hard rock subsea tunnels

B.Nilsen
The Norwegian University of Science and Technology, Trondheim, Norway

A.Palmstrøm
Norconsult as, Sandvika, Norway

ABSTRACT: The many undersea tunnels along the coast of Norway offer excellent opportunities to study the key factors determining stability and water leakage in hard rock subsea tunnels. About 30 such tunnels have been constructed in Norway the last 20 years, all of them excavated by drill and blast. The longest tunnel is 7.9 km with its deepest point 260 metres below sea level. Although all tunnels are located in Precambrian or Palaeozoic rocks, some of them have encountered complex faulting or less competent rocks like shale and schist. The severe tunnelling problems met in these tunnels emphasise the need of a better understanding of the key factors determining stability and water leakage of such projects. This has been discussed based on the experience from several completed projects.

1 INTRODUCTION

In Norway, about 30 subsea tunnels, comprising more than 100 km have been built the last 20 years. Most of these are 2 or 3 lane road tunnels, but some are also for water, sewage, or oil and gas pipelines. All tunnels so far are drill and blast. The locations of some key projects, and tunnels being discussed later in this paper, are shown in Figure 1, and some main figures concerning length and depth are given in Table 1.

The tunnels are located mainly in hard, Precambrian rocks (typically granitic gneisses). This also is the case for the deepest tunnel; the Hitra tunnel (260 meters below sea level at the deepest point). Some of the tunnels are, however, also located in less competent Palaeozoic rocks like shale and schist. This is the case also for one of the longest tunnels; the North Cape tunnel (6.8 km).

The last few years, very complex and difficult ground conditions have been encountered in several subsea tunnels. The problems emphasise the need of a better understanding of the key factors determining stability and water leakage of such projects. In this paper the issue will be discussed based on the experience from completed projects, and with particular reference to a recent study of the Fröya subsea tunnel (Nilsen et al. 1997 and 1999, Palmstrøm et al. 2000), where very difficult rock conditions were encountered.

2 CHARACTERISTICS OF SUBSEA TUNNELS

Compared to conventional tunnels, subsea tunnels are quite special in several ways. Concerning engineering geology and rock engineering, the following factors are the most important (see also Figure 4):

• Most of the project area is covered by water. Hence, special investigation techniques need to be applied, and interpretation of the investigation results is more uncertain than for most other projects.

• The locations of fjords and straits are often defined by major faults or weakness zones in the bedrock. Also in generally good quality rock conditions, the deepest part of the fjord, and hence the most critical part of the tunnel often coincides with weak zones or faults, which may cause difficult excavation conditions.

• The potential of water inflow is indefinite, and all water leakage has to be pumped out of the tunnel due to its geometry.

• The saline character of leakage water represents considerable problems for tunnelling equipment and rock support materials.

The consequences of cave-in or severe water ingress in a subsea tunnel may be disastrous. Thus, despite the fact that Norway is a hard rock province, forming part of the Precambrian Baltic Shield as shown in Figure 1, the subsea tunnels often encounter faults of very poor quality, causing challenging ground conditions.

Figure 1. Locations of some of the main Norwegian subsea tunnel projects.

Table 1. Key data of some of the Norwegian subsea rock tunnels.

Project	Tunnel type	Year completed	Cross section (m²)	Main rock types	Total length (km)	Lowest level (m)
Frierfjord	Gaspipe	1977	16	Limestone/gneiss	3.6	- 252
Vardø	Road	1981	53	Shale/sandstone	2.6	- 68
Karmsund	Gaspipe	1983	27	Gneiss/phyllite	4.8	- 180
Ellingsøy	Road	1987	68	Gneiss	3.5	- 140
Kvalsund	Road	1988	43	Gneiss	1.6	- 56
Godøy	Road	1989	52	Gneiss	3.8	- 153
Freifjord	Road	1992	70	Gneiss	5.2	- 100
Byfjord	Road	1992	70	Phyllite	5.8	- 223
Hitra	Road	1994	70	Gneiss	5.6	- 260
North Cape	Road	1999	50	Shale/sandstone/micaschist	6.8	- 212
Oslofjord	Road	2000	78	Gneiss	7.2	- 130
Frøya	Road	2000	52	Gneiss	5.2	- 157
Bømlafjord	Road	2000	78	Gneiss, greenschist	7.9	- 260

3 FAULTS/ WEAKNESS ZONES

Some of the Norwegian subsea tunnels have been completed without major problems as they have not encountered major weakness zones. This was the case, for instance, for the Kvalsund tunnel in Table1. In all the other tunnels in Table 1, distinct zones with very poor rock quality have, however, been encountered, and the rock support requirement thus has been considerably higher as shown in Table 2. The figures in this table also reflect the traditional Norwegian rock support philosophy that rock support is adjusted or tailored to the actual rock mass conditions, with heavy support like concrete lining applied only in very poor stability conditions. The typical weakness zones have widths of 20-30 meters and more, and consist of heavily crushed and altered rock. Gouge material of swelling type (smectite) is often found in such zones. Swelling pressures of around 1 MPa is common, and in extreme cases swelling pressure of more than 2 MPa has been experienced (clay material with swelling pressure above 0.3 MPa is generally classified as "active"). The particularly high activity of smectite in subsea tunnels reflects the ability of the clay mineral to absorb Na^+ from sea water.

Stability problems due to major weakness zones represent a threat to hard rock subsea tunnel projects. In some cases, severe instability has occurred. In the majority of such cases, the problem has been caused by faulted rock carrying clay minerals and water leakage of relatively high pressure. One such case was the Ellingsøy road tunnel, see Figure 2. Here, despite the fact that continuous probe drilling was carried out, a fault zone containing swelling clay and water-bearing fissures caused a cave-in reaching up to 8-10 meters above the crown before it finally stopped after 24 hours. A concrete plug of approximately 700 m^3 was constructed in order to ensure safe tunnelling through the zone.

Thanks to comprehensive geological preinvestitions and control as well as effective rock support procedures, cave-in of disastrous consequence has never occurred. In some recent cases, the situation has, however, been even more difficult than in the Ellingsøy case:

• The Bjorøy tunnel, where a more than 10 m wide Jurassic, tensional fault zone filled with clay,

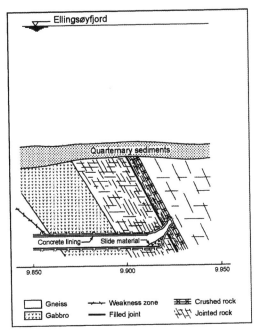

Figure 2. Cave-in situation at the Ellingsøy tunnel (based on Olsen & Blindheim, 1989).

sand and coal fragments quite unexpectedly, due to insufficient refraction seismics, was encountered in the Precambrian bedrock. This was a zone of extremely high permeability and very poor stability, and a very time-consuming procedure involving

Table 2. Extent of rock support in some of the Norwegian subsea tunnels.

| Tunnel | Excavation rate | Rock support | | Concrete lining | Pregrouting Grout consumption | Water leakage | During operation |
| | | Bolts | Shotcrete | | | At opening | |
	m/week	bolts/m tunnel	m³/m tunnel	% of tunnel length	% of tunnel length	kg/m tunnel	l/min/km tunnel	l/min/km tunnel
Vardø	17	6.9	0.95	>50	21	31.7	460	*
Karmsund	34	1.5	0.72	65	15	13.4	*	*
Ellingsøy	28	6.4	0.48	20	3	99.1	300	130
Kvalsund	56	4.0	0.31	*	0	0	320	180
Godøy	*	*	0.40	*	0	265	300	90
Freifjord	45	5.3	1.44	*	2.1	13.7	500	280
Hitra	46	4.2	1.44	*	0.2	11.4	60	*
Frøya	37	5	2.9	**	5	197	8.5	*
Bømlafjord	55	3.8	1.9	**	0	36	< 50	*
Oslofjord	47	4.0	1.7	**	1	165	150	*
North Cape	18/56 ***	3.4	4	**	34	10	60	*

* No data available;   ** The tunnel roof along the entire tunnel has been reinforced by shotcrete;   ***In the shale, sandstone / micaschist parts

stepwise grouting, drainage, spiling and shot-crete arches was necessary to tunnel through it.

- The North Cape tunnel, where flat laying, broken sedimentary rocks (mainly sandstones), often with thin coating of chlorite clay seams, have caused very poor stability. Comprehensive shot-creting and concrete lining at the face was re-quired for rock support, reducing tunnelling progress to about 20 m/week. The difficult con-ditions were not realised from the pre-investigations due to the relatively high seismic velocity of the flat laying layers (4,500-5,500 m/sec).

- The Oslofjord tunnel, where a deep cleft filled with Quaternary soil was encountered, necessi-tating ground freezing to tunnel through. A dis-tinct weakness zone was detected prior to tunnel-ling. Despite very comprehensive pre-investigations including traditional refraction seismics as well as directional core drilling and seismic tomography it was not found that the zone was eroded to a deep cleft.

In the Frøya subsea tunnel project area, a pattern of very distinct, regional faults were identified, see Figure 3. Some of these faults can be followed over a distance of more than 200 km.

In the planning of the Frøya tunnel, great benefit was gained from the construction of the nearby Hitra tunnel, where the major fault in the fjord proved to consist mainly of a mixture of heavily crushed rock and clay minerals (including swelling clay). The water seepage through the zone was minimal, and no concrete lining, but a combination

of short blast rounds, spiling, steel fibre reinforced shotcrete, straps and conventional rock bolts, was used for getting through it.

There were some main differences between the Frøya and Hitra tunnels, making it likely that the former would be more difficult:

- The location closer to the Norwegian Sea, in-creasing the possibility that fault zones may con-tain intercalations of young, high porosity, sedi-mentary rocks.

- The low seismic velocities recorded, indicating rock mass quality far below average.

- The location closer to the main tertiary faults along the Norwegian coast, reducing the normal stress on major faults, and thus increasing the risk of major leakage (see Section 5).

- The less glacial erosion further from the main land (see Section 6).

Very comprehensive pre-investigations were car-ried out for the Frøya tunnel, and all investigation and planning were thoroughly reviewed by two in-dependent panels of experts. Challenging ground conditions were documented, including high perme-ability zones as well as weakness zones containing very loose, sandy material and extremely active swelling clay.

Tunnelling for the Frøya project started in early 1998, and was completed in September 1999. The locations of the main weakness zones are shown in Figure 4, and the technique that was commonly used for excavating through such zones is shown in Figure 5.

4 WATER LEAKAGES

A logical assumption, apparently, would be to ex-pect most of the water inflow in a subsea tunnel to come from major faults or weakness zones. This is, however, seldom the case, probably mainly due to the fact that such zones generally have very low permeability due to a high clay content (as was the case for the main zone encountered in the Hitra tun-nel). The fact that the major zones are in most cases located under the central part of the fjord or strait, with a low permeability soil cover on top of the rock, probably is also a part of the explanation. This hy-pothesis is confirmed by the fact that in some subsea tunnels the major leakage has been encountered un-der land, and not under sea (for instance in the El-lingsøy tunnel, where the maximum inflow of 400 l/min from one single probe drill hole was encoun-tered about 1 km from the sea).

In most cases, the major leakage is encountered at continuous single joints, often near a major fault

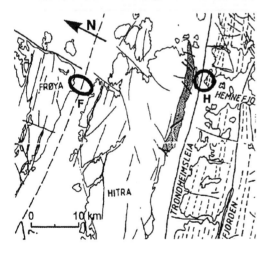

Figure 3. Pattern of regional faults in the area of the Frøya (F) and Hitra (H) subsea tunnels. Shaded area is Palaeozoic rock, the rest Precambrian (after Grønli, 1991).

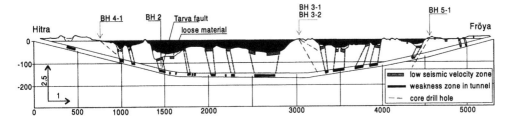

Figure 4. Longitudinal section along the Frøya tunnel.

zone. This situation is very difficult to predict. Table 2 gives an indication of the great variations in leakage and thus grout requirements. During tunnelling, the decisions concerning grouting are based on inflow measurement from probe drill holes ahead of the tunnel face. Attempts have been made to correlate leakage to geological and rock mechanical parameters, but with mixed results so far. To be able to make more reliable prognoses on water leakage, more research definitely is required.

There are a few exceptions to the general trend that the major leakage is normally not connected to a weakness zone. The main one is represented by the water ingress in the tensional zone of the Bjorøy tunnel. To withstand the 0.7 MPa water pressure here, installation of blow out preventers in probe and grout drill holes was required.

As can be seen from Table 2, the water leakage in the subsea tunnels typically is reduced considerably with time.

5 ROCK STRESSES

Problems due to high rock stresses are not considered a major issue for the Norwegian subsea tunnels, which are located mainly in medium high, favourable stress conditions. Low minor principle stress of unfavourable orientation with respect to main discontinuities may, however, increase the water inflow considerably. This is a problem of particular relevance for tunnels located far to the West (close to the main Tertiary fault along the Norwegian coast). The high water inflow (and grout consumption) in the Godøy tunnel (see Table 2) was most likely caused by this effect.

In the Frøya area, the minor principal stress is most likely oriented NE - SW (parallel to the coast line), see Figure 3, and NW - SE oriented discontinuities thus in theory are most likely to give major inflow. Because of the location of the Frøya tunnel so far to the West, some of the same effects as in the Godøy were expected, but (swelling) clay content in most joints and zones resulted in small water leakage.

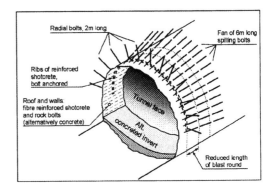

Figure 5. Principles of excavation through poor stability weakness zones applying the spiling technique (revised from Nålsund et al. 1996).

To be able to make reliable prognoses on water inflow based on stress magnitudes and directions, more research is required particularly on the effects of anisotropy and channelling in discontinuities.

6 WEATHERING AND EROSION

In relatively recent geological time, Scandinavia has experienced several glaciations, and is thus in the favourable situation concerning rock engineering, that most weathered material has been removed by the ice. For weakness zones, particularly near the coast (far from the glaciation maximum) this is, however, not necessarily the case, as illustrated in Figure 6.

For the Frøya tunnel, located at a considerable distance from the glaciation maximum, and with a minimum rock cover of about 40 meters, the effect of weathering root is believed to play a significant role. Consequently, the effect has been taken into consideration in the evaluation of investigation results.

501

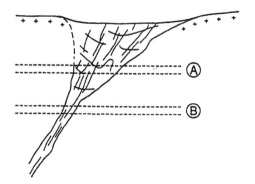

Figure 6. Principle sketch of a weakness zone with weathering root. A shallow tunnel (alternative A) gives a much wider zone crossing than a deeper (alternative B). From Nilsen et al. 1997.

7 CONCLUDING REMARKS

The most unstable conditions in the hard rock subsea tunnels are represented by major faults or weakness zones containing heavily crushed rock and gouge, often including very active swelling clay. Water seepage in such zones may dramatically reduce the stand-up time.

Since the major zones are often located in the middle of the fjord, the most severe problems are generally encountered at a late stage of tunnelling. The tunnelling conditions encountered that far may often have been relatively good, and the problems therefore may come as a surprise if not thorough pre-investigations and following-up during tunnelling are practised. Also, a high degree of readiness for all types of immediate rock support and a continuous quality control of all work are crucial for a successful completion of subsea tunnels.

Major water inflows have been found relatively rarely to be directly connected to the major weakness zones, probably mainly due to the high content of low permeability gouge of such zones. Distinct, continuous single joints apparently are more important. The magnitudes and orientations of rock stresses definitely have influence on water inflow. The effect is, however, complicated by factors like anisotropy and channelling. To be able to make reliable prognoses on water leakage in hard rock subsea tunnels, more research is needed.

REFERENCES

Fejerskov, M. & A. Myrvang 1993. Rock stresses in Norway and on the Norwegian shelf. (In Norwegian). In: *Fjellsprengningsteknikk/Bergmekanikk/*

Geoteknikk - Proc. Annual Norw. Nat. Conf. on Rock Mech., Oslo, Nov. 1993: 25.1-25.17. Tapir.
Grønlie, A. 1991. Joints, faults and breccia systems in outer parts of Trøndelag, central Norway. *Dr. ing. dissertation*, Norwegian Inst. of Techn. (NTH), Dept. of Geol. and Minr. Res. Eng., Trondheim, Norway.
Holmøy K., J.E. Lien and A. Palmstrøm 1999. Going sub-sea on the brink of the continental shelf. *Tunnels & Tunnelling International*, 31(5):25-30.
Melby K. & E. Øvstedal 1999. Daily life of subsea tunnels – construction, operation and maintenance. *Proc. ITA Workshop Strait Crossings – Subsea tunnels*, arranged in connection with 1999 ITA World Tunnel Congress, 14-27. NFF/ITA, Oslo, Norway.
Nilsen, B. 1999. Key factors determining stability and water leakage of hard rock subsea tunnels. *37th US Rock Mech. Symp., Vail 6-9 June 1999:* 593-599. Balkema.
Nilsen, B., A. Palmstrøm & H. Stille 1997. The Frøya tunnel, analysis of excavation and rock support methods as basis for cost calculation, feasibility evaluation and risk estimation. (In Norwegian). *Technical report. Trondheim, Norway.* 50 p.
Nilsen, B., A. Palmstrøm & H. Stille 1999. Quality control of a subsea tunnel project in complex ground conditions. *Proc. ITA World Tunnel Congress, Oslo, Norway.* 137-144.
Nålsund, R., S.Heggstad, A. Mehlum & B. Aagaard 1996. Analysis of excavation methods and rock support . (In Norwegian). *Internal project report, Trondheim, Norway,* 21 p.
Olsen, A.B. & O.T. Blindheim 1989. Prevention is better than cure. *Tunnels & Tunnelling* 20(9):41-44.
Palmstrøm A. 1992. Introduction to Norwegian subsea tunnelling. *Publ. No. 8.* The Norwegian Soil and Rock Engineering Association. 9-12.
Palmstrøm A. & R. Naas 1993. Norwegian subsea tunnelling - rock excavation and support techniques. *Int. Symp. on Technology of bored tunnels under deep waterways,* Copenhagen.
Palmstrøm A. 1994. The challenge of subsea tunnelling. *Tunnelling and Underground Space Technology* 9(2).
Palmstrøm A., H. Stille and B. Nilsen 2000. The Frøya tunnel – a sub-sea road tunnel in complex ground conditions. Proc.*Swedish rock mechanics conference, 2000.* SveBeFo. 19-29.

Modern Tunneling Science and Technology, Adachi et al (eds), © 2001 Swets & Zeitlinger, ISBN 90 2651 860 9

Development of long face reinforcement method with GFRP tubes

Y. Mitarashi, T. Matsuo & T. Okamoto
Kumagai Gumi Co. Ltd., Tokyo, Japan

T. Tsuji, T. Haba & T. Okabe
KFC Ltd., Tokyo, Japan

ABSTRACT: The long face reinforcement method has been developed for the stabilization of mountain tunneling face. This technique uses "casing-injection" type long tubes of glass-fiber reinforced plastics. It is more excellent in ease of work, cost performance and effectiveness than the conventional face piling and long steel pipe forepiling. Demonstrating tests on site indicated a great work efficiency, and measurements of stress generated in tubes showed the effectiveness of the developed method. In addition, by actual construction, measurement and analysis, the effects of the this method were verified, such as control capability of preceding loosening.

1 INTRODUCTION

As auxiliary methods for stabilizing the tunnel face, the injection type long steel pipe forepiling and the face bolting have been widely used in Japan in recent years. However, there is an important difference both in function and effect between long forepiling and forepoling bolt. There is no technique which falls into a category between the two methods from the viewpoint of both effect and economy. In the conventional face reinforcing practice (face stabilizing bolt), it is desirable to install bolts as long as possible, to effectively strengthen the ground ahead of the face influenced by the tunnel excavation. But actually, in a ground with extremely adverse conditions, it is impossible to drive long face bolts because of problems in execution such as collapse of borehole. Also in the case of cable bolt too, the bore hole is required to be intact from drilling till bolt insertion.

As a solution for the difficulties mentioned above, the casing-injection type long face reinforcement has been developed. This technique drives a casing of glass-fiber reinforced plastic tube (GFRP tube) which can be easily cut. This tube serving as reinforcement, with urethane or cement injection agent, forms a "core" (improved ground) in the loosened zone ahead of the face. This paper reports the results of demonstration tests, application to an actual site, and the effects of this reinforcing method verified by analytical study.

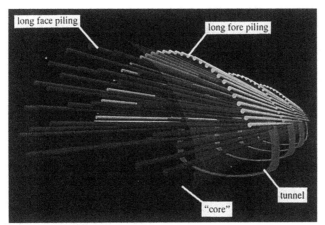

Figure 1. Schematic diagram of the long face reinforcement method with GFRP tubes.

2 CONSTRUCTION SYSTEM

The injection type long face reinforcement method installs long GFRP tubes by drilling, and with these tubes (casings) and injection agent, forms a "core" (improved ground) to reinforce the zone where stresses are redistributed under the influence of tunneling.

The drilling system has a configuration as follows. An outer bit and a casing shoe are mounted on the tip of GFRP tube. A rod provided with an inner bit is inserted into the tube. The rotating rod transmits the revolution of the inner bit to the outer bit. By the use of a drill jumbo or the like, a specified length is drilled. Then, the outer bit and casing shoe are left at the hole end, whereas the rod mounted on the inner bit is retrieved. The GFRP tube is 76 mm in outer diameter and 60 mm in inner diameter. By connecting tubes whose unit length is 3 m, a desired length of reinforcement is made. An urethane or a cement injection agent is used. Two types of injection methods are available, to be selected on the basis of ground conditions and work cycle, that is, valve injection by which the agent is poured through the mouth of hole, and divided injection method using a packer.

The laboratory basic experiments demonstrated that the tube material has a tensile strength of 600 N/mm^2, and the load-carrying capacity of the GFRP tube is 200 kN or more in tension in the coupler portion.

3 FEATURES OF THE FIT METHOD

Since this method utilizes the "casing-injection" technique, it can be applied even to grounds where no stability of boreholes is expected. In addition, the GFRP tubes can be cut. This enables driving in any direction without enlarging the excavation section. When the displacement ahead of the face is small, this method can be used as an alternative for the long forepiles. The excellent features of the method are as follows.

(1)   It can be implemented by a drill jumbo that is generally used in tunneling sites, without requiring any special machines.

(2)   Since no special machines are employed, the method can be easily applied whenever necessary.

(3)   Though the diameter of GFRP tube is small (76 mm OD), it has a high tensile strength.

(4)   The GFRP tube is light in weight, easy to handle. In addition, since its materials are glass fibers and plastics, the portion of tube appearing as the excavation face advances, can be easily cut.

(5)   Less expensive than the conventional long steel pipe forepiling.

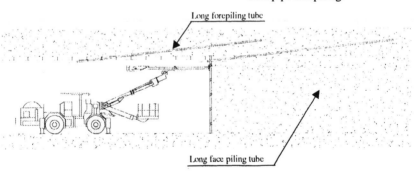

Figure 2. Overview of the long face reinforcement method.

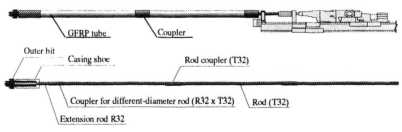

Figure 3. Drilling system.

4 DEMONSTRATION TEST OF THE METHOD AS LONG FOREPILING GFRP TUBE

In the demonstration test in which the method was applied to an actual tunneling site for the first time, two GFRP tubes were driven in addition to the injection type long steel pipe forepiling pattern, for verifying the work efficiency and effect of the method. Measurements were conducted at the same time. The length of the GFRP forepile was 12 m, made of connected four 3-m long tubes.

The test results confirmed that the work efficiency was equal or superior to the injection type long steel pipe forepiling, and that the handling work was very easy, as transport and setting of tubes can be done by one person.

The strain generated in the GFRP tube forepile was measured by plastering type strain gauges. For comparison purpose, measuring instruments were installed on the adjacent steel pipe forepiles. Figure 4 shows the variation of axial strain along with the

advance of face at the main measuring positions. Figure 5 shows the variation of flexural strain.

The axial strain of both steel pipe and GFRP tube significantly changes in the compressive direction just before and after the face passes by. This tendency is especially remarkable with the GFRP tube. The maximum strain was −153 μm (equivalent to 32 MPa) at measuring point 2 (2.5 m deep) of the steel pipe. In the steel pipe, compressive strain occurred in advance before passage of the face, whereas tensile strain was generated in the GFRP tube. Looking at the flexural strain, we know that bending with the outer side (upper side) in tension occurred both in steel pipe and GFRP tube, about 3 to 4 m before the face passed. Then, just before the passage of the face, the strain took on a maximum value, and immediately after the passage, an abrupt change occurred, that is, the tunnel inner side became in tension. This change was more noticeable with the steel pipe.

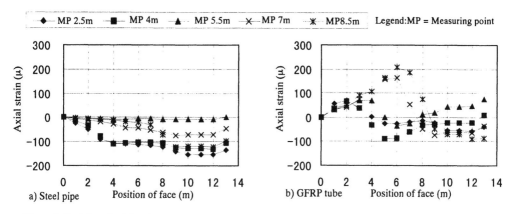

Figure 4. Variation of axial strain.

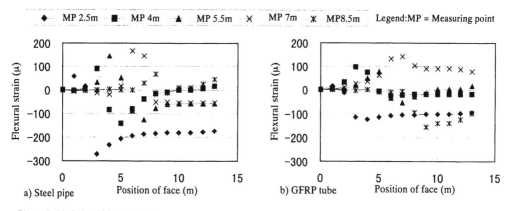

Figure 5. Variation of flexural strain.

505

As discussed above, it was verified that the GFRP tube behaved almost in the same manner as the steel pipe just before and after the passage of the face, though there is a difference in the behavior before the face passage.

5 APPLICATION OF THE METHOD AS LONG FACE PILING GFRP TUBE TO AN ACTUAL TUNNELING SITE

We discuss here the measurement data of a road tunnel where the method was applied as long face piling of top heading section. The sectional area of excavation was about 100 m².
GFRP tubes were driven at a rate of 0.5 per square meter. The face pile is 12 m long, driven at intervals of 7 m, with a lap of 5 m. An urethane injection agent was used. As long forepiling, an injection type was employed with steel pipes of 139 mm in diameter. Since the ground has almost no cohesion, the stability of the face was very poor, and small fragments of rocks in the face fell down. However, a consolidated core was formed around the GFRP tubes, contributing to the stabilization of the face.

For determining the strains generated in the face piling tube, a specific measuring instrument was made, which was of insertion small-diameter pipe type (Figure 6). Since the face piling tubes were cut off while the excavation advanced, the measurement data were taken till the face passed the point concerned. The data obtained by the strain gauge pasted on the measuring point on each tube were stored in the data logger embedded in the tube tip. When the excavation reached the tip, the data logger was collected.

Figure 6(a)(b) shows the measured axial force and bending moment generated along with the face advance during excavation of one shift (7 m long) after the face piles were driven. The axial force was relatively large, whereas the bending moment was small. This result demonstrates the fact that the face piling worked as a tensile member. In addition, the data reveal that the axial force increased while the face advanced. That is, at the measuring points of 5 m, 7 m and 9 m, the axial force significantly increased when the face came at the positions of 3 m, 4 m and 6 m respectively, and it took on a peak value when the face came in close proximity. At each measuring point, the axial and bending force began to occur when the distance from the face was 4 to 5 m. As demonstrated by this result, the tube worked as effective long reinforcement.

6 ANALYSIS OF THE LONG FACE REINFORCEMENT

The effect of the method was verified by numerical analysis considering a weak ground.
The finite difference method (FDM) was used for this purpose, because modeling of large deformations is easier by this method than FEM. A three-dimensional model was created for evaluating the behavior ahead of the face. Three cases were involved in the analysis, that is, forepoling bolt, steel pipe long forepiling and long face reinforcement with GFRP tube (= GFRP tube long forepiling + GFRP tube long face piling).

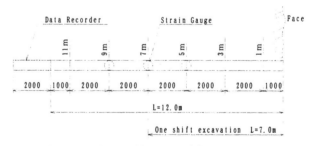

Figure 6.Schematic diagram of long face reinforcement.

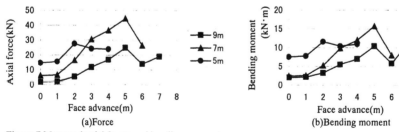

(a)Force
(b)Bending moment
Figure 7.Measured axial forces and bending moments.

The analysis is an elasto-plastic approach, on the basis of the ground properties such as cohesion $C=7kN/mm^2$, internal friction angle $\phi=30$ degrees, and modulus of deformation $E=50MN/mm^2$, with which the ground near the face is prone to becoming plastic as the face approaches. The analytical model is shown in Figure 7.

Out of the analytical results, the crown settlement, the surface settlement and face squeezing of each case are shown in Figures 8, 9 and 10. There is an obvious difference in crown settlement among the three cases; crown settlement is over 10 cm with forepoling bolt while it is 5.5 to 7 cm with forepiling, that is, 50 to 70%. The difference in surface settlement is not noticeable, but its tendency is almost the same as in crown settlement. On the other hand, as for the squeezing, the effect of the GFRP face reinforcement is remarkable. The face squeezing of more than 10 cm was produced with forepoling bolt and with long forepiling, whereas the face squeezing was limited to 3 cm with long forepiling plus long face piling. As indicated by these results, the long forepiling is capable of reinforcing the upper ground, thereby limiting the crown settlement, but under some ground conditions, it is impossible to directly reinforce the ground just in front of the face, where the ground becomes unstable, resulting in loosening ahead of the face. In consequence, the long face reinforcement method which entirely reinforces the ground ahead of the face is optimum.

Figures 10 and 11 show the axial forces and bending moment in each member of long forepiling and the maximum axial forces in long face reinforcement. The axial force induced in the face piling as significant as 130 to 210 kN at maximum varying from portion to portion, whereas the bending moment takes on quite a small value at a maximum of 0.3 kNm. It means that the face piling serves as axial force members, and in this respect, it shows the same tendency as in the measurement discussed in the previous chapter. The long forepiling presents high values in bending moment.

These analytical results cannot be directly compared with the measurement data discussed in Chapter 5, because various conditions were not accurately modeled, such as sectional geometry, geological conditions, and construction conditions including the density of driven reinforcement. However, looking at the generated axial forces and bending moments, it can be said that the analytical results approximately represent the behavior known by measurements. As for quantitative evaluation, the axial force analytically obtained should be evaluated conservatively. In-depth study in the future will offer more accurate evaluation.

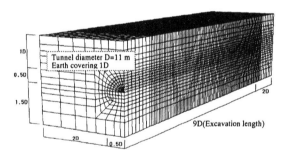

Figure 8. Analytical model.

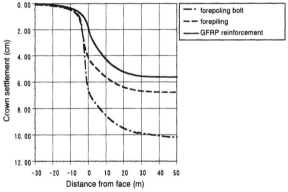

Figure 9. Crown settlement of each case.

507

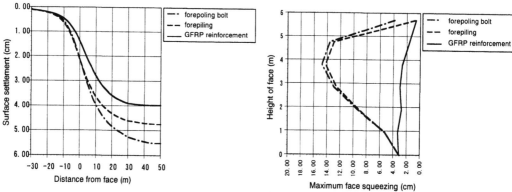

Figure 10. Surface settlement of each case.

Figure 11. Face squeezing of each case.

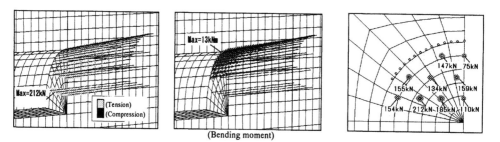

GFRP tube long forepiling + GFRP tube long face piling

Figure 12. Axial forces and bending moments generated in the long face reinforcement

Figure 13. Maximum axial force in each member of the long face reinforcement.

7 CONCLUSIONS

The method presented in this paper provides a sufficient effect of long face reinforcement. If it is suitable for the ground conditions encountered and satisfies the required quality, the method can be also used as an alternative of the conventional injection type long steel pipe forepiling. In addition, it is a powerful solution for providing effects of both face piling and forepiling, because it does not need enlargement of the excavation section, and driving in any direction is practicable from inside the excavated section. In recent years, the injection type long steel pipe forepiling has seen ever increasing use as an auxiliary method for stabilizing the face. Under these circumstances, the application range of the long face reinforcement method with GFRP tubes will be further widened.

Modern Tunneling Science and Technology, Adachi et al (eds), © 2001 Swets & Zeitlinger, ISBN 90 2651 860 9

Study on effect and evaluation of auxiliary methods for mountain tunneling in weak ground

T. Fukui & T. Hirai
Toll Road Corporation by OSAKA Prefectural Government, Osaka, Japan

Y. Kawamura & S. Nishimura
Minoh Tunnel Site Office, Osaka, Japan

Y. Mitarashi, S. Azetaka & H. Tezuka
Kumagai Gumi, CO., LTD., Tokyo, Japan

T. Matsui
Osaka University, Osaka, Japan

ABSTRACT: The Minoh toll road mountain tunneling project (southern lot) includes construction of a main tunnel (2000 meters long), an evacuation tunnel (2009 meters long), a working tunnel (354 meters long) and crosscuts. The south entrance of the main tunnel, which connects with the on-off ramps, has an extremely large section exceeding 300 square meters at maximum. The geology is Osaka stratum group over 100 meters from the south entrance, and Tamba zone in the rest. Actually, the working tunnel construction has been completed, and parts of the main tunnel have been completed including about 500 meters to the north from the working tunnel, and about 150 meters to the south (as of January 2001). Since the geology is complex, the face was extremely unstable during excavation of the working tunnel with a significant volume of water inflow in many places.

This paper evaluates various auxiliary methods used in the working tunnel construction, focusing upon the effect of new injected long face reinforcement, and presents the design principle of this technique.

1 INTRODUCTION

The Shin-Midosuji highway runs northward from the heart of Osaka city. As a part of the project to extend this highway, a bypass of national highway No. 423, to the north, the Minoh toll road tunnel is presently under construction. This road will connect with the Second Meishin Highway in the future. This road tunnel spans about 5.6 km through the 200 to 600 meter high Hokusetsu Mountains in the north of Minoh City.

The south work lot includes tunneling of 2000 meters northward from Bojima-Chinai, Minoh City. At the tunnel entrance of this section, for linking with the on-off ramps, an extremely large-scale tunnel exceeding 300 square meters in section will be constructed in an unconsolidated ground mainly composed of sands and clays.

Before the construction of the main tunnel, a working tunnel (347 m long) parallel to the main tunnel was constructed, applying various auxiliary methods in weak grounds.

In this paper, the work results and applied auxiliary methods in the working tunnel construction, are evaluated referring to the results of measurements and three-dimensional numerical back analysis. The basic concept of the main tunnel project involving an extremely large cross section is also discussed here.

2 OVERVIEW OF THE WORK

Project name: Construction of the Minoh toll road mountain tunnel (south work lot)
Site: Bojima-Chinai, Minoh city, Osaka Prefecture
Main tunnel: 313.0 to 80 m² in section, 2000 m long
Evacuation tunnel: 16 m², 80 to 170 m², 2009 m long
Working tunnel: 40 to 44 m², 347 m long
The plan of the Minoh tunnel in the south work lot is shown in Figure 1.

The construction sequence is as follows. A working tunnel 347 meters long was excavated first; then, the base spaces for the main and evacuation tunnels were excavated. Next, the main tunnel was driven to the south and north; followed by driving the evacuation tunnel. The typical sections of the main and working tunnels and the enlarged sections at the entrance are shown in Figure 2.

3 TOPOGRAPHY AND GEOLOGY

The south work lot of Minoh tunnel is located in a topography with typical reverse faults resulting from the activities of the Arima-Takatsuki tectonic line, where many conspicuous faults have developed, forming a trough at the boundary of hill area between upheaval mountains and subsided low ground.

The geology of this site is composed of a Tamba zone with Mesozoic and Paleozoic strata (shale,

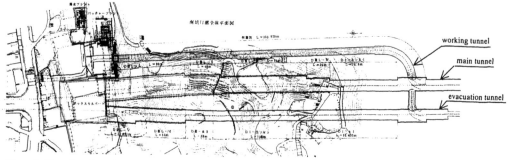

Figure 1. Plan of south work lot of the Minoh tunnel.

Enlarged section of main tunnel

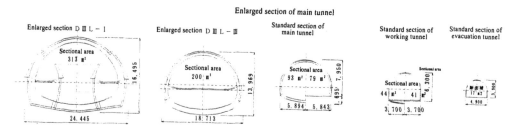

Figure 2. Standard sections of main, working, and evacuation tunnels and enlarged sections of ramp.

sandstone, chert, green rock, etc.), covered with Osaka groups (mainly gravel, sand and clay), which are also irregularly covered with terrace deposits. From the south tunnel entrance, Osaka groups, equivalent Kobe groups and a Tamba zone mainly composed of shale are encountered in order.

4 PROBLEMS IN EXCAVATION OF THE MAIN TUNNEL

In the excavation of the south section of the main tunnel, the following problems were encountered: (1) the instability of the cutting face of a large cross section tunnel in unconsolidated layers of Osaka groups and fractured rocks of a Tamba zone, (2) the unforeseen inflow of groundwater, (3) the impact upon structures on the ground surface, (4) the influence of twin tunnels, and (5) environmental impact (noise, vibration, air pollution, water shortage and pollution, etc.).

In this paper, based on measurements and records of observations in the construction of the working tunnel, the maintenance of face stability and measures for minimizing impact upon the tunnel surroundings, especially countermeasures against surface settlement are discussed.

5 CONSTRUCTION RESULTS OF WORKING TUNNEL

5.1 Work results

The excavation of the working tunnel started on May 17, 1998. In Osaka groups, especially in saturated sand layers, the stability of the cutting face remarkably degraded, being sometimes accompanied by face collapse. As auxiliary methods for reinforcing the face, forepoling was used, and for extremely weak grounds, long forepiling was applied. After the excavation reached the equivalent Kobe groups, small face collapses often occurred. This tendency continued till the excavation reached the Tamba zone located about 300 meters from the tunnel entrance. At around 220 meters from the tunnel entrance, a relatively large scale face collapse occurred, in which finely fractured shale gushed out with water inflow, in spite of applying such auxiliary methods as forepoling, long forepiling and/or face reinforcing bolts. Therefore, it was concluded that it was impossible to maintain the face stability with these measures, and that the combination of long forepiling and injected long face reinforcement (FIT) was applied as an auxiliary method, the pattern of which is shown in Figure 3. As a result of the excavation, as mentioned below, it was confirmed that the ground ahead of the face, could be sufficiently restrained resulting in improved face stability and making the tunnel displacement smaller. Despite

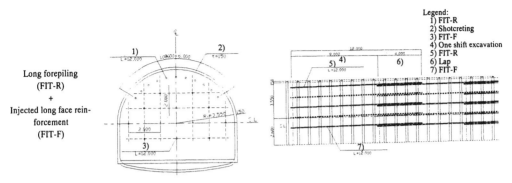

Long forepiling
(FIT-R)
+
Injected long face reinforcement
(FIT-F)

Legend:
1) FIT-R
2) Shotcreting
3) FIT-F
4) One shift excavation
5) FIT-R
6) Lap
7) FIT-F

Figure 3. Pattern of long forepiling and injected long face reinforcement (FIT).

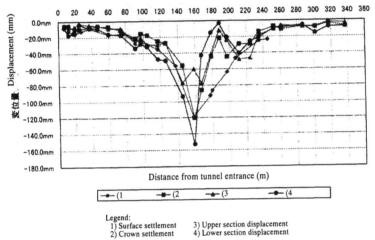

Distance from tunnel entrance (m)

Legend:
1) Surface settlement   3) Upper section displacement
2) Crown settlement   4) Lower section displacement

Figure 4. Measurements of convergence and surface settlement.

severe conditions of geology and face stability with water inflows, the excavation of the working tunnel using various auxiliary methods reached the intersection with the main tunnel in April 2000.

5.2 Measuring results

Measurements made in the working tunnel were daily measurement of surface settlement, and specific measurements (underground settlement measured from the surface, stresses in steel arch support and shotcrete, rockbolt axial force, ground displacement measured from the inside of the tunnel).

Figure 4 shows the measured convergence and surface settlement over the entire length of the working tunnel. This figure shows that, in Osaka groups and the Tamba zone beyond 250 meters from the tunnel entrance, both convergence and surface settlement were limited to a small amount. When sand layers in the Osaka groups contained water, the cutting face became unstable, so forepoling and long

forepiling were applied. As a consequence, the deformation was not significant. In the equivalent Kobe groups, which are located at the middle part of the working tunnel, the convergence and surface settlement were conspicuous, both values were 120 millimeters. This led to a response of pulling the ground surface ahead of the face. When the excavation passed the zone which encountered the maximum deformation, long forepiling and short face reinforcement were applied, but it was impossible to significantly reduce the surface settlement. Figure 5 shows the variation of ground displacement with elapsed time at the point (No.3+26). of 236 meters from the tunnel entrance.

This point is 12 meters ahead of the location where the relatively large-scale face collapse occurred as mentioned in 5-1. When the cutting face reached the position of 45 meters ahead of the measuring point, ground displacement began to occur, and the face collapsed when it advanced 12 meters ahead of the No.3+26. At this location, ap-

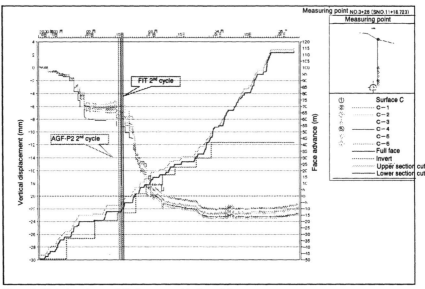

Figure 5. Variation of ground displacement with elapsed time.

plication of long forepiling and injected long face reinforcement (FIT) prevented almost all ground displacement when the face was advanced. But after the face passed the measuring point, the displacement increased abruptly.

This demonstrates that the application of long forepiling with injected long face reinforcement was very effective under such severe conditions that finely fractured shale of the Tamba zone gushed out with water inflow.

6 EVALUATION OF AUXILIARY METHODS

To examine the effects of long forepiling and injected long face reinforcement among various auxiliary methods used in working tunnel construction, back analysis was conducted by the three-dimensional finite difference method (3DFLAC). The analytical model is shown in Figure 6. The input data of physical properties are given in Table 1. The analytical flow is illustrated in Figure 7. For analytical results, Figure 8 shows both the settlements of the tunnel crown with and without auxiliary methods. Figures 9 and 10 show the distributions of the maximum shear strain and the displacement vector diagrams with and without FIT face bolts respectively. The displacement distribution in Figure 8 (with reinforcement) is identical with the measured results, confirming the effectiveness of the back analysis. Through the displacement vector and the maximum shear distributions it is suggested that the auxiliary methods (long forepiling and FIT) restrains the displacement, thus limiting the impact of tunnel exca-

vation. To evaluate the effects of these auxiliary methods, it is essential to implement three-dimensional evaluation. Also in the planning of the main tunnel in the future, three-dimensional numerical analysis, will be applied.

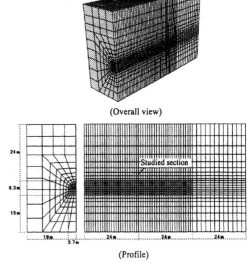

(Overall view)

(Profile)

Figure 6. Analytical model.

512

Table 1. Inputted properties.

Unit weight (γ) KN/m³ t/m³	Elastic modulus (E) MPa (kgf/cm²)	Poisson's ratio (ν)	Cohesion (C) MPa (t/m²)	Internal friction angle (ϕ) degree
22 (2.2)	100 (1,000)	Dead weight analysis 0.45 Excavation analysis 0.35	0.008 (8)	29

Article	Structural element	Unit weight (γ) KN/m³ t/m³	Elastic modulus (E) MPa (kgf/cm²)	Sectional area m²	Thickness m	Moment of inertia m⁴
Shotcrete + Steel support	Shell	23.8 (2.38)	6,600 (66,000)	0.15	0.15	–
GFRP tube (FIT)	Beam	17.5 (1.75)	20,000 (200,000)	1.71E-3	–	1.00E-6

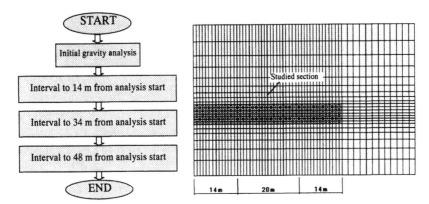

Figure 7. Flow of analysis.

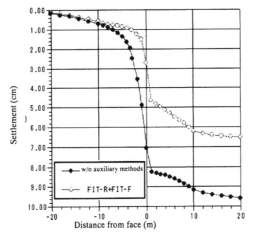

Figure 8. Settlement of tunnel crown.

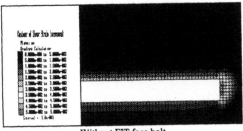
Without FIT face bolt

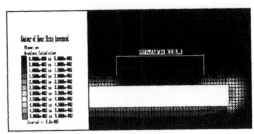
With FIT face bolt

Figure 9. Distribution of maximum shear strain.

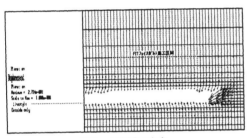
Without FIT face bolt   With FIT face bolt

Figure 10. Displacement vector diagram.

7 FEEDBACK TO PLANNING OF MAIN TUNNEL

For the future construction of the main and evacuation tunnels, it is vital to provide powerful solutions for every difficulty, such as the reduction of impact upon buildings standing 40 meters above the tunnel, the equivalent Kobe groups prone to considerable deformations, and the construction of extremely large cross section exceeding 300 square meters in Osaka groups of soils. Based on the results of the auxiliary methods used in the working tunnel construction, the main tunneling work will be planned and designed.

8 CONCLUSIONS

The authors have been investigating various construction methods in the planning and design of an unprecedented project of Minoh toll road tunnel, that is, tunneling of an extremely large cross section exceeding 300 square meters in Osaka groups of soils. This condition poses the most difficult problem in this project, but effective auxiliary methods were confirmed in such geological conditions through the experience of the working tunnel construction.

The project has been started by the construction of a working tunnel having a small cross section. Then the main tunnels with larger cross-sections will be constructed in order from relatively small to larger. In this sequence, the authors intend to establish a safer and more rational tunnel construction method by feeding back the results of the preceding excavation to the next steps. The construction data of the working tunnel will contribute significantly to the future work of the main tunnel. The authors will present the construction results of the tunnel of extremely large cross section when completed.

Modern Tunneling Science and Technology, Adachi et al (eds), © 2001 Swets & Zeitlinger, ISBN 90 2651 860 9

Numerical evaluation of environmental impact on surrounding ground water by tunnel excavation

Y. Ohnishi & H. Ohtsu
University of Kyoto, Kyoto, Japan
H. Okai & M. Saga
Japan Highway Public Corporation, Takamatsu, Japan
T. Nakai & K. Takahashi
Earth-tech Toyo Corporation, Kyoto, Japan

ABSTRACT: This paper proposes a most suitable method for evaluating an environmental impact on the ground water surroundings at the Houou Tunnel, which crosses the Median Tectonic Line (hereinafter, referred to as M.T.L.), the largest active fault in Japan. The Houou Tunnel is a highway tunnel and consists of two sub-tunnels, the first one of which was constructed in 1990. At the initial evaluation stage we have come to notice a serious environmental problem with the ground water. The previous construction of the study tunnel appears to have given a significant influence on the surface water system around the tunnel. The ground water level has gone down, thus affecting the local residents near the site. The second tunnel was proposed in parallel to the first one. The focal point of the environmental issues is that in the future, the ground water level might be even lowered and the creek water could decrease in amount or might dry out. The three-dimensional finite element ground water analysis is proposed to predict an environmental impact by tunnel excavation on the behavior of the present surrounding ground water in the study mountain area.

1 INSTRUCTIONS

The ground water issues, associated with the mountain tunnel excavation, include the preventative maintenance of tunnel face collapse by inflow water, waste water treatment, safety measures on the groundwater. The construction also generates environmental impact on the surrounding environment such as the pollution of the ground water surroundings, the lowering of water level, thus affecting the existing water resources. To cope with these issues, there has been an increasing demand to evaluate potential environmental impact on surrounding ground water, and to take proper measures.

As for the prediction of the surrounding ground water, the conventional two-dimensional or semi-3-dimensional analysis has gained many good results. However, these analyses have limitations for their application due to the actual complexity of construction methods, the tunnel structure, and the three-dimensionally distributed topography and geology. The construction, in many cases, requires the 3-D analysis of the groundwater behavior, considering environmental impact control in the surrounding natural environment.

This study focuses a series of accurate predictions of the ground water behavior with the application of a large 3-dimensional flow analysis model developed and constructed by the finite element method, using long-term rainfall characteristics and tunnel excavation of the Houou Tunnel. In more concrete term, the groundwater behavior was predicted with the application of the 3-dimensional flow analysis method developed on the basis of the existing first-line construction results. Taking into account the existing first-line tunnel inflows and the localized inflow at the M.T.L. This prediction also includes the evaluation for changes in the ground water level, potential environmental impact on the ground water surroundings, and the necessity of prevention measures on inflows. This paper presents some of the analysis results and advantages of the three-dimensional analysis for practical use.

2 TOPOGRAPHY AND GEOLOGY

The study area (near Kawakami Town, Kawanoe City, Ehime Prefecture) has a first-class active fault (M.T.L.). In the south of the M.T.L. lies the steep mountainous region, mainly comprising the Shimanto metamorphic zone (muddy schist, sandy schist, basic schist). In the north stand hilly districts, which consist of the Izumi strata group - alternated layers of shale and sandstone. Distributed in the end of the Houou mountain range or in the valley wedged between the two of mountainous/hilly districts, talus deposits are thickly accumulated, which cause a series of landslides. In the old talus deposits of schist gravel near the north-side portal entrance, the M.T.L. is distributed with its width of about 50 m and tilted southward at 30-40 degree angle. The geological profile is shown in Table 1 and the location of the tunnel is shown in Fig. 1. The in-situ tests indicate that the M.T.L. has a permeability: K=3.0 ×10⁻⁷ cm/sec, which is highly impermeable. While the field test shows that muddy schist, including fault fracture zone, has K=3.0-8.0×10⁻⁴ cm/sec.

Table.1 Geological profile near the Houou tunnel.

Geological name		Symbols	N-Value Elastic wave velocity (Vp)	Coefficient of Permeability (cm/sec)
Talus deposits		(dt)	N=10~30	-
Classical talus deposits		(Od)	N=50~70	-
Fault		(M.T.L)	N=30~70	3×10^{-7} aquifuge
Muddy schist	Strongly weathered rock	(Psch3)	Vp=2.5~3.0 Km/sec	8×10^{-4} aquifer
	Weathered rock	(Psch4)	Vp=4.0 Km/sec	3×10^{-4}
	Fresh rock	(Psch5)	Vp=4.8~5.0 Km/sec	2×10^{-5}

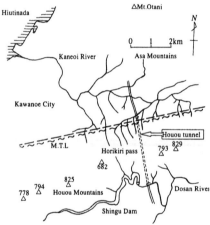

Figure 1. Location of construction site.

3 GROUND WATER CONCERN OF THE 1ST-STAGE TUNNEL

The Houou Tunnel is one of the very few expressways which pass through the M.L.T. The construction of the first line (started in 1987 and ended in 1990) was accompanied with significant volume of localized inflow water at the rate of 4-6 m³/min. Many of tunnel inflows were concentrated at the section 1.2 km from the north-side portal on the M.L.T. Especially through the fault fracture zone of muddy schist, localized inflow occurs. At the completion, total tunnel inflow water volume was approximately 4 m³/min. Up to the present time, the tunnel inflow has drained out at steady rate. In a meantime, as environmental impact by the first-line

construction on the ground water surroundings, there have occurred the drought of the water resources - agricultural-purpose surface water system - within 700-800 m range of the tunnel's one side, and the lowering of the ground water level of the wells for the local households.

4 ANALYSIS METHOD

As the key solution to the construction-related issues, 3-dimensional analysis of the ground water flow is absolutely necessary. With respects to the Houou Tunnel construction, there appear inflow sections or some inflow sites as a result of the previous first-line construction. There is an absolute need to improve the prediction accuracy for the widening of the adjacent second line (a small tunnel was excavated for evacuation. The second-line is made by widening it). A selected model for the analysis, constructed based on the accumulated construction results of the first line, was an accurate method to specify permeability or inflow section distribution and to predict the ground water behavior, associated with the second-line construction, by means of reproduction of tunnel inflow water volume at initial construction stage. Figure 2 shows analysis sequence.

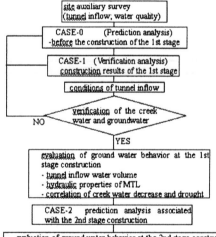

Figure 2. The flow chart of analysis procedure.

5 RESULTS

5.1 Making of analysis model

Due to the significant attributes of the fault fracture zone based on the first-line construction results, a constructed model is to be three-dimensional in order to reproduce the location and width of the zone as closely as possible. Figure 3 shows the outline of the model in the upper figure and shows the location of the tunnel and the fracture zones excluding the surface strata of talus and muddy schist:

a) Scope of analysis: within a range of 2.3 km in the tunnel axis direction and of 2.4 km transverse, applied was the three-dimensional analysis model, using 85,560 nodes, 79,178;

b) Aquifer zones: based on the results of the first-line tunnel and seismic explorations, the model aquifer was partitioned into four sections - talus, weathered-surface, fracture zone including M.L.T. and muddy schist;

c). Hydraulic property: permeability coefficient, volumetric water content, specific storage and unsaturated properties were set based on the verification and analysis results of the site exploration. Rainfall condition was examined in the same way;

d) Tunnel condition: Tunnel shape is to be in rectangular section. The tunnel wall is to be completely drained

e) Boundary condition: Base on the first-line verification and analysis, the water level or flux was fixed and the bottom of the model is impermeable.

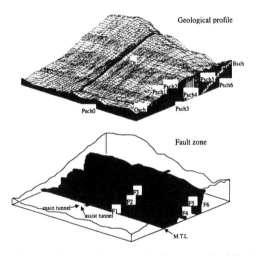

Figure 3. Relationship between the Houou tunnel and distribution of fault zone.

5.2 Verification of analysis model/reproduction of the first line construction

Based on the present tunnel inflow distribution and the first-line construction results, hydraulic properties were verified with the application of the 3-dimensional analysis model. The comparison of the tunnel inflow by actual survey before the second-line construction and by the analysis model indicates that the calculated total drainage volume is almost

matched to be 4.0 m^3/min. Table 2 shows the hydraulic properties of this case.

Table.2 Hydraulic properties of the three-dimensional model..

Symbols	Geology	Coefficient of Permeability (cm/sec)	Volume Water content (%)	Specific Storege (cm^{-1})
Dt	Allvial deposit	1.0×10^{-3}	15	0.01
Osch	Old allvial dep	5.0×10^{-4}	10	0.01
Psch 0 4	Muddy schist	3.0×10^{-3}	5	0.0001
Psch 5 6	Muddy schist	1.0×10^{-5}	5	0.0001
Bsch	Basic schist	1.0×10^{-5}	7	0.0001
F1	Fault(M.T.L)	1.0×10^{-5}	10	0.000001
F2 F6	Fault zone	1.0×10^{-5}	10	0.000001

The reproduction of the first-line construction was verified, assuming on the analysis model that main and evacuation tunnels are excavated using the same previous construction sequence. Figure 4 shows the axial direction of the first-line main tunnel, and Figure 5 shows each pressure head and total head distribution in horizontal direction. Each figure indicates that water level is slowly lowered in the fault fracture zone with the steep hydraulic gradient. Whereas water level is rapidly lowered in muddy schist about 700 m form the north portal entrance, this section takes strong inflow into the tunnel at present. In the deeper tunnel sections the water level is likely to be slowly lowered and almost remained unchanged after the completion of the tunnel.

5.3 Prediction analysis on the widening of the second line

1) Increase/decrease of the tunnel inflow water volume due to the widening

Lowered by the first-line construction, the ground water level around the tunnel has reached at equilibrium state. With this state, inflow water volume by the widening is predicted to become at the rate of approximately 10-100 L/min. This indicates that extraordinary localized constant inflow will very rarely occur. It is predicted that the final increase volume by the widening is about 700-1000L/min. However, The total increase amount by the first and second-line construction is predicted to remain almost unchanged because of the inclusion of the amount of the inflow water by first-line construction. Also, the ground water reservoir type has high permeability in the tunnel transverse direction. Therefore, it is apt to be polluted the area by the muddy water due to widening excavation.

2) Localized inflow volume after the M.L.T. completion

The widening is unlikely to cause additional localized inflow. Due to water pressure at tunnel face and evacuation tunnel construction, ground water level is lowered up to the road base near STA 45. In the fault fraction zone at near STA 47 - 48, auxiliary measures- horizontal boring need to be considered for the widening in the 1st stage

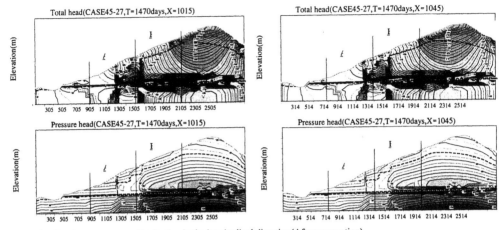

Figure 4. Predicted water pressure distribution in the longitudinal direction(After excavation).

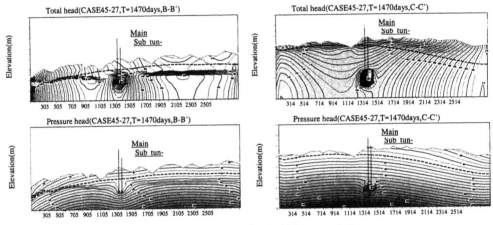

Figure 5. Predicted water pressure distribution in the transverse direction(After excavation).

localized inflow sections. The prediction shows that some section might have high water level.

3) Environmental evaluation by the second-line construction on the surrounding creek water

Drain impact on the ground water surroundings in the first construction is predicted to occur on the surface water system right above the tunnel (Figure 6A and B), and the tributaries has dried up. The validity of the measured water volume were reevaluated, which appear to remain almost unchanged compared to the amount reduced by the first-line construction. It is predicted that impact on the surrounding environment by the second-line construction will not be extensive. The impact on the ground water surroundings, therefore, is not to be referred as long as inflows in the widened area are not extensive.

6 CONCLUSIONS

The ground water issues, associated with the mountain tunnel excavation, has in recent years focused on the scale of measures and its implementation that will definitely be a societal need. To hold a firm understanding of the accurate ground water behavior, it is absolutely necessary to quantitatively approach 3-dimensional unsteady phenomena.

The ground water evaluation method in this study is highly considered taking fully into account the necessity of accurate environmental impact on the ground water flow, and many cases where the 3-dimensional understandings are required to evaluate potential impact on the ground water during and after the second-line construction followed by the first one as of the Houou Tunnel.

Reproduced were the tunnel inflow water volume at the first-line construction, the surface water system before the construction and the ground water

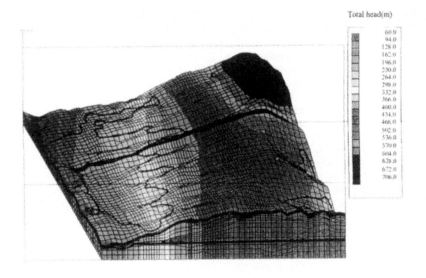

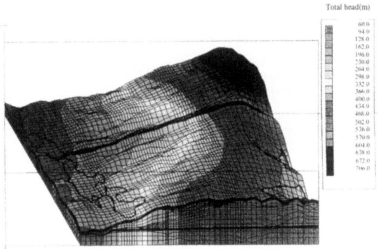

Figure 6A. Result of ground water level change(1).

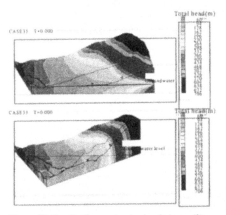

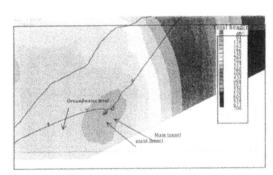

Figure 6B. Result of ground water level change(2).

level at initial stage. At the same time, drain measures as a means of environmental protection, and changes in the ground water behavior over the past nine years since the first-line construction were evaluated. Moreover, possibility of localized inflow in the M.L.T. and the necessity of auxiliary measure were examined through the M.L.T. construction. In the end the ground water flow has reached at equilibrium state due to the first-line construction. It is very unlikely that localized inflow will occur associated with the second-line construction.

Finally it was concluded that the application of this method is tremendous and considered for practical use to solve the ground water issues in other tunnel construction.

REFERENCES

Ohnishi, Y. & M. Tanaka 1996. Evaluation of the Effect by Tunnnel Excavation on Surrounding Groudwater. *Proc. IGCS. Japan.* 31:2137-2138. (Japanese)

Ohnishi, Y., M. Tanaka, T. Yasuda, & K. Takahashi 1997. Evaluation of the Effect by Tunnnel Excavation on Surrounding Ground Water. *Proc. IGCS. Japan.* 32: (Japanese)

Ohnishi, Y., H. Ohtu, T. Yasuda & K. Takahashi 1998. Assessment of Influence on Ground Water Surroundings at Urban Tunnels. *Proc. ITA. World Tunnel Congress* :489-494.

Ohnishi, Y., H. Ohtu, T. Yasuda & K. Takahashi 1999. Analysis of Ground Water Behavior around the Crossing of adjacent Tunnels. *Proc. ITA. World Tunnel Congress* :147-154.

Modern Tunneling Science and Technology, Adachi et al (eds), © 2001 Swets & Zeitlinger, ISBN 90 2651 860 9

Back analysis for the bolted gate by coupling method of FEM and BEM

Tan Yunliang, Han Xianjun, Wang Tongxu & Ma Zhitao
Dept. of Mechanic Engng. Shandong Univ. of Science & Technology, Taian, China

ABSTRACT: The key parameter affecting the deformation of the surrounding rock mass, i.e. elastic modulus E was found by sensitivity analysis, and it was recognized by displacement back analysis based on the coupling method of FEM and BEM. It is shown by an in-situ example that the recognized E is reliable and valuable for the rock mechanics and geotechnical engineering.

1 INTRODUCTION

The development of numerical methods such as FEM and BEM has given powerful passports to solve the intricate geotechnical engineering, but it is difficult to provide reliable parameters by rock sample test because of the complexity of rock mass and changeful geological conditions. As we know, the reliability of parameters of the rock mass is just one key factor determining the reliability of the computation of the geotechnical engineering. In the recent twenty years, more and more attentions have been paid to the back analysis of the parameters of rock mass by in-situ data. Japanese scholar S.Sakurai and K.Takeuchi (1983) proposed the analytical back analyzing answers of even geological stresses and the elastic modulus of rock mass by FEM. Yang Zhifa (1989) contributed to the displacement spectrum method. Yang Linde at el. (1996) have made many efforts to the computational theories and the engineering application of the back analysis. Lu Aizhong (1988) and Jiang Bingsong (1994) introduced the recognize ability condition to the displacement back analysis. However, how to back analyze the equivalent parameters affecting the deformation of bolted surrounding rock is still an unsolved problem. There are many parameters effecting the deformation of the bolted surrounding rock mass. It is difficult and unnecessary in practice to back analyze all of them at the same time. This paper introduces a new coupling method of FEM and BEM to recognize the equivalent mechanical parameters of bolted rock mass.

2 SENSITIVITY ANALYSIS OF THE SURROUNDING ROCK MASS

2.1 Concept of Sensitivity

It is known that displacement is very useful to describe the instability of the surrounding rock mass. There are many factors affecting the displacement of the gate surrounding rock mass, i.e. elastic modulus E, possion ratio μ, cohesion C and internal friction angle φ, etc. For the sake of analyzing the difference of displacement changes which are caused by the changes of different mechanical parameters, we define the concept of sensitivity as follows: note the ratio of the changes of the corresponding displacement u_{xi} under a given mechanical parameter x_i $(i=1,2,\cdots,n)$ as well as the displacement u_0 under a given condition assign its ratio $\eta_{xi} = u_{xi}/u_0$. Moreover, we define a new variable of the ratio of x_i to the primary value x_{i0}, i.e. $\zeta_{xi} = x_i/x_{i0}$ assuming the the other parameters kept unaltered, . Meanwhile within the changes of η_{xi} (viz. $\Delta \eta_{xi}$) affected by that of ζ_{xi} (viz. $\Delta \zeta_{xi}$), the ratio of the $\Delta \eta_{xi}$ to $\Delta \zeta_{xi}$ is called as the sensitivity of the parameter x_i, i.e.

$$\xi_{xi} = \frac{\left|\Delta\eta_{xi}\right|}{\left|\Delta\zeta_{xi}\right|} = \left|\Delta\left(\frac{u_{xi}}{u_0}\right)\right| / \left|\Delta\left(\frac{x_i}{x_{i0}}\right)\right| \quad (1)$$

$$\mu_0 = 0.25\left(1 + e^{\frac{-0.2\sigma_c}{c}}\right) = 0.25 \times \left(1 + e^{-0.2 \times 16}\right) = 0.26 \quad (3)$$

Obviously, by the sensitivity analysis, the key parameters influencing the deformation of the bolted surrounding rock mass can be identified, and the number of the parameters need to be analyzed can be reduced as well.

2.2 Sensitivity analysis

2.2.1 Computational conditions

In order to analyze the sensitivity of the equivalent mechanical parameters of the bolted surrounding rock mass, we choose a typical gate of surrounding rock mass of coal in the analysis process, the bolted surrounding rock mass was considered as an elastic-plastic material and Coulomb-Mohr failure criterion was adopted. The gate is under a uniform stress field of gravity, and the overburden depth is 500m. The average unit weight of the overburden is 0.025MN/m³, the lateral stress ratio is 0.5, The average slope angle of the seam is 5°, the width and height of the gate section are 3.6m and 2.8m respectively. The reinforcing parameters of the bolt are shown in Tab.1. By the laboratory test, the unit weight of the coal is 0.013MN/m³, the compressive strength σ_c =16MPa, cohesion C=6MPa, internal friction angle φ=50°, tensile strength σ_t= 0.8MPa.

Table 1. Bolt reinforcing parameters.

Type	Length /m	Diameter /mm	Space /m×m	Designed anchorage capacity/KN	Drawing force/T
Roof	2.5	20	0.7×0.8	68.6	7
Wall	2.5	20	0.75×0.8	58.8	7

Because the rock mass can be considered as an elastic-plastic material, the main physical parameters affecting the deformation of gate during the elastic-deformation period are elastic modulus E and possion ratio μ, the key parameters affecting the deformation and the failure of the gate also include cohesion C and internal friction angle φ while during the plastic-deformation period. In this paper more attention is paid to do the sensitivity analysis of the above-mentioned parameters. According to the result of Ayden's (1993), the initial values of the elastic parameters are given as follows:

$$E_0 = 80\sigma_c^{1.4} = 80 \times 16^{1.4} = 3880MPa \quad (2)$$

Based on the above given conditions, the gate deformation was simulated separately with the parameter one by one under the condition of other parameters were kept unaltered, the results were shown in Fig. 1.

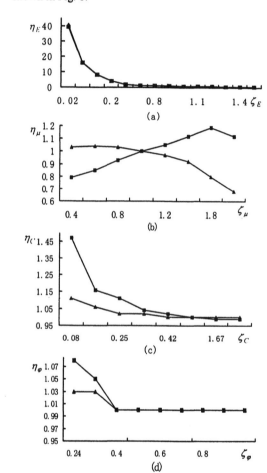

Figure 1. Deformation affected by different parameters.

—■— wall to wall   —▲— roof to floor

2.2.2 Sensitivity of different mechanical parameters

According to the above definition, the sensitivity computation results are shown in figure 1.

It is clear, for the elastic modulus E, when ζ_E changes from 0.02 to 1.1, the deformation of the bolted gate surrounding rock mass changes obviously. From wall to wall, the displacement sensitivity of E is

$$\xi_{E1} = \frac{\left|\Delta\mu_{E1}\right|}{\left|\Delta\zeta_E\right|} = \frac{\left|39.9 - 0.9\right|}{\left|1.1 - 0.02\right|} = 35.56 \qquad (4)$$

From roof to floor, the displacement sensitivity of E is

$$\xi_{E2} = \frac{\left|\Delta\mu_{E2}\right|}{\left|\Delta\zeta_E\right|} = \frac{\left|40.5 - 0.9\right|}{\left|1.1 - 0.02\right|} = 36.67 \qquad (5)$$

With the help of above method, the sensitivity analysis results of all the parameters computed are shown in table 2. From table 2, It is clear that the elastic modulus E is the most key parameter that affects the deformation of the bolted surrounding rock mass.

Table 2. The results of sensitivity analysis.

Parameter	ξ of wall to wall	ξ of roof to floor
E	35.56	36.67
μ	0.29	0.38
C	1.38	0.29
φ	0.50	0.90

3 ANALYSIS AND COMPUTATION FOR ELASTIC MODULUS

3.1 Theory of back analysis

A method of displacement back analysis by the virtual stress BEM was adopted in this paper. Around the gate surface, N boundary line segments or elements are divided, as shown in Fig. 2 and Fig. 3.

For a gate, it can be considered as a plane strain problem, the essential equation is

$$\left.\begin{array}{l} \sum\limits_{j=1}^{N}\left(A_{ss}^{ij}P_j^s + A_{sn}^{ij}P_j^n\right) - \frac{1}{2}\left(\sin 2\beta_i\right)P_x \\[2mm] + \frac{1}{2}\left(\sin 2\beta_i\right)P_z + \left(\cos 2\beta_i\right)P_{xz} = 0 \\[2mm] \sum\limits_{j=1}^{N}\left(A_{ns}^{ij}P_j^s + A_{nn}^{ij}P_j^n\right) - \left(\sin^2\beta_i\right)P_x \\[2mm] + \left(\cos^2\beta_i\right)P_z + \left(\sin 2\beta_i\right)P_{xz} = 0 \end{array}\right\} \quad (6)$$

Where A^{ij} is coefficient of lateral stress, it represents the stress acted at the midpoint of the ith element derived from the jth element on which the uniform

disturbed stress acted. P_j^s and P_j^n are the tangential and normal virtual stress acted on the jth element, P_x, P_z, P_{xz} are components of the initial geological stress vector $\{P\}$.

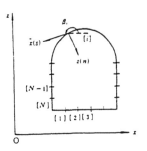

Figure 2. Boundary elements divided.

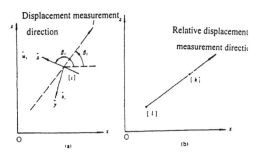

Figure 3. Displacement measurement around a gate.

Given the initial geological stress vector $\{P\}$ and the possion ratio μ of the rock mass. P_j^s, P_j^n $(j = 1,2,\cdots,N)$ can be calculated by formula (6). Furthermore, the elastic modulus E can be made certain as formula (8) by using one measured displacement value to the displacement measurement equation (7).

$$\sum_{k=1}^{N}\left(C_{ls}^{kj}P_j^s + C_{ln}^{kj}P_j^n\right) - \Delta_{lk} = 0 \qquad (7)$$

$$E = \sum_{k=1}^{N}\left(c_{ls}^{kj}P_j^s + c_{ln}^{kj}P_j^n\right)\Big/\Delta_{lk} \qquad (8)$$

Where c^{kj} is the coefficient of boundary relative displacement, $c^{kj} = E \cdot C^{kj}$, and Δ_{lk} is the relative displacement between two elements' midpoints along the measuring direction

3.2 *In-situ example analysis*

The down road gate of the 4303 working face in Xinglongzhuang colliery in China was dug along the floor and its depth is 500m. The average slope angle of the stratum is 5°, the width and height of the gate section are 4.0m and 2.8m respectively. The physical mechanical parameters of the gate surrounding rock mass are shown in table 3. The parameters of the bolt

Table 3. Physical mechanical parameters.

Rock	Far-forth field	Roof	Coal	Floor
Lithological character		siltstone	mudtone	
Thickness/m		0.6	8.3	3.8
Unit weight γ /MN/m^3	0.027	0.026	0.013	0.025
Elastic modulus E/MPa	2000	1400	*	1300
Possion ratio μ	0.3	0.18	0.39	0.3
Tensile strength σ$_t$ /MPa	0.5	2.1	0.2	1.7
Cohesion C/MPa	10.5	12	6	4.6
Internal friction angle φ /°	35	42	38	30

* to be calculated.

supporting are as same as those of table 1. The gate is under a uniform stress field of gravity, and the lateral stress ratio is 0.5. By the in-situ researches, the convergence of the roof to floor is 66mm, and that of the wall to wall is 32mm.

According to the method introduced above, the equivalent elastic modulus E is 540MPa by using the back analysis of the displacement from roof to floor. Moreover, under the calculated elastic modulus E the convergence displacement calculated of the wall to wall is 28-34mm,and its value is consistent to in-situ measured datum.

4 CONCLUSIONS

Complexity of rock mass and the changeful geological conditions make it difficult to use the laboratory sample test results to the computation of geotechnical engineering. This paper firstly analyses the sensitivity of the mechanical parameters affecting the deformation of the bolted surrounding rock mass with the coupling calculation of FEM and BEM. The key parameter affecting the deformation of the surrounding rock mass is elastic modulus E by the sensitivity analysis. It is verified by a living example that the convergence displacement of the gate calculated by using the elastic modulus E back-analyzed well accords with the in situ data.

ACKNOWLEDGEMENT

It is supported by China National Science Foundation and China Education Ministry Foundation.

REFERENCES

S.Sakurai and K.Takeuchi, 1983. Back analysis of measured displacement of tunnels. Rock Mech. and Rock Eng.

Zhou Weiyuan, 1990. Super rock mechanics. Hydro and electric engineering press.

Yang Linde at el., 1996. Back analysis theories and application to geotechnical engineering.

Lu Aizhong, 1988. Discussion on the differentiability of the parameters of surrounding rock mass and the geological stresses during gate's digging. Chinese Journal of Rock Mech. and Eng. 155—164

Jiang Bingsong, 1994. Method of displacement back analysis. Journal of Shandong Inst. of Min. and Tech. 213—217

Yang Zhifa, 1989. Displacement back analysis and its application. Refer to: Advances in Rock Mechanics. Northeastern Technology University Press.

Wu Xingchun, 1995.Integrating computer aid analysis system for the stability of the gate. (Doctorial treatise, Northeastern University).

Chen XiangZhen and Zhang Min, 1985.Back analysis for the deformation of the Ba-yi-lin tunnel in the "Yin Luan Gongcheng" engineering. Chinese Journal of Rock Mech. and Eng. 39—45

Academe of hydro and electric engineering, at el., 1991. Parameters handbook of rock mechanics.

Modern Tunneling Science and Technology, Adachi et al (eds), © 2001 Swets & Zeitlinger, ISBN 90 2651 860 9

Application of back analysis in assessing the stability of an Indian tunnel

Anil Swarup
Scientist, Central Mining Research Institute, India

Shinichi Akutagawa
Associate Professor, Kobe University, Kobe, Japan

ABSTRACT: Prediction of rock mass behavior is one of the most difficult problems of rock engineering in spite of rapid strides made in the recent years in the field of geomechanics. The most promising way to overcome these difficulties seems to be by back analysis of the deformations observed in the tunnel during excavation. Field measurements are a very powerful tool not only for monitoring the stability of underground media, but also for designing structures. An Indian case history was selected for analysis to assess the rock mass behavior using the back analysis approach. The monitored deformations in the tunnel, which were of the order of 8.9% the tunnel diameter, were back analysed. The results of the analysis estimated shear strains upto 15.32% which is much greater than the estimated critical shear strain of 2.9%, indicating severe instability due to excavation. Also the comparison between calculated and measured displacements show good agreement.

1 INTRODUCTION

Review of published literature reveals that the efforts are centered on both empiricism and analytical understanding. Analytical methods cannot lead to a satisfactory solution due to difficulties in simulating the rock mass behavior by a definite mathematical model. Moreover, developments in numerical techniques including finite element method do not provide reliable solutions at the design stage because realistic values of input parameters are rarely available

The most promising way to overcome these difficulties seems to be by back analysis of the deformations observed in the tunnel during excavation. The mechanical properties and initial stresses can be back analyzed from these measured displacements. They can be then used as input data to assess overall stability of the ground by analyzing the stress and strain distributions around the opening. Sakurai (1983) proposed a method for back analysis of measured displacements of tunnel. The technique is based on the concept of strain distributions in the ground. The concept of back analysis has gained acceptance worldwide and Japan in particular. An attempt has been made to apply the concept to Indian tunnels as well and the results have been found encouraging (Swarup 2000).

A 8.56-km long tunnel of 4.75m diameter located in Himalayas was excavated in the rock formations comprising of folded and faulted quartzites and schists. The tunnel passed through high rock covers of upto 900m suggesting development of squeezing pressure. The tunnel encountered problems of cavity formation, support failure and water-inrush.

2 BACK ANALYSIS AND STABILITY ANALYSIS

Sakurai & Takeuchi (1983) proposed an inverse back analysis method. The main objective of the back analysis is to determine the initial stresses and Youngs's modulus from measured displacements and then used as input data for a finite element analysis to determine the strain distribution around the underground opening. Assuming the ground in which a tunnel is bored consists of homogeneous isotropic elastic materials, the following linear relationship between measured displacements and the initial stress can be derived:

$$[A] \{\sigma_0\} = \{u^m\} \tag{1}$$

where $[A]$ is a matrix of only a function of the location of measuring points and Poisson's ratio. $\{u\}$ is a vector of measured displacements, m is the measured displacement vector and $\{\sigma_0\}$ is a normalised initial stress vector defined in a two dimensional state as:

$$\{\sigma_0\} = \{ \sigma_{x0}/E \ \ \sigma_{y0}/E \ \ \tau_{xy0}/E \}^T \tag{2}$$

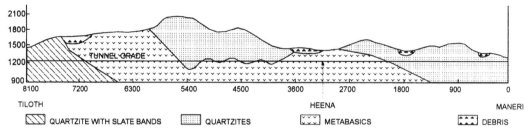

Figure 1. Longitudinal geological cross-section along the tunnel alignment.

where σ_{x0}, σ_{y0}, τ_{xy0} are the components of initial stress existing in the ground before excavation, and E denotes Young's modulus of ground materials.

Normalised initial stress can be uniquely determined from measured displacements by assuming Poisson's ratio. The measurement of relative displacement between two measurement points as well as absolute displacements can be used as input data. The components of initial stress and Young's modulus can be separated from the normalised initial stress by assuming that the vertical stress is equal to overburden pressure.

After determining the normalised initial stress, the displacements at all the nodal points can be calculated by equation (1). The strain in each element, therefore, can be obtained from use of the following relationship between strain and the displacements

$$\{\varepsilon\} = [B] \{u^e\} \tag{3}$$

where matrix $[B]$ is a function of the location of the nodal points alone and e is the element displacement vector.

Sakurai (1981) proposed the concept of critical strain defined as the ratio of uniaxial compressive strength to modulus of elasticity, which may be adopted as an allowable value of strain. The critical strain is always smaller than strain at failure. Thus the stability of tunnel can be monitored by comparing the strain occurring in the ground with the critical strain of the ground material.

3 CASE HISTORY

A 8.56-km long circular tunnel with a finished dia. of 4.75-m has been constructed for Maneri Bhali Hydel Scheme Stage-I on River Bhagirathi located in the Garhwal Himalayas (Fig.1). The tunnel passes through quartzite and metabasic (chlorite schists) rock formations of the young Garhwal Himalayas.

3.1 Regional geology

The rock masses exposed in the area of the tunnel are quartzites, quartzite interbedded with thin bands of slate, chlorite schist, phyllites, metabasics and basic intrusives belonging to the Garhwal group. Towards the northeast and east of the project area, about 4.5-km east of Maneri, the rock formations are affected by a thrust, the Main Central Thrust.

The individual rock units observed along the tunnel alignment between Heena and Tiloth adits are quartzites, metabasics, and quartzites with minor slate bands.

These lithological units are intensely folded and faulted due to tectonic disturbances. The tectonic activity in the area has caused closely spaced jointing, brecciation and shearing even in the quartzites.

3.2 Observed tunnel behaviour

Problems of tunnel face collapse with or without heavy ingress of water, cavity formation and large tunnel closures leading to buckling of steel ribs on account of squeezing ground conditions were encountered (Goel et al. 1995). Tunnelling activities at depth of 700m to 900m between ch. 5550m and 5250m through partially wet and thinly foliated metabasics were beset with severe squeezing problems. The tunnel was supported by ISMB 150x150 mm ribs spaced at 80 to 100cm. The gap between outer flanges of the ribs and the excavated rock surface was filled with concrete. There were no problems during excavation, however, after a period of 5-6 month, the backfill concrete started cracking and the ribs were deformed/buckled due to the squeezing pressures. The project authorities had to resort to forepoling and multi-drift method of excavation and at places the alignment had to be changed to tackle the difficult zones.

3.3 Field measurements

The instrumentation programme for monitoring the adequacy of the supports consisted of measuring hoop loads on steel arches and deformation measurements using tape extensometers. The data analysis shows that the maximum wall closure was of the order of 430 mm (8.9% of the tunnel diameter) in 600 days at the contact of metabasics and quartzites at ch. 5350m (Fig.2) and about 105 mm (2.18%) in over 100 days at ch. 5510m (Fig.3) which was at a depth of 750m. At other locations closure varied from 3 to 20 mm (0.05 to 0.4%).

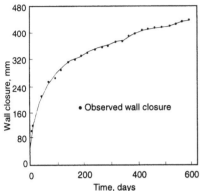

Figure 2. Closure - time relation at ch. 5350 m.

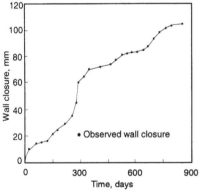

Figure 3. Closure - time relation at ch. 5510 m.

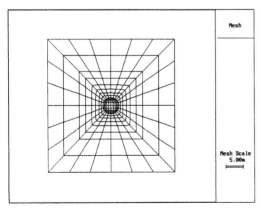

Figure 4. FEM mesh.

3.4 Back analysis of measured displacement

Back analysis was used for obtaining strain distributions from measured displacements. The computer program used for this purpose was DBAP, developed by Sakurai, et al. The finite element mesh used here is shown in Figure 4. The input data used for the analysis is given in Table 1.

Table 1. Input data used in back analysis.

	Quartzite	Metabasic
Unit Weight	2.4 g/cm^3	2.3 g/cm^3
Young's modulus	5 GPa	5.6 GPa
Poisson's ratio	0.25	0.15
Internal friction angle	25°-35°	15°-25°
Compressive strength	230 MPa	100 MPa

In order to verify the accuracy of this back analysis, the displacements were calculated by introducing the initial stress and Young's modulus obtained into the conventional finite element analysis, and compared them with the measured displacements (Fig.5). It can be seen that the back analysis carried out show fairly good accuracy. The results of the analysis are shown in Table 2 and Table 3.

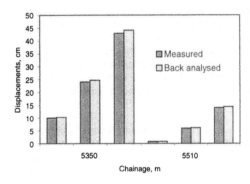

Figure 5. Comparison of measured vs. calculated displacements.

Table 2. Results of back analysis.

Initial Stresses (MPa)	σ_{0x}	σ_{0y}	τ_{0xy}
Chainage (m)			
5350	-14.29	-11.77	0.019
5510	-20.69	-17.65	0.023

Table 3. Maximum shear strains and Young's modulus.

Chainage / Rock Type	No. of Days	Max. Shear Strains (%)	E (MPa)
5350m	0-10	3.56	0.140
Quartzite	100-200	8.55	0.006
	>500	15.32	0.003
5510m	0-10	0.28	0.227
Metabasics	100-200	2.14	0.030
	>500	5.00	0.013

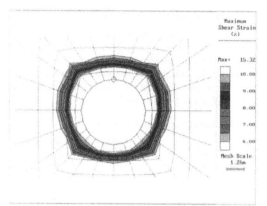

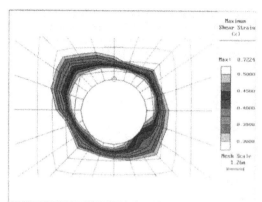

Figure 6. Maximum Shear Strain distribution around the tunnel.

Figure 7. Maximum Shear Strain distribution around the tunnel.

3.5 Stability assessment

The maximum shear strain is often evaluated for monitoring the stability of tunnels during construction. Back analysis method can be used for determining the maximum shear strain distribution from measured displacements (Sakurai 1983). This strain may then be compared with the allowable shear strain or critical shear strain in order to assess the stability of the tunnel or to estimate an unstable zone occurring around a tunnel.

Critical shear strain is defined as the ratio of shear strength to shear modulus of rock material (Sakurai et al. 1993). The critical shear strain can be estimated by using the following expression

$$\gamma_0 = \varepsilon_0 (1+\nu) \tag{4}$$

where ε_0 is the critical strain (Sakurai 1981) and ν is the Poisson's ratio.

The shear strains were estimated through back analysis and compared with the critical shear strains to assess the stability of tunnel. The critical shear strains estimated for quartzite are 2.9% and for metabasics 1.25%. The stability of tunnel is assessed by comparing the maximum shear strain occurring around the tunnel with the allowable values of strain. It can be observed from Table 3 that the back analysed maximum shear strains for chainage 5350m exceeded the value of critical strain soon after excavation indicating that extensive supporting was required. After 400 days the max. shear strains exceeded the critical value by more than 4 times which was indicated by the failure of supports as mentioned above. A plot of maximum shear strains is shown in Figures 6 and 7. The unstable zone is defined as the place where the derived shear strains are greater than the critical shear strains.

Thus it is possible to assess the stability of the tunnel through back analysis by comparing the maximum shear strain with the allowable value.

4 CONCLUSION

In this paper applicability of the back analysis method has been demonstrated for use in tunnels excavated in the Himalayas. The stability of tunnel can be monitored by comparing the strain occurring in the ground with the critical strain of the ground material.

Thus it can be concluded that back analysis can be used as an important tool for analyzing rock deformation and for designing supports for tunnels being excavated in highly tectonic rock formations of Himalayas.

REFERENCES

Dube, A.K., J.L. Jethwa. & B. Singh 1983. Analysis of instrumentation data of Maneri-Uttarkashi Tunnel. *Unpublished report of Central Mining Research Institute, Dhanbad.*

Dube, A.K. etal. 1989. Assessment of rock mass behaviour in the head race tunnel of Maneri Bhali Project Stage-II, Dist. Uttarkashi, Garhwal, India. *Unpublished report of Central Mining Research Institute, Dhanbad.*

Goel, R.K., J.L. Jethwa & A.G. Paithankar 1995. Tunnelling through the young Himalayas - A case history of the Maneri-Uttarkashi power tunnel. *Engineering Geology*,39: 31-44.

Sakurai, S. 1981. Direct strain evaluation technique in construction of underground openings. *Proc. 22nd US Symp. Rock Mech., Cambridge, Massachusetts, M.I.T.:* 278-282.

Sakurai, S. & K. Takeuchi 1983. Back analysis of measured displacements of tunnels. *Rock Mechanics and Rock Engineering*, 16: 173-180.

Sakurai, S., I. Kawashima & T. Otani 1993. A criterion for assessing the stability of tunnels. *EUROCK'93, Lisbon, Portugal:* 969-973. Rotterdam:A.A. Balkema.

Sakurai, S. 1997. Lessons learned from field measurements in tunnelling. *Tunnelling and Underground Space Technology*, Vol.12, No.4: 453-460.

Swarup, A., R.K. Goel & V.V.R. Prasad 2000. Observational approach for stability of tunnels. Tunnelling Asia 2000, New Delhi.

Modern Tunneling Science and Technology, Adachi et al (eds), © 2001 Swets & Zeitlinger, ISBN 90 2651 860 9

Predicting surface settlement at shallow tunnels using Gompertz's curve as a stress release curve

T. Suzuki, T. Domon, & K. Nishimura
Tokyo Metropolitan University, Tokyo, Japan

ABSTRACT: In the convergence-confinement method, the analysis of three-dimensional response of the ground around the tunnel is reduced to a problem of plane strain. A point-symmetrical growth curve is generally adopted as stress release curve to determine the magnitude of deformations, such as surface settlement, displacement at the tunnel crown or the convergence, in the method. For the urban-NATM excavated in soft ground at shallow depth, however, it has been often pointed out that the point-symmetrical growth curve was not fitted to in-situ measured values very well. In this study, the stress release curve is modeled as a non-point symmetrical growth curve, Gompertz's curve, to validate the predicting method of surface settlement above and ahead of the tunnel face under such conditions. We also discuss the method enabling practical application for urban-NATM at shallow depth based on Jeffery elastic solution.

1 INTRODUCTION

The convergence-confinement method has the advantage of being able to take into account the three-dimensional aspect and the statically indeterminate nature of problem in studying tunnel support (Gesta et al. 1986). In generally, to simplify the method, it will be considered the case of a circular section tunnel excavated in homogeneous and isotropic ground in order to satisfy cylindrical symmetry condition. That is, it will be assumed that the initial stresses are isotropic and that the tunnel is sufficiently deep to ignore the variations in the initial hydrostatic stress near the tunnel. With in these restriction hypotheses, we can reduce the three-dimensional problem related to the presence of the tunnel face to a problem of plane deformation. However, some of these hypotheses are unacceptable for shallow tunnels in sandy or soft ground excavated by urban-NATM. The urban-NATM is defined by the ordinary NATM excavated under two conditions as follows (Kansai Branch, JSCE 1988):

1. *Geological condition*: unsolidified natural ground, such as sandy ground.
2. *Geometric and site condition*: relatively shallow depth (approx. less than 50 m) in urban area.

Carranza-Torres & Fairhurst (2000) have been discussed the applicability of the convergence-confinement method for tunnels driven in Hoek-Brown materials under unequal far-field stresses using the Kirsh elastic solution. In their study, there has been renewal of interest in estimating the applicability of the method under non-cylindrical stress field. What seems to be lacking, however, is that the stress release curve or longitudinal deformation profile have still often been modeled by point-symmetrical growth curve. Based on field measurements or 3D numerical analysis, many researchers have pointed out that the point-symmetrical growth curve was not fitted to in-situ measured values.

In this study, we proposed the convergence-confinement method applying the following two modifications:

1. *Stress release curve*: Gompertz's curve is used as non point-symmetric growth curve.
2. *Ground characteristic curve*: it is constructed by the Jeffery elastic solution (Jeffery 1920).

The definition of the stress release curve derived by Gompertz's curve is given in section 2.1. In the Jeffery solution, two approximations are made: the surrounding ground is assumed to be perfectly elastic and the virtual internal pressure, i.e. the cause of the ring deformation, is assumed to be uniformly distributed and radial in direction (Széchy 1973).

In addition, we discuss the applicability in practice for urban-NATM at shallow depth compared with measured value of the surface settlements at actual tunnel construction site.

2 PREDICTION METHOD OF SURFACE SETTLEMENT

2.1 Stress release profile derived by Gompertz's curve

The stress release profile is an important component of the convergence-confinement method. The concept of stress release profile is generally as follows. The profile can be examined that a circular cross section is determined the effect of continuous reduction in the radial stress σ_r acting on the perimeter of the tunnel initial value σ_o to zero. The zero value of the radial stress represents the condition along the tunnel wall excavation, with no support, the tunnel being assumed to be of infinite length. It has been shown to a good degree of accuracy that the nearly front face is equivalent to the application of a virtual support pressure along the wall equal to $(1-\lambda)\sigma_o$, where λ is stress release ratio. The initial state far enough ahead of the front face for it to have no influence correspond to $\lambda=0$. As the face approaches and reaches this section and then recedes from it, λ increases gradually from 0 to 1; the radial displacement of the tunnel wall u_r also increases. This relationship between advancing rate of the face and λ is often modeled by a logistic curve characterized by point-symmetrical in two-dimensional analysis.

However, based on in-situ measurements for the convergence or surface settlement corresponding to λ, the longitudinal deformation profile represents in non point-symmetrical shape. Before the face reaches the measuring section, the measured values are relatively small than calculated values using point-symmetrical curve, on the other hand, after the face passes through the section the measured values tends to become larger than that of calculated values (e.g. Chern et al. 1998).

We adopt the Gomeprtz's curve as the stress release curve enabling to represent the actual behavior of the deformation mentioned above. The Gompertz's curve is defined by following equation.

$$\lambda = k a^{b^t} \tag{1}$$

where k = stress release ratio at the final stage (= 1.0), i.e., the face advances far enough from the measuring section; a = stress release ratio at which the tunnel face reaches the measuring section; b = the gradient of the Gompertz's curve, t = distance to the face. Figure 1 shows the Gompertz's curve comparing with the logistic curve corresponding to the same value of a (=0.27).

2.2 Procedure of the method

We have to determine the parameters in order to construct the Gompertz's curve defined by equation (1). In this study, a and b, and 'equivalent' elastic modulus E set as unknown parameters. Because the parameter k is fixed at 1.0, the effect on tunneling method including auxiliary method represent increasing value on the equivalent elastic modulus E.

Then, we explain the procedure of the prediction method of surface settlement as follows.

STEP 1: *Construction of the face advancing curve.*

The distance from the section measured the surface settlement to the tunnel face are plotted on the vertical axis. On the horizontal axis, the times are usually plotted. To simplicity, we plot the distance on the horizontal axis instead of the times. The vertical axis is equal to the horizontal axis.

STEP 2: *Construction of the surface settlement curve.*

The measured value at the measuring section is plotted on vertical axis, and the distance to the face is plotted on horizontal axis one by one. Negative values on the horizontal axis mean the distance at which the tunnel face does not reach the measuring section. This curve is situated under the curve constructed by STEP 1. In this study, we select the surface settlement as the measured deformation. This is why the surface settlement is one of the most important measuring items for urban-NATM and it can be easily obtained before the tunnel face reaches the measuring section.

STEP 3: *Construction of the stress release curve.*

The stress release curve, that is, Gompertz's curve are determined here. After initial values are given for any parameters, then the parameters are determined by the curve computed by statistical method in which the curve is fitted by least square approximation with nonlinear parameters.

STEP 4: *Construction of the ground characteristic curve.*

The measured value is plotted on vertical axis, and the stress release ratio is plotted on horizontal axis. This curve is situated next to left side of the curve constructed by STEP 1. The internal pressure substituted in the ground characteristic curve

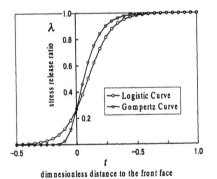

Figure 1. Gompertz's and Logistic curves.

is equal to $(1-\lambda)\,\gamma h_0$, where γ = unit weight of the ground and h_0 = overburden height of the tunnel. The relationship between internal pressure acting on the tunnel wall and the surface settlement is determined by Jeffery elastic solution. In this case using Jeffery solution, the internal pressure with opposite sign of analytical model of Jeffery's is acted on the tunnel wall. Thus, as the face approaches to the measuring section, the internal pressure varies from 0 to $(1-\lambda)\,\gamma h_0$.

STEP 5: *Determine the parameters of Gompertz's curve.*

The measured value is plotted on the curve constructed by STEP 2 corresponding to advancing face. Then, the Gompertz's parameters and equivalent elastic modulus E are determined by minimize the residual difference between the measured value and theoretical value by least square approximation with nonlinear parameters.

The parameters determined by the procedure mentioned above are adopted next excavation stage in order to select a reasonable support system. This method is the convergence-confinement method proposed for shallow tunnels in sandy or soft ground excavated by urban-NATM.

2.3 *Measurement of surface settlement at Tokko Tunnel*

To validate the predicting method of surface settlement above and ahead of the tunnel face, we estimate the theoretical curve particularly concerning surface settlement comparing with measured values for Tokko Tunnel of JR (Japan Railway) Narita-line. In this tunnel, the geological profile of the section adopted by NATM is shown in Figure 2, and the mechanical properties are listed in Table 1. As can be seen, the overburden height and geological condition are approximately invariable along the tunnel. The technical report on this tunnel reported that the ground behaved elastically during the excavation (Terado & Kimura 1982). Though the cross-sectional area of the tunnel is not circular for this case, the shape of tunnel can be regarded as circular with area equal to that of Tokko tunnel.

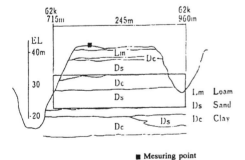

Figure 2. Geological profile of Tokko tunnel.

Table 1. Mechanical properties of the ground.

	D_C	D_S	Mean val.
Elastic modulus E (Mpa)	30	30	30
Poisson's ratio	0.3	0.3	0.3
Unconfined strength q_u (x 10-2 Mpa)	8.9	9.9	9.4
Friction angle ϕ (deg.)	29	32	30

3 EXAMPLE

We discuss the applicability of the proposed method comparing with field measurements for Tokko tunnel. Figure 3 shows the curves constructed by the procedures mentioned above and the results of surface settlements computed by our method. The parameters used in the method are a and b related to the stress release curve and equivalent elastic modulus E. The stress release ratio k at the final stage in which the face advances far away from the measuring section, is fixed at 1.0. The reason why the ground characteristic curve is linear is that the curve are constructed by elastic analysis based on Jeffery's equation. Figure 4 shows the variation of these three parameters including in the dimensionless settlement u/u_{max} along the tunnel line. As we can be seen in Figure 4, the parameter a and b are approximately steady at constant value 0.2 and 0.88, respectively, independent on the surface settlement. From the result, if the ground condition is constant over the tunnel line, the parameters a and b are independent on the amount of surface settlement. On the other hand, the equivalent elastic modulus E varies with the amount of the settlement. The modulus E increases with decreasing the settlement, E decreases with increasing the settlement inversely. That is the amount of the settlement depends on the equivalent elastic modulus considering the effect on the supporting method, but not depends on the stress release ratio, in which the tunneling method is adopted in same excavation procedure overall tunnel line.

4 CONCLUSION

The results obtained from the comparison between the computed vale and the measured value are in good agreement with in-situ measurement of the surface settlements. The parameters of stress release curve a and b are almost constant value, if the tunnel cross-section and geological condition including the overburden height are unchanging. If the parameters

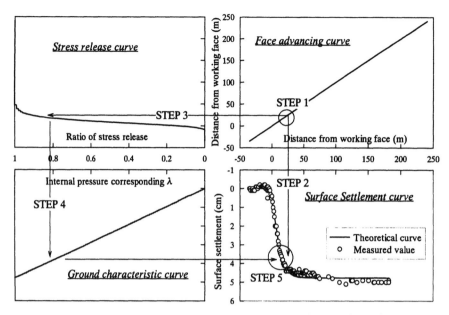

Figure 3. The correlation among the curves and results of the method to predict the surface settlement

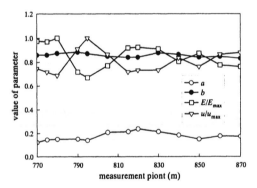

Figure 4. The variation of parameters along the tunnel line.

for this method are too much, the variations of these parameters caused by surface settlement may be not reasonable because of having the much degree of freedom. By the way, the equivalent elastic modulus derived from the method includes the effect of supports, auxiliary methods or division of the tunnel face.

The proposed method can be put in an FEM program. In the case, we can adopt the method to non-circular tunnel or anisotropic and inhomogeneous stress field.

REFERENCES

Carranza-Torres, C. & Fairhurst, C. 2000. Application of the Convergence-Confinement Method of Tunnel Design to Rock Masses That Satisfy the Hoek-Brown Failure Crite-rion, *Tunnelling and Underground Space Technology*, Vol. 15, No. 2, 187-213.

Chern, J. C., Shiao, F. Y. & Yu, C. W. 1998. An empirical safety criterion for tunnel construction, *Proc. Regional Symp. on Sedimentary Rock Engineering*, Taipei, Taiwan, 222-227.

Gesta, P. et al. 1986. Recommendations for use of convergence-confinement method, *Tunnels et ouvrages souter-rain*, No. 73.

Jeffery, G. B. 1920. Plane Stress and Plane Strain in Bipolar Co-ordinates, *Trans. Royal Society, Series A*, Vol.221, 265-293.

Kansai Branch of JSCE. 1988. *Design and construction manual for urban-NATM*, Osaka, Japan.

Széchy, K. 1973. *The Art of Tunnelling*, 2nd edition, Budapest, Hungary.

Terado. Y. & Kimura, K. 1982. Field measurement and consideration on deformability of soft sedimentary, thin overburden soil in tunnel excavation, *Proc. of the 14th Symposium on Rock Mechanics*, 111-115 (in Japanese).

Modern Tunneling Science and Technology, Adachi et al (eds), © 2001 Swets & Zeitlinger, ISBN 90 2651 860 9

Deformation and stability analysis of rectangular tunnel in soft rock ground using a strain softening type elasto-plastic model

T.Adachi, F.Oka, T.Kodaka & J.Takato
Kyoto University, Kyoto, Japan

ABSTRACT: The purpose of this study is to investigate the deformation and stability of rectangular tunnel excavated in the soft rock ground. In order to simulate the excavation of a rectangular tunnel below the underground water level, a series of the soil-water coupled finite element analyses using a strain softening type elasto-plastic constitutive model is carried out. The generally expected negative excess pore water pressure hardly occurs around the rectangular tunnel in the given soft rock ground due to the unloading by the excavation. The unstable behavior caused by the dissipation of the negative pore water pressure cannot be observed. Furthermore, it can be confirmed that the additional settlement of ground surface due to the drainage does not occur.

1 INTRODUCTION

Throughout most of Japan, soft rock ground not only occurs in mountain areas, but also in the relatively shallow subsurface of urban areas. Considering the profuse utilization of underground space in urban areas for expressways, parking lots, and storage areas, the number of excavations using rectangular tunnels is expected to rise. In the present study, both the deformation and stability characteristics of a rectangular tunnel excavated in submerged soft rock ground are discussed. For this purpose, a series of soil-water coupled finite element analyses is carried out. The constitutive model used in the computations is the strain softening, elasto-plastic constitutive model proposed by Oka & Adachi (1985) and Adachi & Oka (1992, 1995). First of all, the derivation of the constitutive model is outlined. Since the stability of submerged underground tunnels is being studied, we conduct a soil-water coupled analysis with a strain softening type of constitutive model. In this present study, the progressive enlargement of the tunnel section into various sizes is made in order to investigate the stability of the rectangular tunnel.

2 STRAIN SOFTENING TYPE ELASTO-PLASTIC CONSTITUTIVE MODEL

Oka and Adachi 1985, Adachi and Oka 1992, 1995, proposed an elasto-plastic constitutive theory for soft rocks by introducing the stress-history tensor with respect to the strain measure. They assumed that the material strength of soft rocks was composed of frictional strength, and a component due to cementation and/or the cohesions of materials as shown in Figure 1. The main factor responsible for strain softening after peak strength is the decrease of cementation and/or cohesion strength component. This can be seen to be due to the degradation of internal structure.

2.1 *Stress history tensor*

Stress history tensor is expressed by

$$\sigma_{ij}{}^* = \int_0^t K(z - z')\sigma_{ij}(z')dz' \qquad (1)$$

$$dz = \sqrt{de_{ij}de_{ij}}$$

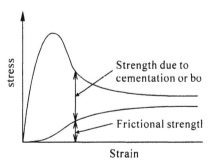

Figure 1. Schematic diagram of stress-strain relation with strain softening.

where z is a strain measure which is the second invariant of deviator strain, and K is the continuous bounded kernel function with the assumption that $\partial K / \partial z < 0$. In this study, the simple exponential function is adopted for soft rock, i.e.

$$K(z) = \exp(-z/\tau)/\tau \qquad (2)$$

where τ is a strain softening material parameter. Substituting Eq.(2) into Eq.(1), the stress history tensor is described as

$$\sigma_{ij}* = \frac{1}{\tau} \int_0^z \exp(-(z-z')/\tau)\sigma_{ij}(z')dz' \qquad (3)$$

2.2 Non-associated flow rule and yield function
The following non-associated flow rule is adopted:

$$d\varepsilon_{ij}^p = H \frac{\partial f_p}{\partial \sigma_{ij}} df_y \qquad (4)$$

where f_p is the plastic potential function, f_y is the yield function and H is the hardening-softening function. The yield function f_y does not directly depend on the current stress state, but depends on stress history ratio and strain hardening-softening parameter κ:

$$f_y = \eta* - \kappa = 0, \quad \eta* = \sqrt{(s_{ij}^* s_{ij}^* / \sigma_m^{*2})} \qquad (5)$$

where s_{ij}^* is the deviatoric part of the stress history tensor σ_{ij}^*, and σ_{ij}^* is the isotropic part of the stress history tensor.

2.3 Strain hardening-softening parameter
The evolution equation of the strain hardening-softening parameter κ is expressed as follows:

$$d\kappa = \frac{G'(M_f^* - \kappa)^2}{M_f^*} d\gamma^p \qquad (6)$$

where M_f^* is the value of $\eta*$ at the residual state. By integration under proportional loading conditions with zero as an initial value of κ, the following hyperbolic function is given:

$$\kappa = \frac{M_f^* G' \gamma^p}{M_f^* + G' \gamma^p} \qquad (7)$$

in which the plastic shear strain is defined in the following equation, i.e.

$$\gamma^p = \int d\gamma^p = \int (d e_{ij}^p d e_{ij}^p)^{1/2} \qquad (8)$$

Then, the strain hardening-softening parameter κ is given by

$$\kappa = \int d\kappa \qquad (9)$$

The strain hardening-softening parameter G' is the initial tangent of the hyperbolic function given in Eq.(7).

2.4 Plastic potential function and overconsolidated boundary surface
The overconsolidated boundary surface f_b is defined as the limit shape of plastic potential function f_p, and denotes as follows:

$$f_b = \bar{\eta} + \bar{M}_m \ln[(\sigma_m + b)/(\sigma_{mb} + b)] = 0 \qquad (10)$$

where $\bar{\eta}$ is the stress ratio defined by

$$\bar{\eta} = \sqrt{s_{ij} s_{ij} /(\sigma_m + b)^2} \qquad (11)$$

The overconsolidated boundary surface parameter $\bar{M}_m$ is the value of $\bar{\eta}$ when maximum volumetric compression occurs. Material parameters b and σ_{mb} describe the material internal structure. Adachi et al. (1998) assumed that σ_{mb} changes with the value of plastic volumetric strain as follows:

$$\sigma_{mb} = \sigma_{mb0} \exp\left(\frac{v^p}{\lambda_p - \kappa_p}\right) \qquad (12)$$

where σ_{mb0} is the initial value of σ_{mb}, and equivalent to the consolidation yield stress. The compression and swelling indexes are denoted by λ_v and κ_v, respectively.

The plastic potential function for soft rock is given by

$$f_p = \bar{\eta} + \bar{M} \ln[(\sigma_m + b)/(\sigma_{mb} + b)] = 0 \qquad (13)$$

where $\bar{M}$ is defined as follows:

In the overconsolidated region $(f_b < 0)$,

$$\bar{M} = -\bar{\eta} / \ln[(\sigma_m + b)/(\sigma_{mb} + b)] \qquad (14)$$

In the normally consolidated region $(f_b \geq 0)$,

$$\bar{M} = \bar{M}_m \qquad (15)$$

2.5 Strain softening type elasto-plastic constitutive equation
The plastic strain increment tensor, $d\varepsilon_{ij}^p$, is derived using Prager's compatibility condition in the following Eq. (16) together with Eqs.(4), (7) and (13), i.e.

$$df_f = d(\eta* - \kappa) = 0 \qquad (16)$$

$$d\varepsilon_{ij}^p = \Lambda\left[\frac{\bar{\eta}_{ij}}{\bar{\eta}} + (\bar{M} - \bar{\eta})\frac{\delta_{ij}}{3}\right]\left[\frac{\eta_{kl}*}{\eta^*} - \eta*\frac{\delta_{kl}}{3}\right]\frac{d\sigma_{kl}*}{\sigma_m^*} \qquad (18)$$

$$\Lambda = \frac{M_f^{*2}}{G'(M_f^* - \kappa)^2} \qquad (19)$$

Consequently, the incremental elasto-plastic constitutive equation can be obtained by adding the plastic strain increment tensor in Eq.(18) to the elastic strain increment tensor.

3 ANALYSIS CONDITIONS

The details of soil-water coupled finite element analysis with a strain-softening elasto-plastic model can be found in Adachi et. al. (2000). The finite element analysis is carried out using four-noded isoparametric elements under the plane strain conditions. At the center of each element, excess pore water pressure (or total head) is defined, while a spatial discretization of continuum equations for the water phase is performed using the finite difference method proposed by Akai and Tamura, 1978.

Table 1 shows the material parameters used in the analysis. These parameters, in fact, correspond to a weathered sandstone that is in the Kansai area in Japan. Figure 2 shows a simulated stress-strain curve for a drained triaxial compression test by the constitutive model with the parameters given in Table 1. In the analysis, the confining pressure dependency of material parameters has not been considered, and constant values were used for each one of them for simplicity. Figure 3 illustrates the finite element mesh used together with the associated boundary conditions. The total numbers of elements and nodes are 1750 and 1836, respectively.

Firstly, an initial stress distribution of ground prior to excavation is calculated. Then equivalent nodal forces for elements to be reduced due to excavation of tunnel are calculated from the initial stress distribution.

Calculations are carried out by the following four stages of analysis

STAGE 1: A standard rectangular section tunnel whose center is located at a depth of 24 m is excavated with a width of 4 m and height of 2 m.

STAGE 2: A 16 m thick upper layer of soft rock ground is removed to decrease the overburden pressure of the tunnel. This stage of analysis is introduced to discuss the stability of the standard sectional rectangular tunnel in the case that the overburden pressure is rapidly decreased.

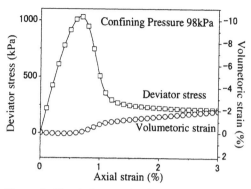

Figure 2. Simulation result of triaxial compression test under drained conditions.

STAGE 3: The tunnel is enlarged to investigate the stability of the standard rectangular section tunnel excavated in STAGE 1. In order to evaluate the stability of the opening, the sectional area of the tunnel is gradually increased. The release rate of stress around the opening is 1% of excavation stress per step with a total number of 100 steps used during the enlarging of the tunnel section. The calculation time period for each step is 1000 seconds. During the enlargement process, it is assumed that the water level is at the ground surface, and the inside of the tunnel is filled with underground water. It is more reasonable to study the stability of the tunnel by considering various sizes of tunnel instead of increasing

Table 1. Material parameters used in the analysis

Young's modulus $E(MN/m^2)$	200.0
Poisson's ratio v	0.33
Submerged unit weight $\gamma'(kN/m^3)$	12.8
Permeability $k(cm/sec.)$	10^{-5}
Strain hardening-softening parameter G'	612.2
Strain hardening-softening parameter M_f^*	1.0
Plastic potential parameter $b(kPa)$	160
Overconsolidation boundary parameter $\overline{M}_m$	0.95
Overconsolidation boundary parameter $\sigma_{mb}(MPa)$	1.47
Stress history parameter τ	0.04

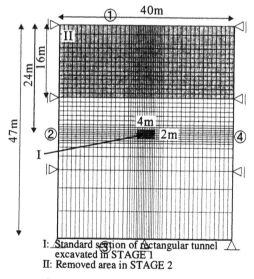

I: Standard section of rectangular tunnel excavated in STAGE 1
II: Removed area in STAGE 2

①: Drained boundary condition
②, ③, ④: Undrained boundary condition

Figure 3. Finite element array and boundary conditions used in the analysis.

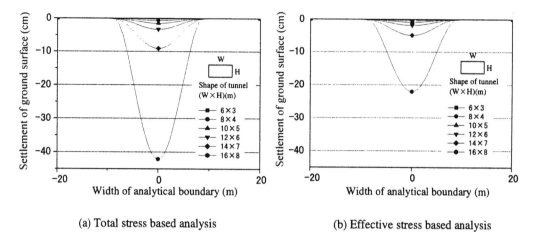

(a) Total stress based analysis   (b) Effective stress based analysis

Figure 4. Settlements of ground surface during the excavation of various rectangular section tunnels.

body forces until collapse is reached for a fixed opening size. Furthermore, a total stress based analysis is also carried out for comparison with an effective stress based analysis.

STAGE4: The underground water that fills the tunnel is forced to drain so that total stress on its boundaries is zero. The stability of the rectangular tunnel is discussed as the internal pressure of the

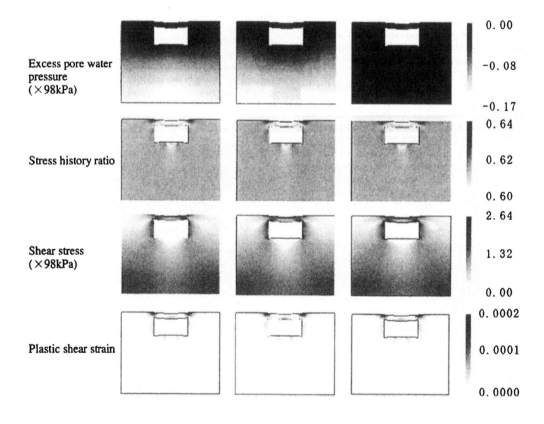

Just after the excavation   1 week after elapsed   10 days elapsed

Figure 5. Simulation results of the excavation of a 14 × 7m sectional rectangular section tunnel.

tunnel is decreased from hydrostatic water pressure to atmospheric pressure levels (zero total stress).

4 DISCUSSION OF RESULTS

Since displacements observed in both STAGE 1 and 2 are small, only computational results obtained in STAGE 3 and 4 are discussed in this section.

Figure 4 shows the settlement of ground surface after the enlargement of the tunnel opening into various sections in STAGE 3. The ground surface is redefined after removing the 16m thick layer in STAGE 2. Figs. 4a and 4b. show the ground surface settlements calculated for various tunnel sections using both total and effective stress analyses. The settlements obtained from the effective stress analysis is smaller than the ones from a total stress analysis because in the former, submerged conditions were assumed. Figure 5 shows the excess pore water pressure, stress history ratio, shear stress and plastic shear strain which are calculated just after the excavation of a 14 m x 17 m rectangular tunnel section in STAGE 3. The results after one week and 10 days are also shown. When the tunnel section is enlarged during excavation, both vertical and lateral stresses are unloaded. In the case that a shear stress due to unloading applies to heavily overconsolidated soft rock, negative excess pore water pressures is expected to occur. However, negative excess pore water pressures hardly develop around both the crown and the sidewall of the tunnel. Also, the changes in shear stress, plastic shear strain, and stress history ratio during the period after excavation are indeed very small. Consequently, the rectangular tunnel assumed in this analysis does not become unstable due to the dissipation of negative pore water pressures. Figure 6 shows the settlements of the ground surface with different sections of the tunnel excavated in STAGE 4. There are hardly any changes in settlements when

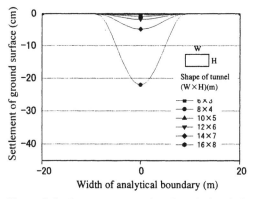

Figure 6. Settlements at ground surface during drainage (STAGE 4)

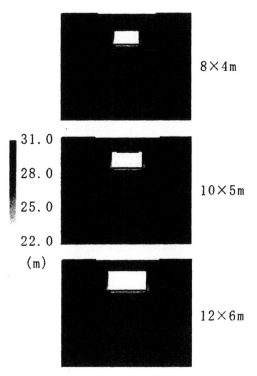

Figure 7. Distributions of total water head with different rectangular section tunnels during drainage (STAGE 4)

comparing Fig.4b with Fig.6, which implies that the drainage of water outside the tunnel does not affect the settlement of the ground surface or the stability of the opening. Figure 7 illustrates the total water head distribution after the drainage. The drop in total water head is observed only around the tunnel. It should be noted that this result is limited to the special condition under which the underground water is considered abundant and there is no water level drawdown. In the case where there is water level drawdown due to drainage, results may be different.

5 CONCLUSIONS

In order to study the stability of rectangular tunnels excavated in soft rock ground, a series of soil-water coupled finite element analyses using a strain softening elasto-plastic constitutive model has been performed. It was found that generally expected negative pore water pressures due to excavation hardly develop around the rectangular tunnel in the given soft rock ground. Hence no tunnel instability was observed. Furthermore, it was confirmed that no additional surface settlement due to drainage occurred. The relative length of lateral boundaries against an opening of tunnel may influence the obtained results.

However, it can be expected that the results obtained with the shorter lateral boundaries are severer against a collapse of tunnel.

REFERENCES

Adachi, T. and F. Oka 1992. An elasto-plastic constitutive model for soft rock with strain softening. *Proc. of JSCE, Journal of Geotechnical Engineering*. 445/III-18: 9-16. (in Japanese)

Adachi, T. and F. Oka 1995. An elasto-plastic constitutive model for soft rock with strain softening. *Int. J. Numerical and Analytical Methods in Geomechanics*: 19, 233-247.

Adachi, T., F. Oka, A. Koike and M. Koike 1998. Modified Adachi-Oka's elasto-plastic constitutive model for soft rock. *Proc. of JSCE, Journal of Geotechnical Engineering*. 589/III-42: 31-40. (in Japanese)

Adachi, T., F. Oka, H. Osaki and F. Zhang 2000, Soil-water coupling analysis of progressive failure in cuts with a strain softening model, Proc. Constitutive Modelling of Granular Materials: 471-490.

Akai, K. and T. Tamura 1978. Elasto-plastic numerical analysis of soil-water coupling problem, *Journal of JSCE*: 269: 95-104. (in Japanese)

Oka, F. and T. Adachi 1985. A constitutive equation of geologic materials with memory. *Proc. of Int. Conf. on Numerical Methods in Geomechanics*, 1, 293-300.

Modern Tunneling Science and Technology, Adachi et al (eds), © 2001 Swets & Zeitlinger, ISBN 90 2651 860 9

Study on tunnel design and construction method for New Tomei-Meishin Expressway

Ikuo Suzuki, Hiromichi Shiroma, Tetsuo Ito
Expressway Research Institute, Japan Highway Public Corporation

Shinobu Kaise
Head Office, Highway Engineering Division, Engineering Department,

Japan Highway Public Corporation

ABSTRACT: The New Tomei-Meishin expressway is a high-speed highway facility having six lanes (three lanes for each direction) throughout its total length of 502km. The expressway is planned to form a new transportation backbone for Japan in the 21st Century by connecting between Tokyo and Kobe. This expressway is functioning as a supplemental expressway of the existing Tomei and Meishin expressways. The tunnels on the New Tomei-Meishin expressway have a very large, flattened cross-section of approximately 190m^2 excavation area, while the excavation width is approximately 18m. There is no previous example of a large number of tunnels having a huge cross-section in the world, which can be used as references. Thus, there are various problems appearing regarding to determine the technical configuration and support structure, in terms of construction such as how to excavate a large group of tunnels safely and efficiently. The study and investigation related to design and construction method for the large cross-section tunnels located in various ground conditions is described in this paper.

1 INTRODUCTION

The whole service of Tomei-Meishin expressway started in 1969. This expressway has accomplished great contribution to the development of the industry and economy of Japan. The expressway has functioned as the national highway that connected three metropolitan areas, Tokyo-Nagoya-Osaka, of Japan. However, the maintenance works and the rehabilitation works on damaged by deterioration are increasing because of the enlargement of the vehicles and the increase of traffic density, in the progress in about 30 years. The expressway is in a situation that is not able to secure the high speed and regular time traveling which is expected having the function of the original expressway.

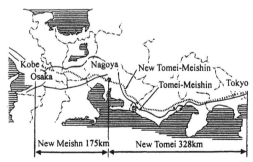

Figure 1. Location of New Tomei-Meishin expressway.

The New Tomei-Meishin expressway, so called by "the 2nd Tomei-Meishin Expressway" has been planned serving as the existing road of Tomei-Meishin expressway, which connects three metropolitan areas. Furthermore, this new expressway should become the root of the whole country network formation. The New Tomei-Meishin expressway consists of the New Tomei expressway (Tokyo-Nagoya, length of approximately 328 km) and the New Meishin Expressway (Nagoya-Kobe, length of approximately 175km), as shown in Figure 1. The length of tunnel constructions in a ratio reaches approximately 25% of the length of the whole route plan. The route plan is going across the Itoigawa-Shizuoka tectonic line and the Median tectonic line, and these accompanied lines as well. These tectonic lines have played an important role to the formation of Japanese Islands. The tunnels constructions on the New Tomei-Meishin expressway have three lanes and full shoulder for each direction.

In this paper, the authors describe the design and the construction method for large cross-section tunnels located in various ground conditions.

2 STUDY OF THE CROSS SECTION

2.1 Construction gauge

The construction gauge of the tunnels has about 15m in width and 4.7m in height. The crossing slope is 2% in principle.

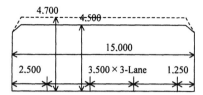

Figure 2. Construction gauge of New Tomei-Meishin expressway tunnels.

Figure 2 shows the standard construction gauge of the New Tomei-Meishin expressway tunnels.

2.2 Cross Section

There are little achievements of tunnel constructions, which have over 200m² large cross-section in Japan. Accordingly, we have to create new standard of tunnel cross-section, which match to the New Tomei-Meishin expressway. We may decide the minimum cross-section of the tunnels by considering mechanical stability and economical efficiency to cover the construction gauge. Generally, three-centered arch or five-centered arch cross-section has adopted for road tunnels in Japan. However, we examined the most suitable arch figure from a single-centered arch, three-centered arch and five-centered arch.

Figure 3 illustrates the relation between the stability and the excavation volume to the cross-section of tunnels. Based on this relationship, we designed the cross-section of tunnels for the New Tomei-Meishin expressway after conducting a further examination for the balance of construction on the stability and the economical efficiency.

2.2.1 The experience of the performance on the excavation for a large cross section tunnel
Figure 4 presents the achievement in a ratio of the height to the width of tunnel cross-section, which has over 100m²-excavation area in Japan. Here, we defined the ratio as follows;

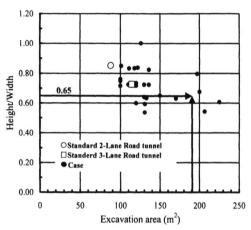

Figure 4. The ratio of height to width of the large section tunnels in Japan.

(Height): The distance from the end of invert to the crown.

(Width): The width of cross-section.

As shown in this figure, there are many cases with 0.65 of the ratio value for the large cross-section tunnels in the past construction. By considering to this phenomenon, we judged that the ratio value of 0.65 might be possible for the constructions of large cross-section tunnels of the New Tomei-Meishin expressway. Then, the ratio was adopted as a tentative standard of the large cross-section tunnels.

2.2.2 Standard cross section
We analyzed the shape of cross-section such as the upper section, lower section, and radius of inverted arch by using frame analysis. The two-dimensional FEM was used to check the stability of the tunnel

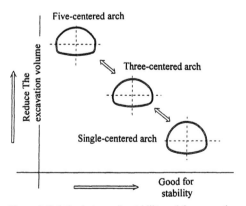

Figure 3. Relation between the stability and the excavation volume of the tunnel.

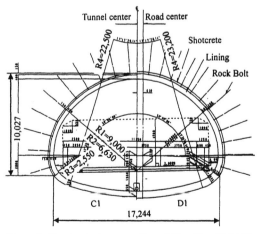

Figure 5. The standard cross-section of New Tomei-Meishin expressway tunnels.

540

lining and ground. The standard cross-section then was determined by evaluating the stability and the economical efficiency. The standard cross-section of New Tomei-Meishin expressway can be seen in Figure 5.

3 DEVELOPMENT OF A NEW SUPPORT MATERIAL

The quality of support of the tunnel measured by the conventional standard specification becomes down by the decline of construction efficiency, by the increase of the cycle time of the excavation phase, and the decline of the quality of the construction work. To cope with these difficulties, we have developed the new specification of the support material.

3.1 *New support materials and the purpose of development*

Table 1 shows the comparison between the standard specification of support members that newly made (high strength support) for New Tomei-Meishin expressway tunnels and the usual one that used for the normal tunnel constructions of two or three-lanes.

The purposes of development of the support materials in each support member are;
1) Shotcrete
The purposes of development of the shotcrete are to obtain first, the reduction of design thickness by using the high strength concrete and second, the improvement of the support effect by growing up the initial strength of shotcrete.
2) Steel fiber reinforced shotcrete (SFRS)
The purposes of development of the SFRS are to obtain the reduction of the design thickness and the replacement of the steel arch support in good ground condition by the improvement of the toughness of shotcrete.
3) Steel arch support
The purpose of development of steel arch support is to achieve the economical efficiency and the excavation efficiency by using the lightweight steel support of the high strength steel.
4) Rock bolt

Table 1. Standard specifications of support members.

Support member	Specification	
	Usually use(2-Lane or 3-Lane tunnels)	New Tomei-Meishin tunnels
Shotcrete	$\sigma_{28day}=18N/mm^2$ $\sigma_{1day}=5N/mm^2$	$\sigma_{28day}=36N/mm^2$ $\sigma_{1day}=10N/mm^2$ $\sigma_{3hr}=2N/mm^2$
Shotcrete	Plain Shotcrete	Steel Fiber Reinforced Shotcrete (Bace concrete is same as the top)
Steel arch support	SS400 Tensile strength 450-510N/mm^2 Yield point 245N/mm^2	SS540/HT590 Tensile strength 590N/mm^2 Yield point 440N/mm^2
Rock Bolt	Yield strength,steel 120-180kN	Yield strength,steel 180-290kN
Lining	$\sigma_{28day}=18N/mm^2$	$\sigma_{28day}=30N/mm^2$

The purposes of development of rock bolt are to obtain the reduction of set number in one cycle and the improvement of the tunnel stability by using the high strength bolt.
5) Lining
The purpose of development of lining is to obtain the reduction of design thickness by using the high strength concrete.

The performance of these support members has been confirmed by an indoor test. Moreover, we developed them through the field tests in two-lane tunnels that it is now under constructing by JH.

4 SITE EXAMINATION

4.1 *Excavation method*

4.1.1 *Trial construction to determine the excavation method*
In the previous plan, the excavation method of New Tomei-Meishin expressway would be taken by considering three conditions. First, the conventional top heading excavation method would be applied at the comparatively good ground condition. Second, the side excavation method would be established at the

Table 2. Standard support patterns of New Tomei-Meishin tunnels at the planning.

Ground Class	Cut Per Advance (Top Heading) (m)	Rock Bolt			Steel arch support	Thickness of Shotcrete (cm)	Thickness of Lining (cm)	
		Length (m)	Spacing					
		Upper section and Bench	Peripheral (m)	Longitudinal (m)	Upper and Bench		Archi/side wall	Invert
B	1.5	4.0	1.5	1.5	H-150	15	50	50
C I			1.2			20		
C II	1.2	6.0	1.0	1.2	H-200	25		
D I	1.0			1.0		30	60	80

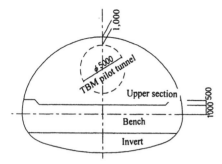

Figure 6. Position of TBM pilot tunnel.

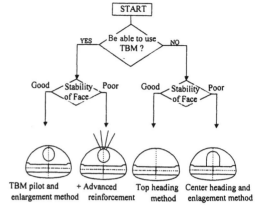

Figure 7. The selecting flowchart of excavation method.

portal department for a condition if the ground lacked its bearing stress. Third is the center dia-phragm excavation method, which would be used at poor ground condition as the unsettled face.

In addition, it will become necessary to take coun-termeasures for the stability of the tunnel face, the water inflow, and the settlement of the foot, on large cross-section tunnel such as on this new expressway. To do the countermeasures efficiently, it is important to know the ground condition. Especially, it has a great influence in terms of construction and cost of construction that can be considered in order to choose the method of the countermeasures for a long tunnel. It is also necessary to develop the additional excavation method after the excavation method, which has mentioned above.

We performed the center drift advanced method by tunnel boring machine (TBM) for a new excavation method. The center drift is excavated by the high speed TBM. The diameter of the drift was 5.0m. The aim of using this method is to confirm the ground stability, the face stability, and the improvement of excavation efficiency by the existence of the pilot tunnel in the upper section, as shown in Figure 6.

The Japan Highway Public Corporation (JH) has carried out this excavation method with two-lane tunnel on Yuda Daini tunnel on 1993; the diameter

of the drift was 3.5m. Then the execution test of this method was carried out on Shimizu Daisan tunnel and Ritto tunnel in the new expressway.

4.1.2 Summary of the tunnel excavation method in new expressway

Table 3 shows the summary of the tunnel excavation method by TBM pilot and the enlargement method. Based on the result of trial excavation, the flow chart of the selection of the excavation method for the general part of the new expressway may be drawn as in Figure 7. Basically, the excavation method of TBM pilot is applicable to the tunnel that is satisfied with the following conditions;
1) The length of tunnel is about 1.5km more.
2) The ground condition is not so bad.
3) The tunnel location is suitable for the installation of TBM.

The tunnel excavation method by means of ad-vanced center drift without TBM is applicable for the following conditions.
1) Fault zone as geology condition
2) Large water inflow

Table 3. The effect of the TBM pilot tunnel at the enlargement.

Effect	Contents
Confirmation of the grund condition	· The ground condition is able to confirm in the pilot tunnel.
Stability of the face	· The face is stabilized by the pilot tunnel. · The face is stabilized by drainage from the pilot tunnel. · The face is stabilized by the advanced reinforcement in the pilot tunnel.
Improvement of efficiency at excavation	· The cut per advance is extended by effect of center cut with pilot tunnel. · Explosion is economized by the pilot tunnel. · Selection of the support becomes appropriate by the confirmation of the ground condition.
Others	· The ventilation becomes efficient by using the pilot tunnel.

Table 4. The trial construction of support patterns.

Trial Case		Standard	Case4	Case6
Cross section		1,500 200	1,500 150	2,000 150
Shotcrete		Nomal	High-strength	High-strength(SFRS)
		σck= 18N/mm², t= 200mm	σck= 36N/mm², t= 150mm	σck= 36N/mm², t= 150mm
Steel support		H=200, Standard (SS400)	H= 150, High-strength (SS540)	None
Cut per advance		1.5m	1.5m	2.0m
Rock bolt	Top Heading	L= 6.0m , Strength-180kN	L= 6.0m , Strength:270kN	L= 6.0m, Strength: 300kN
		Peripheral P= 1.2m	Peripheral P= 1.6m	Peripheral P= 1.5m
		Units N (6.0m)= 17	Units N (6.0m)= 13-14	Units N (6.0m)= 15-16
	Bench	L= 6.0m, Strength: 180kN	L= 6.0m, Strength: 180kN	L= 4.0m, Strength: 180kN
		Circumferential P= 1.2m	Circumferential P= 1.6m	Circumferential P= 1.5m
		Units N (6.0m)= 8	Units N (6.0m)= 5	Units N (4.0m)= 4

4.2 Trial construction to determine support pattern

4.2.1 The trial construction

We carried out the trial construction of the support pattern by using the developed new support material that mentioned above. In this paper, we described the result of the trial construction of support pattern at Shimizu Daisan tunnel. The ground classes in this case are called B or C1 (it is 1.0-2.2 in Q system). At the trial construction about the consideration of support pattern, we changed the bolt spacing or the cut per advance or exclusion of steel arch support. Table 4 shows the representative trial construction patterns.

Figure 8 shows the value of ground condition (by JH system, RMR system, and Q system) that evaluated at tunnel faces, and it shows result of measurements in each trial construction case. The result of the trial construction can be seen below.

a) The results of measurements
1) Convergence and crown settle measurements were approximately 10mm in any case. And we have not admitted the differences in any case.
2) The force acting on the steel ribs was almost normal axial forces. The bending moment was little in all cases.
3) We had not admitted the difference of stress of shotcrete in any case.
4) The difference of the thickness of shotcrete had no influence on the measurement of convergence and stress, and the existence of steel support had not given any difference on the measurement as well as without the steel support.

5) Axial force acting on the rock bolt for the lower section (Bench) gave a satisfied value for an allowable axial force.

b) Efficiency of excavation
1) There was a difference in the cycle time by the number and length of rock bolt, and it influenced the excavation.
2) Steel arch support that usually used was bad handling because of its weight.

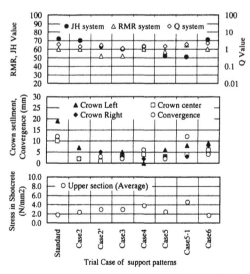

Figure 8. Result of measurements at the trial construction cases.

543

Table 5. The Standard support patterns of New Tomei-Meishin expressway.

| Ground Class | Cut Per Advance (Top heading) (m) | Rock Bolt | | | | | Steel arch support | | Thickness of Shotcrete (cm) | Thickness of Lining (cm) | |
| | | Length (m) | | Peripheral | | Longitud-inal | | | | | |
		Upper section	Bench	(m)	(m²/piece)	(m)	Upper section	Bench		Archi/side wall	Invert
B	2.5 (2.0)	4	4	3.0 (2.0)	3.8	2.5 (2.0)	-	-	10*	40	-
C I	2.0 (1.5)	6**	4	1.5 (2.0)	2.9	2.0 (1.5)	- (H-154)	- (-)	15* (15)	40	55
C II	1.5 (1.2)	6**	4	1.3 (1.6)	1.8	1.5 (1.2)	H-154	H-154	15	40	55
D I	1.2 (1.0)	6**	6	1.3 (1.6)	1.4	1.2 (1.0)	H-154	H-154	20	50	70

1) ( ) is without TBM pilot advance

2) Shotcrete is high strength type , * Upper section uses SFRS (Steel Fiber Reinforced Shotcrete)

3) Rock bolt: ** has strength of 290kN , Others has strength of 180kN

4) Steel arch support is high grade type

3) Cutting per advance of 2.0m is effective on a cycle but 2.5m was not good, although it depend on the machine.

4.2.2 *Standard support pattern of New Tomei-Meishin tunnels*

The standard support pattern is effective in support design of the road tunnels as a liner structure. And it is the premise naturally that is changed by the evaluation of the ground condition in construction. We proposed the standard support pattern of the New Tomei-Meishin expressway tunnels as listed in Table 5. There were resulted by the tests on the trial constructions and of indoor tests. As shown in the table, it has reduced the cost of construction that made up the high-strength support members as compare to the cost listed in Table 2.

5 CONCLUSIONS

The study of rational design and construction method for New Tomei-Meishin expressway with the trial constructions on the sites and indoor tests may be concluded as follows:

(1) In terms of cross-section, the three-centered arch cross-section may be a stable and economic shape for the New Tomei-Meishin Expressway tunnels.

(2) It may be able to construct the tunnels that achieve the requirements in terms of economical and stability by the development of new support members in comparison on the usual one.

(3) The excavation method, which is corresponding to various ground conditions, may be selected from the flow chart in Figure 7.

(4) The TBM pilot and enlargement method are the effective method of excavations for long tunnels with a good ground condition, in order to keep the stability of face and improve the efficiency of the excavation.

The results are written above then will be used to design and construct the New Tomei-Meishin Expressway tunnels with total length of about 200km.

In order to get better results, the subjects of the future research are considered to be taken as follows:

(1) To study the support pattern which will be applied on the special ground condition.

(2) To study the TBM machine data that evaluatesthe ground condition for the time of the excavation of enlargement.

REFERENCES

2000. THE 3rd design manual (tunnel) the 2nd Tomei-Meishin Expressway. Japan highway public corporation (JH). (in Japanese)

Takeuchi, J & Mitani, K & Nakata, M. 1997. Development of New High-Strength Shotcrete Experimental Study. *Nihon Doro Kodan Research Institute Report* Vol. 34:123-130. (in Japanese)

Mitani, K & Takeuchi, J. 1998. A Design Method and a Specification of the Fiber Reinforced Shotcrete. *Nihon Doro Kodan Research Institute Report* Vol. 35:56-63. (in Japanese)

Yoshizuka, M & Mitani, K & Shiroma, H. 1999. A Study on the Properties of Various Rock Bolts. *Nihon Doro Kodan Research Institute Report* Vol. 36:104-110. (in Japanese)

Modern Tunneling Science and Technology, Adachi et al (eds), © 2001 Swets & Zeitlinger, ISBN 90 2651 860 9

Construction Work of Yamba Tunnel TBM Test Section

K.Takagi, M.Tsukada & Y.Watabe
Joshin-etsu Construction Office, East Japan Railway Company, Japan

H.Iizuka
Construction Department, East Japan Railway Company, Japan

ABSTRACT: For a full face excavation of a single-track railway tunnel, a TBM construction method has not been used in Japan so far. But we will apply this method to the Yamba tunnel for fast excavation. In order to check propriety of this method, we plan to set a test section at start of excavation for 300m-long. We report and introduce our forward exploration methods and setting of primary support criteria in this paper.

1 INTRODUCTION

To adjust flood waters and develop water resources of the Tone river, the Ministry of Land, Infrastructure and Trasport of JAPAN is now constructing the Yamba dam on the Agatsuma river, which is an affluent of the Tone river, that runs through Naganohara Town, Agatsuma Country, Gumma Prefecture. As the dam will submerge a 6km-long section of JR Agatsuma Line where the Kawarayu Onsen station is

located, it is planned to construct a 10.4km-long rerouting line through the Kawarayu hot spring area that will be newly developed on the right bank of the Agatsuma river. Figure 1 shows the outline of the rerouting section of Agatsuma Line.

This paper reports a test plan to discuss the forward exploration methods that will be applied to the construction work of the Yamba tunnel on the rerouting start point side and introduces the setting of execution control criteria for primary supports.

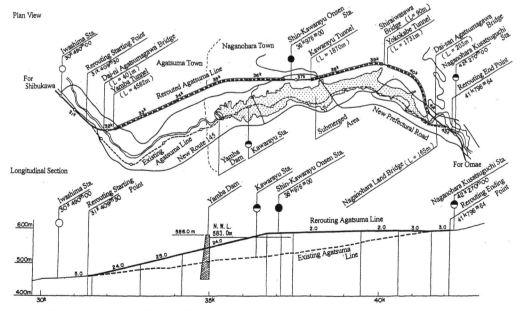

Figure 1. Outline of the Rerouting Section of Agatsuma Line.

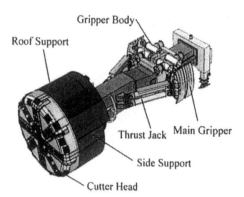

Figure 2. Outline of TBM.

2 YAMBA TUNNEL

2.1 *Topography and Geometry*

The Yamba tunnel that is located on near the rerouting start point passes a slope of northern aspect sandwiched by the Agatsuma river and a ridge of about 1,000m above sea level on the right bank of the river.

Regarding the topography and geometry of the Yamba tunnel, the following are estimated by exploration boring, elastic wave exploration and resistivity exploration methods.

1) The ground is under good condition to excavate as a whole, though there are some zones with low elastic wave velocity.
2) The rock types are mainly andesite, volcanic breccia and tuff breccia of medium hardness.

2.2 *Conditions for Execution of the Construction*

The conditions for execution of the construction work are as follows.

1) The tunnel is 4,582m long.
2) The construction work shall be completed in a short period of time, as it makes a critical path in the rerouting work.
3) Noise and vibration shall be minimized as there is a hot spring area on the end side of the tunnel.

2.3 *Method of Construction Work*

Based on the conditions for execution and the results of geological survey, we selected the full face excavation method by using a TBM for the following reasons.

1) High-speed excavation is possible.
2) Noise and vibration are comparatively small.
3) Most of the route passes a ground of medium hardness.

Figure 2 shows the outline of TBM which is going to be used at the Yamba tunnel.

2.4 *Support*

In order to reduce the NATM standard support patterns, we determined the design of support for the tunnel as shown in Table 1 by the following reasons.

The full face excavation by a TBM:

1) makes the cut section circular and dynamically stable.
2) less relaxes the ground when compared with blasting excavation.

For railway tunnels, a standard support pattern is prescribed for each ground class based on the results of NATM execution in the past.

2.5 *Control of Execution*

High-speed excavation, which is a feature of the TBM construction method, hasn't been executed much for railway tunnel construction work in Japan, presumably because it will cause collapse or sudden water inflow, if the conditions of the forward ground are not assessed sufficiently or measures are not taken for bad ground conditions. The subjects to be addressed in detail to execute high-speed excavation are:

1) prior assessment of bad ground conditions ahead, and
2) adoption of reasonable support execution control criteria.

Therefore, we have set a test section (for about 300m from the start position) at the initial stage of

Table 1. Comparison between set as the NATM standard and the support patterns used in this project.

Support Member / Support Pattern		Rock Bolt			Shotcrete		Support Category
		Arrangement	Length * Quantity	Interval	Material	Thickness(cm)(Arch/Wall)	
IV$_{NP}$	Standard	------	------	------	N	5 (Average)	------
	Yamba Tunnel	------	------	------	F	Optimal	------
III$_{NP}$	Standard	Arch	2m * 0~4	Optimal	N	10(Average)	------
	Yamba Tunnel	Optimal	Optimal	Optimal	F	2 (Minimum, Upper half)	------
II$_{NP}$	Standard	Arch	2m * 6	1.5 m	N	10(Average)	------
	Yamba Tunnel	Arch	1.5m * 6	1.5 m	F	2 (Minimum, Except at Invert)	------
I$_{NP}$	Standard	Arch / Side Wall	3m * 10	1.0 m	N	10(Minimum)	100H
	Yamba Tunnel	Arch / Side Wall	1.5m * 10	1.2 m	F	3 (Minimum)	125H
I$_{LP}$	Standard	Arch / Side Wall	3m * 8	1.0 m	N	15(Minimum)	100H
	Yamba Tunnel	Arch / Side Wall	1.5m * 8	1.2 m	F	3 (Minimum)	125H

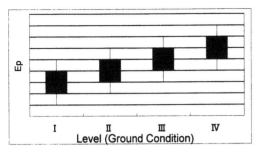

Figure3. Example of the image of the Box Bar Chart.

TBM advancement, where we will test various forward exploration methods to verify their applicability (effect on the cycle and precision of exploration), for the purpose of prior assessment of ground conditions. We will also set primary support execution control criteria by drawing a box bar chart based on the TBM data and ground conditions including those of cutting face and excavated circumference (arch side wall). Figure 3 shows an example of the image of the Box Bar Chart [1]. The vertical axis means a gripper coefficient, an energy of excavation or a destructive energy and so on. And the horizontal axis means the ground condition. We can evaluate the ground condition by use of this figure.

3 FORWARD EXPLORATION METHODS

Table 2 shows an example of forward exploration method. Although advancing boring (that is, core boring) is the most reliable, it considerably affects the construction cost and cycle, if it is executed for the entire length of the tunnel. Therefore, we will effectively combine different forward exploration methods based on the results of prior geological survey. We are now considering to use the following methods. Namely, we will:

1) adopt exploration boring in the cycle for the entire length of the tunnel. (however, exploration boring is not core boring.)
2) start exploring ground conditions by the elastic wave reflection method at a point about 100m before the elastic wave low speed zone predicted by the prior geological survey to check the positions of multi-crack zone.

Table 2. Example of Forward Exploration Method.

Exploration Method	Elastic Wave Reflection Method	Exploration Boring	Advancing Boring (Short)
Outline	*We predict the existence of reflection plane (stratum boundaries) around the tunnel by emitting an elastic wave by a small-scale blasting in the tunnel pit.	*We judge the exsitence of water inflow and ground conditions ahead based on the drilling conditions through forward boring by a drilling machine. *The rock conditions can be assessed by the drilling speed and measurement of driving energy.	*We collect cores by using a hydraulic percussion drill for advancing drilling.
Capability of Forward Exploration	*100m ~ 150m *Medium Hardness or Over ($V_p>2.5$km/s) *Stratum Boundary	*~ 40m *There are no limits due to geological conditions. *Conditions of drilling and underground water.	*~ 100m *There are no limits due to geological conditions *Collection of cores, rock hardness, crack and ground-water.
Reliability	*Errors will be contained in the distance to the forward reflection plane assumed from the V_p at the cutting face. *The reflection plane may be running forward or inclined. *The existence of underground water cannot be explored.	*We judge the geological conditions based on the conditions of drilling and slime. *The underground water conditions can be explored. *The lower half part cannot be explored depending on the position where the drilling machine is installed.	*We judge the geological conditions based on the conditions of slime and cores. (More reliable than the exploring boring and physical boring methods.)
Easiness of Execution	*It takes one day from preparation to analysis. *The work in the pit must be suspended for about two hours. *There is little effect on the execution cycle as the exploration can be performed on holidays.	*It can be implemented during the rest periods everyday. *Drilling speed (about 10km/h) *There is little effect on the execution cycle, if exploration workers are assigned separately.	*As it takes 2 or 3 days from preparation until drilling completes, the exploration affects the execution cycle.
Past Applications	*About 15 companies in Japan experienced this method.	*Has been applied frequently. *The method of judgement based on the measurement of drilling speed and driving energy is under development.	*Has been applied frequently. *We have experienced the method of judgement by the measurement of drilling speed and driving energy.

547

3) execute advancing boring after checking the conspicuous geological boundaries where a multi-crack zone is expected in the step 2.

As the elastic wave reflection method, we will perform the three-component HSP (Ashida laboratory, University of Kyoto) and TSP (Amberg Co.) for comparison.

The HSP method is to emit an elastic wave at the advancing point set on the tunnel wall and receive it by a few receivers simultaneously to determine the existence of reflection plane, which is the same as the TSP exploration method in principle. The three-component HSP that will be applied to the test section receives and processes the reflection wave divided into three components to eliminate virtual images.

4 SETTING OF PRIMARY SUPPORT EXECUTION CONTROL CRITERIA

Figure 4 shows the flow chart to set primary support execution control criteria for the test section, that will be performed in the following three aspects.

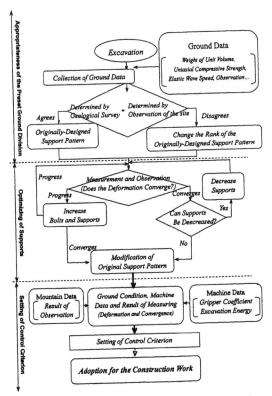

Figure 4. Flow to set a primary support execution control criterion.

4.1 Verification of the Appropriateness of Preset Ground Classification

Based on the ground data (weight of unit volume, uniaxial compressive strength, elastic wave speed in the pit, cutting face observation results) obtained from the excavation work, we evaluate the ground classification at the excavation point and verify the applicability of the classification determined through the geological survey. We will use the designed support pattern when the actual ground falls in the preset classification, or change it to an appropriate design pattern otherwise.

4.2 Optimization of Support

We measure the steel support stresses, axial force of rock bolts, tunnel space displacement and displacement of the ground in the vicinity of the tunnel after executing the primary support according to the design support pattern. If the displacement is converging, we will reduce supports while considering the safety factor. If the displacement is increasing or deformation or cracks emerge, we will increase bolts, shotcrete and frames, and perform measurement and observation until the displacement and deformation converges.

We will review the design support pattern based on the ground conditions obtained by the above procedure, results of primary support measurement and analysis of their relationship with the number of supports.

4.3 Setting of Control Criteria

In this project, we will collect the data of TBM and ground conditions of the test section and evaluate it together with measurement results, in order to set control criteria. According to the geological survey, there are only ground categories I_N and III_N in the test section. We will set an execution control criterion for the II_N ground through the interpolation between the I_N and III_N data, and more, estimate the criterion for the IV_N. We will modify these criteria when judged necessary based on the results of excavation as the tunnel construction progresses.

5 CONCLUSIONS

We have reported above the forward exploration methods for the test section of the Yamba tunnel and setting of primary support execution criteria. By effectively using the results of the comparison between different forward exploration methods and the set primary support execution control criteria, we would like to safely complete the Yamba tunnel construction work.

6 REFERENCES

Report on the "Evaluation and Review of the TBM Construction Work for the Koshibakama and Shirohata Tunnel Evacuation Pits, Tokai Hokuriku Highway", Japan Highway Public Corporation Niigata Branch Office, June 1996.

Modern Tunneling Science and Technology, Adachi et al (eds), © 2001 Swets & Zeitlinger, ISBN 90 2651 860 9

External water pressure on lining of tunnels in mountain area

ZHANG Youtian
China Institute of Water Resources and Hydropower Research (IWHR), Beijing, China

ABSTRACT: The external water pressure acting on tunnel lining is discussed in the paper. The load due to ground water on tunnel is generally seepage body force. In condition that normal stress on the interface of lining and rock is tensile, the lining is under the action of surface load due to ground water, so called external water pressure, $p_0 = \beta_1 \beta_2 \beta_3 \gamma h_0$, where γ is specific weight of water, h_0 is the height of water column between ground water table and axis of tunnel, β_1 is the correction coefficient considering the initial seepage field of groundwater, β_2 is a reduction coefficient, ranging from 0 to 1.0, considering the permeability difference of surrounding rocks and lining in condition of $\beta_1=1.0$ and β_3 is a coefficient considering the effectiveness of drainage system, specially designed for reducing the external water pressure on tunnel lining in condition of $\beta_1 = \beta_2 =1.0$. Detailed theoretical and practical discussions and suggestion of the ranging values of β_1, β_2 and β_3 are given. The proposed approach can help tunnel designer avoid the subjective estimation of external water pressure on lining and therefore to make the design more safe and reasonable.

1 INTRODUCTION

China is rich in water power resources, and the exploitable hydroelectric capacity reaches 3.78×10^9 kW, which ranks the top position all over the world. In China, hydropower resources are mainly concentrated in mountainous areas of the Southwest due to topographic and climatic conditions. Therefore, the abundant rainfall and narrow river valleys are suitable for construction of hydropower projects with high-head and large capacity in the area. In order to obtain concentrated high-head drop, arrangement of long diversion tunnels and high dams are always required. According to statistics, more than 40 hydraulic tunnels with the length of over 5 km have been constructed or under construction (ZHANG 2000) .In Table 1 19 hydraulic tunnels with length over 8 km are listed.

More and more long hydraulic tunnels will be constructed in China. For example, There are two hydropower tunnels diameter in Jingping Project with diameter of 9,5 m and each length of 18.7 km. The maximum overburden of the tunnels is up to 2400 m. The west line of the water diversion project from the Upper Yalong River, the tributary of the Yangtze River, to the Yellow River passes through the Bayankela Mountain. The length of any of one single tunnel for different alternatives is more than 100 km and the longest one is 243.8 km.

Most of these tunnels, as well as a lot of long railway and highway tunnels traverse the mountain area with high overburden. There are many serious engineering problems, especially the problem due to high ground water level, to be solved in tunnel design and construction. High ground water level definitely produces high ground water load acting on the lining and surrounding rock. The load due to ground water on tunnel is generally seepage body force. In condition that normal stress on the interface of lining and rock is tensile, the lining probably separates from surrounding rock and bears the seepage body force distributed in domain of lining itself. As the concrete lining is relative thin, the seepage body force on lining can be treated as the surface load acting on the external surface of the lining due to ground water, so called external water pressure. Generally the external water pressure on lining is not equal to static water head pressure produced by ground water level. How to determine the external water pressure is very difficulty due to the complexity of geological and hydro-geological as well as topographical conditions. Based on the research and consulting work in hydraulic tunnels, an empirical approach to determine the external water pressure on lining is given in the paper by using three coefficients to correct the static pressure due to ground water surface.

Table 1. Hydraulic tunnels completed and under construction in China (length over 8km).

No.	Projects (province)	Surrounding rock	Length (km)	Type/ size of section (m)
1	Tunnel 7# SM,YRDP(Shanxi)	Sandstone	42.9	F/D=4.2
2	Tunnel 5# SM,YRDP(Shanxi)	Limestone	26.4	F/D=4.2
3	Yindaruqing(Gansu)	Sandstone	15.7	F/4.2×4.4
4	Yiner(Xinjiang)	Gravel stone	15.3	F/D=6.2
5	Tunnel 6# SM,YRDP(Shanxi)	Limestone	16.4	F/D=4.2
6	Tunnel 7# CF,YRDP(Shanxi)	Limestone	13.4	F/D=4.8
7	Tunnel 8# GM,YRDP(Shanxi)	Limestone	12.2	F/D=5.46
8	Yindaruqing(Gansu)	Limestone Sandstone	11.7	F/D=4.8
9	Yinluanrujin(Tianjin)	Gneiss	11.4	F/5.7×6.45
10	Taipingyi(sichuan)	Granite	10.6	P/D=9.0
11	Tianshengqiao II (Guizhou)	Limestone	10.0	P/D=9.8
12	Tunel 11# GM,YRDP(Shanxi)	Limestone	10.0	F/5.0×5.36
13	Suojing Mountain (Hubei)	Limestone shale	9.8	F/2.6×3.7
14	Yinfengjiyong		9.7	F/
15	Baoquan Mountain (III) (Jilin)		9.5	F/4.2×4.4
16	Lubuge (Yunnan)	Dolomite, Limestone	9.4	P/D=8.0
17	Fuer River (Liaoning)	Granite	8.8	F/4.4×4.2
18	Yuzixi I (Sichuan)	Granodiorite	8.4	P/D=5.0
19	Xier River I (Yunnan)	Biotite gneissic rock	8.2	P/D=4.3

*: YRDP: Yellow River Diversion Project, GM: General Main, SM: South Main, CF: Connect Faction, F: Free flow, P: Pressure.

2 EXTERNAL WATER LOAD AND EXTERNAL WATER PRESSURE ACTING ON TUNNEL LINING

External water load induced by ground water flow is the most important one for design of tunnels in mountain area. Strictly speaking, only if the lining is impermeable, such as steel lining, the water load acting on the lining is the surface force, which is so called water pressure. Concrete lining and surrounding rock are permeable materials. The action of groundwater on the tunnel and its lining is a body force due to seepage. As the tunnel is composed of rock and lining, study of a tunnel not only lies in lining, but mainly in surrounding rock. Therefore, the water load is the seepage body force acting in the media under the groundwater table, which includes both surrounding rock and lining.

There are joints and fissures produced due to different reasons in rock body. Water transferred through joints and fissures is much larger than that through the pore of intact rock. For studying steady seepage field the equivalent pore media by averaging the conductivity of joints to rock body can be used. Concrete is a kind of pore media. If permeability coefficients of rock body and lining are known, the seepage field H (x, y, z) can be solved by finite element method from boundary conditions. Let

$$H = P / \gamma + Z , \qquad (1)$$

where P is pore pressure, γ is the unit weight of water and z is vertical coordinate of calculating region (upward is positive of z).

If the seepage field is solved, the seepage body force F is also obtained, its components are as follows:

$$\left. \begin{array}{l} X = \partial P / \partial x = -\gamma \, \partial H / \partial x \\ Y = \partial P / \partial y = -\gamma \, \partial H / \partial y \\ Z = \partial P / \partial z = -\gamma \, \partial H / \partial z + \gamma \end{array} \right\} \qquad (2)$$

Known the seepage body force, the stress distribution produced in the surrounding rock and lining may be obtained by FEM according to following procedure.

3 INCREMENT THEORY OF TUNNEL STRESS ANALYSIS

There exists an initial seepage field H_0 of the groundwater prior to tunnel construction. After the excavation of a tunnel, the internal surface, pore pressure of which equals 0, is formed. The primary seepage field H_0 changes to H_1. After lining, the seepage field changes to H_2 affected by lining. Therefore, the stress of lining and surrounding rock should be analyzed according to accumulative stress by increment of the seepage body force step by step. The detailed procedures are as follows (ZHANG & ZHANG 1980, ZHANG & ZHANG 1982, ZHANG & WANG 1985).

(1) Calculate the initial seepage field H_0 prior to tunnel excavation, the body force F_0 corresponding to H_0 by formula (2). Calculate the primary stress field σ_0^R by back analysis from the measured values.

(2) Calculate the seepage field H_1 and F_1 after excavation. The increment stress of surrounding rock due to increment seepage body force $\Delta F_1 = F_1 - F_0$ and the effect of excavation will be $\Delta \sigma_1^R$. The stress of surrounding rock after excavation is

$$\sigma_1^R = \Delta \sigma_1^R + \sigma_0^R \qquad (3)$$

(3) Calculating the seepage field H_2 and F_2 of tunnels after lining, the increment stress of the surrounding rock $\Delta \sigma_2^R$ and also the increment stress of lining $\Delta \sigma_2^L$ due to the increment of the seepage body force $\Delta F_2 = F_2 - F_1$ and the action of other load increment (dead load of lining). The stresses of sur-

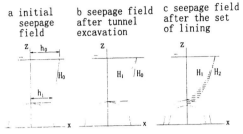

a initial seepage field   b seepage field after tunnel excavation   c seepage field after the set of lining

Figure1. Increment theory of water load for tunnel analysis.

Table 2. Reduction Coefficients of external water pressure.

Classification	Activity status of groundwater	Suggested value of β in formula (6)
1. no	Tunnel wall is dry or damp	0
2. weak	Seepage or drip water along structure surfaces	0.00~0.40
3. significant	Lined water or water spray along the cracks or faults	0.25~0.60
4. strong	Serious flowing water like strand	0.40~0.80
5. serious	Serious drip or flowing water along the faults	0.65~1.00

rounding rock and lining after the lining completed are

$$\left.\begin{array}{l}\sigma_2^R = \sigma_1^R + \Delta\sigma_2^R \\ \sigma_2^L = \Delta\sigma_2^L\end{array}\right\} \tag{4}$$

(4) Calculating the seepage field H_3 and F_3 for pressure tunnels after its impounding, and the increment stress of the surrounding rock $\Delta\sigma_3^R$ and also the increment stress of lining $\Delta\sigma_3^L$ due to the increment of the seepage body force $\Delta F_3 = F_3 - F_2$. The stresses of rock and lining after tunnel impounding are

$$\begin{array}{l}\sigma_3^R = \sigma_2^R + \Delta\sigma_3^R \\ \sigma_3^L = \sigma_2^L + \Delta\sigma_3^L\end{array} \tag{5}$$

It is seen from above that the stress state of tunnels depends on a historical process. This procedure is shown ideally in Fig. 1. It can be seen in formula (4), That the lining bears only the increment seepage force ΔF_2. If the permeability of lining is more larger than that of the surrounding rock, the seepage field keeps unchanged after the set of lining. There is no stress in lining produced by ground water in spite how high the ground water level is.

If the lining is relative impermeable, the seepage field changes a lot from H_1 to H_2 after the set of lining and the increment seepage force ΔF_2 is produced (shaded area in Fig.1c). ΔF_2 is centripetal in lining area and centrifugal in rock area. The tensile stress is obviously produced in normal of the lining/rock interface. The lining probably separates from surrounding rock because of the low lining/rock bonding strength. In this case the lining bears only the part of distributed in its area, which approximately equals the pore pressure at the rock/lining interface, so called the external water pressure.

4 GENERAL SITUATION OF PRACTICE TO DETERMINE EXTERNAL WATER PRESSURE IN TUNNEL DESIGN

It was suggested in the Design Code for Hydraulic Tunnels issued in China in 1985 that the external water pressure P should be determined by formula

$$P = \beta\gamma h_0, \tag{6}$$

Where h_o is static water head due to groundwater level, β is a reduction coefficient ranging from 0.0 to 1.0 considering the relative permeability of lining and surrounding rock and may be obtained by seepage analysis. According to the practice of tunnel design in China the reduction coefficient is suggested to be adopted from Table 2 according to the condition of hydrogeology.

Such a method of reduction coefficient is also used in Australia, United States and Japan. From statistical data of tunnel projects in these countries the reduction coefficients are ranging from 0.15 to 0.90 (EPRI 1987).

In design of tunnels in German and sometimes in Japan the external water pressure is adopted as static head due to groundwater surface i.e. the reduction coefficient β =1.0.

The ground water rises in rain seasons. After a heavy and long time rain, ground water level rises to some extent. For safety consideration the maximum water level is assumed up to that of ground surface. Static water head pressure due to water level up to the ground surface is adopted as the external water pressure for design of tunnel lining in some countries, such as Brazil, Canada, and also United States [6].

From the situation mentioned above there is no common rule to determine the value of external water pressure in tunnel lining design. Different designer may adopt quite different values in designing the same tunnel. Tunnel in mountain area often has a relative high ground water level, which is the critical load in tunnel lining design. If the external water pressure used in tunnel design has an unallowable error, the safety of tunnel has no guarantee or extra money should be invested.

5 EXTERNAL WATER PRESSURE ON LINING OF TUNNEL IN MOUNTAIN AREA

5.1 *The empirical formula to determine the external water pressure on lining of tunnel in mountain area*

Formula (6) means that the initial field of groundwater is distributed hydro-statically. This is correct only if groundwater in initial state is not flowing. In mountain area groundwater always flows and the pore pressure at a point is usually not equal to the hydrostatic induced by water level on vertical line of this point. Usually the pore pressure at a point in mountain area is smaller than hydrostatic value due to ground water level. While tunnel passes underneath the river valley, where the groundwater is always in a confined state, the pore pressure beneath the valley should be higher than static head due to the ground water level or the surface of river water. To determine the external water pressure on lining should primarily to know the water pressure P_1 at the tunnel position before its excavation. For this reason a correction coefficient β_1 should be defined to correct the water pressure in initial state.

$$P_1 = \beta_1 \gamma h_0 \qquad (7)$$

External water pressure after tunnel constructed should reduced from its initial state, thus

$$P_2 = \beta_2 P_1 \qquad (8)$$

where β_2 is the same of β in formula (6).

The drainage system is often designed further to reduce the external water pressure for tunnels with high ground water level by reduction coefficient β_3

$$P_3 = \beta_3 P_2 \qquad (9)$$

Finally, the formula estimating the external water pressure on lining can be expressed as

$$P = \beta_1 \beta_2 \beta_3 \gamma h_0 \qquad (10)$$

Therefore the determination of design external

water pressure acting on lining by a single reduction coefficient β is not consistent with the actual conditions. The initial seepage field should be analyzed according to the hydro-geological conditions, then the initial pore pressure at the position of tunnel P_1 is obtained. Then the P_1 multiplies β_2, the same as the β in formula (6) or in Table 2 is the real external water pressure for lining design. Certainly the value of β in Table 2 is given in condition of β_1 =1. While the external water pressure is extremely high, the drainage measures should be taken. External pressure should be further reduced by multiplying a reduction coefficient β_3.

1.1 *The value of the correction coefficient of initial water pressure β_1*

1.1.1 *Tunnel located in slope of a big mountain*
Fig.2 shows a slope area of a big mountain with ground water surface and isopotential lines obtained by seepage analysis. The correction coefficients of water pressure in initial state are different in different points in the area. Let a tunnel located at point A in slope area (Fig.2). The elevation of the point is 160 m. The ground water level at the vertical line above point A is El. 290 m. The static water head at the point h_0 should be 290-160=130 m. The water head at point A is 272 m, and so the head of the pore pressure of the point is 272-160=112 m in initial state. Thus β_1 =112/130=0.86 at point A. 0.86 can be assumed as an average value of correction coefficient for tunnel located in slope area of a mountain. β_1 is approximately ranging from 0.75~1.00 obtained by using the finite element method of seepage analysis for different tunnel location in the slope area and also the different topography the mountain.

1.1.2 *Tunnel located under the ridge of a mountain*
Point B in Fig.2 is located under the ridge of a mountain. There ground water flows downward from top. Here the losses of the water head reach its maximum value. The elevation of the point is 160 m. The ground water level is about 376 m. The water head at the point is 316 m. Therefor β_1 =(316-160)/(376-160)=0.72. Higher the mountain and dipper the slope, bigger the water losses should be. In consideration of this condition the range of β_1 is estimated as 0.6 to 0.8 for tunnel under the ridge of mountain.

1.1.3 *Tunnel located under the narrow valley*
Point C in Fig.2 is located under the narrow river valley with an elevation of 116 m. Ground water flows upward beneath narrow valley and the water head there should be higher than that of river water surface. For example, the water head at point C is 206 m. The level of the river water is 200 m. β_1= 1.07 can be easily got. The correction coefficient of water pressure under the narrow valley for initial

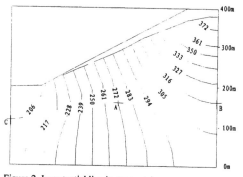

Figure.2. Isopotential line in a mountain area.

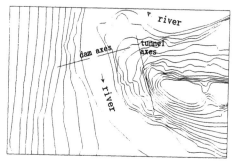

Figure 3. The semi-isolated mountain in left bank of Manwan Project.

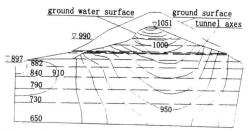

Figure 4. The initial seepage field of the profile along the diversion tunnel axes of Manwan Project.

state is ranging 1.00~1.20 depending on how narrow and cliff of the valley.

1.1.4 *Tunnel in isolated or semi-isolated small mountain*

Isolated or semi-isolated small mountain has a special hydro-geological condition. The rainfall infiltrates into ground surface and percolates radially to the hillside around. The water losses beneath the top of the mountain are greater than that of a big mountain where the ground water flows only to the one side. The left bank of the dam of Manwan Hydropower Project is a semi-isolated small mountain showed in Fig.3, where the river flows around the three sides of the mountain. A diversion tunnel passes through the mountain from upstream to downstream of the dam. The isopotential lines of initial seepage field are shown in Fig.4. The water pressure at the tunnel under the top of the mountain is much smaller than its static pressure due to the ground water level. There the β_1 is only 0.445. β_1 would be smaller value in case of an isolated mountain.

1.1.5 *Tunnel located in mountain with karst formation*

The karst is often developed in deep of the mountain with carbonate formation. The karst tubes even karst rivers form a very complicated underground drainage system, which remarkably reduces the pore pressure around the karst tubes and rivers. But the permeability of the rock (possibly the limestone), a little bit far from the karst zone, is usually small, thus a relative high ground water level is frequently observed in mountain with karst zone in deep area of it. Tunnel in Tianshengqiao Second Hydropower Project is an example. There are three pressure tunnels with each length of 10 km and diameter of 9.6 m through the mountain of limestone formation. Along the tunnel axis the karst is quite developed. Fig.5 shows the profile of hydro-geological condition along the one of tunnel axes. The water pressure near the tunnel position is very low, but the ground water surface is relative high. Obviously the values of correction coefficient β_1 are very small with the ranging of about 0.0~0.5.

5.3 *The reduction coefficient accounting the designed drainage system β_3*

The drainage system is often designed for reducing the external water pressure on lining while the tunnel

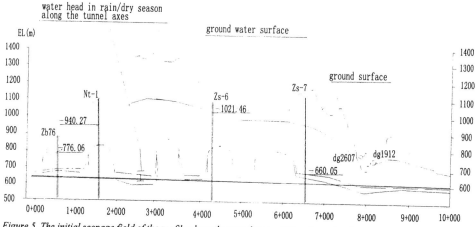

Figure 5. The initial seepage field of the profile along the tunnel axes of Tianshengqiao Project

Table 3. Suggested value of reduction coefficients β_3 due to drainage hole system.

Hole spacing along tunnel axis (D is diameter of tunnel)	D/3	D	2D
Holes only along the crown line	0.30~0.50	0.50~0.70	0.70~0.90
Holes in four lines evenly distributed on perimeter	0.15~0.25	0.25~0.35	0.35~0.45

is located in the mountain with high ground water level. The drainage holes from lining into surrounding rock with the depth of 2~3 m in radius direction are mostly used for railway tunnels, highway tunnels and hydraulic free flow tunnels and also the pressure tunnels with ground water pressure much higher than internal water pressure, According to the practice in China, for example, the 15.7 km long free flow water supply tunnel from the Luanhe river to Tianjin, the tunnel No.7 with the length of 43 km in Yellow River Diversion Project in Shanxi Province, reduction coefficient β_3 due to the drainage hole system in condition of $\beta_1 = \beta_2 = 1$ can be summarized in Table 3 depending on its spacing (IWHR & TIDI 1997). The big value can be adopted for tunnel with large diameter.

6 CONCLUSION

The load produced by ground water with high water level is the serious one for lining design of tunnels in mountain area. Concrete lining and surrounding rock are permeable materials. The action of groundwater on the tunnel and its lining is a body force due to seepage. After the excavation of a tunnel, the seepage field around the tunnel is changed from its initial state. It is changed also after setting the lining. The increment theory should be used in numerical analysis for tunnel stresses. As the complexity of geological, hydro-geological and topographical conditions in mountain area, the mathematical model and also the parameters are difficult to determine. A lot of research work proves that the lining probably separates from rock because the normal stresses of the rock/lining contact surface are often tensile due to the increment seepage force. The load of the ground water can be treated as external water pressure acting on the external surface of lining. External water pressure is not equal to the pressure of static water head due to ground water level. It should be corrected by three different coefficients β_1, β_2, and β_3. Detailed theoretical and practical discussions and suggestion of the ranging values of β_1, β_2 and β_3 are given. The proposed approach can help tunnel designer avoid the subjective estimation of external water pressure on lining and therefore to make the design more safe and reasonable.

REFERENCES

Design Code for Hydraulic Tunnels in China (SD134/84), 1985, Beijing.
EPRI. 1987. Design guideline for pressure tunnels and shafts, University of California at Berkeley.
IWHR & TIDI 1997. No. 7 Tunnel Detrital Rock Area of YRDP South Main. Special Report on Research of Initial Seepage Field and External Water Pressure on Lining.
Zhang Youtian 2000. Hydraulic underground structure, Chapter 11 of *Large dams in China, A Fifty-year Review*, edited by Jiazheng Pan & Jing He, CHINA WATERPOWER PRESS, Beijing, 615-653.
Zhang Youtian & Zhang Wugong 1980. Static calculation for tunnels due to water load, *Hydraulic Journal*, 3, 52-62 (in Chinese).
Zhang, Y. T., & Zhang, W. G. 1982. The influence of history of water load on tunnel stress, *ISMR Symposium*, 753-758, Aachen
Zhang, Y. T. & Wang, L. 1985. Finite element analysis of tunnel stress due to water load, *Proc. 5th Int. Conf. Num. Method in Geomech.*, 1221-1224. Nagoya.

Modern Tunneling Science and Technology, Adachi et al (eds), © 2001 Swets & Zeitlinger, ISBN 90 2651 860 9

Slope failure at tunnel entrance due to excavation and its countermeasure

A.Yashima
Gifu University, Gifu, Japan

A.Matsumoto & K.Tanabe
SUNCOH Consultants Co. Ltd., Nagoya, Japan

ABSTRACT: A large slope failure occurred at the arrival entrance of A-tunnel and the tunnel support linings were totally destroyed. The geological profile and the reason of the failure are discussed in detail. To complete the tunnel construction, firstly the fallen soil was removed and then a re-cut of the slope with a more stable gradient was carried out to assure the long-term stability of the cutting slope with rock bolts and shotcrete. The slope was cut with 15 stages and the total amount of the excavated soil reached 240,000 m^3. Furthermore, the entrance of the tunnel is redesigned and extended 40m using opening-concrete structure to avoid an eccentric earth pressure. As a counter weight, a retaining structure was also constructed to keep the stability of the riverbank that goes paralleled to the tunnel.

1 INTRODUCTION

A-tunnel considered in this paper has a length of 543m. Its construction started in Oct. 1996 and the tunnel was excavated through in Nov. 1997. However, the slope at the ending point of the tunnel collapsed in a region with a width of 65m and a length of 80m on Nov. 10, 1997, resulting in the failure of the tunnel. In this paper, the geologic condition of the failed slope is firstly made clear and then the mechanism of the failure is investigated. Based on the results of the field survey and the countermeasure against the large-scale failure of the slope, a method containing the removal of the failed soils and a slope cutting by which a safety gradient of the slope can be achieved is adopted in the remediation. The slope cutting contains 15 stages and its volume is about 240,000 m^3. Furthermore, the ending mouth of the tunnel is redesigned and extended 40m using opening-concrete structure to avoid an eccentric earth pressure. As a counter weight, a retaining structure was also constructed to keep the stability of the riverbank that goes paralleled to the tunnel.

2 TOPOGRAPHIC AND GEOLOGIC PROFILE

The topographic features of the ending mouth of the tunnel obtained from aerial photographs and field survey are listed as follow:
a) A trace of large-scale slope failure with 30m×30m along the riverside of the tunnel is

observed. The failed soils moved towards the river.
b) The failed slope locates at the corner of the river, undergoing the erosion of the water flow.
c) A clear evidence of the thrust of the ground due to slope-failure or landslide can be observed at the erosive slope.
d) The traces of old landslide and slope failure can be observed on upper part of the slope.
e) The failed slope has a steeper gradient as the slope goes down.

Based on above observations, it can be concluded that the location of the failed slope is just a typical candidate of slope failure due to its unstable condition where the top of the slope is big and its foot is erased by the flow of river.

The geologic features of the ending mouth of the tunnel obtained from the field survey are listed as follow:
a) The ground of the slope from the tunnel to the utmost of the failed slope is a granite diorite.
b) The green-colored metamorphic clayey stratum distributed in the failed zone is in a large cross angle with the slope.
c) The weathering condition of the rock mass is different at the upper and the lower parts whose boundary is the metamorphic stratum. The lower part is rather intact on whole with a few non-opened cracks, while the upper part has many opened cracks and is outstandingly weathered.

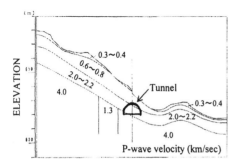

Figure 1. P-wave velocity distribution around A-tunnel.

d) By seismic prospecting (see Figure 1), a weak zone where the P-wave velocity is only about 0.6~0.8 km/s is found.

3 MECHANISM OF THE SLOPE FAILURE

From above surveys on the geologic and topographic features, a retaining wall and a repressing banking supported by soil-cement materials were originally designed before the tunnel is excavated through so that the excavation continues only after the ground becomes stable. The design cross section of the structure at the ending mouth of the tunnel is shown in Figure 2. In real excavation, however, the ground at the ending mouth of the tunnel is much better than expected. Therefore, engineers made a decision that the designed retaining structure was not constructed after the tunnel was cut through and continued the excavation at the foot of the slope (see Figure 3), resulting in the failure of the slope whose process is listed in Table 1. Figure 4 shows the geologic profile, the position of the failed slope and the line of the tunnel.

3.1 Factors of the failure

The factors of the failure can be listed as follow:

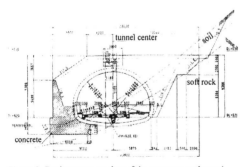

Figure 2. Design cross section of the structure at the ending mouth of A-tunnel.

Topography: The lower part of the slope is located at the corner of a river and its gradient is the steepest in the area.

Geology: A few weakened clayed strata existed in the slope and weathered rocks heavily distribute at the upper part of the slope

Because of the geologic and topographical features, the slope failure is ignited by a series of small-scale failures happened at the lower parts of the slope and finally failed totally due to the progressive failure.

3.2 Ignition of the slope failure

According the weather data at the time before the slope failed, there is no evidence of heavy rain or a quick arise of underwater level which is often related to the failure of a slope. On the other hand, it is evident that there is a rather long time log between the failure and the excavation of the tunnel and the construction of the lining. It is estimated that the tunnel was excavated through the potential sliding surface on Nov. 4, 1997 and is excavated through in the same day. After then, a key-stone plate was set up along the river side and excavation was continued. The excavation was conducted with excavation machines and explosion, by inserting 6m-long extra

Figure 3. Excavation work at the foot of the slope.

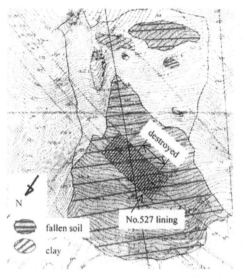

Figure 4. Geologic profile, the position of the failed slope and the line of the tunnel.

Figure 5. Crack in tunnel lining.

rockbolt into the ground in the mountainside. The slope failed at the afternoon of Nov.11, 1997. Before that time, no abnormal change in the inner convergence, which is often regarded as a warning of a failure, was observed. Such kind of time log is usually considered as the main reason of the slope failure due to the progressive failure caused by loosening of the surrounding ground. Based on above discussion, it is concluded that the main reasons of the slope failure are the excavation through the potential sliding surface and the loosening of the surrounding ground due to tunnel excavation (Okuzono, 1997).

3.3 Relation between tunnel excavation and failure

In evaluating the safety factor of the slope, there are several assumption of the sliding surfaces among which the one shown in Figure 7 is used. From the deformation pattern of the steel arches after the failed soils are removed, the mechanism of the shear force acting on the arches due to the sliding can be comprehensively understood from Figure 8 and Figure 9, by which the sliding surface can be clearly assumed in the safety calculation. The cohesion and the internal frictional angle of the assumed sliding surface can be back analyzed by assuming that the safety factor of the slope before the excavation of the tunnel is 1.05. Its results are c=10.0 kPa and tan ϕ=0.64. Based on the results, the difference of the safety factor caused by the existence the supporting embankment and the passive retaining wall is estimated, assuming that the expected sliding surface is a complex sliding surface. Furthermore, in the case of the supporting embankment and the passive

Table 1. Sequence of the slope failure at the ending mouth of A-tunnel.

Nov. 4, 1997	cut through of the tunnel
Nov. 9, 1997	no abnormal change in inner convergence or other observation
Nov. 11, 1997	the day of the failure
7:30	no abnormal observation
13:15	completion of the 527th steel arch (see Figure 4)
13:30	no abnormal change in the steel arch
	some 2~3m cracks are found in the crown and shoulder (60°) between 523th ~527th steel arches (see Figure 5)
14:30	construction of a bank within the tunnel between 523th ~527th steel arches, expecting some enhancement of the earth bearing
15:15	broken of the rockbolt plate in 527th ~528th steel arches (see Figure 6)
15:40	extending of the cracks within the shotcrete and then fallen down
15:50	outstanding deformation of the steel arches and continuously fallen down of rocks at the two sides of the mouth
15:55	
16:00	men withdraw from the construction site
16:15	failure happened

Figure 6. Failure of rockbolt plate.

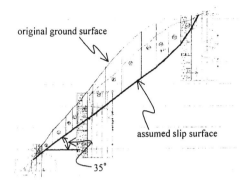

Figure 7. Slope stability analysis for an assumed slip surface.

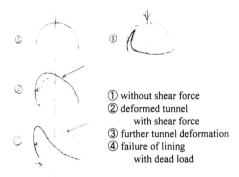

① without shear force
② deformed tunnel
    with shear force
③ further tunnel deformation
④ failure of lining
    with dead load

Figure 9. Failure mechanism of tunnel lining.

retaining wall, the reacting forces against the sliding from the effect of the structures or the weight of themselves are discussed respectively. Table 2 shows the calculated results. It is found that the supporting embankment and the passive retaining wall are very effective in repressing the sliding of the slope. With the existence of these earth structures, the reduction of the safety factor of the slope due to the tunnel excavation is only about 0.2%.

Usually in the case when a tunnel is excavated in an oblique direction with a slope, a supporting embankment or a passive retaining wall is often used as

Table 2. Result of slope stability analysis .

without embankment and wall	
no reacting force	Fs=0.886
with embankment and wall	
no reacting force	Fs=0.976
reacting force=68tf	Fs=0.886

Figure 10. Cutting work of the failed slope.

a countermeasure against the eccentric earth pressure acting on the tunnel. It is, however, not designed for the earth retaining structures that can be effectively used to prevent large-scale slope failure as discussed in this paper. In spite of this, it is still regarded to be effective and its repressing effect is shown in Table 2.

4 REMIDIAL WORK OF THE FAILED SLOPE

The remedial works of the failed slope are conducted in the following steps:

The failed soils are removed at first and then the slope is re-cut to a safety gradient. Figure 10 shows the cutting work of the failed slope. Secondly the destroyed tunnel due the failure of the slope is excavated out and reconstructed with concrete up to the initially designed ending mouth of the tunnel (Sometime it is also called as opening lining).

At the end of the excavation mentioned above,

Figure 8. Total failure of tunnel lining due to sliding force.

however, some wedge-shaped failures in the cutting surface of the slope happened. For this reason, there exists a danger that the slope may fail again in the excavation.

The reason of the failure is regarded as the clayed strata existed in the cut slope that rapture during the excavation of the slope. The gradient of the cut slope at the upper part of the tunnel is 1:1.3~1:1.5. Therefore the gradient of the discontinuous strata is larger than those of the slope, resulting in a safety cut of slope. In the final cutting stage, in order to make a space for a passive retaining wall, the cut slope was excavated in a gradient of 1:0.5, in which the gradient of the discontinuous strata is less than those of the slope, resulting in an unstable cutting.

For this reason, in the final cutting stage mentioned above, rockbolt and shotcrete were designed for the cutting at the beginning. However, because the cutting line is just about the same level of the spring line, it is necessary to cut more deep so that the tunnel can be constructed to the invert. In this case, there is a possibility of further cut slope failure due to the excavation through unknown weakened zone. For this reason, the following enhanced measures were taken to prevent the potential failure:

a) The ground of one span of the steel arch is cut down in the direction from the ending to the starting of the tunnel.
b) A steel arch is then installed at the excavated place.
c) A key-stone plate is setup to the steel arch and then shotcrete is constructed in the inside of the arch.
d) Concrete that used to fix the ground and the steel arch is constructed to prevent the deformation of the surrounding ground.
e) Repeat a)~d) and finally connect the steel arches to existed tunnel that does not fail yet.
f) After the completion of the shelter with steel

arches and shotcrete, a RC second lining is constructed.

The main feature of the method is that because it takes a long time to construct the RC second lining, a shelter with steel arches and shotcrete are constructed at first to ensure that the ground to be cut is in stable condition. After the completion of the RC second lining, the mouth structure is constructed and then the construction of the supporting embankment, as shown in Figure 11 which shows the completion of the remedial work.

5 LESSON FROM THE CASE STUDY IN DESIGN AND CONSTRUCTION OF A MOUTH STRUCTURE OF TUNNEL

In present case study, the designed countermeasures against the slope failure due to the excavation of tunnel were not used during the excavation, resulting in a total failure of the slope. Therefore, something must be learned from the accident.

5.1 Identification of dangerous slopes by advance survey

In design the line of tunnel, particular attention should be paid to the dangerous slope. Mouth structure should be avoided at such dangerous slope because it will cost a lot of time and money to survey and to construct. By picking up and analyzing the dangerous slopes, it is possible to avoid them or in the case of being impossible to avoid them, to find out a applicable countermeasure against the slope failure and the evaluate the time and the money needed for the construction.

5.2 Effect of supporting embankment at mouth of the tunnel

In the present case, the designed supporting embankment was not used during the excavation. The only thing that has been done is that after the construction of the tunnel openly, a loose soil is banked in the surrounding of the tunnel to serve as an earth pressure balance. For this reason, the toe of the slope is in an overhang situation temporarily, resulting in a loosen state of mountainside. Then the slide of the slope develops and its shear force made the tunnel deformed. The development of the deformation in the tunnel is mainly caused by the insufficient stiffness of the banking ground against the normal deformation of the tunnel. In other words, the reacting force against the sectional deformation of the tunnel is not sufficient. If the supporting embankment and

Figure 11. Completion of remedial work.

the passive retaining wall were constructed before the excavation of the tunnel, and if the stiffness of the repressing banked ground is enough, the tunnel would not fail even if the shear force acts on the tunnel due to the loosing of the ground caused by the tunnel excavation.

In the paper, the ground, the supporting embankment and passive retaining wall are assumed as rigid in the calculation of evaluating the effectiveness of enhance the safety of the slope. It is found, in this case, that the decrease of the safety due to the excavation is very small. The problem is that if the stiffness of the commonly used materials such as soil-cement and foaming cement is enough to sufficient to support the shear force that may occur. Therefore, not only the strength, but also the unit weight and the stiffness of the banking materials are very important factors of design.

6 CONCLUSION

In the present paper, a large-scale slope failure due to tunnel construction was studied. The topographic and geologic condition of the failed slope was firstly made clear and then the mechanism of the failure was investigated. The designed supporting embankment was not used during the excavation work. For this reason, the toe of the slope was in an overhang situation temporarily, resulting in a loosen state of mountainside. Therefore, the slide of the slope developed and its shear force made the tunnel deformed.

Based on the results of the stability analysis, the remedial method for further construction was proposed. The method containing the removal of the failed soils and a slope cutting by which a safety gradient of the slope can be achieved was firstly adopted in the remediation. Furthermore, the ending mouth of the tunnel was redesigned and extended 40m using opening-concrete structure to avoid an eccentric earth pressure.

7 REFERENCE

S.Okuzono (1997) : Stability analysis condition for slope failure at the mouth of the tunnel, Landslide Technology, 23-3, pp.15-21 (in Japanese).

Modern Tunneling Science and Technology, Adachi et al (eds), © 2001 Swets & Zeitlinger, ISBN 90 2651 860 9

Design and construction of tunnel through counterweight fill

Hong-Gyu LEE & T. SUZUKI
Fuji Research Institute Corporation, Tokyo, Japan

K. OOKUBO
Japan Highway Public Corporation, Shizuoka, Japan

ABSTRACT: In an area where located on the toe part of landslide, and also where alluvium is distributed under toe part of landslide, a two-lane expressway was constructed by a soil cement tunneling method through counterweight fill. In this method, after composing a part of counterweight fill with strengthened soil stabilized by cement as a treatment during filling, and then the inner part of strengthened zone is excavated. In this paper, the authors introduce the investigation contents and monitoring results for soil cement tunneling method, from its design to construction.

1 INTRODUCTION

Two-lane expressway has been designed and constructed to the area in which following restrictions lies in about 100m-long nearby portal.

1. Located in the toe part of landslide.
2. Beneath toe part of landslide is distributed alluvium of soft ground overlying bedrock.
3. Since passing by natural park and national remains, present topography has to be kept to the utmost.

In original design, it was planned to construct a tunnel by cut and cover method in counterweight fill, after conducting counterweight fill work (including preloading) and ground water drainage work by drainage well and horizontal drainage drilling as a countermeasure against landslide and residual settlement.

However, in case of cut and cover method, the stability of landslide could not be ensured during temporary cut, and there was a technical problem of differential settlement.

Therefore, we have investigated a method, which excavates a tunnel through counterweight fill; after composing a part of counterweight fill with "strengthened soil" stabilized by cement as a treatment during the filling, excavate the part within strengthened soil zone.

This method is named as "soil cement tunneling method" was employed because it is superior economically, and for safety and construction time than other construction method.

"Soil cement tunneling method" was designed after investigating mainly with two-dimensional elastic FEM analysis, and the tunnel construction

was done after strengthening external part (3.5m) of tunnel excavation section using soil cement within counterweight fill.

In this paper, as a new challenge, we introduce the investigation contents and monitoring result for soil cement tunneling method of this site, from its design to construction in order to support future similar cases.

2 GEOLOGY

Within landslide area located in western part of the site, old landslide scarp and traces of tension cracks are observed (Figure 1). The geology of bedrock consists of mainly mudstone and partly alternation of strata of sandstone and mudstone. As shown in Figure 1, Landslide mass is 100m-wide and 300m-long colluvium deposit over bedrock and composed from 2 blocks, generally varying 10 to 15m in thickness and SPT N-values less than 10. In addition, alluvium of soft ground is distributed downward from right under the toe part, and forms mainly gravelly clay with SPT N-value less than 10 (Figure 2).

3 TECHNICAL PROBLEM IN ORIGIN DESIGN

In origin design, in a cross-section shown in Figure 2, stability analysis have been carried out with each condition of groundwater table for following three cases, after counterweight fill, after temporary cut, and after tunnel construction. The conditions of groundwater water are at present, at dry season, at thaw and at 2m below than present. As a result, in

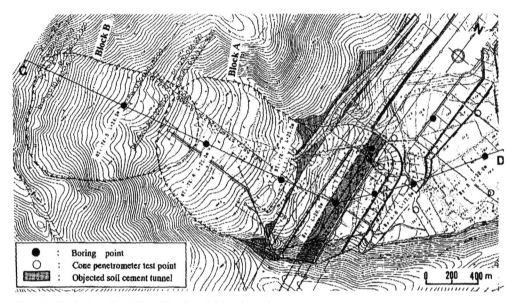

Figure 1. Distribution of landslide and location of objected tunnel.

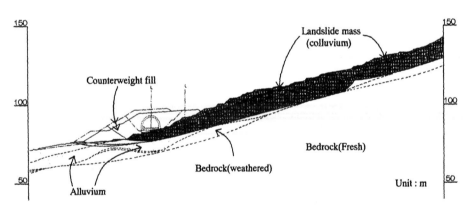

Figure 2. Geologic section according to CD line in Figure 1.

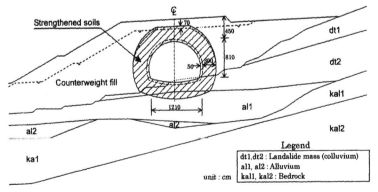

Figure 3. Example of geologic section around soil cement tunnel.

the present condition of groundwater table, safety factor can not be ensured at temporary cut. Then, the present groundwater table is lowered to 2m by drainage well and horizontal drainage drilling in order to ensure the safety factor at temporary cut.

Moreover, following technical problems have been raised for tunnel constructions with cut and cover method in the counterweight fill.

1. In order to ensure safety factor during temporary cut operation, lowering 2m of groundwater table by ground water drainage works is set as a prerequisite. But there remains a question for lowing groundwater table in overall landslide mass and its prevention effect from a geological point of view.

2. During the actual construction period, temporary cut will experience thaw. In this case, lowering water level is also indispensable, and the risk will be considerable if failing the lowering.

3. Temporary cut will potentially require countermeasure for cut slope during temporary cut.

4. Because alluvium of soft ground (about 10m in depth) is deposited under counterweight fill, it is predicted that consolidation settlement will be generated by counterweight fill (about 12~14m of height), rebound by temporary cut, and resettlement after tunnel construction and cover.

From technical problems above, it was concerned that stability of landslide mass can not be ensured during the temporary cut operation, and differential settlement occurs on cut and cover method. Therefore, we have investigated a method, which excavates a tunnel through counterweight fill. After comparative investigation on tunnel construction method, "soil cement tunneling method" was employed because it is superior economically, and for safety and construction time than other construction method.

4 DESIGN

4.1 Investigation on counterweight fill on alluvium

For counterweight fill as landslide countermeasure, due to constructing on alluvium of soft ground and constructing a tunnel through the counterweight fill, it was required to investigate on following problems.

1. consolidation settlement of soft ground
2. stability of landslide
3. rebound and resettlement.

In original design, those problems were investigated based on available laboratory test result, and shape of counterweight fill, height of preload, and consolidation period were decided. However, in order to feed back those investigated in practical point of view to design and construction, test embankment has been carried out with speed of 30cm/day in field. In order to grasp real strength increase of soft ground by consolidation, monitoring using settle

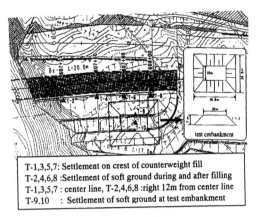

T-1,3,5,7: Settlement on crest of counterweight fill
T-2,4,6,8 :Settlement of soft ground during and after filling
T-1,3,5,7 : center line, T-2,4,6,8 :right 12m from center line
T-9.10   : Settlement of soft ground at test embankment

Figure 4. Monitoring location of settlement.

ment plate and borehole inclinometer are carried out. Locations of test embankment and monitoring are shown in Figure 4. And, an example of settlement observation is shown in Figure 5. Also, check boring has been done before and after the test embankment. As a result of trial embankment, it was proved that consolidation velocity is very fast, and that real settlement is greater than assumed one with e-log p curve in laboratory test.

Then, coefficient of consolidation, e-log p curves and drains distance for design was reviewed. And the relation among degree of consolidation during planned consolidation period of about 180 days, fill height and settlement were re-investigated. As a result, it was proved that counterweight fill height should be 14m in order to make degree of consolidation nearly 100% after consolidation period of 180 days and that the final settlement is 67cm at that time. Moreover, it was confirmed that there is no problem in landslide stability with counterweight fill, and it is assumed that the stability is secured even with 30cm/day in construction speed of counterweight fill.

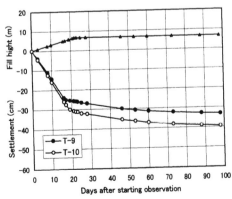

Figure 5. Monitoring result of settlement plates during test Embankment.

4.2 Investigation on soil cement tunneling method

Design conditions of this method are for two-lane tunnels with width 11.2m (excavation width 12.1m) and tunnel length is 115m, and special for small scale in which overburden is 4.5m(Figure 3).

Since tunnel design imposes a earth pressure of fill on strengthened soil constructed around a tunnel, it is aimed to make economical strengthened soil with enough strength against stress generated by excavation. So it becomes significant point to comprehend the relation between reformed thickness and reformed strength of economical strengthened soil. Accordingly, mix proportion test of some stabilized soil was carried out with field generated soil to obtain basic properties, and two-dimensional FEM elastic analysis was done based on those valued. A example of FEM analysis results is shown in Figure 6. As a result of mechanical and economical point of view, it was determined for strengthened soil within counterweight fill that reformed thickness and reformed strength of strengthened soil are 3.0m and q_u=0.5N/m² respectively.

For construction extent of strengthened soil at real fillings, as shown in Figure 3, considering allowance of deformation (50cm) at tunnel excavation and predicted maximum settlement (67cm) of alluvium, reformed zone of strengthened soil was 50cm thicker and 70cm was added on the crown.

For the section design of secondary lining, its thickness and unconfined compressive strength after 28 days are 35cm and 18N/m² respectively.

Besides, for the case of tunnel covered by colluvium of landslide mass, due to an anxiety of differential settlement at right and left legs, it was required to replace those colluvia to strengthened soil.

4.3 Investigation on combination of Soil Cement and filling method

4.3.1 Laboratory test results
Laboratory test was carried out in order to grasp the physical and mechanical properties of fill material and strengthened soil, and to obtain basic information for determining specifications of rolling test.

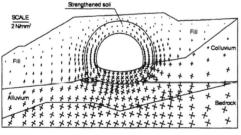

Figure 6. Principal stress distribution.

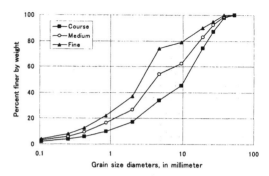

Figure 7. Grain size distribution curves (in laboratory).

Figure 7 shows representative grain size distribution curves of fill materials, which is excavation muck of neighboring tunnel. Cement content of 1, 3, 5, 7, and 9% were employed for the laboratory tests. Sample was made to be 93% of maximum dry density in optimum water content, by compaction. The scale of sample is ϕ 10cm × h 20cm and its curing period is 7 days and 28 days. Unconfined compression test, slaking test, wetting and drying tests of compacted soil-cement mixtures, and triaxial compression test etc were conducted. Figure 8 shows an example of test result. This figure is a result of unconfined compression test after curing time of 7days. q_u is 0.2 – 1.5N/m² in cement contents of 1 - 9%. In the relation of material to grain size distribution, finer grain size distribution obtains higher strength.

Further, following test results are obtained.
1. Because mudstone muck of fill material is easy to be slaking, it is required to secure resistance against weathering by stabilization by cement. With cement contents of 5%, it was thought to have some resistance against weathering from the result of wetting and drying tests of compacted soil-cement mixtures.
2. According to relation between water content and unconfined compressive strength, in case of water content exceeding ±3% of optimum water content, strength of strengthened soil extremely decreased.

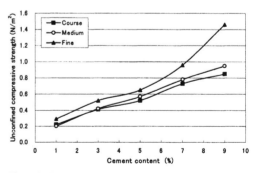

Figure 8. Result of unconfined compression test (curing time : 7 days).

Table 1. Control criteria of counterweight fill.

Control item	Controlled value	Frequency of measurement
Grain size distribution	53±20% passing 5mm	1 / day
Water content	14.3±2.5%	1 / day
Spreading depth	30cm	--
Number of rolling compaction	8 times	--
Degree of Compaction（RI）	> 92%	15 points / 1 layer·day
Unconfined compressive strength	> 0.5 MPa	5points/1000m³

Therefore, it was required to control water content within ±2.5% of optimum water content during construction.

3. It was expected to have about 20% of strength increase from curing time of 7 days to 28 days.

4.3.2 *Improvement method of fill material*
Tunnel muck had variation in grain size distribution. There were many existence of boulders larger than 100mm, and sometimes larger than 1000mm. Therefore, crushing method and stirring/mixing method was problems for improvement of fill materials. As an economical method with uniformity of grain size distribution and improvement effect, mixed in place type stabilizer method was employed.

4.3.3 *Rolling test*
Rolling test was subjected to obtain basic information needed for determining specifications of design and construction method in real works. Fill material for rolling test is one type with grain distribution close to medium gradation in laboratory test. Cement contents were set 3 types, 3, 5, and 7% referred to the result of laboratory tests.

Preliminary crushing, crushing/stirring/mixing, spreading, and rolling compaction were carried out with bulldozer, stabilizer, bulldozer, and vibration compaction roller, respectively. Spreading depth was 30cm, number of rolling compaction has 5 patterns of number of times; 2, 4, 6, 8, 10. Also, physical and mechanical properties of each condition were examined.

As a result, as well as deciding cement content of 5.3% at strengthened soil, control criteria has been settled as shown in Table 1.

5 CONSTRUCTION OF COUNTERWEIGHT FILL

Construction extent of strengthened soil in counterweight fill was divided into following three construction standard section, according to positional

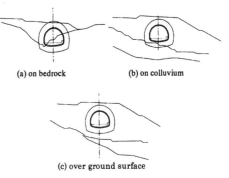

(a) on bedrock   (b) on colluvium

(c) over ground surface

Figure 9. Positional relation between ground surface and soil cement tunnel.

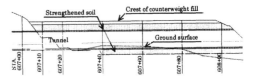

Figure 10. Longitudinal section of soil cement tunnel section.

relation between ground surface after clearing and grubbing and soil cement tunnel (Figure 9, 10).

1. STA.607+4.2~STA.607+40.0 : external part of strengthened soil is on bedrock (Figure 9(a))

2. STA.607+40.0~STA.607+87.2 : external part of strengthened soil is on colluvium (Figure 9(b))

3. STA.607+87.2~STA.608+20.2 : external part of strengthened soil is over ground surface(Figure 9(c))

For the filling of counterweight fill, where the external part of strengthened soil is on colluvium of landslide mass, cut was done with least excavation length (slope of 1.2H: 1.0V), and filling was replaced by strengthened soil because of differential settlement. And, for the internal part of strengthened soil, in order to secure stability of the tunnel face, cement content of 3% which is equal to about 1/2 of cement content at external part was employed.

Monitoring result during and after filling is shown in Figure 11. Monitoring locations were

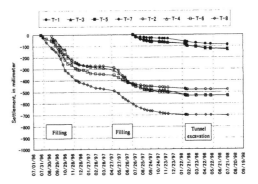

Figure 11. Monitoring results of settlement.

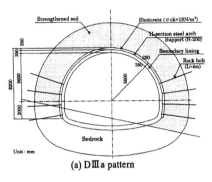

(a) DⅢa pattern

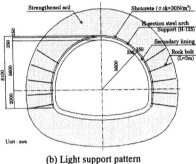

(b) Light support pattern

Figure 12. Standard section.

shown in Figure 4. Almost same result with one by review in test embankment was obtained.

6 TUNNEL EXCAVATION

Top heading method was employed as excavation method. Excavation speed was 1m/1cycle length, and advanced 2m in a day. Back hoe or twin header was used to excavate internal part of tunnel which cement content is 3%, and finishing by ring cut was employed for strengthened soil of cement content 5% encroaching inside pay line. Because the internal part of tunnel is also strengthened soil with cement,

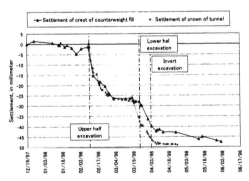

Figure 13. Monitoring results of settlement in crest of counter-weight fill and crown of tunnel (in STA.607+60).

Figure 14. View after completion of soil cement tunnel.

so stand-up of face was fine, and only a little fall of rocks is proved. Also, it was proved by needle penetrometer test that improved strength at the face was expressed more than design strength.

Figure 12 shows standard section in real construction. At the design stage of this construction method, it is not considered to employ rock bolt and H-section steel arch support as primary support. However, because of anxiety of differential settlement and unsymmetrical pressure, and in expectation of sewing effect in strengthened soil, those supports were added.

Monitoring has been carried out inside tunnel as well as surface of counterweight fill. Following tendencies are obtained from the monitoring result in soil cement tunnel.

1. Pre-deformation was extent of 23-30%.

2. Deformation mode was obtained tendency that crown is sinking and convergence measure of upper half become longer, which is equal to tendency in FEM analysis.

3. After tunnel face advanced 3 times of tunnel diameter, deformation rut an end.

Figure 13 shows an example of monitoring result during tunnel excavation.

7 CONCLUSION

Soil cement tunneling method has advantages; it is possible to construct not only on the site in which arch culvert was constructed, but also on the site with unique conditions such as landslide block and soft ground, and there is no restrain of construction after filling in snow cover zone.

In the future, for the opportunity of discussing soil cement tunneling method by comparing to existing arch culvert method etc, please use this case by referring to not only this construction result but also the circumstance from design to construction.

Modern Tunneling Science and Technology, Adachi et al (eds), © 2001 Swets & Zeitlinger, ISBN 90 2651 860 9

Causes of Primary Crusher Conveyor Tunnel Failure in Sar Cheshmeh Copper Mine in Iran

M. M. Toufigh
Associated Professor, Department of Civil Engineering, Shahid Bahonar University of Kerman, Kerman, Iran

M. E. Mirabedini
Graduate Student, Department of Civil Engineering, Shahid Bahonar University of Kerman, Kerman, Iran

ABSTRACT: The primary crusher and its conveyor tunnel are built in 1976 for processing of copper in Sar Cheshmeh near Kerman, Iran. The conveyor tunnel approximately 5 meters by 2.5 meters by 115 meters long will extend horizontally from the bottom of the crusher north ward to the toe of the fill slope. Right after the construction of the tunnel a few horizontal and one diagonal cracks were appeared inside the tunnel. The slope stability analysis with static and dynamic loads shows factor of safety above 1.5. The analysis of the tunnel shows there is no deficiency in the design of the structural part of the tunnel. The analysis shows that the rock foundations under tunnel with RQD less tan 50 and washing of fines through drainage system not support these loads and required to stabilized these highly weathered rocks. This can be done by the grouting cement under the tunnel.

1 INTRODUCTION

The primary crusher and its conveyor tunnel are built in 1976 for processing of copper in Sar Cheshmeh near Kerman, Iran. The primary crusher is about 24 meters by 25 meters in plan with conveyor tunnel approximately 5 meters by 2.5 meters by 115 meters long will extend almost horizontally with slope of 5.4% from the bottom of the crusher north ward to the toe of the fill slope. The excavated rocks for the primary crusher is about 40 meters and the height of excavation decreases to zero for first 65 meters and the rest of the tunnel is resting on 97% compacted soil and rock fill up to height of 7 meters. The tunnels constructed by reinforced concrete with various thickness from 60 cm up to 200 cm with different reinforcement depend on overburden pressure. Since this tunnel was constructed on half rock and half soil, they left two constructed joints almost at border of soil and rock. Figure 1 shows the plan and section view of the primary crusher and conveyor tunnel.

2 PAST HISTORY AND ANALYSIS OF SETTLEMENT

Half of the built tunnel in 1976 was on rock foundation and the other half on soil embankment. There are two constructed joints existed at intersection of the soil and rock foundation. After the construction of the tunnel a few verticals and one diagonal crack

were appeared inside of the tunnel. Some of the cracks occur within constructed joints resulted in more opening, and some cracks outside of these joints.

Right after the construction and development of cracks, the surveying group of the tunnel started to evaluate and recording the settlements and movement of the tunnel. This information is available up

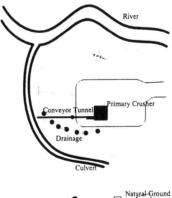
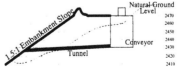

Figure 1. Plan and section view of the primary crusher and conveyor tunnel.

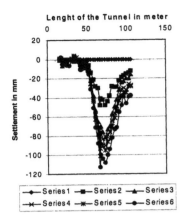

Length of the Tunnel in meter

series1 = 1977, series2 = 1982, series3 = 1986
series4 = 1990, series5 = 1994, series6 = 1999

Figure 2. Variation of settlement in the tunnel with time (year).

to now for interval of 6 months to 2 years. Result of surveying shows that settlement of tunnel and opening of cracks increases every year. The results are shown in Figure 2.

In the past years the owner with help of consulting engineers started various studies on the tunnel. They came up with three major solutions, two in 1982 and the other one in 1990. In 1982 the performed an incomplete grouting under the tunnel in the rock section and also stabilize the slope of the overburden soil. Right after this process the speed of settlements suddenly decreased but never stopped. By the assumption of the owner that this method is not successful, in 1990 they hired another consulting engineer and they designed and constructed a drainage system by the assumption that the existed one is not performing well.

The intersecting result is that this method actually at the beginning deceased but later increased the speed of settlements of the tunnel. The above would discuss in detail latter in this paper. It should be noted that the recorded settlement of tunnel on the rock part is greater than that the part on the 7-meter embankments soil with maximum of 115 mm and 40 mm respectively. The measured rotation of the tunnel is very small and is about 3 to 4 mm.

The boring data is available exactly under the tunnel with depth of 15 meter. Most of the rock is highly weathered with some portion of RQD less than 50 and the strength classified as weak to very weak. The joint system in the bedding is very irregular and the materials are highly broken.

3 STRUCTURE OF THE TUNNEL

The 115-meter long tunnel of the primary crusher started at the elevation of 2429 m and finished at the elevation of 2424 with drop of 5 meter, which is equivalent to slope of 5.4%

The internal section of the tunnel is a constant rectangular with width of 5 meter and height of 2.5 meter . The thickness of tunnel varies from 0.6 to 2 meter. The variation of the tunnel thickness depend on excavation surface and overburden material weight on the tunnel, which varies from zero at the outlet of the tunnel to about 40 meter next to the crusher.

This constructed by reinforced concrete with compressive strength of 280 kg/cm^2. In the concrete design of the tunnel the slabs at the roof and bottom are assumed to behave as one way slabs which resulted in mostly transverse reinforcement and much less longitudinal reinforcement, mainly for shrinkage and temperature of concrete. The same methodology was assumed for design of side walls. At the construction joints perpendicular to the director of the tunnel the longitudinal reinforcement are discontinued and waterstop are installed.

Analysis of loads acting on the tunnel were done by FLAC program and modeled by structure mode based on Mohr Columb behavior. According to available soil properties and by applying all the influence loads on the roof and walls of the tunnel, the results of analysis by FLAC program are given in the Table 1.

It should be noted that only alluvial material from explosion are overburdens the tunnel. The results are showing that designed structure can easily handle all of the flexural and share stresses influenced by overburden material on the tunnel.

Table 1. Results of analysis by FLAC program.

Max y Displacement	2.4 cm
Min y Displacement	0 cm
Max x Displacement	7 cm
Min x Displacement	2 cm
Min shear stress (xx)	-2.5 Mpa
Max shear stress (xx)	0 Mpa
Min shear stress (yy)	-2.5 Mpa
Max shear stress (yy)	0 Mpa
Max principal stress	3.5 Mpa
Max major principal compressive stress	1 Mpa
Min major principal compressive stress	-3.3 Mpa
Max minor principal compressive stress	0.8 Mpa
Min minor principal compressive stress	-1 Mpa

4 SLOPE STABILITY AND IT'S EFFECT

Referring to Fig 1 the existed slope above the tunnel is about 33.5 degrees with the maximum height of about 40 meter. The material are classified as GW with friction angle of 32 degrees and small cohesion of 0.5 gr/cm2, due to low compaction and irregular surface of the slope at the early years the water migrates in to the slope and a few surface cracks was appeared.

At that time the owner with consulting engineer suggestion regulates and smoothen the surface of the slope and compacted the material in a few layers (just at the surface) up to 97% modified compaction effort. It 's not clear that whether these cracks were only at the surface at that time or had any influence on the tunnel settlement. After the stabilization of the slope, the settlements of the tunnel did not stopped but they have not seen any new cracks on the surface. the question rise is that still at the some portion of the tunnel due to non-stability of the slope material, these material loads slices may cause and extra loads on the system. In the other words at some portions of the tunnel, the loads are not actually equal to overburden pressure and because of the slope failure something similar to stress concentration increase the influence of the load on the tunnel. This effect may causes the cracks inside the tunnel. For the above reason for slope stability analysis two cases were considered in this investigations. In the first case the under layer rock material at the level of bottom tunnel were considered as rigid and solid material and only failure surface in the analysis can pass through only overburden material (soils). For the second case these failure surfaces can pass through under layer rocks. As explained previously most rocks are highly broken and therefore the failure surface does not have to pass through certain joint paths.

For stability analysis computer program STABL was adopted with material density of 1.85 ton/m^3 and other material propriety as explained previously in this paper and owner documents. Since this tunnel is near mining explosion with the closest distance of 1.5 km, therefore lateral pressure of 0.1g also considered in the calculations. The circular and random failure surfaces with and without explosion effect were considered in the calculation. For all of the above cases the minimum 1.5 factor of safety were computed.

5 DARINAGE SYSTEM AND IT'S EFFECT

In the original design of the drainage system three drainage wells were constructed next to the tunnel with discharging pipes to the nearby stream. These wells were destroyed by tracks and heavy equipments at the site. Addition to the above system two

Table 2. Chemistry test results on discharged water from drainage system.

PH	P Alk	M Alk
6	0	10
TH	Ca H	SiO2
610	560	18.5
SiO4	Cl	Na
775	16	19
K	Cu	Zn
30	88	150
TDS	TSS	Cond
581	10	1197

horizontal drainage system were constructed next to the outside walls of the tunnel. The later case is still working well. As explained previously, in 1990 the consulting engineer designed and constructed seven drainage wells with variable depths but all of them lower than bottom of the tunnel and all are connected together and finally to the nearby stream. Right after construction of the new system the settlement decrease but later increases. After looking backward to the variation in the settlement it can be assumed that, this adopted new system actually washing away or solve the fine material under the tunnel.

Therefore an analysis of the discharge water has been done and the results are shown in Table 2.

The results are stated that the levels of the total hardness solved in the discharge water are very high. Chemistry analysis of the water are shown that existing of free ions such as Cl, Zn, Cu, and SiO4 with pH of 6 (acidic) indicate that due to weathering process in the rocks causes this effect. Finally it can be concluded that washing of the rocks occurring and resulting the creation of the voids or settlement in the tunnel.

6 RESULT AND CONCLUSIONS

Slope stability analysis for material above and bellow the tunnel gave the minimum factor of safety 1.5. Therefore the slope above the tunnel may not have any influence on the tunnel settlement. Structural analysis of the tunnel show that the tunnel is highly resistible against the external loads, which computed by computer program, and therefore due to overburden pressure the tunnel structure is stable, and can not causes the structural failure. Due to low RQD of the rocks under the tunnel and water analysis it can be concluded that the main cause of the settlements is because of originally high void in the system and additionally to the above is washing and solving of the material fines beneath the tunnel.

Therefore this existing void and creature of the new voids would introduce the settlement of the system. Since the settlements under longitudinal direction of the tunnel are not equal, therefore the cracks would appear in the tunnel. In order to solve this problem at this stage with knowledge of the problem, probably two simultaneously methods should be considered

1) Stop drainage any water bellow the bottom level of the tunnel

2) Adopting a grouting process under the tunnel to filling the exciting voids which may reduce any type of settlements.

7 ACKNOWLEDGMENT

The first author should like to express his gratitude to Mr. Ahmadi, Mr. Mirzaiee, Mr. Afghani and Mr. Etminan and also National Iranian Copper Industries for their continuos help, geotechnical test result, and for their valuable comments and discussion.

REFERENCES

Bieniawski, Z.T. 1989. *Engineering rock mass Classifications*. John Wily & Sons New York.

Bieniawski, Z.T. 1984. *Rock mechanics design in mining and tunneling*. A.A Balkema.

Brown, E.T. 1981. *Rock characterization testing and monitoring*. ISRM suggested methods, British library.

Modern Tunneling Science and Technology, Adachi et al (eds), © 2001 Swets & Zeitlinger, ISBN 90 2651 860 9

Study on the construction and design method of bolting support in coal roadway

N. Zhang, J. Bai, J. Zhou
China University of Mining and Technology, Xuzhou, jiangsu China
C. Min, S. hai
Xuzhou Mining administration, Xuzhou city, China

Abstract: The bolting technology of coal roadway has been widely applied in China with obvious economic benefits and technical effects. This paper aims to give a general introduction of it from the following two aspects. Firstly it presents two commonly used bolting design methods, i.e. engineering analogy method based on the classification of coal roadway and system design method with computer numerical simulation. Secondly it deals with the analysis of bolting stability and rational bolting technology of gob edge entry based on the "Big-small structure" theory of surrounding rock.

Bolting is a kind of efficient, economical and safe supporting method of coal roadway. It has many obvious advantages and has become a main support technology for coal roadway all over the world. In resent years, China University of Mining and Technology, Central coal mining Research institute, Xintai Mining Administration, Xinwen Mining Administration, and other related organizations have cooperated with each other in solving several key problems of bolting support for coal roadway with breaking-through progress.

1 BOLTING DESIGN METHODS

Here two commonly used bolting design methods are to be briefly introduced.

1.1 engineering analogy method based on the stability classification of coal roadway

China University of Mining and Technology, together with China Central Coal Mining Research Institute, adopts a method that combines theoretical study with extensive field monitoring and numerical analysis, and then puts forward corresponding bolting pattern and selecting scheme of main parameters of coal roadway. Fuzzy cluster analysis

is used to classify the stability of surrounding rock of roadway into 5 categories according to 7 parameters. (which are rock strength of roof, two sides and bottom, burying depth of roadway, width of pillar, coefficient of mining activity and integrated coefficient of rock surrounding roadway respectively.) The five categories are I extremely stable, II stable, III medium stable, IV unstable, and V extremely unstable. Bolting pattern and parameters vary from category to category, as shown in Table1.

1.2 System bolting design with computer numerical simulation

The system bolting design method is characterized by computer numerical simulation. Geological mechanic evaluation, design, construction, monitoring and information feedback are taken as an integral system in this method, and design is within constant adjustment in practice. A set of software for bolting design of coal roadway, which includes pre- and post- handling programs, has been developed. It can be used to input and transform original parameters, form and compare schemes, select best scheme, output results and so on. Once main parameters of roadway are input in practical use, various possible schemes will be automatically

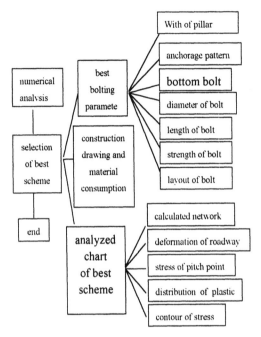

Figure1. Exported result of of numerical analysis

formed and further analyzed via a computer. The best scheme will be selected according to the principle of low cost and small displacement, as shown in Figure1. Information will be fed back from construction and field monitoring. And five parameters (relative displacement of two side ribs, incoherent deformations of in- and out-bolting rock zone, the number of yield points of testing bolts and anchorage power of bolts) are input into the computer in order to decide whether and how design should be revised.

2 THE BOLTING SUPPORT TECHNOLOGY OF GOB EDGE ENTRY

The gob edge entry is a kind of gateway which is driven along gob of next face with a narrow chain pillar of 3~5 meters wide. One side of which is the relative integrated coal seam and the other side is the narrow pillar in which plastic deformation has occurred. Whether can the bolting support be used in the gob edge entry? Whether can the structure of bolting support keep stable? What about is the safety of bolting structure? Experts at home and abroad have different views concerning these questions. The scholars of Australia and Britain strongly argue that it is unsuitable to support gob edge entry by bolting for its two small safety coefficient. But there has been considerable successful engineering practices in China with the "big-small structure" theory as guidance.

2.1 *the stability analysis of gob edge entry by bolting*

According to the key stratum theory and movement pattern of the overlying strata on gob side of coalface, the main stratum will fracture and form an articulated beam on gob side, which is called 'big structure'. The structure is relatively stable until the next face is pushed forward. The concentration stress will occur on the coal seam around working face, meanwhile the stress-reducing zone of the gob edge will occur, roadway will be driven in the zone, shown as Figure 2, where H stands for depth of the roadway below the surface. γ stands for the average density of the rock above the floor, k_1, k_2 and k_3 stand for stress concentration factors. When the roadway is excavated along the gob edge, it can be shielded by the 'big structure'; thus the supporting is relatively easy. But the coal seam along the next gob is subjected to the big structure, the fracture and plastic zone of 2~4 meters wide will occur at the edge of coal seam before driving roadway. The width of chain pillar is only 3~5 meters, therefore the chain pillar will basically be in plastic status after the roadway is driven. It can be expressed in other words that the gob edge entry has two properties 1) in low stress state, 2) under the broken rock condition. The deep surrounding rock has formed stable "big structure", So only the shallow surrounding rock of roadway needs to be supported. The modern newly bolting can reinforce the broken rock and create stable "small structure".

In the course of extraction, the stress equilibrium of the big structure will be disturbed by advancing abutment pressure, whose block B will be rotated again, the narrow pillar will wholly be in fracture status, its load-bearing capacity will be further reduced. Meanwhile, because of superposition of the advancing abutment pressure and the concentration stress of the gob side on

Table1. Basic patterns and main parameters of roof bolting of coal roadway

Roadway state	Basic pattern of support		Main parameter
Extremely Stable	Integrated sandstone、limestone strata: unsupported		/
	Other strata: single bolts		Head anchored, rod diameter 16~18mm,
Stable	Relatively integrated roof: single bolts		bolt length 1.6~1.8mm, interval 0.8~1.2mm, design anchorage power: 64~80KN
	Relatively broken roof: bolts + mesh		Head anchored, rod diameter 16~18mm,
Medium stable	Relatively integrated roof: bolts + steel reinforcement		bolt length 1.6~2.0mm, interval 0.8~1.0mm, design anchorage power 64~80KN
	Relatively broken roof: bolts + W steel band(steel reinforcement) +mesh or added cable anchor; truss mesh or added cable anchor;		Head anchored, rod diameter 16~18mm, bolt length 1.8~2.2mm, interval 0.6~1.0mm, design anchorage power 64~80KN; Fully anchored, Rod diameter 18~22mm, bolt length 1.8~2.4mm, interval 0.6~1.0mm;
Unstable	Bolts + W steel band(steel reinforcement) +mesh or added cable anchor; truss mesh or added cable anchor;		Fully anchored, rod diameter 18~22mm, bolt length 1.8~2.4mm, Interval:0.6~1.0mm
Extremely Unstable	Relatively integrated roof: bolts + yieldable steel set or added cable anchor; relatively broken roof: bolts + mesh + yieldable steel set or added cable anchor; strongly floor heave: bolts +steel circular rigid support		Fully anchored, rod diameter 18~24mm, bolt length 1.8~2.6mm, Interval:0.6~1.0mm

Figure2. big-small structure of surrounding rock of gob edge entry

integrated coal beside the roadway, the stress on the integrated coal will increase. In general, the stress concentration factor of the integrated coal and the narrow pillar are 2.5~3 and 0.25~0.45 respectively. So the original small structure will also be disturbed, then the distance is not rather long, the strengthening support may be used to keep the roadway safe in the last period, such as prop-support, cable anchor, grouting reinforcement and so on.

3 RATIONAL BOLTING SUPPORTING OF GOB EDGE ENTRY

Firstly, it is vertically important to keep two sides of roadway stable. Bolts should provide strong mean load density in order to control the deformation of coal side ribs, meanwhile they should have proper elongation for rock deformation. Enhanced extensible bolts can satisfy these demands, whose strength of heat-treated thread part will be exceed that of the rod, then the drawing breakage of bolt occurs in the rod instead of at the thread and it will leave enough room for elongation. Usually the anchorage power should be more than 40kN, and the bolts should be more than 1.8m long.

Secondly, the integral sinking of two sides is bound to cause the floor heave and incoherent deformations in and out of roof anchorage zone, so it is necessary to keep the roof stable. Supporting with full-length-resin-grouted bolts can decrease deformation of roof and pressure of bolt head obviously, with few bolts losing effectiveness. These factors justify the choice of high function bolts with full-length resin grouted. There are two basic bolting patterns: one is bolt-steel reinforcement-mesh, the other is truss bolting.

Thirdly, bolting should reinforce stress concentrated area in two floor-corners.

In case of too thick coal or immediate roof, cable anchor of small diameter may be applied to keep roof stable. Each cable anchor is 28mm in diameter and 6~9m in length. They should be installed in stable strata.

4 ENGINEERING INSTANCE:

The bolting technology is illustrated by actual application in gob edge roadway of 2445 working face in Pangzhuang coalmine. The test roadway is 3.6m wide, 2.8m high, and has a 5.0-meter-wide pillar along gob edge. The soft compound roof is made of carbonaceous mudstone containing two coal rider seams with well-developed joints and fractures. The coal seam is loose and broken, and will easily to fall off as soon as been driven. It must be repaired at least three times when the I-steel support is used.

Support parameters are as follows:

1) Newly bolts with extra-length-resin-grouted (Φ20*2200mm) for roof;

2) Enhanced extensible bolts with ended-length resin-grouted (Φ18*1800mm) for two side ribs;

3) Small diameter cable anchor for the thicker soft roof in every 10m distance.

Support effect: In comparison to I-steel support, displacement of two sides can be decreased by 60%, advancing rate of roadway can be increased by 40%, while the cost of support and maintenance is decreased by 45%. In a word, the need of safe produce can be totally satisfied by bolting.

5 CONCLUSIONS

1. Two commonly used bolting design methods, engineering analogy method and system bolting design method are briefly introduced.

2. The big-small structure theory is applied to expound the stability of surrounding of gob edge entry.

3. Enhanced flexible bolts and high-function bolts have been successfully applied in gob edge entry supporting.

REFERENCES

Hou Chaojiong et al.1996.Development of bolting of coal roadway in China. *Journal of China Coal Society*(2):113~118.

Ma Lianjie et al. 1997.Design method based ground stress for bolting of coal roadway. *Ground Pressure and Strata Control* (3,4):195~197.

Hou Chaojiong et al.1997.High strength bolt. *Ground Pressure and Strata Control* (3,4):176~179.

Chaojiong Hou, Nong Zhang & Jianbiao Bai, Development of bolting technology for coal roadway in China, Proceedings of the mining science and technology'99: 447-450, A.A. balkema /rotterdam /brookfield/ 1999..

Modern Tunneling Science and Technology, Adachi et al (eds), © 2001 Swets & Zeitlinger, ISBN 90 2651 860 9

Rapid excavation for small section tunnel using TBM

H. Namura, H. Imaoka, T. Takamichi & Y. Kobayashi
Sato Kogyo Co., Ltd., Tokyo, Japan

ABSTRACT: In smaller tunnel projects for construction of, for example, headrace tunnels for hydropower plants or pilot bores of large-section tunnels, use of tunnel boring machines (TBM), which enables safer, faster tunneling than in conventional methods, has grown recently. When applied to complex ground conditions in Japan or to small-section tunnels, however, the TBM approach poses problems concerning supports, hauling and excavation, and an efficient and systematic method has not yet been established. Before headrace tunneling at the Shin-Onagatani No. 1 Hydropower Plant, problems likely to hamper expeditious TBM excavation were identified, and the Combined Boring and Lining TBM Method, an automatic shotcreting system, a double-track mucking system and a new ground evaluation system using TBM machine data were developed and adopted. As a result, an average monthly advance of 381 m and a maximum monthly advance of 785 m, both of which are a new record in Japan, were achieved.

1 INTRODUCTION

In order to develop a small or medium-sized hydroelectric power plant economically within a short period of time, it is important to reduce the cost of headrace tunnels, which accounts for a significant percentage of the total construction cost of such a plant. In line with this, there have been a growing number of cases involving the use of a TBM(tunnel boring machine)-based excavation method, which enables safer and faster excavation than in conventional excavation methods.

Because of the factors that have contributed to the formation of its archipelago, geology in Japan is complicated and varied to the extent of being almost unique in the world. If a TBM is to be used for tunnel excavation, therefore, it is necessary to allow for soft formations and a large number of accompanying faults and fracture zones. In the case of a small section tunnel, there are many problems yet to be addressed, such as the difficulty of shotcreting immediately after TBM excavation because of space limitation and relatively low efficiency in muck removal.

In the construction of the headrace tunnel for Shin-Onagatani No. 1 Hydropower Plant, a number of new approaches were adopted to enhance efficiency and expedite work in small-section tunnel excavation, such as a TBM for excavation and lining, an automatic shotcreting system, a double-track mucking system and a ground evaluation system using fed-back data from the TBM. This paper outlines the various new systems employed for the construction.

2 PROJECT OVERVIEW

The construction of the Shin-Onagatani No. 1 Hydropower Plant was planned and ordered by the Public Enterprise Bureau of the Toyama Prefectural Government. This project involves the replacement of the existing Onagatani No. 1 Hydropower Plant, which is getting obsolete. It also aims to make more effective use of rivers. Under a power generation plan, a new water intake facility is to be built on the Onagatani River at Yatsuo-machi (Nei County, Toyama Prefecture) to take in up to 6.0 m^3/s of water and generate electricity of up to 7500 kW using an effective head of 152 m induced by a 5.3 kilometer long headrace channel.

The work involved the construction of a headrace tunnel. Most (4925 m) of the total tunnel length of 5281 m was driven with a double-shield type TBM for both boring and lining with an outside diameter of 2.8 m.

Figure 1 shows a geological profile of the construction site. The geology of the site is composed mainly of highly wet tuff (maximum rate of water inflow: 10 t/h), highly swelling (with high slaking potential) silty rock, and granite prone to rockfalls. A single-shell lining was formed by spraying fiber-

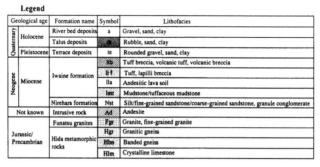

Legend

Geological age		Formation name	Symbol	Lithofacies
Quaternary	Holocene	River bed deposits	a	Gravel, sand, clay
		Talus deposits	dt	Rubble, sand, clay
	Pleistocene	Terrace deposits	te	Rounded gravel, sand, clay
Neogene	Miocene	Iwaine formation	Itb	Tuff breccia, volcanic tuff, volcanic breccia
			Itf	Tuff, lapilli breccia
			Ila	Andesitic lava soil
			Imt	Mudstone/tuffaceous mudstone
		Nirehara formation	Nst	Silt/fine-grained sandstone/coarse-grained sandstone, granule conglomerate
Not known		Intrusive rock	Ad	Andesite
Jurassic/ Precambrian		Funatsu granites	Pgr	Granite, fine-grained granite
		Hida metamorphic rocks	Hgr	Granitic gneiss
			Hbn	Banded gneiss
			Hlm	Crystalline limestone

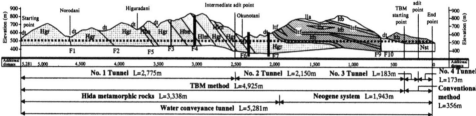

Figure 1. Geological profile.

reinforced mortar (standard thickness: 2 cm to 4 cm) that is suitable for application to wet or rockfall-prone ground.

3 COMBINED BORING AND LINING TBM METHOD

3.1 *Challenges associated with conventional TBM methods*

In conventional TBM methods of excavation, it is difficult to install shotcreting equipment in a limited working space, and it is inefficient to work in a dust-laden environment by shotcreting. Common practice, therefore, is to construct lining after the completion of TBM excavation.

The excavated bore, however, is usually left to stand unlined for a long period of time. Conse-

quently, depending on the geological conditions, weathering of ground or rockfalls may unstabilize the ground with the elapse of time. In addition, lining work carried out as a separate process may lengthen the time required for construction, and cost reduction cannot be achieved eventually.

3.2 *Overview of the new TBM method*

The Combined Boring and Lining TBM Method is a method of construction developed specifically to overcome the problems of conventional TBM methods described above. In the newly developed method, shotcreting is carried out immediately after TBM excavation. The Agency of Natural Resources and Energy of the Ministry of International Trade and Industry (now reorganized as the Ministry of Economy, Trade and Industry) commissioned New Energy Foundation (NEF), is conducting research

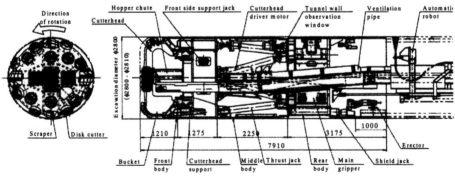

Figure 2. TBM structure.

Table 1. Main specifications of TBM.

Type	Combined Boring & lining TBM
Machine structure	Double-shield type (manufactured by Komatsu)
Excavation diameter	2.80 m (2.79-2.81 m)
Main body length	7.91 m
Total length	Approx. 100 m (to rear end of trailing carriage)
Total weight	Approx. 150 t
Total motor output	480 kw
Power supply	AC6,600 V 3ϕ · 60 Hz
Cutter	13 inches × 22 cutters (die steel)
Cutterhead torque	max 358 kN (37 tf-m)
Cutterhead speed	3.4 rpm-10.0 rpm (4 steps)
Main thrust	max 3,822 kN (390 tf)
Auxiliary thrust	max 4,704 kN (480 tf)
Main stroke	1.20 m
Auxiliary stroke	1.15m
Gripper thrust	max 8,232 kN (840 tf)
Internal mucking capacity	400 mm × 77 m³/h (belt conveyor type)
Spray material	Fiber-reinforced mortar (premixed type)
Spraying capacity	2 m³/h
Shotcreting system	Automatic shotcreting robot

and studies for the development of the new TBM technologies. As a part of this technology development, the Combined Boring and Lining TBM Method was employed for the construction of the hydropower plant mentioned above with the aims of early stabilization of excavated ground, improvement of the working environment and reduction of the construction period in small and medium-sized tunnel projects. Figure 2 shows the TBM's main structure, and Table 1 shows its specifications.

4 AUTOMATIC SHOTCRETING SYSTEM

4.1 *Problems of conventional shotcreting methods*

In conventional shotcreting methods used in mountain tunnel construction, it is common practice to use remote-controlled robots for large-section tunnels and manual work for small-section tunnels. Both methods, however, require human involvement, and the finishing of surfaces and determination of shotcrete thickness depend much on skilled workers' ex-

perience and intuition. From the viewpoint of hygiene, dust and shotcrete rebound in a limited working space pose serious problems in the conventional shotcreting methods.

As for shotcrete thickness control, measurement with measuring pins and holes drilled into applied shotcrete is not very accurate and is time consuming.

4.2 *Outline of automatic shotcreting system*

The newly developed automatic shotcreting system consists of an automatic shotcreting robot, an automatic section measurement system, an automatic shotcrete thickness control system and an automatic surveying system. The automatic shotcreting robot shown in Figure 3 automates shotcreting (lining) work to be carried out concurrently with excavation, even in a small section tunnel TBM, by data-linking an auto tracking survey instrument, section measuring equipment, a shotcrete pump and a TBM. The

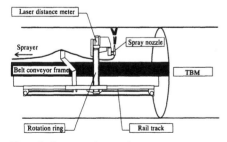

Figure 3. Structure of automatic shotcreting robot.

Figure 4. Automatic shotcreting robot.

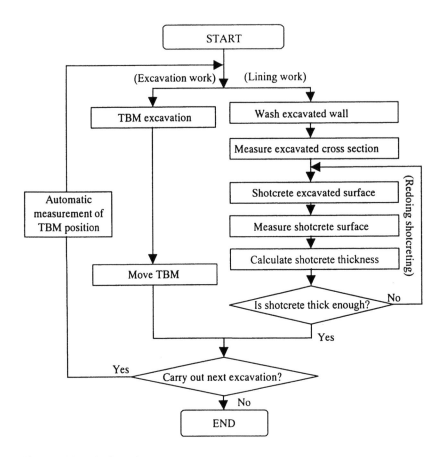

Figure 5. Automatic shotcreting process.

development of this automatic shotcreting system has made the Combined Boring and Lining TBM Method a reality.

Figure 4 is a photograph of the automatic shotcreting robot installed to the TBM. Figure 5 is a flowchart for the automatic shotcreting process. The system has the following characteristics:

(1) The shotcreting robot is located directly behind the TBM tail and comprises a rotation ring mounted on a belt conveyor and a shotcreting nozzle. The robot is capable of self traveling and rotation, and the swinging of the nozzle. They can be activated with a touch on a button.

(2) Research and development of nozzle movement patterns taking account of shotcrete finish has made uniform thickness and smooth shotcreting possible.

(3) The amount of extra spray of shotcrete and shotcrete thickness required are automatically calculated without interrupting excavation work. Thus, highly accurate control has now become possible on a personal computer.

5 DOUBLE-TRACK MUCKING METHOD

5.1 Problems of conventional small section tunnel excavation

In order to construct a small section tunnel expeditiously, it is necessary to remove muck and carry in materials efficiently. Limited working space in a small section tunnel makes it difficult to adopt an efficient belt conveyor method, necessitating the use of muck cars on a single track. In a single track, however, there is interference between mucking and the hauling of equipment and materials, and of efficiency of tunneling declines as the hauling distance increases. Providing turnouts at certain intervals through partial widening is one way to address this problem; however, partial widening of excavation is time consuming and the goal of shortening the construction period may not be attained.

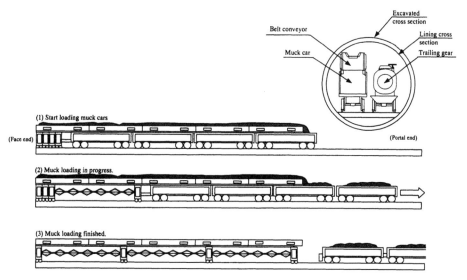

■ Procedure

(1) Connect the muck cars carried in from outside to the belt conveyor carriage in the state described in (3), and move the muck cars toward the TBM while supporting the belt conveyor with the muck cars.

(2) As excavation proceeds, move the muck cars toward the portal and load them with muck, distributing it uniformly among the cars.

(3) Disconnect the muck cars from the belt conveyor and carry the muck cars our of the tunnel for muck removal.

Figure 6. Muck removal method.

5.2 *Newly developed technologies for double-track mucking*

A belt conveyor located in the right or left half of the tunnel is supported by conveyor bogies or small-size muck cars designed for use in a small-section tunnel, located in the other half of the tunnel (i.e. supported temporarily without fixed supports), and muck is loaded into muck cars from the belt conveyor. This method enables the use of a double-track method with concurrent and continuous mucking and materials-and-equipment hauling. The use of the double-track system also facilitates probing ahead of the working face in pilot boring and Tunnel Seismic Prediction (TSP). Appropriate measures can be taken by detecting poor geological conditions in advance and to carry in materials and equipment quickly. Thus, it has now become possible to construct a small-section tunnel safely in a much shorter period. Figure 6 illustrates the mucking method.

6 GROUND EVALUATION SYSTEM USING TBM DATA

6.1 *Problems with ground evaluation by conventional TBM methods*

In the TBM method, the geological conditions at the working face cannot be determined accurately because direct observation is not possible during exca-vation. This can cause problems, such as a delay in mobilization of temporary support materials or in selecting appropriate types of support, that may halt excavation work for a long period of time and hamper speedy tunnel construction, canceling the advantage of speediness of TBM tunneling.

6.2 *Outline of ground evaluation system*

Evaluating the ground conditions ahead of the working face on the basis of TBM machine data, which can be fed back in real time during excavation, was considered. Analysis of actual TBM machine data showed that determination of ground conditions at the working face, which is generally believed to be difficult, can be made in real time by use of excavation energy values (Equation 1) obtainable from TBM data (e.g. thrust, cutter torque).

Selection of appropriate support patterns is made possible by these data and analysis.

$$E = \{(F - F_0) \times L + 2\pi \times (T - T_0) \times N \} / V$$
$$N = n \times L / v \tag{1}$$

where

E : specific energy of excavation (MJ/m^3)
F : thrust (N)
F_0 : initial thrust (N)
T : cutter torque (N·m)

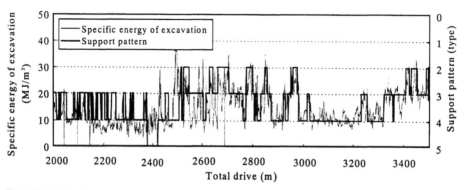

Figure 7. Relationship between specific energy of excavation and support pattern.

T_0 : initial cutter torque (N·m)
N : number of cutterhead rotations
n : speed per minute (rpm)
v : thrust velocity (m/s)
V : excavated volume (m^3)
L : advance per stroke (m)

Figure 7 shows the relationship between excavation energy and support pattern.

7 ACCOMPLISHMENTS OF NEW SYSTEM

The adoption of the Combined Boring and Lining TBM Method, the automatic shotcreting system, the double-track mucking system and the new ground evaluation system based on TBM data has brought about a number of accomplishments in terms of quality improvement, cost reduction, shorter construction period, working environment improvement, and enhanced safety:

Quality: Early completion of linings helped stabilize the ground. A sufficient design shotcrete thickness, a smooth finish of shotcrete surface, and improved roughness coefficients and quality of the walls were also attained.

Cost: Total construction cost was reduced through reduction in extra application of shotcrete, expeditious excavation, and concurrent execution of excavation and lining work.

Construction period: A shorter construction period was achieved through quick support selection, expeditious excavation, and concurrent execution of excavation and lining work.

Safety: Automation of shotcreting has freed workers from work in a dusty environment and from arduous tasks, improved the working conditions and enhanced safety.

8 CONCLUSION

The TBM work reported in this paper started in mid-May of 1999 by two shifts, and was completed in early July, 2000, despite a number of hardships including repeated incidents of considerable water inflow (at a maximum flow rate of 10 t/min), a large amount of sludge accumulated at the excavation bottom, squeeze of the TBM in complex ground conditions on three occasions, falling of rocks from the tunnel crown caused by unexpected heavy water inflows, and continual collapses of ground. After overcoming the complex geological conditions, a headrace tunnel with a total length of 4925 m was driven through in about 13 months while constructing the permanent lining concurrently. In this project, the maximum monthly advance of 785 m was achieved, setting a new record in Japan.

Hydropower is a clean and environment-friendly kind of energy. It is also invaluable as energy produced purely domestically. It is hoped that the achievements accomplished in this project will help build safe and inexpensive small and medium-sized hydropower plants.

Modern Tunneling Science and Technology, Adachi et al (eds), © 2001 Swets & Zeitlinger, ISBN 90 2651 860 9

The use of neurofuzzy modeling for performance prediction of tunnel boring machines

P.A.Bruines

Faculty of Civil Engineering and Geosciences, Department of Applied Earth Sciences, Section of Engineering Geology, Delft University of Technology, The Netherlands. Current address: Département de Génie Civil, Institut de Sols, Roches et Fondations, Laboratoire de Mécanique des Roches, École Polytechnique Fédérale de Lausanne, ISRF-LMR, DGC, EPFL, 1015 Lausanne, Switzerland, email: patrick.bruines@epfl.ch

ABSTRACT: This paper presents the results of a study into the application of neurofuzzy techniques to model performance of tunnel boring machines. It is shown that this method gives better results than other, more conventional modelling approaches. Fuzzy set theory, fuzzy logic and neural network techniques seem very well suited for typical engineering geological applications. In conjunction with statistics and conventional mathematical methods, hybrid models can be developed that may prove a step forward in engineering geology.

1 INTRODUCTION

The excavation industry has a need for good working models describing the performance of excavation equipment, since it determines the total cost to complete a project. However modelling rock excavation machine performance of tunnel boring machines (TBMs), as well as roadheaders, dredgers, bulldozers, rippers and trenchers is difficult, since the processes involved are complex.

This paper presents an alternative modelling approach to assist the prediction of complex real-world engineering geological problems. The principal constituents of the modelling approach are fuzzy sets, fuzzy logic, approximate reasoning, neural networks and data clustering. They are combined into a so-called hybrid intelligent synergism. We focus on how to construct a model from data, how to use this framework to interpret the results and who to assess its reliability. We present in this paper the application of the modelling approach to the performance prediction of TBMs in terms of penetration rate (PR). For that purpose a database consisting of about 640 TBM projects worldwide was used (Courtesy of Texas University & MIT, USA). In the database, information on TBM performance is recorded and presented at different levels of information detail (Nelson et al. 1994). In this study we used data set levels 1 and 2 only.

To assess the significance of the findings, and to show the added value of the approach, the results are compared with empirically developed formulas and conventional statistical methods (e.g. multi-linear regression analysis). Moreover, the results are validated using the data and theoretical principles.

This paper is organized as follows: First a general overview of the neurofuzzy modelling method adopted in this study is presented. Next, the modelling approach is applied to real world cases within the framework of TBM performance. Finally, conclusions are given and further directions are pointed out. In addition, we briefly mention other possible areas of applications of this type of modelling strategy in the field of Engineering Geology.

2 NEUROFUZZY MODELLING APPROACH - AN OVERVIEW

Neurofuzzy modelling is an emerging computational tool, which combines fuzzy logic (FL) and neural network (NN) methods. NNs recognize patterns and adapt themselves to cope with changing environments, while FL allows to incorporate human knowledge, to deal with uncertainty and imprecision, and to perform inference and decision making. Several different ways exist to construct neurofuzzy models (see Nauk et al. 1997). The modelling approach adopted in this study is shown in figure 1. It consists of a logical and a numerical component.

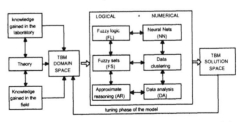

Figure 1. Neurofuzzy modelling approach adopted in this study.

The logical component comprises fuzzy sets, fuzzy logic and approximate reasoning; the numerical component comprises NN, data clustering and data analysis. We will briefly explain why we have adopted this modelling strategy.

During the operations of rock excavation machines many factors are involved (Alvarez Grima 2000, Bruines 1998, Verhoef 1997, Deketh 1995). Among these factors are: rock mass characteristics, intact rock properties, geological setting, machine characteristics, operator skills and expert knowledge. The interaction between machines and rock masses is dynamic, uncertain, complex, nonlinear, and ill defined. In most cases rock masses behave anisotropic, nonlinear and discontinuous. Under these circumstances it is difficult to model the machine performance using 'purely' mathematical modelling techniques based entirely on first principles (i.e. white box models) or using statistics or probability methods solely.

A discussion about the disadvantages of conventional methods frequently used in Engineering Geology such as statistics and probability is beyond the scope of this paper. Suffice to mention here that statistical models are not robust enough to describe complex, multivariable and nonlinear systems accurately (Alvarez Grima 2000, Verhoef 1997). Furthermore, their generalization capability is poor in the presence of outliers and extreme values in the data. Probability theory can deal with random events but is limited when vague and fuzzy concepts as used in our daily language, have to be represented. The solution of complex real-world engineering geological problems therefore demands an alternative modelling strategy. It should be noted that we do not ignore conventional methods; we rather seek for a proper combination to approach the overall solution of the problem at hand. In the following, the logical and numerical components of the modelling approach adopted in this study are explained.

2.1 The logical component

The logical component consists of fuzzy sets, fuzzy logic and approximate reasoning (Figure 1).

2.1.1 Fuzzy sets
Fuzzy sets were proposed by Zadeh (1965) in order to mathematically represent, linguistic expressions, and fuzzy and vague concepts (non-probabilistic in nature). Fuzzy sets can be seen as an extension of the classical set theory, such that the membership of an element ranges over the unit interval [0, 1]. The ultimate goal of fuzzy sets is to mimic human approximate reasoning.

2.1.2 Fuzzy logic and approximate reasoning
Fuzzy logic can be considered an extension of the multi-valued logic developed by Lukasiewicz in 1930. An important concept in fuzzy logic is a fuzzy proposition. Fuzzy propositions are statements that possess fuzzy variables. Fuzzy propositions allow describing input-output relationships via fuzzy If-Then rules. Fuzzy If-Then rules can take, for example, the following form: 'If the strength of the rock is high Then the PR is low', where the terms high and low can be represented by fuzzy sets (Zadeh 1965, Zadeh 1979).

2.2 The numerical component

The numerical component consists of NN, data clustering and data analysis (Figure 1). Basic knowledge of data analysis is assumed.

2.2.1 Neural networks
NNs are mathematical models inspired by the biological structure and functioning of the brain. NN models consist of elementary processing units called neurons. The neurons are interconnected in a predefined topology called layers. Usually the neurons operate in parallel layers. A typical NN topology consists of the input layer, one or more hidden layers and the output layer (e.g. Figure 6). On the basis of the type of learning, NN are classified into a supervised mode and an unsupervised mode. The supervised learning mode is used in this study. The major advantage of NN is the ability to learn, recall, and generalize from training data by assigning or adjusting the connection weights. The major disadvantage of a NN is, that it is usually difficult to explain via its connection weights, what exactly the NN has learned (Menhrotra et al. 1997).

2.2.2 Data clustering
The ultimate goal of data clustering is to partition the data into similar subgroups by employing some similarity measures (e.g. the Euclidean distance). In this paper, data clustering is used to derive membership functions from measured data (i.e. knowledge discovery), which in turn determines the number of fuzzy If-Then rules in the model (i.e. rule induction). Several clustering methods exist. The method employed in this study is the subtractive clustering method (Chiu 1994). For more details concerning the use of data clustering within the framework of fuzzy model identification the readers is referred to Babuška (1996).

3 APPLICATION

This section presents the application of the neurofuzzy modelling approach to predict the PR of TBMs. For a more detailed description of this study see Bruines (1998) and Alvarez Grima et al.(2000).

Figure 2 depicts the modelling strategy used to construct the PR model using neurofuzzy tech-

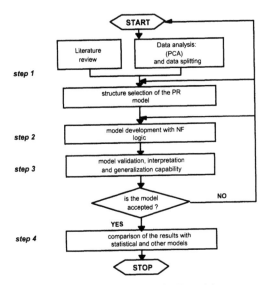

Figure 2. Modelling strategy used for the PR model.

niques. The data set used, corresponds to data from level 1 and 2 from Nelson et al. (1994).

3.1 Structure selection of the model and data reduction

The selection of the structure of the model was based on literature study and a thorough analysis of the available data. The literature study shows that the most relevant factors influencing the PR can be grouped into rock mass properties, machine characteristics and tunnel geometry (Figure 3).

3.1.1 Rock mass properties
Rock mass properties are determined by the intact rock and the discontinuity structure of the rock mass. The most influential intact rock parameter in the database is the unconfined compressive strength (UCS). Higher rock strength leads to lower PR (Figure 4).

Discontinuities weaken the rock mass, which gives rise to a higher PR. The data set includes data on core fracture frequency that can easily be related to a RQD value. Other factors such as discontinuity orientation and degree of weathering might also be of importance on the overall TBM performance.

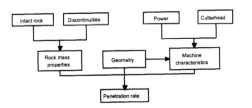

Figure 3. Main factors influencing the PR (Bruines 1998).

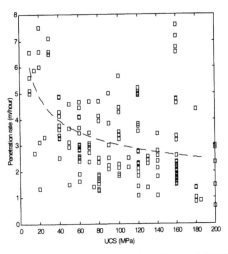

Figure 4. Penetration rate versus the rock strength (UCS).

This information, however, is not provided in the available data sets.

3.1.2 Machine characteristics
The most important machine characteristic is its propulsion. The number of cutters mounted on the TBM allows the calculation of the maximal thrust per cutter, a parameter used in most existing models. Increasing the thrust per cutter has the most advantageous effect on the PR. Also related to the propulsion are power, torque and RPM (revolutions per minute). These parameters however also depend on the diameter of the tunnel (Figure 5).

Another important parameter is the type of cutters used. Larger and stronger cutters allow for more normal pressure to be applied and more RPM.

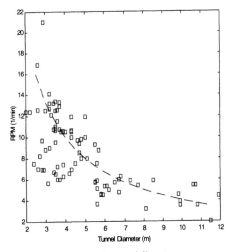

Figure 5. RPM versus the tunnel diameter.

3.1.3 Geometry

The geometry of the tunnel resulted to be a very important parameter. Many parameters like RPM, torque and total power consumption are influenced by the tunnel diameter, as is the stability of the rock mass. In general it can be said that the PR per length decreases with increasing diameter.

3.1.4 Result of literature review and data analysis

After careful evaluation of the literature and the analysis of the data available to us for this study, the following input parameters were selected for further modelling:

- Unconfined compressive strength (UCS),
- Core fracture frequency (CFF),
- Tunnel diameter,
- Hole-through year,
- Thrust per cutter,
- Revolutions per minute (RPM),
- Torque,
- Diameter of disc-cutters.

In Bruines (1998) the same strategy was used to analyze the parameters affecting advance rate, wear rate and utilization. Utilization is the percentage of the shift time the machine is effectively used.

The literature review also revealed a multitude of existing models to assess TBM performance (Nelson 1994, Bruines 1998), however only a single model was compatible with the database.

3.1.5 Data reduction using principle component analysis (PCA)

To further reduce the number of input variables, PCA was used. Although the data are not linearly related, the PCA solution gave interesting results. The most important result being that the input variables can be subdivided into machine characteristics, tunnel dimension and rock mass properties when three components are chosen (Table 1). A further split in UCS and CFF, and cutter technology and thrust per cutter occurs when four respectively five components are used. This agrees with engineering expectations and the results of the literature survey (Figure 3).

Table 1. Rotated component matrix with three components (Variance accounted for in %).

Parameters	Factor 1 (29.9 %)	Factor 2 (29.3 %)	Factor 3 (16.2 %)
CFF	.213	-.134	-.774
UCS	.256	.133	.748
Tunnel diameter	.074	.947	.198
Year of completion	.827	.012	-.158
Thrust per cutter	.803	-.078	.129
RPM	.477	-.759	.055
Torque	.212	.906	.216
Cutter diameter	.823	.098	.062

Principal components were rotated using the Varimax method.

Table 2. Two passes in the hybrid learning procedure for ANFIS (Jang, 1993).

	Forward pass	Backward pass
Premise parameters	Fixed	Gradient descent
Consequent parameters	Least-squares estimator	Fixed
Signal	Node output	Error signals

3.2 Neurofuzzy modelling: Takagi - Sugeno method

The neurofuzzy method used to construct the PR model evolved from the classical Takagi-Sugeno (TS) method (Takagi & Sugeno 1985). The method combines the adaptive network-based fuzzy inference system, ANFIS (Jang 1993), least square estimation and subtractive clustering (Chiu, 1994). Without loss of generality, a TS fuzzy rule takes the following form: If x is A and y is B then z = f (x, y); where A and B are fuzzy sets, or more specifically, membership functions in the antecedent of the rule (i.e. If part) and z is a crisp function in the consequent of the rule (i.e. Then part). In our case the consequent part of the fuzzy rule is a first-order polynomial of the input variables, called a first-order TS fuzzy model.

Table 2 summarizes the hybrid learning process for ANFIS. In the forward pass the consequent parameters are identified by least square estimate, while in the backward pass, the error rates propagate backward and the premise parameters (i.e. membership functions) are updated by means of a gradient-based optimization method. The optimization method employed in the original ANFIS algorithm is the so-called back propagation algorithm. Note that other learning methods exist and can also be applied. Figure 6 shows the ANFIS structure of the final PR model.

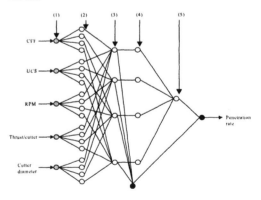

Figure 6. ANFIS structure for the PR model with five input parameters (CFF, UCS, RPM, thrust per cutter, cutter diameter) and four rules. The numbers (1)-(5) at the top indicate the layers number. (1) Inputs, (2) input membership function, (3) rules, (4) output membership function, (5) weighted sum output, and (6) output. The black dots represent a normalization factor for the rules.

With the eight initial input variables selected, different combinations were attempted. The groups found with PCA were used as a starting point in the modelling exercise. The final model was obtained by using five parameters, being CFF, UCS, RPM, thrust per cutter, and cutter diameter (Figure 6).

3.3 Results, interpretation, validation and generalization capabilities

To validate the model and to check its generalization capability ten different checking sets (filters) were used. The model using CFF, UCS, RPM, thrust per cutter, and cutter diameter not only has a low error in the training set, but also yielded good results for the checking set. This indicates the good generalization capability of the neurofuzzy model approach. Four rules were used in the final model. This model has the best interpretability and best validity over the solution space. Another model that included the year of completion gave good numerical results but was not valid outside the range of years included in the data, making such a model useless for future predictions.

Figure 7 graphically shows the TS fuzzy rules and the corresponding membership functions extracted from the data using the clustering algorithm and optimized with the gradient descent method. Figure 8 shows a surface view of the PR model. Note that the PR decreases as the UCS increases and that the PR increases as RPM increases. The results found, are in agreement with the literature study and the data analysis. This indicates the excellent identification capability of the ANFIS algorithm.

Figure 9 shows the model output evaluated with training and checking data. The average modelling error was 0.8884 for the training set and 0.8999 for

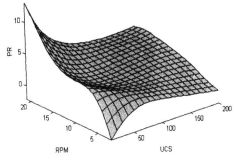

Figure 8 Surface plot describing the relationship between the PR, the UCS and RPM.

the checking set, and indicates that the model generalizes well.

3.4 Comparison of different methods

To show the added value of the neurofuzzy approach, the results were compared with two models developed using multi-linear regression analysis, an empirical model developed by Graham and an adapted Graham model (Bruines 1998). The Graham models use thrust per cutter, UCS and RPM. One multi-linear model is developed using stepwise regression analysis and includes RPM, UCS and thrust per cutter as independent variables. A second model developed with a backward regression analysis includes RPM, UCS, thrust per cutter, tunnel diameter and torque.

Table 3 lists training and checking error for the different models. The lowest errors were obtained for the neurofuzzy model.

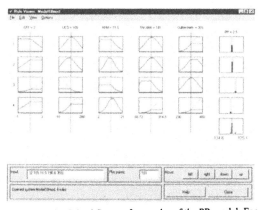

Figure 7. The Takagi-Sugeno fuzzy rules of the PR model. For a certain set of rock and TBM parameters each rule leads to a PR calculated by the polynomials. The average and range of the calculated values indicates the PR predicted by the model.

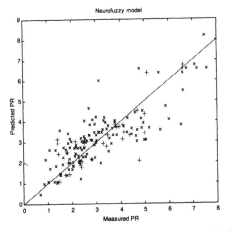

Figure 9. Plot of the measured and calculated PR with the neurofuzzy model. Training data (x), checking data (+).

Table 3. Different models and their performance as described by the root mean square error (RMSE).

Models	RMSE (trn)	RMSE (chk)
MLR: stepwise	1.0724	1.8098
MLR: backward	1.0569	1.7845
Graham	*	11.828
Graham (>80MPa)	*	1.7123
Adj. Graham (>80MPa)	1.2737	1.4532
Neurofuzzy	0.8884	0.8999

* Existing model so no training and all data was used for checking

4 CONCLUSIONS AND FURTHER DIRECTIONS

This paper presented the results of the modelling of mechanical rock excavation by TBM, using the MIT-University of Texas database by neurofuzzy methods. The models were developed by a combination of general knowledge of mechanical rock excavation, careful data-analysis and selection of the appropriate computational modelling technique. The model thus developed performs better than traditional statistical methods based on regression analysis. With the modelling methods used here, it is possible to make better use of the often hard to get geological, geotechnical and machine performance data gathered during projects.

In a next step of this research project we would like to develop models using the data sets of levels 3 and 4 of the MIT-Texas University database and to improve the models presented here. These data sets have a more detailed and complete information of the rock mass and the machine performance.

The importance of the neurofuzzy modelling techniques available for the Engineering Geology discipline is obvious. It allows using also vague and imprecise (fuzzy) information on the rock or soil present in the subsurface and allows us to use large amounts of data of which the physical meaning is not obvious (e.g. rebound test data on rock cores, or geophysical well logging parameters). By using the techniques described, relationships of such data with geotechnical significant information can be established and used. Perhaps the most interesting feature of this approach is that we can cope scientifically with subjectivity and uncertainty in the knowledge domain, rather that blindly avoiding them.

We would like to encourage others in the field of applied earth sciences to start using the computational methods as described in this paper. The advantages are numerous: uncertainty, non-linearity and expert opinion are much better dealt with. Other promising fields where this type of modelling approach might be of great help are: decision making, mapping, rock mass classification systems, underground construction, monitoring, hydrogeology, laboratory experiments and so forth.

5 ACKNOWLEDGEMENTS

I would like to thank Dr. Mario Alvarez Grima and Dr. P.N.W. Verhoef from the Section of Engineering Geology at the TU Delft for their support and advise. The University of Texas and MIT are acknowledged for providing the data sets.

REFERENCES

Alvarez Grima, M. 2000. *Neuro-fuzzy modeling in engineering geology*, Balkema, Rotterdam, 244 pp.

Alvarez Grima, M., P.A. Bruines & P.N.W. Verhoef 2000. Modeling tunnel boring machine performance by neuro-fuzzy methods, *Tunneling and underground space technology*, 15 (3), pp. 259-269.

Babuška, R. 1996. *Fuzzy Modelling and Identification*. PhD dissertation, Delft University of Technology, Delft, The Netherlands. ISBN 90-9010153-5, 294pp.

Bruines, P. 1998. *Neurofuzzy modelling of TBM performance with emphasis on the PR*, Memoirs of the Centre of Engineering Geology in The Netherlands no. 173, Delft, 202 pp.

Chiu, S. 1994. Fuzzy model identification based on cluster estimation, *Journal of intelligent & fuzzy systems*, Vol. 2, No. 3, Sept. 1994.

Deketh, H. J. R. 1995. *Wear of rock cutting tools. Laboratory experiments on the abrasivity of rock*. Balkema, Rotterdam, 144 pp.

Jang, J.S.R. 1993. ANFIS: Adaptive Network-based Fuzzy Inference System. *IEEE Transactions on Systems, Man, and Cybernetics*, Vol. 23, pp.665-685.

Menhrotra, K, C. K. Mohan. & S. Ranka. 1997. *Elements of artificial neural networks*. The MIT Press. 344 pp.

Nauck, D, F. Klawonn. & R. Kruse. 1997. *Foundations of neuro-fuzzy systems*. John Wiley & Sons. 305 pp.

Nelson, P.P, A. A. Yousof. P.E. Laughton. 1994. *Tunnel Boring Machine project data. Bases and construction simulation*. Geotechnical Engineering Report GR94-4, Geotechnical Engineering Center, Department Civil Engineering, The University of Texas at Austin, 78712.

Tagaki, T. & M. Sugeno 1985. Fuzzy identification of systems and ist application to modelling and control. *IEEE Transactions on systems, man, and cybernetics*, vol SMC-15, no. 1, pp. 116-132.

Verhoef, P.N.W. 1997. *Wear of Rock Cutting Tools: Implications for the site investigation of rock dredging projects*. Balkema, Rotterdam, 327 pp.

Zadeh, L. A. 1965. Fuzzy sets. *IEEE Information and Control*, Vol. 8: 338-353

Zadeh, L. A. 1979. A Theory of approximate reasoning. in *Machine Intelligence* 9, ed. by J. E. Hayes, D. Michie and L. I. Mikulich, New York: John Wiley & Sons.

Modern Tunneling Science and Technology, Adachi et al (eds), © 2001 Swets & Zeitlinger, ISBN 90 2651 860 9

Construction of Kamosaka Tunnel by the NARAI Excavation System

H.Haga, N.Takahashi, &S.Morishima
Kamosaka Tunneling Site Office of Kumagai Gumi Co.Ltd, Yamagata, Japan

H.Kamiyama
Kumagai Gumi Co.Ltd, Tokyo, Japan

ABSTRACT: The "NARAI excavation system" has been developed, that performs optimum excavation with minimized overbreak. The system combines excavator position/posture measurement instruments with a cutter boom numerical control function. The cutter drum position relative to the preset cutting section is recognized automatically and in real time, and displayed on a screen. The control system outputs signals to the boom hydraulic circuit to prevent the cutter drum from exceeding the preset section. This function enables highly accurate cutting, regardless of operator's skill. Long period continuous operation of this system was implemented for the first time, resulting in satisfactory excavation. This paper discusses the overview of the system and application results.

1 INTRODUCTION

Mechanical excavation in NATM tunneling has most critical elements in the work process. Since its accuracy which influences the profitability of work depends upon operator's skill, it gives operators extreme strains both mentally and physically. The mechanical excavation using a partial cutting machine, which is highly efficient, cannot free them from poor work environment with vibration, noise and dust. It inevitably leads to lack of skilled operators, and to the necessity for a system ensuring the excavation accuracy regardless of operator's skill.

As a "streamlined system" for tunnel excavation, this system has been designed to provide a satisfactory finish precision ensuring optimized excavation with minimized overbreak, and systematic operation that is not affected by the operator's skill level. The system integrates an excavator position/posture measuring system, with cutter boom numerical control, displaying in real time the cutting position against the planned section, thereby controlling the cutting range. Before introduction to an actual work site, the system was tested by the use of simulated bedrock. The test demonstrated its cutting accuracy of ±50 mm as expected. In addition, the system performance was verified in three tunneling projects. In every projects, the system offered a highly accurate working form, though its use spanned a short period. This paper reports the results of its long-term use in the whole length of Kamosaka tunnel project (776.4 m long). (This

system was developed jointly by Chiba National Road Project Office of Kanto Regional Construction Bureau of Ministry of Construction, Advanced Construction Technology Center, Kumagai Gumi, Hazama-Gumi, and Mitsui Miike Machinery)

2 OVERVIEW OF THE SYSTEM

Figure 1 shows the overall view of the system. Figure 2 illustrates the functional flow of the system. The reference position and posture of the excavator unit in the tunnel are detected by two independent automatic follow-up total stations (referred to as "follow-up system") and an inclinometer. On the basis of the measured values of these instruments, the posture of the cutter boom (referred to as "boom") is converted from the in-machine relative coordinates to external absolute coordinates. From the magnitude of each movement of the boom, the cutting drum position relative to the preset cutting section is recognized automatically and in real time. The recognized results are displayed on the screen of the control system. When the cutting drum is on the verge of exceeding the preset cutting section (or just enters the "overbreak area"), a signal is sent to the hydraulic circuit to automatically stop the boom, and operation of the boom into the preset cutting section only becomes available. With this control function, no high level skill of operator is required for ensuring accurate excavation conforming to the preset cut section. In addition, this system can be operated like a conventional machine to achieve accurate cutting precisely following the planned excavation section.

The grounds of the mountain tunneling sites in Japan are extremely diversified not only in rock properties, but also in stratification, with cracks, and so on. Consequently, with standardized automatic control of the partial cutting machine boom, it is impossible to achieve effective excavation well suited to the ground conditions encountered. Keeping this point in mind, the NARAI system was designed around the following concept ; the excavation inside the cutting section is carried out by comprehensive judgment of the operator as in the conventional practice, whereas the control system assists the operator's judgment in cutting the circumference of the section.

The Kamosaka tunnel project was designed with steel support patterns to be installed over the whole length of the tunnel. So, a laser marking function was added to one of the follow-up devices of overbreak limiting system, which provided the reference for erecting the steel supports. The system was modified so that change-over between automatic follow-up for NARAI system and laser marking can

be done in the tunnel station. This modification aimed at integration of multifunction and cost reduction, which greatly contributed to rational tunneling work.

3 WORK RESULTS

The ground to be excavated of the Kamosaka tunneling site is mainly composed of mudstone and tuff of Neogene. Its strength is in general 20 to 30 MPa. Except for the tunnel entrance (starting point), cracking is generally not so remarkable, so the ground was relatively suitable for mechanical cutting. As the tunnel section is relatively large, with two lanes and sidewalk (3 m wide), the length of one excavation cycle is 1.2 m at maximum with CII pattern, and steel supports (H-150, H-200) were installed over the whole length of tunnel.

We will discuss below the work results of the Kamosaka tunnel project, from the viewpoint of

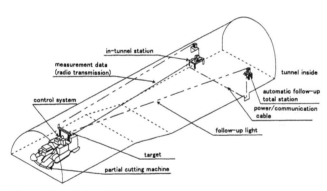

Figure 1. Overall view of the system.

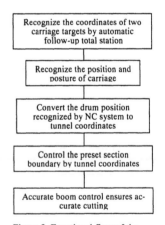

Figure 2. Functional flow of the system.

Figure 3. View of cutting.

Figure 4. Monitor display.

both excavation accuracy and system operation status.

(I) Excavation accuracy

Figure 5 and 6 show the data of the working form with CII pattern (excavation cycle length 1.2 m with steel support) by manual cutting and system-controlled cutting. These data were taken at locations with minimum cracks. The average error is 3.8 cm with system-controlled cutting, whereas 6.1 cm with manual cutting. The standard deviation is 2.2 cm with system-controlled cutting, whereas 3.2 cm with manual cutting. These results demonstrate the remarkable excellence of system-controlled cutting both for average and variance. For ensuring the specified dimensions of the working form, the fabricated steel supports were 5 cm larger than the designed size, to cope with the settlement, convergence and work error, including also erection margin and overbreak. As for the accuracy at different positions of tunnel section, the system-controlled cutting shows little variance in any points, whereas with the manual cutting, the cutting tends to be excessive near the crown, because it is difficult for the operator to visually confirm this location. The cutting accuracy was confirmed by the actual tunneling work, that was more excellent than cutting practice using, as work reference, steel supports installed at 1.2 m intervals. It greatly contributed to reduction of overbreak, consumption of spraying material and lining concrete. This has a great significance for further development of the system.

(II) System operation status

A length more than 700 m was excavated by a partial cutting machine in the Kamosaka tunnel. The NARAI system worked continuously in such a middle scale tunnel, and inconveniences with the system were made clear, for which an effective improvement was made. These are fruits of a great significance. Furthermore, the vibration and dust proof devices were improved for avoiding adverse influence of vibration and dust produced in cutting operation, and almost all troubles with the software were also removed. Problems to be tackled for further development are reduction of time when the control capability is not available because of "target lost" (the automatic follow-up device fails to recognize the target), and saving of labor for rearrangement of the system (move toward the face).

(1) Target lost

When hard rock in the face was encountered, dense dust was produced, causing bad visibility which led to frequent target lost. It is impossible for the current equipment alone to keep the dust density low only by water sprinkling from the drum of the partial cutting machine. A solution may be the use of a compact large capacity dust collector near the dust source. This is a vital problem from the viewpoint of work environment too.

In usual practice of mechanical cutting of road tunnel, by top heading short bench method, the upper and lower halves are frequently cut at a time. Also in the site of Kamosaka tunneling, the upper and lower halves were simultaneously excavated. Consequently, the construction machine for the lower half cutting frequently obstructed the follow-up function. In addition, when mucking was started in the latter half of upper section cutting, the mucking machine (side dump type wheel loader) often shut off the follow-up function. The position of the excavator itself is not changed so often. However, if

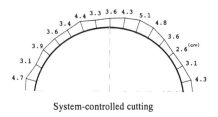

System-controlled cutting

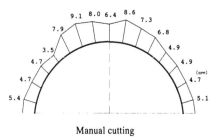

Manual cutting

Figure 5. Working form section.

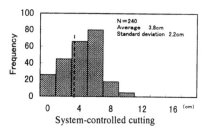

System-controlled cutting

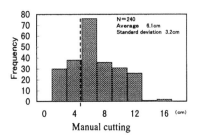

Manual cutting

Figure 6. Histogram of working form difference.

its position is moved even slightly under the effect of cutting vibration, influence is inevitably exerted upon the control capability. Possible solutions are ingenious selection of installation position of the automatic follow-up system that can minimize the risk of target lost, and a maximized distance of muck loading from the excavator, with a relatively upper section bottom. Another way for shortening the loss time is to move the two targets toward back and forth respectively, at a higher level on the excavator.

(2)   Labor saving of the rearrangement of the follow-up system

Labor of two men-one day is necessary to rearrange the NARAI system, moving it toward the face. It must be done on a holiday, so further labor saving of displacement of the system will be required. By the use of more lightweight and integrated machines in the displacement work, the rearrangement must be carried out easily. In addition, remote control function shall be added to the follow-up system, by which maintenance can be done without aerial work vehicle.

4 CONCLUSIONS

At the initial stage of introduction of the NARAI system, because of the laser marking function integrated as an additional function, troubles occurred in the software for shifting from follow-up to laser marking. Other troubles occurred due to aging of the machine which had been used from the initial development stage. Through improvements of these points, we have achieved a highly reliable integrated system.

The merits of the use of NARAI system is indeed more remarkable in the following cases:
- in lots without steel supports than lots with them,
- as the excavation cycle becomes longer,
- as the distance from the operator's eyes to the cutting surface becomes longer, and
- in full face cutting than in top heading short bench cutting.

We hope, on the basis of the accumulated know-how, to construct a two-lane road tunnel by full face cutting, using the NARAI system with a one-class larger partial cutting machine.

Modern Tunneling Science and Technology, Adachi et al (eds), © 2001 Swets & Zeitlinger, ISBN 90 2651 860 9

Development of low noise and vibration tunneling methods using slot by single hole continuous drilling

T.Noma & T.Tsuchiya
FUJITA CORPORATION, Tokyo, Japan

ABSTRACT: The authors have newly developed a continuous hole drilling method as a rock fracturing method in hard rock tunnel excavation and as a free face forming method to be used at the time of controlled blasting excavation. Here, they overview the continuous hole drilling method and explain the tunnel drilling by the rock fracturing method and the controlled blasting by using this method.

1 INTRODUCTION

It goes without saying that the adoption of the blasting method is most efficient and very economical for drilling bedrock. However, the blasting method brings about large noises and vibrations and would not be suitable for excavating tunnels in suburbs of cities. Despite, examples of this excavation have increased in Japan.

Machine drilling of hard rock is classified into two kinds, i.e., single machine drilling and rock fracturing method. The single machine drilling has a few problems: drilling in short distance is not economical and compressive strength for drilling is limited. So, the rock fracturing method is considered most suitable for machine drilling of hard rock in short distance at portal and so on. For the rock fracturing method, free face is artificially formed at working face. Then, the free face is used to crush bedrock.

Though various free face forming methods were proposed (Hagimori et al, 1991), they have such problems as necessity of special forming device, forming efficiency and accuracy in continuity. So, the authors have developed the method to use the general-purpose drill jumbo more efficiently and higher in continuity than conventional methods. In addition, this method is effective for reducing vibrations even in controlled blasting if free face is formed on the periphery of tunnel.

2 SIGNIFICANCE OF FREE FACE

A free face means a face that does not restrict rupture at the time of crushing. When a tunnel is excavated by blasting, only one free face is formed at the working face. In case of blasting method, the crushing force is strong and excavation has become possible even with a free face as the center cut blasting has been devised.

On the other hand, the crushing capacity of the rock fracturing method is less powerful than blasting, and crushing with one free face is impossible. In the other words, though cracks are caused at the working face by some method, it is difficult to cause large cracks when only one free face is formed. Two or more free faces enable to secure bedrock-moving spaces and to make cracks larger. So, it is necessary to artificially form free faces at the working faces of tunnels.

Figure1 shows the conceptual diagram of free face formation in excavating tunnel. When a continuous hole is drilled like a groove as shown in the Figurer, the working face is divided into blocks. This brings about multiple free faces and makes crushing bedrocks easily. When natural ground becomes hard, it is difficult to crack it and many free faces should be formed. Continuity and widths of free faces to be formed at the time influence rock fracturing efficiency. If there remain rock bridges that are shown in the Figurer and that interrupt continuity of free faces, a large compressive force is necessary for cracking. Also, the wider free faces to be formed are, the larger bedrock moving spaces can be secured and the easier rocks come to be cut.

However, it takes considerable labor and time to form free faces on hard rocks. So, it is very important for excavating tunnels by the rock fracturing method to form continuous free faces efficiently and economically. As mentioned above, the usual method requires a special forming device and has some problems in forming efficiency and accuracy of continuity. To solve these problems, they have

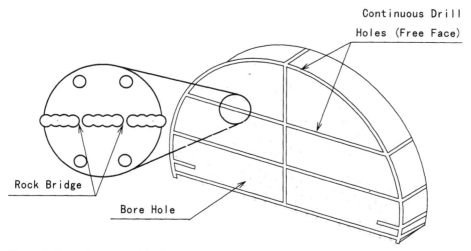

Figure 1. General concept of the free face formation at the tunnel.

developed the method to drill single holes more efficiently and continuously than by the conventional method.

3 OVERVIEW OF SINGLE HOLE CONTINUOUS DRILLING METHOD

As mentioned above, they paid special attention to the following points in developing the new free face forming method: (1) Use of general purpose drill jumbo, (2) Superiority in free face continuity and (3) High free face forming efficiency. From this point of view, they have developed the single hole continuous drilling method.

When a single hole is drilled continuously, the bit always turns to an existing hole in the vicinity. In view of this property, an SAB (Spinning Antibed) rod in the existing hole so that a free face is formed when the bit contacts or strikes the SAB rod in drilling a continuous hole. When the bit contacts or strikes the SAB rod, there is no clearance between the bit and the SAB rod, and this secures continuity of free face. Figure 2 shows the concept of the SAB rod, and Figure 3 shows the free face forming procedures. The procedures are as follows:

① Drill a hole in which the SAB rod is inserted and then insert it in the hole.

② Start drilling a continuous hole. The bit drills a hole while contacting or striking the SAB rod. The SAB rotates as the bit rotates, and this makes high speed drilling possible.

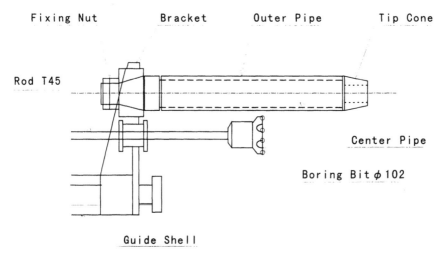

Figure 2. Concept of SAB rod.

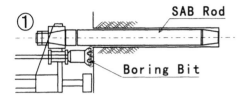

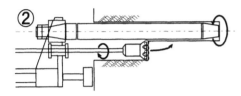

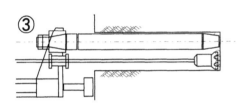

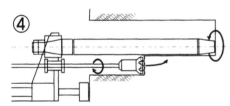

Figure 3. Procedure of continuous hole drilling.

③ Keep drilling to the specified position.
④ Insert SAB rods one after another and keep drilling to form a continuous hole (free face).

Since SAB rods can rotate, both the rods and the bit come to be worn less, and the SAB rods are worn equally. So, they can be used for a long time. The SAB rod is fitted to the bracket on the tip of the guide cell, and the guide cell is slid to insert and pull out the SAB rod. The continuous hole drilling capacity allows drilling hard granite 1.1m deep with a bit 102mm across at the speed of 3.5~4.0m²/h. Figure 4 shows relations between bedrock strength and the continuous hole drilling capacity.

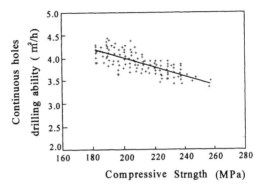

Figure 4. Relationship between continuous hole drilling ability and rock strength.

Figure 5. Continuous hole drilling.

Features of this method are easy mounting and detaching of SAB rod as well as high drilling capacity, and when a machine can drill holes by rock fracturing and fix rock bolts, the number of machines used in tunnel can be reduced. Also, since the consumables of the SAB rod are thick pipes only, this method is very economical.

Figure 5 shows the situation of continuous hole drilling.

4 APPLICATION TO ROCK FRACTURING METHOD

The Kaminiko Tunnel in Hiroshima Prefecture is a 550m long tow-lane road tunnel. This tunnel was restricted as follows:
①There were many houses around the construction site.
② There were many boulders along the tunnel route, they were supposed to fall down due to blasting vibration even if protection measures were provided.

Therefore, the blasting method could not be adopted along the whole tunnel route. In addition, the stratum along the tunnel route was granite of

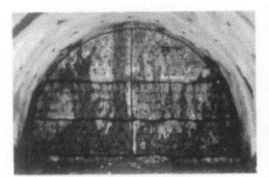

Figure 6. Tunnel face after forming free face.

Figure8. Secondary fracturing.

compressive strength of more than 100MPa, and single machine excavation with a free section excavator was not possible. Therefore, free faces were formed on the tunnel working face, and the free faces were used to excavate the tunnel by the rock fracturing method.

As mentioned above, the stronger the bedrock is, the more difficult it is to drill the bedrock, and the number of free faces should be increased. Therefore, they prepared the free face forming pattern for the strength for the construction. Figure 6 shows the free face forming patter for the hardest working face of the compressive strengths of 200~250MPa. 2 each of free faces were formed in the peripheral, 1 free face was formed in the vertical direction and 2 free faces were formed at the both corners. For this tunnel, 1 progress head was 1m, the depth of formed free face was 1.1m and the total free face forming area was about 88m2. Though the drill jumbo used for the construction was of the 3-boom drifter type weighing 150kgf, the free face forming efficiency was very high and the forming time per 1 progress was less then 6 hours.

After the continuous hole drilling, the hydraulic wedge shown in Figure 7 was used for rock fracturing. The hydraulic wedge has the following features:① High crushing capacity, ② Causing wide cracks and ③Control of crack causing direction.

After these processes, breaker was used for the secondary crushing to end the rock fracturing. Figure 8 shows the secondary fracturing situation.

5 APPLICATION TO CONTROLLED BLASTING

In case of Yasumiyama Tunnel, the portal was located in the center of a city and there were many houses around. As the tunnel was excavated, the bedrock became harder, and there spread granite of the uniaxial compressive strength of more than 150MPa beyond 500m from the portal. Therefore, the single machine excavation with a a partial face machine became impossible for the construction. They examined if the blasting method was applicable, but there were houses 70m above the construction site. Also, they had to consider the environment around the portal.

There, they tested various controlled blasting, compared and examined them to find the most applicable controlled blasting method for the following process. Concretely speaking, they examined control with peripheral free face, control with electronic detonator and combination of the both. They conducted usual blasting by the above methods and with ordinary detonator, measured vibrations and noises at the times of blasting, compared their control effects and examined economy of these methods.

Table 1 summarizes the tests conducted this time, Figure 9 shows various excavation patterns except division blasting, and Table 2 shows vibration and noise measurements and measuring position. Table 3 shows the results of noise and vibration measurements. They evaluated these methods as follows from the results of these tests:

①The measured noise values show that the controlled blasting with free face and electronic detonator caused the least noise as they expected and that controlled blasting with electronic detonator,

Figure 7. Rock fracturing by hydraulic wedge.

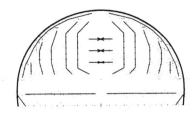

Normal blasting

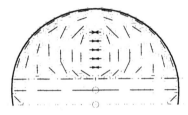

Electric detonator

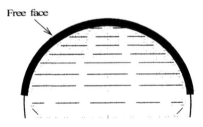

Free face + normal or electric

Figure 9. Drilling pattern for each test.

Table 1. Testing pattern of each method.

Testing method	Detonator	Free face
Electric detonator only	Electric	×
Electric detonator + free face	Electric	○
Normal detonator + free face	Normal	○
Division	Normal	×
Normal blasting	Normal	×

Table 2. Monitoring site, monitoring item and distance from tunnel face.

Monitoring site	Distance from tunnel face (m)	Monitoring item
House above tunnel face	Just upsaide 70	Displacement speed Vibration level meter Noise meter
	Forward 100	Displacement speed Vibration level meter
	Backward 100	Displacement speed Vibration level meter
Portal	500	Noise meter Low frequency noise meter
House beside portal	From portal about 40	Low frequency noise meter

controlled blasting with free face and ordinary detonator and division blasting caused noises at the similar levels.   And the ordinary blasting caused the largest noises.

② In case of ordinary blasting, the vibration value reduced gradually from the peak at the time of center cut.   When electronic detonator was used, the vibration peak value was lower than that of ordinary blasting.   However, the blasting with electronic detonator took more time, and vibration of certain level continued for a longer time (4~5 seconds).   When both free face and ordinary

Table 3. Results of vibration and noise monitoring.

Test pattern	Vibration (dB)			Low frequency noise (dB)		Noise (dB)	
	-100m	Just upside	+100m	Portal	House beside portal	Portal	Just upside
Electric	50	52	54	116	101	77	47
Electric + free face	43	48	47	113	100	75	53
Normal + free face	50	52	53	118	111	79	52
Division (max value)	52	50	53	122	112	79	51
Normal blasting	56	57	57	122	117	82	51

detonator were used, the vibration value came to the peak at the beginning and reduced quickly.

③ For low frequency noise, there were significant difference between electronic detonator and ordinary detonator irrespective of free face, and remarkable noise reduction was recognized in blasting with electronic detonator.

④ For economy, blasting with free face and electric detonator was most expensive, and division blasting, blasting with free face and ordinary detonator, blasting electronic detonator and ordinary blasting followed in this order.

6 CONCLUSION

We have overviewed our new free face forming method and now we will explain features of this method hereunder:

①No special machine is necessary for forming free faces, and the general drill jumbo will do all drilling works.

②Even in case of bedrocks of compressive strength of more than 200MPa, the free face forming capacity of more than $3.5m^2/h$ is available.

③This method is superior in continuity of free faces to be formed and allows efficient rock cutting fracturing and controlled blasting.

In view of these features, we will gather various construction data further to establish this method as the efficient low vibration tunnel excavation method.

REFERENCE
Hagimori, K., Furukawa, K., Nakagawa, K., Yokozeki, Y 1991: Study of non-blasting tunneling by slot drilling method, *Proc. 7th ISRM Int. Congr. on Rock Mechanics*, Aachen, 1001-1004.

Modern Tunneling Science and Technology, Adachi et al (eds), © 2001 Swets & Zeitlinger, ISBN 90 2651 860 9

Experimental Study on Rock Cutting by Use of Actual Size Disc Cutter with Round Tip

H. Takahashi, T. Sato, H. Yamanaka
Dept. of Geoscience and Technology, Graduate School of Engineering
Tohoku University
Sendai 980-8579, JAPAN

K. Kaneko
Dept. of Environment and Resource Engineering, Graduate School of Engineering
Hokkaido University

K. Sugawara
Dept. of Architecture and Civil Engineering, Faculty of Engineering
Kumamoto University

ABSTRACT: TBM(Tunnel Boring Machine) is an useful machine to construct the tunnel. In the excavation by TBM, the rock is broken by the disc cutter installed on the cutting face of TBM. Recently, the round tip disc cutter has been widely used. This paper describes the rock cutting by use of actual size disc cutter with round tip. It was confirmed that hard rock can be broken by the round tip disc cutter. Furthermore, the internal of the rock was visualized by X ray CT equipment. It was observed that the fracture was generated horizontally from the dip made by the rolling of the disc cutter.

1 INTRODUCTION

TBM(Tunnel Boring Machine) is an useful machine to construct the tunnel. It is often said that the uniaxial compressive strength of the rocks for the effective rock cutting by TBM is about 10 MPa – 100MPa (Muro et al. 1997). However, recently TBM needs to cut the rock whose strength is higher than 100MPa. For example, in the construction of Maiko Tunnel in Japan, the hard rock of more than 200MPa was cut by TBM. Therefore, it is considered that TBM in the near future is required to have an ability to cut both soft rocks and hard rocks.

By the way, the performance of TBM strongly depends on the cutting performance of disc cutter installed on the cutting face. Therefore, many studies on rock cutting by the disc cutter have been carried out. However, most of these studies dealt with the small disc cutter of approximately 100 mm i.d. and few works have been done by using the actual size disc cutter. Furthermore, the previous studies reported the cutting performance by wedge type disc cutter. However, the reduction of the cutting performance due to the wear is significant for the wedge type disc cutter. In result, disc cutters have to be replaced frequently. However, Li et al.(1997) reported that the cost of install and replacement of the disc cutter is up to half of total construction cost of the tunnel. Therefore, if the number of install and replacement of the disc cutter can be reduced, it is possible to save the tunnel construction cost. Due to the above-mentioned background, the development of the disc cutter with round tip has been receiving considerable attentions. The round tip disc cutter is strong for the wear, but it is not made clear as yet that the hard rock whose uniaxial compressive strength is more than 200MPa is cut effectively by the round tip disc cutter.

On the other hand, it has already reported that the fractures in the rock affect the excavation rate. This means that if the fractures can be generated effectively in the rock because of the device of the shape and arrangement of the disc cutter, effective rock cutting will be possible. However, it is not made clear how the fractures are generated by the rolling of the disc cutter. Therefore, the purposes of this study are to collect the basic data of the cutting performance by the actual size disc cutter with round tip and to investigate the fracture mechanism by visualizing the internal of the rock by X-ray CT equipment.

2 EXPERIMENTAL APPARATUS AND ROCK SAMPLES

2.1 Experimental Apparatus

For the simplicity of experiments, the trajectory of the disc cutter was limited to the linear rolling. The schematic diagram of the experimental apparatus is shown in Fig.1. The size of the experimental apparatus is 2.3m width, 2.3m length, and 3.0m height. As shown in this figure, the rock sample was fixed in the table. The disc cutter was penetrated in the rock by the hydraulic cylinder.

Three hydraulic cylinders were used. One is the cylinder used to penetrate the disc cutter into the rock sample(400kN). The long stroke cylinder (1.1m) was used to roll the disc cutter. The rolling velocity was changeable to control the hydraulic pressure and maximum rolling velocity was about 0.25m/sec. in this apparatus. The third hydraulic cylinder was used to move the rock sample to the cutting position. The potentiometer and transducer of displacement were installed on the thrust hydraulic cylinder, and the rolling distance and penetration depth were measure by these sensors, respectively.

Figure 2 shows the schematic diagram of the dynamometer. The dynamometer was used to measure both the thrust force and rolling force. The several strain gauges were attached on the dynamometer. The signals from the strain gauges and some sensors were transmitted to the personal computer through strain meters and A/D converter. The sampling frequency was 100Hz.

The actual size disc cutter was used in this experiment. The size and shape were shown in Fig.3. The diameter was 360mm and the tip was round having 8mm radius.

2.2 Rock Samples

In this experiment, Iidate Granite and Shirakawa Welded Tuff were used. The size of the rock sample was 0.8m width, 1.2m length and 0.2m thickness. The physical properties of the rock sample were listed in Table 1. In this experiment, the side of the rock sample was covered with iron board of 15mm thickness to realize the side pressure and to avoid the easy broken due to the absence of the side pressure.

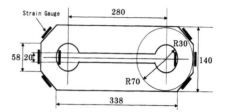

Figure 2. Schematic diagram of dynamometer.

2.3 Experiment Procedure

In this experiment, 5 Runs of experiments were carried out with changing the parameters of cutting width, rolling velocity and the kind of the rock. The rolling length was 0.8m. The parameters of each run were listed in Table 2. The cutting width was defined by the distance between the ditch of the rolling of the disc cutter as shown in Fig.4. The rolling velocity was defined as the velocity in the horizontal direction of the disc cutter. If the cutting width is small, the fracture generated from the ditch is considered to connect with the fracture generated from the neighbor ditch, and the chipping will be occurred. However, the cutting width in Run 5 is 0.2m and it is considered to be large enough to avoid the chipping. In this experiment, the experimental condition which the chipping is expected was defined as "adjacent", and the experimental condition which the chipping is not expected was defined as "non-adjacent".

The thrust and rolling force, penetration depth

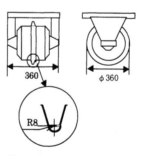

Figure 3. Schematic diagram of disc cutter used in this experiment (Radius of the cutter tip is 8mm.).

Table 1. Physical properties of rock samples.

Rock Sample	Density [kg/m^3]	Uniaxial Compressive Strength σ_c[MPa]	Poisson Ratio [−]
Iidate Granite	2634	259.9	0.239
Shirakawa Welded Tuff	1998	46.8	0.229

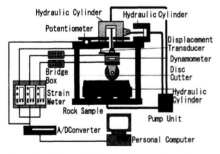

Figure 1. Schematic diagram of experimental apparatus.

Table 2. Experimental conditions of each Run.

Run No.	Rock Sample	Rolling Velocity V[m/s]	Cutting Width S [mm]	Remarks
1		0.25	65	General Cutting Width
2	Granite		40, 50, 65, 80	Effect of Cutting Width
3		0.125	40, 50, 65, 80	Effect of Rolling Velocity
4	Welded Tuff	0.25	65	Effect of Rock Sample
5	Granite		200	Non-Adjacent

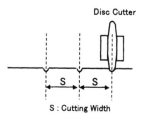

Figure 4. Definition of the cutting width.

were measured in each cutting. Furthermore, the rock fragments were collected in each cutting and the weight of them were measured in order to obtain the specific energy of cutting.

3 EXPERIMENTAL RESULTS

Figure 5 shows an example of the measured data. In Fig.5, the relationship between the thrust force, rolling force, penetration depth against the rolling distance was shown. Large fluctuations are seen in thrust force and penetration depth. This fluctuation is considered to be due to the chipping or uneven on the cutting ditch. The rolling force is much smaller compared with the thrust force, and it was

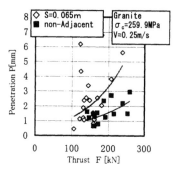

Figure 6. Relationship between penetration and thrust. (Run1+Run5)

about 1-3% of the thrust force for the granite and 5-8% for the tuff. In this experiment, the penetration depth was not so large. Therefore, it is inferred that the rolling force was small because of the small penetration depth. The penetration depth increased with increase the rolling distance at the beginning of each experiment and approached to the constant value. This result coincides with the result in the previous research (Gong et al. 1992).

Figure 6 shows the comparison between the results of Run1(adjacent) and Run5(non-adjacent). S and V in this figure indicate the cutting width and rolling velocity, respectively. Black and white symbols indicate the results of Run5 and Run1, respectively. Although the scatter of the data is large, it is confirmed that the penetration depth became large if the adjacent breakage is occurred. The solid line in this figure shows the empirical equation described latter.

Figure 7 shows the comparison between the results of Run1(Granite) and Run4(Tuff). The rolling velocity was 0.25m/sec. and cutting width was 0.065m. As the uniaxial compressive strength of the tuff is less than one fifth of the one of granite, it is found that larger penetration depth was obtained with less thrust force.

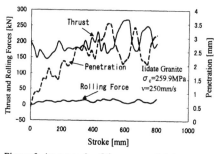

Figure 5. An example of the measured data.

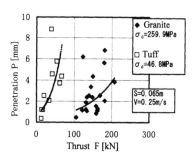

Figure 7. Relationship between penetration and thrust. (Run1+Run4)

601

Table 3. Empirical equations obtained in this experiment.

Run No.	Empirical Equations
1	$F = 0.369 \cdot \sigma_c \cdot P^{0.641}$
2	$F = 0.198 \cdot \sigma_c \cdot P^{0.757}$
3	$F = 0.201 \cdot \sigma_c \cdot P^{0.707}$
4	$F = 0.319 \cdot \sigma_c \cdot P^{0.789}$
5	$F = 0.530 \cdot \sigma_c \cdot P^{0.741}$
Penetration	$F = 0.114 \cdot \sigma_c \cdot P^{1.124}$

The relationship between the thrust force and penetration depth of the disc cutter is generally expressed by the following equation in terms of the uniaxial compressive strength of the rock.

$$F = a\sigma_c P^n \tag{1}$$

By using the experimental results, the coefficient, a and power index, n were calculated. The obtained empirical equations were shown in Table 3. The unit of thrust force F, penetration depth P and uniaxial compressive strength of the rock σ_c are [kN], [mm] and [MPa], respectively. The power index is almost the same regardless of the experimental condition and it was about 0.65-0.8.

Figure 8 shows the relationship between the thrust force and specific energy of cutting for adjacent breakage and non-adjacent breakage. The specific energy of cutting is defined by the required work to generate the fragment per unit volume by the disc cutter. That is, smaller specific energy of cutting means the effective rock cutting. In this study, the specific energy of cutting was calculated by Eq.(2) after Nishimatsu et al.(1975).

$$S_E = F_R L / V_0 \tag{2}$$

S_E is the specific energy of cutting[MJ/m³], F_R is the rolling force[N], L is the cutting distance[m] and V_0 is the volume of fragment cut by disc cutter[m³]. In Fig.8, black and white symbols

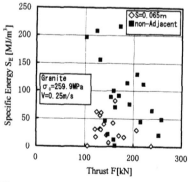

Figure 8. Relationship between specific energy and thrust. (Run1+Run5)

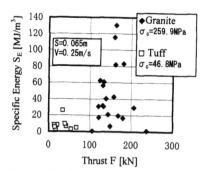

Figure 9. Relationship between specific energy and thrust. (Run1+Run4)

indicate the results of non-adjacent breakage and adjacent breakage, respectively. It seems that the specific energy of cutting for adjacent breakage is smaller compared with the specific energy of cutting for non-adjacent breakage. This means that the hard rock is cut effectively by the effect of adjacent breakage, and this result shows that the disc cutter should be arranged on the cutter face so that the adjacent breakage will easily be occurred.

Figure 9 shows the relationship between the thrust force and specific energy of cutting for granite and tuff. Since the tuff is soft and is broken largely, the specific energy of cutting is very small and one tenth of the one for granite.

We have already carried out the static penetration experiment of the disc cutter into the rock by using the same size and shape of the disc cutter (Takahashi et al. 1997). The specific energy of cutting for granite was 100-1000 times as much as the one for tuff in the static penetration experiment. However, it was confirmed by the visualization using X-ray CT equipment that some fractures exist inside of the granite and it was pointed out that effective rock cutting would be possible if the fractures are connected each other by the rolling of the disc cutter. In this experiment, the difference of specific energy of cutting for granite and tuff decreased and it is considered that this is due to the effect of rolling of the disc cutter. In order to confirm this effect, the core sample was obtained after the cutting experiment was over and the inside of the core sample was visualized by X-ray CT equipment. In the next chapter, the mechanism of rock cutting is discussed based on the visualization results.

4 VISUALIZATION OF FRACTURES IN ROCKS BY X-RAY CT EQUIPMENT

Core samples were obtained after the rock cutting experiment was finished. The diameter of the core

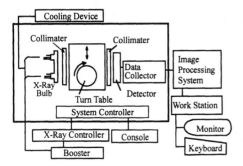

Figure 10. Outline of the X-ray CT equipment.

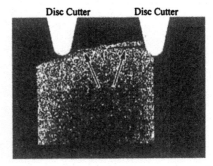

Figure 11. An example of the image visualized by X-ray CT equipment.

sample was 0.1m. Figure 10 shows the schematic diagram of the X-ray CT equipment. X-ray bulb and 176 detectors were fixed on the same horizontal plane, which is movable in up and down direction. The core sample was fixed on the turntable. The table turns and moves horizontally.

In this experiment, the core sample was set so that the X-ray was shot to the perpendicular to the rolling direction.

Figure 11 shows an example of the image of visualization. In order to help the comprehension,

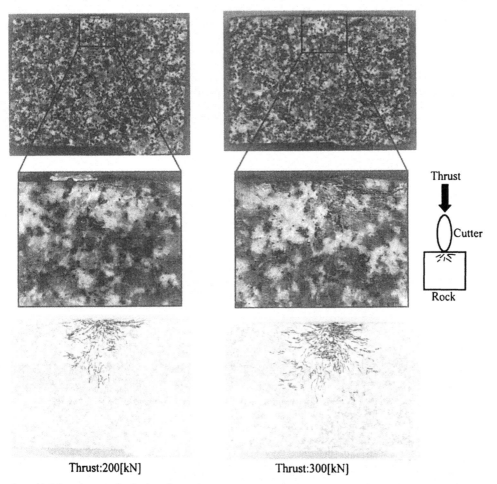

Thrust:200[kN]   Thrust:300[kN]

Figure 12. Micro fracture distribution after static penetration of the round tip disc cutter into the rock (Granite).

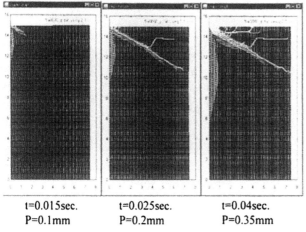

| t=0.015sec. | t=0.025sec. | t=0.04sec. |
| P=0.1mm | P=0.2mm | P=0.35mm |

Figure 13. Calculation results on the fracture growth by penetration of the disc cutter by use of DEM: White parts indicate the fractures.

the disc cutters were shown at rolling position with the same scale compared to the core sample. In Fig.11, white parts are corresponded to the rock, and black parts are corresponded to the porosity or fractures. As shown in Fig.11, the fractures are generated horizontally and they are parallel to the rock surface. The clear fractures are not observed in the depth direction and it is confirmed that the fractures are generated horizontally at the 2-3mm below from the rock surface. That is, the feature of the fractures generated by the rolling of the disc cutter is that the fractures are generated horizontally and the chipping is occurred if the fracture is connected with the fracture from the neighbor cutting ditch. In Fig.11, as two fractures are not connected, the chipping is not occurred. After the fractures grow because of the rolling of the disc cutter and they are connected, it can be considered that chipping is occurred and the rock fragment is removed.

5 INVESTIGATIONS ON MICRO FACTURES IN ROCKS GENERATED BY PENETRATION OF THE DISC CUTTER

In order to investigate the detail mechanism of rock cutting by the disc cutter, the static penetration experiment was carried out, and the micro fractures which were generated by the disc cutter were observed. Fig. 12 shows the images of the cross section of the core sample after the disc cutter was statically penetrated into the rock. Micro fractures were detected by the visual observation for the surface of the cross section. A

large horizontal fracture is significant, but many micro fractures in the radial direction were observed. Furthermore, numerical simulations were carried out by Distinct Element Method (DEM), and an example of the calculated results is shown in Fig.13. Horizontal fractures were well simulated, but there are some discrepancies between the simulated results and experimental ones. Therefore, further investigation on the numerical simulations will be still necessary.

REFERENCES

Gong F., Sato K. & Asai H. 1992. Optimal Cutting Condition and Maximal Tool Force in Circular Rock Cutting, J. of Mining and Material Processing Institute of Japan, Vol.108, No.12, pp.849-854.

Li X.S. Gurgenci, H. & Guan Z. 1997. Experimental Study of Disc Cutter Temperatures, Proc. of 4th Int. Symposium on Mine Mechanization and Automation, Vol.1, pp.A5-25-A5-34.

Muro T., Uematsu M & Iwatani, N 1997. Study on the Excavation Properties of a Roller Cutter Bits, Terra-mechanics, pp.111-115, 1997

Nishimatsu Y., Okuno N. & Hirasawa Y. 1975. The Rock Cutting with Roller, J. of Mining and Metallurgical Institute of Japan, Vol.91, No.1052, pp.653-658.

Takahashi H., Suzuoki S., Hatakeyama N. Nunomura S. & Shimizu Y. 1997. Penetration Experiment of Rocks by Use of TBM Roller Bit with Aiming the Fast Execution, Proc. of Int. Symposium on Rock Stress, pp.333-338.

Modern Tunneling Science and Technology, Adachi et al (eds), © 2001 Swets & Zeitlinger, ISBN 90 2651 860 9

Establishment of ventilation design method for New Tomei-Meishin Expressway Tunnels

T.Iwasaki
Consultant, Engineering Department, Japan Highway Public Corporation

K.Takekuni
Assistant Section Manager, Road Engineering Section, Engineering Department, Japan Highway Public Corporation

T.Otsu
Deputy Councilor, Road Engineering Section, Engineering Department, Japan Highway Public Corporation

M.Yamada
President, EchoPlan Inc

ABSTRACT: The 500 km-long six-lane Tomei and Meishin expressway No. 2 (hereinafter referred to as " New Tomei and Meishin expressway") will be a trunk line for the 21st century, connecting Tokyo, Nagoya and Kobe, along with the existing Tomei and Meishin expressway. The expressway will be constructed to the highest standards, designed for an optimal speed of 140 km, in light of improved vehicle performance and an increasing demand for highways that can accommodate high-speed driving. To ensure safe and comfortable driving conditions, the highway is horizontally aligned, with a large width and gentle longitudinal gradients. As it passes through mountainous areas, about 25% of the highway's total length will be in tunnels. This paper discusses the maximum exhaust concentration for a safe driving environment, a ventilation system to accommodate the emission control standards which will be intensified, and a ventilation system design that utilizes the space near the tunnel ceiling.

1 EXHAUST SMOKE

New Tomei and Meishin expressway will be a heavily-traveled route, and we assume that large vehicles will represent a proportion of about 60% of total traffic. When the proportion of large diesel vehicles is high, the main object of tunnel ventilation will normally be smoke from vehicle exhaust. The smoke ventilation system is designed to improve the visibility in the tunnel, and its scale will be determined by the estimated volume and concentration of exhaust gases.

1.1 Survey of Tunnel Conditions

To assess the volume of exhaust gas and to check the effect of emission controls on smoke production, we surveyed the conditions of the Ena tunnel (Lr = 8,625 m: equipped with a shaft air supply and exhaust-type longitudinal ventilation system and electrostatic precipitators) in 1997, and determined the following.

The survey method was as follows: The visible concentration in the tunnel was measured using two visibility (VI) meters; the weight concentration of particles was measured using two Tapered element oscillating microbalances (TEOM), with one of them set to measure SPM(Suspended Particulate Matter) 2.5, neglecting concentrations of 2.5 μ m or higher, and the other set to measure SPM10, neglecting concentrations of 10 μ m or higher; and, through an

analysis of the differences between these two measurements, the volume of air based on the measured wind velocity, and the traffic volume, the volume of exhaust smoke per vehicle, the relationship between the visible concentration and weight concentration, and the particle-size distribution were determined. As a result, we obtained the facts discussed below.

(1) Relation between particulates and grits
(2) Effect of emission control

We obtained the following particle size distribution in the tunnel (Fig. 2), which can be separated into two types, each having a peak, with a particle size of 2.5 μm as the boundary between the types. The dust in tunnels is comprised of those emitted from exhaust pipes and those on the road surface

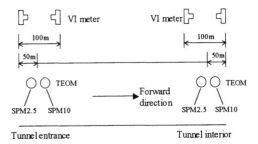

Figure 1. Basic Configuration of Measuring Instruments.

that are disturbed by moving vehicles. Normally, the particulate matter emitted from exhaust pipes is of fine diameter. The smaller particles (2.5 μm or less) in this survey can be attributed chiefly to carbon from burned light oil. The larger particles consisted mostly of silicon and calcium. We regarded the particles smaller than 2.5 μm (the assumed boundary between the two types of particles), as having been emitted from exhaust pipes, and considered the larger particles as having been disturbed by moving vehicles. These assumptions were based on the fact that the size of particles emitted from exhaust pipes is very small, and we then examined the effect of emission controls on exhaust pipe particles of grain sizes smaller than 2.5 μm.

As the safety of the driving environment is evaluated based on visible smoke concentration, we measured the visible smoke concentration with a smoke transmissometer, and compared the results to with analysis of the SPM weight concentration.

Table 1 shows the relation between the volume of exhaust smoke obtained from a visible smoke concentration with a volume of particulate as X and that of grits as Y and SPM weight concentration, obtained by multiple regression analysis.

Given the above, the following equation holds:

$$K = \alpha X + \beta Y$$

When the following symbols are used:
K: mixed volume based on visible concentration (m³/km/vehicle)
α: Coefficient for particulates
β: Coefficient for grits
X: mixed volume based on the weight concentration of particulates (g/km/vehicle)
Y: mixed volume based on the weight concentration of grits (g/km/vehicle)

The result of analysis proves that the effect of particulates on the visibility is, by weight, 1.3 (or more) times that of the effect of grits.

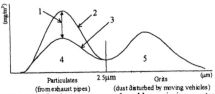

1 Volume of exhaust gas, as reduced by emission controls
2 Current volume of exhaust gas
3 Future volume of exhaust gas
4 Decreases resulting from emission controls.
5 Does not decrease further to emission controls.
Note: A component analysis of particulates shows that 80 to 98% of particulates, and 40 to 90% of grits are emitted from exhaust pipes.

Figure 2. Representation of the volume of exhaust smoke, as decreased further to emission controls.

Table 1. Values of coefficients α and β.

Coefficient α	2.92*
Coefficient β	2.19*

*: Values are given in grams.

In this analysis, we assumed that the grits are dust disturbed from the road surface. According to the survey results, the concentrations of particulates and grits are in high correlation.

1.2 Effect of Emission Controls

To determine the present design value for exhaust gas volume, the Ena tunnel conditions were surveyed in 1988, before the new emission controls were implemented. To evaluate the effects of emission controls, the survey results of 1988 and 1997 were compared, i.e., before and after implementation of the controls.

The 1997 survey also considered the proportion and types of vehicles covered by the controls. Based

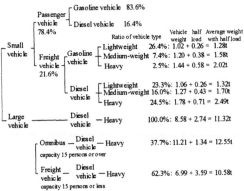

*: Two vehicles for which the fuel type is unknown are excluded.
**: Gasoline passenger vehicles include an LPG vehicle.

Figure 3. Vehicle types summarized based on the survey results.

Table 2. Ratio of PM-controlled vehicles (%).
(Vehicles subjected to the 1993 to 1994 controls)

Small vehicle	Passenger vehicle		37.1
	Freight vehicle	Lightweight	65.8
		Medium-weight	69.2
		Heavy	35.0
Large vehicle		Heavy	27.4
	Omnibus	Heavy	16.8
	Freight vehicle	Heavy	33.8

*: The proportion of small lightweight and medium-weight vehicles subjected to emission controls are as high as 60%, while the proportion of heavy-freight vehicles is only about 30%.

Table 3. Estimated future volume of smoke (m²/km/vehicle).

Type	PM division	Past (Estimated)	1988 (Measured)	Present (Estimated)	1997 (Measured)
Small vehicle	Emitted from exhaust pipes	0.51	---	0.42	---
	Disturbed from the road surface	0.31	---	0.31	---
	Total	0.82	---	0.73	0.88
Large vehicle	Emitted from exhaust pipes	7.77	---	6.84	---
	Disturbed from the road surface	1.63	---	1.63	---
	Total	9.40	9.28	8.47	8.45

Table 4. Control of diesel exhaust smoke.

		Short-term target	Long-term target	New short-term target	New long-term target
Fiscal year		1993 1994	1997 1998 1999	2002 2003 2004	2007
Month		10 10	10 10 10		
Passenger car (g/km)	1.25t or less	0.2 0.08		0.052	0.026
	Over 1.25t	0.2	0.08	0.056	0.028
Lightweight car (g/km)		0.2	0.08	0.052	0.026
Medium-Weight car (g/km)	MT	0.25	0.09	0.06	0.03
	AT	0.25	0.09	0.06	0.03
Heavy car (g/kWh)	2.5t<GVW ≤3.5t	0.7 0.25		0.18	0.09
	3.5t<GVW ≤12t	0.7	0.25	0.18	0.09
	12t<GVW	0.7	0.25	0.18	0.09

on the vehicle type configuration and data on exhaust smoke obtained, we calculated the volume of exhaust smoke for 1988. Fig. 3 and Table 2 show the vehicle type configuration and the proportion of vehicles covered by the controls, respectively, based on survey results. For large vehicles that contribute a large proportion of the exhaust emitted, the estimates calculated for 1997 levels agree closely with corresponding measurements, and those of 1988 levels agree with values obtained in the 1988 survey (see Table 3).

2 EXHAUST SMOKE AND VENTILATION VOLUMES

The volume of ventilation (i.e., fresh air) required to maintain a good driving environment is calculated based on the volume of exhaust smoke, with corrections for velocity gradient and altitude. In the past, the volume of exhaust smoke was determined through site surveys, and the basic data thus obtained was used for ventilation planning.

In 1993 the government decided to tighten controls over annual diesel exhaust emissions. The volume of exhaust smoke is expected to decrease to less than 40% of its present level after the implementation of the 1999 controls, and to less than 10% of present; levels after implementation of the 2007 controls. For New Tomei and Meishin expressway tunnels, we set the estimated volume of exhaust gas until fiscal 2007 to satisfy the emission control that will be implemented in that year as a new long-term target, in consideration of the replacement of vehicles with those that conform to the emission control.

We assumed the following in our calculations of the future volumes of exhaust smoke.

(1) The control of the diesel smoke will be implemented as scheduled.

(2) Vehicles will be comprised of the types set forth in Table 5 (manual of the assessment law).
(3) There is a fixed relation between the visible concentration of particles and the concentration of particles by weight.

Given the above assumptions, Table 6 shows the ratio of the required volume of ventilation to the volume of the exhaust smoke per unit of tunnel length, at the projected large vehicle proportion of 60%, after reduction of the emission of fine particles from exhaust pipes further to emission control standards.

Table 5. Total vehicle types (in different years).

Type	Base year	1 year earlier	2 years earlier	3 years earlier	4 years earlier	5 years earlier
Small vehicle	15.99	14.79	13.40	12.10	10.60	9.17
Large vehicle	17.53	15.83	14.14	12.45	10.73	9.02

Type	6 year earlier	7 years earlier	8 years earlier	9 years earlier	10 years earlier	More than 10 years earlier
Small vehicle	7.63	6.06	4.46	2.93	1.84	1.03
Large vehicle	7.32	5.57	3.83	2.27	1.08	0.23

Table 6. Value relative to the currently adopted value.

Year	Value relative to the currently adopted value
2005 ~ 2008	0.6
2009 ~	0.5

∗: A_r = 116m^2 τ = 50%
　　Gradient ± 0%
　　Driving speed 80km/h

3 DESIGN CONCENTRATION (TRANSMISSIVITY)

The variables to be controlled in order to ensure a safe driving environment in tunnels are smoke, and carbon monoxide, affecting visibility and health (respectively) of drivers and passengers in the tunnels.

As the estimated proportion of large vehicles is high, smoke removal has been selected as the major ventilation objective for New Tomei and Meishin expressway tunnels. As ventilation to combat smoke is designed to improve visibility, the design value (concentration) is determined as a function of driving speeds. In Japan, the following minimum concentration (smoke transmissivity) is used, applicable to a design driving speed of 100 km/h.

As the design speed is 140 km/h for New Tomei and Meishin expressway, we sought to determine the design smoke transmissivity given driving speeds of 100 to 140 km/h.

As design transmissivity was to be determined to ensure driving safety, we set the obstacle-detection distance on an asphalt-paved road (black) to 170 m, based on an experimentally-obtained relationship between obstacles on a road (e.g., dropped objects) with reflectivity 20% and height 20 cm × approx. lane width) and an obstacle-detection distance while

Table 7. Driving speed versus design concentration. (transmissivity)

Driving speed	Design transmissivity (%)
60 km/hr or less	40
80 to 100 km/hr	50

Per 100 m
Proportional from 60 km/h to 80 km/h
∗ : Table 8 gives an evaluation scale for transmissivity (PIARC 1979 Vienna Conference)

Table 8. Smoke transmissivity and discomfort levels.

Transmissivity (%)	Condition
100 ~ 60	Clean
60 ~ 50	Smoke visible
50 ~ 40	Uncomfortable
40 ~ 30	Very uncomfortable
30 or less	Intolerable

(PIARC 1979 Vienna Conference)

Table 9. Relation between pavement type, transmissivity and obstacle-detection distance (standard road surface).

Pavement type	Smoke transmissivity	Obstacle-detection distance (m)
Concrete (White)	τ = 50%	λ ≤ 140 ~ 170
	τ = 60%	λ ≤ 170 ~ 220
	(τ = 65%)	(λ ≤ 220 ~ 260)
	τ = 70%	λ ≤ 260 ~ 320
Asphalt (Black)	τ = 50%	λ ≤ 80 ~ 110
	τ = 60%	λ ≤ 110 ~ 140
	(τ = 65%)	(λ ≤ 140 ~ 170)
	τ = 70%	λ ≤ 170 ~ 200

Table 10. Relation between pavement type, required road surface illuminance and smoke transmissivity.

Pavement type (cd/m^2)	Smoke transmissivity (%)	Road surface illuminance
Concrete (White)	60	17
	65	15
	70	13
High-function pavement (Black)	65	15
	70	13

driving at 140 km/h (see Table 9). The obstacle-detection distance depends on the pavement and illumination in the background of the obstacle. As it is planned to adopt high-function pavement (black) to improve the visual environment for New Tomei and Meishin expressway, we set the design transmissivity at 70% in consideration of the road illuminance set in the illumination plan. Based on these results, Fig. 4 shows the relationship between driving speed and design transmissivity in tunnels in Japan.

The design transmissivity is set at 50% (per 100 m) as a standard for a visibility distance of 100 m at speeds of 80 to 100 km/h. To reproduce these conditions for an obstacle-detection distance of 170 m at the design speed of 140 km/h, the transmissivity level should theoretically be set at 67%. It will

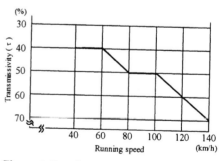

Figure 4. Running speed versus design transmissivity.

therefore be sufficient to set the transmissivity at 70% for a design speed of 140 km/h.

4 APPLICABLE TUNNELS AND THE INTRODUCTION OF CEILING-INSTALLED DUST COLLECTOR SYSTEM

New Tomei and Meishin expressway has 99 tunnels. By length the 99 tunnels can be classified as shown in Fig. 5. Twenty-two tunnels are longer than 2 km. Thirty-seven tunnels require ventilation systems.

As New Tomei and Meishin expressway tunnels are designed for high-speed one-way traffic, with a cross-sectional area as large as 116 m^2(Fig. 6), we plan a longitudinal ventilation system that will effectively utilize the ventilation force of traffic, which will combine with jet fans, electrostatic precipitators, as well as other air-supply and exhaust devices. Of the tunnels requiring ventilation, twenty-five will use electrostatic precipitators. In an initial plan, the dust-collecting chamber of an electrostatic precipitator was to be have been built independently (Fig. 7 shows the bypass method), wherein the chamber diverts a part of the air flow into the tunnel wall, with the branched air subsequently treated by the precipitator and returned to the road. However, the tunnels in New Tomei and Meishin expressway have ample space above the road, because the tunnels have large cross-sections. The space should be put into good use. Hence, we decided to use a design in which the dust collector system is installed on the tunnel ceiling, since this appeared an economical solution to expressway management.

The three-channel system maximizes the volume of treated air for a given cross-sectional area of the tunnel. The first channel collects the air in the tunnel flowing in a longitudinal direction, while the second and third channels collect the air immediately above the traffic lanes. The treated air is sent back to the traffic lanes by fans.

As shown in the Fig. 8, electrostatic precipitators and precipitator fans are installed on the ceiling, and the ancillary high-voltage generating boards, control

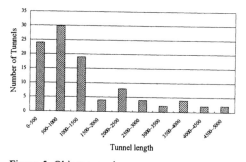

Figure 5. Object tunnels.

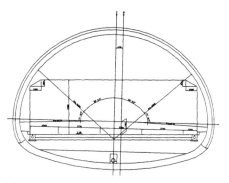

Figure 6. Standard cross-sectional area.

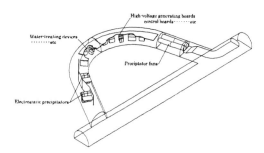

Figure 7. The construction of precipitator room (bypass method).

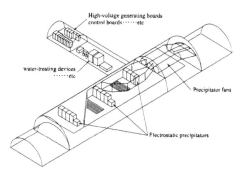

Figure 8. The construction of precipitator room.

boards and water-treating devices are located in the horizontal shaft of each precipitator room.

5 CONCLUSION

We have established a method to design a ventilation system for tunnels with a large cross-sectional area designed for high-speed traffic, for which we use a novel means of estimating the volume of exhaust smoke. The ventilation systems of the Tokyo Harbor Aqua-Line and the Maiko tunnel demonstrate the first large-scale uses in Japan of the ceiling-installed dust collector system that uses free

Table 11. Effect of emission control.

	At the present value	At the controlled value
Natural ventilation	62 tunnels	86 tunnels
Jet-fan system	2 tunnels (5 sets)	3 tunnels (18 sets)
Electrostatic precipitator system	25 pieces (40 sets) (volume of treated air 23,000m³/s)	10 piece (14 sets) (volume of treated air 7000m³/s)
Saccardo nozzle and shaft system	10 tunnels	—

Jet fan : Nominal diameter 1500

space above the traffic lanes. As a result of efforts to meet emission standards, the number of tunnels in which a ventilation system needs to be installed fell from 87 to 13, as shown in Table 11, with subsequent elimination of the need for conventional air feed and exhaust methods, which require gigantic equipment and a shaft, upright or inclined.

By making efforts to meet emission controls and introducing a ceiling-installed dust collector system in the tunnels of New Tomei and Meishin expressway, the scale of the ventilation system becomes smaller, and the cost is cut by 60% or more. This means that the system minimizes costs while at the same time ensuring driving safety and comfort.

For Product Safety Concerns and Information please contact our EU
representative GPSR@taylorandfrancis.com
Taylor & Francis Verlag GmbH, Kaufingerstraße 24, 80331 München, Germany